D1336409

INSTRUMENTAL ANALYSIS OF INTRINSICALLY DISORDERED PROTEINS

WILEY SERIES ON PROTEIN AND PEPTIDE SCIENCE

Vladimir N. Uversky, Series Editor

Metalloproteomics · Eugene Permyakov

Protein Misfolding Diseases: Current and Emerging Principles and Therapies · Edited by Marina Ramirez-Alvarado, Jeffery W. Kelly, and Christopher M. Dobson

Instrumental Analysis of Intrinsically Disordered Proteins: Assessing Structure and Conformation · Edited by Vladimir Uversky and Sonia Longhi

INSTRUMENTAL ANALYSIS OF INTRINSICALLY DISORDERED PROTEINS

Assessing Structure and Conformation

Edited by

VLADIMIR N. UVERSKY AND SONIA LONGHI

A John Wiley & Sons, Inc., Publication

Published by John Wiley & Sons, Inc., Hoboken, New Jersey
Published simultaneously in Canada

Library of Congress Cataloging-in-Publication Data

ISBN 978-0-470-34341-8

Printed in the United States of America

10 9 8 7 6 5 4 3 2 1

CONTENTS

PREFACE

The notion that protein function relies on a precise three-dimensional (3D) structure constitutes one of the central paradigms of biochemistry. According to this concept, a protein can perform its biological function(s) only after folding into a unique 3D structure, and all the information necessary for a protein to fold into this unique 3D structure is encoded in the amino acid sequence. Only recently has the validity of this structure–function paradigm been seriously challenged, primarily through the wealth of counterexamples that have gradually accumulated over the past 20–25 years. These counterexamples demonstrated that many functional proteins or protein parts exist in an entirely or partly disordered state. Intrinsically disordered proteins (IDPs), also referred to as natively unfolded proteins, lack a unique, stable 3D structure in solution, existing instead as a dynamic ensemble of conformations. Many IDPs function without a prerequisite stably folded structure. The protein flexibility that is inherent to disorder confers functional advantages. Their increased plasticity (1) enables the binding of or to numerous structurally distinct targets; (2) provides the ability to overcome steric restrictions by enabling larger surfaces of interaction in protein complexes than those obtained with rigid partners; and (3) allows protein interactions to occur with both high specificity and low affinity.

IDPs possess peculiar sequence properties that allow them to be distinguished from globular proteins. In particular, (1) they are generally enriched in amino acids preferred at the surface of globular proteins (i.e., A, R, G, Q, S, P, E, and K) and depleted in core-forming residues (W, C, F, I, Y, V, L, and N); (2) they possess a distinct combination of a high content of charged residues and of a low content of hydrophobic residues; (3) they typically possess a low predicted secondary structure content; (4) they tend to have a low sequence complexity (i.e., they make use of fewer types of amino acids);

and (5) they often have a high sequence variability. The peculiar sequence features of IDPs have led to the development of various disorder predictions, which have allowed an estimation of the occurrence of disorder in biological systems. Computational studies have shown that the frequency and length of disordered regions increases with increasing an organism's complexity. For example, long intrinsically disordered regions have been predicted to occur in 33% of eukaryotic proteins, with 12% of these latter being fully disordered. Furthermore, viruses and eukaryota were predicted to have 10 times more conserved disorder (roughly 1%) than archaea and bacteria (0.1%). Beyond these computational studies, an increasing amount of experimental evidence has been gathered in the last years pointing out the large abundance of intrinsic disorder within the living world: More than 523 proteins containing 1195 disordered regions have been annotated so far in the Disprot database (http://www.disprot.org).

Regions lacking specific 3D structure have been so far associated with a number of distinct biological functions, including nucleic acid and protein binding, interaction with small molecules, display of posttranslational modification sites, display of proteolysis sites, and prevention of interactions by means of excluded volume effects. Most IDPs are involved in functions that imply multiple partner interactions (e.g., one-to-many and many-to-one binding scenarios), such as molecular recognition, molecular assembly (and amyloidogenesis), cell cycle regulation, signal transduction, and transcription. As such, IDPs are implicated in the development of several pathological conditions, including cancer and cardiovascular diseases. They have been shown to be promising targets for drug development.

Intrinsic disorder is a distinctive and common feature of "hub" proteins, with disorder serving as a determinant of protein promiscuity. A recent study indicates that the size of macromolecular assemblies is related to the abundance of disorder of the protein components. Intrinsic disorder also serves as a determinant of the transient nature of the interactions that IDPs can establish, by virtue of the presumed rather low affinity that typifies interactions involving IDPs. Indeed, the ability of "social hubs" to establish transient interactions has been associated to the overall flexibility of the hub proteins. The relationship between structural disorder and regulation provides a plausible explanation for the prevalence of disorder in higher organisms, which have more complex signaling and regulatory pathways. However, the abundance of disorder within viruses likely reflects the need for genetic compaction, where a single disordered protein can establish multiple interactions and hence exert multiple concomitant biological effects. In addition, structural disorder might endow viral proteins with broader ability to interact with the components of the host and may also be related to high adaptability levels and mutation rates observed in viruses, thus representing a unique strategy for buffering the deleterious effects of mutations.

Although there are IDPs that carry out their function while remaining disordered all the time (e.g., entropic chains), many of them undergo a disorder-to-order transition upon binding to their physiological partner(s), a process termed induced folding. IDPs show an extremely wide diversity in their structural properties: They can attain extended conformations (random coil like) or remain globally collapsed (molten globule like), where the latter possess regions of fluctuating secondary structure. Conformational and spectroscopic analyses showed that random coil-like IDPs can further be divided into two major groups. While the first group consists of proteins with extended maximum dimensions typical of random coils with no (or little) secondary structure, the second group comprises the so-called premolten globules, which are more compact (but still less compact than globular or molten globule proteins) and contain some residual secondary structure. The residual intramolecular interactions that typify the premolten globule state may enable a more efficient start of the folding process induced by a partner.

Many IDPs bind to their target(s) through "molecular recognition elements" (MoREs) or "molecular recognition features" (MoRFs). MoRFs are interaction-prone short segments with an increased foldability, which are embedded within long disordered regions and which become ordered upon binding to a specific partner. The conformation of MoRFs in isolation can be either disordered or partially preformed, thus reflecting an inherent conformational preference. In this latter case, a transiently populated folded state would exist even in the absence of the partner, thus implying that the folding induced by the partner would rely on conformer selection (i.e., selection by the partner of a preexisting conformation) rather than on a "fly-casting" mechanism. It has been proposed that the restriction in the conformational space of MoRFs in the unbound state could reduce the entropic cost of binding thereby enhancing affinity. IDPs can bind their target(s) with a high extent of conformational polymorphism, with binding generally involving larger normalized interface areas than those found between rigid partners, with protein interfaces being enriched in hydrophobic residues. Thus, protein–protein interactions established by IDPs rely more on hydrophobic–hydrophobic than on polar–polar contacts.

Although a large body of experimental evidence of the abundance (and biological relevance) of disorder in the living world has been gathered in the last decade, the notion of a tight dependence of protein function on a precise 3D structure is still deeply anchored in many structuralists' mind. The reasons for this lack of awareness or even "resistance" to the concept of protein intrinsic disorder are multiple. First, the growing numbers of protein structures determined by X-ray crystallography and by NMR in the last three decades has shifted the attention of structuralists away from the numerous examples of IDPs. Second, IDPs have been long unnoticed because researchers encountering examples of structural disorder mainly ascribed it to errors and artifacts, and as such, purged them from papers and reports. Third, structural disorder is hard to conceive and classify. Fourth, IDPs have been neglected because of the perception that a limited amount of mechanistic data could be derived from their study.

The evidence that IDPs exist both *in vitro* and *in vivo* is compelling and justifies considering them as a separate class within the protein realm. Furthermore, their distinguishing interaction abilities justify designating the protein–protein interactions they establish as "close encounters of the third kind." Key questions that have to be addressed to unravel the way IDPs exert their biological functions concern how they can be correctly recognized by their partner proteins in the absence of a folded, stable structure, and assessing the extent to which these proteins are structurally preconfigured prior to partner protein binding.

Answering these questions requires appropriate methods for the description of conformational ensembles in the free, pre-recognition states, as well as in complex with their partner proteins. IDPs inherently escape structural characterization by conventional high-resolution techniques. In fact, since IDPs exist as a dynamic continuum of conformations, their crystallization is generally precluded, and in those rare cases where it is successful, it only leads to a snapshot of poorly representative conformations. As such, the structural characterization of IDPs requires the use of combined, complementary physicochemical approaches. In this book, a thorough description of the principal experimental approaches that can be used to assess structural disorder and induced folding is provided. We highlight the type of information that can be derived by the different approaches and point out how information can be broadened by using different methods.

The book has seven sections dedicated to (1) assessment of IDPs in the living cell (chapters 1 and 2), (2) spectroscopic techniques for the analysis of IDPs (chapters 3–12), (3) single-molecule techniques applied to the study of IDPs (chapters 13 and 14), (4) assessment of IDP size and shape (chapters 15–18), (5) tools for the analysis of IDP conformational stability (chapters 19 and 20), (6) mass spectrometry (chapter 21), and (7) approaches for expression and purification of IDPs (chapters 22–24). A brief outline of the corresponding chapters is presented below.

In chapter 1, Yosef Shaul, Peter Tsvetkov, and Nina Reuven represent an overview of the mechanisms underlining the process of intracellular IDP degradation and describe experimental techniques used for the analysis of IDP degradation *in vivo*. The authors also describe various mechanisms used by IDPs for escaping degradation by default and show that several human diseases, including certain types of cancers and neurological disorders, can result from aberrant protein degradation.

In chapter 2, Philipp Selenko introduces a set of in-cell NMR methods that are used to extend the structural understanding of IDPs in a cellular environment. It is emphasized that these techniques are comparable to an atomic resolution microscope and, therefore, can provide high-resolution structural information on proteins, including IDPs, inside prokaryotic and eukaryotic cells.

Chapter 3, authored by Frans A. A. Mulder, Martin Lundqvist, and Ruud M. Scheek, is dedicated to the description of the peculiarities of IDP analysis by NMR spectroscopy, which is believed to be the most suitable tool to investigate the behavior of these highly dynamic proteins. The authors emphasize that many NMR parameters, including chemical shifts, line widths, spin-spin multiplet patterns, relaxation rates, and residual dipolar couplings, are atom specific and harbor information about the local conformations that polypeptide chains adopt, including record of their dynamic behavior. As such, they provide unique information on the peculiarities of IDP structure.

In chapter 4, Martin Blackledge, Pau Bernadó, and Malene Ringkjøbing Jensen show how IDP ensembles and local structural propensity in IDPs can be characterized at the atomic level using NMR residual dipolar couplings that report on time- and ensemble-averaged conformations up to the millisecond timescale. The authors conclude that the combination of appropriate ensemble descriptions allows the extraction of unique and important information on the conformational propensities of IDPs.

In chapter 5, Gary W. Daughdrill describes approaches for the determination of realistic structural ensembles of IDPs and emphasizes that this task requires the development of suitable experimental and computational methods. Very well-suited experimental techniques are small-angle X-ray scattering (SAXS), which can be used for the determination of a Boltzmann-weighted distribution of gyration radii, and NMR spectroscopy, which provides ensemble-averaged angular and interatomic distance information. The usefulness of this approach is illustrated in the case of the intrinsically disordered transactivation domain of the human tumor suppressor, p53.

Valérie Belle, Sabrina Rouger, Stéphanie Costanzo, Sonia Longhi, and André Fournel, in chapter 6, introduce site-directed spin labeling electron paramagnetic resonance (EPR) spectroscopy as a useful and unique tool for the structural analysis of IDPs. The theoretical principles of this approach are introduced and an illustrative example is provided highlighting the adequacy of this method to investigate the structural properties and the induced folding of IDPs.

In chapter 7, Reinhard Schweitzer-Stenner, Thomas J. Measey, Andrew M. Hagarman, and Isabelle C. Dragomir consider the application of vibrational spectroscopy in the structural analysis of IDPs and unfolded peptides. This chapter introduces the basic physical concepts and reviews utilization of UV resonance Raman, visible nonresonance Raman, vibrational circular dichroism (CD), Raman optical activity, and electronic CD spectroscopies for the exploration of IDPs and disordered peptides.

Chapter 8, by Antonino Natalello and Silvia Maria Doglia, is focused on applications of Fourier transform infrared spectroscopy in the analysis of IDPs and induced folding. This chapter represents a general survey of the standard experimental methods to obtain the infrared absorption spectrum of a protein, together with the data analysis that enables evaluation of the secondary structure content. The potential of infrared spectroscopy to document induced folding is illustrated through examples of IDPs that either fold in the presence of different effectors, such as DNA, partner proteins, and osmolytes, or undergo amyloid aggregation.

Chapter 9 by Natalya I. Topilina, Vitali Sikirzhytski, Seiichiro Higashiya, Vladimir V. Ermolenkov, John T. Welch, and Igor K. Lednev describes the analysis of IDP structure and aggregation by deep UV resonance Raman spectroscopy (DUVRR). DUVRR is a novel method for acquisition of quantitative information on the peptide backbone conformation. In particular, the authors focused on the use of genetic engineering in the study of the mechanism of fibrillation of large biopolymers that are excellent models for IDPs. They believe that the selective design of the polypeptide sequence by genetic engineering represents a great tool for studies of the relationship between the sequence and the cross-β-sheet structure.

In chapter 10, Robert W. Woody describes application of CD to analysis of IDPs. As IDPs are readily recognized by CD, a large number of IDPs have been identified and characterized using this technique. Complex (e.g., heterogeneous) IDPs containing folded domains interspersed with extensive unordered regions are more difficult to diagnose. These IDPs can be analyzed by combining limited proteolysis with CD, where CD analysis of the individual domains identified by proteolysis can be used to decipher their modular organization.

Fluorescence spectroscopy of IDPs is discussed by Eugene A. Permyakov and Vladimir N. Uversky in chapter 11. As IDPs are typically characterized by surface location of tryptophan residues, their intrinsic fluorescence is red-shifted and is readily quenched by external fluorescence quenchers. However, IDP interactions with binding partners transfer tryptophan residues into a more hydrophobic or more rigid environment, resulting in a measurable blue-shift of the fluorescence maximum position and a noticeable decrease in the quencher accessibility. Furthermore, spectral probes and labels are widely used for IDP analysis.

Chapter 12 (by Kálmán Tompa, Monika Bokor, and Peter Tompa) is dedicated to the analysis of structural and dynamical properties of interfacial water at the protein surface by wide-line NMR spectroscopy and nuclear relaxation time measurements. The authors provide a detailed description of the theoretical background and practice of this approach, followed by the description of its implementation on ordered and disordered proteins.

Benjamin Schuler in chapter 13 introduces methods for single-molecule spectroscopy of IDPs. Recent years have seen a remarkable development of single-molecule spectroscopy, especially single-molecule Förster resonance energy transfer (FRET), which has evolved into a versatile method to probe distances, distance distributions, and dynamics of unfolded proteins, including IDPs.

The theme of single-molecule spectroscopy is further addressed in chapter 14, where Massimo Sandal, Marco Brucale, and Bruno Samorì describe the technique of single-molecule force spectroscopy (SMFS), a description of the various SMFS experimental approaches applicable for protein conformational analysis, and application of these techniques for the evaluation of the conformational equilibria and structural diversity of IDPs.

Chapter 15 starts the section dedicated to various hydrodynamic techniques. This chapter, authored by Florence Manon and Christine Ebel, introduces analytical ultracentrifugation (AUC) and describes how this technique can be used to probe and characterize the size and oligomeric state of IDPs. The authors provide basic information on theory of AUC and describe the AUC instrumentation. They show how the sedimentation phenomenon can be used to evaluate various hydrodynamic parameters of IDPs and how two usual types of AUC experiments, sedimentation equilibrium (SE) and velocity (SV), are used for IDP characterization.

Pau Bernadó and Dmitri I. Svergun in chapter 16 describe the type of information on IDPs that can be extracted by SAXS, which is one of the very few techniques yielding low-resolution structural information about flexible macromolecules. This chapter introduces the technical and experimental details of SAXS, including classical approaches based on the analysis of overall parameters and a recent development, the ensemble optimization method. The latter approach, in combination with other biophysical techniques such as NMR, FRET, and molecular simulations, provides unique insights into the analysis of IDP structure and structural perturbations induced by the environmental changes or binding to biological partners.

Chapter 17, authored by Klaus Gast, describes dynamic and static light scattering (SLS) and shows how these techniques can be used to evaluate various molecular parameters of IDPs, such as size, molar mass, and intermolecular interactions. The physical bases of light scattering, experimental techniques, sample treatment, and data evaluation schemes are outlined, with special emphasis on studies on IDPs. It is also emphasized that since SLS and dynamic light scattering (DLS) yield different physical quantities of macromolecules, the combined application of these techniques improves the accuracy of the molar mass evaluation and is essential when changes in the molecular dimensions and molecular association/dissociation take place simultaneously.

Chapter 18 by Vladimir N. Uversky is focused on size exclusion chromatography (SEC) and its applications for the analysis of IDPs, allowing estimation of the hydrodynamic dimensions, evaluation of the association state, analysis of IDP interactions with binding partners, and assessment of induced folding events. As SEC can physically separate IDP conformers based on their hydrodynamic dimensions, this method provides a unique possibility for the independent analysis of the physicochemical properties of these conformers.

The conformational behavior of IDPs is characterized by the low cooperativity of the denaturant-induced unfolding, lack of the measurable excess heat absorption peak(s) in calorimetric analysis, "turned out" response to heat and changes in pH, the ability to gain structure in the presence of various counterions, and the unique response to macromolecular crowding. Chapter 19 (by Vladimir N. Uversky) describes these unique features of the conformational behavior of IDPs.

Angelo Fontana, Patrizia Polverino de Laureto, Barbara Spolaore, Erica Frare, and Marcello Zambonin in chapter 20 show how limited proteolysis can be used to analyze protein structure and dynamics, and to identify disordered sites or regions within otherwise folded globular proteins. Often, sites of limited proteolysis coincide with sites of enhanced flexibility of the polypeptide chain, that is, with regions characterized by high values of the crystallographic temperature factor (B-factor) or with regions of missing electron density. As limited proteolysis can be used to probe protein structure and dynamics, and to detect sites of disorder in proteins, this technique complements other physicochemical and computational approaches.

Mária Šamalíková, Carlo Santambrogio, and Rita Grandori dedicate chapter 21 to the description of mass spectrometry tools, which recently became central to structural biology, and their applications for the investigation of the structure and the dynamics of protein conformations and protein assemblies. The authors focus on methods based on maintenance of noncovalent interactions under electrospray conditions, such as charge-state distribution analysis and ion mobility, and show that these methods offer useful structural information, which are complementary to that gained by other biophysical techniques.

Dmitri Tolkatchev, Josee Plamondon, Richard Gingras, Zhengding Su, and Feng Ni dedicate chapter 22 to the description of a procedure for recombinant expression of IDPs in *Escherichia coli*, which leads to yields comparable to and even higher than those typically obtained for well-folded proteins. The method is based on the high-level expression of IDPs fused to a carrier protein derived from the N-terminal oligonucleotide binding domain of staphylococcal nuclease, leading to the accumulation of inclusion bodies in the cell.

In chapter 23, Vladimir N. Uversky, Marc S. Cortese, Peter Tompa, Veronika Csizmok, and A. Keith Dunker argue that the unique conformational behavior of IDPs (such as their high temperature and acid stability, and apparent insensitivity to chemical denaturation) can be exploited for the large-scale identification of IDPs and isolation of these proteins from cell extracts.

Aviv Paz, Tzviya Zeev-Ben-Mordehai, Joel L. Sussman, and Israel Silman continue the theme of IDP purification in chapter 24. The authors survey some of the specific procedures reported in the literature for purifying recombinant IDPs that are sensitive to the degradation machinery of the host cell in which they are being overexpressed. They also describe approaches elaborated to stabilize IDPs in the course of purification.

Sonia Longhi
A. Keith Dunker
Vladimir N. Uversky

INTRODUCTION TO THE WILEY SERIES ON PROTEIN AND PEPTIDE SCIENCE

Proteins and peptides are the major functional components of the living cell. They are involved in all aspects of the maintenance of life. Their structural and functional repertoires are endless. They may act alone or in conjunction with other proteins, peptides, nucleic acids, membranes, small molecules and ions during various stages of life. Dysfunction of proteins and peptides may result in the development of various pathological conditions and diseases. Therefore, the protein/peptide structure-function relationship is a key scientific problem lying at the junction point of modern biochemistry, biophysics, genetics, physiology, molecular and cellular biology, proteomics, and medicine.

The *Wiley Series on Protein and Peptide Science* is designed to supply a complementary perspective from current publications by focusing each volume on a specific protein- or peptide-associated question and endowing it with the broadest possible context and outlook. The volumes in this series should be considered required reading for biochemists, biophysicists, molecular biologists, geneticists, cell biologists and physiologists as well as those specialists in drug design and development, proteomics and molecular medicine with an interest in proteins and peptides. I hope that each reader will find in the volumes within this book series interesting and useful information.

First and foremost I would like to acknowledge the assistance of Anita Lekhwani of John Wiley & Sons, Inc. throughout this project. She has guided me through countless difficulties in the preparation of this book series and her enthusiasm, input, suggestions and efforts were indispensable in bringing the *Wiley Series on Protein and Peptide Science* into existence. I would like to take this opportunity to thank everybody whose contribution in one way or another has helped and supported this project. Finally, special thank you goes to my wife, sons and mother for their constant support, invaluable assistance, and continuous encouragement.

LIST OF CONTRIBUTORS

Valérie Belle, Bioénergétique et Ingénierie des Protéines, UPR 9036 CNRS and Universités Aix-Marseille I et II, Marseille, France

Pau Bernadó, Institute for Research in Biomedicine, Parc Científic de Barcelona, Barcelona, Spain

Martin Blackledge, Protein Dynamics and Flexibility, Institut de Biologie Structurale, Grenoble, France

Monika Bokor, Research Institute for Solid State Physics and Optics, Hungarian Academy of Sciences, Budapest, Hungary

Marco Brucale, Department of Biochemistry "G. Moruzzi," Bologna, Italy

Marc S. Cortese, Institute for Intrinsically Disordered Protein Research, Center for Computational Biology and Bioinformatics, and Department of Biochemistry and Molecular Biology, Indiana University School of Medicine, Indianapolis, IN

Stéphanie Costanzo, Architecture et Fonction des Macromolécules Biologiques, UMR 6098 CNRS and Universités Aix-Marseille I et II, Campus de Luminy, Marseille, France

Veronika Csizmok, Institute of Enzymology, Biological Research Center, Hungarian Academy of Sciences, Budapest, Hungary

Gary W. Daughdrill, Division of Cell Biology, Microbiology, and Molecular Biology, and the Center for Biomolecular Identification and Targeted Therapeutics, University of South Florida, Tampa, FL

Silvia Maria Doglia, Department of Biotechnology and Biosciences, University of Milano-Bicocca, Milan, Italy

Isabelle C. Dragomir, Department of Chemistry, Drexel University, Philadelphia, PA

A. Keith Dunker, Institute for Intrinsically Disordered Protein Research, Center for Computational Biology and Bioinformatics, and Department of Biochemistry and Molecular Biology, Indiana University School of Medicine, Indianapolis, IN

Christine Ebel, Commissariat à l'Energie Atomique, Institut de Biologie Structurale, Grenoble, France; Centre National de la Recherche Scientifique, Grenoble, France; Université Joseph Fourier, Grenoble, France

Vladimir V. Ermolenkov, Department of Chemistry, University at Albany, State University of New York, Albany, NY

Angelo Fontana, CRIBI Biotechnology Centre, University of Padua, Padua, Italy

André Fournel, Bioénergétique et Ingénierie des Protéines, UPR 9036 CNRS and Universités Aix-Marseille I et II, Marseille, France

Erica Frare, CRIBI Biotechnology Centre, University of Padua, Padua, Italy

Klaus Gast, Universität Potsdam, Institut für Biochemie und Biologie, Physikalische Biochemie, Potsdam-Golm, Germany

Richard Gingras, Biomolecular NMR and Protein Research Group, Biotechnology Research Institute, National Research Council of Canada, Montreal, Quebec, Canada

Rita Grandori, Department of Biotechnology and Biosciences, University of Milano-Bicocca, Milan, Italy

Andrew M. Hagarman, Department of Chemistry, Drexel University, Philadelphia, PA

Seiichiro Higashiya, Department of Chemistry, University at Albany, State University of New York, Albany, NY

Igor K. Lednev, Department of Chemistry, University at Albany, State University of New York, Albany, NY

Sonia Longhi, Architecture et Fonction des Macromolécules Biologiques, UMR 6098 CNRS and Universités Aix-Marseille I et II, Campus de Luminy, Marseille, France

Martin Lundqvist, Department of Biophysical Chemistry, University of Groningen, Groningen, The Netherlands

Florence Manon, Commissariat à l'Energie Atomique, Institut de Biologie Structurale, Grenoble, France. Centre National de la Recherche Scientifique, Grenoble, France; Université Joseph Fourier, Grenoble, France

Thomas J. Measey, Department of Chemistry, Drexel University, Philadelphia, PA

Frans A. A. Mulder, Department of Biophysical Chemistry, University of Groningen, Groningen, The Netherlands

Antonino Natalello, Department of Biotechnology and Biosciences, University of Milano-Bicocca, Milan, Italy

Feng Ni, Biomolecular NMR and Protein Research Group, Biotechnology Research Institute, National Research Council of Canada, Montreal, Quebec, Canada

Aviv Paz, Departments of Neurobiology and Structural Biology, Weizmann Institute of Science, Rehovot, Israel

Eugene A. Permyakov, Institute for Biological Instrumentation of the Russian Academy of Sciences, Pushchino, Moscow, Russia

Josee Plamondon, Biomolecular NMR and Protein Research Group, Biotechnology Research Institute, National Research Council of Canada, Montreal, Quebec, Canada

Patrizia Polverino de Laureto, CRIBI Biotechnology Centre, University of Padua, Padua, Italy

Nina Reuven, Department of Molecular Genetics, Weizmann Institute of Science, Rehovot, Israel

Malene Ringkjøbing Jensen, Protein Dynamics and Flexibility, Institut de Biologie Structurale UMR 5075 CEA-CNRS-UJF, Grenoble, France

Sabrina Rouger, Architecture et Fonction des Macromolécules Biologiques, UMR 6098 CNRS and Universités Aix-Marseille I et II, Campus de Luminy, Marseille, France

Mária Šamalíková, Department of Biotechnology and Biosciences, University of Milano-Bicocca, Milan, Italy

Bruno Samorì, Department of Biochemistry "G. Moruzzi," Bologna, Italy

Massimo Sandal, Department of Biochemistry "G. Moruzzi," Bologna, Italy

Carlo Santambrogio, Department of Biotechnology and Biosciences, University of Milano-Bicocca, Milan, Italy

Ruud M. Scheek, Department of Biophysical Chemistry, University of Groningen, Groningen, The Netherlands

Benjamin Schuler, Biochemisches Institut, Universität Zürich, Zürich, Switzerland

Reinhard Schweitzer-Stenner, Department of Chemistry, Drexel University, Philadelphia, PA

Philipp Selenko, Leibniz Institute of Molecular Pharmacology (FMP), Berlin, Germany

Yosef Shaul, Department of Molecular Genetics, Weizmann Institute of Science, Rehovo, Israel

Vitali Sikirzhytski, Department of Chemistry, University at Albany, State University of New York, Albany, NY

Israel Silman, Department of Neurobiology, Weizmann Institute of Science, Rehovot, Israel

Barbara Spolaore, CRIBI Biotechnology Centre, University of Padua, Padua, Italy

Zhengding Su, Biomolecular NMR and Protein Research Group, Biotechnology Research Institute, National Research Council of Canada, Montreal, Quebec, Canada

Joel L. Sussman, Department of Structural Biology, Weizmann Institute of Science, Rehovot, Israel

Dmitri I. Svergun, European Molecular Biology Laboratory, Hamburg Outstation, Hamburg, Germany

Dmitri Tolkatchev, Biomolecular NMR and Protein Research Group, Biotechnology Research Institute, National Research Council of Canada, Montreal, Quebec, Canada

Kálmán Tompa, Research Institute for Solid State Physics and Optics, Hungarian Academy of Sciences, Budapest, Hungary

Peter Tompa, Institute of Enzymology, Biological Research Center, Hungarian Academy of Sciences, Budapest, Hungary

Natalya I. Topilina, Department of Chemistry, University at Albany, State University of New York, Albany, NY

Peter Tsvetkov, Department of Molecular Genetics, Weizmann Institute of Science, Rehovot, Israel

Vladimir N. Uversky, Institute for Intrinsically Disordered Protein Research, Center for Computational Biology and Bioinformatics, and Department of Biochemistry and Molecular Biology, Indiana University School of Medicine, Indianapolis, IN; and Institute for Biological Instrumentation, Russian Academy of Sciences, Pushchino, Moscow Region, Russia

John T. Welch, Department of Chemistry, University at Albany, State University of New York, Albany, NY

Robert W. Woody, Department of Biochemistry and Molecular Biology, Colorado State University, Fort Collins, CO

Marcello Zambonin, CRIBI Biotechnology Centre, University of Padua, Padua, Italy

Tzviya Zeev-Ben-Mordehai, Departments of Neurobiology and Structural Biology, Weizmann Institute of Science, Rehovot, Israel

LIST OF ABBREVIATIONS

2D-DUVRR	two-dimensional correlation deep UV resonance Raman spectroscopy
A	acid-denatured protein
A9	AcYEAAAKEAAAKEAAAKANH_2
α-LA	α-lactalbumin
α-syn	α-synuclein
Aβ	amyloid beta peptide
AD	Alzheimer's disease
AFA	abstract factor analysis
AFM	atomic force microscopy
ALS	alternating least square
Am I	amide I vibration
Am II	amide II vibration
Am III	amide III vibration
ANS	1-anilino-8-naphthalenesulfonate
APD	avalanche photo diode
apo-LA	α-LA depleted of the protein-bound calcium ion
apoMb	apomyoglobin
A(r)	total signal (either absorbance or interference fringe)
ATR	attenuated total reflection
AUC	analytical ultracentrifugation
β-I	β-rich protein with a CD spectrum-like model β-polypeptides
β-II	β-rich protein with a CD spectrum-like unordered polypeptides
BSA	bovine serum albumin
C	cold-denatured protein
c	concentration (g/L)
CaM	calmodulin
CD	circular dichroism

Cdk	cyclin-dependent kinase
CH1	carboxy-terminal peptide of Histone H1
CLAM	cholinesterase-like adhesion molecules
CPMG echo	Carr–Purcell–Meiboom–Gill echo
CRL1	*Candida rugosa* lipase 1
Cro-WT	wild type λ Cro repressor
CSD	charge state distribution
CSD1	calpastatin
CTD	C-terminal domain
CTRD	C-terminal regulatory domain
CW	continuous wave
cyt	cytoplasmic domain
DBD	DNA-binding domain
DDW	doubly-distilled water
DF31	Drosophila protein of chromatin decondensation
DLA	diffusion limited aggregation
DLS	dynamic light scattering
DSC	differential scanning calorimetry
DT	diphtheria toxin
DTGS	deuterated triglycine sulphate
DUVRR	deep UV resonance Raman spectroscopy
$E_{0.1\%}$	extinction coefficient (mg/mL)/cm
EBNA	Epstein Barr virus nuclear antigen
EBV	Epstein Barr virus
EDTA	ethylenediaminetetraacetic acid
EFA	evolving factor analysis
EM	electron microscopy
EPR	electron paramagnetic resonance
ERD10	early responsive to dehydration 10
E:S	enzyme to substrate ratio
ESI	electrospray ionization
ESI-MS	electrospray ionization mass spectrometry
f	frictional coefficient
f/f_{min}	frictional ratio
FCS	fluorescence correlation spectroscopy
FFF	field flow fractionation
FID	free induction decay
FSD	Fourier self-deconvolution
FTIR	Fourier transform infrared
fwhm	full width at half maximum
GAPDH	glyceraldehyde 3-phosphate dehydrogenase
GdnHCl	guanidinium hydrochloride
GFP	green fluorescent protein
GlpO	α-glycerophosphate oxidase
GuCl	guanidinium chloride

H	heat-denatured protein
HBV	hepatitis B virus
HCV	hepatitis C virus
H/D	hydrogen-deuterium
HECT	homologous to E6-AP C-terminus
HEWL	hen egg white lysozyme
HFIP	hexafluoroisopropanol
Hfip	hexafluoro-2-propanol
hGH	human growth hormone
hNL3	human neuroligin 3
HN-NMR	heteronuclear NMR
HOPG	highly oriented pyrolytic graphite
HPLC	high-performance liquid chromatography
Hsp72	heat shock protein 72
ID	intrinsically disordered
IDP	intrinsically disordered protein
IUP	intrinsically unstructured protein
L	large protein
LEA	late embriogenesis abundant
L-P	polymerase complex
M	matrix protein
M	molar mass (g/mol)
MALDI	matrix-assisted laser-desorption ionization
MAP2c	microtubule-associated protein 2c
MBD	methylcytosine-binding domain
MCT	mercury cadmium telluride
MDS	molecular dynamics simulation
MeCP2	methylCpG-binding protein 2
MeOH	methanol
MES	2-(N-morpholino)ethanesulfonic acid
metMb	metmyoglobin (ferrimyoglobin)
MeV	measles virus
MoRE	molecular recognition element
MoRF	molecular recognition feature
MS	mass spectrometry
MTSL	methane thiosulfonate
η	viscosity
N	native protein
N	nucleoprotein
N_A	Avogadro's number
Na-TCA	sodium trichloroacetate
N_{CORE}	structured N-terminal domain of N
nm	nanometer
NMR	nuclear magnetic resonance
NMRD	nuclear magnetic resonance dispersion

Nrt	neurotactin
NOE	nuclear Overhauser effect
NQO1	NAD(P)H quinone oxidoreductase 1
NR	nucleoprotein receptor
Ntail	unstructured C-terminal domain of N
NTD	N-terminal domain
ODC	ornithine decarboxylase
ρ	density
P	phosphoprotein
P_{II}	poly(L-Pro)II conformation or helix
P9	$AcYEAAAKEAPAKEAAAKANH_2$
PABP	poly(A)-binding protein
PAGE	polyacrylamide gel electrophoresis
Paip2	PABP-interacting protein 2
PBS	phosphate buffer saline
PCNA	proliferating cell nuclear antigen
PCT	C-terminal domain of P
PDB	Protein Data Bank
PDC	pyruvate decarboxylase
PFG NMR	pulsed field gradient NMR
PGK	phosphoglycerate kinase
PMG	premolten globule
PNT	N-terminal domain of P
ProTα	prothymosin α
PrP82-146	prion peptide sequence from 82 to 146
PVM	pure variable method
r	radial position
R	gas constant
RdRp	RNA-dependent RNA-polymerase
REG	red blood cell regulator
R_H	hydrodynamic radius
RLA	reaction limited aggregation
R_{min}	minimum theoretical hydrodynamic radius
RNase A	ribonuclease A
RNase T1	ribonuclease T1
RNP	ribonucleoproteic complex
ROA	Raman optical activity
RP	reversed phase
rpm	revolution per minute
R_s	Stoke radius
SASA	solvent-accessible surface area
SAXS	small-angle X-ray scattering
SDSL	site directed spin labeling
SE	sedimentation equilibrium
SEC	size exclusion chromatography

SLS	static light scattering
SN	staphylococcal nuclease
S/N	signal to noise ratio
SPM	scanning probe microscopy
SSNMR	solid state NMR
SV	sedimentation velocity
t	time
T	absolute temperature
TAD	transcription activation domain
Tat	transactivator of transcription
TD	tetramerization domain
TEM	transmission electron microscopy
TFA	target factor analysis
TFA	trifluoroacetic acid
TFE	trifluoroethanol
ThDP	thiamine diphosphate
TM-AFM	tapping-mode atomic force microscopy
TMAO	trimethylamine-N-oxide
TOF	time-of-flight
TRD	transcription repression domain
Tris	tris(hydroxymethyl)-aminoethane
trx	thioredoxin
U	urea- or GuCl-denatured protein
V	volume
$\bar{v}$	partial specific volume
VCD	vibrational circular dichroism
wt	wild type
XD	X domain of P
YB-1	Y-box binding protein-1

PART I

ASSESSING IDPs IN THE LIVING CELL

1

IDPs AND PROTEIN DEGRADATION IN THE CELL

Yosef Shaul, Peter Tsvetkov, and Nina Reuven
Department of Molecular Genetics, Weizmann Institute of Science, Rehovot, Israel

ABSTRACT

The degradation of the majority of cellular proteins is mediated by the proteasomes. Ubiquitin-dependent proteasomal protein degradation is executed by a number of enzymes that interact to modify the substrates prior to their engagement with the 26S proteasomes. The 26S proteasome is made of two complexes, the 20S and the 19S. The role of 19S is to unfold the proteins to gain entry into the 20S particle, where the protein is cleaved into short peptides. Thus, some of the functions of the 19S complex are expected to be dispensable for degradation of intrinsically disordered proteins, or IDPs. Indeed, in cell-free systems, at least some of the IDPs are digested by 20S particles in the absence of the 19S. In fact, it appears that susceptibility to the 20S proteasome may represent a hallmark of IDPs. Recent evidence suggests that IDPs are susceptible to degradation *in vivo* by the 20S proteasome as well. The process is ubiquitin independent and takes place by default. However, the process of default degradation can be regulated by different strategies. The described process of IDP degradation provokes new predictions and explanations in the field of protein regulation and functionality.

Instrumental Analysis of Intrinsically Disordered Proteins: Assessing Structure and Conformation, Edited by Vladimir Uversky and Sonia Longhi

1.1 INTRODUCTION

The proteins that provide the machinery for growth and maintenance of living cells turn over by complex processes of synthesis and degradation. Thus, the extended polypeptide chain synthesized on the ribosome collapses to a compact molten globule (MG; 80), which subsequently acquires the three-dimensional (3D) structure of the functional protein, with or without the aid of molecular chaperones (49). However, many proteins contain extensive disordered regions, and some are even completely disordered under physiological conditions (42). These natively unfolded (136) or intrinsically disordered proteins (IDPs) are involved in many cellular processes, including transcription regulation and signal transduction (44, 132). Thus, as well formulated by Dyson and Wright (44), functional proteins fall on a structural continuum, ranging from tightly packed proteins, which are almost completely ordered, thus displaying well-defined tertiary structures, through proteins that contain both large-ordered and large-disordered stretches, a good example being the p53 tumor suppressor (15, 52), on to whole proteins such as the ~440-residue tau protein, which appears to be ID from its NH_2-terminus to its COOH-terminus (123). Furthermore, in multidomain proteins, tightly packed and ordered domains may be connected by ID flexible linker sequences, as is the case for a construct that contains the first three zinc fingers of transcription factor-IIIA (24).

Protein degradation plays an important role in almost every basic cellular process. The selective degradation of many short-lived proteins in the cell is mediated via the ubiquitin-26S proteasomal degradation pathway (54, 60). Ubiquitin, a 76-amino acid residue protein, is covalently conjugated in a highly regulated multistep process to the substrate protein, marking it for degradation by the 26S proteasomes. Recent discoveries in the field of protein degradation have revealed that several proteins are also susceptible to ubiquitin-independent degradation that is mediated via the core 20S proteasomes (10). New evidence raises the interesting possibility that IDPs are subjected to this pathway of degradation that does not require prior modification, a process we refer to as "degradation by default."

In this chapter, the evidence for IDPs degradation by the default mechanism, both *in vitro* and in the cells, is summarized. In addition, the biological outcomes and the unique contributions of IDPs' susceptibility to degradation by default are addressed. Given the fact that this concept is in its infancy, some of the proposed models call for additional experimental data.

1.2 PROTEASOMAL DEGRADATION

Protein degradation, a pivotal biochemical process, regulates many basic aspects of cellular behavior. Degradation of the majority of cellular proteins is carried out by the proteasomes (section 1.3). In the ubiquitin–proteasome degradation pathway, substrates are marked by covalent linkage to ubiquitin chains to be recognized and degraded by the 26S proteasome (54, 61, 63). Ubiquitination generally requires three proteins: the ubiquitin-activating enzyme (E1), the ubiquitin-carrier or conjugating enzyme (E2), and the ubiquitin ligase (E3). The E1 enzyme transfers the activated ubiquitin to one of many E2 carrier enzymes, which, in combination with one of many E3 ubiquitin ligases, transfer the ubiquitin monomer to the substrate. The E3 ubiquitin ligase binds to both the substrate and the E2 enzyme, assembling them in the correct position to allow efficient ubiquitin transfer. E3 ubiquitin ligases can either be single proteins or protein complexes. Ubiquitin is conjugated to its target protein in most cases by the formation of an isopeptide bond between its carboxyl group of glycine 76 (the carboxy terminus) and the epsilon amino group of one or more lysine residues or N-terminal amino group of the target protein (34, 61). Additional ubiquitin molecules are added to the lysine 48 residue of the previous ubiquitin in the chain. Specificity of the ubiquitination reaction is determined by the E3 ubiquitin ligase (61, 104).

Many cellular proteins are substrates of ubiquitin-mediated proteasomal degradation. The tumor suppressor p53 is one of the best studied substrates of the ubiquitin proteasome degradation pathway. Wild-type p53 is a short-lived protein that accumulates following various types of stress and induces growth arrest or apoptosis. The level of p53 is tightly regulated by the rate of its proteasomal degradation (138). A major regulator of p53 is Mdm2, an E3 really interesting new gene (RING) finger ubiquitin ligase that binds to the amino terminal transactivation domain of p53 and ubiquitinates p53 (59, 79). p53 is further polyubiquitinated either by p300 or Mdm2, and the polyubiquitinated p53 is eventually degraded by the 26S proteasomes. Mdm2 is a RING domain protein that binds and ubiquitinates p53, directing p53 to 26S proteasomal degradation. Mdm2 is transcriptionally upregulated by p53, thus establishing an autoregulatory negative feedback loop. p53 is also ubiquitinated by several other independent E3 ligases. In addition, it is ubiquitinated following its recruitment into a complex containing the viral E6 protein of the human papilloma viruses (HPVs) and the E6-associated protein, which serves as an E3 HECT domain ubiquitin ligase (117).

While ubiquitin-dependent proteasomal degradation is likely to represent the primary means of proteasomal degradation for most proteins, there are a number of proteins that have been demonstrated to undergo proteasomal degradation that is ubiquitin independent (reviewed recently by Jariel-Encontre et al. [65]). Ornithine decarboxylase (ODC) was the first protein found to undergo ubiquitin-independent degradation by the proteasome, and it is the most well studied. ODC is the first rate-limiting enzyme in polyamine biosynthesis that catalyzes the decarboxylation of ornithine to form putrescine. In its active and stable form, ODC is a homodimer with two enzymatic active sites. Binding of antizyme, a polyamine-induced protein, to ODC disrupts ODC homodimers and exposes the C-terminal region of ODC, inducing ubiquitin-independent 26S proteasomal degradation of ODC (37). The mammalian ODC carboxy-terminal 37 amino acids constitute a degron necessary and sufficient to destabilize ODC. This region is a molecular mimic of a polyubiquitin chain (143) and also provides an unstructured or loosely structured region from which degradation can initiate (126).

1.3 THE VARIOUS COMPLEXES OF THE PROTEASOMES

In eukaryotic cells, most proteins are degraded by the 26S proteasome, a huge molecular machine designed for controlled proteolysis. It degrades the majority of cellular proteins in eukaryotes. It controls the levels of various regulatory proteins and prevents the accumulation of misfolded mutant and damaged proteins (61). The 26S proteasome is comprised of a 19S regulatory complex and the 20S protealytic core. The symmetrical core 20S proteasome consists of 28 subunits arranged in a cylindrical form of four rings. Seven different alpha and beta subunits are assembled as four heptameric rings to form the 20S catalytic core particle. The rings are stacked on top of each other to form a hollow cylindrical structure (57). Beta subunits in the middle of the complex provide the proteolytic activity with little specificity for their substrates. Alpha subunits form the external rings of the cylinder to generate the entry gates (56). The 20S proteasome is in a latent state, since entry is gated, and therefore, the substrate entry is the rate-limiting step for hydrolysis.

The 19S regulatory complex is composed of about 18 polypeptides, six of which are ATPases. It associates with the small ends of the 20S core particle at the entrance to the degradation channel. The 19S complex can be dissociated into two subcomplexes, the “lid” and the “base.” The lid possesses the subunits for recognition of polyubiquitin chains. The base, containing the ATPase activity, binds and unfolds proteins, the “reverse chaperone” activity, to unfold substrates in preparation for their degradation by the 20S core (150). The base also controls the gating to the degradation channel (124). Together, the lid and the base of the 19S regulatory complex recognize the polyubiquitin chains and catalyze substrate deubiquitination, denaturation, and translocation into the 20S catalytic core for degradation.

Correctly folded globular proteins in their “native” conformation should not be readily susceptible to 20S proteasomal attack unless denatured by the regulatory 19S subunit (68, 140). Interestingly, the proteolysis of tightly folded proteins by the 26S proteasome is greatly accelerated when an unstructured region is attached to the substrate (107). It has been proposed that the unstructured region acts as a site for initiation of unfolding. This region, which may lie at the terminus of the protein or constitute a surface loop, is the first portion of the substrate to enter the proteasome, a process that improves substrate unfolding. Thus, a ubiquitin chain is sufficient for proteasome docking, but unfolding cannot be initiated efficiently in the absence of an unstructured initiation site.

The proteasome 20S core particle in isolation is active in degradation of certain proteins. It has been proposed that unfolded protein substrates are fed into the 20S proteasome via their gates as extended chains and are degraded progressively (93, 139). Liu and coworkers, however, have provided evidence for susceptibility of internal disordered regions to endoproteolytic activity of the proteasome (84), suggesting that a disordered protein can be degraded by the proteasome in a unique manner as discussed below.

There are two additional and smaller protein complexes, PA28α/β (REGα/β) and PA28γ (REGγ), that belong to the REG or 11S family of proteasome activators that have been shown to bind and activate the 20S proteasome based on their ability to promote the degradation of model peptide substrates. The REG complex is an ATP-independent and ubiquitin-independent proteasome activator that strongly enhances the catalytic activity of proteasomes *in vitro* (110). REGs are rings composed of seven subunits that bind the alpha rings (the gates) of the 20S proteasome and activate peptide hydrolysis. Of the three REG family members, REGα and REGβ subunits form a heteroheptameric complex, whereas REGγ forms a homoheptameric complex with a molecular mass of about 200 kDa (111). REGγ has a nuclear-restricted expression pattern and can be found independently or associated with 20S proteasomes. REGγ is found in worms, insects, and higher animals, whereas REGα and REGβ are found only in vertebrates (86). Originally, REGγ had been only characterized in terms of its ability to degrade small peptide model substrates *in vitro*, and it was initially thought that REGγ is not involved in the destruction of intact endogenous proteins. As discussed below, REGγ directs degradation of proteins by the 20S proteasome in an ATP-independent and ubiquitin-independent manner.

1.4 IDPs AND DEGRADATION BY 20S PROTEASOME *IN VITRO*

The architecture of the proteasome suggests that correctly folded globular proteins in their "native" conformation are too big to enter the gates of the 20S proteasome and would not be readily susceptible to 20S proteasomal attack unless denatured by the regulatory 19S subunit. It has been proposed that structured substrates are denatured prior to being fed into the 20S gates as extended chains to progressively undergo degradation (93, 139). However, IDPs override this requirement. For example, p21 and α-synuclein are unstructured proteins and are directly susceptible to uncapped 20S proteasomal degradation (21, 131), possibly by partial opening of the 20S gates.

The differential behavior between IDPs and the globular proteins can be easily demonstrated *in vitro* where only IDPs are susceptible to the 20S proteasome digestion (135). This is also the case with partially disordered proteins, such as p53 (see section 1.7). IDPs are completely degraded to smaller fragments. However, certain motifs may be resistant to degradation as was reported for the polyglutamine chain (137).

Well-defined preparations of proteins in a long-lived MG state, in which the protein is devoid of most of its tertiary structure, but retains its secondary structure, and is still quite compact, are also resistant to the 20S proteasome. The MG species are significantly more flexible than the corresponding native species, based on their capacity to undergo disulfide reshuffling, as well as their being highly susceptible to proteinase K. Nevertheless, they apparently do not present sufficiently extended unstructured sequences to the 20S proteasome to permit their penetration into its cavity so as to undergo peptide bond cleavage (135). Given that folded proteins are resistant to proteinase K and 20S digestion, that IDPs are sensitive to both 20S and to proteinase K digestion, and that proteins in the MG state are sensitive to proteinase K degradation only, the utilization of these two digestion protocols in tandem provides a convenient operational tool for discriminating between the three states (135). The susceptibility to 20S proteasomal degradation is a simple, rapid, and reliable method of operationally defining IDPs and unstructured sequences. Furthermore, utilization of this method could lead to the characterization of motifs that sensitize IDPs to proteasomal degradation.

Certain structured proteins contain one or more disordered regions. The question is whether there is a specific requirement for the position of the disordered region in order to be sensitized to degradation *in vitro* by the 20S proteasome. Location of a disordered region either at the N-terminus or the C-terminus facilitates proteasomal degradation in general (107). Interestingly, however, the susceptibility to 20S proteasomal digestion was maintained even when these IDPs were hooked at both N and C-termini to GFP, a highly structured protein, suggesting that internal disordered regions are susceptible to endoproteolytic activity of the proteasome (84). Thus, the location of the disordered region seems not to follow a simple rule.

The ability of the core 20S proteasomes to degrade primarily unstructured proteins and protein regions can be utilized as an assay for the identification of such protein regions. The protein of interest can be incubated *in vitro* with purified 20S proteasomes, and the degradation products can be analyzed with the expectation that the ID region will be degraded first. Furthermore, in cases when a protein folds upon binding to its biological counterpart, such an assay can be a valuable tool for identifying the interacting target protein or molecule as these are expected to rescue the protein from the 20S proteasomal degradation. More importantly, such a biological assay enables us to examine whether a suspected candidate protein is degraded by the 20S proteasomes and protected upon binding to its biological counterpart.

1.5 *IN VITRO* STABILIZATION OF IDPs

It has been estimated that 36–63% of the proteins in eukaryotic proteomes contain long sequences that are ID *in vitro* (43). This does not imply that these IDPs, or protein sequences, are necessarily disordered *in vivo*. On the contrary, in many cases, convincing evidence has been presented that they form functional complexes within the cell, adapting a well-defined conformation upon interaction with their partner(s). Based on this observation, it was predicted that protein–protein interactions can protect IDPs from 20S proteasomal degradation. This may be achieved either by masking the unstructured domain or by folding it. We have recently demonstrated the validity of this principle experimentally (135). It has further been argued that their ID character may enable IDPs to form relatively low-affinity complexes, thus permitting plasticity and rapid turnover, and, consequently, to serve as hub proteins interacting with multiple partners (41). These kinds of interactions may be enough to protect IDPs from 20S proteasomal degradation.

The Cdk inhibitor p21, a naturally unstructured protein, is a critical regulator of cell division and DNA replication (21). In the cells, p21 proteins are rarely found in a "free" form and are almost always detected as a part of a complex, raising the possibility that the "free" p21 is rapidly degraded in the cells. *In vitro* degradation studies with p21 show that this protein is degraded by the 20S proteasomes in a ubiquitin-independent manner. Degradation of p21 is inhibited upon binding of p21 to the proliferating cell nuclear antigen (PCNA) or in the presence of the cyclin E and Cdk2 complex (20, 38, 134). p21 is just one example of an IDP that is stabilized by its binding partners. More examples exist, as in the case of p53 and other proteins discussed below and specified in Table 1.1.

1.6 UBIQUITIN-INDEPENDENT PROTEASOMAL DEGRADATION *IN VIVO*

As discussed in section 1.2, recognition of substrates by the 26S proteasome is typically accomplished through ligation of a polyubiquitin chain to one or more internal lysine residues or to the N-terminal amino group of a target substrate. Once bound to the proteasome, the polyubiquitinated substrates must be unfolded and inserted into the proteasome's catalytic chamber, where proteolysis occurs. As discussed in this chapter, a growing number of IDPs are subjected to proteasomal degradation without prior ubiquitination. Experimental strategies were formulated to demonstrate that proteasomal degradation of a given protein is neither ubiquitinated nor dependent on ubiquitination. The key experiments are:

1. Biochemical analysis ruling out the possibility that ubiquitin moieties are attached to the substrate.
2. Demonstration that polyubiquitinated substrates do not accumulate in cells treated with known chemicals that inhibit proteasomes. However, certain proteins are subjected to both ubiquitin-dependent and ubiquitin-independent pathways; therefore the fact that protein is ubiquitinated does not rule out the possibility that it is subjected to the other

TABLE 1.1 IDPs Suspected to Undergo Degradation by Default

Protein	Ubiquitin-Independent Proteasomal Degradation	26S/20S *In Vitro* Degradation	Binding Stabilizes	Unstructured
α-Catenin	(64)	—	—	FoldIndex† (255 disordered residues, 11 segments)
α-Synuclein	(131)	(84, 131)	—	DisProt (completely)
Calmodulin	(128)	(16, 48, 128)	Ca^{2+} (16, 48, 128)	DisProt (77–81)
Casein	—	(73)	—	DisProt (completely)
C/EBPδ	(149)	—	(149)	FoldIndex (207 disordered residues, 4 segments)
c-Fos	(22)	—	—	DisProt (216–380)
Cholera toxin	—	(98)	CTA2 (98)	FoldIndex (132 residues, 4 segments)
c-Jun	(66)	(66)	—	DisProt (185–235)
DHFR	—	(2)	MTX (2)	DisProt (43 residues, 5 segments)
eIF4F and eIF3	(14)	(14)	—	DisProt (393–490)
Fra-1	(13)	—	—	High homology to c-Fos
GRK2	—	(102)	—	FoldIndex (367 residues, 6 segments)
HIF-1α	(74)	(74)	—	(115)
IκBα	(76, 87, 103)	(1, 78, 87)	p65 (1, 76, 87, 94)	DisProt 1–66, 276–317)
NF-κB (p65)	(91)	(91)	—	FoldIndex (167 residues, 8 segments)
ODC	(114)	(4, 17, 18, 36)	NQO1 (4)	DisProt (1–35, 158–165, 297–311, 412–425)
$p14^{ARF}$	—	(105)	TBP-1 (105)	DisProt $p19^{ARF}$ (1–37)
p16	(30)	(30)	—	(127)
p21	(119) REGγ (30, 82)	(30, 82, 84, 134)	PCNA (134) Cdk/cyclin (82) cyclinE/ Cdk2 (30)	DisProt (completely)

Table 1.1 Continued

Protein	Ubiquitin-Independent Proteasomal Degradation	26S/20S *In Vitro* Degradation	Binding Stabilizes	Unstructured
p33[ING1b]	(53)	(53)	NQO1 (53)	FoldIndex (165 disordered residues, 2 segments)
p53	(7, 27)	(11, 55)	NQO1 (11) NQO2 (55)	(15)
p73	(11)	(11)	NQO1 (11)	DisProt* (549–564)
Pertussis toxin S1	—	(97, 98)	NAD^+(97)	FoldIndex (70 residues, 6 segments)
PrP	—	(129)	—	DisProt (23–120)
Rb	(70, 118)	(118)	—	FoldIndex (319 disordered residues, 12 disordered segments)
RPN4	(69)	—	—	FoldIndex (394 disordered residues, 5 segments)
SRC-3	(83)	(83)	—	FoldIndex (831 residues, 30 segments)
Tau	—	(28, 39, 106, 145)	Hsc70, BAG-1 (45)	DisProt (completely)
Thymidylate synthase	(51)	—	—	(51)
Troponin C	—	(16)	Ca^{2+} (16)	DisProt (103–158)
YB-1	(125)	(125)	RNA (125)	FoldIndex (246 residues, 2 segments)

Note: This table provides a partial list of proteins suspected to undergo degradation by default as defined by parameters in section 1.10. References are provided for: UI degradation *in vivo*, 20S/26S proteasomal degradation *in vitro*, stabilization by binding to a protein or co-factor and intrinsic disorder. In addition to specific citations regarding structure, the DisProt database and the FoldIndex program were used to identify proteins listed here as disordered.

*The Database of Protein Disorder (DisProt) is a curated database that provides information about proteins that lack fixed 3D structure in their putatively native states, either in their entirety or in part. DisProt is a collaborative effort between the Center for Computational Biology and Bioinformatics at Indiana University School of Medicine and the Center for Information Science and Technology at Temple University (122).

† FoldIndex©: a tool to predict whether a given protein sequence is intrinsically unfolded (108).

CTA2, cholera toxin A2 polypeptide; DHFR, dihydrofolate reductase; GRK2, G protein-coupled receptor kinase 2; HIF, hypoxia-inducible factor; MTX, methotrexate; RPN4, Regulatory Particle Non-ATPase.

mode of degradation. Also, since lysine is the site of ubiquitin ligation in most molecules targeted for proteasomal degradation, "lysine-less" substrates, usually with lysine residues replaced by arginine, are used to show that degradation is not compromised. However, the maintenance of normal protein degradation in "lysine-less" mutants does not rule out the possibility that ubiquitin ligation occurs at the N-terminal amino group (20). Therefore, it is important to show, often by MS measurements, that the N-terminal end is blocked by N-acetylmethionine. For these reasons, these methods are insufficient to show ubiquitin-independent degradation for proteins that are susceptible to both pathways.

3. Genetic manipulation offers another method to demonstrate the ubiquitin-independent nature of the proteasomal degradation. A conventional tool is a cell line that bears a temperature-sensitive E1 mutant, such as the A31N-ts20 cell line (33). E1 is the first enzyme in the consecutive steps of protein ubiquitination, and under the restrictive condition (i.e., high temperature), the primary vertebrate E1, encoded by Ube1, is inactivated, and therefore, proteasomal degradation under this condition provides biological evidence for ubiquitin-independent degradation of the substrate. Using these approaches, an increasing number of proteins have been shown to be degraded by the proteasome in a ubiquitin-independent manner, including ODC (4), inhibitory κBα (IκBα) (76), p53 (7), retinoblastoma protein (Rb) (70), T-cell receptor α (142), c-Jun (66), calmodulin (128), and thymidylate synthase (TS) (51, 101). For some of these, degradation is completely independent of ubiquitination, while for others, it occurs through both ubiquitin-dependent and ubiquitin-independent pathways. Recently, an alternate ubiquitin-activating E1, called Ube1L2, Uba6, or E1-L2, has been discovered (32, 67, 100). Although this E1 appears to interact with a different subset of E2s than Ube1, this finding demonstrates the complexity of the ubiquitin-proteasome system and the need to use several experimental systems to prove that a particular protein is being degraded in a ubiquitin-independent manner.

TS is an S-phase enzyme involved in thymidine 5′-monophosphate synthesis. Recently, it was shown that degradation of human TS is carried out by proteasomes in a ubiquitin-independent fashion (51, 101). The N-terminal region of the polypeptide was identified as a primary determinant of this degradation (51). The region is disordered in X-ray crystallographic structures (29). Deletion of as few as six amino acids from the N-terminal end of the molecule elicits marked stabilization of the enzyme, with further deletions resulting in varying degrees of stability (51). Thus, the disordered N-terminal domain seems to mediate recognition by the proteasome and to target the polypeptide for proteolytic destruction. This domain is sufficient to destabilize an evolutionarily distinct TS molecule, indicating that it functions as an independent degradation signal (101).

As exemplified by ODC degradation, a ubiquitin-independent pathway is executed by the 26S proteasome and not necessarily by the 20S proteasome. However, in that case, a second protein, antizyme, is required for targeting of ODC to the 26S particle. However, the ODC case appears to be an exception, since the emerging picture is that the 26S proteasome is not the catalytic particle in many other cases of ubiquitin-independent degradation. As discussed below, other particles, such as 20S either in isolation or in association with REG or other proteins, may provide the catalytic domain in the process of ubiquitin-independent degradation.

1.7 *IN VIVO* DEGRADATION OF IDPs

An important question is whether IDPs are present as such within the living cell. By analogy with the *in vitro* scenario, it might be possible to address this question by the mode of degradation to which IDPs are subjected. A growing number of proteins bearing disordered regions are reported to undergo ubiquitin-independent degradation. However, we are not yet in the position to say that a given protein that undergoes ubiquitin-independent degradation has to contain disordered regions. As discussed below, another angle is susceptibility to the 20S proteasome or related particles that lack the 19S particle with the "reverse chaperone" activity. Susceptibility to the 20S proteasome *in vitro* may be extended in the future to address the issue of the status of IDPs *in vivo*, either by inducing the 20S proteasomal degradation or by selectively blocking 26S proteasome activity, but not that of the uncapped 20S proteasome. Proteins found to undergo proteasomal degradation under such conditions are likely to have retained their ID status *in vivo*. In addition, IDPs are protected from 20S proteasome digestion *in vitro* when they are in association with a partner. This principle of "interact to protect" may be applied for *in vivo* analysis as well to further demonstrate the ID nature of a given protein. Furthermore, in the future, this kind of assay should enable identification of novel binding partners of a given IDP.

An example is the tumor suppressor p53, which is a short-lived protein that accumulates following exposure to different types of stress and induces cell cycle arrest or apoptosis. Regulation of p53 stability plays a central role in the control of proper function of p53. Degradation of p53 has been intensively studied, and indeed, p53 has become a hallmark for ubiquitin-dependent 26S proteasomal degradation. Several specific E3 ubiquitin ligases were reported to bind and polyubiquitinate p53, marking it for degradation by the 26S proteasomes (see references 7–10 in Asher et al. [10]; 40, 59, 79, 81). Both the p53 N-terminal transactivation domain and the C-terminal regulatory domain were identified as unstructured regions (15). These unstructured regions may facilitate and serve as a signal for degradation by the core 20S proteasomes. Interestingly, p53 is also susceptible to degradation in a ubiquitin-independent manner in the cells that are mediated via the core 20S proteasomes (5–9, 11).

The c-Fos proto-oncogene is a component of activator protein-1 (AP-1) transcription factor regulating numerous cellular processes. c-Fos acts as an obligatory heterodimer with members of basic region leucine zipper proteins from the JUN, ATF, and MAF families (31, 62), and binds to specific DNA motifs via a basic domain N-terminally adjacent to the leucine zipper domain. The C-terminal half of c-Fos is ID (26), and *in vitro* c-Fos is susceptible to 20S proteasome digestion (see Jariel-Encontre et al. [65]). This region is involved in ubiquitin-independent degradation of c-Fos (22). Fra-1, a related protein, appears to behave in a similar manner. Indeed, enhanced GFP is destabilized by a 40-amino acid C-terminal fragment of Fra-1 (13). Alternative pathways may contribute to c-Fos degradation in different subcellular compartments, in the nucleus ubiquitin-independent but in the cytoplasm ubiquitin-dependent (116). It will be interesting to investigate whether or not ubiquitin independence in the nucleus may be attributed to the nuclear REGγ-20S complex.

These examples, and the others listed in Table 1.1, suggest that two distinct proteasomal pathways can degrade IDPs *in vivo*. The ubiquitin-dependent pathway is a common feature of many proteins including long-lived structured proteins in addition to IDPs. In contrast, ubiquitin-independent proteasomal degradation, with the exception of ODC, has been mainly shown to degrade IDPs, raising the possibility that the ubiquitin-independent degradation is a specific degradation pathway common to IDPs.

1.8 IDPs AND *IN VIVO* EVIDENCE FOR DEGRADATION BY THE 20S PROTEASOME

IDPs are subjected to degradation by the conventional ubiquitination pathway involving ATP-dependent substrate denaturation by the 19S cap particle. However, as summarized above, a growing number of proteins undergo degradation in a ubiquitin-independent manner. Given the fact that IDPs are unstructured, ATP-dependent substrate denaturation by the 19S particle is dispensable. Therefore, in principle, IDPs may undergo degradation by the 20S proteasome. To date, it is rather difficult to distinguish the activity of the 26S from that of the 20S in the cells. A major problem is that the known chemical inhibitors of the 26S proteasome inhibit 20S as well. This introduces confusion in the field since many reports have demonstrated the involvement of the 26S proteasome in their system by simple utilization of these inhibitors, ignoring the possibility of coinhibition of the "free" and uncapped 20S proteasome as well. Therefore, in this chapter, we prefer to refer to these cases as 26S/20S proteasome. This will include any proteasome containing the 20S core particle regardless of the presence and the nature of the cap particle. One strategy to determine whether the uncapped 20S proteasome is responsible for IDPs degradation is to eliminate subunits from the 19S particle and, therefore, to reduce or eliminate 26S proteasomes in the cells. This can be achieved by knocking down crucial subunits of the 19S regulatory complex that would prevent the association of the 19S complex to the 20S proteasome. In addition, proteasomal ATPase-associated factor 1 (PAAF1) protein has been shown to associate with the ATPases of the 19S regulatory complex, and overexpression of PAAF1 interferes with the association of the 19S with the 20S proteasomes (99). Thus, PAAF1 can increase the level of 20S proteasomes, and susceptibility to PAAF1 would provide an additional way to determine the involvement of the uncapped 20S proteasome in the process. We anticipate that these strategies will help answer the question as to whether the 20S proteasome degrades IDPs *in vivo*.

Another possible strategy would be to find the proper physiological settings to alter the 20S/26S ratio. For example, starvation to glucose increases the relative amount of the uncapped 20S proteasome in yeast cells and changes proteasomal distribution in mammalian cells (12, 96). Oxidative stress was also shown to alter the activity of the 20S/26S proteasomes (112, 113). However, at the moment, it is not clear what physiological settings elevate the 20S/26S proteasome ratio in mammalian cells.

There are several examples of IDPs that are degraded *in vitro* by 20S proteasomes and *in vivo* in a ubiquitin-independent manner that does not require interaction with E3 ubiquitin ligase or any other tagging such as polyubiquitination. For example, the degradation of p53 and its related family member p73 both exhibit these characteristics (11). Because susceptibility to this degradation is determined by the inherent characteristics of the protein, we refer to it as degradation by default (see section 1.10). By analogy with the *in vitro* system, our model is that *in vivo* degradation by default is carried out by the uncapped 20S proteasomes. The methods delineated above will be useful in definitively establishing that the 20S proteasomes are responsible for default degradation. The present *in vitro* and *in vivo* data at hand suggest that this is the case.

ODC is an interesting example since it undergoes ubiquitin-independent degradation via both 26S and 20S proteasomes. The former is true for the homodimer complex, whereas the latter is true for the ODC monomer (4). Degradation of ODC monomers by the 20S proteasomes does not require tagging (such as polyubiquitination) or protein–protein interaction, suggesting that these monomers are inherently unstable and are degraded by the 20S proteasomes in the cells. In contrast, ODC dimers are resistant to degradation by the 20S proteasomes. Remarkably, antizyme that binds ODC also sensitizes ODC to degradation by the 26S proteasome but inhibits it from degradation by the 20S proteasome (4). This switching mechanism suggests a cross talk between these distinct pathways.

1.9 IDPs AND *IN VIVO* DEGRADATION, THE ROLE OF REGγ

As detailed in section 1.3, in the cells, some of the 20S particles are capped with two smaller protein complexes, PA28α/β (REGα/β) or PA28γ (REGγ), both members of the 11S family of proteasome activators. The REG complex is an ATP-independent and ubiquitin-independent proteasome activator that strongly enhances the catalytic activity of proteasomes *in vitro* (110). REGγ has been only characterized in terms of its ability to degrade small peptide model substrates *in vitro*, but recently, evidence was provided for REGγ to direct degradation of proteins by the 20S proteasome in an ATP-independent and ubiquitin-independent manner.

REGγ interacts with and enhances the proteolysis of several proteins such as HCV core protein (92), steroid receptor coactivator-3 (SRC-3) protein (83), p21 (30, 82), $p19^{ARF}$, and p16 (30). REGγ enhances 20S proteasome-mediated degradation of these proteins in a ubiquitin-independent and ATP-independent manner. Given the fact that this process does not require the "reverse chaperone" activity of the 19S, it has been speculated that IDPs are the natural substrates of the REGγ-activated 20S proteasome (83). $p19^{ARF}$, p21, and p16 are all IDPs when not associated with specific binding partners (such as cyclins and Cdks, for p21 and p16, and nucleophosmin in the case of $p19^{ARF}$). Moreover, although monomeric p21 is unstructured, the great majority of p21 *in vivo* is highly structured within p21–cyclin–Cdk complexes. Therefore, the susceptibility to degradation by this pathway is likely to be confined to the monomer. Certain truncation mutants of p21, which are predicted to be unstructured like the full-length p21, are not substrates of the REGγ pathway. Therefore, unfolded structure might be necessary but not sufficient.

1.10 IDPs AND DEGRADATION BY DEFAULT

In eukaryotes, the conventional model of ubiquitin-dependent proteasomal degradation argues that the regulatory step is in the hands of the E3 ligase group of enzymes (see above). A given protein is stable unless polyubiquitinated by a specific E3 ligase, a modification that targets the modified protein to the 26S proteasome, the major protein destructive machine. In addition to this conventional model of "modification to degradation," the IDPs also seem to undergo degradation without prior modification. Furthermore, unlike the ODC case, IDPs do not require an antizyme-like entity to target the substrate to the proteasome. Therefore, IDPs appear to be inherently susceptible to uncapped 20S degradation. We regard this mechanism as degradation by default, that is, the protein is digested by the 20S proteasome unless it actively escapes this process, such as through interaction with a partner. To reach the conclusion that a protein is subjected to proteasomal degradation by default, the following observations have to be made:

1. The protein is digested *in vitro* by the 20S proteasome.
2. The *in vitro* degradation is blocked in the presence of an interacting protein.

3. In the cells, the protein is degraded by proteasomes in the absence of a functional ubiquitin system.
4. The half-life of the protein is increased under high expression of an interacting protein.
5. Knocking down the partner protein destabilizes the substrate.

The definition is operational at the moment, and in the future, new strategies are expected to be developed. For example, at least some of the proteins, such as p53, are stabilized by NQO1, an NADH-dependent enzyme that is in association with the 20S proteasomes (see below). We have been able to show that the p53 degradation process meets all the experimental criteria of degradation by default. There are many other proteins that meet at least some of these criteria (as listed in Table 1.1), and additional experiments are needed for their better classification. Degradation by default is a regulated process, and some of the 20S-associated proteins are suspected candidates for accelerating or inhibiting the process. In addition, when our knowledge of the molecular basis of substrate targeting to the uncapped 20S proteasome is improved, we will be in a better position to identify the substrates of the process of degradation by default.

This operational definition can be used to reevaluate some of the published observations. For example, the protein α-catenin undergoes ubiquitin-independent proteasomal degradation, but this process is blocked by the interacting β-catenin (64). Therefore, in accordance with our experimental criteria, α-catenin is likely to undergo degradation by default. To fulfill all the criteria, we have to demonstrate that α-catenin is a substrate of the uncapped 20S proteasome. Although no experiments were performed to meet these criteria, some hints are provided by sequence analysis. Using different algorithms to predict the structure of α-catenin, we found that this protein is expected to contain several unstructured regions and therefore might be a putative substrate of the uncapped 20S proteasomes in the cells. This prediction can be easily challenged by using the *in vitro* assay described above.

1.11 PROTEASOME TARGETING OF IDPs

Proteasome targeting of proteins is a crucial step in degradation. In the ubiquitin-dependent proteasomal degradation pathway, the 19S cap particle binds to the polyubiquitin chains and catalyzes substrate deubiquitination, denaturation, and translocation of the unfolded substrate into the 20S catalytic core for degradation. A major role of polyubiquitination is to target proteins to proteasomes. The actual degradation step requires that ubiquitinated proteins first be denatured and deubiquitinated, to allow them to fit into the narrow channel of the core particle. The polyubiquitin chain is not crucial for proteasomal degradation per se as has been first demonstrated for ODC (18). Thus, the substrates for the actual proteolysis are unfolded and nonubiquitinated.

Therefore, if a protein can be delivered to the proteasome in a denatured or partially unfolded state, ubiquitination might be required for substrate targeting but not for its degradation.

Direct binding of the substrates to the uncapped 20S via its different subunits could provide a possible mechanism for targeting of IDPs for proteasomal degradation. Consistent with this model, it has been shown that p21, a substrate of uncapped 20S proteasome, binds α7, a 20S subunit (134). The authors proposed that this physical interaction is responsible for p21 targeting to the proteasome for degradation. However, although p21 can bind to isolated α7, it has not been possible to detect p21 in complex with the uncapped 20S proteasome (134). More importantly, a p21 (1–82) mutant lacking the binding to α7 is also readily degraded by purified proteasomes (20). Interestingly, the same six amino acids in p21 that are necessary for binding the α7 subunit are required for binding REGγ (30). Although the relationship between the binding of p21 to both α7 and REGγ remains to be defined, it is possible that these reflect concerted steps in the pathway of p21 degradation, namely, targeting and entry to the catalytic chamber.

Interestingly, interaction with the 20S α7 subunit is important for degradation of other proteins as well. The Rb plays a critical role in the development of human malignancies. Based on the available prediction algorithms, Rb is a highly unstructured protein. Mdm2 promotes Rb proteasomal degradation in a ubiquitin-independent manner by promoting Rb interaction with the α7 subunit of the 20S proteasome (118). This model attributes to Mdm2 a role of adaptor protein that acts to target Rb to the 20S proteasome. However, the human cytomegalovirus has developed an alternative mechanism to target Rb to the proteasome (see below). The 20S proteasome α7 subunit seems to be a preferred subunit for substrate targeting. Whether this subunit facilitates degradation of the 20S proteasome client proteins is an important possibility that would improve our understanding of the mechanism of 20S proteasome targeting and activation.

As discussed below, another emerging mechanism of 20S proteasome targeting was recently described. The enzyme NQO1 selectively associates with the uncapped 20S proteasome. Furthermore, NQO1 interacts with a number of IDPs. It is therefore likely that at least certain IDPs are targeted to the 20S via NQO1. As it is clarified below, NQO1 also protects the substrate from the 20S proteasomal digestion (4, 11).

1.12 IDPs STABILIZATION VIA PROTEIN–PROTEIN INTERACTION

An intuitive conclusion from the model that IDPs are prone to degradation by default is that IDPs are, in general, less stable proteins. However, there are mechanisms to protect IDPs from degradation by default. A large body of evidence suggests that IDPs escape degradation by default when they are in association with a partner, possibly by masking the disordered region. This process might provide a mechanism for supporting the formation of functional complexes and at the same time, for eliminating excess of "free" subunits that might interfere with the function of these complexes. A number of cases are described here, but the literature is loaded with many more cases.

The p21 IDP is rarely found in a "free" form and is often detected as a part of a complex. A likely explanation is derived from the fact that degradation of p21 is inhibited upon binding of p21 to the PCNA or in the presence of the cyclin E and Cdk2 complex pushing the equilibrium toward complex formation (20, 38, 134). How p21 is protected is an important issue. It is very likely that p21 is in a more defined 3D structure while engaging into complexes, but some alternative explanations should also be considered. For example, p21 is targeted to the 20S proteasome via interaction with the α7 subunit. Thus, inhibition of p21 degradation via masking of the α7-binding site may represent an alternative mechanism of p21 stabilization. Consistent with this possibility is the fact that PCNA binds p21 at the region very close to that considered to be critical for α7 binding (134). A similar mechanism was proposed for cyclin D1-dependent p21 stabilization (38). This may also explain how another p21-interacting protein, CCAAT/enhancer binding protein alpha (C/EBPα), stabilizes p21 (130).

Another example is the case of IκBα, a major regulator of the NF-κB transcription factor. IκBα is unstructured at its N and C-termini (summarized in the DisProt database [122]). Following exposure to different types of stimuli, IκBα is phosphorylated and rapidly degraded via the ubiquitin 26S proteasome degradation pathway (76). Independent of its rapid stimulation-induced breakdown, IκBα is inherently unstable and undergoes continuous turnover. Analysis of both the basal degradation and the stimulation-induced breakdown of IκBα revealed that although in both cases the degradation is mediated via the proteasome, these two pathways of degradation are distinct. In contrast to the stimulation-induced breakdown, the basal degradation of IκBα does not require polyubiquitination of IκBα (76). Interestingly, expression of p65, which interacts with IκBα, significantly reduces the basal turnover of IκBα. Several studies have shown that IκBα is degraded by the uncapped 20S proteasomes in a ubiquitin-independent manner and that p65 can protect IκBα from this degradation (1, 78, 87).

In these last two examples, the basal degradation of both p21 and IκBα is ubiquitin independent and mediated via the core 20S proteasomes, whereas ubiquitin-dependent degradation occurs only upon exposure to certain physiological stimuli. These examples further suggest that the degradation of certain proteins through the ubiquitin-independent 20S degradation pathway depends upon the inherent quality of the substrate protein, whereas ubiquitin-dependent degradation is an active process that promotes increased levels of degradation. It is our assumption that assembly of substrate proteins into large protein complexes protects these substrates only from the 20S proteasomal degradation.

The $p14^{ARF}$ tumor suppressor is a key regulator of cellular proliferation, frequently inactivated in human cancer. ARF can activate the p53 tumor surveillance pathway by interacting with and inhibiting the p53 antagonist Mdm2 (120). $p14^{ARF}$ is dynamically disordered in aqueous solution and becomes structured upon binding to Mdm2 (23). $p14^{ARF}$ is degraded *in vitro* by the 20S proteasome in the absence of ubiquitination. $p14^{ARF}$ also binds transactivator of transcription (tat) binding protein 1 (TBP-1). TBP-1 inhibits ubiquitin-independent $p14^{ARF}$ degradation both *in vitro* and *in vivo* (105). By the experimental parameters outlined in this chapter, it is very likely that the IDP $p14^{ARF}$ undergoes degradation by default.

1.13 NQO1 REGULATES IDPS DEGRADATION BY DEFAULT

The process of degradation by default must be blocked under certain conditions to allow protein accumulation on demand. As discussed above, large protein complexes are not susceptible to 20S proteasomal degradation; however, this is not the only means of protecting proteins from degradation by default. A novel mechanism of protection is regulated by NQO1, a ubiquitous enzyme that utilizes NADH to catalyze the reduction of various quinones. NQO1 binds a subset of short-lived proteins including p53 and p73α (11), ODC (4), $p33^{ING1b}$ (53), p21, and Fos (unpublished observations). All these proteins bear ID regions and are subjected to degradation *in vitro* by the 20S proteasome and *in vivo* to degradation by both ubiquitin-dependent and ubiquitin-independent mechanisms. This set of proteins that undergo degradation by default is protected by NQO1, with the exception of p73β, which does not bind NQO1 (11). Binding of NQO1 to at least some of these proteins is augmented in the presence of NADH and inhibited by dicoumarol, an inhibitor of NQO1, which competes with NADH (11). In the case of the tumor suppressor $p33^{ING1b}$, phosphorylation of the protein in response to genotoxic stress increases its association with NQO1 and its half-life (53). Inhibition of NQO1 by dicoumarol or NQO1 knockdown with specific NQO1 siRNA induces ubiquitin-independent degradation of these proteins. *In vitro* degradation assays further confirmed that NQO1 together with NADH selectively protects p53, p73α, and ODC from 20S proteasomal degradation, and dicoumarol reduces the protection. The binding of NQO1 to the 20S proteasomes and the ability of NQO1 to bind and protect a subset of short-lived proteins from uncapped 20S proteasomal degradation suggest that NQO1 is an important regulator of degradation by default of IDPs.

1.14 NQO1 AS A 20S PROTEASOME GATEKEEPER

NQO1 is in association with the 20S core particle where it plays the function of "gatekeeper" by regulating the degradation of certain substrates (11).

Purification of the 20S proteasomes from mice livers (11) and human red blood cells (unpublished data) shows that NQO1 is physically associated with the 20S proteasomes but not with the 26S proteasomes. As described above, NQO1 interacts with certain 20S substrates such as p73α, p53, and ODC. Interestingly, NADH regulates the association of NQO1 with the substrates but not with the 20S proteasome. At high levels of NADH, the substrates are protected and do not enter the 20S catalytic chamber. At low levels of NADH, the substrates are not effectively protected and are degraded by the proteasome. Certain small drugs, like dicoumarol, by competing with NADH, abolish the NQO1–client substrate interaction and, therefore, sensitize these proteins to degradation by the 20S proteasomes. Dicoumarol does not alter the binding of NQO1 to the 20S proteasomes. Indeed, treatment of the 20S–NQO1–p53 ternary complexes with dicoumarol resulted in dissociation only of p53 but not of NQO1 and 20S.

NQO1, the gatekeeper, can be switched to an "on" or "off" position by NADH to allow or to block protein degradation. NADH level in turn is determined by the energy and oxidative state of the cells. The balance between these conditions, and the overall cell metabolism and catabolism processes must be properly maintained. Thus, NQO1 and NADH may provide a novel means of keeping this balance. Consistent with this hypothesis, the interaction of NQO1 with p53, p73, and ODC is modulated under oxidative state and ionizing radiation (4, 11).

The finding that an enzyme regulates the 20S proteasome in the context of degradation of IDPs may be the early bird of more regulators to be discovered. There are a number of additional roles that NQO1 may play in this context. NQO1 may be involved in targeting a subset of ID-containing proteins to the 20S proteasome, a possibility that is discussed below. Also, NQO1 may be functioning in regulating the latent state of the 20S catalytic core particle, with yet unknown mechanism.

1.15 OTHER MECHANISMS OF ESCAPING IDPs DEGRADATION BY DEFAULT

As discussed so far, the ID region, upon interacting with a partner, either acquires a defined structured or is sequestered. In either case, this would protect the IDPs from degradation by the 20S proteasome *in vitro*. By the same rationale, we explained the fact that IDPs in the cells are stabilized in the presence of a binding partner or destabilized when the level of the partner is decreased. For example, IκB proteins bind tightly to NF-κB dimers to ensure their nuclear exclusion. Free IκBα undergoes degradation by default, a process that depends on the C-terminal sequence. NF-κB binding to IκBα masks the C-terminus domain from proteasomal recognition, precluding its degradation by default, and as a result, degradation of free IκBα is about three orders of magnitude slower (94). As discussed above, bound IκBα undergoes degradation by a second pathway that requires its N-terminal phosphorylation by the IκB kinase.

Protein interaction with other macromolecules may provide another possible mechanism of escaping degradation by default. Y-box binding protein-1 (YB-1) exemplifies a case of protein–RNA interaction. Based on the FoldIndex prediction program, YB-1 is a highly unstructured protein. YB-1 is a nucleocytoplasmic shuttling protein involved in many DNA-dependent and RNA-dependent events. In the nucleus, YB-1 regulates transcription of many genes involved in cell proliferation and differentiation. Recently, it was demonstrated that YB-1 undergoes a specific proteolytic cleavage by the 20S proteasome. Cleavage of YB-1 by the 20S proteasome is ubiquitin independent and is inhibited following association of YB-1 with messenger RNA (125).

Protein modification may provide another mechanism of escaping degradation by default. For example, c-Fos degradation by default (ubiquitin independent) is blocked when c-Fos heterodimerizes with c-Jun. However, c-Fos may escape degradation by default by modification. The C-terminus of c-Fos is an ID region. Under certain conditions, phosphorylation at this region (S362, S374) stabilizes c-Fos (47). Fra-1, a Fos-related protein, behaves in a similar manner in response to phosphorylation at the C-terminus (13).

Another emerging mechanism is interaction with small molecules. Some small molecules that serve as protein substrates or cofactors stabilize IDPs, probably by reducing their disordered conformation. Calmodulin becomes partially resistant to proteasomal degradation upon binding to Ca^{2+} (16, 48, 128), and the same was shown for troponin C (16). Other small molecules are NAD^+ and NADH. These may regulate IDP stability either indirectly via NQO1, as described above, or by direct binding to 20S substrates. For example, the binding of NAD^+ prevents pertussis toxin S1 degradation by the 20S proteasome (97). Another case is p53 that binds NAD^+ and NADH, and undergoes conformational changes (89); however, it is not known whether NAD^+ and NADH alone, in the absence of NQO1, may regulate p53 stability. NADH and NAD(P)H regulate some other protein–protein interactions that may indirectly lead to protein stabilization. These include the increased binding of the transcriptional repressor C-terminal binding protein (CtBP) to the transcription factor E1A in the presence of NADH (50, 144) and the increased binding of the metabolic enzyme glyceraldehyde 3-phosphate dehydrogenase (GAPDH) to the POU domain of the transcription factor OCT-1 by NAD^+. The NAD^+-dependent binding of GAPDH to OCT-1 does not require the enzymatic activity of GAPDH (148). A CtBP mutant that cannot bind NADH was also shown to be less stable (85). Stabilization of the lens quinone oxidoreductase ζ-crystallin and its binding to the chaperone ζ-crystallin require NADPH, which induces a conformational change in ζ-crystallin (109). Changes in metabolism that result in altered NAD^+/NADH ratio may thus alter the stability of IDPs through the default degradation pathway. This may occur due to NADH-induced interaction of NQO1 with its client proteins, or due to direct interaction of NAD^+ or NAD(P)H with IDPs, that affects their conformation and susceptibility to default degradation.

1.16 THE BIOLOGICAL MEANING OF IDPs DEGRADATION BY DEFAULT

An important question is how degradation by default of IDPs would serve the cell's needs. An interesting case is activation of NF-κB in response to UV radiation. Inflammatory signaling induces rapid degradation of NF-κB-bound IκBα, but it does not induce the degradation of free IκBα (87, 94). In contrast, UV irradiation causes the depletion of both free and NF-κB-bound pools of IκBα. This is possible because free IκB undergoes degradation by default and because UV-induces IκB translation inhibition. The unstable nature of free IκBα demands its high synthesis rate to accumulate to the level needed to bind NF-κB to saturation. Inhibition of the synthesis of highly unstable free IκB, the inhibitor of NF-κB, rapidly shifts the equilibrium toward active NF-κB (95). Thus, the IκBα degradation by default is essential to obtain NF-κB activation in response to UV radiation.

Because many unstructured regions are folded upon binding to their biological targets in the cells, it seems that nature has developed an elegant way to link protein stability with biological functionality. Such a mechanism is extremely important in cases of functional multiprotein complexes, in which any excess of one of the subunits might interfere with the proper biological function of the complex. A theoretical analysis of gene circuits in bacteria presents a model for how selective degradation of monomers could be advantageous to the cells (25). This study analyzes the nonlinear dependence of degradation rate on the protein concentration. The degradation rate of proteins at lower concentrations is elevated compared with the degradation rate at higher protein concentrations. Since monomers predominate at low protein concentrations and dimerization is favored at higher concentrations, the model proposes that multimeric complexes support protein stability and terms it "cooperative stability." The authors of the "cooperative stability" model suggest that it improves the functioning of genetic networks and provides the means for evolutionary changes. This is based on the premise that genetic networks can be better regulated if the proteins involved are at higher concentrations. For example, if a protein dimer is the active species, then a low protein concentration would require high affinity for the dimer to be formed, and also would demand very stringent control of the protein expression. Higher protein concentrations offer the cell more flexibility in the regulation of expression, allow for lower affinity interactions, and allow for the variability that is essential for evolution. However, high protein concentrations and lower affinity mean that excess protein monomers may interfere with reactions catalyzed by the active dimers. Therefore, there must be a means to selectively degrade the excess monomers. The degradation by default model explains how excess monomers are selectively degraded without requiring any additional molecular components.

One example of this type of regulation is the degradation of poly(A)-binding protein (PABP). PABP binds to the mRNA 3′ poly(A) tail and stimulates recruitment of the ribosome to the mRNA at the 5′ end. PABP activity is tightly controlled by the PABP-interacting protein 2 (Paip2), which inhibits translation by displacing PABP from the mRNA. The carboxy-terminal third of PABP contains an unstructured region (75). An E3 ubiquitin ligase, EDD (E3 isolated by differential display), targets Paip2 for degradation. Interestingly, EDD-mediated degradation of Paip2 gives rise to the IDP PABP degradation as well possibly by the default mechanism. Upon PABP knockdown, Paip2 interacts with EDD, which leads to Paip2 ubiquitination and degradation. Thus, an IDP protein (PABP) interacts with a regulatory-structured protein (Paip2). The PABP/Paip2 complex resists proteasomal degradation by sequestering the structured protein from interacting with the E3 ligase on the one hand, and protecting the IDP from default degradation on the other hand. This interesting coregulatory mechanism serves as a homeostatic feedback to control the activity of PABP in cells (141).

The nonhomologous DNA end-joining pathway is used in animal cells to repair double-strand DNA breaks. To this aim, the heterodimer Ku70–Ku80 proteins, which form a DNA end-binding complex, play an important role. The central region of the Ku80 C-terminal domain has a well-defined structure with disordered N- and C-termini (146). Ku70 is unstructured in part as well (147). Not much is known about the mechanisms of degradation of these proteins. Interestingly, knockout of one results in dramatic reduction in the level of the other (46, 58). Therefore, the Ku70 and Ku80 proteins appear to play reciprocal roles in stabilizing each other. The simplest explanation is that these proteins are subjected to degradation by default but protected upon their interaction. Having said so, the data on knockout of a gene encoding an IDP should be treated cautiously while attributing to a gene a given function. Based on this principle, the coreduced protein level under a knockout background may provide a simple way to identify the potential interactors.

1.17 DEGRADATION OF IDPs AND VIRUSES

HCV capsid is, by large, a disordered protein based on prediction by FoldIndex. This protein undergoes degradation via the 20S proteasome and the 20S activator REGγ (92). This degradation process is ubiquitin independent and initiated by direct contact between the capsid protein and REGγ. The fact that the monomers of the capsid protein that form the highly structured capsid are partially ID is not unique to HCV and appears to be the case with other viruses such as HBV and human immunodeficiency virus type 1 (HIV) (based on in silico analysis). However, in the case of HCV, the capsid has many regulatory roles as well (88), and therefore, its ID nature may be important to this function.

Tat is a small RNA-binding protein that plays a central role in the regulation of HIV replication and in approaches to treating latently infected cells. The tat amino acid sequence has a low overall hydrophobicity and a high net positive charge, and analyses by several algorithms and experimental data suggest that it is a natively unfolded protein (121). Tat has been shown to interact with a number of 20S subunit including $\alpha 7$ (3). EBNA, an EBV-coded protein, also targets the 20S $\alpha 7$ subunit (133). Whether this subunit facilitates degradation of its client proteins is an important possibility that may explain how substrates are targeted to the uncapped 20S proteasome.

DNA tumor viruses often induce degradation of the major tumor suppressors p53 and Rb, both of which are, in part, unstructured or predicted to be so. Recent findings suggest that this process can take place by both ubiquitin-dependent and ubiquitin-independent pathways. The human papilloma virus (HPV) E6 protein interacts with two regions of p53, namely, the DNA-binding region and the C-terminus unstructured region. The former enhances p53 degradation via the ubiquitin pathway, whereas the latter enhances via a ubiquitin-independent mechanism that is likely to be 20S proteasomal default degradation (27). Consistent with this possibility is the finding that the HPV E6 protein was ineffective in destabilizing p53 under overexpression of NQO1, which blocks degradation by default (6). E7, another regulatory protein encoded by HPV, interacts to destroy Rb. Interestingly, E7 also binds to the 19S subunit of the proteasome and might therefore direct the degradation of the Rb protein (19). However, proteasome binding has not yet been shown to be required for degradation, and the dependence of the degradation on polyubiquitination has not been examined. The pp71 protein of cytomegalovirus interacts with and facilitates the ubiquitin-independent proteasomal degradation of Rb, p107, and p130 (70). The question of whether pp71 targets the 20S proteasome was not addressed.

Viruses, therefore, encode IDPs and exploit the process of IDP degradation to satisfy their needs. These studies are in their infancy, and it is expected that many additional examples will be reported in the future. Nevertheless, already at this initial stage, one can think of future directions of how to manipulate IDPs roles in virus–host interactions in order to block virus infection and the associated diseases.

1.18 IDPs AND DISEASE

Protein degradation plays a role in many cellular processes, such as cell cycle regulation, antigen presentation, and the disposal of denatured, unfolded, or oxidized proteins. Several diseases, including certain types of cancers and neurological disorders, can result from aberrant protein degradation. In many neurodegenerative diseases, the basis for the pathology is the cellular inability to degrade misfolded and damaged proteins, resulting in the formation of cytotoxic aggregates such as seen in AD, Parkinson's Huntington's, and prion diseases. Parkinson's is caused by the accumulation of aggregated proteins in the neuronal or glial cytoplasm. The inclusions of aggregated and filamentous proteins, called Lewy bodies, are comprised primarily of α-synuclein, a protein of unknown function that is highly expressed in the human brain. Monomeric α-synuclein is a natively unfolded protein without defined structure that undergoes ubiquitin-independent degradation by the 20S proteasomes (131). α-Synuclein spontaneously multimerizes to form highly stable insoluble fibrils that are resistant to degradation. Inhibition of the proteasomes in cells leads to an accumulation of α-synuclein that is phenotypically similar to the inclusions found in diseased tissue. Furthermore, inherited forms of Parkinson's disease include mutations that increase the copy number of α-synuclein, mutations that impair proteasome function, and mutations in α-synuclein that promote oligomerization of the protein and resistance to proteasomal degradation (90). One of the characteristics in Alzheimer's (AD) is the formation of neurofibrillary tangles that are intracellular and rich in tau, a structural protein that is normally associated with the microtubules. In several studies, it has been suggested that in AD, the proteasomal machinery is impaired (71, 72). Furthermore, it has been shown that the disordered tau protein can undergo ubiquitin-independent degradation by the proteasome, whereas the induction of conformational change by sodium dodecyl sulfate (SDS) or the binding to heat shock cognate 70 (Hsc70) or BCL2-associated athanogene (BAG-1) can prevent tau degradation (39, 45). Prion diseases are fatal neurodegenerative disorders whose pathogenesis is associated with a conformational rearrangement of the normal cellular prion protein (PrP^C) to abnormal conformers (PrP^{Sc}). PrP^C can undergo ubiquitin-independent degradation (129), whereas not only is the PrP^{Sc} conformation resistant to proteasomal degradation but it also inhibits the 20S proteasome catalytic activity (77).

These neurodegenerative diseases illustrate the importance of functioning 20S degradation. Interestingly, the proteins that are discussed are IDPs (see Table 1.1), highlighting the importance of degradation by default. Although some have pointed out that the ubiquitin system is important in the pathology of these diseases (as reviewed [35]), this does not contradict the possibility that ubiquitin-independent proteasomal degradation by default might play a significant role in these abnormalities. Monomeric proteins, such as α-synuclein and tau, if not forming their functional complex, need to be degraded to prevent their accumulation and subsequent spontaneous oligomerization that leads to the protein inclusions associated with these diseases. The degradation by default pathway can prevent the accumulation of uncomplexed ID proteins, suggesting that if this pathway is impaired, it could lead to the associated diseases.

1.19 IDPs: ARE THEY DISORDERED OR HIGHER ORDERED?

We do not think that the terms IUPs or IDPs adequately describe the functional significance of this category of proteins. Both terms are inaccurate, inasmuch as these proteins may become structured/ordered at given points in time and space, both in the presence and absence of partners. Furthermore, the use of the terms IDP and IUP may also give the impression that these proteins are in some way "inferior" to structured proteins. This is completely wrong, given the fact that such proteins are mainly associated with higher organisms and functions, as can be seen, for example, by their prevalence at synapses in the nervous system. We proposed that they be called "4D proteins," based on the fact that their structures are not fixed, as is generally the case for "3D proteins," but are rather defined by time and space (135). IDPs, therefore, are more highly ordered than the 3D proteins, and the term 4D proteins would more accurately define this group of proteins.

ACKNOWLEDGMENTS

This work was supported by grants from the Samuel Waxman Cancer Research Foundation, the Israel Science Foundation, and the Minerva Foundation with funding from the Federal German Ministry for Education and Research. Y. S. is the Oscar and Emma Getz Professor.

REFERENCES

1. Alvarez-Castelao, B., and J. G. Castano. 2005. Mechanism of direct degradation of IkappaBalpha by 20S proteasome. FEBS Lett **579**:4797–802.
2. Amici, M., D. Sagratini, A. Pettinari, S. Pucciarelli, M. Angeletti, and A. M. Eleuteri. 2004. 20S proteasome mediated degradation of DHFR: implications in neurodegenerative disorders. Arch Biochem Biophys **422**:168–74.
3. Apcher, G. S., S. Heink, D. Zantopf, P. M. Kloetzel, H. P. Schmid, R. J. Mayer, and E. Kruger. 2003. Human immunodeficiency virus-1 Tat protein interacts with distinct proteasomal alpha and beta subunits. FEBS Lett **553**:200–4.
4. Asher, G., Z. Bercovich, P. Tsvetkov, Y. Shaul, and C. Kahana. 2005. 20S proteasomal degradation of ornithine decarboxylase is regulated by NQO1. Mol Cell **17**:645–55.
5. Asher, G., J. Lotem, B. Cohen, L. Sachs, and Y. Shaul. 2001. Regulation of p53 stability and p53-dependent apoptosis by NADH quinone oxidoreductase 1. Proc Natl Acad Sci U S A **98**:1188–93.
6. Asher, G., J. Lotem, R. Kama, L. Sachs, and Y. Shaul. 2002. NQO1 stabilizes p53 through a distinct pathway. Proc Natl Acad Sci U S A **99**:3099–104.
7. Asher, G., J. Lotem, L. Sachs, C. Kahana, and Y. Shaul. 2002. Mdm-2 and ubiquitin-independent p53 proteasomal degradation regulated by NQO1. Proc Natl Acad Sci U S A **99**:13125–30.
8. Asher, G., J. Lotem, L. Sachs, and Y. Shaul. 2004. p53-dependent apoptosis and NAD(P)H:quinone oxidoreductase 1. Methods Enzymol **382**:278–93.
9. Asher, G., J. Lotem, P. Tsvetkov, V. Reiss, L. Sachs, and Y. Shaul. 2003. P53 hot-spot mutants are resistant to ubiquitin-independent degradation by increased

binding to NAD(P)H:quinone oxidoreductase 1. Proc Natl Acad Sci U S A **100**: 15065–70.

10. Asher, G., N. Reuven, and Y. Shaul. 2006. 20S proteasomes and protein degradation "by default". Bioessays **28**:844–9.
11. Asher, G., P. Tsvetkov, C. Kahana, and Y. Shaul. 2005. A mechanism of ubiquitin-independent proteasomal degradation of the tumor suppressors p53 and p73. Genes Dev **19**:316–21.
12. Bajorek, M., D. Finley, and M. H. Glickman. 2003. Proteasome disassembly and downregulation is correlated with viability during stationary phase. Curr Biol **13**:1140–4.
13. Basbous, J., D. Chalbos, R. Hipskind, I. Jariel-Encontre, and M. Piechaczyk. 2007. Ubiquitin-independent proteasomal degradation of Fra-1 is antagonized by Erk1/2 pathway-mediated phosphorylation of a unique C-terminal destabilizer. Mol Cell Biol **27**:3936–50.
14. Baugh, J. M., and E. V. Pilipenko. 2004. 20S proteasome differentially alters translation of different mRNAs via the cleavage of eIF4F and eIF3. Mol Cell **16**:575–86.
15. Bell, S., C. Klein, L. Muller, S. Hansen, and J. Buchner. 2002. p53 contains large unstructured regions in its native state. J Mol Biol **322**:917–27.
16. Benaroudj, N., E. Tarcsa, P. Cascio, and A. L. Goldberg. 2001. The unfolding of substrates and ubiquitin-independent protein degradation by proteasomes. Biochimie **83**:311–8.
17. Bercovich, Z., and C. Kahana. 1993. Involvement of the 20S proteasome in the degradation of ornithine decarboxylase. Eur J Biochem **213**:205–10.
18. Bercovich, Z., Y. Rosenberg-Hasson, A. Ciechanover, and C. Kahana. 1989. Degradation of ornithine decarboxylase in reticulocyte lysate is ATP-dependent but ubiquitin-independent. J Biol Chem **264**:15949–52.
19. Berezutskaya, E., and S. Bagchi. 1997. The human papillomavirus E7 oncoprotein functionally interacts with the S4 subunit of the 26 S proteasome. J Biol Chem **272**:30135–40.
20. Bloom, J., V. Amador, F. Bartolini, G. DeMartino, and M. Pagano. 2003. Proteasome-mediated degradation of p21 via N-terminal ubiquitinylation. Cell **115**:71–82.
21. Bloom, J., and M. Pagano. 2004. To be or not to be ubiquitinated? Cell Cycle **3**:138–40.
22. Bossis, G., P. Ferrara, C. Acquaviva, I. Jariel-Encontre, and M. Piechaczyk. 2003. c-Fos proto-oncoprotein is degraded by the proteasome independently of its own ubiquitinylation in vivo. Mol Cell Biol **23**:7425–36.
23. Bothner, B., W. S. Lewis, E. L. DiGiammarino, J. D. Weber, S. J. Bothner, and R. W. Kriwacki. 2001. Defining the molecular basis of Arf and Hdm2 interactions. J Mol Biol **314**:263–77.
24. Bruschweiler, R., X. Liao, and P. E. Wright. 1995. Long-range motional restrictions in a multidomain zinc-finger protein from anisotropic tumbling. Science **268**:886–9.
25. Buchler, N. E., U. Gerland, and T. Hwa. 2005. Nonlinear protein degradation and the function of genetic circuits. Proc Natl Acad Sci U S A **102**:9559–64.

26. Campbell, K. M., A. R. Terrell, P. J. Laybourn, and K. J. Lumb. 2000. Intrinsic structural disorder of the C-terminal activation domain from the bZIP transcription factor Fos. Biochemistry **39**:2708–13.
27. Camus, S., S. Menendez, C. F. Cheok, L. F. Stevenson, S. Lain, and D. P. Lane. 2007. Ubiquitin-independent degradation of p53 mediated by high-risk human papillomavirus protein E6. Oncogene **26**:4059–70.
28. Cardozo, C., and C. Michaud. 2002. Proteasome-mediated degradation of tau proteins occurs independently of the chymotrypsin-like activity by a nonprocessive pathway. Arch Biochem Biophys **408**:103–10.
29. Carreras, C. W., and D. V. Santi. 1995. The catalytic mechanism and structure of thymidylate synthase. Annu Rev Biochem **64**:721–62.
30. Chen, X., L. F. Barton, Y. Chi, B. E. Clurman, and J. M. Roberts. 2007. Ubiquitin-independent degradation of cell-cycle inhibitors by the REGgamma proteasome. Mol Cell **26**:843–52.
31. Chinenov, Y., and T. K. Kerppola. 2001. Close encounters of many kinds: Fos-Jun interactions that mediate transcription regulatory specificity. Oncogene **20**: 2438–52.
32. Chiu, Y. H., Q. Sun, and Z. J. Chen. 2007. E1-L2 activates both ubiquitin and FAT10. Mol Cell **27**:1014–23.
33. Chowdary, D. R., J. J. Dermody, K. K. Jha, and H. L. Ozer. 1994. Accumulation of p53 in a mutant cell line defective in the ubiquitin pathway. Mol Cell Biol **14**:1997–2003.
34. Ciechanover, A., and R. Ben-Saadon. 2004. N-terminal ubiquitination: more protein substrates join in. Trends Cell Biol **14**:103–6.
35. Ciechanover, A., and P. Brundin. 2003. The ubiquitin proteasome system in neurodegenerative diseases: sometimes the chicken, sometimes the egg. Neuron **40**:427–46.
36. Coffino, P. 2001. Antizyme, a mediator of ubiquitin-independent proteasomal degradation. Biochimie **83**:319–23.
37. Coffino, P. 2001. Regulation of cellular polyamines by antizyme. Nat Rev Mol Cell Biol **2**:188–94.
38. Coleman, M. L., C. J. Marshall, and M. F. Olson. 2003. Ras promotes p21(Waf1/Cip1) protein stability via a cyclin D1-imposed block in proteasome-mediated degradation. EMBO J **22**:2036–46.
39. David, D. C., R. Layfield, L. Serpell, Y. Narain, M. Goedert, and M. G. Spillantini. 2002. Proteasomal degradation of tau protein. J Neurochem **83**:176–85.
40. Dornan, D., I. Wertz, H. Shimizu, D. Arnott, G. D. Frantz, P. Dowd, K. O'Rourke, H. Koeppen, and V. M. Dixit. 2004. The ubiquitin ligase COP1 is a critical negative regulator of p53. Nature **429**:86–92.
41. Dosztanyi, Z., J. Chen, A. K. Dunker, I. Simon, and P. Tompa. 2006. Disorder and sequence repeats in hub proteins and their implications for network evolution. J Proteome Res **5**:2985–95.
42. Dunker, A. K., C. J. Brown, J. D. Lawson, L. M. Iakoucheva, and Z. Obradovic. 2002. Intrinsic disorder and protein function. Biochemistry **41**:6573–82.
43. Dunker, A. K., Z. Obradovic, P. Romero, E. C. Garner, and C. J. Brown. 2000. Intrinsic protein disorder in complete genomes. Genome Inform Ser Workshop Genome Inform **11**:161–71.

44. Dyson, H. J., and P. E. Wright. 2005. Intrinsically unstructured proteins and their functions. Nat Rev Mol Cell Biol **6**:197–208.

45. Elliott, E., P. Tsvetkov, and I. Ginzburg. 2007. BAG-1 associates with Hsc70.Tau complex and regulates the proteasomal degradation of Tau protein. J Biol Chem **282**:37276–84.

46. Errami, A., V. Smider, W. K. Rathmell, D. M. He, E. A. Hendrickson, M. Z. Zdzienicka, and G. Chu. 1996. Ku86 defines the genetic defect and restores X-ray resistance and V(D)J recombination to complementation group 5 hamster cell mutants. Mol Cell Biol **16**:1519–26.

47. Ferrara, P., E. Andermarcher, G. Bossis, C. Acquaviva, F. Brockly, I. Jariel-Encontre, and M. Piechaczyk. 2003. The structural determinants responsible for c-Fos protein proteasomal degradation differ according to the conditions of expression. Oncogene **22**:1461–74.

48. Ferrington, D. A., H. Sun, K. K. Murray, J. Costa, T. D. Williams, D. J. Bigelow, and T. C. Squier. 2001. Selective degradation of oxidized calmodulin by the 20 S proteasome. J Biol Chem **276**:937–43.

49. Fink, A. L. 1999. Chaperone-mediated protein folding. Physiol Rev **79**:425–49.

50. Fjeld, C. C., W. T. Birdsong, and R. H. Goodman. 2003. Differential binding of NAD+ and NADH allows the transcriptional corepressor carboxyl-terminal binding protein to serve as a metabolic sensor. Proc Natl Acad Sci U S A **100**: 9202–7.

51. Forsthoefel, A. M., M. M. Pena, Y. Y. Xing, Z. Rafique, and F. G. Berger. 2004. Structural determinants for the intracellular degradation of human thymidylate synthase. Biochemistry **43**:1972–9.

52. Friedler, A., D. B. Veprintsev, S. M. Freund, K. I. von Glos, and A. R. Fersht. 2005. Modulation of binding of DNA to the C-terminal domain of p53 by acetylation. Structure **13**:629–36.

53. Garate, M., R. P. Wong, E. I. Campos, Y. Wang, and G. Li. 2008. NAD(P)H quinone oxidoreductase 1 inhibits the proteasomal degradation of the tumour suppressor p33(ING1b). EMBO Rep **9**:576–81.

54. Glickman, M. H., and A. Ciechanover. 2002. The ubiquitin-proteasome proteolytic pathway: destruction for the sake of construction. Physiol Rev **82**:373–428.

55. Gong, X., L. Kole, K. Iskander, and A. K. Jaiswal. 2007. NRH:quinone oxidoreductase 2 and NAD(P)H:quinone oxidoreductase 1 protect tumor suppressor p53 against 20s proteasomal degradation leading to stabilization and activation of p53. Cancer Res **67**:5380–8.

56. Groll, M., M. Bajorek, A. Kohler, L. Moroder, D. M. Rubin, R. Huber, M. H. Glickman, and D. Finley. 2000. A gated channel into the proteasome core particle. Nat Struct Biol **7**:1062–7.

57. Groll, M., L. Ditzel, J. Lowe, D. Stock, M. Bochtler, H. D. Bartunik, and R. Huber. 1997. Structure of 20S proteasome from yeast at 2.4 A resolution. Nature **386**:463–71.

58. Gu, Y., S. Jin, Y. Gao, D. T. Weaver, and F. W. Alt. 1997. Ku70-deficient embryonic stem cells have increased ionizing radiosensitivity, defective DNA end-binding activity, and inability to support V(D)J recombination. Proc Natl Acad Sci U S A **94**:8076–81.

59. Haupt, Y., R. Maya, A. Kazaz, and M. Oren. 1997. Mdm2 promotes the rapid degradation of p53. Nature **387**:296–9.
60. Hershko, A. 1996. Lessons from the discovery of the ubiquitin system. Trends Biochem Sci **21**:445–9.
61. Hershko, A., and A. Ciechanover. 1998. The ubiquitin system. Annu Rev Biochem **67**:425–79.
62. Hess, J., P. Angel, and M. Schorpp-Kistner. 2004. AP-1 subunits: quarrel and harmony among siblings. J Cell Sci **117**:5965–73.
63. Hochstrasser, M. 1996. Ubiquitin-dependent protein degradation. Annu Rev Genet **30**:405–39.
64. Hwang, S. G., S. S. Yu, J. H. Ryu, H. B. Jeon, Y. J. Yoo, S. H. Eom, and J. S. Chun. 2005. Regulation of beta-catenin signaling and maintenance of chondrocyte differentiation by ubiquitin-independent proteasomal degradation of alpha-catenin. J Biol Chem **280**:12758–65.
65. Jariel-Encontre, I., G. Bossis, and M. Piechaczyk. 2008. Ubiquitin-independent degradation of proteins by the proteasome. Biochim Biophys Acta **1786**:153–177.
66. Jariel-Encontre, I., M. Pariat, F. Martin, S. Carillo, C. Salvat, and M. Piechaczyk. 1995. Ubiquitinylation is not an absolute requirement for degradation of c-Jun protein by the 26 S proteasome. J Biol Chem **270**:11623–7.
67. Jin, J., X. Li, S. P. Gygi, and J. W. Harper. 2007. Dual E1 activation systems for ubiquitin differentially regulate E2 enzyme charging. Nature **447**:1135–8.
68. Johnston, J. A., E. S. Johnson, P. R. Waller, and A. Varshavsky. 1995. Methotrexate inhibits proteolysis of dihydrofolate reductase by the N-end rule pathway. J Biol Chem **270**:8172–8.
69. Ju, D., and Y. Xie. 2004. Proteasomal degradation of RPN4 via two distinct mechanisms, ubiquitin-dependent and -independent. J Biol Chem **279**:23851–4.
70. Kalejta, R. F., and T. Shenk. 2003. Proteasome-dependent, ubiquitin-independent degradation of the Rb family of tumor suppressors by the human cytomegalovirus pp71 protein. Proc Natl Acad Sci U S A **100**:3263–8.
71. Keck, S., R. Nitsch, T. Grune, and O. Ullrich. 2003. Proteasome inhibition by paired helical filament-tau in brains of patients with Alzheimer's disease. J Neurochem **85**:115–22.
72. Keller, J. N., K. B. Hanni, and W. R. Markesbery. 2000. Impaired proteasome function in Alzheimer's disease. J Neurochem **75**:436–9.
73. Kisselev, A. F., T. N. Akopian, K. M. Woo, and A. L. Goldberg. 1999. The sizes of peptides generated from protein by mammalian 26 and 20 S proteasomes. Implications for understanding the degradative mechanism and antigen presentation. J Biol Chem **274**:3363–71.
74. Kong, X., B. Alvarez-Castelao, Z. Lin, J. G. Castano, and J. Caro. 2007. Constitutive/hypoxic degradation of HIF-alpha proteins by the proteasome is independent of von Hippel Lindau protein ubiquitylation and the transactivation activity of the protein. J Biol Chem **282**:15498–505.
75. Kozlov, G., J. F. Trempe, K. Khaleghpour, A. Kahvejian, I. Ekiel, and K. Gehring. 2001. Structure and function of the C-terminal PABC domain of human poly(A)-binding protein. Proc Natl Acad Sci U S A **98**:4409–13.

76. Krappmann, D., F. G. Wulczyn, and C. Scheidereit. 1996. Different mechanisms control signal-induced degradation and basal turnover of the NF-kappaB inhibitor IkappaB alpha in vivo. EMBO J **15**:6716–26.
77. Kristiansen, M., P. Deriziotis, D. E. Dimcheff, G. S. Jackson, H. Ovaa, H. Naumann, A. R. Clarke, F. W. van Leeuwen, V. Menendez-Benito, N. P. Dantuma, J. L. Portis, J. Collinge, and S. J. Tabrizi. 2007. Disease-associated prion protein oligomers inhibit the 26S proteasome. Mol Cell **26**:175–88.
78. Kroll, M., M. Conconi, M. J. Desterro, A. Marin, D. Thomas, B. Friguet, R. T. Hay, J. L. Virelizier, F. Arenzana-Seisdedos, and M. S. Rodriguez. 1997. The carboxy-terminus of I kappaB alpha determines susceptibility to degradation by the catalytic core of the proteasome. Oncogene **15**:1841–50.
79. Kubbutat, M. H., S. N. Jones, and K. H. Vousden. 1997. Regulation of p53 stability by Mdm2. Nature **387**:299–303.
80. Kuwajima, K. 1989. The molten globule state as a clue for understanding the folding and cooperativity of globular-protein structure. Proteins **6**:87–103.
81. Leng, R. P., Y. Lin, W. Ma, H. Wu, B. Lemmers, S. Chung, J. M. Parant, G. Lozano, R. Hakem, and S. Benchimol. 2003. Pirh2, a p53-induced ubiquitin-protein ligase, promotes p53 degradation. Cell **112**:779–91.
82. Li, X., L. Amazit, W. Long, D. M. Lonard, J. J. Monaco, and B. W. O'Malley. 2007. Ubiquitin- and ATP-independent proteolytic turnover of p21 by the REGgamma-proteasome pathway. Mol Cell **26**:831–42.
83. Li, X., D. M. Lonard, S. Y. Jung, A. Malovannaya, Q. Feng, J. Qin, S. Y. Tsai, M. J. Tsai, and B. W. O'Malley. 2006. The SRC-3/AIB1 coactivator is degraded in a ubiquitin- and ATP-independent manner by the REGgamma proteasome. Cell **124**:381–92.
84. Liu, C. W., M. J. Corboy, G. N. DeMartino, and P. J. Thomas. 2003. Endoproteolytic activity of the proteasome. Science **299**:408–11.
85. Mani-Telang, P., M. Sutrias-Grau, G. Williams, and D. N. Arnosti. 2007. Role of NAD binding and catalytic residues in the C-terminal binding protein corepressor. FEBS Lett **581**:5241–6.
86. Masson, P., O. Andersson, U. M. Petersen, and P. Young. 2001. Identification and characterization of a Drosophila nuclear proteasome regulator. A homolog of human 11 S REGgamma (PA28gamma). J Biol Chem **276**:1383–90.
87. Mathes, E., E. L. O'Dea, A. Hoffmann, and G. Ghosh. 2008. NF-kappaB dictates the degradation pathway of IkappaBalpha. EMBO J **27**:1357–67.
88. McLauchlan, J. 2000. Properties of the hepatitis C virus core protein: a structural protein that modulates cellular processes. J Viral Hepat **7**:2–14.
89. McLure, K. G., M. Takagi, and M. B. Kastan. 2004. NAD+ modulates p53 DNA binding specificity and function. Mol Cell Biol **24**:9958–67.
90. Moore, D. J., A. B. West, V. L. Dawson, and T. M. Dawson. 2005. Molecular pathophysiology of Parkinson's disease. Annu Rev Neurosci **28**:57–87.
91. Moorthy, A. K., O. V. Savinova, J. Q. Ho, V. Y. Wang, D. Vu, and G. Ghosh. 2006. The 20S proteasome processes NF-kappaB1 p105 into p50 in a translation-independent manner. EMBO J **25**:1945–56.
92. Moriishi, K., T. Okabayashi, K. Nakai, K. Moriya, K. Koike, S. Murata, T. Chiba, K. Tanaka, R. Suzuki, T. Suzuki, T. Miyamura, and Y. Matsuura. 2003. Proteasome

activator PA28gamma-dependent nuclear retention and degradation of hepatitis C virus core protein. J Virol **77**:10237–49.

93. Navon, A., and A. L. Goldberg. 2001. Proteins are unfolded on the surface of the ATPase ring before transport into the proteasome. Mol Cell **8**:1339–49.
94. O'Dea, E. L., D. Barken, R. Q. Peralta, K. T. Tran, S. L. Werner, J. D. Kearns, A. Levchenko, and A. Hoffmann. 2007. A homeostatic model of IkappaB metabolism to control constitutive NF-kappaB activity. Mol Syst Biol **3**:111.
95. O'Dea, E. L., J. D. Kearns, and A. Hoffmann. 2008. UV as an amplifier rather than inducer of NF-kappaB activity. Mol Cell **30**:632–41.
96. Ogiso, Y., A. Tomida, H. D. Kim, and T. Tsuruo. 1999. Glucose starvation and hypoxia induce nuclear accumulation of proteasome in cancer cells. Biochem Biophys Res Commun **258**:448–52.
97. Pande, A. H., D. Moe, M. Jamnadas, S. A. Tatulian, and K. Teter. 2006. The pertussis toxin S1 subunit is a thermally unstable protein susceptible to degradation by the 20S proteasome. Biochemistry **45**:13734–40.
98. Pande, A. H., P. Scaglione, M. Taylor, K. N. Nemec, S. Tuthill, D. Moe, R. K. Holmes, S. A. Tatulian, and K. Teter. 2007. Conformational instability of the cholera toxin A1 polypeptide. J Mol Biol **374**:1114–28.
99. Park, Y., Y. P. Hwang, J. S. Lee, S. H. Seo, S. K. Yoon, and J. B. Yoon. 2005. Proteasomal ATPase-associated factor 1 negatively regulates proteasome activity by interacting with proteasomal ATPases. Mol Cell Biol **25**:3842–53.
100. Pelzer, C., I. Kassner, K. Matentzoglu, R. K. Singh, H. P. Wollscheid, M. Scheffner, G. Schmidtke, and M. Groettrup. 2007. UBE1L2, a novel E1 enzyme specific for ubiquitin. J Biol Chem **282**:23010–4.
101. Pena, M. M., Y. Y. Xing, S. Koli, and F. G. Berger. 2006. Role of N-terminal residues in the ubiquitin-independent degradation of human thymidylate synthase. Biochem J **394**:355–63.
102. Penela, P., A. Ruiz-Gomez, J. G. Castano, and F. Mayor, Jr. 1998. Degradation of the G protein-coupled receptor kinase 2 by the proteasome pathway. J Biol Chem **273**:35238–44.
103. Petropoulos, L., and J. Hiscott. 1998. Association between HTLV-1 Tax and I kappa B alpha is dependent on the I kappa B alpha phosphorylation state. Virology **252**:189–99.
104. Pickart, C. M. 2001. Mechanisms underlying ubiquitination. Annu Rev Biochem **70**:503–33.
105. Pollice, A., M. Sepe, V. R. Villella, F. Tolino, M. Vivo, V. Calabro, and G. La Mantia. 2007. TBP-1 protects the human oncosuppressor p14ARF from proteasomal degradation. Oncogene **26**:5154–62.
106. Poppek, D., S. Keck, G. Ermak, T. Jung, A. Stolzing, O. Ullrich, K. J. Davies, and T. Grune. 2006. Phosphorylation inhibits turnover of the tau protein by the proteasome: influence of RCAN1 and oxidative stress. Biochem J **400**:511–20.
107. Prakash, S., L. Tian, K. S. Ratliff, R. E. Lehotzky, and A. Matouschek. 2004. An unstructured initiation site is required for efficient proteasome-mediated degradation. Nat Struct Mol Biol **11**:830–7.
108. Prilusky, J., C. E. Felder, T. Zeev-Ben-Mordehai, E. H. Rydberg, O. Man, J. S. Beckmann, I. Silman, and J. L. Sussman. 2005. FoldIndex: a simple tool to

predict whether a given protein sequence is intrinsically unfolded. Bioinformatics **21**:3435–8.

109. Rao, P. V., J. Horwitz, and J. S. Zigler, Jr. 1994. Chaperone-like activity of alpha-crystallin. The effect of NADPH on its interaction with zeta-crystallin. J Biol Chem **269**:13266–72.
110. Realini, C., C. C. Jensen, Z. Zhang, S. C. Johnston, J. R. Knowlton, C. P. Hill, and M. Rechsteiner. 1997. Characterization of recombinant REGalpha, REGbeta, and REGgamma proteasome activators. J Biol Chem **272**:25483–92.
111. Rechsteiner, M., and C. P. Hill. 2005. Mobilizing the proteolytic machine: cell biological roles of proteasome activators and inhibitors. Trends Cell Biol **15**: 27–33.
112. Reinheckel, T., N. Sitte, O. Ullrich, U. Kuckelkorn, K. J. Davies, and T. Grune. 1998. Comparative resistance of the 20S and 26S proteasome to oxidative stress. Biochem J **335** (Pt 3):637–42.
113. Reinheckel, T., O. Ullrich, N. Sitte, and T. Grune. 2000. Differential impairment of 20S and 26S proteasome activities in human hematopoietic K562 cells during oxidative stress. Arch Biochem Biophys **377**:65–8.
114. Rosenberg-Hasson, Y., Z. Bercovich, A. Ciechanover, and C. Kahana. 1989. Degradation of ornithine decarboxylase in mammalian cells is ATP dependent but ubiquitin independent. Eur J Biochem **185**:469–74.
115. Sanchez-Puig, N., D. B. Veprintsev, and A. R. Fersht. 2005. Binding of natively unfolded HIF-1alpha ODD domain to p53. Mol Cell **17**:11–21.
116. Sasaki, T., H. Kojima, R. Kishimoto, A. Ikeda, H. Kunimoto, and K. Nakajima. 2006. Spatiotemporal regulation of c-Fos by ERK5 and the E3 ubiquitin ligase UBR1, and its biological role. Mol Cell **24**:63–75.
117. Scheffner, M., J. M. Huibregtse, R. D. Vierstra, and P. M. Howley. 1993. The HPV-16 E6 and E6-AP complex functions as a ubiquitin-protein ligase in the ubiquitination of p53. Cell **75**:495–505.
118. Sdek, P., H. Ying, D. L. Chang, W. Qiu, H. Zheng, R. Touitou, M. J. Allday, and Z. X. Xiao. 2005. MDM2 promotes proteasome-dependent ubiquitin-independent degradation of retinoblastoma protein. Mol Cell **20**:699–708.
119. Sheaff, R. J., J. D. Singer, J. Swanger, M. Smitherman, J. M. Roberts, and B. E. Clurman. 2000. Proteasomal turnover of p21Cip1 does not require p21Cip1 ubiquitination. Mol Cell **5**:403–10.
120. Sherr, C. J. 2001. The INK4a/ARF network in tumour suppression. Nat Rev Mol Cell Biol **2**:731–7.
121. Shojania, S., and J. D. O'Neil. 2006. HIV-1 Tat is a natively unfolded protein: the solution conformation and dynamics of reduced HIV-1 Tat-(1-72) by NMR spectroscopy. J Biol Chem **281**:8347–56.
122. Sickmeier, M., J. A. Hamilton, T. LeGall, V. Vacic, M. S. Cortese, A. Tantos, B. Szabo, P. Tompa, J. Chen, V. N. Uversky, Z. Obradovic, and A. K. Dunker. 2007. DisProt: the Database of Disordered Proteins. Nucleic Acids Res **35**:D786–93.
123. Smet, C., A. Leroy, A. Sillen, J. M. Wieruszeski, I. Landrieu, and G. Lippens. 2004. Accepting its random coil nature allows a partial NMR assignment of the neuronal Tau protein. Chembiochem **5**:1639–46.

124. Smith, D. M., S. C. Chang, S. Park, D. Finley, Y. Cheng, and A. L. Goldberg. 2007. Docking of the proteasomal ATPases' carboxyl termini in the 20S proteasome's alpha ring opens the gate for substrate entry. Mol Cell **27**:731–44.

125. Sorokin, A. V., A. A. Selyutina, M. A. Skabkin, S. G. Guryanov, I. V. Nazimov, C. Richard, J. Th'ng, J. Yau, P. H. Sorensen, L. P. Ovchinnikov, and V. Evdokimova. 2005. Proteasome-mediated cleavage of the Y-box-binding protein 1 is linked to DNA-damage stress response. EMBO J **24**:3602–12.

126. Takeuchi, J., H. Chen, and P. Coffino. 2007. Proteasome substrate degradation requires association plus extended peptide. EMBO J **26**:123–31.

127. Tang, K. S., A. R. Fersht, and L. S. Itzhaki. 2003. Sequential unfolding of ankyrin repeats in tumor suppressor p16. Structure **11**:67–73.

128. Tarcsa, E., G. Szymanska, S. Lecker, C. M. O'Connor, and A. L. Goldberg. 2000. Ca2+-free calmodulin and calmodulin damaged by in vitro aging are selectively degraded by 26 S proteasomes without ubiquitination. J Biol Chem **275**: 20295–301.

129. Tenzer, S., L. Stoltze, B. Schonfisch, J. Dengjel, M. Muller, S. Stevanovic, H. G. Rammensee, and H. Schild. 2004. Quantitative analysis of prion-protein degradation by constitutive and immuno-20S proteasomes indicates differences correlated with disease susceptibility. J Immunol **172**:1083–91.

130. Timchenko, N. A., T. E. Harris, M. Wilde, T. A. Bilyeu, B. L. Burgess-Beusse, M. J. Finegold, and G. J. Darlington. 1997. CCAAT/enhancer binding protein alpha regulates p21 protein and hepatocyte proliferation in newborn mice. Mol Cell Biol **17**:7353–61.

131. Tofaris, G. K., R. Layfield, and M. G. Spillantini. 2001. alpha-synuclein metabolism and aggregation is linked to ubiquitin-independent degradation by the proteasome. FEBS Lett **509**:22–6.

132. Tompa, P. 2005. The interplay between structure and function in intrinsically unstructured proteins. FEBS Lett **579**:3346–54.

133. Touitou, R., J. O'Nions, J. Heaney, and M. J. Allday. 2005. Epstein-Barr virus EBNA3 proteins bind to the C8/alpha7 subunit of the 20S proteasome and are degraded by 20S proteasomes in vitro, but are very stable in latently infected B cells. J Gen Virol **86**:1269–77.

134. Touitou, R., J. Richardson, S. Bose, M. Nakanishi, J. Rivett, and M. J. Allday. 2001. A degradation signal located in the C-terminus of p21WAF1/CIP1 is a binding site for the C8 alpha-subunit of the 20S proteasome. EMBO J **20**: 2367–75.

135. Tsvetkov, P., G. Asher, A. Paz, N. Reuven, J. L. Sussman, I. Silman, and Y. Shaul. 2007. Operational definition of intrinsically unstructured protein sequences based on susceptibility to the 20S proteasome. Proteins **70**:1357–1366.

136. Uversky, V. N., J. R. Gillespie, and A. L. Fink. 2000. Why are "natively unfolded" proteins unstructured under physiologic conditions? Proteins **41**:415–27.

137. Venkatraman, P., R. Wetzel, M. Tanaka, N. Nukina, and A. L. Goldberg. 2004. Eukaryotic proteasomes cannot digest polyglutamine sequences and release them during degradation of polyglutamine-containing proteins. Mol Cell **14**:95–104.

138. Vogelstein, B., D. Lane, and A. J. Levine. 2000. Surfing the p53 network. Nature **408**:307–10.

139. Voges, D., P. Zwickl, and W. Baumeister. 1999. The 26S proteasome: a molecular machine designed for controlled proteolysis. Annu Rev Biochem **68**:1015–68.
140. Wenzel, T., and W. Baumeister. 1995. Conformational constraints in protein degradation by the 20S proteasome. Nat Struct Biol **2**:199–204.
141. Yoshida, M., K. Yoshida, G. Kozlov, N. S. Lim, G. De Crescenzo, Z. Pang, J. J. Berlanga, A. Kahvejian, K. Gehring, S. S. Wing, and N. Sonenberg. 2006. Poly(A) binding protein (PABP) homeostasis is mediated by the stability of its inhibitor, Paip2. EMBO J **25**:1934–44.
142. Yu, H., G. Kaung, S. Kobayashi, and R. R. Kopito. 1997. Cytosolic degradation of T-cell receptor alpha chains by the proteasome. J Biol Chem **272**:20800–4.
143. Zhang, M., C. M. Pickart, and P. Coffino. 2003. Determinants of proteasome recognition of ornithine decarboxylase, a ubiquitin-independent substrate. EMBO J **22**:1488–96.
144. Zhang, Q., D. W. Piston, and R. H. Goodman. 2002. Regulation of corepressor function by nuclear NADH. Science **295**:1895–7.
145. Zhang, Y. J., Y. F. Xu, Y. H. Liu, J. Yin, H. L. Li, Q. Wang, and J. Z. Wang. 2006. Peroxynitrite induces Alzheimer-like tau modifications and accumulation in rat brain and its underlying mechanisms. FASEB J **20**:1431–42.
146. Zhang, Z., W. Hu, L. Cano, T. D. Lee, D. J. Chen, and Y. Chen. 2004. Solution structure of the C-terminal domain of Ku80 suggests important sites for protein-protein interactions. Structure **12**:495–502.
147. Zhang, Z., L. Zhu, D. Lin, F. Chen, D. J. Chen, and Y. Chen. 2001. The three-dimensional structure of the C-terminal DNA-binding domain of human Ku70. J Biol Chem **276**:38231–6.
148. Zheng, L., R. G. Roeder, and Y. Luo. 2003. S phase activation of the histone H2B promoter by OCA-S, a coactivator complex that contains GAPDH as a key component. Cell **114**:255–66.
149. Zhou, S., and J. W. Dewille. 2007. Proteasome-mediated CCAAT/enhancer-binding protein delta (C/EBPdelta) degradation is ubiquitin-independent. Biochem J **405**:341–9.
150. Zwickl, P., and W. Baumeister. 1999. AAA-ATPases at the crossroads of protein life and death. Nat Cell Biol **1**:E97–8.

2

THE STRUCTURAL BIOLOGY OF IDPs INSIDE CELLS

PHILIPP SELENKO
Leibniz Institute of Molecular Pharmacology (FMP), Berlin, Germany

ABSTRACT

It has long been axiomatic that the function of a protein is directly related to its three-dimensional structure. Intrinsically disordered proteins (IDPs) defy this structure–function paradigm. Although many IDPs have been well characterized *in vitro*, we know little about their structural *in vivo* properties. What kind of conformational features do natively unfolded proteins exhibit inside live cells? How does the crowded intracellular environment affect their disordered states? Do cellular IDPs exist as isolated monomers, or are they always bound to interacting proteins? Is the disordered state sustainable under native *in vivo* conditions at all? Until recently, we had no means of "looking" into cells with high enough resolution to address these questions. The advent of in-cell NMR spectroscopy has changed this notion. Comparable to an atomic resolution microscope, this technique provides high-resolution structural information on proteins inside prokaryotic and eukaryotic cells. In this chapter, I will outline how in-cell NMR methods can advance our structural understanding of natively unfolded proteins in cellular *in vivo* environments.

2.1 INTRODUCTION

Many gene sequences encode entire proteins or large segments of proteins that lack any well-defined secondary and/or tertiary structure in their pure

Instrumental Analysis of Intrinsically Disordered Proteins: Assessing Structure and Conformation, Edited by Vladimir Uversky and Sonia Longhi

states and under isolated *in vitro* conditions. These intrinsically disordered proteins (IDPs) are estimated to account for up to 30% of all human proteins (23) and have been shown to exert important regulatory functions in key cellular processes (22). IDPs can be highly conserved between species, both in composition and sequence, and, contrary to the traditional view that protein function equates with a stable three-dimensional structure, that is, the structure–function paradigm (102), IDPs are often functional in ways that we are only beginning to understand (22, 38). Many disordered protein segments fold upon binding to their biological targets (i.e., coupled folding and binding) (24), whereas others have constitutively flexible linkers that function in the assembly of macromolecular arrays (95) or as favorable environments for posttranslational protein modifications (39). Collectively, IDPs appear to accommodate many of the functional characteristics of folded proteins without actually being folded in the "classical" way.

One reason for the growing interest in IDPs stems from the fact that many of them are implicated in human diseases. IDPs are overrepresented in cell signaling and cancer-associated proteins (38), and correlate with prevalent human neurodegenerative disorders, like Alzheimer's (AD), Parkinson's (PD), Huntington's (HD), and Creutzfeld–Jakob's disease (CJD). The Aβ peptide and tau protein of AD, α-synuclein (α-syn) of PD, are classified IDPs that completely lack any defined secondary structure over their entire protein sequence (100). Mammalian PrP proteins of the transmissible spongiform encephalopathies (prion or CJD) are equally divided into intrinsically unstructured N-terminal halves and globular C-terminal portions. HD is caused by trinucleotide expansions in the *huntingtin* gene, which similarly introduces large segments of polyQ repeats in this otherwise folded protein. All of the proteins above are implicated in the formation of highly ordered, insoluble, fibrillar deposits (or β-aggregates), which serve as hallmarks of the respective neuropathological disorders.

2.1.1 Physical and Structural Properties of IDPs *In Vitro*

IDPs exhibit amino acid sequence propensities that make their *in silicio* predictions highly accurate (67). They typically display high compositional bias toward a low content of bulky hydrophobic amino acids, which would normally form the core of a folded globular protein, and a high proportion of polar and charged residues. These properties normally render purified IDPs highly soluble and reluctant to spontaneous aggregation (98), which is in contrast to their cellular tendency for pathological fibrillation (see above). Surprisingly, IDPs also exhibit a lower frequency of "β-aggregation nucleating segments" than globular proteins or unstructured regions within globular proteins (50). They score low in sequence complexity and high in the occurrence of repeat expansions (96).

From a structural point of view, many IDPs display features of preferred segmental orientations and overall conformations that are often closer to globular folds than to fully extended polypeptide chains. In that sense, it is important to emphasize that the absence of "classical" secondary or tertiary structure does not necessarily imply that IDPs adopt random conformations. Or, for that matter, that molecules of a certain IDP sample random conformational spaces. It rather seems that IDPs display preferred segmental orientations that are populated to varying degrees. *In vitro*, many IDPs exhibit residual helical propensities, whereas β-sheet properties are much less common. This is surprising given the fact that the conformational states of IDPs in all known amyloid aggregates are β-sheet. In addition, a helical conversion is the preferred structural transition of IDPs that adopt a folded conformation upon binding to other proteins.

2.1.2 IDPs inside Cells

While we appreciate many of the physical *in vitro* properties of IDPs, we know very little about their structural *in vivo* characteristics. The obvious question to ask is: What do IDPs actually look like inside live cells? Before I introduce the appropriate biophysical tools to address this question, I will try to justify the apparent naivety of my inquiry by outlining some general considerations about the intracellular space that make it far from trivial to answer. These considerations can be divided into two main parts: physical parameters of the intracellular environment that affect the thermodynamic, kinetic, and conformational properties of IDPs inside live cells, and biological activities that modulate their cellular *in vivo* behavior.

2.1.2.1 Macromolecular Crowding A general feature of the intracellular space is the high content of biological macromolecules. A typical cell contains ~25% of protein by volume, of which ~10% are cytoskeletal filaments and ~90% are soluble globular proteins, along with substantial amounts of RNA and DNA, and other biopolymers like lipids. Collectively, the term macromolecular crowding denotes the combined effects that the cellular interior exerts on any of its suspended components. Macromolecular crowding represents an important functional aspect of cellular complexity, and its influence on biological reactions has been addressed experimentally (26, 27) and quantitatively described with NMR spectroscopy as the method of choice (4, 90). Macromolecules occupy about 30% of the cell volume, thus making this space unavailable to other macromolecules. This property largely affects intracellular diffusion, which in turn has direct consequences on the kinetic and thermodynamic properties of macromolecular binding events. On the one hand, macromolecular crowding reduces the overall reaction kinetics of macromolecular association processes by decreasing diffusion-controlled on-rates (77). On the other hand, the "excluded volume effect" eventually increases the effective local concentration of ligands at protein binding sites and hence decreases the energy barrier for the formation of "activated state" intermediates. The net result of these opposing effects largely depends on the precise nature of the respective reactants and the specific macromolecular environment. In general, the main contribution of crowding on biochemical equilibria is to favor the association of macromolecules. Equilibrium constants for binding events can be increased by as much as two to three orders of magnitude, depending on the relative sizes and shapes of the reactants and products, and on those of the background molecules (60). For IDPs, it has been proposed that macromolecular crowding enhances amyloid formation in protein deposition diseases (59). The validity of this assumption has been experimentally confirmed by the increased fibrillation of human apolipoprotein C-II (35) and α-syn (86, 99) under crowded *in vitro* conditions. In addition, it has been shown that crowding can induce the formation of protein structure in the intrinsically unfolded FglM protein (19). This observation has raised the question whether some IDPs can generally display properties of folded proteins under native *in vivo* conditions. Although it is unlikely that this is true as a general feature of IDPs, especially when considering their bias in amino acid composition (see above), it emphasizes the importance of cellular crowding in understanding the native *in vivo* characteristics of natively unfolded polypeptides.

2.1.2.2 Biological Activities There is culminating evidence that certain biological activities can alter the conformational properties of IDPs inside cells. As has been mentioned earlier, IDP folding upon cognate protein binding is likely to constitute a frequent cellular event (24). Clearly, the intracellular environment provides a plethora of potential interacting partners, which can include proteins, membranes, or RNA/DNA molecules. All of the above can potentially serve as scaffolding surfaces for structural transitions that could not be populated under isolated *in vitro* conditions. In that sense, IDP folding upon protein binding may not necessarily depend on a single designated biological interaction inside a cell. It could rather constitute a random and unspecific event that exerts a catalytic function in allowing the IDP to escape a certain structural entrapment. The degree of macromolecular crowding of the intracellular space might then be sufficient to stabilize the folded conformation or to make global unfolding less favorable. Although there does not exist, to my knowledge, a single experimental study that specifically addresses this question, some data hint to unspecific priming steps in the course of IDP folding upon protein binding (93). The wealth of possible encounters in the intracellular space seems to make such scenarios plausible. Thus, suggesting that initial priming interactions could possibly be facilitated by many cellular proteins and might not stringently depend on a defined biological binding event.

Other types of biological activities have been shown to exert a strong influence on the cellular behavior of IDPs, both from a structural and functional point of view. Most prominently among these are posttranslational protein modifications (73). Sites of protein phosphorylation, for example, can be found in regions of structural disorder, as well as within regions of well-ordered structure. However, the similarity in sequence complexity, amino acid composition, and flexibility parameters between protein phosphorylation sites and disordered protein regions suggests that intrinsic disorder in and around modification sites constitutes a common recognition feature for many eukaryotic serine, threonine, and tyrosine kinases (39). With regard to the variety of structural consequences of protein phosphorylation events, all sorts of conformational responses have been observed: maintenance of disorder, disorder-to-order transitions, as well as order-to-disorder transitions, and the preservation of the ordered state (42). One must therefore take into account that posttranslational protein modifications might indeed alter the conformational characteristics of IDPs, especially in a cellular context.

When considering cellular biological activities and their effects on the *in vivo* fate of IDPs, some other questions intuitively arise. What about cellular chaperones? How do these proteins affect the conformational properties of IDPs in live cells? What about the *in vivo* stability of IDPs? Should IDPs not function as preferred substrates for cellular proteases? Surprisingly, there is no evidence that IDPs constitute preferred targets for cellular chaperones or for endogenous protein proteases. As a matter of fact, a recent computational survey identified a negative correlation between the degree of protein disorder and a proteins' propensity to serve as a chaperone substrate (36). Moreover, this anticorrelation behavior appears to be preserved across multiple species, ranging from prokaryotes, to unicellular eukaryotes like yeast, to metazoans. Surprisingly, some IDPs, rather than serving as substrates for cellular chaperones, even display chaperone-like activities themselves (45). In a second instance, structural disorder appears to serve as a rather weak signal for intracellular protein degradation (97). Only among very short-lived proteins, this study found significant, although small, correlations between the levels of protein disorder (i.e., number of disordered residues, length of disordered segments) and the *in vivo* protein half-lives. Taken together, IDPs, in comparison to folded proteins, do not appear to be more rapidly degraded inside live cells. In addition, this behavior does not seem to be attributed to elevated levels of protection by intracellular chaperones.

2.2 IDPs AND NMR SPECTROSCOPY

The biophysical characterization of IDPs is tightly connected to spectroscopic techniques in aqueous solutions (see chapters 3–12). Because IDPs do not form ordered crystals, as would be required for high-resolution X-ray analyses, or yield suitable specimens for EM experiments, both methods cannot be employed to investigate the structural *in vitro* properties of IDPs. Indeed, the method of choice for atomic resolution IDP characterization has been solution-state NMR spectroscopy (25). Numerous NMR methods have been specifically developed, or adapted, to analyze the conformational properties of unfolded proteins *in vitro* (8). NMR pulse sequences have been tailored for degenerate chemical shift analyses and for studying small nuclear Overhauser effects in disordered protein regions (14, 44, 94, 105). Complementary, a variety of paramagnetic spin labeling techniques have been devised to observe residual long-range interactions in unfolded polypeptide chains (31, 49). IDPs have been dynamically characterized by spin relaxation experiments (7, 9, 12) and by residual dipolar coupling measurements (53, 85). Chapters 3 and 4 of this book will outline some of the *in vitro* NMR techniques that can be employed for the characterization of isolated IDPs in solution. For the remainder of this chapter, I will focus on a special form of solution-state NMR spectroscopy that enables the *in vivo* characterization of IDPs inside live cells: high-resolution in-cell NMR spectroscopy.

2.2.1 In-Cell NMR Spectroscopy

Most structural investigations of biomolecules are typically confined to artificial and isolated *in vitro* experimental setups. It is beyond any doubt that *in vitro* approaches have proven extremely powerful in shaping our structural knowledge about many biomolecules, including IDPs. It is nevertheless apparent that our knowledge about the cellular states of proteins in general, and IDPs in particular, is only fractional.

To study biomolecules in their natural environments (i.e., within cells), recent attempts have aimed at the development of novel *in vivo* techniques for high-resolution structural biology (76). As mentioned before, X-ray crystallography and electron microscopy are intrinsically restricted from *in vivo* methods due to their requirement for pure samples and crystalline or vitrified specimens. NMR spectroscopy, the only other biophysical method for structural investigations at the atomic level, allows for the direct observation of NMR-active nuclei within any NMR-inactive environment and can thus be employed to structurally investigate labeled proteins inside cells (80).

The overall concept of in-cell NMR spectroscopy (82) represents the extended application of a basic magneto resonance principle. Most nuclei in natural substances are NMR inactive and, hence, not detectable by NMR methods. Biomolecular NMR experiments therefore make it necessary to substitute NMR-inactive nuclei with NMR-active stable isotopes in order to make proteins and other biomolecules "visible" to the spectroscopic eye. Today, isotope labeling constitutes a routine procedure in biomolecular NMR spectroscopy (33). In addition to labeling all residues of a protein, elaborate laboratory protocols have been devised to introduce NMR-observable isotopes at specific protein sites (i.e., site-selective labeling) or at subsets of amino acid residues (i.e., residue specific labeling). Together, these techniques allow us to either visualize whole proteins or to engineer NMR-observable sites within proteins of interest. For in-cell NMR applications, the isotope effect is exploited as a selective visualization filter. By analyzing isotope-labeled proteins in complex but NMR-inactive environments, that is, inside cells, researchers can specifically study the influence of the intracellular environment on a protein's structure and function.

2.2.1.1 NMR Parameters for In-Cell NMR We have recently reviewed some of the most important NMR parameters with regard to in-cell NMR applications (80). In the following section, I will reiterate some of the points made in this paper.

Two-dimensional heteronuclear correlation experiments of ^{1}H and ^{15}N or ^{13}C-labeled proteins serve as the primary NMR techniques employed for high-resolution in-cell measurements. The correlation of these NMR-active atomic nuclei, by means of specifically tailored NMR pulse sequences, yields individual NMR signals (or resonance cross-peaks) at the respective resonance frequencies (or chemical shift values, $\delta[^{1}H]$, $\delta[^{15}N]$, and/or $\delta[^{13}C]$) of the observable spin systems. For a folded protein, the exact chemical shift of any NMR-active nuclei is a direct consequence of its immediate chemical environment, which, in turn, is defined by a proteins' unique three-dimensional structure. This leads to a characteristic pattern (or "fingerprint") of resonance cross-peaks in the two-dimensional correlation spectrum, which reflects the specific conformational state of the labeled compound under investigation. Local changes in the chemical environment of labeled molecules, either by unlabeled ligand binding or other conformational rearrangements, result in different resonance frequencies of the involved residues. The complexity of the initial correlation spectrum remains unchanged because both reactions do not introduce observable spin pairs. Differences in resonance frequencies of the labeled specimen are measured as "chemical shift changes" ($\Delta\delta_{total} = \Delta\delta[^{1}H] + \Delta\delta[^{15}N$ or $^{13}C]$) and indirectly translated into three-dimensional information about localized structural alterations. Hence, chemical shift values, and changes thereof, serve as unique indicators of a protein's structure, conformational modulations, or localized binding events both *in vitro* and *in vivo* (106).

In addition, the characteristic appearance of each NMR signal, the NMR line shape, contains information about the spin system under investigation. A prerequisite for liquid-state NMR spectroscopy is that the molecule of interest tumbles freely in solution. The resulting overall tumbling rate depends on the size of the molecule, and the temperature and the viscosity of the sample solution. These parameters determine the overall line shape of respective NMR resonance signals. Yet, NMR resonances from one protein or from one protein domain do not all exhibit identical peak intensities or uniform NMR line shapes. Properties like internal mobility, solvent, and conformational exchange differentially affect the relaxation properties of individual spins and, conversely, the appearance of the respective NMR signals (29, 70, 71). Residues of secondary structure elements often exhibit faster relaxation rates than spin systems within unstructured, loop regions. This typically results in narrower, slightly more intense resonance signals in the latter case. These dynamic properties may undergo differential alterations in a cellular environment, and comparative analyses of the resulting changes of these peak parameters can provide information about the modulation of internal dynamics and exchange in a cellular context. In general, small proteins, or intrinsically unfolded polypeptides, display large tumbling rates, which lead to slow overall relaxation and conversely narrow peak shapes. Molecules of larger size tumble more slowly, relax faster, and exhibit broader resonance signals. Because the overall rotational tumbling rate is a direct function of the viscosity of the medium in which the macromolecule is dissolved, intracellular viscosity becomes a crucial parameter for in-cell NMR experiments. Any molecule in a cellular context will exhibit a reduced overall tumbling rate due to intracellular viscosity and hence display broader NMR line shapes. In the absence of sample binding to endogenous cellular factors, a direct comparison of the appearance of individual protein NMR signals in buffer and in in-cell experiments will readily yield a qualitative estimate about the contribution of viscosity and intracellular crowding to a proteins' in-cell behavior. Upon interaction with cellular components, the resulting protein complexes can either display a tumbling rate that corresponds to the sum of their individual masses, in the case of tight overall binding, or exhibit individual contributions to a mixed set of rates, when binding is restricted to a subset of residues. The latter results in residue-specific differential line broadening, which yields additional information on the dynamics and localization of the interaction. Interactions with quasi-static cellular structures such as organelles or membranes result in severe signal broadening, which can still serve as qualitative indicators of the biological nature of these interactions. Additionally, such binding events are often reversible and modulated by cellular signaling events. Therefore, some of them might become, at least partially, amenable to structural investigations.

It is apparent that any of these aforementioned scenarios, and superpositions thereof, are possible for a biologically active protein in a cellular context. This may result in increasingly complicated in-cell NMR spectra or in a poor overall quality of the spectroscopic data. In such cases, the researcher needs to reduce the complexity of the system under investigation. This can be achieved by "chopping up" full-length proteins into individual domains to selectively probe differential cellular contributions or by introducing mutations that abolish certain functional characteristics and similarly enable to discriminate between specific biological activities.

2.2.1.2 In-Cell NMR of IDPs in Prokaryotic Cells Historically, the first high-resolution in-cell NMR experiments have been performed in bacterial cells (83). Indeed, most in-cell NMR applications today still employ bacterial cells and rely on the same rationale of sample overexpression, labeling and measuring within the same cell type (81). While most prokaryotic in-cell NMR studies describe the *in vivo* behavior of folded proteins, a selected few specifically investigate the cellular properties of IDPs.

The *in vivo* NMR analysis of the bacterial FglM protein is one such example (19). FglM is a 97-residue polypeptide from *Salmonella typhimurium*, which regulates flagellar synthesis by binding to the transcription factor δ^{28} (37). Free FglM is mostly unstructured in dilute solution, but its C-terminal half can form a transient α-helix (16). Upon interaction with δ^{28} *in vitro*, this portion of FglM becomes structured, which is manifested by the disappearance of a set of C-terminal NMR resonance signals (15). In their study, Dedmon et al. have exploited this behavior to investigate the conformational properties of FglM in different *in vitro* and *in vivo* solutions. Inside live *Escherichia coli* cells, the same set of NMR resonance signals was absent that would also disappear upon δ^{28} binding *in vitro*. The authors reasoned that this could possibly be explained by a structural rearrangement that was similar to the one observed with δ^{28} (15). Dedmon et al. also argued that binding of FglM to an endogenous homologue of δ^{28} could not explain the observed effect because δ^{28} does not naturally occur in *E. coli*. Further analyses in glucose- (450 g/L), BSA- (400 g/L), or ovalbumin-crowded solutions (450 g/L) revealed that the observed transition did indeed, and exclusively, depend on the absolute amount of the respective crowding agent and, hence, on the effective degree of macromolecular crowding (19). Thus, providing a fine example of the influence of intracellular crowding on the structural *in vivo* properties of an IDP.

In another recent study from the Pielak laboratory, Li et al. compared the effects of macromolecular crowding on a folded versus an intrinsically unfolded protein (47). The authors found that NMR resonances of the intrinsically unfolded human protein α-syn (~14 kD), the polypeptide that forms insoluble aggregates (amyloid fibrils) in PD patients, were detectable inside live *E. coli* cells, whereas NMR signals of the much smaller but folded chymotrypsin inhibitor 2 (CI2, ~7 kD) were not. Li et al. speculated that this difference could possibly be explained by a differential protein response to intracellular viscosity and macromolecular crowding. Indeed, measuring NMR relaxation parameters in artificially crowded solutions for both proteins indicated that these changed more unfavorably for the folded CI2 than for the intrinsically unfolded α-syn protein. What exactly were those changes and how could they be explained on an atomistic level? In brief, Li et al. comparatively measured two sets of NMR parameters that reflect the dynamic properties of CI2 and α-syn: the ^{15}N longitudinal relaxation rate (R_1 or $1/T_1$) and the in-phase ^{15}N transverse relaxation rate (R_2 or $1/T_2$). Both are inherently connected to the overall signal quality of NMR experiments. R_1 and R_2 can very loosely be described as reflecting the global motion or the internal mobility of a protein, respectively (40). Whereas R_1 values of α-syn were quite insensitive to the overall increase in viscosity of the employed *in vitro* experimental setup (~50 times that of a dilute solution), values for CI2 decreased threefold to fourfold. At the same time, the individual R_2s of α-syn only increased between 1.5-fold and 6-fold. R_2s of CI2 were collectively increased up to 40-fold. Taken together, these results demonstrated that the dynamic properties of folded proteins appeared to be more sensitive toward changes in the macromolecular environment than those of IDPs. Concurrently, NMR detection of IDPs inside live cells, in the absence of other biological interactions, could generally be more favorable than the observation of folded proteins. One of the reasons for this behavior could lie in the differences in global and local motions for ordered and disordered proteins. Because of their rigidity, the relaxation rates of globular proteins are most sensitive to global motions, which are usually described by a single rotational correlation time (τ_c) (40). Disordered proteins are flexible, and their motions are best described by an ensemble of interconverting conformers. Therefore, every IDP residue can exhibit a different effective correlation time (43). It thus seems plausible that the overall flexibility of IDPs lessens the negative effects of intracellular viscosity on the quality of their in-cell NMR spectra.

In a preceding in-cell NMR study about the prokaryotic *in vivo* properties of human α-syn, McNulty et al. also showed that this classical IDP remained soluble, monomeric, and, in contrast to FglM, completely unfolded in the periplasm of live *E. coli* cells (58). This finding was surprising especially given the fact that macromolecular crowding had been shown to function as a strong enhancer of α-syn aggregation *in vitro* (86, 99). Considering the high degree of crowding and intracellular viscosity, >10 times that of water in the cytoplasm of *E. coli* (65), one would have expected a much greater tendency of α-syn to aggregate inside those cells. With regard to the structural *in vivo* properties of α-syn, however, the situation is more complicated.

Whereas isolated α-syn (140aa) is disordered in solution, it exhibits structural *in vitro* properties that deviate from a completely extended, random coil conformation (3, 18). Long-range tertiary interactions, mediated primarily by the acidic C-terminal tail of α-syn and by a central region of the protein, result in ensembles of compact monomeric conformers with hydrodynamic radii that are, on average, ~10 Å smaller than would be expected for a completely denatured 140-residue polypeptide (61). Hence, isolated α-syn adopts an unfolded but compact shape *in vitro*. In this form, its N-terminal ~100 residues show experimentally detectable helical propensities (10). In the presence of lipid vesicles or detergent micelles, these helical segments are stabilized and function in "membrane" anchoring (17). In the membrane-bound state, only the C-terminal ~40 residues of α-syn remain in a disordered and extended conformation, solvent exposed, and with sufficient internal mobility to be detected by NMR (10). Thus, in the presence of adequate membrane or membrane-like environments, the globular conformation of α-syn is lost and the protein forms an overall extended structure. In addition, the unfolded characteristics of isolated α-syn are highly sensitive to both temperature and pH (57, 100). Significant amounts of partially folded α-syn intermediates, predominantly of monomeric β-structures, are being formed with increasing temperature (>25°C) or upon lowering the pH (<7.5) (100). This could be the reason for the poor quality *in vitro* NMR spectra of isolated α-syn at physiological temperature (37°C) and at neutral pH (7.5) (57). Under these conditions, the first ~100 residues of α-syn are severely line broadened and effectively undetectable by *in vitro* NMR experiments (which could reflect intermediate conformational exchange between unfolded and partially folded forms of the protein). In the cytoplasm of live *E. coli* cells and at the "physiological" temperature of 35°C, α-syn adopted overall structural features that were reminiscent of the unfolded, isolated protein state at 10°C (i.e., also the first 100 residues were detectable by in-cell NMR measurements) (58). This behavior was interpreted as α-syn remaining preferentially disordered under crowded cellular conditions without forming partially folded intermediates. (A similar effect was observed with artificially crowded solutions *in vitro*.) Whereas this is a valid interpretation of the provided NMR data, McNulty et al. did not specifically address the possibility that differential chemical exchange effects *in vitro* and *in vivo* could also cause the observed differences in NMR signal qualities. Indeed, a recent study by Croke et al. identified chemical exchange as the likely source for the observed line-broadening effects of α-syn, both *in vitro* and *in vivo* (13). Our own data strongly support this notion (unpublished).

What are the biological implications of these prokaryotic in-cell NMR experiments? Foremost, they indicate the overall feasibility of high-resolution, structural studies of IDPs under native cellular *in vivo* conditions. They prove that different *in vivo* protein conformations can be readily identified by simple NMR experiments and that changes in protein structures can be detected in a straightforward manner. Furthermore, they underscore the universal importance of cellular crowding in determining the *in vivo* features of intracellular protein structures (26, 27). In this regard, the relevance of the FglM study is readily appreciated. The biological significance of the α-syn investigation is more difficult to assess. Above all, α-syn is a human protein that was assayed in a bacterial cellular context. With regard to α-syn's biological role in associating with cellular membranes (41), it is of particular importance to keep in mind that the lipid composition of *E. coli* differs greatly from mammalian cells (even more so from human neuronal cells) (68). In that sense, the intracellular space of *E. coli* constitutes a vastly different biological environment than α-syn's native cellular context. The point to make is that in-cell NMR measurements in bacterial cells might not faithfully reflect the true native environments of proteins from higher organisms. The development of eukaryotic cellular model systems for in-cell NMR applications thus constituted an important step toward analyzing eukaryotic proteins under more physiologically relevant conditions.

2.2.1.3 In-Cell NMR of IDPs in Eukaryotic Cells In-cell NMR applications in eukaryotic cells are fairly new additions to the rapidly growing field of high-resolution *in vivo* NMR studies. The first in-cell NMR measurements in eukaryotic cells were reported for folded proteins inside *Xenopus laevis* oocytes (75, 79). These amphibian cells have long served as important laboratory tools in the disciplines of developmental and cellular biology (52, 66). Mature oocyte cells (stage VI) arrest in prophase at the G2/M boundary of the first meiotic division and contain large cell volumes (~1 μL, compared with a few picoliter (pL) as for most somatic cells), 20% of which comprises the nuclear organelle (or germinal vesicle). Most importantly, *Xenopus* oocytes are large enough to be conveniently manipulated by microinjection. In this way, defined amounts of isotope-labeled NMR-active proteins can easily be deposited in the cytoplasm of these cells (84). The general rationale for introducing labeled specimens into the otherwise unlabeled environment of live cells thus differs from bacterial in-cell NMR applications. Proteins are recombinantly produced in *E. coli*, purified to homogeneity, and then introduced into the eukaryotic cellular environment for in-cell NMR analyses (80). This procedure offers several advantages over the bacterial in-cell NMR approach of sample overproduction and measurement within the same cell type. Foremost, background-labeling artifacts that often affect the quality of prokaryotic in-cell NMR measurements (74) are effectively eliminated.

With regard to eukaryotic in-cell NMR studies of IDPs, only one such analysis has been reported to date (6). In this paper, Bodart et al. investigate the conformational *in vivo* properties of the human tau protein inside *X. laevis* oocytes. Tau is one of the largest known IDP. It exhibits disordered features over the entireness of its 441 amino acid residues (~45 kD) (51). In contrast to other IDPs, only weak propensities for transient secondary structure (turn conformations and β-structures) have been observed (63, 64). Tau binds to, and stabilizes, microtubules (28, 62, 87). At the same time, tau constitutes a major part of the intracellular tangles that form inside neurons of AD patients and displays high tendencies for self-aggregation (32). The tau protein thus presents, by any means, a veritable challenge for NMR spectroscopy (51, 63, 88, 89): first, because of its size; second, because of its relatively low signal-to-noise behavior in NMR experiments; third, because of its congesting degree of spectral overlap (the protein is completely disordered after all); and fourth, because of its biological function to bind to one of the largest macromolecular structures that exist inside every living cell (i.e., microtubules), and which are typically present at high natural abundance (>10 μM in general).

With their in-cell NMR project, Bodart et al. took upon them the truly remarkable effort to study tau at close to physiological protein concentrations. At an intracellular concentration of 5 μM, in-cell NMR spectra revealed a predominantly unfolded conformation of tau (6). Due to the large degree of viscosity-driven line broadening and the resulting additional increase in NMR signal overlap, it was difficult to assess whether some portions of tau were adopting novel structural properties that were not present in the pure state. In general, intracellular crowding did not appear to induce a drastic conformational rearrangement of the protein. Several features of the in-cell NMR spectra did, however, indicate that large portions of intracellular tau could be bound to endogenous microtubules. This conclusion was drawn based upon some striking similarities between *in vitro* NMR spectra of tubulin-bound tau (87) and the observed *in vivo* NMR data. Indeed, given the endogenous concentration of tubulin in *Xenopus* oocytes (~20 μM [30]) and the known stochiometry of tau binding to preformed microtubules (1:3 [55]), such a scenario is well possible. In addition, the in-cell NMR spectra displayed several resonance signals that strongly suggested that residues of tau became posttranslationally phosphorylated by endogenous *Xenopus* kinases. Comparing the new NMR resonance signals with those of *in vitro* phosphorylated tau (46) indeed confirmed that *in vivo* protein phosphorylation was likely to take place.

This observation highlights an important feature of eukaryotic in-cell NMR measurements that I have not yet elaborated on: the ability to directly observe the establishment of posttranslational protein modifications inside live cells (80). We have recently investigated this propensity in a systematic manner *in vitro* and *in vivo* (78). In the next paragraph, I will briefly outline why cellular protein modifications are "naturally" detected by in-cell NMR measurements in eukaryotic cells, and then discuss the relevance of this property for *in vivo* IDP analyses.

2.2.2 In-Cell NMR and Posttranslational Protein Modifications

In 2007, we set out to determine the suitability of time-resolved, high-resolution NMR measurements for observing posttranslational protein phosphorylation *in vitro* and *in vivo*. We reasoned that NMR spectroscopy is ideally suited to detect de novo protein phosphorylation, because the resonance signals of individual amino acids are highly sensitive to phosphorylation-induced changes in the environment of the respectively modified atomic nuclei (5). Therefore, site-specific protein phosphorylation can be detected at atomic resolution and in a quantitative manner. Another advantage of analyzing protein phosphorylation events by NMR spectroscopy is that the progressive establishment of these amendments can be monitored in a noninvasive and nondisruptive way. Protein phosphorylation reactions can therefore be studied in a continuous fashion, and stepwise modification events can be dissected in a time-dependent manner. Furthermore, the modification of substrate residues by phosphorylation constitutes a covalent chemical addition that is not subject to exchange behaviors, unlike protein–protein interactions, for example. Finally, phospho-modifications result in negligible mass increases so that unmodified and modified substrate species are equally well detected. These features, we reasoned, predestine NMR spectroscopy as a novel tool for posttranslational modification research. To put it bluntly, NMR theory predicts that posttranslational protein modifications of isotope-labeled NMR-active proteins cannot escape detection by NMR spectroscopy, irrespective of whether the specimen is dispensed in an *in vitro* solution or localized inside a live cell. The experimental proof of this concept, with regard to protein phosphorylation, has, in the meantime, been provided by our (78) and other laboratories (6, 46, 69). What about other types of posttranslational protein modifications? Again, NMR theory predicts that any form of residue-specific modification will yield changes in the observable resonance signals, irrespective of the chemical nature of the alteration. I therefore expect that high-resolution NMR spectroscopy will turn out to be equally well suited to similarly detect other types of posttranslational modifications, both *in vitro* and *in vivo*.

What are the implications of this notion for the NMR analysis of IDPs? Foremost, it adds another level of structural and functional information that can, in principle, be obtained by in-cell NMR studies. As has been discussed in section 2.1.2.2, IDPs often undergo posttranslational protein modifications. As also mentioned earlier, these can have profound effects on the structural and functional characteristics of IDPs. In that sense, in-cell NMR spectroscopy represents a high-resolution method that is capable of jointly reporting the structural and functional characteristics of IDPs inside live cells. Who would argue that this does not sound like a promising new tool for future *in vivo* IDP research?

2.3 FUTURE PERSPECTIVES

Without a doubt, in-cell NMR spectroscopy will continue to constitute a most versatile tool for analyzing the conformational properties of IDPs inside live cells. After all, this technique is still in its infancy and a great deal of methodological advancements is to be expected for the near future. One of the drawbacks of the currently available cellular model systems, both prokaryotic and eukaryotic, is the limited number of cellular environments that they represent. Ideally, of course, we would like to analyze IDPs that occur in a specific subset of human brain cells, like many of the amyloid disease proteins, for example, in their truly native biological contexts. Only in-cell NMR analyses in these environments could genuinely be considered physiologically relevant. Unfortunately, these types of experiments are currently not possible. What is missing are other cellular model systems for eukaryotic in-cell NMR applications and alternative means to introduce isotope-labeled proteins into these cells. Intracellular sample deposition by microinjection, like with *X. laevis* oocytes, is clearly not feasible for many other cell types, as they are often several orders of magnitude smaller than amphibian oocytes. In order to extend the general applicability of in-cell NMR measurements toward mammalian cells, for example, the development of alternative protocols for intracellular sample delivery is much needed. Ideally, such procedures should be applicable to a wide range of different cell types and easy to perform on a large number of cells.

One solution to this problem, suggested by us in a recent publication (80), could be the application of engineered cell-permeable protein transduction constructs. I am particularly intrigued by the so-called Trojan peptide tags, which confer efficient cell membrane transduction activities to a wide range of fused protein substrates (20, 21). These internalization sequences are composed of short, positively charged amino acid residues, which can be genetically engineered to be part of virtually any recombinant polypeptide (48). Upon labeled expression and purification of tagged fusion proteins from *E. coli*, these substrates are simply added to the growth medium of a variety of cultured laboratory cell lines and then readily internalized into the cytoplasms of these cells. In theory, this method should be generally applicable to a wide range of eukaryotic cell types and quantitatively accomplishable for a large number of cells.

Another approach could involve sample overexpression and in-cell NMR measurements in cells other than *E. coli*. Recent advances in structural genomics have also led to a more thorough investigation of possible alternatives to bacterial recombinant protein production and isotope-labeling schemes (33, 104). Among these, a few exotic approaches in mammalian CHO and HEK cells have been reported for *in vitro* NMR sample preparations (34, 54, 103). More prominent systems include the yeast *Pichia pastoris* and baculovirus-infected insect cells (11, 72, 91, 92). All of these eukaryotic cells have been employed to prepare labeled NMR samples for *in vitro* NMR analyses, and I can therefore only speculate about their experimental suitability for in-cell NMR measurements. The major obstacles for the selective labeling of recombinant proteins with NMR-active isotopes in these cells have been the difficulty to achieve adequate levels of protein overexpression, sufficient isotope incorporation rates, and the costs of isotope-enriched growth media. Growth media for NMR labeling in *E. coli* are simple in their composition, easily prepared, and, depending on the type of labeling, relatively cheap. Bacteria will also incorporate isotopes with high efficiency (~98%). Labeling media for eukaryotic cells are sophisticated; they must often be obtained commercially for satisfactory results; and they are expensive. The yeast *P. pastoris* might represent an exception to this notion since cells can be grown in glycerol/glucose and labeled in ^{15}N-ammoniumchloride and ^{13}C-methanol (101). Due to the complexity of most other eukaryotic metabolisms, isotope incorporation is typically less than 90%. With regard to in-cell NMR measurements, induction times for recombinant protein expression are in the range of days rather than hours, which is likely to increase the amount of adverse background labeling artifacts. In summary, I believe that in-cell NMR measurements in these eukaryotic cells could possibly be accomplished but are unlikely to constitute practicable routine approaches in the near future.

One of the most fundamental problems of any form of NMR spectroscopy is the inherent low sensitivity of this biophysical method. For in-cell NMR applications, this is especially detrimental because it makes *in vivo* measurements at physiological protein concentrations virtually impossible. The NMR-knowledgeable reader will now rightfully comment that overcoming the sensitivity problem of NMR spectroscopy is not going to be a trivial pursuit (to say the least). There are, however, new advancements in NMR technology that might, at least in principle, be also applicable to in-cell NMR measurements at some point in the future. Dynamic nuclear polarization, for example, has become a highly useful addition to the repertoire of novel NMR methods (2, 56). This technique holds the promise to increase the sensitivity of certain NMR experiments by up to a factor of 10,000 (1), which would clearly enable in-cell NMR measurements at concentrations far below those of most naturally occurring cellular proteins.

In conclusion, I hope to have convinced the valued reader that high-resolution in-cell NMR spectroscopy can greatly advance our structural understanding of IDPs inside live cells. For such a young technique, in-cell NMR spectroscopy has already contributed an impressive wealth of novel insights into the nature of the intracellular environment and its effects on the *in vivo* properties of IDPs. I am sure that this method will continue to do so in the years to come.

REFERENCES

1. Ardenkjaer-Larsen, J. H., B. Fridlund, A. Gram, G. Hansson, L. Hansson, M. H. Lerche, R. Servin, M. Thaning, and K. Golman. 2003. Increase in signal-to-noise ratio of >10,000 times in liquid-state NMR. Proc Natl Acad Sci U S A **100**:10158–63.
2. Bajaj, V. S., M. K. Hornstein, K. E. Kreischer, J. R. Sirigiri, P. P. Woskov, M. L. Mak-Jurkauskas, J. Herzfeld, R. J. Temkin, and R. G. Griffin. 2007. 250 GHz CW gyrotron oscillator for dynamic nuclear polarization in biological solid state NMR. J Magn Reson **189**:251–79.
3. Bernado, P., C. W. Bertoncini, C. Griesinger, M. Zweckstetter, and M. Blackledge. 2005. Defining long-range order and local disorder in native alpha-synuclein using residual dipolar couplings. J Am Chem Soc **127**:17968–9.
4. Bernado, P., J. Garcia de la Torre, and M. Pons. 2004. Macromolecular crowding in biological systems: hydrodynamics and NMR methods. J Mol Recognit **17**: 397–407.
5. Bienkiewicz, E. A., and K. J. Lumb. 1999. Random-coil chemical shifts of phosphorylated amino acids. J Biomol NMR **15**:203–6.
6. Bodart, J. F., J. M. Wieruszeski, L. Amniai, A. Leroy, I. Landrieu, A. Rousseau-Lescuyer, J. P. Vilain, and G. Lippens. 2008. NMR observation of Tau in *Xenopus* oocytes. J Magn Reson **192**:252–7.
7. Bracken, C. 2001. NMR spin relaxation methods for characterization of disorder and folding in proteins. J Mol Graph Model **19**:3–12.
8. Bracken, C., L. M. Iakoucheva, P. R. Romero, and A. K. Dunker. 2004. Combining prediction, computation and experiment for the characterization of protein disorder. Curr Opin Struct Biol **14**:570–6.
9. Buevich, A. V., U. P. Shinde, M. Inouye, and J. Baum. 2001. Backbone dynamics of the natively unfolded pro-peptide of subtilisin by heteronuclear NMR relaxation studies. J Biomol NMR **20**:233–49.
10. Bussell, R., Jr., and D. Eliezer. 2001. Residual structure and dynamics in Parkinson's disease-associated mutants of alpha-synuclein. J Biol Chem **276**:45996–6003.
11. Chen, C. Y., C. H. Cheng, Y. C. Chen, J. C. Lee, S. H. Chou, W. Huang, and W. J. Chuang. 2006. Preparation of amino-acid-type selective isotope labeling of protein expressed in *Pichia pastoris*. Proteins **62**:279–87.
12. Choy, W. Y., and L. E. Kay. 2003. Probing residual interactions in unfolded protein states using NMR spin relaxation techniques: an application to delta131delta. J Am Chem Soc **125**:11988–92.
13. Croke, R. L., C. O. Sallum, E. Watson, E. D. Watt, and A. T. Alexandrescu. 2008. Hydrogen exchange of monomeric alpha-synuclein shows unfolded structure per-

sists at physiological temperature and is independent of molecular crowding in *Escherichia coli*. Protein Sci **17**:1434–45.

14. Crowhurst, K. A., and J. D. Forman-Kay. 2003. Aromatic and methyl NOEs highlight hydrophobic clustering in the unfolded state of an SH3 domain. Biochemistry **42**:8687–95.
15. Daughdrill, G. W., M. S. Chadsey, J. E. Karlinsey, K. T. Hughes, and F. W. Dahlquist. 1997. The C-terminal half of the anti-sigma factor, FlgM, becomes structured when bound to its target, sigma 28. Nat Struct Biol **4**:285–91.
16. Daughdrill, G. W., L. J. Hanely, and F. W. Dahlquist. 1998. The C-terminal half of the anti-sigma factor FlgM contains a dynamic equilibrium solution structure favoring helical conformations. Biochemistry **37**:1076–82.
17. Davidson, W. S., A. Jonas, D. F. Clayton, and J. M. George. 1998. Stabilization of alpha-synuclein secondary structure upon binding to synthetic membranes. J Biol Chem **273**:9443–9.
18. Dedmon, M. M., K. Lindorff-Larsen, J. Christodoulou, M. Vendruscolo, and C. M. Dobson. 2005. Mapping long-range interactions in alpha-synuclein using spin-label NMR and ensemble molecular dynamics simulations. J Am Chem Soc **127**:476–7.
19. Dedmon, M. M., C. N. Patel, G. B. Young, and G. J. Pielak. 2002. FlgM gains structure in living cells. Proc Natl Acad Sci U S A **99**:12681–4.
20. Derossi, D., G. Chassaing, and A. Prochiantz. 1998. Trojan peptides: the penetratin system for intracellular delivery. Trends Cell Biol **8**:84–7.
21. Dietz, G. P., and M. Bahr. 2004. Delivery of bioactive molecules into the cell: the Trojan horse approach. Mol Cell Neurosci **27**:85–131.
22. Dunker, A. K., C. J. Brown, J. D. Lawson, L. M. Iakoucheva, and Z. Obradovic. 2002. Intrinsic disorder and protein function. Biochemistry **41**:6573–82.
23. Dunker, A. K., Z. Obradovic, P. Romero, E. C. Garner, and C. J. Brown. 2000. Intrinsic protein disorder in complete genomes. Genome Inform Ser Workshop Genome Inform **11**:161–71.
24. Dyson, H. J., and P. E. Wright. 2002. Coupling of folding and binding for unstructured proteins. Curr Opin Struct Biol **12**:54–60.
25. Dyson, H. J., and P. E. Wright. 2001. Nuclear magnetic resonance methods for elucidation of structure and dynamics in disordered states. Methods Enzymol **339**:258–70.
26. Ellis, R. J. 2001. Macromolecular crowding: an important but neglected aspect of the intracellular environment. Curr Opin Struct Biol **11**:114–9.
27. Ellis, R. J. 2001. Macromolecular crowding: obvious but underappreciated. Trends Biochem Sci **26**:597–604.
28. Fischer, D., M. D. Mukrasch, M. von Bergen, A. Klos-Witkowska, J. Biernat, C. Griesinger, E. Mandelkow, and M. Zweckstetter. 2007. Structural and microtubule binding properties of tau mutants of frontotemporal dementias. Biochemistry **46**:2574–82.
29. Fischer, M. W., L. Zeng, A. Majumdar, and E. R. Zuiderweg. 1998. Characterizing semilocal motions in proteins by NMR relaxation studies. Proc Natl Acad Sci U S A **95**:8016–9.

30. Gard, D. L., and M. W. Kirschner. 1987. Microtubule assembly in cytoplasmic extracts of *Xenopus* oocytes and eggs. J Cell Biol **105**:2191–201.
31. Gillespie, J. R., and D. Shortle. 1997. Characterization of long-range structure in the denatured state of staphylococcal nuclease. II. Distance restraints from paramagnetic relaxation and calculation of an ensemble of structures. J Mol Biol **268**:170–84.
32. Goedert, M., and M. G. Spillantini. 2006. A century of Alzheimer's disease. Science **314**:777–81.
33. Goto, N. K., and L. E. Kay. 2000. New developments in isotope labeling strategies for protein solution NMR spectroscopy. Curr Opin Struct Biol **10**:585–92.
34. Hansen, A. P., A. M. Petros, A. P. Mazar, T. M. Pederson, A. Rueter, and S. W. Fesik. 1992. A practical method for uniform isotopic labeling of recombinant proteins in mammalian cells. Biochemistry **31**:12713–8.
35. Hatters, D. M., A. P. Minton, and G. J. Howlett. 2002. Macromolecular crowding accelerates amyloid formation by human apolipoprotein C-II. J Biol Chem **277**:7824–30.
36. Hegyi, H., and P. Tompa. 2008. Intrinsically disordered proteins display no preference for chaperone binding in vivo. PLoS Comput Biol **4**:e1000017.
37. Hughes, K. T., K. L. Gillen, M. J. Semon, and J. E. Karlinsey. 1993. Sensing structural intermediates in bacterial flagellar assembly by export of a negative regulator. Science **262**:1277–80.
38. Iakoucheva, L. M., C. J. Brown, J. D. Lawson, Z. Obradovic, and A. K. Dunker. 2002. Intrinsic disorder in cell-signaling and cancer-associated proteins. J Mol Biol **323**:573–84.
39. Iakoucheva, L. M., P. Radivojac, C. J. Brown, T. R. O'Connor, J. G. Sikes, Z. Obradovic, and A. K. Dunker. 2004. The importance of intrinsic disorder for protein phosphorylation. Nucleic Acids Res **32**:1037–49.
40. Jarymowycz, V. A., and M. J. Stone. 2006. Fast time scale dynamics of protein backbones: NMR relaxation methods, applications, and functional consequences. Chem Rev **106**:1624–71.
41. Jo, E., J. McLaurin, C. M. Yip, P. St George-Hyslop, and P. E. Fraser. 2000. alpha-Synuclein membrane interactions and lipid specificity. J Biol Chem **275**:34328–34.
42. Johnson, L. N., and R. J. Lewis. 2001. Structural basis for control by phosphorylation. Chem Rev **101**:2209–42.
43. Kim, S., C. Bracken, and J. Baum. 1999. Characterization of millisecond timescale dynamics in the molten globule state of alpha-lactalbumin by NMR. J Mol Biol **294**:551–60.
44. Klein-Seetharaman, J., M. Oikawa, S. B. Grimshaw, J. Wirmer, E. Duchardt, T. Ueda, T. Imoto, L. J. Smith, C. M. Dobson, and H. Schwalbe. 2002. Long-range interactions within a nonnative protein. Science **295**:1719–22.
45. Kovacs, D., E. Kalmar, Z. Torok, and P. Tompa. 2008. Chaperone activity of ERD10 and ERD14, two disordered stress-related plant proteins. Plant Physiol **147**:381–90.
46. Landrieu, I., L. Lacosse, A. Leroy, J. M. Wieruszeski, X. Trivelli, A. Sillen, N. Sibille, H. Schwalbe, K. Saxena, T. Langer, and G. Lippens. 2006. NMR analysis of a Tau phosphorylation pattern. J Am Chem Soc **128**:3575–83.

47. Li, C., L. M. Charlton, A. Lakkavaram, C. Seagle, G. Wang, G. B. Young, J. M. Macdonald, and G. J. Pielak. 2008. Differential dynamical effects of macromolecular crowding on an intrinsically disordered protein and a globular protein: implications for in-cell NMR spectroscopy. J Am Chem Soc **130**:6310–1.
48. Li, Y., R. V. Rosal, P. W. Brandt-Rauf, and R. L. Fine. 2002. Correlation between hydrophobic properties and efficiency of carrier-mediated membrane transduction and apoptosis of a p53 C-terminal peptide. Biochem Biophys Res Commun **298**:439–49.
49. Lietzow, M. A., M. Jamin, H. J. Jane Dyson, and P. E. Wright. 2002. Mapping long-range contacts in a highly unfolded protein. J Mol Biol **322**:655–62.
50. Linding, R., J. Schymkowitz, F. Rousseau, F. Diella, and L. Serrano. 2004. A comparative study of the relationship between protein structure and beta-aggregation in globular and intrinsically disordered proteins. J Mol Biol **342**: 345–53.
51. Lippens, G., A. Sillen, C. Smet, J. M. Wieruszeski, A. Leroy, L. Buee, and I. Landrieu. 2006. Studying the natively unfolded neuronal Tau protein by solution NMR spectroscopy. Protein Pept Lett **13**:235–46.
52. Liu, J. X. (ed.) 2006. *Xenopus Protocols: Cell Biology and Signal Transduction*. Totowa, NJ: Humana Press.
53. Louhivuori, M., K. Paakkonen, K. Fredriksson, P. Permi, J. Lounila, and A. Annila. 2003. On the origin of residual dipolar couplings from denatured proteins. J Am Chem Soc **125**:15647–50.
54. Lustbader, J. W., S. Birken, S. Pollak, A. Pound, B. T. Chait, U. A. Mirza, S. Ramnarain, R. E. Canfield, and J. M. Brown. 1996. Expression of human chorionic gonadotropin uniformly labeled with NMR isotopes in Chinese hamster ovary cells: an advance toward rapid determination of glycoprotein structures. J Biomol NMR **7**:295–304.
55. Makrides, V., M. R. Massie, S. C. Feinstein, and J. Lew. 2004. Evidence for two distinct binding sites for tau on microtubules. Proc Natl Acad Sci U S A **101**: 6746–51.
56. Maly, T., G. T. Debelouchina, V. S. Bajaj, K. N. Hu, C. G. Joo, M. L. Mak-Jurkauskas, J. R. Sirigiri, P. C. van der Wel, J. Herzfeld, R. J. Temkin, and R. G. Griffin. 2008. Dynamic nuclear polarization at high magnetic fields. J Chem Phys **128**:052211.
57. McNulty, B. C., A. Tripathy, G. B. Young, L. M. Charlton, J. Orans, and G. J. Pielak. 2006. Temperature-induced reversible conformational change in the first 100 residues of alpha-synuclein. Protein Sci **15**:602–8.
58. McNulty, B. C., G. B. Young, and G. J. Pielak. 2006. Macromolecular crowding in the *Escherichia coli* periplasm maintains alpha-synuclein disorder. J Mol Biol **355**:893–7.
59. Minton, A. P. 2000. Implications of macromolecular crowding for protein assembly. Curr Opin Struct Biol **10**:34–9.
60. Minton, A. P. 2001. The influence of macromolecular crowding and macromolecular confinement on biochemical reactions in physiological media. J Biol Chem **276**:10577–80.
61. Morar, A. S., A. Olteanu, G. B. Young, and G. J. Pielak. 2001. Solvent-induced collapse of alpha-synuclein and acid-denatured cytochrome c. Protein Sci **10**:2195–9.

62. Mukrasch, M. D., J. Biernat, M. von Bergen, C. Griesinger, E. Mandelkow, and M. Zweckstetter. 2005. Sites of tau important for aggregation populate {beta}-structure and bind to microtubules and polyanions. J Biol Chem **280**:24978–86.

63. Mukrasch, M. D., P. Markwick, J. Biernat, M. Bergen, P. Bernado, C. Griesinger, E. Mandelkow, M. Zweckstetter, and M. Blackledge. 2007. Highly populated turn conformations in natively unfolded tau protein identified from residual dipolar couplings and molecular simulation. J Am Chem Soc **129**:5235–43.

64. Mukrasch, M. D., M. von Bergen, J. Biernat, D. Fischer, C. Griesinger, E. Mandelkow, and M. Zweckstetter. 2007. The "jaws" of the tau-microtubule interaction. J Biol Chem **282**:12230–9.

65. Mullineaux, C. W., A. Nenninger, N. Ray, and C. Robinson. 2006. Diffusion of green fluorescent protein in three cell environments in *Escherichia coli*. J Bacteriol **188**:3442–8.

66. Murray, A. W. 1991. *Xenopus laevis*: practical uses in cell and molecular biology. Methods Cell Biol **36**:1–718.

67. Obradovic, Z., K. Peng, S. Vucetic, P. Radivojac, C. J. Brown, and A. K. Dunker. 2003. Predicting intrinsic disorder from amino acid sequence. Proteins **53** (Suppl 6):566–72.

68. Opekarova, M., and W. Tanner. 2003. Specific lipid requirements of membrane proteins—a putative bottleneck in heterologous expression. Biochim Biophys Acta **1610**:11–22.

69. Paleologou, K. E., A. W. Schmid, C. C. Rospigliosi, H. Y. Kim, G. R. Lamberto, R. A. Fredenburg, P. T. Lansbury, Jr., C. O. Fernandez, D. Eliezer, M. Zweckstetter, and H. A. Lashuel. 2008. Phosphorylation at Ser-129 but not the phosphomimics S129E/D inhibits the fibrillation of alpha-synuclein. J Biol Chem **283**:16895–905.

70. Palmer, A. G., 3rd. 2001. Nmr probes of molecular dynamics: overview and comparison with other techniques. Annu Rev Biophys Biomol Struct **30**:129–55.

71. Peng, J. W., and G. Wagner. 1994. Investigation of protein motions via relaxation measurements. Methods Enzymol **239**:563–96.

72. Pickford, A. R., and J. M. O'Leary. 2004. Isotopic labeling of recombinant proteins from the methylotrophic yeast *Pichia pastoris*. Methods Mol Biol **278**:17–33.

73. Radivojac, P., L. M. Iakoucheva, C. J. Oldfield, Z. Obradovic, V. N. Uversky, and A. K. Dunker. 2007. Intrinsic disorder and functional proteomics. Biophys J **92**: 1439–56.

74. Rajagopalan, S., C. Chow, V. Raghunathan, C. G. Fry, and S. Cavagnero. 2004. NMR spectroscopic filtration of polypeptides and proteins in complex mixtures. J Biomol NMR **29**:505–16.

75. Sakai, T., H. Tochio, T. Tenno, Y. Ito, T. Kokubo, H. Hiroaki, and M. Shirakawa. 2006. In-cell NMR spectroscopy of proteins inside *Xenopus laevis* oocytes. J Biomol NMR **36**:179–88.

76. Sali, A., R. Glaeser, T. Earnest, and W. Baumeister. 2003. From words to literature in structural proteomics. Nature **422**:216–25.

77. Schnell, S., and T. E. Turner. 2004. Reaction kinetics in intracellular environments with macromolecular crowding: simulations and rate laws. Prog Biophys Mol Biol **85**:235–60.

78. Selenko, P., D. P. Frueh, S. J. Elsaesser, W. Haas, S. P. Gygi, and G. Wagner. 2008. In situ observation of protein phosphorylation by high-resolution NMR spectroscopy. Nat Struct Mol Biol **15**:321–9.

79. Selenko, P., Z. Serber, B. Gadea, J. Ruderman, and G. Wagner. 2006. Quantitative NMR analysis of the protein G B1 domain in *Xenopus laevis* egg extracts and intact oocytes. Proc Natl Acad Sci U S A **103**:11904–9.

80. Selenko, P., and G. Wagner. 2007. Looking into live cells with in-cell NMR spectroscopy. J Struct Biol **158**:244–53.

81. Serber, Z., L. Corsini, F. Durst, and V. Dotsch. 2005. In-cell NMR spectroscopy. Methods Enzymol **394**:17–41.

82. Serber, Z., and V. Dotsch. 2001. In-cell NMR spectroscopy. Biochemistry **40**:14317–23.

83. Serber, Z., A. T. Keatinge-Clay, R. Ledwidge, A. E. Kelly, S. M. Miller, and V. Dotsch. 2001. High-resolution macromolecular NMR spectroscopy inside living cells. J Am Chem Soc **123**:2446–7.

84. Serber, Z., P. Selenko, R. Hansel, S. Reckel, F. Lohr, J. E. Ferrell, Jr., G. Wagner, and V. Dotsch. 2006. Investigating macromolecules inside cultured and injected cells by in-cell NMR spectroscopy. Nat Protoc **1**:2701–9.

85. Shortle, D., and M. S. Ackerman. 2001. Persistence of native-like topology in a denatured protein in 8 M urea. Science **293**:487–9.

86. Shtilerman, M. D., T. T. Ding, and P. T. Lansbury, Jr. 2002. Molecular crowding accelerates fibrillization of alpha-synuclein: could an increase in the cytoplasmic protein concentration induce Parkinson's disease? Biochemistry **41**:3855–60.

87. Sillen, A., P. Barbier, I. Landrieu, S. Lefebvre, J. M. Wieruszeski, A. Leroy, V. Peyrot, and G. Lippens. 2007. NMR investigation of the interaction between the neuronal protein tau and the microtubules. Biochemistry **46**:3055–64.

88. Sillen, A., J. M. Wieruszeski, A. Leroy, A. B. Younes, I. Landrieu, and G. Lippens. 2005. High-resolution magic angle spinning NMR of the neuronal tau protein integrated in Alzheimer's-like paired helical fragments. J Am Chem Soc **127**:10138–9.

89. Smet, C., A. Leroy, A. Sillen, J. M. Wieruszeski, I. Landrieu, and G. Lippens. 2004. Accepting its random coil nature allows a partial NMR assignment of the neuronal Tau protein. Chembiochem **5**:1639–46.

90. Snoussi, K., and B. Halle. 2005. Protein self-association induced by macromolecular crowding: a quantitative analysis by magnetic relaxation dispersion. Biophys J **88**:2855–66.

91. Strauss, A., F. Bitsch, B. Cutting, G. Fendrich, P. Graff, J. Liebetanz, M. Zurini, and W. Jahnke. 2003. Amino-acid-type selective isotope labeling of proteins expressed in baculovirus-infected insect cells useful for NMR studies. J Biomol NMR **26**:367–72.

92. Strauss, A., F. Bitsch, G. Fendrich, P. Graff, R. Knecht, B. Meyhack, and W. Jahnke. 2005. Efficient uniform isotope labeling of Abl kinase expressed in baculovirus-infected insect cells. J Biomol NMR **31**:343–9.

93. Sugase, K., H. J. Dyson, and P. E. Wright. 2007. Mechanism of coupled folding and binding of an intrinsically disordered protein. Nature **447**:1021–5.

94. Tollinger, M., J. D. Forman-Kay, and L. E. Kay. 2002. Measurement of side-chain carboxyl pK(a) values of glutamate and aspartate residues in an unfolded protein by multinuclear NMR spectroscopy. J Am Chem Soc **124**:5714–7.

95. Tompa, P. 2002. Intrinsically unstructured proteins. Trends Biochem Sci **27**: 527–33.

96. Tompa, P. 2003. Intrinsically unstructured proteins evolve by repeat expansion. Bioessays **25**:847–55.

97. Tompa, P., J. Prilusky, I. Silman, and J. L. Sussman. 2008. Structural disorder serves as a weak signal for intracellular protein degradation. Proteins **71**:903–9.

98. Uversky, V. N. 2002. What does it mean to be natively unfolded? Eur J Biochem **269**:2–12.

99. Uversky, V. N., E. Cooper M., K. S. Bower, J. Li, and A. L. Fink. 2002. Accelerated alpha-synuclein fibrillation in crowded milieu. FEBS Lett **515**:99–103.

100. Uversky, V. N., J. Li, and A. L. Fink. 2001. Evidence for a partially folded intermediate in alpha-synuclein fibril formation. J Biol Chem **276**:10737–44.

101. Wood, M. J., and E. A. Komives. 1999. Production of large quantities of isotopically labeled protein in *Pichia pastoris* by fermentation. J Biomol NMR **13**: 149–59.

102. Wright, P. E., and H. J. Dyson. 1999. Intrinsically unstructured proteins: re-assessing the protein structure-function paradigm. J Mol Biol **293**:321–31.

103. Wyss, D. F., J. S. Choi, J. Li, M. H. Knoppers, K. J. Willis, A. R. Arulanandam, A. Smolyar, E. L. Reinherz, and G. Wagner. 1995. Conformation and function of the N-linked glycan in the adhesion domain of human CD2. Science **269**: 1273–8.

104. Yokoyama, S. 2003. Protein expression systems for structural genomics and proteomics. Curr Opin Chem Biol **7**:39–43.

105. Zhang, O., J. D. Forman-Kay, D. Shortle, and L. E. Kay. 1997. Triple-resonance NOESY-based experiments with improved spectral resolution: applications to structural characterization of unfolded, partially folded and folded proteins. J Biomol NMR **9**:181–200.

106. Zuiderweg, E. R. 2002. Mapping protein-protein interactions in solution by NMR spectroscopy. Biochemistry **41**:1–7.

PART II

SPECTROSCOPIC TECHNIQUES

3

NUCLEAR MAGNETIC RESONANCE SPECTROSCOPY APPLIED TO (INTRINSICALLY) DISORDERED PROTEINS

Frans A. A. Mulder, Martin Lundqvist, and Ruud M. Scheek

Department of Biophysical Chemistry, University of Groningen, Groningen, The Netherlands

ABSTRACT

NMR spectroscopy is undoubtedly the most suitable tool for investigating the details of IDPs. Many parameters in NMR, such as chemical shifts, line widths, spin-spin multiplet patterns, relaxation rates, and residual dipolar couplings (RDCs), are atom specific and harbor information about the local conformations that polypeptide chains adopt, including record of their dynamic behavior.

3.1 INTRODUCTION AND SCOPE

Disordered proteins have hitherto deserved little attention in the scientific literature, and most studies that have dealt with unfolded proteins have focused on the "other half of the folding equation" where the folded state is no longer attained, under conditions such as extremes of pH, temperature,

Instrumental Analysis of Intrinsically Disordered Proteins: Assessing Structure and Conformation, Edited by Vladimir Uversky and Sonia Longhi

and denaturants. Several excellent reviews on the subject of NMR spectroscopy of unfolded proteins exist, mostly with the focus on unfolded natively folded proteins. We mention here those by Shortle (31), Dyson and Wright (11–14, 39), Bracken et al. (4), Chatterjee and coworkers (9), Wirmer, Schlörb, and Schwalbe (36), and Meier et al. (23). The purpose of this chapter is to review various NMR observables and discuss their relation to the properties of unfolded protein chains. Much of what will be described for intrinsically disordered proteins (IDPs) applies also to unfolded natively folded proteins, as well as to shorter peptides. This chapter starts, at the beginning, describing the theory of interactions of the nuclear spins with magnetic fields of various forms. Subsequent sections focus on the application of NMR spectroscopy to IDPs, with examples taken from our own work and the scientific literature. We have chosen here not to attempt to review the many interesting publications in the literature, which may be found in the papers mentioned above. For this reason, the list of references has remained succinct, with suggestions for further reading mainly pointing to books and reviews. Finally, in the beginning of this chapter, we will go a little in depth on the fundamentals of NMR spectroscopy. Although a proper description of NMR theory can be cumbersome, simplistic pictures may sometimes be misleading and are inadequate to describe present-day multinuclear multidimensional biomolecular NMR spectroscopy. Since space is limited and the descriptions are short, it is necessary to refer the reader already here to a number of excellent books on NMR spectroscopy and its application to proteins, including those by Levitt (21), Abragam (1), Ernst et al. (15), Cavanagh et al. (8), van de Ven (32), Keeler (18), and Wüthrich (40, 41). It is hoped that this chapter may provide a foothold for other parts of this book that describe various developments of NMR techniques for and applications to IDPs.

3.2 NMR SPECTROSCOPY: MANIPULATING HAMILTONIANS

Spectroscopy aims to employ the interaction of matter with electromagnetic waves. Various forms of spectroscopy monitor the absorption or emission of photons with particular energies to derive some information about molecular structure. In NMR spectroscopy radiofrequency (RF), waves in the MHz range are applied to samples immersed in a strong and homogeneous magnetic field (B_0). The magnets for this purpose are made of cryogenically cooled superconducting materials and reach fields of several tesla. In the presence of the strong external B_0 field, the nuclear spin states attain different energies, an effect named after the Dutch physicist Zeeman. Due to the small energy difference between the nuclear spin states, the NMR phenomenon arises from a small ($\sim 10^{-6}$) macroscopic excess nuclear spin magnetization in the sample. The NMR signal is registered as a tiny current induced in a receiver coil, due to stimulated emission from the nuclear spin system. This signal, known as the free induction decay (FID), is then Fourier transformed to resolve the

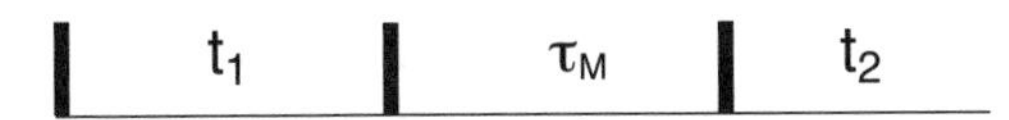

Figure 3.1 Timing diagram of the NOESY pulse sequence: The vertical bars represent 90° pulses; the time periods in between pulses are delays, and are further explained in the text.

frequency components that are present. The pulse–acquire scheme has paved the way to chemical structure determination. The strength of NMR is only truly unleashed when manipulating the nuclear spins between the excitation pulse and detection of the signal. This is the area of nuclear spin Hamiltonian design, better known as pulse sequence development. Pulse sequences are a series of timed events, including delays and RF pulses, during which the nuclear spin ensemble, described by the density operator $\sigma(t)$, evolves into a new state: $\sigma(t) \rightarrow \sigma(t + \tau)$.

As an example, one may consider the timing diagram of the popular nuclear Overhauser spectroscopy (NOESY) pulse sequence (Fig. 3.1). Here, each vertical bar represents a short RF pulse, and the empty regions in between are periods of chemical shift encoding, here marked with "t_1" and "t_2," or fixed delays to build up a new state, such as the delay τ_M for transfer of population (nuclear Overhauser enhancement; NOE) due to short-range dipole–dipole interactions between spins in the molecule.

The Hamiltonian $\hat{H}$ describes the energies of interaction of the nuclear spins with internal and external electric and magnetic fields. For protein NMR—which is focused on spin ½ nuclei ^1H, ^{13}C, and ^{15}N—electric interactions are not relevant, and the strongest interaction takes place with the external magnetic field B_0. Relevant to the application of NMR spectroscopy are the much smaller terms due to the indirect magnetic interaction of the nuclear spin with the external magnetic field through the involvement of the electrons (chemical shift), the indirect magnetic interaction of the nuclear spins with each other through the intervening electrons (scalar or J coupling), and the magnetic interactions of nuclear spins with each other through space (direct dipole–dipole [DD] coupling). The energy operator, or Hamiltonian of spin j, $\hat{H}_j$, due to interaction with its environment, then takes the following form:

$$\hat{H}_j = \hat{H}_j^{static} + \hat{H}_j^{CS} + \hat{H}_{jk}^{J} + \hat{H}_{jl}^{DD} + \hat{H}_j^{RF} + \hat{H}_j^{other}.$$

The terms on the right result from interactions with the static field (*static*), the electrons around the nucleus (*CS*), the scalar coupling with nucleus k (J), nuclear spins l in the direct environment (*DD*), external radiofrequency fields (*RF*), and any remaining fields (*other*), respectively, and a summation over all nuclei k and l is implicit in the above equation. In isotropic liquids—of relevance to everything described in this chapter, with the exception of RDCs—rapid molecular reorientation will average out any orientation dependence of the interactions, and only the isotropic parts of this Hamiltonian remain:

$$\text{chemical shift:} \quad \hat{H}_j^{static} + \hat{H}_j^{CS} = \omega_{0,j}\hat{I}_{jz} = -\gamma_j B_0 (1+\delta_j)\hat{I}_{jz}$$
$$\text{J coupling:} \quad \hat{H}_{jk}^{J} = \pi J_{jk} 2\hat{I}_{jz}\hat{I}_{kz}$$
$$\text{dipole-dipole coupling:} \quad \hat{H}_{jl}^{DD} = 0,$$

where ω_0 is the Larmor precession frequency of the spins about the B_0 field (with $\upsilon_0 = \omega_0/2\pi$ in the MHz range), γ is the magnetogyric ratio (a physical constant for a particular nuclear species), and δ is the small deviation due to the shielding of the nucleus by the electrons, known as the chemical shift (in ppm $\times 10^{-6}$, relative to a reference compound). Note that in the usual convention, $\hat{H}_j^{static}$ is not just the interaction of a bare nucleus with the magnetic field, but also that of a proton in a particular reference compound, defined to resonate at 0 ppm. The Hamiltonian describing the interaction with weak rotating RF fields is not described in detail here but encompasses, for example, pulses, isotropic mixing schemes, and decoupling. $\hat{H}_j^{other}$ may contain further interactions that can be introduced into the system, such as those involving unpaired electrons.

In addition to all of the above, there is another manifestation of the interaction between the nuclei and their surroundings, operating through the time dependence of the energetic interactions. As molecules tumble in solution, the Hamiltonians will vary in time. These variations then become a means for the nuclear spin system to relax back to thermal equilibrium. Nuclear spin relaxation may arise from the anisotropic part of the chemical shift interaction (CSA) or from DD coupling. Depending on the state of the spin density operator, the relaxation rates have different names. Most well known and popular to measure are the return of longitudinal magnetization back to the Boltzmann equilibrium value and the decay of transverse magnetization, known as T_1 and T_2 relaxation, respectively. Since the time dependence originates from the modulation of the interaction energies due to reorientation, the connection to dynamic properties of the molecules becomes obvious. We will return to relaxation in sections 3.10–3.12.

Through the application of clever RF irradiation schemes, parts of the Hamiltonian can be suppressed or isolated, yielding information only about a subset of interactions. One such example is the transfer of the large spin polarization of protons to the attached heteronuclei to increase the sensitivity of their detection. Another such application is the continuous irradiation at the proton frequencies during detection of carbon nuclei to collapse the splitting due to the scalar coupling (hence the term decoupling). The fact that the nuclear spin Hamiltonian is so complex has allowed for the continuous development of new schemes to focus on a particular aspect of molecular structure or dynamics.

Despite the complexity of the interactions at the core of NMR, the spectrum of a molecule dissolved in a low-viscous solvent is well resolved, demonstrating a collection of signals for all chemically distinct atoms, augmented by possible splitting patterns due to scalar couplings. As an example, Figure

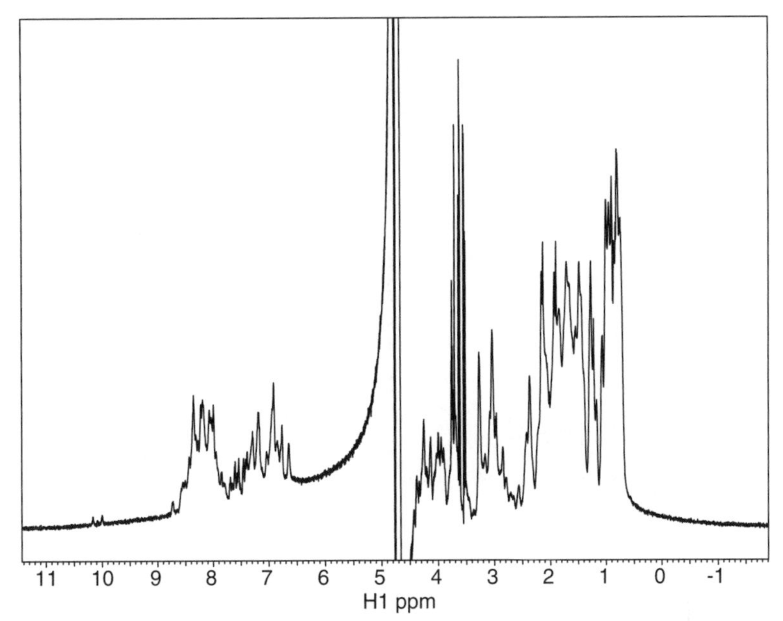

Figure 3.2 Proton 1D spectrum of an IDP: The 139-residue intrinsically disordered cytoplasmic domain of human neuroligin-3 (Nlg3cyt). The protein was ^{13}C and ^{15}N labeled, giving rise to doublet structure in the ^{1}H spectrum as a result of large one-bond couplings ($^{1}J_{NH}$ ~ 90 Hz; $^{1}J_{CH}$ ~ 140 Hz).

3.2 shows the one-dimensional (1D) proton NMR spectrum of the intrinsically disordered cytoplasmic domain (cyt) from the human neuronal adhesion protein neuroligin-3 (Nlg3). The protein construct contains 139 amino acids, and, as a consequence, the proton NMR spectrum consists of roughly 1000 lines! So why does the spectrum not appear to show that many signals? This is due to the fact that—in the absence of structure—the chemical environment of protons in the same chemical group will result in very similar chemical shifts. Take, for example, the methyl protons of the *pro-R* methyl group of the 16 leucine residues in Nlg3cyt: These all overlap in a very narrow interval between 0.86 and 0.91 ppm.

For comparison, the proton 1D spectrum of a small folded protein (bovine calbindin D_{9k}) is shown in Figure 3.3. Despite containing only 75 amino acids, the signals are spread over a much wider range and display a more idiosyncratic pattern. The *pro-R* methyl proton signals of the 12 leucine residues in this small folded protein are spread over a range of 0.72–1.09 ppm, with a mean of 0.86 ppm and a standard deviation of 0.11 ppm. This much larger variation is due to the fact that—in a folded protein—the local magnetic environments are very different. Thus, the methyl region of a protein 1D

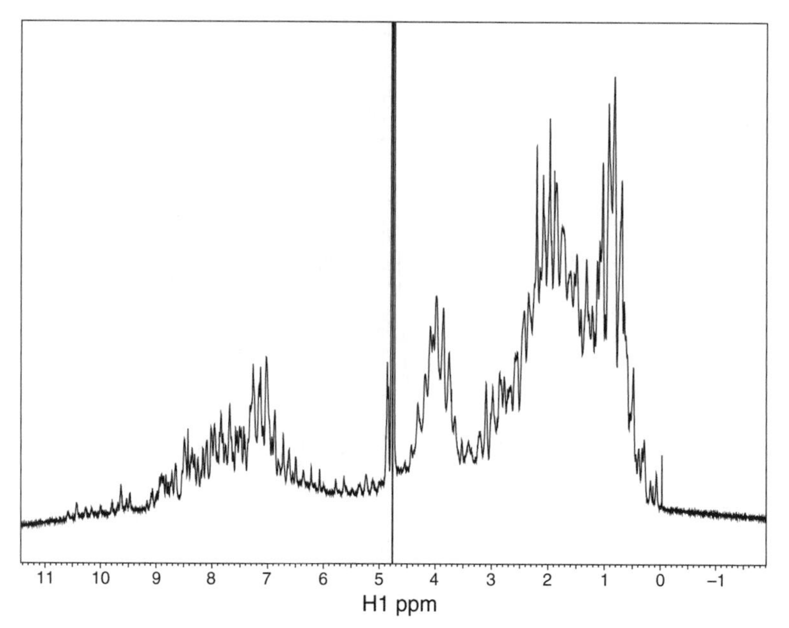

Figure 3.3 Proton 1D spectrum of a stable folded protein: the 75 amino acid intestinal calcium binding protein calbindin D_{9k} from cow. The sample was ^{13}C and ^{15}N labeled, giving rise to doublet structure in the ^{1}H spectrum as a result of large one-bond couplings. The solvent signal was suppressed by irradiation with a weak RF field.

proton spectrum is a very sensitive tool to gauge the presence of stable *tertiary* structure.

Besides the methyl region (0–2 ppm), protein proton 1D spectra also contain a well-separated signature region for the presence of stable *secondary* structure, the amide region (7–11 ppm). In proteins devoid of stable secondary structure, the amide ^{1}H chemical shifts are restricted to a narrow (~1 ppm) interval. For Nlg3cyt, we find all amide signals between 7.7 and 8.7 ppm, with average ± standard deviation being 8.23 ± 0.19. For calbindin D_{9k}, the signals are spread from 6.9 to 10.5 ppm, with a distribution 8.45 ± 0.98. Thus, 1D proton NMR can be used to directly and unambiguously arrive at an operational definition of an IDP, based simply on the lack of secondary and/or tertiary structure gauged from a single spectrum recorded in a few seconds. However, the 1D spectrum is just the beginning: We can now continue and get information about the *local structural, electrostatic, and dynamic environment of any atom in the molecule*. This is what makes NMR spectroscopy the single most valuable technique for the study of IDPs. In order to have access to many probes in the molecule, we need two things: spectral resolution and the assignment of each resonance to a particular atom. These requirements are discussed below.

3.3 MULTIDIMENSIONAL NMR AND ISOTOPIC ENRICHMENT

Even the 1D spectrum of a small folded protein contains an overwhelming number of signals, and their assignment is a daunting task. Early protein NMR results were based on a few well-resolved peaks, but it was the advent of two-dimensional (2D) Fourier transform (FT) NMR that made highly detailed studies of proteins possible. The pivotal point is that a correlation spectrum can be produced with two frequency domains, connected via a period of "mixing." A *homonuclear* 2D spectrum would show the 1D spectrum on the diagonal and display off-diagonal correlations (cross-peaks), due to transfer from spin A (with chemical shift δ_A) to spin B (with chemical shift δ_B) under the influence of a "mixing" Hamiltonian. The mixing Hamiltonian is designed or chosen by the experimenter, to effect a desired change $\sigma(t) \rightarrow \sigma(t + \tau)$ of the nuclear spin system. For example, it is possible to correlate the nuclear spin frequencies belonging to a scalar coupled network, by correlation spectroscopy (COSY) or total correlation spectroscopy (TOCSY). It is also possible to transfer polarization directly between nuclei close in space, such as in NOESY. Homonuclear 2D NMR was the *modus operandi* for many years, predicated on the fact that protons have high natural abundance (~100%) and high sensitivity. With the solution of three-dimensional (3D) structures of many small proteins, NMR became a principal technique for structural biology, second only to X-ray diffraction off crystals. For most of this time, little attention was focused on unfolded proteins, as their relevance to biology seemed secondary. In addition, the lack of dispersion of the NMR lines for even a small, unfolded protein is prohibitive in assigning resonances to specific atoms, as underscored by Figures 3.2 and 3.3. The breakthrough forced for folded proteins—isotopic enrichment with ^{13}C and ^{15}N, coupled with triple resonance NMR pulse sequence development—was instrumental for structural biology and has been crucial for the structural investigation of unfolded proteins. First, the possibility to correlate two, three, four, or maybe even more chemical shifts in a single experiment mitigates the overlap problem by spreading the peaks in additional dimensions. Second, heteronuclear backbone spins such as ^{15}N and $^{13}C'$ are very sensitive to the side-chain identities on both sides, giving rise to much better resolved spectra than those based on $^1H\alpha$ or ^1H_N. This is illustrated further in the following.

3.4 THE SEQUENTIAL RESONANCE ASSIGNMENT PROBLEM

The first, obligatory, step in a high-resolution NMR study of a protein is solving the *resonance assignment problem*. This is a big puzzle, where each peak in the spectrum is assigned to a specific atom in the molecule. With a complete assignment in hand, it is now possible to study various aspects of that atom in the protein: the dynamics it undergoes, its distance to other groups, the angle between the bond it is part of, and other bonds, and so on.

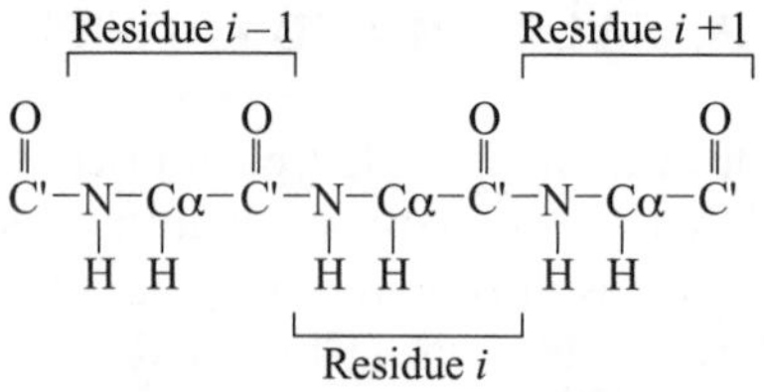

```
      Residue i-1            Residue i+1
O         O          O          O
‖         ‖          ‖          ‖
C'-N-Cα-C'-N-Cα-C'-N-Cα-C'
   |  |      |  |      |  |
   H  H      H  H      H  H
          Residue i
```

Figure 3.4 Schematic representation of a portion of the polypeptide backbone. A 3D HN(CA)CO experiment produces a 3D spectrum with signals that correlate the carbonyl chemical shifts left and right of each amide group, i.e. produces cross-peaks with coordinates (ω_H[i], ω_N[i], ω_C[i–1]) and (ω_H[i], ω_N[i], ω_C[i]), respectively. Identical correlations can be detected with a (HCA)CO(CA)NH experiment, although the magnetization transfer pathway is different.

Many advances have made this problem solvable in a limited period of time for *well-behaved* small structured proteins.

To understand the sequential assignment strategy by triple resonance NMR spectroscopy, consider the polypeptide chain in Figure 3.4. All ^{1}H, ^{15}N, and ^{13}C nuclei are NMR active, so that magnetization can be transferred between them. The dominant design factors for a pulse sequence that correlates chemical shift are (1) the size of the J coupling constant between the nuclei, as the transfer time needed is *inversely* proportional to this; and (2) the rate of decay during each transfer period. For example, transfer from ^{15}N to $^{1}H_N$ is very favorable as the one-bond coupling constant is large (~93 Hz), and amide nitrogen-15 transverse relaxation is relatively slow. A pulse sequence directed at the transfer from ^{13}Cα to ^{15}N is expected to have a much lower sensitivity, because of the much smaller (7–11 Hz) coupling constant, exacerbated by rapid carbon-13 relaxation due to the strong dipolar interaction with the attached proton. Since relaxation rates are very sensitive to molecular reorientation times, the sensitivity of a pulse sequence is generally lower for higher molecular weight systems. For small proteins, the following subset of 3D experiments has been popular and successful: HNCA and HN(CO)CA; HNCACB and CBCA(CO)NH; and HNCO and HN(CA)CO. The acronyms indicate which nuclei are being correlated and which coupling pathways are followed. For example, HN(CO)CA refers to successive transfers $^{1}H_N \rightarrow {}^{15}N \rightarrow {}^{13}C' \rightarrow {}^{13}C\alpha$ through the mutual one-bond couplings and correlates the chemical shifts of ^{13}Cα(i – 1), ^{15}N(i), and $^{1}H_N$ (i). The brackets around CO indicate that the ^{13}C′ is not frequency labeled in the experiment but only passed *in transit*. As you may have noted, the experiments were presented in pairs, and the HN(CO)CA is coupled with the HNCA experiment. This is because they share one or more common correlations. In the HNCA, cross-peaks are observed for ^{13}Cα(i), ^{15}N(i), and $^{1}H_N$(i), and, often with less intensity, also for ^{13}Cα(i – 1), ^{15}N(i), and $^{1}H_N$(i). Thus, given a certain cross-peak in the 2D (^{15}N – ^{1}H) – Heteronuclear Single Quantum Correlation (HSQC) spectrum, we retrieve the ^{13}Cα chemical shift of residue (i – 1) in the HN(CO)CA. Should this shift be unique, then we can look in the HNCA for the correlation that would connect it with the $^{15}N - ^{1}H$ pair belonging to residue ($i - 1$). This procedure can be repeated until no cross-peak is observed, the newly found $^{13}C\alpha$ chemical shift is not unique (but degenerate), or when overlap occurs in the $^{15}N - ^{1}H$ spectrum. In practice, all situations are found to occur, and a "divide and conquer" strategy is used, by simultaneously looking for sequential connectivities through $^{13}C\alpha$, $^{13}C\beta$, and $^{13}C'$, using the experiment couples mentioned above. Looking at Figure 3.4, one can easily think of a host of alternative pulse sequences, some of which surely exist. However, the design criterion—to transfer quickly through nuclei with favorable relaxation properties—limits the number of experiments that offer good sensitivity. For further insightful discussion and detailed descriptions of the experiments described above, the interested reader is referred to the books by Cavanagh and coauthors (8) and van de Ven (32), and to review articles by Kay (17), Wider (35), and Sattler et al. (28).

Now, as mentioned before, the NMR spectrum of an IDP is extremely congested. This has consequences for the choice of experiments: Those pulse sequences that are the crux for resonance assignment of folded proteins, HN(CO)CA, HNCA, CBCA(CO)NH, and HNCACB, show no or limited dispersion for the carbon chemical shift of the same amino acid in a different place in the sequence, and therefore yield highly ambiguous data. Instead, experiments that correlate neighboring $^{13}C'$ (like HN[CA]CO and [HCA]CO[CA]NH) or adjacent ^{15}N frequencies (such as HNN, HN[C]N, or HN[COCA]NH) meet with higher success. In addition, pulse sequences that would be too insensitive for folded proteins can now be used, as the decay of spin coherences is slowed down in highly flexible systems. For example, TOCSY mixing through the small (~1 Hz) coupling constant between $^{13}C'$ of adjacent residues now becomes feasible.

To illustrate the overlap problem for unfolded proteins, we now turn to an example, using the unstructured cytoplasmic domain of Nlg3. In order to make clear the advantages and disadvantages of detecting the various backbone chemical shift correlations in a multidimensional experiment, we focus here on a five-dimensional (5D) HACACONH pulse sequence that establishes a connectivity between the following nuclei: $^{1}H\alpha(i-1)$, $^{13}C\alpha(i-1)$, $^{13}C'(i-1)$, $^{15}N(i)$, and $^{1}H_N(i)$. Note that, although nuclei that belong to adjacent residues are connected in this spectrum, the 5D correlation peaks do not *share* any chemical shifts that are necessary to sequentially link them (in contrast to the simultaneous presence of cross-peaks for $^{13}C\alpha[i-1]$ and $^{13}C\alpha[i]$ in an HNCA spectrum, for example). Of the 5D experiment, we can record 2D, 3D, or four-dimensional (4D) subspectra, which all have the amide proton domain in common (as this is "directly" recorded, in the form of an FID). 3D spectra can be recorded by traditional nested time incrementation of the indirect evolution times, but combined chemical shift evolution in two or more indirect domains is necessary to retain high resolution while maintaining manageable acquisition times (3, 20, 22).

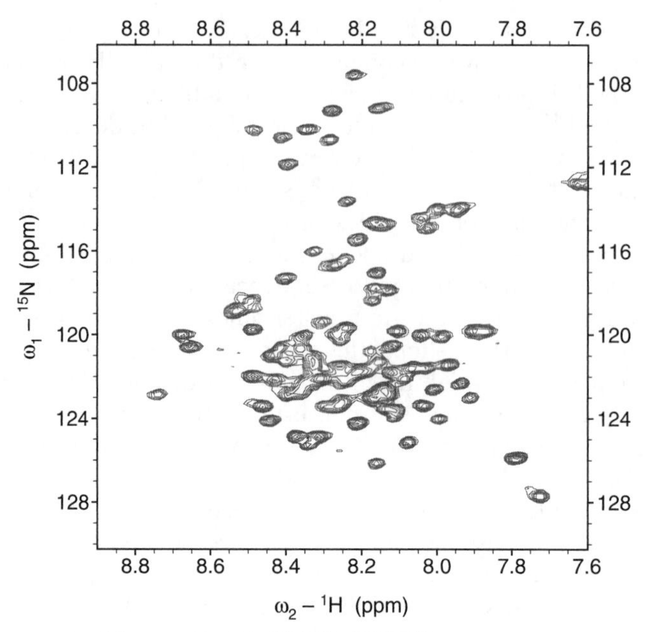

Figure 3.5 2D (HACACO)NH spectrum.

Figure 3.5 shows the 2D ($^{15}N - {}^{1}H$) plane for the 5D HACACONH experiment, that is, a 2D (HACACO)NH spectrum, giving a peak for each amide group in the protein (observe its similarity to a 2D $^{15}N - {}^{1}H$ HSQC spectrum). Although many peaks are resolved, there are also severely congested regions. Since triple resonance assignment strategies for folded proteins rely heavily on connecting the amide pairs via intervening carbon atoms, it can be readily foreseen that many ambiguities arise for an unfolded protein and call for alternatives. Figure 3.6 shows that correlating the amide proton chemical shift with that of the carbonyl of the preceding residue may be a particularly fruitful strategy. Even better resolution can be obtained if one correlates $^{13}C'(i - 1)$ and $^{15}N(i)$. This is visualized by projecting a 3D (HACA)CONH subspectrum along the proton axis, shown in Figure 3.7.

Next, we show the 2D (HA)CA(CON)H plane in Figure 3.8. Very little chemical shift variation is observed for the $^{13}C\alpha$ of any particular amino acid type, so that the identity of several residues can be deduced from this spectrum directly. For example, the seven glycine residues in Nlg3cyt are observed in a very narrow range $\delta(^{13}C\alpha) = 45.22 \pm 0.08$ ppm. In this calculation, $\delta(^{13}C\alpha) = 44.67$ ppm for Gly60 was excluded, as it experiences a large direct effect from proline at the next position in the sequence. Note that this correlation is not visible in the spectrum, Figure 3.8, as Pro has no amide proton that is required for detecting the signal.

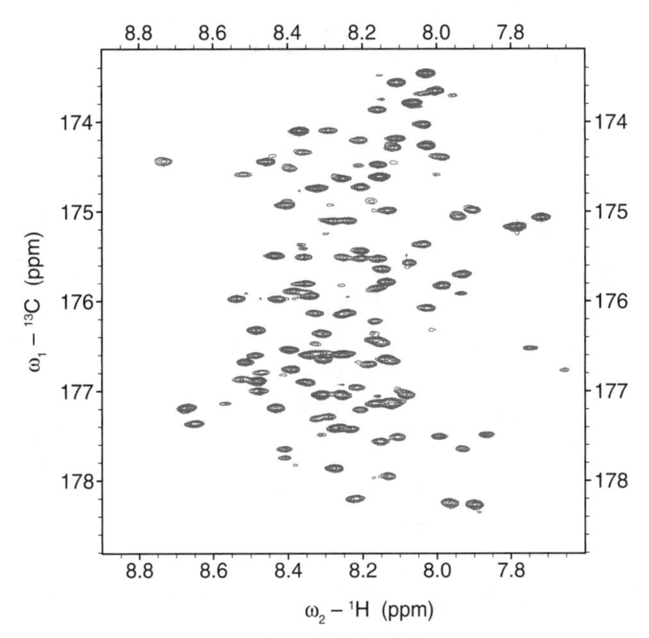

Figure 3.6 2D (HACA)CO(N)H spectrum.

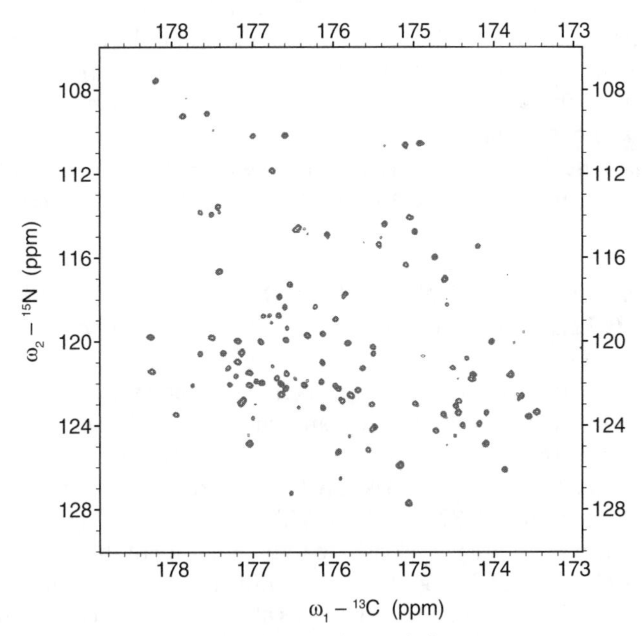

Figure 3.7 2D (HACA)CON(H) spectrum.

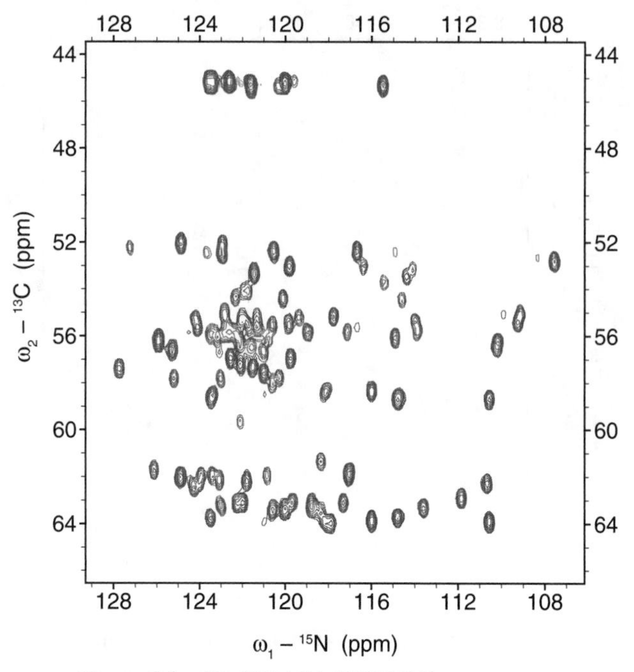

Figure 3.8 2D (HA)CA(CO)N(H) spectrum.

Finally, we show in Figure 3.9 strips from a 3D CBCA(CO)NH spectrum, yielding correlations between the ^{13}Cα/^{13}Cβ pairs and the amide proton of the following residue for 14 leucine residues. Again, it can be appreciated that the narrow region over which the signals are dispersed makes matching ^{13}Cα/^{13}Cβ pairs to backbone amides highly ambiguous.

In the next section, we will discuss one possible avenue to solve the assignment problem, although there are several alternatives in the literature.

3.5 USING HIGH-DIMENSIONAL NMR DATA FOR THE ASSIGNMENT

Because of spectral congestion and likeness of chemical shifts for similar chemical structures, correlating as many chemical shifts as possible offers a route to resolve the overlap problem. Excellent spectral resolution with little overlap is expected in the 5D HACACONH spectrum, as a result of the following factors (5, 30, 34, 37, 38, 42): (1) The backbone ^{13}C′(i – 1), ^{15}N(i), and ^{1}H$_N$(i) chemical shifts are very sensitive to the identity of the side-chain groups of residues (i – 1), i, and (i + 1), with each of them showing good dispersion (*vide supra*). The 20 × 20 × 20 = 8000 possible side-chain combinations yield a unique set of chemical shifts for any tripeptide in the protein sequence. (2)

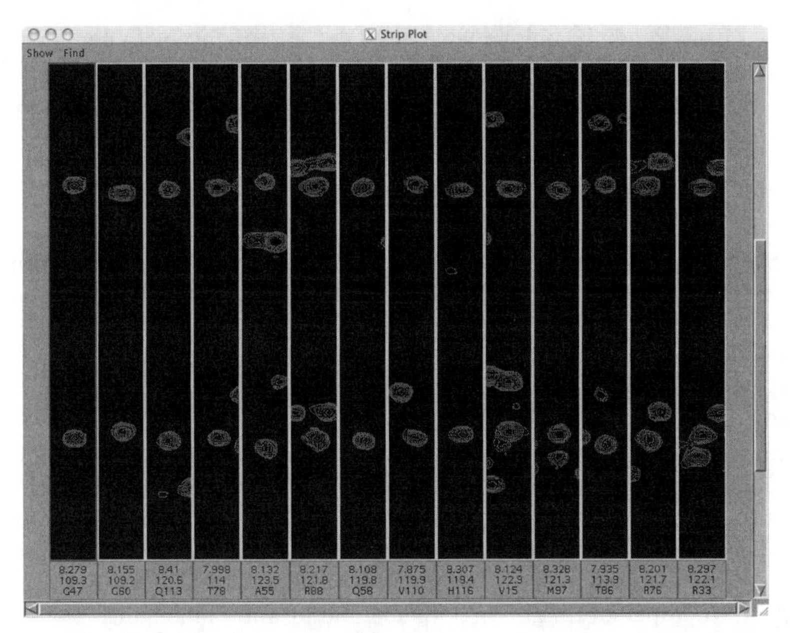

Figure 3.9 Strip plots for 14 of 16 leucine residues of Nlg3cyt, correlating $^{13}C\alpha/^{13}C\beta$ pairs with the backbone ^{1}H of the next residue, indicated at the bottom. Leucine residues that are followed by proline do not yield any cross-peaks. In case where (^{15}N,^{1}H) correlations (nearly) overlap, cross-peaks of other spin systems are observable in addition. See color insert.

$^{1}H\alpha(i-1)$ and $^{13}C\alpha(i-1)$ offer the unique signature of the identity of the preceding amino acid. (3) All the chemical shifts are sensitive to backbone conformation, with $^{1}H\alpha$, $^{13}C'$, and $^{13}C\alpha$ showing the highest sensitivity to alpha-helix formation, while $^{1}H\alpha$ and $^{13}C\beta$ are most useful to detect the propensity for extended structures (β-sheet and polyproline-II conformation). A synergistic approach is taken by measuring as many chemical shifts at once in a single high-dimensional experiment: Labeling many chemical shifts on either side of the amide moiety gives little overlap, and thereby reduces ambiguity in the assignment process, while at the same time providing a large set of chemical shifts per residue, exposing any structural preferences that are present. The 5D HACACONH meets these criteria and has high inherent sensitivity. An HACACO(CA)NH companion experiment provides the necessary sequential connectivities, albeit with lower S/N.

As long as each combination of frequency correlations ($^{15}N[i]$, $^{1}H_N[i]$) and ($^{1}H\alpha[i-1]$, $^{13}C\alpha(i-1)$, $^{13}C'[i-1]$) is unique, no ambiguity arises when making the "sequential walk" from residue to residue in the sequence through the data, and the assignment can be readily completed. Naturally, prolines break the sequential walk and form the borders of the assigned stretches. Given the typical chemical shift signatures of most amino acids, stretches of only a few residues can already be uniquely matched to the sequence with high probability. A distinctive pattern is much sooner obtained for IDPs than for folded proteins, precisely due to the absence of variation in the local surroundings.

3.6 THE CHEMICAL SHIFT (δ)

The chemical shift is easily measured and is extremely rich in information about primary, secondary, and tertiary structure. It is also subject to averaging on timescales from milliseconds to picoseconds. At the same time, it remains the most underused observable of current-day applications in protein NMR spectroscopy. The reason for this paradox is that calculating chemical shifts to sufficient accuracy to be used as structural parameters (<1 ppm for ^{13}C, for example) has remained difficult. However, state-of-the-art *ab initio* calculations using density functionals are increasingly successful for folded proteins and promising for structure determination and refinement in the future. For unfolded proteins, the inherent flexibility of most bonds in the molecule introduces averaging over multiple conformations, the populations of which are difficult to establish, and renders structure determination both demanding and challenging.

As an alternative to *ab initio* calculations, empirical relationships have been established between chemical shifts and conformational preferences in model systems. In one approach, homologous series of peptides are used, such as Gly-Gly-Xxx-Gly-Gly, where Xxx is any of the 20 naturally occurring amino acids. One possible uncertainty in the chemical shift values arises due to the possibilities that small peptides present conformational preferences, and thereby bias the shifts. As an alternative, typical chemical shifts for amino acids in disordered regions may be gauged from an analysis of "coil libraries," containing those amino acids that are distant from regular secondary structure, like helices, strands, and regular turns. One potential problem here is that the spatial confinement leaves its mark on the backbone angular distributions and introduces uncertainty in the correspondence of these shifts with the conformational distributions found in truly unfolded proteins, such as IDPs. As a third alternative, one may use the distributions of chemical shifts that are actually observed in IDPs. Although a database analysis of IDP chemical shifts is beyond the scope here, we present the random coil chemical shifts obtained from the Nlg3 assignments. These values are averages taken over the data for each amino acid type, and are presented in Table 3.1. As may be expected, the values are similar to the chemical shifts obtained from the other two approaches.

3.7 THE SCALAR COUPLING CONSTANT (J)

Whereas the interpretation of chemical shift in terms of molecular geometry and configuration is very difficult, the dependence of three-bond scalar

TABLE 3.1 Average Residue-Specific Amino Acid Backbone and Side-Chain Chemical Shifts for Nlg3cyt in ppm*†

Amino Acid	$^{1}H_{N}$	^{15}N	$^{13}C'$	$^{13}C\alpha$	$^{1}H\alpha$	Side-chain $^{13}C/^{1}H$
A	8.10	124.4	177.6	52.6	4.29	19.1/1.32 (β)
C‡	8.49	119.7	174.5	58.5	4.46	28.1/2.92 (β)
D	8.22	120.0	176.2	54.2	4.56	41.1/2.59/2.67 (β)
E	8.52	121.9	176.7	57.0	4.26	30.0/1.98/2.04 (β) 36.2/2.26 (γ)
F	8.08	120.9	175.6	57.9	4.54	39.8/3.00/3.08 (β)
G	8.32	110.0	173.9	45.2	3.93	
H	8.42	119.5	174.7	55.7	4.69	29.7/3.11/3.13 (β)
I	8.10	123.2	176.2	61.3	4.33	38.6/1.86 (β) 27.3/1.14/n.d. (γ_1) 17.3/0.91 (γ_2) 12.9/0.84 (δ_1)
K	8.31	122.4	176.4	56.6	4.23	32.9/1.72/1.79 (β) 24.8/1.40 (γ) 29.3/1.67 (δ) 42.1/2.98 (ε)
L	8.11	122.6	177.4	55.4	4.33	42.2/1.56/1.62 (β) 27.0/1.56 (γ) 24.7/0.88 (δ_1)§ 23.6/0.84 (δ_2)§
M	8.27	121.8	175.7	55.3	4.45	32.9/1.94/2.11 (β) 32.0/2.45/2.49 (γ) 17.0/2.03 (ε)
N	8.41	119.5	175.3	53.3	4.65	38.8/2.73/2.77 (β)
P	–	n.d.	176.8	63.2	4.36	32.1/1.85/2.24 (β) 27.4/1.96/1.97 (γ) 50.6/3.64/3.70 (δ)
Q	8.33	121.0	175.8	55.8	4.35	29.3/1.95/2.08 (β) 33.9/2.34 (γ)
R	8.31	122.5	176.1	56.0	4.31	30.9/1.74/1.81 (β) 27.1/1.59/1.60 (γ) 43.3/3.16 (δ)
S	8.34	116.9	174.7	58.5	4.52	63.7/3.83/3.86 (β)
T	8.11	115.3	174.4	62.1	4.29	69.7/4.18 (β) 21.7/1.16 (γ_2)
V	8.00	121.4	176.8	62.9	4.21	32.6/2.04 (β) 20.9/0.91 (γ_1) 20.7/0.91 (γ_2)
W	8.12	120.6	175.1	57.3	4.55	29.7/3.20 (β)
Y	8.20	121.0	175.9	58.0	4.54	38.5/2.96/3.05 (β)

* ^{1}H and ^{13}C chemical shifts of residues followed by Pro, or those at the N and C-termini, are not included in the table.

† Atoms that do not exist, and for which chemical shifts were not determined, are marked "–" and "n.d.," respectively. For all side chains, only the aliphatic ^{1}H and ^{13}C chemical shifts are given. When methylene proton shifts are degenerate, only a single value is given.

‡ Reduced form (cysteine).

§ Stereospecifically assigned.

couplings on intervening bond angles has a long and fruitful history. Based on 3D structures and coupling data obtained for several folded proteins, parameterizations have been obtained. Thanks to this, coupling constants are nowadays routinely measured and used as input restraints in structure calculation protocols. For unfolded proteins, many ^3J measurements suffer from the lack of dispersion in the NMR spectrum, making their application limited. In particular, HNHA couplings have been obtained for several proteins and peptides, as they can even be obtained for unlabeled systems. The values obtained are intermediate, leading to an interesting irresolution in their interpretation: Do the intermediate values correspond to well-defined and persistent backbone structure, or are they averaged between many possible structures? Both views have been expressed in the literature, and a lot of data are required to support or dismiss either, given the number of unknowns in the system.

3.8 HYDROGEN EXCHANGE RATES (k_{ex}) AND pKa VALUES

Use can be made of exchanging stable isotopes between solvent molecules and labile sites in the protein. The rate at which a particular hydrogen exchanges with the solvent depends on its pKa, which is determined by the extent of bond polarization. For folded proteins, hydrogen bonding in cooperative structural elements lengthens the lifetimes of amide protons by many orders of magnitude, in some cases, to months. For unfolded proteins, however, the *intrinsic* exchange rates are on the order of several per second at ambient temperature and neutral pH. The sequence-dependent variations can be calculated from the primary sequence: Acid groups will prolong the life time, whereas positive charges at the N-terminus or in the side chain will facilitate exchange. Bond polarization effects depend on the side-chain identities on both sides of the amide.

In addition, side-chain protonation equilibria of basic (His, Arg, Lys, N-terminus), and acidic (Asp, Glu, Cys, Tyr, C-terminus) groups can be used to gauge locally persistent electrostatic interactions, involving charges or dipoles. As chemical shifts and protonation constants both critically depend on electron density distributions, pKa values are exquisitely sensitive reporters. As Coulomb forces act over long distances because of their 1/r dependence, polarization-induced chemical shifts constitute unique long-range NMR parameters. The shifts are, however, often difficult to relate to specific interactions, although screening by salt can be used to modulate the interaction strength. At present, this area appears to be largely underdeveloped.

3.9 MOLECULAR ALIGNMENT AND RDCs

It was mentioned in section 3.2 that the dipolar interaction is averaged out to zero in isotropic solutions, such that narrow resonance lines result at the isotropic value of the chemical shift, and the splittings observed in the spectra derive exclusively from the *indirect* dipolar coupling, or J coupling. This situation can be radically changed when anisotropy is introduced into the molecular environment, and the dipolar coupling for two nuclei, i and j, at a fixed distance r becomes averaged in time over all angles θ relative to the external field (27):

$$D_{ij} = d_{ij} \langle P_2(\cos(\theta(t))) \rangle \quad \text{with} \quad d_{ij} = \left(\frac{\mu_0}{4\pi}\right) \times \left(\frac{h\gamma_i\gamma_j}{2\pi^2 r^3}\right).$$

Here, $P_2(\cos\theta) \equiv (3\cos^2\theta - 1)/2$, h is Planck's constant ($6.626 \times 10^{-34}\,\mathrm{kg\,m^2\,s^{-1}}$), μ_0 is the permeability of free space ($4\pi \times 10^{-7}\,\mathrm{kg\,m\,s^{-2}\,A^{-2}}$), γ_i and γ_j are the magnetogyric ratio of species i and j, respectively, and r is the vibrationally averaged internuclear distance.

One situation of magnetic anisotropy we are obviously familiar with: The magnetic field in which the sample is placed is pointing in one direction. Since molecules generally possess anisotropic magnetic susceptibilities, there will be a tendency of the molecules to align and minimize their interaction energy. In this case, an alignment tensor A relates the orientation of the protein molecule in the magnetic field. Since the magnetic susceptibility anisotropy is small in diamagnetic proteins, use can be made of paramagnetic tags, and this yields a measure of the direct alignment of the chain itself in the field. In addition, there exist many ways to introduce a small net preference of molecular orientation by intermolecular interactions (steric, electrostatic, hydrophobic) with the environment (27). For example, dilute solutions of large particles that strongly align in the field will generate an environment at the nanoscale in which a net directionality of the dissolved protein is obtained by simple excluded volume effects and/or transient interactions. Although alignment produced by the medium is the method of choice for folded proteins, the intermolecular interactions responsible for external alignment potentially introduce bias in the conformational ensemble itself.

Experimentally, RDCs are obtained from the difference between the sum of *direct* (D) and *indirect* (J) dipolar splittings between aligned and nonaligned samples. Since the average dipolar coupling is zero in the absence of alignment, the difference yields D directly. For example, in a rigid protein molecule, the one-bond coupling between the amide nitrogen and proton D_{NH} can be related to the polar angles (ϑ, φ) of the bond vector in the frame of the alignment tensor as follows:

$$D_{NH}(\vartheta, \varphi) = d_{NH} A_{zz} \left[P_2(\cos\vartheta) + \frac{1}{2}\eta \sin^2\vartheta \cos 2\varphi \right],$$

where A_{zz} is the z-component of the alignment tensor in its principal axis frame, and the rhombicity is $\eta = (A_{xx} - A_{yy})/A_{zz}$. Hence, in the absence of internal dynamics, each dipolar coupling can be related to its orientation relative to a common molecular frame. Through this property, RDCs offer one of the most direct and sensitive probes of molecular structure. This is of particular value for IDPs, where such restraints are few. However, how the dipolar interaction is averaged over time and over the molecular ensemble in flexible systems needs careful reevaluation. For an insightful discussion on this matter, the interested reader is referred to the recent review by Meier et al. (23). The use of RDCs for modeling IDPs is further discussed by M. Blackledge et al. in this book.

3.10 NUCLEAR SPIN *AUTO*RELAXATION (T_1, T_2) AND *CROSS*-RELAXATION (NOE)

As alluded to in section 3.2, nuclear spin relaxation describes the phenomenon of return to equilibrium of the nuclear spins with their surrounding after a perturbation. In the NMR sample, relaxation is caused by local magnetic fields, which vary in direction and amplitude. The most important interactions were already mentioned, such as the dipole–dipole coupling between nuclei close in space and nuclear shielding by electrons. The kinetics is derived from nuclear spin relaxation theory, which describes the energetic coupling with the environment (lattice) and the rate of fluctuations in this environment. The decay rate of a particular system state, for example, corresponding to bulk longitudinal or transverse magnetization is described by an *auto*relaxation rate. When two or more spins are involved, then magnetization can also be exchanged between them. In that case, we speak of *cross*-relaxation. Both will be treated in this section. However, before we can arrive at a structural and dynamic description of IDPs from experimental relaxation data, the concepts of time correlation function (TCF) and spectral density function (SDF) require some introduction (7).

There exists a direct connection between the nuclear spin relaxation rates and the molecular motions that supply the rapidly varying magnetic fields that are necessary for it, through the power SDF, $J(\omega)$: The SDF is mathematically connected (via the real part of the FT) to the TCF, which describes the motion of the relaxation interaction in a reference frame connected to the molecule. For example, a rigid globular object will rotate in solution due to collisions with its neighbors, that is, Brownian diffusion. It is obvious that small spheres will tumble more rapidly than larger ones. Suppose now, as illustrated in Figure 3.10, that two nuclei would be present in this globular object at a distance from each other, and one would have a small magnetic moment associated with it. In such a case, the local field B_{loc} felt by the second nucleus would change with time, as their relative orientation and distance changes. In the case of two directly bonded atoms, the (vibrationally averaged) distance is a constant, and we only need to consider the rotational part of the TCF: The relative orientation of the two nuclei after some time τ will be slightly different, and after many periods τ, the final orientation will no longer be correlated with that at the outset. If we take the dot product of the vectors at time zero, and after a time τ, and average many dot products over a large ensemble, we

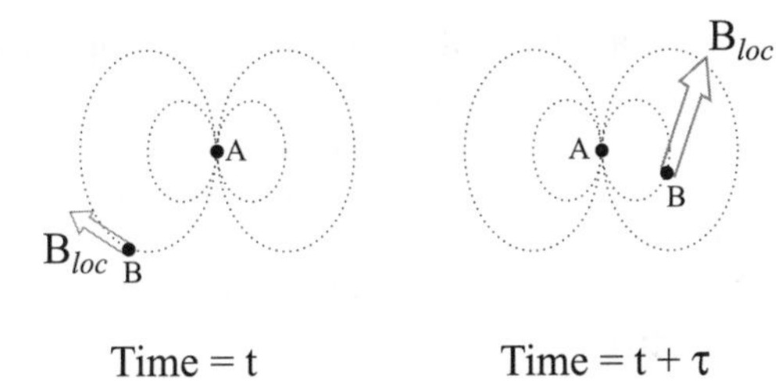

Figure 3.10 Dipole–dipole interaction of two spins A and B. The magnetic moment of B will generate a field at position A. As the relative distance and orientation changes, so do the magnitude and direction of the local field B_{loc} at B.

get a feel for how this function decays, and the correlation is lost with increasing τ. In NMR relaxation, the relevant mathematics are slightly more complicated as the spin interactions are actually rank-2 tensors, and the relevant TCF is given by the following definition:

$$C(\tau) \equiv \langle P_2[\mu(t) \cdot \mu(t+\tau)] \rangle,$$

where μ is the (time-dependent) unit vector between the two spins, $P_2(\cos\theta) \equiv (3\cos^2\theta - 1)/2$, and the brackets indicate a time and ensemble average over all conformations in the system. In the case of isotropic molecular tumbling, when the dipole–dipole vector is rigidly attached to the molecular frame, the TCF is an exponentially decaying function, with decay constant τ_C being the overall or isotropic rotational correlation time. In this case, the SDF will be described by a Lorentzian function. When reorientation is rapid, the TCF decays fast, and the SDF is broad. When tumbling is slow, the TCF is longer lived, and the SDF now falls off more rapidly. Since the spin relaxation rates are proportional to the SDF at specific frequencies, they harbor the information about the exact shape of the SDF, and hence of the details of the underlying molecular motions. As these concepts may be new to the reader, the following statement confers some more physical insight into the TCF and to the question which dynamics will contribute to relaxation: *Motion on any timescale may contribute to the decay of the TCF and to relaxation, as long as the TCF has not already decayed to zero*. For example, small globular proteins sample all directions within a few nanoseconds (ns). This means that the TCF has died away after a few tens of nanoseconds, and any motions that are slower than that do not affect the relaxation rates. On the contrary, for IDPs, rapid jumps about dihedral angles and diffusive dynamics around the average structure take place within the span of roughly a nanosecond, and are therefore manifested in their nuclear spin relaxation rates.

Relaxation experiments can roughly be divided in two groups: those where the dynamics is known or assumed to be understood such that the coupling parameter can be determined. In this class, we find back the NOESY experiment: We assume that two protons are tumbling together and are rigidly attached to the macromolecule. Under these conditions, the cross-relaxation rate will yield the internuclear distance. Alternatively, we may select a fragment of well-defined geometry, such as a directly bonded ^{15}N–^{1}H pair in the polypeptide backbone, and use the relaxation rates to unravel the details of molecular motion. Practical issues pertaining to IDPs are described in sections 3.10.1 and 3.10.2, respectively.

3.10.1 Homonuclear *Cross*-Relaxation and NOESY

The cornerstone of protein structure determination by NMR spectroscopy is 2D ^{1}H NOESY spectroscopy; two protons that have distinct chemical shifts give rise to an off-diagonal cross-peak, the intensity of which is a measure of the internuclear distance. For a structured protein, NOESY cross-peaks are observable out to 5–6 Å. Several assumptions are often implicit in the analysis of NOESY data, such as (1) the distance between the two protons is fixed, (2) the internuclear vector reorients together with the entire molecule, (3) tumbling is isotropic, and (4) each ensemble member displays the same dynamics. In the process of structure calculation, these effects can be mitigated, by applying generous distance boundaries.

For IDPs, NOESY spectra often display poor sensitivity. This is due to the fact that the limitations pointed out above for folded proteins become rather acute for flexible molecules: The ensemble of molecules exists of many diverse copies of the protein molecule, and any relative motion of protons in the protein includes both translational and rotational components, which are all entangled into a single cross-peak intensity. Hence, satisfying an NOE can be done in many different ways, and it is impossible to derive any unique description of an IDP from NOESY data alone. In addition, the fact that NOEs are averaged as $\langle r^{-6}\rangle^{-1/6}$ means that fleeting attractions at short distance strongly dominate the result. An observed cross-peak, therefore, does not by itself prove a dominant or persistent type of secondary structure, although the lack of a NOESY cross-peak can be a strong indicator for the absence of a particular type of structure. As an example, the observation of a stretch of sequential $H_N(i) - H_N(i + 1)$ cross-peaks would indicate that the α-helical region of Φ/Ψ space is sampled at each backbone position at least in some structures for part of the time, whereas the absence of medium-range NOEs (e.g., $H_N[i] - H_N[i + 3]$) would indicate that amide protons of residues i and $i + 3$ are never in close proximity. If the above situations hold simultaneously, then they argue against stable, cooperatively folded helical segments in the particular IDP. Ideally, this hypothesis would be tested by additional experiments.

3.10.2 Heteronuclear ^{15}N–^{1}H Relaxation

The relaxation rates most readily and often measured for the investigation of protein dynamics are those of the backbone ^{15}N nuclei. The relaxation rates for longitudinal magnetization, $R\{N_Z\}$, and transverse magnetization, $R\{N_{X,Y}\}$, and the ^{15}N–^{1}H cross-relaxation rate, $R\{N_Z \leftrightarrow H_Z\}$, are given by (33):

$$R_1 = R\{N_Z\} = \rho = d^2[3J(\omega_N) + J(\omega_H - \omega_N) + 6J(\omega_H + \omega_N)] + c^2 J(\omega_n)$$

$$R_2 = R\{N_{X,Y}\} = \lambda = \frac{d^2}{2}[4J(0) + 3J(\omega_N) + 6J(\omega_H) + J(\omega_H - \omega_N) + 6J(\omega_H + \omega_N)]$$
$$+ \frac{c^2}{6}[4J(0) + 3J(\omega_n)] + R_{ex}$$

$$R\{N_Z \leftrightarrow H_Z\} = \sigma = d^2[-J(\omega_H - \omega_N) + 6J(\omega_H + \omega_N)],$$

where $d = \left(\frac{\mu_0}{4\pi}\right) \times \left(\frac{h\gamma_H\gamma_N}{4\pi r^3}\right)$ and $c = \frac{(\Delta\sigma\gamma_N B_0)}{\sqrt{3}}$, h is Planck's constant (6.626 $\times 10^{-34}$ kg m^2 s^{-1}), μ_0 is the permeability of free space ($4\pi \times 10^{-7}$ kg m – s^2 A^{-2}), γ_N and γ_H are the magnetogyric ratio of ^{15}N and ^{1}H, -2.713×10^7 rad T^{-1} s^{-1} and 26.75 × 107 rad T^{-1} s^{-1}, respectively (1T = 1 kg s^{-2} A^{-1}), r is the vibrationally averaged ^{15}N–^{1}H internuclear distance (1.04 Å = 1.04×10^{-10} m), and $\Delta\sigma$ is the ^{15}N chemical shift anisotropy (–160 to –170 ppm). The SDFs in the above expressions are normalized such that J(0) = (2/5) × τ_C for a rigid sphere. R_{ex} refers to excess transverse relaxation (line broadening) caused by chemical exchange. This topic is discussed further in the next section.

The *auto*relaxation rates for longitudinal magnetization, $R_1 = R\{N_Z\}$, and transverse magnetization, $R_2 = R\{N_Z\}$ are measured by monitoring the recovery (R_1) and decay (R_2) of N_Z and $N_{X,Y}$, respectively, as a function of a parametric relaxation delay in the pulse sequence. The heteronuclear {^{1}H}^{15}N NOE

$$\text{NOE} = 1 + \frac{\gamma_H}{\gamma_N}\frac{\sigma}{\rho}$$

is obtained by comparison of the magnitude of the ^{15}N Boltzmann magnetization with the steady-state value obtained upon constant saturation of the proton transitions. The {^{1}H}^{15}N NOE is smaller than 1, since the magnetogyric ratios of proton and nitrogen are of opposite sign. For rigid segments in folded proteins, the NOE is close to one, and negative values only result for very flexible loops. For IDPs, the NOE measured at a 600 MHz spectrometer is typically close to 0, emulating an effective rotational diffusion time for the peptide segment of about 1 ns. Since the NOE depends on the fall off between $J(\omega_H + \omega_N)$ and $J(\omega_H - \omega_N)$, its value is very sensitive to the exact timescale of rapid (ps) high amplitude motions, as well as to the field strength of the instrument.

Unfortunately, the sampling frequencies of the SDF—given by the transition frequencies of the nuclear spin system—are only a handful. This means that even with measurements done at several spectrometer fields, a rather sketchy picture of a system's dynamics emerges. For folded proteins, where the backbone is mostly quite rigid, this crude approach is very successful, as the motions of the N–H bond vectors of the backbone can often be described by a small number of parameters, such as the overall tumbling time of the molecule, and the timescale and amplitude of restricted local motion. However, for unfolded proteins, the motions of the backbone are much more complex. There is no rigid scaffold that holds the atoms locked in place, and dynamics result from diffusion within local energetic potential energy wells, as well as due to transitions between rotameric states about backbone dihedral angles. In addition, each peptide segment is also spatially restrained through the physical presence of the remainder of the polypeptide chain. This includes "hard sphere" exclusion from certain regions of space, but also limitations in the movement of the covalent chain in certain directions, simply by being tethered together. For IDPs, a proper description of their flexibility needs to consider the fact that the molecular ensemble is heterogeneous, with specific reorientation behavior for different members of the molecular ensemble, as well as to encompass the internal dynamics that occur on ps to ns timescales, resulting from multiple degrees of freedom. It is therefore fair to say that our understanding of dynamics in IDPs is still rather rudimentary, with some important initial insight provided by the studies on chemically denatured proteins by Schwalbe et al. (29), Buevich and Baum (6), Ochsenbein et al. (25), and, more recently, by Modig and Poulsen (24).

Since only approximate insight can be gleaned from NMR relaxation experiments, an alternative would be to investigate the microscopic details of flexibility using molecular dynamics (MD) simulations, provided that accurate force fields are available and satisfactory agreement with the experimental data can be obtained.

3.11 SPIN ECHOES AND CONFORMATIONAL EXCHANGE BROADENING (R_{ex})

In the previous section, we have seen that very rapid movement manifests itself in the nuclear spin relaxation rates. For a small protein in water, molecular tumbling will lead to decay of the TCF in nanoseconds, and nuclear relaxation is therewith "blind" to slower motions. There is, however, an exception, known as chemical exchange, where the nuclei change chemical environment and experience different shielding in the two environments. The resultant time-dependent Hamiltonian will then add a contribution R_{ex} to the relaxation of transverse magnetization. The time dependence can originate from a real physical exchange process, such as the exchange of amide protons between the amide group of a protein and the solvent water, or can result from conformational interconversion. For IDPs, exchange with the solvent is an important issue: Since amides are exposed to the solvent, they exchange readily, and OH^--assisted catalysis leads to an increase in the exchange rate by an order of magnitude with each pH unit. As the pH increases to above 6.0, those residues with the largest intrinsic exchange rates (see section 3.8) start showing broadening at room temperature. A further increase in the pH will gradually lead to overall broadening and the disappearance of resonance lines. In those cases where the protein tends to aggregate at lower pH values, it is then essential to perform NMR experiments at low temperature, if the observation of amide proton signals is desired.

In the literature about chemical exchange (26), three regimes are generally considered: slow, intermediate, and fast. In the *slow limit*, a separate resonance is observed for each species, and the exchange broadening is proportional to the rate of leaving that particular site. In the above example of solvent exchange, this is what happens around pH 6–7: The resonance difference between the proton on the protein and in the water is about 3 ppm, corresponding to a few kHz in resonance frequency, whereas the exchange rate is only several per second. In this situation, the amide proton signal will progressively broaden and disappear, with no or only a very small movement in its spectral position in the direction of the water signal. As exchange depends predominantly on kinetics, the spectral appearance will be the same on spectrometers with different field strength. *Intermediate exchange* refers to the regime where the total exchange rate constant $k_{ex} = k_{A \to B} + k_{B \to A}$ matches closely the angular resonance frequency difference $\delta\omega = \omega_A - \omega_B$ in the assumed two sites, and broadening is maximal. Intermediate exchange spectra are a nightmare, barring resonance assignment and subsequent analyses. Intermediate in this case refers to exchange taking place on the millisecond timescale. Molten globule proteins appear to often undergo these kinds of dynamics, making them very difficult to work with. It should be emphasized that the dynamics in this case are due to interconverting molecular conformations with sufficiently different chemical shifts between the members of the ensemble. Finally, *fast exchange* refers to the regime where k_{ex} exceeds $\delta\omega$. In this regime, the exchange broadening scales with the square of the field strength, such that the resonance lines will appear broader when using a more expensive magnet. In the above examples, two-site exchange has been considered for convenience, but in proteins, conformational exchange may take place between many states, and it is even possible to be in slow and fast exchange at the same time.

The study of conformational exchange broadening in the presence of Carr–Purcell–Meiboom–Gill echo pulse trains or continuous wave RF irradiation provides a powerful tool for the study of slower dynamics in the ensemble. Over the past decade, these long-known techniques have been transformed into a powerful suite of *transverse relaxation dispersion* experiments (26), which have been applied to diverse systems (2), including the first studies of unfolded states and folding intermediates. As exchange broadening is ubiquitous in loosely packed structures where dynamics are present at all timescales, we may expect numerous applications to IDPs in years to come.

3.12 PARAMAGNETIC RELAXATION ENHANCEMENT

When an unpaired electron (radical) is present or introduced into the molecular system, a number of new interactions emerge between the electron spin and the nuclei. For simplicity, we only consider the relaxation brought about by the dipolar interaction of a proton spin on the protein and the electron spin of a nitroxide radical, covalently attached to a single site on the IDP (16,19). In this case, the nuclear spin relaxation rates will show additional contributions R_1^{para} and R_2^{para} to the longitudinal and transverse relaxation rates, respectively:

$$R_1^{para} = \frac{2K}{r^6}\left[3\tau_C/\left(1+\omega_H^2\tau_C^2\right)\right]$$

$$R_2^{para} = \frac{K}{r^6}\left[4\tau_C + 3\tau_C/\left(1+\omega_H^2\tau_C^2\right)\right],$$

where $K = 1.23 \times 10^{-44}\,m^6\,s^{-2}$ for the case of a nitroxide radical considered here, and r is the distance between the proton and the unpaired electron. The correlation time, τ_C, depends on the properties of the radical through the electron spin relaxation time, but for nitroxide radicals in proteins, it is essentially dominated by the proton–electron reorientation correlation time, which is on the order of several nanoseconds. The excess relaxation is most easily obtained from a comparison of the proton line width in the paramagnetic and diamagnetic form of the protein adduct, that is, before and after reduction of the radical. Filling in the numbers in the above equations, we observe that the nitroxide labels are useful probes for distances of 10–20 Å from the unpaired electron.

The application of paramagnetic relaxation enhancement (PRE) restraints to modeling conformational ensembles to represent the native state of an IDP is treated in the chapter by G. Daughdrill.

3.13 GRADIENT ECHOES AND TRANSLATIONAL DIFFUSION

The observables alluded to so far mainly contain information about the geometry, distances, or dynamics of one or a small collection of atoms. NMR diffusion measurements, however, contain information about the shape and size of the molecule, or the size distribution of the molecular ensemble. These techniques make use of small (mT) pulsed magnetic field gradients that can be generated by coils placed around the sample. For simplicity, we will consider here only diffusion along the z-axis (assumed to be the long axis of the NMR sample tube, in the direction of the field gradient). Transverse nuclear spin magnetization in a small slice at position z of the sample will experience an additional field due to the external gradient, such that the nuclear Larmor frequency is temporarily altered. Application of a linear gradient $G_z = \Delta B_0/\Delta z$ for a period τ to nuclei with a magnetogyric ratio γ leads to a phase build up $\Phi = \gamma \times \tau \times G_z \times z$. After a certain time interval T, a second gradient pulse is given of the same magnitude but with opposite polarity. This reverses the accrued phase buildup and creates a so-called gradient echo. However, molecules that have diffused out of the slice during the period T will experience an imperfect gradient echo, as the decoding phase is slightly mismatched to the encoding phase. This will lead to an attenuation of the signal intensity, from which the diffusion constant can be obtained. For IDPs, the "molecular diffusion constant" can subsequently be used to put limits on the gyration radii of the members of the rapidly interconverting ensemble, each with a unique shape. Since there exist no unique solution to this problem, minimal ensembles or clusters of conformations are derived from statistical libraries or MD simulations, and are simultaneously required to satisfy additional experimental data, such as chemical shifts, coupling constants, or restraints derived from other biophysical techniques, such as SAXS (10).

ACKNOWLEDGMENTS

We wish to thank Aviv Paz and Joel Sussman for the sample of Nlg3cyt, and Eva Thulin and Mikael Akke for the sample of calbindin D_{9k}. Katy Wood is acknowledged for preparing Table 3.1. We thank Klaas Dijkstra for many important contributions to the software and hardware of our NMR spectrometer. NMR spectra were analyzed and graphed using the programs NMRPipe (NIH) and Sparky (UCSF). We thank Frank Delaglio and Thomas Goddard for continued development of these program suites and keeping them available to the NMR community. Financial support was provided by EU research training network HPRN-CT-2002-00241. F. A. A. M. is the recipient of a VIDI grant from the Netherlands Organization for Scientific Research (NWO).

REFERENCES

1. Abragam, A. 1994. Principles of Nuclear Magnetism. Oxford University Press, Oxford.
2. Akke, M. 2002. NMR methods for characterizing microsecond to millisecond dynamics in recognition and catalysis. Curr Opin Struct Biol **12**:642–7.
3. Atreya, H. S., and T. Szyperski. 2005. Rapid NMR data collection. Methods Enzymol **394**:78–108.
4. Bracken, C., L. M. Iakoucheva, P. R. Rorner, and A. K. Dunker. 2004. Combining prediction, computation and experiment for the characterization of protein disorder. Curr Opin Struct Biol **14**:570–6.
5. Braun, D., G. Wider, and K. Wuthrich. 1994. Sequence-corrected N-15 random coil chemical-shifts. J Am Chem Soc **116**:8466–9.
6. Buevich, A. V., and J. Baum. 1999. Dynamics of unfolded proteins: incorporation of distributions of correlation times in the model free analysis of NMR relaxation data. J Am Chem Soc **121**:8671–2.
7. Case, D. A. 2002. Molecular dynamics and NMR spin relaxation in proteins. Acc Chem Res **35**:325–31.

8. Cavanagh, J., W. J. Fairbrother, A. G. Palmer, III, M. Rance, and N. J. Skelton. 2007. Protein NMR Spectroscopy. Principles and Practice. Academic Press, London.
9. Chatterjee, A., A. Kumar, J. Chugh, S. Srivastava, N. S. Bhavesh, and R. V. Hosur. 2005. NMR of unfolded proteins. J Chem Sci **117**:3–21.
10. Choy, W. Y., F. A. A. Mulder, K. A. Crowhurst, D. R. Muhandiram, I. S. Millett, S. Doniach, J. D. Forman-Kay, and L. E. Kay. 2002. Distribution of molecular size within an unfolded state ensemble using small-angle X-ray scattering and pulse field gradient NMR techniques. J Mol Biol **316**:101–12.
11. Dyson, H. J., and P. E. Wright. 1996. Insights into protein folding from NMR. Annu Rev Phys Chem **47**:369–95.
12. Dyson, H. J., and P. E. Wright. 2002. Insights into the structure and dynamics of unfolded proteins from nuclear magnetic resonance. Adv Protein Chem **62**: 311–40.
13. Dyson, H. J., and P. E. Wright. 2001. Nuclear magnetic resonance methods for elucidation of structure and dynamics in disordered states. Methods Enzymol **339**:258–70.
14. Dyson, H. J., and P. E. Wright. 2004. Unfolded proteins and protein folding studied by NMR. Chem Rev **104**:3607–22.
15. Ernst, R. R., G. Bodenhausen, and A. Wokaun. 1994. Principles of Nuclear Magnetic Resonance in One and Two Dimensions. Oxford University Press, Oxford.
16. Gillespie, J. R., and D. Shortle. 1997. Characterization of long-range structure in the denatured state of staphylococcal nuclease. 1. Paramagnetic relaxation enhancement by nitroxide spin labels. J Mol Biol **268**:158–69.
17. Kay, L. E. 1995. Pulsed field gradient multi-dimensional NMR methods for the study of protein structure and dynamics in solution. Prog Biophys Mol Biol **63**:277–99.
18. Keeler, J. 2007. Understanding NMR Spectroscopy. John Wiley & Sons, Chichester, UK.
19. Kosen, P. A. 1989. Spin labeling of proteins. Methods Enzymol **177**:86–121.
20. Kupce, E., and R. Freeman. 2008. Hyperdimensional NMR spectroscopy. Prog Nucl Magn Reson Spectrosc **52**:22–30.
21. Levitt, M. H. 2001. Spin Dynamics. Basics of Nuclear Magnetic Resonance. John Wiley & Sons, Chichester, UK.
22. Malmodin, D., and M. Billeter. 2005. High-throughput analysis of protein NMR spectra. Prog Nucl Magn Reson Spectrosc **46**:109–29.
23. Meier, S., M. Blackledge, and S. Grzesiek. 2008. Conformational distributions of unfolded polypeptides from novel NMR techniques. J Chem Phys **128**:052204.
24. Modig, K., and F. M. Poulsen. 2008. Model-independent interpretation of NMR relaxation data for unfolded proteins: the acid-denatured state of ACBP. J Biomol NMR **42**:163–77.
25. Ochsenbein, F., J. M. Neumann, E. Guittet, and C. Van Heijenoort. 2002. Dynamical characterization of residual and non-native structures in a partially folded protein by N-15 NMR relaxation using a model based on a distribution of correlation times. Protein Sci **11**:957–64.

26. Palmer, A. G., C. D. Kroenke, and J. P. Loria. 2001. Nuclear magnetic resonance methods for quantifying microsecond-to-millisecond motions in biological macromolecules. Methods Enzymol **339**:204–38.
27. Prestegard, J. H., C. M. Bougault, and A. I. Kishore. 2004. Residual dipolar couplings in structure determination of biomolecules. Chem Rev **104**:3519–40.
28. Sattler, M., J. Schleucher, and C. Griesinger. 1999. Heteronuclear multidimensional NMR experiments for the structure determination of proteins in solution employing pulsed field gradients. Prog Nucl Magn Reson Spectrosc **34**:93–158.
29. Schwalbe, H., K. M. Fiebig, M. Buck, J. A. Jones, S. B. Grimshaw, A. Spencer, S. J. Glaser, L. J. Smith, and C. M. Dobson. 1997. Structural and dynamical properties of a denatured protein. Heteronuclear 3D NMR experiments and theoretical simulations of lysozyme in 8 M urea. Biochemistry **36**:8977–91.
30. Schwarzinger, S., G. J. A. Kroon, T. R. Foss, J. Chung, P. E. Wright, and H. J. Dyson. 2001. Sequence-dependent correction of random coil NMR chemical shifts. J Am Chem Soc **123**:2970–8.
31. Shortle, D. R. 1996. Structural analysis of non-native states of proteins by NMR methods. Curr Opin Struct Biol **6**:24–30.
32. van de Ven, F. J. M. 1995. Multidimensional NMR in Liquids. VCH, New York.
33. Wagner, G. 1993. Nmr relaxation and protein mobility. Curr Opin Struct Biol **3**: 748–54.
34. Wang, Y. J., and O. Jardetzky. 2002. Investigation of the neighboring residue effects on protein chemical shifts. J Am Chem Soc **124**:14075–84.
35. Wider, G. 1998. Technical aspects of NMR spectroscopy with biological macromolecules and studies of hydration in solution. Prog Nucl Magn Reson Spectrosc **32**:193–275.
36. Wirmer, J., C. Schlörb, and H. Schwalbe. 2005. Conformation and dynamics of nonnative states of proteins studied by NMR spectroscopy, pp. 737–808. In J. Buchner and T. Kiefhaber (eds.), Protein Folding Handbook. Part I. Wiley-VCH, Weinheim, Germany.
37. Wishart, D. S., C. G. Bigam, A. Holm, R. S. Hodges, and B. D. Sykes. 1995. H-1, C-13 and N-15 random coil NMR chemical-shifts of the common amino-acids. 1. Investigations of nearest-neighbor effects. J Biomol NMR **5**:67–81.
38. Wishart, D. S., and B. D. Sykes. 1994. Chemical-shifts as a tool for structure determination. Nucl Magn Reson **239**:363–92.
39. Wright, P. E., and H. J. Dyson. 1999. Intrinsically unstructured proteins: re-assessing the protein structure-function paradigm. J Mol Biol **293**:321–31.
40. Wüthrich, K. 1976. NMR in Biological Research: Peptides and Proteins. North-Holland, Amsterdam.
41. Wüthrich, K. 1986. NMR of Proteins and Nucleic Acids. John Wiley & Sons, New York.
42. Yao, J., H. J. Dyson, and P. E. Wright. 1997. Chemical shift dispersion and secondary structure prediction in unfolded and partly folded proteins. FEBS Lett **419**:285–9.

4

ATOMIC-LEVEL CHARACTERIZATION OF DISORDERED PROTEIN ENSEMBLES USING NMR RESIDUAL DIPOLAR COUPLINGS

MARTIN BLACKLEDGE,[1] PAU BERNADÓ,[2] AND MALENE RINGKJØBING JENSEN[1]

[1]*Protein Dynamics and Flexibility, Institut de Biologie Structurale UMR 5075 CEA-CNRS-UJF, Grenoble, France*
[2]*Institute for Research in Biomedicine, Parc Científic de Barcelona, Barcelona, Spain*

ABSTRACT

NMR spectroscopy is an extremely powerful tool for studying the conformational behavior of intrinsically disordered proteins in solution at atomic resolution. Over the last decade, it has become clear that the measurement of residual dipolar couplings (RDCs), induced by the incomplete averaging of anisotropic interactions that become measurable under conditions of weak alignment of the protein, provides particularly sensitive probes of both local and long-range structure in these highly flexible systems. In this chapter, we present a review of recent developments in the interpretation of RDCs in terms of explicit ensemble descriptions of the protein and demonstrate the first steps toward a quantitative description of the unfolded state. We address both theoretical and practical aspects of the approach and illustrate its suc-

Instrumental Analysis of Intrinsically Disordered Proteins: Assessing Structure and Conformation, Edited by Vladimir Uversky and Sonia Longhi

cessful application to the characterization of nascent structure in the proteins tau, α-synuclein, and the C-terminal domain of the nucleoprotein of Sendai virus.

4.1 INTRODUCTION

Over the past two decades, the field of structural biology has concentrated on the elucidation of the three-dimensional structures of soluble proteins with stable three-dimensional folds. However, proteins also adopt very different structural states in their native physiological environment, as diverse as membrane proteins whose active sites occupy the interface between solid and liquid state, and amyloid-forming proteins that undergo functionally devastating phase transitions. As an example of the importance of understanding the behavior of proteins in their different molecular phases, the onset of serious disease is known to accompany transformation of prion-like proteins from highly flexible physiological conformations to aggregated, pathological forms of the same protein (13). Similarly the mechanisms of transition from unfolded to folded forms of the same protein are now understood to represent a significant fraction of the interactome but remain beyond the reach of most experimental techniques. The mechanisms of these kinds of dynamic transitions are not yet understood, and a fundamental description of the dynamic and thermodynamic properties of these different phases will undoubtedly help to elucidate these transition processes.

Perhaps the most intriguing domain of structural biology concerns the remarkably prevalent family of proteins that are either fully or partially disordered in their functional form (17, 57, 60). Intrinsically disordered proteins (IDPs) represent over 40% of the human proteome, playing key roles in many physiological and pathological processes, including signaling, cell cycle control, molecular recognition, transcription and replication, as well as endocytosis and amyloidogenesis. This figure rises to 80% for proteins involved in carcinogenesis. IDPs pose an entirely new set of questions concerning the molecular description of functional biology, with many of these proteins folding only upon binding (2, 14, 19, 58, 61, 62). Despite their obvious importance, this vast family of proteins is impossible to characterize using the standard tools of classical structural biology, due to their highly heterogeneous conformational behavior. It is clear that a significant amount of work will be required in order to develop methods to understand the behavior of IDPs and, eventually, to solve important biological and medical paradigms involving these proteins.

As described in the overview provided in chapter 3, nuclear magnetic resonance (NMR) spectroscopy is without any doubt the most powerful tool for studying IDPs (15). The principal reason for this lies in the coincidence between the averaging properties of NMR observables and the necessity for an appropriate ensemble description of the unfolded state. The dynamic properties of disordered proteins allow for the use of multidimensional NMR experiments that at least partially compensate the comparative crowding experienced in the amide region of the proton spectrum, allowing complete assignment of ^{1}H, ^{15}N, and ^{13}C resonances in $^{13}C/^{15}N$-labeled proteins.

In recent years, NMR spectroscopy has thus regularly and successfully provided a significant volume of very important experimental information on the unfolded state (1, 39, 47, 49). The simplicity of chemical shift measurements should not mask their importance for the characterization of the unfolded state. The measured average shift depends on all local conformations sampled by the ensemble in rapid exchange on timescales up to the millisecond. Similarly, three-bond scalar couplings depend on backbone dihedral angles (48, 53) and have similar averaging properties to the chemical shift. More detailed structural information can, in theory, be derived from interproton nuclear Overhauser interactions (NOE) reporting on distance information (30). However, the quantitative interpretation of NOEs is complicated by the strong dependence on the angular correlation function, as well as the fluctuating distance, and cannot therefore be simply interpreted in terms of a distance distribution function.

Paramagnetic relaxation enhancements (PREs) (20) also present a measurement of distance distributions, offering certain clear advantages over NOE measurements, as they are significantly stronger interactions, and therefore provide longer range information. These data can either be interpreted in terms of average distances or statistically interpreted in terms of probability distributions.

Residual dipolar couplings (RDCs) report on time and ensemble-averaged conformations up to the millisecond timescale and can therefore be used to characterize both the structure and dynamics of unfolded proteins (27, 50). RDCs are highly sensitive to the atomic resolution details of structure in unfolded proteins at levels of population and precision that are difficult to detect using other techniques and, like chemical shift and scalar coupling, represent straightforward time and ensemble averages, providing a direct tool to determine quantitative levels of order in unfolded proteins (5). In combination, RDCs and PREs have made it possible to describe angular and distance fluctuations of the unfolded state at unprecedented detail. Not surprisingly, the interpretation of these data has required a certain amount of lateral thinking, in particular, a reinvention of the structure/function paradigm and a significant change of perspective with respect to classical methods of structural biology.

Over the last 5 years, significant progress has been made in developing a clearer understanding of the nature of RDCs in the unfolded state, either using analytical random chain descriptions of the chain derived from polymer physics or using explicit conformational ensemble descriptions of the protein. These novel approaches to the description of structural biology in terms of heterogeneous molecular ensembles are highly adapted to the unfolded state. In this chapter, we will discuss the basis assumptions of standard interpretation of RDCs in terms of explicit ensembles, presenting both the advantages and disadvantages of these approaches.

4.2 RDCs

As described in chapter 3, dipolar couplings between two spins *i* and *j* depend on the geometry of the internuclear spin vector in the following way (8, 55):

$$D_{ij} = -\frac{\gamma_i \gamma_j \hbar \mu_0}{4\pi^2 r^3} \left\langle \frac{(3\cos^2\theta(t) - 1)}{2} \right\rangle = D_{\max} \langle P_2(\cos\theta(t)) \rangle \tag{4.1a}$$

with

$$D_{\max} = -\frac{\gamma_i \gamma_j \hbar \mu_0}{4\pi^2 r^3}. \tag{4.1b}$$

Here, θ is the instantaneous angle of the internuclear vector relative to the static magnetic field, and r is the internuclear distance. For the case of RDCs measured between covalently bound nuclei, r is assumed to be fixed and represents a vibrationally averaged distance. The angular parentheses represent an average over all conformations sampled until a characteristic time in the millisecond range. The physical constants are described in chapter 3.

The average described in Equation 4.1 depends on the second-order polynomial $P_2(\cos\theta)$, such that if all possible orientations θ are sampled with equal probabilities relative to the magnetic field, the value of the coupling averages very efficiently to zero. This is the case in free solution, where dipolar couplings, despite their intrinsic strength (around 20 kHz for a covalently bound $^{13}C^{\alpha}$–$^{1}H^{\alpha}$ spin pair), are vanishingly small unless the isotropic angular sampling is somehow perturbed. Such a perturbation can exist naturally due to intrinsic diamagnetic or paramagnetic anisotropic magnetic susceptibility of the protein itself or can be induced by dissolving the protein in a dilute liquid crystal.

Numerous alignment media have been proposed, and the practical aspects of their preparation and application have recently been extensively reviewed (43). Common media applied to the study of IDPs are lipid bicelles (55), filamentous bacteriophages (21), lyotropic ethylene glycol/alcohol phases (44), and polyacrylamide gels that have been strained either laterally or longitudinally to produce anisotropic cavities (46, 59). In most of these cases, alignment results from a steric interaction with the medium, while in the case of bacteriophage or charged forms of the other media, electrostatic interactions exert additional electric forces that combine with the steric exclusion. In the latter case, the interpretation of dipolar couplings in partially or completely unfolded proteins in terms of local structure and dynamics, although feasible (52), is significantly complicated, such that sterically aligning media are the most commonly used.

4.3 INTERPRETATION OF RDCs IN TERMS OF LOCAL STRUCTURE

The interpretation of RDCs measured in folded proteins is relatively straightforward. The average residual coupling given in Equation 4.1 can be recast in terms of the orientation of the internuclear vector relative to a common molecular alignment tensor, where the alignment tensor describes the net alignment of the protein relative to the molecule in terms of a second rank order matrix:

$$D = D_{\max} A_{zz}[P_2(\cos\vartheta) + \eta/2 \sin^2\vartheta \cos 2\varphi], \tag{4.2}$$

where A_{zz} is the longitudinal component of the alignment tensor, η is the rhombicity defined as $\eta = (A_{xx} - A_{yy})/A_{zz}$, and (ϑ, φ) are expressed as polar coordinates of the internuclear vector. If one assumes that the global alignment is not strongly coupled to local fluctuations, the measurement of different RDCs for different spin pairs can therefore be interpreted in terms of different orientations of the internuclear vectors relative to the molecular frame. The validity of this assumption has been tested quite stringently for folded proteins and appears reasonable (22). RDCs measured in folded molecules with a stable three-dimensional conformation have thus been shown to provide valuable constraints for structure refinement (3), for the study of long-range order in extended molecules (56) or protein complexes (10), or for the characterization of local dynamics in proteins on a slower timescale (9, 36, 42).

4.4 INTERPRETATION OF RDCs IN HIGHLY FLEXIBLE SYSTEMS

For the case of IDPs, the average given in Equation 4.1 cannot be expressed in such simple terms. The alignment of each molecule in the time and ensemble conformational average can be expected to vary significantly, as a function of the shape of the conformation in the case of steric alignment or electrostatic charge superimposed on this shape in the case of electrostatically aligning media. In this case, the measured coupling should rather be expressed in terms of the average of the specific time averages for each N molecule in the ensemble:

$$D = D_{\max} \frac{1}{\mathrm{N}} \sum_{k=1}^{\mathrm{N}} \frac{1}{t_{\max}} \int_{t=0}^{t_{\max}} P_2(\cos\theta_k(t))\, dt. \tag{4.3}$$

As described elsewhere (32), we can assume that each copy of the protein samples the conformational space of the ensemble, thereby removing the explicit description of the ensemble from Equation 4.3, such that:

$$D = D_{\max} \frac{1}{t_{\max}} \int_{t=0}^{t_{\max}} P_2(\cos\theta(t))\, dt. \tag{4.4}$$

Despite this simplification, the average requires the development of specific procedures in order to extract relevant information concerning the characterization of the unfolded state.

In view of the high flexibility of IDPs, it may appear, at first sight, unlikely that RDCs could provide any useful information in these systems or even that the dipolar coupling would be measurable in the unfolded state. Early examples of the measurement of RDCs in chemically denatured proteins, however, demonstrated convincingly that the orientational sampling of internuclear vectors was not isotropic in these proteins (16, 34, 37, 50), but that measured ^{15}N–^{1}H RDCs were, in general, negative, with maximal values measured in the center of the protein, tailing off via a so-called bell-shaped distribution to zero at the extremities.

N–H^N RDCs were thus measured in the native and the Δ131Δ mutant of staphylococcal nuclease in 8 M urea (50), eglin C (41), protein GB1 (12), apomyoglobin (37) and acyl-CoA-binding protein (ACBP) (16) under diverse denaturing conditions. The presence of residual secondary structure that was postulated to occur in partially unfolded proteins on the basis of chemical shift measurements was also found to coincide with deviations of measured RDCs from the expected bell-shaped distributions observed in apparently fully denatured proteins. Thus, changes in magnitude and sign of $^{1}D_{HN}$ residues in regions of acid-denatured states of apomyoglobin and ACBP were linked to helical propensities for regions which form helices in the native state and rationalized in terms of a change in sign that would be expected for α-helical segments that orient parallel to the principal direction of the chain. The N–H^N vector would then be parallel to this orientation axis and not perpendicular as in extended conformations (see above). Unfolding of a β-hairpin structure in the fibritin foldon domain was also followed by observing the diminution of the structure of the experimental RDC profile with increasing temperature (34). Similarly, a gradual decrease of the size of RDCs indicative of helix destabilization was observed with increasing temperatures or decreasing salt concentration in α-helical ribonuclease S-peptide (45), while Ding and coworkers studied thermal unfolding of GB1 using RDCs (12).

Less work was initially applied to IDPs, until RDCs were used to study the amino acid specificity, dictating local conformational sampling in short peptides (11), and to identify the level of local and long-range structure in the proteins, such as tau protein (38, 51) and α-synuclein (4, 7) (see below).

In order to understand the origin of these characteristics, the key problem is then to develop a framework that can usefully describe, and interpret, the time and ensemble average represented in Equation 4.4. Significant progress was made in this direction by Fredriksson and coworkers (18) who described the unfolded protein in terms of a series of connected segments of equal length experiencing a restricted random walk. Integration of Equation 4.4 over all available orientations of each segment leads to the observation that in the presence of an obstacle, orientational sampling is more restricted in the center of the chain than at the termini and would indeed be expected to lead to nonvanishing RDCs, even when the torsion angles along the polymer chain adopt random conformations. The result is that larger RDCs are measured in the center and that the effectively more flexible ends (with fewer neighbors) have smaller RDCs, leading to the experimentally observed bell-shaped distribution. For the N–H^N vector, the average in Equation 4.4 is also predicted to be negative in the case where the protein is aligned parallel to the magnetic field in an elongated cavity, indicating that on average, the amide vector is more likely to be perpendicular to the field (and therefore, the average direction of the chain) than parallel to it. This model has recently been extended, revised, and, to an extent, corrected by Obolensky and coworkers, confirming the overall observations and changing none of the main conclusions (40).

These models therefore elegantly describe many aspects of the physical alignment of the unfolded polypeptide and present a relatively simple conceptual framework for the qualitative understanding of the experimental observations. However, the description of a natural amino acid sequence as a homopolymer is unrealistic, and although these approaches laid the foundations for the basic physical understanding of RDCs in unfolded proteins, these analytical models could not be used to interpret data derived from complex heteropolymeric systems such as proteins. The importance of site-specific variation of nonaveraged RDCs was indeed demonstrated by Louhivuori and coworkers in their work (28), where the incorporation of glycine and proline residues into simulations of random homopolymers resulted in prediction of RDC profiles that increased or decreased due to the change in local flexibility of the amino acid. Experimentally, the significant site-to-site variation of observed RDCs along an unfolded peptide chain compared with the overall bell-shaped prediction from the random flight models suggested that it would be essential to somehow introduce amino acid-specific conformational behavior into a more detailed interpretation.

4.5 MODELING RDCs IN HIGHLY FLEXIBLE SYSTEMS USING EXPLICIT ENSEMBLE MODELS

More direct approaches to the interpretation of RDCs from unfolded proteins were therefore proposed that rely on the development of explicit ensemble descriptions of the unfolded state (5, 25). In analogy to the case of the folded protein (Eq. 4.2, the measured couplings are therefore expressed in terms of a discrete average over all sampled conformers:

$$D = D_{\max}\frac{1}{M}\sum_{k=1}^{M} A_{k,zz}[P_2(\cos\vartheta_k)+\eta_k/2\sin^2\vartheta_k\cos 2\varphi_k], \tag{4.5}$$

where the alignment characteristics of each conformer are explicitly predicted, normally, on the basis of the molecular shape or on the basis of electrostatic charge distribution in the case of electrostatic alignment. The dipolar couplings are then averaged over a sufficient number of conformers to fully represent the conformational average and to achieve convergence. This kind of approach was discussed in the Introduction and has been applied to the prediction of scalar couplings and chemical shifts in unfolded chains. Two very similar approaches were developed essentially simultaneously, explicitly accounting for the heteropolymeric nature of the peptide chain and sampling amino acid-specific (Φ/Ψ) propensities in a random way to construct the conformational ensemble of the protein of interest (5, 25). The amino acid-specific conformational energy basins were, in the case of Bernado et al., selected from a recent compilation of 500 high-resolution crystal structures (29). α-Helical and β-sheet conformations were removed from the sampled structures, although turn conformations were retained. Additional pseudo-amino acids were developed that account for the changes induced in specific backbone dihedral angle sampling in, for example, residues preceding prolines. Rudimentary van der Waals-type interactions were accounted for between amino acid side chains by removing structures when a steric clash occurred between residue-specific spheres centered on the β-carbon atoms of each amino acid (α-proton in the case of glycines). Conformers are constructed by randomly sampling these amino acid-specific conformational basins, and RDCs were predicted from the ensemble of structures by averaging RDCs (Eq. 4.5 predicted for each copy of the ensemble using a shape-based alignment algorithm.

An example of the latter approach, termed flexible Meccano or FM, is shown in Figure 4.1, where experimental RDCs from a two-domain viral protein, protein X, from the Sendai virus phosphoprotein, comprising an apparently entirely unfolded and a folded domain, are predicted using the FM approach. ${}^{1}D_{NH}$ and ${}^{2}D_{C'NH}$ RDCs are evidently accurately reproduced from throughout the protein. Note that the relative level of alignment of the two interdependent domains provides a quantitative test of the validity of the approach.

The FM approach also accurately reproduced experimentally measured ${}^{1}D_{NH}$ dipolar couplings measured in the Δ131Δ mutant of nuclease (50) simply on the basis of local conformational propensities, without invoking residual tertiary fold. Numerous further examples of the reproduction of experimental RDC profiles, for example, ${}^{1}D_{NH}$ measured in apomyoglobin (37), established the statistical coil sampling approach, and in particular FM, as a method of choice for predicting baseline values that result directly from the conformational properties of the primary sequence and can therefore be considered as "random coil."

4.6 INTERPRETING DEVIATIONS OF EXPERIMENTAL RDCs FROM RANDOM COIL BEHAVIOR—TAU K18 AND A-SYNUCLEIN

Having established an apparently reasonable description of the unfolded state, the next step is to develop techniques whereby a departure from baseline values can then be interpreted in terms of specific local or long-range confor-

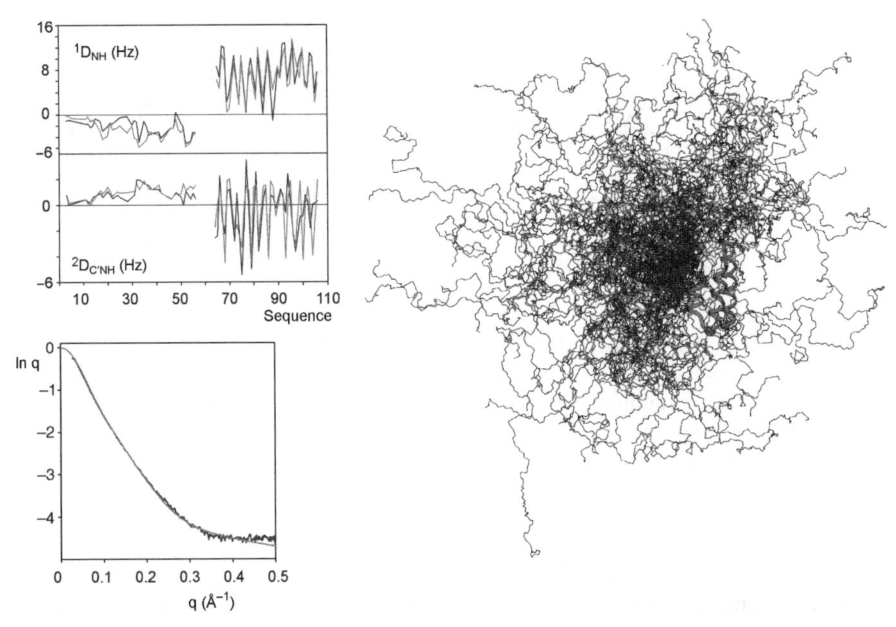

Figure 4.1 Top left: experimental $^1D_{NH}$ and $^2D_{C'NH}$ from the two-domain protein, PX, from the Sendai virus. $^1D_{NH}$ and $^2D_{C'NH}$ RDCs are evidently accurately reproduced from throughout the protein using the explicit ensemble flexible-Meccano approach. Experimental values are shown in blue. Small-angle X-ray scattering data are reproduced from the same ensemble (from Reference 5). See color insert.

mational behavior. In a recent study of K18, $^1D_{NH}$ couplings were measured from throughout a 130-amino acid construct of the natively unfolded protein tau that physiologically controls microtubule dynamics and stability, and represents a significant fraction of the proteins found in tangles in Alzheimer's disease (31). This construct comprises four repeat regions (R1–R4) and contains the hexapeptide segments identified as the interaction sites of tau with microtubules, as well as the sites involved in self-association, formation of paired helical filaments, and eventual aggregation. Local inversion of $^1D_{NH}$ RDCs in the four repeated domains was not reproduced by the simple coil FM approach (38). In this case, extended molecular dynamics (MD) simulations revealed strong tendencies to form βI-turns for repeats R1–R3. The model from simulation could be validated by replacing the intrinsic statistical coil backbone dihedral angles for the four amino acids involved in each turn region with the backbone dihedral sampling resulting from the MD simulations. This model reproduced experimental RDC values very closely, indicating that the local βI-turn conformations and, importantly, their predicted populations were in good agreement with the experimental data.

FM has also been able to demonstrate the sensitivity of RDCs in IDPs to the presence of transient long-range contacts. Simple application of the

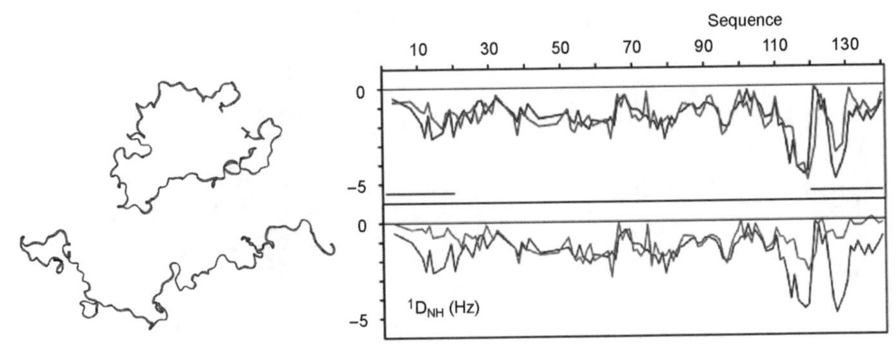

Figure 4.2 $^1D_{NH}$ RDCs from α-synuclein are better reproduced in the presence (top) of long-range contacts between the C-terminal and N-terminal domains of the protein than in their absence (bottom). Experimental values are shown in light gray (from Reference 4).

random coil FM approach did not reproduce experimental RDCs from the protein α-synuclein, with strong variations in the RDC profile at the termini of the protein (Fig. 4.2) (4). Using extensive sampling of the possible presence of long-range order in the ensemble, it could be shown that the deviations from random coil RDCs were due to transient contacts between the N-terminal and C-terminal regions of the protein. These interactions were also detected using PREs (see chapter by G. Daughdrill) and were shown to disappear upon addition of denaturant, probably due to a screening of the long-range interactions. The interactions also disappear at high temperature and upon polyamine binding, in both cases, conditions that favor aggregation of α-synuclein *in vitro*. Most importantly, α-synuclein mutants, which are associated with increased neurotoxicity (A30P and A53T) and occur in early-onset familial forms of Parkinson's disease, do not show the presence of this contact, suggesting a protective role of the long-range interactions in α-synuclein against misfolding and aggregation (6).

Long-range order can also be detected from RDCs by the measurement of $^1H^N$–$^1H^N$ RDCs in highly deuterated proteins. This was demonstrated in a study of urea and acid-denatured ubiquitin where long-range RDCs clearly demonstrated the presence of significantly populated native-like local structure in the N-terminal β-hairpin (35). In a study of the same protein, up to seven RDCs per peptide unit were measured, including $^1H^N$–H^N and $^1H^N$–$^1H^\alpha$ RDCs. The FM approach was used to predict expected one-bond RDCs ($^1D_{NH}$, $^1D_{C\alpha H\alpha}$, and $^1D_{C\alpha C'}$) from the unfolded chain, and the conformational sampling behavior was refined to achieve simultaneous reproduction of all different types of RDCs ($^1D_{NH}$, $^1D_{C\alpha H\alpha}$, and $^1D_{C\alpha C'}$). The required modification of the sampling of the coil database indicated that a larger population of extended conformations would be necessary to reproduce all data that were present in the standard coil library. The conclusion was verified by a similar comparison of calculated and experimental sequential, interproton ($^1H^N_i$–$^1H^N_{i+1}$, $^1H^N_i$–$^1H^N_{i+2}$, and $^1H^N_i$–$^1H^{\alpha}_{i-1}$) RDCs. A simultaneous analysis of $^3J_{HNH\alpha}$ scalar couplings measured under the same conditions suggested that neither polyproline II nor extended β-regions dominate the additional sampling of extended conformations. The requirement of more extended conformations for urea-denatured proteins than for IDPs appears to demonstrate that binding of urea to the polypeptide chain extends the unfolded amino acid chain, an observation that would be in agreement with local binding of urea to the backbone, inducing restricted and more extended sampling of backbone dihedral angles (32, 33).

4.7 QUANTITATIVE ANALYSIS OF LOCAL CONFORMATIONAL PROPENSITIES FORM RDCs—APPLICATION TO SENDAI VIRUS N-TAIL

As described in the Introduction, the extent to which regions of the protein that contribute to binding and function are preconfigured prior to interaction remains an important question that can prove difficult to resolve in highly flexible systems. Recently, RDCs have been combined with *FM* to study the structural properties of the partially ordered recognition element of N-tail from the Sendai virus, whose interaction site exhibits helical propensity and folds into a complete helix upon binding. Using a minimum ensemble representation analysis of $^1D_{HN}$, $^1D_{CC}$, $^2D_{CHN}$, and $^1D_{C\alpha H\alpha}$, RDCs in terms of an interconverting structural ensemble demonstrated that rather than fraying randomly, the molecular recognition sequence of N-tail preferentially populates specific helical conformers (23, 24). The two most populated conformers were found to differ by one helical turn in length at both termini and to enclose the recognition site amino acids (Fig. 4.3). Interestingly, the limits of the preferentially populated helical segments were found to be preceded by commonly observed N-capping motifs, suggesting that the preferential helices are stabilized by these motifs and that the selected conformations are encoded in the primary sequence of the molecular recognition element. Equally importantly, the direction in which the unfolded strand adjacent to the helix is projected via these interactions is controlled by the capping motif, indicating that the termination of the helices selectively controls for specific directionality in the unfolded strands via the primary sequence. This study therefore provided strong evidence for the molecular basis of nascent helix formation in partially folded chains, with clear implications for our understanding of the early steps of protein folding.

The mechanisms governing protein folding upon interaction has been greatly enhanced by prediction (26) and observation (54) of weakly binding encounter complexes formed between partially folded proteins and their partners. In this study, the authors have therefore identified a mechanism by which the partially folded form of the protein could project the unfolded strand in the most functionally useful direction to achieve these efficient "fly-casting" interactions.

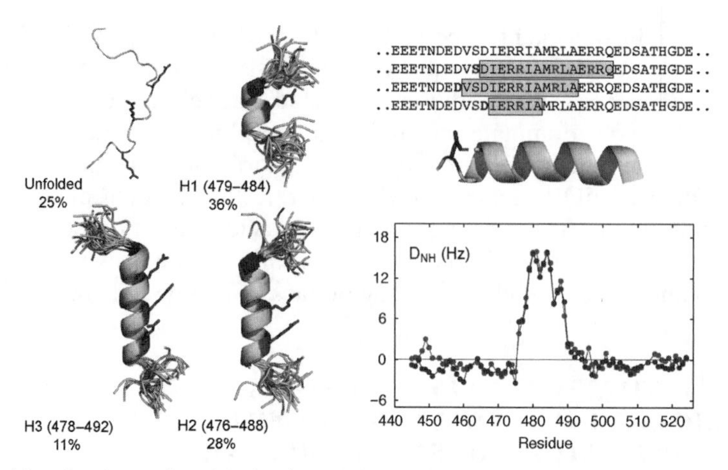

Figure 4.3 Conformational behavior of the molecular recognition element of N-tail protein from the Sendai virus. This protein samples helical conformations in the free state before folding upon binding. Using multiple RDCs and a modified FM approach to derive minimum interconverting conformational ensemble descriptions of the protein, it was possible to determine the details of preferential helical sampling and the associated populations. Importantly, the preferentially populated helices comprising the conformational equilibrium are all stabilized by identifiable N-capping interactions that precede the helices (shown in blue) (from References 23 and 24). See color insert.

4.8 COMBINATION OF RDCs WITH COMPLEMENTARY BIOPHYSICAL APPROACHES—APPLICATION TO THE TUMOR SUPPRESSOR p53

As we have seen, RDCs offer both precise local structural detail and important long-range structural information, and thereby provide powerful additional tools with which to map the vast conformational space available to IDPs. In order to fully understand unfolded proteins, it is important to combine data from complementary biophysical approaches as possible in order to attempt to understand the nature of the unfolded state. A particularly pertinent example of the combination of RDCs with complementary biophysical tools was illustrated in the recent development of the first explicit ensemble description of the tumor suppressor p53 (63). p53 plays crucial roles in maintaining the integrity of the human genome and controlling apoptosis, cell-cycle arrest, and DNA repair, and exists as a homotetramer, with folded tetramerization and core domains that are linked together and flanked by intrinsically disordered (or natively unfolded) domains at the N and C-termini. As such, with 37% of its structure intrinsically disordered, p53 is actually typical of the structural content of the human proteome. Having characterized the structure and quaternary geometry of the folded domains of p53, Wells et al then went on to study the intrinsically disordered N-terminal transactivation domain

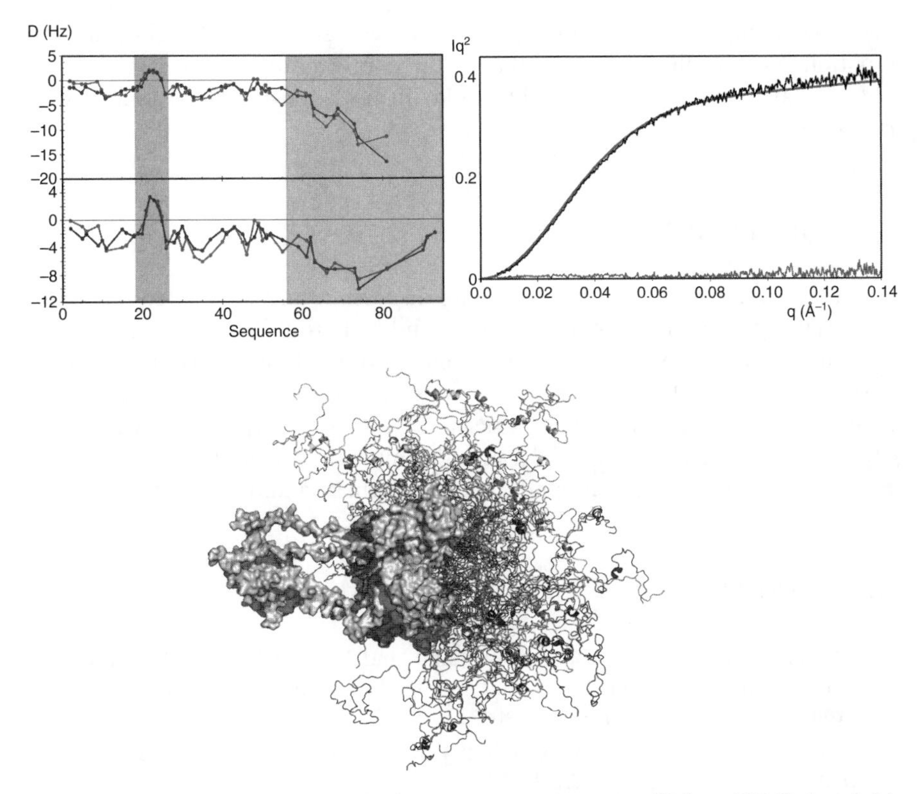

Figure 4.4 Ensemble model of the human tumor repressor p53 from RDCs (top left), SAXS (top right), MD simulation, and *FM* (bottom). RDCs were measured in the isolated TAD and in the full-length DNA-bound and free forms of the protein. Nascent helical propensity was clearly visible in the binding site of the TAD prior to interaction with Mdm2 (red shading) and enhanced rigidity in the proline-rich region that projects the TAD from the surface of the folded core domain of the protein (blue shading) (from Reference 63). See color insert.

(TAD) of p53 using RDCs and small-angle scattering, and the domain in the full-length form of the protein bound to DNA and in the unbound form.

Combining RDCs, SAXS, MD simulation, and FM, this study identified differential flexibility in the N-terminal unfolded domain of p53 (Fig. 4.4). The proline-rich region, attached to the folded core domain, exhibited enhanced stiffness relative to the TAD, projecting the important Mdm2 interaction site away from the surface of the protein. Within the TAD, where chain flexibility is higher overall, the presence of a single helix turn was found to be populated to approximately 30%, consistent with earlier suggestion of a nascent helix that fully folds upon interaction with Mdm2. In this case, the dynamic properties of the unfolded chain allow the measurement of RDCs from the flexible domain region in the presence of the full-length protein, both in the free and the DNA-complexed forms (a particle of molecular mass of 240 kD). In combination, the results indicated that local conformational sampling of the N-terminal domain is remarkably similar in both full-length protein and in isolation.

4.9 CONCLUSIONS

It is therefore clear from the preceding discussion and associated examples that RDCs provide remarkably sensitive and agile tools for the study of local structural propensity in IDPs. The combination of appropriate ensemble descriptions allows the extraction of unique and important information on the conformational propensities of IDPs and will hopefully contribute significantly to our understanding of the relationship between conformational preferences in the unfolded state and molecular function and malfunction.

REFERENCES

1. Alexandrescu, A. T., C. Abeygunawardana, and D. Shortle. 1994. Structure and dynamics of a denatured 131-residue fragment of staphylococcal nuclease: a heteronuclear NMR study. Biochemistry **33**:1063–72.
2. Aloy, P., and R. B. Russell. 2004. Ten thousand interactions for the molecular biologist. Nat Biotechnol **22**:1317–21.
3. Bax, A. 2003. Weak alignment offers new NMR opportunities to study protein structure and dynamics. Protein Sci **12**:1–18.
4. Bernado, P., C. W. Bertoncini, C. Griesinger, M. Zweckstetter, and M. Blackledge. 2005. Defining long-range order and local disorder in native α-synuclein using residual dipolar couplings. J Am Chem Soc **127**:17968–9.
5. Bernado, P., L. Blanchard, P. Timmins, D. Marion, R. W. Ruigrok, and M. Blackledge. 2005. A structural model for unfolded proteins from residual dipolar couplings and small-angle x-ray scattering. Proc Natl Acad Sci U S A **102**: 17002–7.
6. Bertoncini, C., C. Fernandez, C. Griesinger, T. Jovin, and M. Zweckstetter. 2005. Familial mutants of alpha-synuclein with increased neurotoxicity have a destabilized conformation. J Biol Chem **280**:30649–52.
7. Bertoncini, C. W., Y. S. Jung, C. O. Fernandez, W. Hoyer, C. Griesinger, T. M. Jovin, and M. Zweckstetter. 2005. Release of long-range tertiary interactions potentiates aggregation of natively unstructured α-synuclein. Proc Natl Acad Sci U S A **102**:1430–5.
8. Blackledge, M. 2005. Recent advances in the use of residual dipolar couplings for the study of biomolecular structure and dynamics in solution. Prog Nucl Magn Reson Spectrosc **46**:23–61.
9. Bouvignies, G., P. Bernado, S. Meier, K. Cho, S. Grzesiek, R. Brüschweiler, and M. Blackledge. 2005. Identification of slow correlated motions in proteins using

residual dipolar and hydrogen-bond scalar couplings. Proc Natl Acad Sci U S A **102**:13885–90.

10. Clore, G. M. 2000. Accurate and rapid docking of protein–protein complexes on the basis of intermolecular nuclear Overhauser enhancement data and dipolar couplings by rigid body minimization. Proc Natl Acad Sci U S A **97**:9021–5.
11. Dames, S. A., R. Aregger, N. Vajpai, P. Bernado, M. Blackledge, and S. Grzesiek. 2006. Residual dipolar couplings in short peptides reveal systematic conformational preferences of individual amino acids. J Am Chem Soc **128**:13508–14.
12. Ding, K., J. M. Louis, and A. M. Gronenborn. 2004. Insights into conformation and dynamics of protein GB1 during folding and unfolding by NMR. J Mol Biol **335**:1299–1307.
13. Dobson, C. M. 2003. Protein folding and misfolding. Nature **426**:884–90.
14. Dyson, H. J., and P. E. Wright. 2002. Coupling of folding and binding for unstructured proteins. Curr Opin Struct Biol **12**:54–60.
15. Dyson, H. J., and P. E. Wright. 2004. Intrinsically unstructured proteins and their functions. Chem Rev **104**:3607–22.
16. Fieber, W., S. Kristjansdottir, and F. M. Poulsen. 2004. Short-range, long-range and transition state interactions in the denatured state of ACBP from residual dipolar couplings. J Mol Biol **339**:1191–9.
17. Fink, A. L. 2005. Natively unfolded proteins. Curr Opin Struct Biol **15**:35–41.
18. Fredriksson, K., M. Louhivuori, P. Permi, and A. Annila. 2004. On the interpretation of residual dipolar couplings as reporters of molecular dynamics. J Am Chem Soc **126**:12646–50.
19. Fuxreiter, M., I. Simon, P. Friedrich, and P. Tompa. 2004. Preformed structural elements feature inpartner recognition by intrinsically unstructured proteins. J Mol Biol **338**:1015–26.
20. Gillespie, J. R., and D. Shortle. 1997. Characterization of long-range structure in the denatured state of staphylococcal nuclease. I. Paramagnetic relaxation enhancement by nitroxide spin labels. J Mol Biol **268**:158–69.
21. Hansen, M. R., L. Mueller, and A. Pardi. 1998. Tunable alignment of macromolecules by filamentous phage yields dipolar coupling interactions. Nat Struct Biol **5**:1065–74.
22. Hus, J.-C., and R. Brüschweiler. 2002. Principal component method for assessing structural heterogeneity across multiple alignment media. J Biomol NMR **24**: 123–32.
23. Jensen, M. R., and M. Blackledge. 2008. On the origin of NMR dipolar waves in transient helical elements of partially folded proteins. J Am Chem Soc **130**:11266–7.
24. Jensen, M. R., K. Houben, E. Lescop, L. Blanchard, R. W. H. Ruigrok, and M. Blackledge. 2008. Quantitative conformational analysis of partially folded proteins from residual dipolar couplings: application to the molecular recognition element of Sendai virus nucleoprotein. J Am Chem Soc **130**:8055–61.
25. Jha, A. K., A. Colubri, K. F. Freed, and T. R. Sosnick. 2005. Statistical coil model of the unfolded state: resolving the reconciliation problem. Proc Natl Acad Sci U S A **102**:13099–104.

26. Levy, Y., S. S. Cho, J. N. Onuchic, and P. G. Wolynes. 2005. A survey of flexible protein binding mechanisms and their transition states using native topology based energy landscapes. J Mol Biol **346**:1121–45.
27. Louhivouri, M., K. Pääkkönen, K. Fredriksson, P. Permi, J. Lounila, and A. Annila. 2003. On the origin of residual dipolar couplings from denatured proteins. J Am Chem Soc **125**:15647–50.
28. Louhivuori, M., K. Fredriksson, K. Paakkonen, P. Permi, and A. Annila. 2004. Alignment of chain-like molecules. J Biomol NMR **29**:517–24.
29. Lovell, S. C., I. W. Davis, W. B. Arendall, III, P. I. W. de Bakker, J. M. Word, M. G. Prisant, J. S. Richardson, and D. C. Richardson. 2003. Structure validation by C alpha geometry: phi, psi and C beta deviation. Proteins **50**:437–50.
30. Macura, S., and R. R. Ernst. 1980. Elucidation of cross relaxation in liquids by two-dimensional NMR-spectroscopy. Mol Phys **41**:95–117.
31. Mandelkow, E.-M., and E. Mandelkow. 1998. Tau in Alzheimer's disease. Trends Cell Biol **8**:425–7.
32. Meier, S., M. Blackledge, and S. Grzesiek. 2008. Conformational distributions of unfolded polypeptides from novel NMR techniques. J Chem Phys **128**:052204.
33. Meier, S., S. Grzesiek, and M. Blackledge. 2007. Mapping the conformational landscape of urea-denatured ubiquitin using residual dipolar couplings. J Am Chem Soc **129**:9799–807.
34. Meier, S., S. Guthe, T. Kiefhaber, and S. Grzesiek. 2004. Foldon, the natural trimerization domain of T4 fibritin, dissociates into a monomeric A-state form containing a stable beta-hairpin: atomic details of trimer dissociation and local beta-hairpin stability from residual dipolar couplings. J Mol Biol **344**:1051–69.
35. Meier, S., M. Strohmeier, M. Blackledge, and S. Grzesiek. 2007. Direct observation of dipolar couplings and hydrogen bonds across a β-hairpin in 8 M urea. J Am Chem Soc **129**:754–5.
36. Meiler, J., J. Prompers, C. Griesinger, and R. Brüschweiler. 2001. Model-free approach to the dynamic interpretation of residual dipolar couplings in globular proteins. J Am Chem Soc **123**:6098–107.
37. Mohana-Borges, R., N. K. Goto, G. J. Kroon, H. J. Dyson, and P. E. Wright. 2004. Structural characterization of unfolded states of apomyoglobin using residual dipolar couplings. J Mol Biol **340**:1131–42.
38. Mukrasch, M. D., P. Markwick, J. Biernat, M. V. Bergen, P. Bernado, C. Griesinger, E. Mandelkow, M. Zweckstetter, and M. Blackledge. 2007. Highly populated turn conformations in natively unfolded tau protein identified from residual dipolar couplings and molecular simulation. J Am Chem Soc **129**:5235–43.
39. Neri, D., M. Billeter, G. Wider, and K. Wuthrich. 1992. NMR determination of residual structure in a urea-denatured protein, the 434 repressor. Science **257**:1559–63.
40. Obolensky, O. I., K. Schlepckow, H. Schwalbe, and A. V. Solov'yov. 2007. Theoretical framework for NMR residual dipolar couplings in unfolded proteins. J Biomol NMR **39**:1–16.
41. Ohnishi, S., A. L. Lee, M. H. Edgell, and D. Shortle. 2004. Direct demonstration of structural similarity between native and denatured eglin C. Biochemistry **43**:4064–70.

42. Prestegard, J. H., H. M. Al-Hashimi, and J. R. Tolman. 2000. NMR structures of biomolecules using field oriented media and residual dipolar couplings. Q Rev Biophys **33**:371–95.

43. Prestegard, J. H., C. M. Bougault, and A. I. Kishore. 2004. Residual dipolar couplings in structure determination of biomolecules. Chem Rev **104**:3519–40.

44. Ruckert, M., and G. Otting. 2000. Alignment of biological macromolecules in novel nonionic liquid crystalline media for NMR experiments. J Am Chem Soc **122**: 7793–7.

45. Sallum, C. O., D. M. Martel, R. S. Fournier, W. M. Matousek, and A. T. Alexandrescu. 2005. Sensitivity of NMR residual dipolar couplings to perturbations in folded and denatured staphylococcal nuclease. Biochemistry **44**: 6392–403.

46. Sass, H. J., G. Musco, S. J. Stahl, P. T. Wingfield, and S. Grzesiek. 2000. Solution NMR of proteins within polyacrylamide gels: diffusional properties and residual alignment by mechanical stress or embedding of oriented purple membranes. J Biomol NMR **18**:303–9.

47. Schwalbe, H., K. M. Fiebig, M. Buck, J. A. Jones, S. B. Grimshaw, A. Spencer, S. J. Glaser, L. J. Smith, and C. M. Dobson. 1997. Structural and dynamical properties of a denatured protein. Heteronuclear 3D NMR experiments and theoretical simulations of lysozyme in 8M urea. Biochemistry **36**:8977–91.

48. Serrano, L. 1995. Comparison between the φ distribution of the amino acids in the protein database and NMR data indicates that amino acids have various φ propensities in the random coil conformation. J Mol Biol **254**:322–33.

49. Shortle, D. 1996. The denatured state (the other half of the folding equation) and its role in protein stability. FASEB J **10**:27–34.

50. Shortle, S., and M. S. Ackerman. 2001. Persistence of native-like topology in a denatured protein in 8 M urea. Science **293**:487–9.

51. Sibille, N., A. Sillen, A. Leroy, J. M. Wieruszeski, B. Mulloy, I. Landrieu, and G. Lippens. 2006. NMR investigation of the interaction between the neuronal protein tau and the microtubules. Biochemistry **45**:12560–72.

52. Skora, L., M. K. Cho, H. Y. Kim, S. Becker, C. O. Fernandez, M. Blackledge, and M. Zweckstetter. 2006. Charge-induced molecular alignment of intrinsically disordered proteins. Angew Chem Int Ed Engl **45**:7012–5.

53. Smith, L. J., K. A. Bolin, H. Schwalbe, M. W. MacArthur, J. M. Thornton, and C. M. Dobson. 1996. Analysis of main chain torsion angles in proteins: prediction of NMR coupling constants for native and random coil conformations. J Mol Biol **255**:494–506.

54. Sugase, K., H. J. Dyson, and P. E. Wright. 2007. Mechanism of coupled folding and binding of an intrinsically unstructured protein. Nature **447**:1021–5.

55. Tjandra, N., and A. Bax. 1997 Direct measurement of distances and angles in biomolecules by NMR in a dilute liquid crystalline medium. Science **278**:1111–4.

56. Tjandra, N., J. Omichinski, A. M. Gronenborn, G. M. Clore, and A. Bax. 1997. Use of dipolar 1H-15N and 1H-13C couplings in the structure determination of magnetically oriented macromolecules in solution. Nat Struct Biol **4**:732–8.

57. Tompa, P. 2002. Intrinsically unstructured proteins. Trends Biochem Sci **27**: 527–33.

58. Tompa, P., and M. Fuxreiter. 2007. Fuzzy complexes: polymorphism and structural disorder in protein-protein interactions. Trends Biochem Sci **33**:2–8.
59. Tycko, R., F. J. Blanco, and Y. Ishii. 2000. Alignment of biopolymers in strained gels: a new way to create detectable dipole-dipole couplings in high-resolution biomolecular NMR. J Am Chem Soc **122**:9340–1.
60. Uversky, V. N. 2002. Natively unfolded proteins: a point where biology waits for physics. Protein Sci **11**:739–56.
61. Vacic, V., C. J. Oldfield, A. Mohan, P. Radivojac, M. S. Cortese, V. N. Uversky, and A. K. Dunker. 2007. Characterization of molecular recognition features, MoRFs, and their binding partners. J Proteome Res **6**:2351–66.
62. Vucetic, S., Z. Obradovic, V. Vacic, P. Radivojac, K. Peng, L. M. Iakoucheva, M. S. Cortese, J. D. Lawson, C. J. Brown, J. G. Sikes, C. D. Newton, and A. K. Dunker. 2005. DisProt: a database of protein disorder. Bioinformatics **21**:137–40.
63. Wells, M., H. Tidow, T. J. Rutherford, P. Markwick, M. R. Jensen, E. Mylonas, D. I. Svergun, M. Blackledge, and A. R. Fersht. 2008. Structure of tumor suppressor p53 and its intrinsically disordered N-terminal transactivation domain. Proc Natl Acad Sci U S A **105**:5762–7.

5

DETERMINING STRUCTURAL ENSEMBLES FOR INTRINSICALLY DISORDERED PROTEINS

GARY W. DAUGHDRILL

Division of Cell Biology, Microbiology, and Molecular Biology and the Center for Biomolecular Identification and Targeted Therapeutics, University of South Florida, Tampa, FL

ABSTRACT

Determining realistic structural ensembles for intrinsically disordered proteins (IDPs) requires the development of experimental and computational methods that will accurately model a heterogeneous ensemble using mostly ensemble-averaged and some Boltzmann-weighted data. Two promising experimental techniques for determining accurate structural ensembles of IDPs are small-angle X-ray scattering (SAXS) and nuclear magnetic resonance (NMR) spectroscopy. SAXS can be used to determine a Boltzmann-weighted distribution of gyration radii. Residual dipolar coupling and paramagnetic relaxation enhancement measurements from NMR experiments provide ensemble-averaged angular and interatomic distance information, respectively. The focus of this chapter will be using distance restraints from paramagnetic relaxation enhancement experiments to determine a structural ensemble for the intrinsically disordered transactivation domain of the human tumor suppressor, p53. Using SAXS and residual dipolar couplings to determine structural ensembles of IDPs are covered in other chapters in this series.

Instrumental Analysis of Intrinsically Disordered Proteins: Assessing Structure and Conformation, Edited by Vladimir Uversky and Sonia Longhi

5.1 INTRODUCTION

5.1.1 Current Model for the Structure and Dynamics of Intrinsically Disordered Proteins

Intrinsically disordered proteins (IDPs) form ensembles of structures that can have a broad range of compactness, secondary structure content, and conformational dynamics (13, 18, 23, 24, 51, 52, 61). A method that could characterize the structure of individual conformations in the equilibrium ensemble would be ideal. Unfortunately, determining atomic structures for the individual conformations in the equilibrium ensembles of both folded and IDPs remains a challenge (55). Two groups have proposed general models that are based on local interactions to describe the structural ensembles for chemically denatured and IDPs (11, 34, 59). In both studies, ensembles were generated using a database of dihedral angles that are observed in the loop regions of folded proteins. Ensembles generated by this method were used to predict residual dipolar couplings (RDCs) and small-angle X-ray scattering (SAXS) data for chemically denatured proteins and at least two IDPs with high accuracy. The results from both of these studies suggest that any conformational biasing observed in the unfolded state is defined locally. Interestingly, one of the same groups had to modify the method described above to accurately model intrinsically disordered α-synuclein (10). The approach was modified to account for the presence of long-range structure by selecting models that satisfied long-range distance restraints. Transient long-range structure was also detected for the intrinsically disordered transactivation domain of the tumor suppressor protein, p53 (42, 56). The presence of transient long-range structure for some IDPs is consistent with the notion that IDPs have been selected to perform functions that require them to maintain a specific structural ensemble. It would be very surprising if all of the different selective pressures acting on IDPs resulted in a single category of structural ensemble.

Since it is likely that different functional classes of IDPs will form different classes of structural ensembles, it is important to develop and test new methodologies that are capable of determining the structural ensembles of IDPs even if the resolution of these methods is less than atomic. In addition to the iterative fitting method described above, a method using distance restraints derived from paramagnetic relaxation enhancement (PRE) data is being used to determine structural ensembles for IDPs (17). PRE data can be used to estimate internuclear distances in a dynamic ensemble of structures at distances from 5 to 30 Å, depending on the spin label that is used. This makes it a suitable method for structural studies of IDPs (17, 21, 28, 29). This approach was recently used to determine a structural ensemble for the p53 transactivation domain (p53TAD) (42, 56). In the initial manuscripts describing this work, relevant biological conclusions and procedural details were discussed. In this chapter, relevant procedural details will be reiterated to help the editors provide a compendium for the structural analysis of IDPs. Some important technical considerations will also be discussed and additional details will be provided on the analysis of IDPs using a new method we developed to identify persistent structural features in a heterogeneous ensemble (41).

5.1.2 Description of the Model System

We are currently investigating the structure, function, and dynamics of the human p53TAD. Residues 1–73 of p53 form a transactivation domain (p53TAD) that is responsible for regulating the transcriptional activity and the cellular stability of p53 (5, 35, 60). This domain contains binding sites for the ubiquitin ligase, Mdm2, and the 70 kDa subunit of replication protein A, RPA70. When bound to Mdm2, p53 becomes ubiquinated and targeted for proteosome-mediated degradation. When bound to RPA70, p53 may be stabilized and available to amplify the cellular response to DNA damage (57).

Several studies have shown that human p53TAD is intrinsically disordered with some minimal preferences for local secondary structure (9, 19, 37, 57). More recently, my group used PRE to show that human p53TAD maintains some transient long-range structure, resulting in an ensemble that deviates significantly from a random coil (42, 56). However, long-range NOEs have not been observed and only a small number of medium-range NOEs have been observed (37).

We are investigating how conserved the structural ensembles are for IDP families, and the availability of homologous sequences is one reason that p53TAD was chosen as a model system for studying the structure and dynamics of IDPs. Figure 5.1 shows a protein sequence alignment for p53TAD from seven mammals. Sequence identity for these homologues ranges from 91% between human and macaque to 42% between dog and mouse. Previous work on the evolutionary relationships between compact globular proteins has shown that amino acid sequence identity of greater than 40% leads to nearly identical protein folds that often have identical functions (3, 4, 15, 16, 38, 39, 44). It is unclear whether a similar relationship between sequence and structure applies to IDPs. We are currently testing this hypothesis by determining structural ensembles for the p53TAD homologues shown in Figure 5.1 using data from PRE, RDC, and SAXS experiments. To determine whether any structural differences are expected for the homologues in Figure 5.1, their disorder probabilities were predicted using IUPred (22). According to the predictions shown in Figure 5.1, a broad range of structural variation is expected between the seven homologues. Based on the disorder prediction shown in Figure 5.1, human p53TAD is the most disordered homologue. If the predictions are accurate, then partially collapsed structures should be detectable for the other homologues.

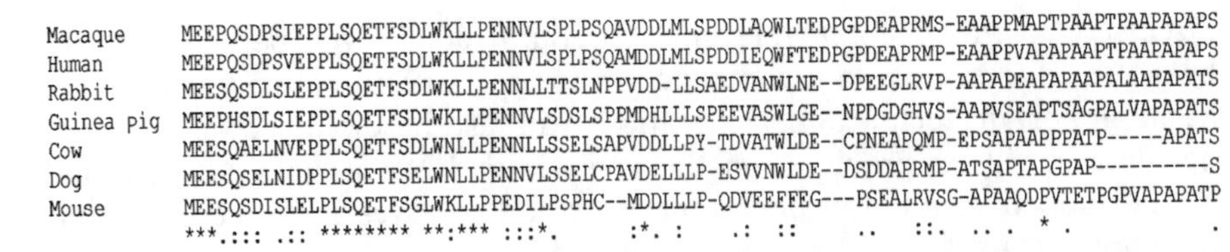

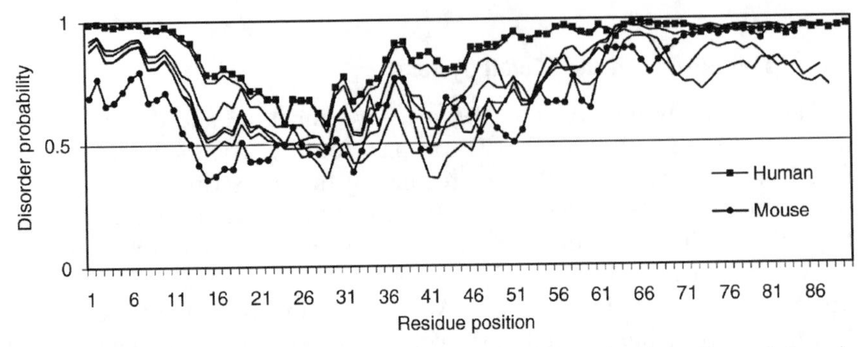

Figure 5.1 Protein sequence alignment of the p53TAD and polyproline subdomain from the seven mammals. The protein sequences correspond to residues 1–91 of human p53. Percent identity ranges from 91% (between human and macaque) to 42% (between dog and mouse). The graph shows the IUPred disorder probabilities. For clarity, only the sequences with the highest (human) and lowest (mouse) average disorder probabilities are labeled.

5.2 MUTAGENESIS AND SPIN LABELING

5.2.1 Criteria for Site Selection of Cysteine Mutants

To determine a structural ensemble for human p53TAD using data from PRE experiments, the spin label MTSL was covalently attached to cysteine mutants. Four mutants were cloned that substitute cysteine for D7, E28, A39, and D61. Three criteria were considered important for site selection of cysteine mutants. First, it is probably most important to choose a set of sites that will not disrupt the function of the protein. Unfortunately, this may not be possible if one is to satisfy the other two criteria. However, as is shown below, it should be possible to choose sites that have a minimal effect on the structure of the ensemble. Second, one should try and choose sites that provide a relatively uniform coverage of the sequence. Placing all of the mutants at one end of the sequence may bias your ability to detect the presence of long-range interactions at the other end of the sequence. Third, it is also important to choose sites that are structurally similar to the cysteine side chain. Replacement of bulky, aromatic, or nonpolar side chains with a cysteine may have a detrimental effect on function and on any long-range structure present. Of course, these are just guidelines and in-depth knowledge about functionally relevant sites, and other structural data should be taken into consideration during site selection. In the case of p53TAD, a fourth criterion that may not be valid was considered. This criterion is based on a general assumption about the location of charged residues in folded proteins. Charged residues in folded proteins are typically located on the surface, and when selecting these sites for mutagenesis, one can usually assume that no significant changes in the structure will occur. It is unclear whether structural ensembles for IDPs will follow a similar principle, since water may not be excluded from any transient hydrophobic core. For the particular case of p53TAD, it is important to recognize that the high net negative charge (−13 at pH 7) is a consequence of evolutionary selection and essential for function (1). For instance, all of the p53TAD homologues in Figure 5.1 have pKa values between 3 and 4. It is reasonable to expect that reducing the net charge will have an effect on both the structural ensemble of p53TAD and its function. In particular, the molecule could become more collapsed as the net charge is reduced. In section 5.5.2, SEC is used to show that this was not the case for the p53TAD cysteine mutants.

5.2.2 MTSL Labeling and NMR Data Analysis

The paramagnetic spin label MTSL is covalently attached to the free thiol of the cysteine residues. The cysteines are maintained in their reduced form using DTT. Prior to the addition of MTSL, the dithiothreiotol (DTT) is removed using a gravity flow desalting column. The MTSL is then added directly to the protein fractions and incubated at room temperature for 1 h. It may be necessary to empirically determine incubation times for specific IDPs since the transient burial of the cysteine residue could influence its availability to form a covalent bond with the MTSL. Excess MTSL is removed using another desalting column, and the MTSL-labeled protein is concentrated for NMR analysis. The extent of the MTSL labeling should be greater than 95%. This can be verified using mass spectrometry.

It is likely that the addition of MTSL to the cysteine residues, as well as the mutagenesis itself, will have some local effects on the chemical shifts of backbone nuclei. For the case of p53TAD, only small chemical shift changes were observed for the resonances in the ^{15}N heteronuclear single-quantum coherence (HSQC) spectra of the D7C, A39C, and D61C variants when compared with the wild-type (wt) resonance assignments, which were previously determined using standard triple resonance methods (57). If this is the case, the amide nitrogen and proton resonance assignments can be made by inference, assuming a minimal perturbation in the ^{15}N HSQC spectrum. Whenever there is any ambiguity in the assignment of resonances using this principle, as was the case for E28C, the resonance assignments are made using the standard triple resonance methods that were applied to wt p53TAD (57). Analysis of C_α and C_β chemical shifts can be used to determine whether significant changes in the ^{15}N HSQC spectrum are accompanied by structural changes. This was not the case for E28C.

5.3 DETECTING THE EFFECTS OF MUTAGENESIS AND SPIN LABELING ON THE STRUCTURAL ENSEMBLE

5.3.1 Using NMR to Determine the Effects of Mutagenesis and Spin Labeling on the Structural Ensemble

It is recommended that a number of tests be performed to ensure that labeling the cysteine mutants with a hydrophobic molecule like MTSL does not induce or stabilize any long-range structure that may be present in the protein. The tests described below probably do not represent an exhaustive list, which underscores the need for a discussion in the structural biology community on a general protocol for determining structural ensembles of IDPs. The first test determines whether self-association between the protein molecules in solution is responsible for the observed PRE. To perform this test, a ^{1}H–^{15}N HSQC spectrum is collected on a ^{15}N-labeled sample of p53TAD that is independently titrated with MTSL-labeled samples of the cysteine mutants. The cysteine mutant samples are not ^{15}N labeled and therefore do not give any direct signals in the ^{1}H–^{15}N HSQC spectrum. However, self-association between the ^{15}N-labeled sample of p53TAD and the MTSL-labeled cysteine mutants will be detected as PRE. For human p53TAD, the results from these experiments showed no evidence for self-association at the working concentrations for NMR experiments (0.3–0.6 mM).

It is also important to determine whether labeling the cysteine mutants with MTSL stabilizes any transient structure that is present in the ensemble. To perform this test, free MTSL was titrated into a ^{15}N-labeled sample of wt p53TAD. Following the addition of free MTSL, a ^{1}H–^{15}N HSQC spectrum was collected to detect any chemical shift and resonance intensity changes. The presence of free MTSL induced mostly global and some local intensity changes (data not shown). The free MTSL was subsequently reduced using an excess of ascorbic acid, and another ^{1}H–^{15}N HSQC spectrum was collected. In this spectrum, some local chemical shift changes were observed. A plot of the combined amide ^{1}H and ^{15}N chemical shift changes are shown in Figure 5.2A. The amide ^{1}H and ^{15}N chemical shift changes were combined and scaled using the equation shown in the figure legend. Combined chemical shift changes >0.02 ppm are considered significant. A comparison of the combined chemical shift changes in Figure 5.2A with hydrophobicity values for p53TAD residues showed no significant correlation. The result shown in Figure 5.2A suggests that free MTSL may be binding to multiple sites on human p53TAD. To ensure that these potential interactions were not the result of tight binding between MTSL and p53TAD, the sample was dialyzed back into the original buffer and a final ^{1}H–^{15}N HSQC spectra was collected. In this spectrum, all chemical shifts returned to their original positions.

A similar analysis of chemical shift changes should be performed on the cysteine mutants with and without MTSL. In the case of p53TAD, the largest combined chemical shift changes were observed for the E28C mutant. The

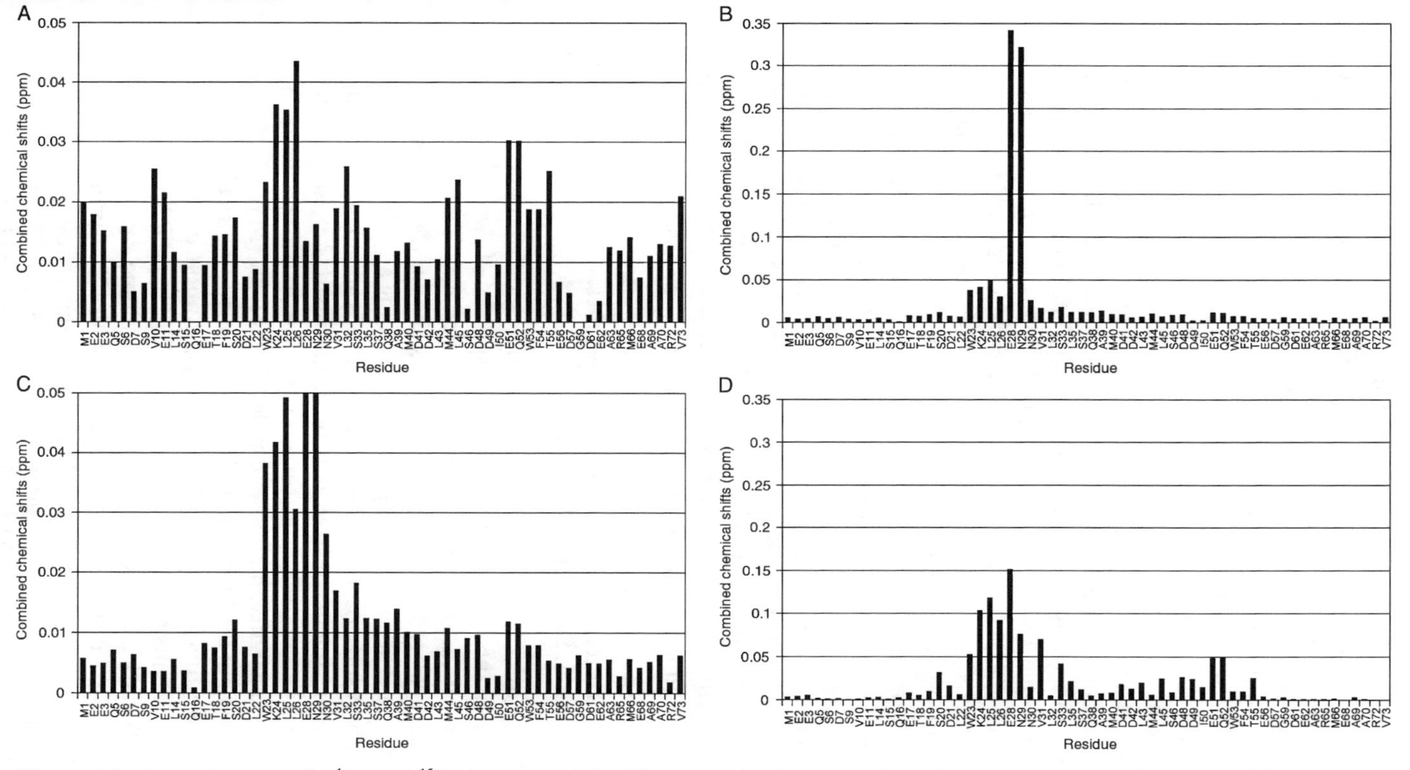

Figure 5.2 Combined amide ^{1}H and ^{15}N chemical shift differences for human p53TAD. Combined chemical shift differences are calculated using the following relationship:

$$\left[\left((H1-H2)^2+((N1-N2)/5)^2\right)/2\right]^{1/2},$$

where H1 and H2 correspond to the amide proton chemical shifts in two different conditions, and N1 and N2 correspond to the amide nitrogen chemical shifts in two different conditions. (A) Combined chemical shift differences for p53TAD before and after the addition of an equivalent amount of MTSL. The MTSL was reduced with ascorbic acid to detect the chemical shift change. (B) Combined chemical shift differences between p53TAD and E28C shown on a full scale. (C) Combined chemical shift differences between p53TAD and E28C shown on an expanded scale. (D) Combined chemical shift differences for E28C with and without the MTSL label.

combined amide ^{1}H and ^{15}N chemical shift changes between wt p53TAD and E28C are shown on two different scales in Figure 5.2B,C. As expected, all of the significant chemical changes are near the mutation site. Figure 5.2D shows the combined amide ^{1}H and ^{15}N chemical shift differences between E28C with and without MTSL. In this experiment, the largest chemical shift changes are localized near the cysteine, but there are also significant chemical shift changes observed for residues that are outside of the expected range. In particular, E51 and Q52 have combined chemical shifts in the range of 0.05 ppm. This observation can be interpreted in one of two ways, either the mutation site at E28 is already in close proximity to E51 and Q52, or the presence of the MTSL label is inducing a transient collapse between these two regions. In the next section, SEC is used to show that the Stokes radius of p53TAD is identical to E28C, with and without the MTSL. A close inspection of Figure 5.2A,C also suggests that E28 is in close proximity to E51 and Q52. This is because the correlation coefficient between the combined chemical shifts in Figure 5.2A,C is 0.31, whereas the correlation coefficient between Figure 5.2C,D is 0.84. To perform this correlation analysis, the obvious outliers at E28 and N29 were removed and the fit was forced to go through the origin. The higher correlation coefficient between Figure 5.2C,D means that the pattern of chemical shift changes observed between p53TAD and E28C is more similar to the pattern of chemical shift changes observed for E28C when the MTSL is attached. If the MTSL was inducing a transient collapse, then the pattern of chemical shift changes in Figure 5.2C should be more similar to that observed in Figure 5.2A.

5.3.2 Using SEC to Determine the Effects of Mutagenesis and Spin Labeling on the Structural Ensemble

Due to the patterns of chemical shift changes observed for some of the p53TAD cysteine mutants with and without the spin label, we felt it was prudent to develop an independent approach to asses whether any global conformational changes are being induced. SEC was chosen for convenience and because we were particularly interested in determining any changes in the Stokes radius that may have occurred due to mutagenesis and/or spin labeling. Details of the SEC experiments have already been published (42). In summary, three replicate experiments were run for each sample and the average V_e values were used to calculate an apparent molecular weight (MW_{app}). MW_{app} was then used to calculate the Stokes radii (R_s) based on an empirical relationship between the two quantities (53). The averaged elution profiles for the three replicate experiments are shown for p53TAD, E28C, and E28C + MTSL in Figure 5.3A,B. The standard deviations in elution volume were similar for all samples, with a value of 0.14 mL. Average R_s for all three proteins were also similar but did show a small trend toward compactness going from p53TAD to E28C. The R_s values in nanometer were 2.38 ± 0.02 for p53TAD, 2.36 ± 0.02 for E28C, and 2.36 ± 0.02 for E28C labeled with MTSL. Based on these results, it was concluded that there were no large changes in R_s due to

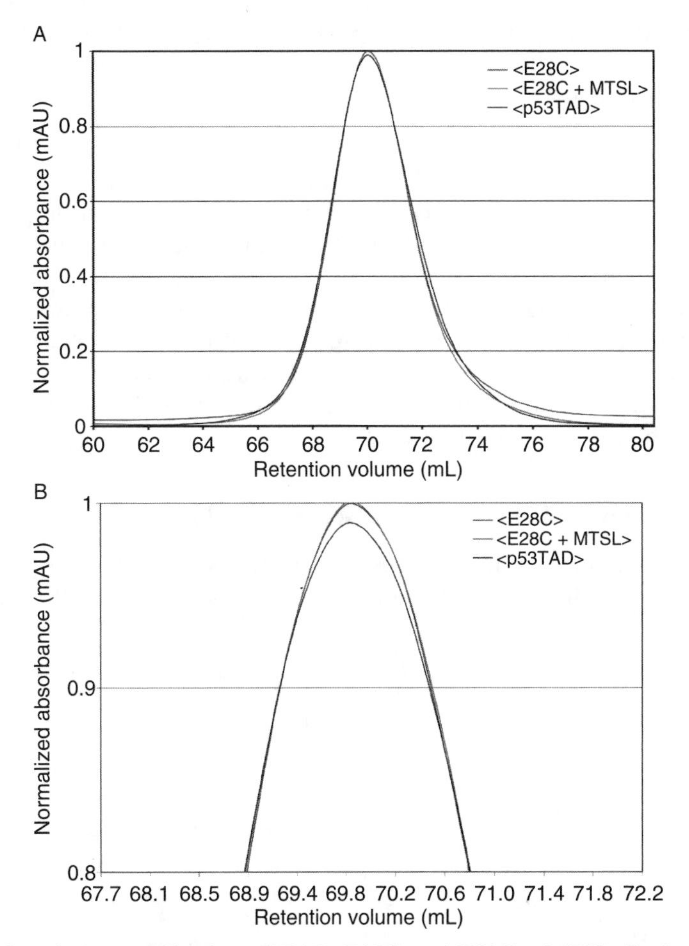

Figure 5.3 Elution profiles for p53TAD, E28C, and E28C + MTSL. Each trace represents the average of three consecutive runs. The vertical axis shows the normalized absorbance at 280 nm, and the horizontal axis shows the elution volume. (A) Shows the full chromatograms. (B) Shows an expanded view.

mutation or MTSL labeling. Similar results were obtained for the other cysteine mutants, with and without the spin label (unpublished observation).

The R_s values presented above were calculated using a relationship between MW_{app} and R_s that was developed for ordered proteins (53). More recently, Uversky proposed relationships between MW_{app} and R_s for the molten globule (MG) and pre-MG (PMG) forms of ordered proteins as well as the coil and PMG forms of disordered proteins (54). Using the MG relationship gives an R_s value for p53TAD of 2.65 nm. Interestingly, this value is very close to R_s and R_g values of 2.55 and 2.83 nm, determined using pulsed field gradient NMR and SAXS, respectively (unpublished observation).

5.4 PRE MEASUREMENTS AND STRUCTURE CALCULATIONS

5.4.1 Collecting PRE Data

^{15}N HSQC spectra are collected on the spin labeled samples. The spin label is then reduced with ascorbic acid, and additional spectra are collected on the mutant proteins. We typically collect two ascorbic acid titration points at a 5-fold and 10-fold excess. It is important to make sure that the addition of ascorbic acid does not change the sample pH. Figure 5.4 shows selected regions of the ^{15}N HSQC spectra for two of p53TAD cysteine mutants, showing resonances affected by PRE and the recovery of these resonances following addition of ascorbic acid. A resonance intensity quotient is determined by measuring the peak intensities from the two spectra and dividing the MTSL intensities by the reduced intensities.

5.4.2 Calculation of Distances and Generation of Restraints

The intensity reduction that is observed for H_N resonances in the presence of the spin label is directly proportional to the enhancement in the transverse

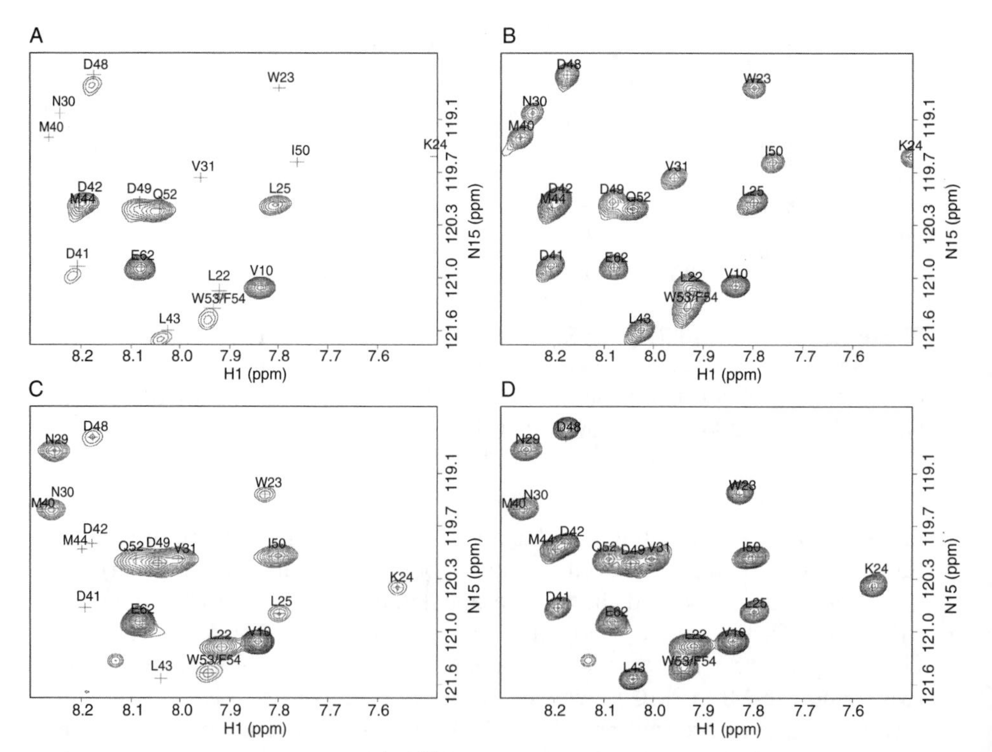

Figure 5.4 Selected regions of ^{1}H–^{15}N correlation spectra for p53TAD cysteine variants showing PRE. (A) E28C + MTSL, (B) E28C + MTSL + vitamin C, (C) A39C + MTSL, and (D) A39C + MTSL + vitamin C.

relaxation rate. To calculate the electron–proton distance using the intensity quotient, ^{15}N spin relaxation, longitudinal ^{1}H relaxation, and the difference between in-phase and antiphase transverse ^{1}H magnetization can be ignored because these effects have a small effect on the calculation. These approximations, together with a measurement of the intrinsic proton transverse relaxation rate in the reduced state, allow a fit of the contribution of the spin label to the proton transverse relaxation rate (6). In our study of human p53TAD, the intrinsic proton transverse relaxation rate was measured by fitting resolved HSQC resonances to a Lorentzian function and calculating the average value. In the HSQC experiments, the proton was transverse during the INEPT delays for a total of 9.8 ms.

Once the paramagnetic contribution to proton transverse relaxation is determined, the electron–proton distance is calculated using standard relaxation theory, assuming that the electron–proton correlation time is similar to the average overall rotational correlation time of the proton–nitrogen vectors (28). An estimate of the overall rotational correlation time for the proton–nitrogen vectors can be obtained from model-free analysis or reduced spectral density mapping of the ^{15}N relaxation data (57). For p53TAD, data from reduced spectral density mapping were used. This analysis showed that the overall correlation time for individual residues varied from 3 to 5 ns. However, this variation is not a problem since the calculation of proton–electron distances is insensitive to changes in correlation times of up to 50% (29). The average rotational correlation time from the reduced spectral density mapping was estimated to be 3.3 ns, and this value was used in the calculation of the electron–proton distances (57).

The PRE data from the four cysteine mutants was used to calculate 207 long-range distances. Two classes of restraints were generated for structure calculations using these distances: (1) a square-well potential centered on the target distance and 10 Å in width was used for residues that were less than 20 Å from the spin label, and (2) a step potential was used with a step at 20 Å and no upper bound for residues greater than 20 Å. The distances were either restrained from the C_α or the C_β of the corresponding spin labeled cysteine to the amide proton of residues that experience PRE. The C_α–H_N restraint was used with a backbone model of the protein during the simulated annealing step of the molecular dynamics, and the C_β–H_N was used with an all-atom model.

5.4.3 Calculation of Structures Using Torsion Angle Dynamics and Ensemble-Averaged Restraints

The PRE-based distance measurements were used as restraints to calculate an ensemble of structures for human p53TAD. Structures were calculated with the program XPLOR-NIH 2.14 on a cluster supercomputer running 128 parallel processors (48, 49). A 1000-member ensemble was calculated starting from random extended structures. Torsion angle dynamics were initiated at

3500 K and annealed to 100 K in 15,000 steps. For the models restrained at the backbone, the energy term during annealing consisted of bond lengths, bond angles, impropers, steric repulsion, and PRE-derived distance restraints. For the models restrained at the side chain, all of these restraints were used as well as the Ramachandran basin sampling option provided in XPLOR. The annealing is followed by 10,000 steps of torsion angle minimization and all-atom Cartesian minimization. In the final ensembles of 1000 structures, restrained either at the backbone or the side chain, there were no violations of the PRE-derived distance restraints. This statement does not reflect the quality of the structures but rather underscores the ±5-Å uncertainty in the distance measurement.

To generate the structural ensemble for human p53TAD, torsion angle dynamics was combined with ensemble averaging of the distance restraints. Ensemble averaging is used because the distance estimates based on PRE are biased toward shorter distances (29, 32, 40). This inherent distance bias leads to overrestrained structures, and ensemble averaging helps to alleviate this problem (55). For human p53TAD, an ensemble of noninteracting replicas was simulated in parallel. For each restraint, an ensemble average distance is calculated from the replicate structures. These averaged restraints are then compared with the PRE-derived distance restraints using a penalty function with a square-well potential. This function is designed to account for the 5-Å uncertainty in the PRE-derived distance restraint. The optimal size of the ensemble was determined by a series of structure calculations with ensemble sizes from 2 to 20, with a total number of 128 structures in each run. The distribution of gyration radii for each run was compared to find a limit in ensemble size where the distribution mean does not significantly change (40). An ensemble size of 5 was used in the calculations described above for human p53TAD.

5.4.4 Comparing Structural Ensembles Determined Using Restraints on Backbone or Side-Chain Atoms

Figure 5.5A shows two distributions of R_g values for 1000-member ensembles of human p53TAD. In one ensemble, the 207 PRE-derived distance restraints were applied between C_α of the cysteine mutants and the H_N where PRE was observed. In the other ensemble, restraints were applied between the C_β of cysteine mutants and the H_N where PRE was observed. The average R_g value for the distribution restrained at the backbfone is 2.0 nm and the values for the distribution restrained at the side chain is 2.2 nm. The shift in the R_g distribution for the ensemble restrained at the side chain is probably due to the increased degrees of freedom available during the simulated annealing for models restrained at C_β.

A structural ensemble was also calculated without the PRE-derived distance restraints to generate a random coil model for p53TAD (Fig. 5.5B, unrestrained). The average R_g value for the unrestrained distribution was

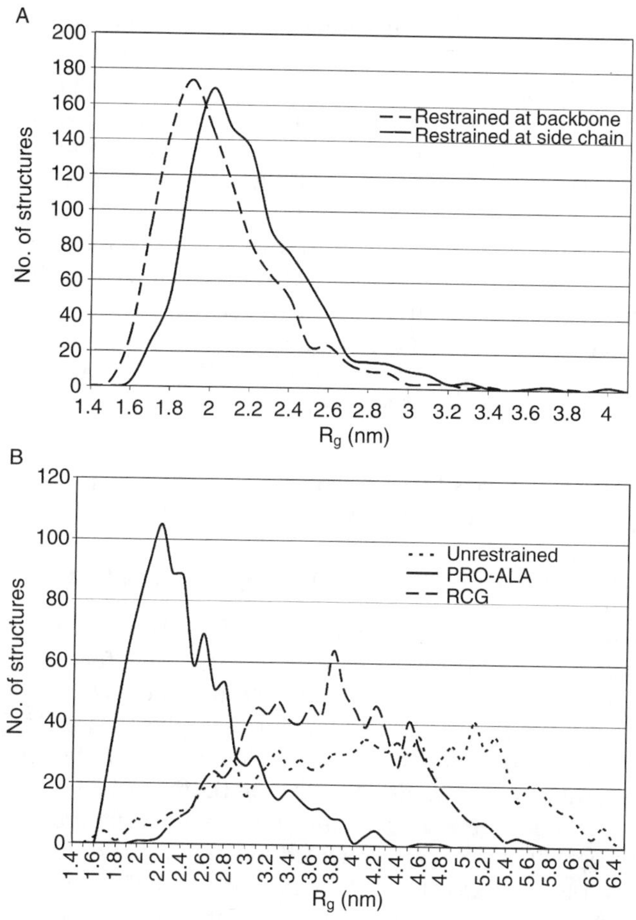

Figure 5.5 Smoothed histograms showing the distribution of R_g values for 1000 p53TAD structures. (A) R_g values for structures restrained either at the backbone or the side chains. (B) R_g values for unrestrained structures.

4.1 nm, and this ensemble is broader and has a larger average value than expected for a random coil based on theoretical considerations (20, 27). The unrestrained distribution is abnormal because human p53TAD contains 13 prolines. It has two PXP motifs, one PXXP motif, one PP, and five single prolines. Prolines reduce flexibility and extend the polypeptide chain because of the ring structure formed between the side chain and backbone. To test how the presence of prolines influenced the distribution of R_g values for the random coil model of p53TAD, unrestrained torsion angle dynamics was performed on a sequence identical to p53TAD except all of the prolines were changed to alanines (Figure 5.5B, PRO–ALA). The distribution of R_g values for the random coil model of p53TAD without prolines has an average value of 2.44 nm, which is much closer to the expected value for a theoretical random coil the size of p53TAD. We also wanted to determine if one of the new models for unfolded proteins that combine side-chain-excluded volume and amino acid-specific backbone dihedral angle propensities defined by a restricted library of loop fragments that were taken from a database of folded proteins generated an ensemble for p53TAD that was closer to the PRE-restrained or the unrestrained ensemble (11, 34). The Random Coil Generator (RCG) program provided by Sosnik's group was used for the calculations (34). The results are shown in Figure 5.5B (RCG) and shows that the model fails to recover the long-range structure identified by PRE. It does, however, produce an ensemble that is more similar to the unrestrained ensemble than flexible Meccano, the program developed by the Blackledge group for modeling the structural ensembles of IDPs (59).

5.5 PROSPECTIVE: VALIDATION, ANALYSIS, AND INTERPRETATION OF STRUCTURAL ENSEMBLES

5.5.1 Validating Structural Ensembles Using RDCs and SAXS

RDCs provide distance-independent information about the angular relationships between bond vectors in pairs of amino acids and, for compact globular proteins, provide an unequivocal demonstration of the presence of long-range structure (7, 45). The picture seems a bit more complicated for RDCs observed in IDPs (26, 34, 43, 50). While the current debate is not settled, it does appear that the observation of nonvanishing RDCs in IDPs provides both information about local sampling of ϕ/ψ space and long-range structure (2, 10, 26, 46). Despite the rich structural data provided by RDCs, it is difficult to use them to determine de novo structures. However, it is relatively simple to predict RDCs from a given structure, making them particularly useful for cross-validation of structural ensembles determined using distance restraints from PRE data (8). In an effort to develop such a cross-validation scheme, one-bond D_{NH} dipolar couplings were measured in anionic liquid crystalline media for human p53TAD (21). For this study, D_{NH} dipolar couplings were measured by recording ^{1}H–^{15}N HSQC spectra without the regular ^{1}H 180° decoupling pulse that is normally applied during the t_1 evolution period. Figure 5.6 shows the results for human p53TAD. In Figure 5.6, the quadropolar splitting is shown for D_2O in 5 wt% anionic liquid crystals at 25°C. Anionic liquid crystals were prepared using the method described by Ruckert and Otting (46). Couplings are shown for selected residues from human p53TAD in both isotropic and anisotropic ($\perp$) conditions. Both positive and negative RDCs are observed in anisotropic conditions. Large couplings like those observed for S18 may indicate an interaction with the liquid crystals, and RDCs will be measured in additional alignment media to test this idea. We are currently using the structural ensembles determined from PRE-derived distance restraints, as well as structural ensembles based on a restricted library of loop fragments from the PDB to predict

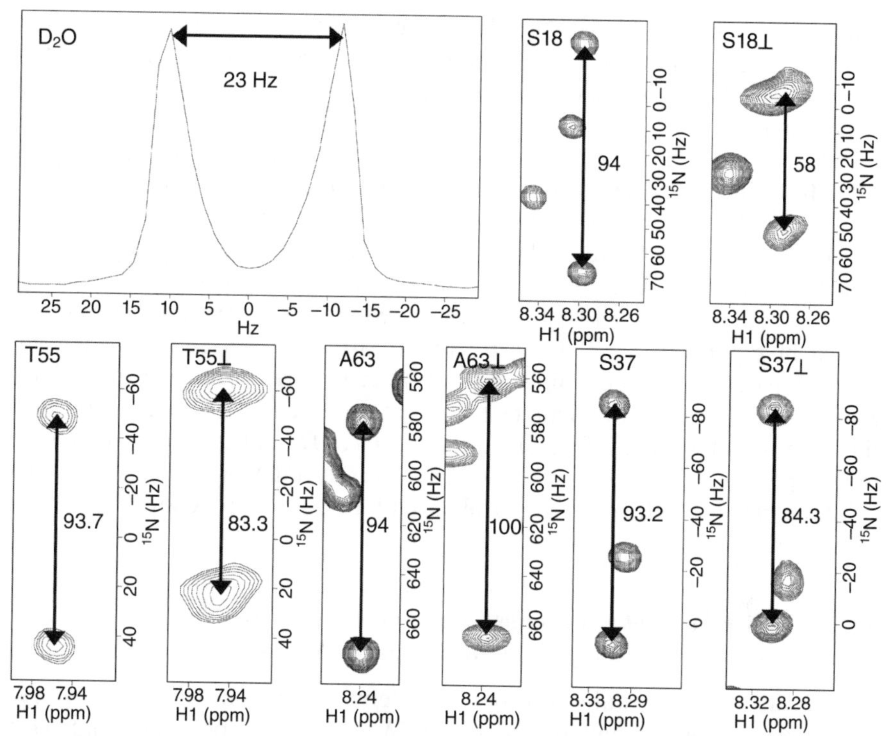

Figure 5.6 RDCs for p53TAD. Quadropolar splitting is shown for D_2O in 5 wt% anionic liquid crystals at 25°C. RDCs are shown for selected residues in both isotropic and anisotropic conditions (⊥). Both positive and negative couplings are observed. Large couplings like those observed for S18 may indicate an interaction with the crystals.

RDCs for comparison with the measured values. This will provide an important cross-validation of the structural ensembles determined using the PRE-derived distance restraints.

SAXS is also being used to investigate the equilibrium distribution of structures for human p53TAD and the other homologues presented in Figure 5.1. This work is being conducted in collaboration with Dr. Veronique Receveur-Brechot at the Architecture et Fonction des Macromolécules Biologiques in Marseilles, France. We are preparing a manuscript describing this work as well as denaturation studies of human p53TAD, so I will only briefly describe our progress. Scattering data for human p53TAD was collected at the small-angle scattering station at the European Molecular Biology Laboratory (EMBL)-Hamburg. The D_{max} extracted from probability distribution for all internuclear distances P(r) is 10.3 ± 0.5 nm (unpublished observation). This value is in reasonable agreement with the largest R_g values observed for the unrestrained distribution of structures. The R_g for p53TAD was determined from the scattering profile using the Guinier approximation and gave a value of 2.8 ± 1 nm (unpublished observation). The R_g value determined using the Guinier approximation is larger than the Stokes radius determined using SEC, and the average R_g value from the distribution of PRE-restrained structures is shown in Figure 5.5A.

While the R_g value determined using SAXS is greater than the average value from the restrained distribution of structures shown in Figure 5.5A (2.8 vs. 2.2 nm), it is much less than the average R_g value from the unrestrained distribution (2.8 vs. 4.1 nm). The SAXS data are capturing the equilibrium distribution of structures, and this distribution appears to contain structures that have R_g values from both the restrained and unrestrained distributions shown in Figure 5.5. In the future, structural ensembles determined using PRE-derived distance restraints, no restraints, or a restricted library of loop fragments will be used to calculate X-ray scattering curves for comparison with the experimental curves (58). Given the values of D_{max} and R_g determined for human p53TAD from the SAXS data, the best fit of a weighted distance distribution calculated from a structural ensemble may require combining structures from the three ensembles.

5.5.2 Analysis of Heterogeneous Structural Ensembles Using Eigenvector Decomposition of Distance Matrices

One challenge in the study of IDPs is to identify common structural features in a heterogeneous ensemble of structures. We have developed a new approach that takes advantage of the information provided by atomic distance matrices or contact maps (41). The variance between these matrices can be rapidly analyzed for large, heterogeneous ensembles using principal component analysis (PCA). This permits the identification of structures with common features. We performed this analysis on the p53TAD ensemble and observed a number of interesting results that are summarized in Figure 5.7. Figure 5.7 shows the average contact maps and surface representations for all of the clusters identified following PCA of the 1000 contact maps calculated from the p53TAD ensemble. Four surface representations are shown for the aligned structures within a given cluster. In the top row, the clusters are oriented to show the side with the maximum negative charge potential. The second row is in the same orientation as the first row, but the cluster surface is now colored to show the location of the Mdm2 and RPA70 binding sites, in gold and green, respectively. The third and fourth rows show the cluster surfaces rotated by 180° around the vertical axis. Row 3 shows the electrostatic potential, and row 4 shows the Mdm2 and RPA70 binding sites. In every cluster, the side with the maximum negative charge potential is opposite the side with the lowest potential. It also appears that the Mdm2 and RPA70 binding sites are roughly organized on opposite sides of the clusters, with the Mdm2 binding site on the least negative side and the RPA70 binding on the most negative side. This is consistent with the important role of electrostatics for the interaction between p53TAD and RPA70 (1, 12, 33, 57).

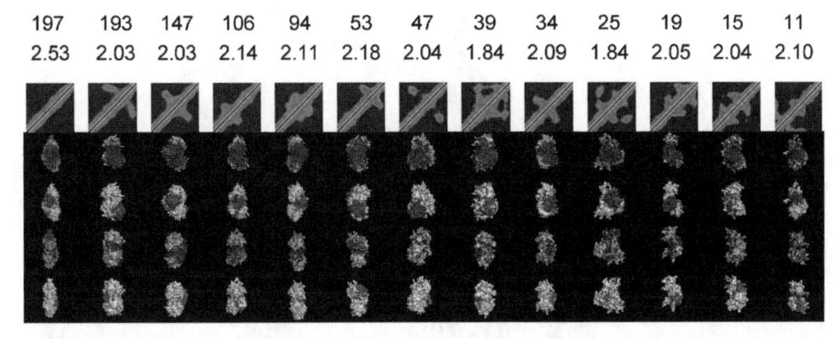

Figure 5.7 Average contact maps and surface representations for clusters of structures identified using PCA. The first row of numbers gives the cluster population, and the second row gives the average R_g for a cluster in nanometer. The contact maps are averaged over all cluster members, and the surface representations are averages for the cluster following spatial alignment. The first row of surface representations shows the side of the cluster with the highest negative electrostatic potential. The second row shows the same view as the first row with the Mdm2 and RPA70 binding sites colored orange and green, respectively. The third and fourth rows are colored like rows 1 and 2, respectively, and show 180° rotations around the vertical axis. See color insert.

The 13 clusters shown in Figure 5.7 were identified by hierarchical clustering of the first 21 principal components of the contact maps and together include 98% of the total structures in the ensemble. The number of structures and average R_g for each cluster is shown at the top of the figure. A mean-centered Pearson's correlation coefficient of 0.4 was used as the similarity metric between contact maps. This value was chosen to balance cluster size and the similarity of individual structures within a cluster. Once the clusters are identified, spatial alignments are more effective. The approach appears to be general, and we encourage other investigators to try this out on their ensembles.

5.5.3 The Importance of Biological Context for Interpreting Structural Ensembles

Given the fact that rigorous cross-validation procedures have not been developed for the structural ensembles of IDPs, it is important that these ensembles are consistent with what is known about the biology of the IDP. In the case of p53TAD, we were surprised at how strong the correlations were. Whether examining individual structures or the entire ensemble, we were able to recover biologically significant features. Since the individual structures were not calculated with any torsion angle restraints, we are reluctant to draw strong conclusions from them. However, the approach described in section 5.5.2 for analyzing the entire ensemble was rich in information and corroborated previous structural studies. In particular, the asymmetry in negative charge distribution observed for all the clusters shown in Figure 5.7 is preserved in the structure of

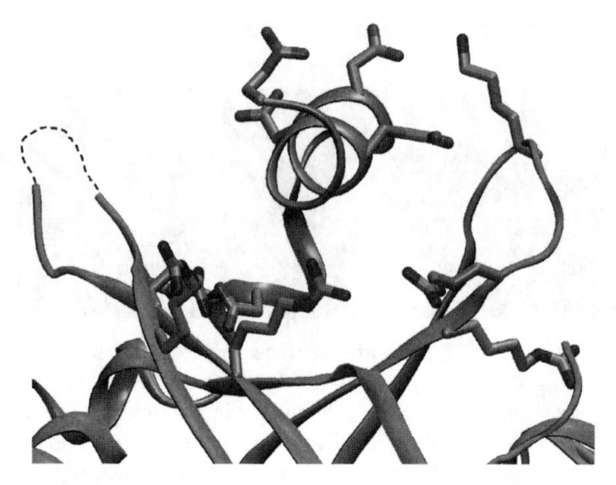

Figure 5.8 Ribbon diagram showing asymmetric distribution of charge for p53TAD when bound to RPA70. This figure was made using coordinates in the PDB file 2B3G and the visual molecular dynamics molecular graphics package (12, 31). Side chains show GLU and ASP residues in p53TAD, and ARG and LYS residues in RPA70. See color insert.

the p53TAD–RPA70 complex. This structure was determined using the minimal binding domains of p53TAD and RPA70 (12). These minimal binding domains included residues 33–56 of p53TAD and residues 1–120 of RPA70. A ribbon diagram of this structure is shown in Figure 5.8. In Figure 5.8, p53TAD is colored orange and RPA70 is colored green. Human p53TAD is acidic with a negative charge of −13 at pH 7. The side chains for ASP and GLU residues are shown in the figure, and the majority of these residues are asymmetrically arranged on the structure. For RPA70, the side chains for ARG and LYS residues that form a prominent basic cleft are shown. Interestingly, a number of potential electrostatic interactions between negative residues from p53TAD and positive residues from RPA70 are not satisfied. We think these residues may play a role in the formation of a weak encounter complex that was shown to be the first step in the coupled folding/binding mechanism for the Gal4 and VP16 transactivation domains (25, 30).

Another approach we used prior to developing the method for PCA of contact maps was to examine one-dimensional distance distributions. As mentioned above, structural ensembles calculated using PRE-derived distance restraints are not expected to contain regular secondary since the minimum length scale for the PRE interaction is too long (i.e., 1 nm). However, the PRE restraints should define collapsed or extended regions over a six-residue length scale. For typical secondary structure, turns and helices result in the amide nitrogen of residues 1 and 6 of a six-residue segment being between 0.5 and 1.0 nm apart. Extended regions or beta strands result in residues 1 and 6 being about 1.5 nm apart. Histograms showing the distribution of amide nitrogen

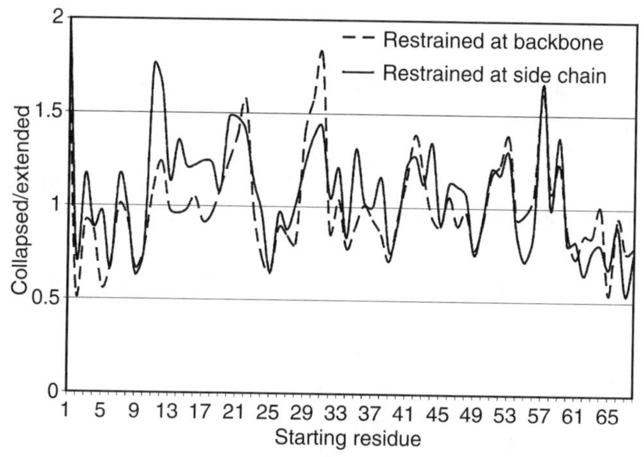

Figure 5.9 Graph showing the ratio of collapsed and extended structures in the p53TAD ensembles calculated with distance restraints between backbone atoms or distance restraints between side-chain and backbone atoms.

distances for all amino acid pairs that are four residues apart were constructed for the 1000-member ensemble of p53TAD structures. An inspection of these distance distributions showed that all six residue segments had some probability of being both collapsed and extended, but the observed distances did not match the values expected for canonical secondary structures, with peaks in the distribution centered around 1.1 and 1.4 nm. It was also observed that different regions had a higher probability of being either collapsed or extended.

The propensity of different regions to be more collapsed or extended was examined by taking the ratio of the peak heights at 1.1 and 1.4 nm from the amide nitrogen distance distributions and plotting these ratios as a function of residue *i*. In this analysis, a value of 1 means that there is an equal probability of collapsed versus extended structures in the distribution. Values that are greater and less than 1 correspond to a probability of more collapsed and extended structures, respectively. The results of this analysis are shown for the p53TAD ensemble restrained at the backbone and the side chain in Figure 5.9. Using this approach, we are able to identify collapsed regions that overlap with the nascent helix that forms the primary Mdm2 binding site and the two nascent turns that form the primary RPA70 binding site, which overlaps with two secondary Mdm2 binding sites (14, 36, 37, 47, 57). The region with the highest probability for collapse includes a PXP motif at residues 34–36.

ACKNOWLEDGMENTS

Amber Stancik and David Lowry are gratefully acknowledged for their previous work on this project. The work is funded by the American Cancer Society (RSG-07-289-01-GMC) and the National Science Foundation (0744839).

REFERENCES

1. Abramova, N. A., J. Russell, M. Botchan, and R. Li. 1997. Interaction between replication protein A and p53 is disrupted after UV damage in a DNA repair-dependent manner. Proc Natl Acad Sci U S A **94**:7186–91.
2. Ackerman, M. S., and D. Shortle. 2002. Molecular alignment of denatured states of staphylococcal nuclease with strained polyacrylamide gels and surfactant liquid crystalline phases. Biochemistry **41**:3089–95.
3. Al-Lazikani, B., J. Jung, Z. Xiang, and B. Honig. 2001. Protein structure prediction. Curr Opin Chem Biol **5**:51–6.
4. Baker, D., and A. Sali. 2001. Protein structure prediction and structural genomics. Science **294**:93–6.
5. Bargonetti, J., and J. J. Manfredi. 2002. Multiple roles of the tumor suppressor p53. Curr Opin Oncol **14**:86–91.
6. Battiste, J. L., and G. Wagner. 2000. Utilization of site-directed spin labeling and high-resolution heteronuclear nuclear magnetic resonance for global fold determination of large proteins with limited nuclear overhauser effect data. Biochemistry **39**:5355–65.
7. Bax, A. 2003. Weak alignment offers new NMR opportunities to study protein structure and dynamics. Protein Sci **12**:1–16.
8. Bax, A., G. Kontaxis, and N. Tjandra. 2001. Dipolar couplings in macromolecular structure determination. Methods Enzymol **339**:127–74.
9. Bell, S., C. Klein, L. Muller, S. Hansen, and J. Buchner. 2002. p53 contains large unstructured regions in its native state. J Mol Biol **322**:917–27.
10. Bernado, P., C. W. Bertoncini, C. Griesinger, M. Zweckstetter, and M. Blackledge. 2005. Defining long-range order and local disorder in native alpha-synuclein using residual dipolar couplings. J Am Chem Soc **127**:17968–9.
11. Bernado, P., L. Blanchard, P. Timmins, D. Marion, R. W. Ruigrok, and M. Blackledge. 2005. A structural model for unfolded proteins from residual dipolar couplings and small-angle x-ray scattering. Proc Natl Acad Sci U S A **102**:17002–7.
12. Bochkareva, E., L. Kaustov, A. Ayed, G. S. Yi, Y. Lu, A. Pineda-Lucena, J. C. Liao, A. L. Okorokov, J. Milner, C. H. Arrowsmith, and A. Bochkarev. 2005. Single-stranded DNA mimicry in the p53 transactivation domain interaction with replication protein A. Proc Natl Acad Sci U S A **102**:15412–7.
13. Buevich, A. V., U. P. Shinde, M. Inouye, and J. Baum. 2001. Backbone dynamics of the natively unfolded pro-peptide of subtilisin by heteronuclear NMR relaxation studies. J Biomol NMR **20**:233–49.
14. Chi, S. W., S. H. Lee, D. H. Kim, M. J. Ahn, J. S. Kim, J. Y. Woo, T. Torizawa, M. Kainosho, and K. H. Han. 2005. Structural details on mdm2-p53 interaction. J Biol Chem **280**:38795–802.
15. Chothia, C., and A. M. Lesk. 1987. The evolution of protein structures. Cold Spring Harb Symp Quant Biol **52**:399–405.
16. Chothia, C., and A. M. Lesk. 1986. The relation between the divergence of sequence and structure in proteins. EMBO J **5**:823–6.

17. Clore, G. M., C. Tang, and J. Iwahara. 2007. Elucidating transient macromolecular interactions using paramagnetic relaxation enhancement. Curr Opin Struct Biol **17**:603–16.
18. Daughdrill, G. W., G. J. Pielak, V. N. Uversky, M. S. Cortese, and A. K. Dunker. 2005. Natively disordered proteins, pp. 275–357. In J. Buchner and T. Kiefhaber (eds.), Protein Folding Handbook, vol. **3**. Wiley-VCH, Darmstadt, Germany.
19. Dawson, R., L. Muller, A. Dehner, C. Klein, H. Kessler, and J. Buchner. 2003. The N-terminal domain of p53 is natively unfolded. J Mol Biol **332**:1131–41.
20. De Gennes, P. G. 1979. Scaling Concepts in Polymer Physics. Cornell University Press, Ithaca, NY.
21. Dedmon, M. M., K. Lindorff-Larsen, J. Christodoulou, M. Vendruscolo, and C. M. Dobson. 2005. Mapping long-range interactions in alpha-synuclein using spin-label NMR and ensemble molecular dynamics simulations. J Am Chem Soc **127**:476–7.
22. Dosztanyi, Z., V. Csizmok, P. Tompa, and I. Simon. 2005. The pairwise energy content estimated from amino acid composition discriminates between folded and intrinsically unstructured proteins. J Mol Biol **347**:827–39.
23. Dunker, A. K., J. D. Lawson, C. J. Brown, R. M. Williams, P. Romero, J. S. Oh, C. J. Oldfield, A. M. Campen, C. M. Ratliff, K. W. Hipps, J. Ausio, M. S. Nissen, R. Reeves, C. Kang, C. R. Kissinger, R. W. Bailey, M. D. Griswold, W. Chiu, E. C. Garner, and Z. Obradovic. 2001. Intrinsically disordered protein. J Mol Graph Model **19**:26–59.
24. Dyson, H. J., and P. E. Wright. 2005. Intrinsically unstructured proteins and their functions. Nat Rev Mol Cell Biol **6**:197–208.
25. Ferreira, M. E., S. Hermann, P. Prochasson, J. L. Workman, K. D. Berndt, and A. P. Wright. 2005. Mechanism of transcription factor recruitment by acidic activators. J Biol Chem **280**:21779–84.
26. Fieber, W., S. Kristjansdottir, and F. M. Poulsen. 2004. Short-range, long-range and transition state interactions in the denatured state of ACBP from residual dipolar couplings. J Mol Biol **339**:1191–9.
27. Flory, P. J. 1953. Principles of Polymer Chemistry. Cornell University Press, Ithaca, NY.
28. Gillespie, J. R., and D. Shortle. 1997. Characterization of long-range structure in the denatured state of staphylococcal nuclease. I. Paramagnetic relaxation enhancement by nitroxide spin labels. J Mol Biol **268**:158–69.
29. Gillespie, J. R., and D. Shortle. 1997. Characterization of long-range structure in the denatured state of staphylococcal nuclease. II. Distance restraints from paramagnetic relaxation and calculation of an ensemble of structures. J Mol Biol **268**:170–84.
30. Hermann, S., K. D. Berndt, and A. P. Wright. 2001. How transcriptional activators bind target proteins. J Biol Chem **276**:40127–32.
31. Humphrey, W., A. Dalke, and K. Schulten. 1996. VMD: visual molecular dynamics. J Mol Graph **14**:27–8, 33–8.
32. Iwahara, J., C. D. Schwieters, and G. M. Clore. 2004. Ensemble approach for NMR structure refinement against (1)H paramagnetic relaxation enhancement data arising from a flexible paramagnetic group attached to a macromolecule. J Am Chem Soc **126**:5879–96.

33. Jacobs, D. M., A. S. Lipton, N. G. Isern, G. W. Daughdrill, D. F. Lowry, X. Gomes, and M. S. Wold. 1999. Human replication protein A: global fold of the N-terminal RPA-70 domain reveals a basic cleft and flexible C-terminal linker. J Biomol NMR **14**:321–31.

34. Jha, A. K., A. Colubri, K. F. Freed, and T. R. Sosnick. 2005. Statistical coil model of the unfolded state: resolving the reconciliation problem. Proc Natl Acad Sci U S A **102**:13099–104.

35. Kaustov, L., G. S. Yi, A. Ayed, E. Bochkareva, A. Bochkarev, and C. H. Arrowsmith. 2006. p53 transcriptional activation domain: a molecular chameleon? Cell Cycle **5**:489–94.

36. Kussie, P. H., S. Gorina, V. Marechal, B. Elenbaas, J. Moreau, A. J. Levine, and N. P. Pavletich. 1996. Structure of the MDM2 oncoprotein bound to the p53 tumor suppressor transactivation domain. Science **274**:948–53.

37. Lee, H., K. H. Mok, R. Muhandiram, K. H. Park, J. E. Suk, D. H. Kim, J. Chang, Y. C. Sung, K. Y. Choi, and K. H. Han. 2000. Local structural elements in the mostly unstructured transcriptional activation domain of human p53. J Biol Chem **275**:29426–32.

38. Lesk, A. M., and C. Chothia. 1980. How different amino acid sequences determine similar protein structures: the structure and evolutionary dynamics of the globins. J Mol Biol **136**:225–70.

39. Lesk, A. M., M. Levitt, and C. Chothia. 1986. Alignment of the amino acid sequences of distantly related proteins using variable gap penalties. Protein Eng **1**:77–8.

40. Lindorff-Larsen, K., S. Kristjansdottir, K. Teilum, W. Fieber, C. M. Dobson, F. M. Poulsen, and M. Vendruscolo. 2004. Determination of an ensemble of structures representing the denatured state of the bovine acyl-coenzyme a binding protein. J Am Chem Soc **126**:3291–9.

41. Lowry, D. F., A. C. Hausrath, and G. W. Daughdrill. 2008. A robust approach for analyzing a heterogeneous structural ensemble. Proteins (in press).

42. Lowry, D. F., A. Stancik, R. M. Shresta, and G. W. Daughdrill. 2007. Modeling the accessible conformations of the intrinsically unstructured transactivation domain of p53. Proteins (in press).

43. Mohana-Borges, R., N. K. Goto, G. J. Kroon, H. J. Dyson, and P. E. Wright. 2004. Structural characterization of unfolded states of apomyoglobin using residual dipolar couplings. J Mol Biol **340**:1131–42.

44. Petsko, G. A., and D. Ringe. 2004. Protein Structure and Function. New Science Press Ltd., London.

45. Prestegard, J. H., H. M. al-Hashimi, and J. R. Tolman. 2000. NMR structures of biomolecules using field oriented media and residual dipolar couplings. Q Rev Biophys **33**:371–424.

46. Ruckert, M., and G. Otting. 2000. Alignment of biological macromolecules in novel nonionic liquid crystalline media for NMR experiments. J Am Chem Soc **122**:7793–7.

47. Schon, O., A. Friedler, M. Bycroft, S. M. Freund, and A. R. Fersht. 2002. Molecular mechanism of the interaction between MDM2 and p53. J Mol Biol **323**:491–501.

48. Schwieters, C. D., Kuszewski, J. J., N. Tjandra, N., and Clore, G. M. 2003. The Xplor-NIH NMR Molecular Structure Determination Package. J Magn Res **160**:66–74.

49. Schwieters, C. D., Kuszewski, J. J., and Clore, G. M. 2006. Using Xplor-NIH for NMR molecular structure determination. Progr NMR Spectroscopy **48**:47–62.

50. Shortle, D., and M. S. Ackerman. 2001. Persistence of native-like topology in a denatured protein in 8 M urea. Science **293**:487–9.

51. Tompa, P. 2002. Intrinsically unstructured proteins. Trends Biochem Sci **27**: 527–33.

52. Uversky, V. N. 2002. Natively unfolded proteins: a point where biology waits for physics. Protein Sci **11**:739–56.

53. Uversky, V. N. 1993. Use of fast protein size-exclusion liquid chromatography to study the unfolding of proteins which denature through the molten globule. Biochemistry **32**:13288–98.

54. Uversky, V. N. 2002. What does it mean to be natively unfolded? Eur J Biochem **269**:2–12.

55. Vendruscolo, M. 2007. Determination of conformationally heterogeneous states of proteins. Curr Opin Struct Biol **17**:15–20.

56. Vise, P., B. Baral, A. Stancik, D. F. Lowry, and G. W. Daughdrill. 2007. Identifying long-range structure in the intrinsically unstructured transactivation domain of p53. Proteins **67**:526–30.

57. Vise, P. D., B. Baral, A. J. Latos, and G. W. Daughdrill. 2005. NMR chemical shift and relaxation measurements provide evidence for the coupled folding and binding of the p53 transactivation domain. Nucleic Acids Res **33**:2061–77.

58. von Ossowski, I., J. T. Eaton, M. Czjzek, S. J. Perkins, T. P. Frandsen, M. Schulein, P. Panine, B. Henrissat, and V. Receveur-Brechot. 2005. Protein disorder: conformational distribution of the flexible linker in a chimeric double cellulase. Biophys J **88**:2823–32.

59. Wells, M., H. Tidow, T. J. Rutherford, P. Markwick, M. R. Jensen, E. Mylonas, D. I. Svergun, M. Blackledge, and A. R. Fersht. 2008. Structure of tumor suppressor p53 and its intrinsically disordered N-terminal transactivation domain. Proc Natl Acad Sci U S A **105**:5762–7.

60. Woods, D. B., and K. H. Vousden. 2001. Regulation of p53 function. Exp Cell Res **264**:56–66.

61. Wright, P. E., and H. J. Dyson. 1999. Intrinsically unstructured proteins: re-assessing the protein structure-function paradigm. J Mol Biol **293**:321–31.

6

SITE-DIRECTED SPIN LABELING EPR SPECTROSCOPY

Valérie Belle,[1] Sabrina Rouger,[2] Stéphanie Costanzo,[2] Sonia Longhi,[2] and André Fournel[1]

[1]*Bioénergétique et Ingénierie des Protéines, UPR 9036 CNRS, and Universités Aix-Marseille I et II, Marseille, France*
[2]*Architecture et Fonction des Macromolécules Biologiques, UMR 6098 CNRS and Universités Aix-Marseille I et II, Campus de Luminy, Marseille, France*

ABSTRACT

Electron paramagnetic resonance (EPR) spectroscopy is a technique that specifically detects unpaired electrons. EPR-sensitive reporter groups (spin labels or spin probes) can be introduced into biological systems via site-directed spin labeling (SDSL). The basic strategy of SDSL involves the introduction of a paramagnetic group at a selected protein site. This is usually accomplished by cysteine substitution mutagenesis, followed by covalent modification of the unique sulfydryl group with a selective nitroxide reagent, such as the 1-oxyl-2,2,5,5-tetramethyl-δ3-pyrroline-3-methyl methane thiosulfonate. SDSL followed by EPR spectroscopy has been extensively used to study conformational changes within structured proteins, as well as folding and unfolding processes of structured proteins in the presence of denaturing agents, such as urea and guanidium chloride. In this chapter, we review the theoretical principles of this approach and illustrate how we successfully applied it to investigate the structural properties and the induced folding of

Instrumental Analysis of Intrinsically Disordered Proteins: Assessing Structure and Conformation, Edited by Vladimir Uversky and Sonia Longhi

the intrinsically disordered C-terminal domain of the measles virus nucleoprotein (N-tail, aa 401–525) upon binding to the X domain (aa 459–507) of the viral phosphoprotein (P).

6.1 INTRODUCTION

Electron paramagnetic resonance (EPR) spectroscopy is a technique dedicated to the study of species containing paramagnetic centers. It is based on the observation of the energy absorbed by a paramagnetic system in a homogeneous magnetic field. In biological systems, these centers are mainly free radicals or some ions of transition metals such as iron, copper, and nickel. Site-directed spin labeling (SDSL) combined with EPR is a powerful technique for detecting structural changes in proteins that are devoid of such paramagnetic centers. The strategy of SDSL involves the insertion of a paramagnetic label at a selected site of a protein and its observation by EPR spectroscopy. The paramagnetic label is usually a stable nitroxide reagent that is introduced at the desired position via cysteine substitution mutagenesis, followed by chemical modification of the sulfydryl group. SDSL EPR spectroscopy has emerged as a valuable tool for mapping elements of secondary structure in a wide range of proteins, including proteins that could not be studied by classical structural techniques, such as nuclear magnetic resonance (NMR) and X-ray crystallography (16, 37–39).

A variety of EPR approaches can be used depending on the system under study. The common approach consists in the analysis of the influence of the nitroxide motion on EPR spectral shapes. This approach is particularly well suited to monitor conformational changes induced by either protein–protein or protein–ligand interactions (4, 37) since such interactions modify the mobility of the label and, consequently, its EPR spectrum. Another approach, which has been developed mainly in order to study membrane proteins where label motion is often damped, is based on the study of the relaxation properties of the label by using various fast-relaxing agents, such as paramagnetic metal complexes and molecular oxygen. Analysis of the relaxation properties allows determination of solvent accessibility of the label that, in some cases, may be more informative in characterizing the overall protein structure (1). Finally, SDSL can be used to measure distances in biological systems. The method requires the introduction of two spin probes in the protein of interest. Then, the distance between the two probes can be measured through spin-spin interaction analysis. This technique has been applied to a wide range of structural systems, including peptides, soluble proteins, and membrane proteins (40, 58, 69).

In this chapter, the basics of EPR spectroscopy of nitroxide radicals will be presented. After a detailed description of the approach based on the study of spin label motion, the extraction of parameters of interest from an EPR spectrum and the structural information that can be derived from these parameters will be described. Finally, a case study concerning the induced folding of an intrinsically disordered protein (IDP) domain, namely, the C-terminal domain of the nucleoprotein of the measles virus (MeV), will be treated in details.

6.2 EPR SPECTROSCOPY OF NITROXIDE RADICALS

6.2.1 General Principles

Nitroxide radicals are anisotropic paramagnetic centers characterized by the interaction between an electronic spin S = ½ and a nuclear spin I = 1 arising from the magnetism of the ^{14}N nucleus located in the vicinity of the unpaired electron. For the sake of simplicity, we will first present the case of a center with isotropic magnetic properties before describing the effect of the magnetic anisotropy.

6.2.1.1 EPR Spectrum of an Isotropic Center S = ½ and I = 1 For an electronic spin system S = ½, the two possible orientations of the spin determine two energy states (Ms = +½; Ms = −½), which are degenerated in the absence of an external magnetic field $\vec{B}$. The interaction of $\vec{B}$ with this system (called *electronic Zeeman interaction*) induces a splitting of the two energy states. The energy difference ΔE between the two states is proportional to the value of the magnetic field (Fig. 6.1). Taking into account the *hyperfine interaction* arising from the interaction of the electronic spin with the nuclear spin I = 1, each level (Ms = +½; Ms = −½) is subdivided into three levels corresponding to each possible value of M_I ($M_I = -1$; $M_I = 0$; $M_I = +1$). The energetic diagram shown in Figure 6.1 gives an overview of the position of the different energetic levels. An EPR experiment is based on the application of an

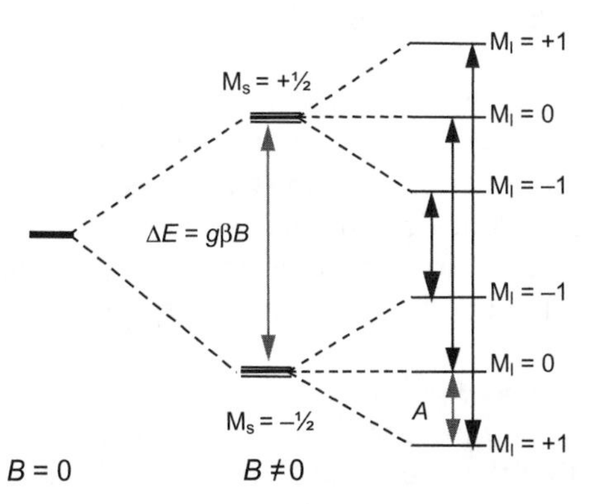

Figure 6.1 Diagram of the energy levels for a system S = ½ and I = 1. ΔE is the energy splitting caused by the Zeeman interaction between the spin system S = ½ and the external magnetic field. A is the hyperfine coupling constant. Gray arrows show the three possible transitions.

electromagnetic radiation at a suitable frequency υ. An EPR spectrum is recorded by keeping υ constant and by varying continuously the external magnetic field. The absorption of the electromagnetic wave (called *resonance phenomenon*) takes place when the energy $h\upsilon$, with h being the Planck constant, fits exactly the energy difference ΔE between two levels and when $\Delta M_s = \pm 1$ and $\Delta M_s = 0$ for these two levels.

The determination of the energies of a spin system necessitates a quantum mechanical description using spin operators and Hamiltonian. For an isotropic spin system (S = ½; I = 1), the spin Hamiltonian describing the magnetic interactions of this system with an external magnetic field is:

$$\mathrm{H} = \mathrm{H}_{\mathrm{Zeeman}} + \mathrm{H}_{\mathrm{Hyperfine}}$$

$$\mathrm{H} = g\beta\vec{B}\vec{S} + A\vec{S}\vec{I},$$

where $g = 2.0023$ is the electronic Zeeman interaction factor for a free electron, β is the Bohr electron magneton (9.274×10^{-24} J/T), A is the hyperfine coupling constant, $\vec{B}$ is the external magnetic field, and $\vec{S}$ and $\vec{I}$ are the electron and nuclear spin operators, respectively. The hyperfine interaction is small compared with the Zeeman one and can thus be considered as a perturbation of the latter. The energy values E corresponding to this Hamiltonian are given by the expression:

$$E = \mathrm{M_S}(g\beta B + A\mathrm{M_I}) \text{ with } \mathrm{M_S} = \left(-\frac{1}{2}, +\frac{1}{2}\right) \text{ and } \mathrm{M_I} = (-1, 0, +1).$$

The resonance conditions are fulfilled for three different values of the external magnetic field, called resonant field B_{MI}, given by:

$$B_{M_I} = (h\upsilon - A\mathrm{M_I})/g\beta$$

$$B_{+1} = (h\upsilon - A)/g\beta;\ B_0 = h\upsilon/g\beta;\ B_{-1} = (h\upsilon + A)/g\beta.$$

As g is close to 2.0, this leads to an EPR spectrum centered at around $B = 340\,\mathrm{mT}$ with an X-band spectrometer working at $\upsilon \approx 9\,\mathrm{GHz}$. The EPR spectrum is composed of three equidistant lines separated by the quantity $\Delta B = A/g\beta$ (Fig. 6.2A). A device producing a modulation of the external magnetic field is used in order to improve the sensitivity, leading to the acquisition of the signal as the first derivative of the absorption lines (Fig. 6.2B).

6.2.1.2 EPR Spectrum of an Anisotropic Center S = ½ and I = 1 A nitroxide radical is an anisotropic center and the magnetic axes attached to this molecule are such that the x-axis lies along the N–O bond, z is parallel to the nitrogen $2p_z$ orbital, and y is perpendicular to the x, z plane (Fig. 6.3). In this case, the g factor and the hyperfine coupling are tensorial entities noted $\tilde{g}$ and $\tilde{A}$. $\tilde{A}$ is nearly axial, and conventionally, the principal values of the g and A tensors follow the relations $g_z < g_y < g_x$ and $A_x \approx A_y < A_z$. The spin Hamiltonian

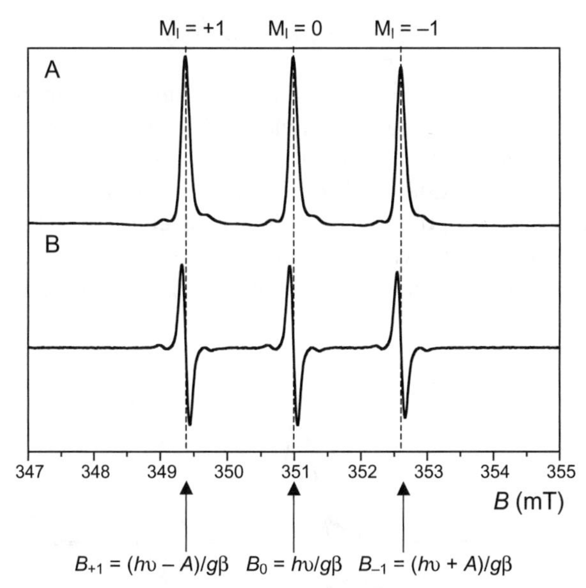

Figure 6.2 (A) EPR absorption lines of an isotropic center with S = ½ and I = 1; each line corresponds to a value of a resonant magnetic field $B_{\mathrm{M_I}}$. (B) First derivative of the absorption lines corresponding to the EPR spectrum as it is acquired experimentally. mT, milliTesla.

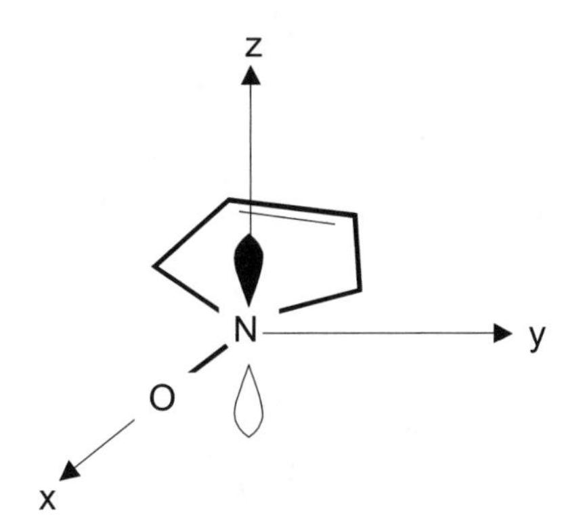

Figure 6.3 Diagram of the nitroxide radical inserted into the pyrrolinyl group with the representation of the axes attached to the molecule.

describing the magnetic interactions of this system with an external magnetic field is:

$$\mathrm{H} = \beta \vec{B}\tilde{g}\vec{S} + \vec{S}\tilde{A}\vec{I}.$$

The energy values E corresponding to this Hamiltonian are given by the expression:

$$E = M_S(g_{eff}\beta B + A_{eff}M_I),$$

where g_{eff} and A_{eff} are the so-called effective values that depend both on the orientation of the external magnetic field, with respect to the magnetic axes, and on the principal values of $\tilde{g}$ and $\tilde{A}$ ($g_z < g_{eff} < g_x$ and $A_x \approx A_y < A_{eff} < A_z$). For a collection of nitroxides with uniform random orientations, as it is the case in a frozen solution, the resonant fields for each M_I values are not discrete as described in the previous section but lie in between a range of values (Fig. 6.4A). For nitroxide radicals, the anisotropy of the g tensor is weak

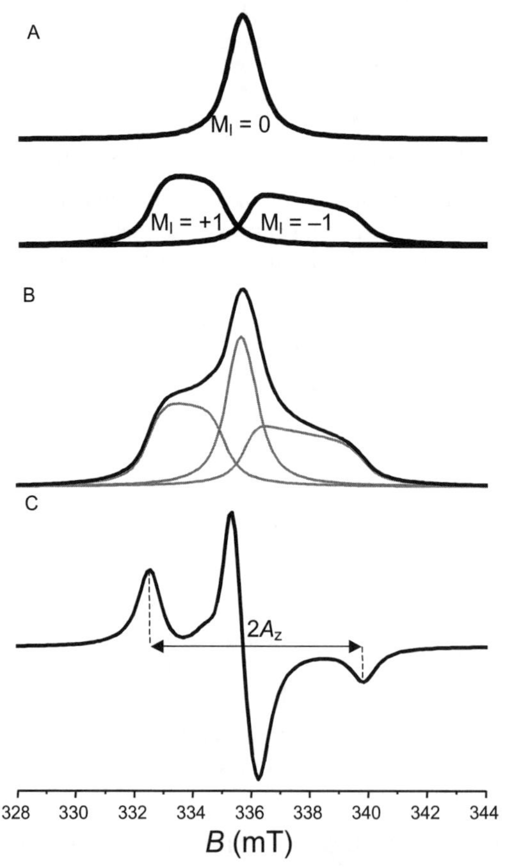

Figure 6.4 Simulated spectra of a frozen solution or a polycrystalline sample of nitroxides obtained with the following values:

$$\tilde{g} = (g_x, g_y, g_z) = (2.0089, 2.0064, 2.0027)$$
$$\tilde{A}/g\beta = (A_x, A_y, A_z)/g\beta = (0.49, 0.49, 3.51)\,\text{mT}.$$

The absorption spectrum (B) is decomposed in three components for each value of M_I (A). The EPR spectrum (C) is the first derivative of the envelop (B).

(g_x, g_y, and g_z are very close to 2.0023, the g factor of the free electron), whereas the hyperfine tensor anisotropy is relatively large and axial: While $A_x/g\beta$ and $A_y/g\beta$ values are almost equivalent (≈0.5 mT), A_z is much larger (≈3.5 mT). In this condition, the typical EPR spectrum of a nitroxide obtained at X-band frequency is shown in Figure 6.4C. Since the anisotropy of the ^{14}N hyperfine interaction is relatively large, the contribution of the components $M_I = +1$ and $M_I = -1$ is well separated from the central feature of the spectrum, the width of which depending only on the g anisotropy ($M_I = 0$) (see Fig. 6.4A,B). The value of the larger hyperfine component A_z can be deduced from the spectrum by measuring the outer line splitting as indicated in Figure 6.4C.

6.2.1.3 Dynamic Line Shape Effects In the previous section, we have described the EPR spectrum of a nitroxide radical in a frozen solution, when the spin labels are totally immobilized. EPR spectra are very sensitive to the motion of nitroxide spin labels. In the case of a free spin label in solution, this motion is expressed by an isotropic rotational correlation time τ_r. Figure 6.5 shows a set of characteristic EPR spectra for various motional regimes. The power of the SDSL technique is based on the fact that the g and A tensors are anisotropic, making EPR spectra critically dependent on the mobility of the label. When nitroxide spin labels are allowed to tumble rapidly in an isotropic way, as in the case of a liquid, magnetic interactions are completely averaged, and the system becomes similar to an isotropic center, the values of g and A tensors being averaged to:

$$g_{av} = \frac{1}{3}(g_x + g_y + g_z)$$

$$A_{av} = \frac{1}{3}(A_x + A_y + A_z).$$

This arises for $\tau_r \leq 10^{-11}$ s at X-band frequency, and the EPR spectrum is similar to the one described in section 6.2.1.1 with three narrow lines. As the motion becomes progressively slower, the magnetic anisotropy is no longer totally averaged, and this results in a differential broadening of lines in the spectrum, while line positions remain constant. This is the so-called *fast motional regime*, valid until $\tau_r = 10^{-9}$ s. For values of $\tau_r > 10^{-9}$ s up to 10^{-6} s, the averaging of tensor components becomes less and less efficient, leading to shape distortions of the EPR spectrum. This range of τ_r values defines the *intermediate motional regime*. When τ_r becomes superior to 10^{-6} s, the *slow motional regime* is reached, in which the full effects of the anisotropy of g and A tensors are observed. The EPR spectrum corresponds to the one obtained from a frozen solution described in the previous section.

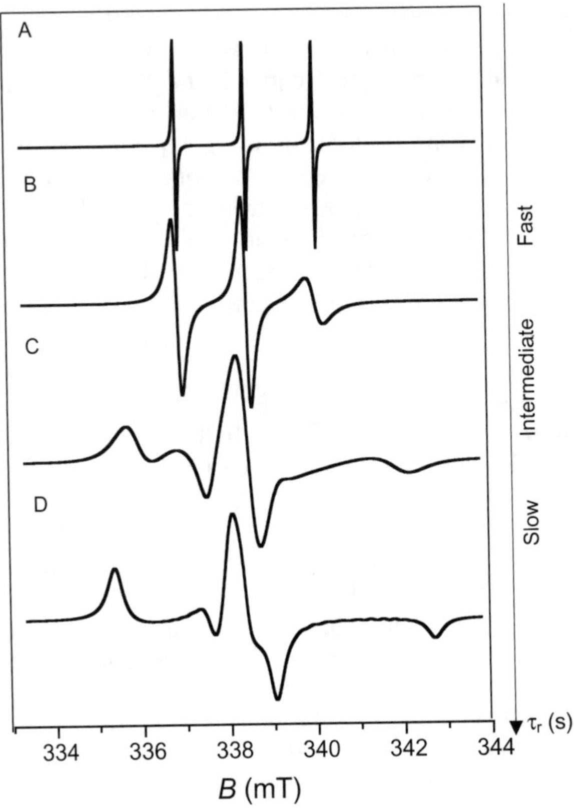

Figure 6.5 Set of various motional regimes and the corresponding characteristic EPR spectra for isotropic motion of a nitroxide spin label described by the rotational correlation time τ_r. The spectra have been simulated by the EasySpin software (70) for different values of τ_r: (A) 1.10^{-12} s, (B) 1.10^{-9} s, (C) 1.10^{-8} s, and (D) 1.10^{-5} s.

6.3 EPR SPECTROSCOPY OF NITROXIDE RADICALS GRAFTED ON PROTEINS

6.3.1 SDSL Methodology

SDSL consists in manipulating the biological system of interest so that it contains a single reactive amino acid residue, usually a cysteine that may be selectively modified with an appropriate spin label. Cysteine residues can be modified with a variety of reagents, but the most commonly used are either methane thiosulfonate (MTSL) or maleimide derivatives. The spin label (1-oxyl-2,2,5,5-tetramethyl-δ3-pyrroline-3-methyl) (MTSL) (see Fig. 6.6A) has the advantage of being specific for cysteine residues as compared with maleimide derivatives that can also modify the amino group of lysines. Moreover, MTSL has a relatively small molecular volume, similar to natural

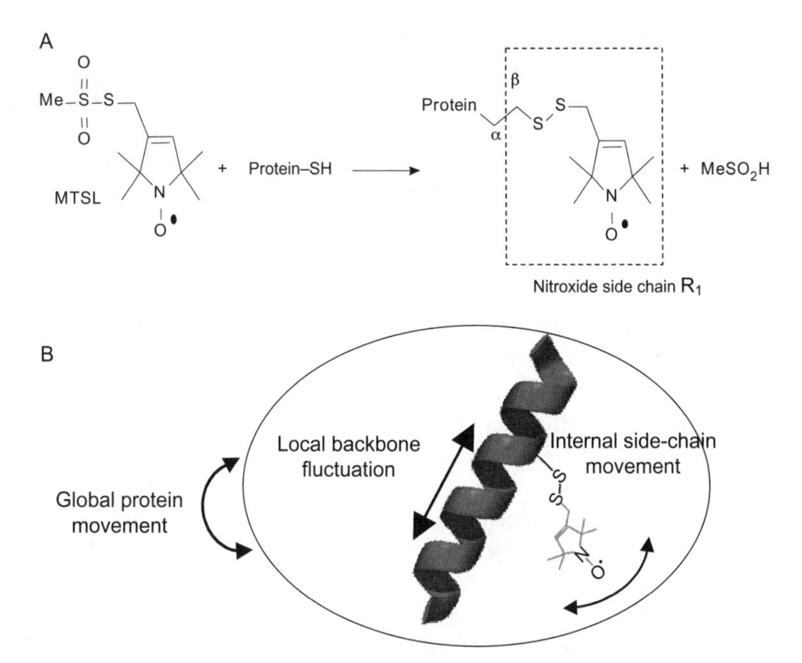

Figure 6.6 (A) Spin labeling reaction with MTSL producing the nitroxide side chain, referred to as R_1. (B) Schematic representation of the different contributions of mobility that influence the EPR spectral shapes of the spin label. SH, sulfydryl group of a cysteine.

tryptophan or phenylalanine side chains, thus leading to a minimal perturbation in the protein. The methodology requires the control of the functional viability of the cysteine variant protein, preferably before and after spin labeling. Then, a panel of EPR approaches can be applied, the classical one consisting in the study of spin label motion.

6.3.2 Mobility of Spin Labels Grafted on Proteins

The mobility of a spin label grafted on a protein is due to the movement of the entire protein, to the local backbone fluctuations, to the internal dynamics of the spin labeled lateral chain, and to possible tertiary contacts (Fig. 6.6B). This mobility can be described in terms of rotational correlation times by:

$$\frac{1}{\tau} = \frac{1}{\tau_{prot}} + \frac{1}{\tau_{local}},$$

where τ_{prot} is the rotational correlation time of the protein, and τ_{local} contains all the other contributions. As the goal of SDSL is to observe local modifications, the influence of the movement of the entire protein on the EPR spectra must be taken into account. This influence is negligible when $\tau_{prot} \gg \tau_{local}$. This is the case for a protein with a molar mass ≥50 kDa ($\tau_{prot} \geq 20$ ns) and when τ_{local} is in the nanosecond timescale. For smaller proteins, the influence of the movement can be reduced by increasing the viscosity of the solution by adding up to 30% w/v sucrose. This induces an increase of τ_{prot} by a factor of about three, while τ_{local} remains unchanged (73).

6.3.3 How to Extract Mobility Information from EPR Spectra?

One way to process the data in a quantitative way is to simulate EPR spectra (14, 70). However, the large number of parameters used for such simulations may lead to an ambiguous determination of the spin label mobility. A better extraction of these parameters from an EPR spectrum can be done by recording spectra at different frequencies, but this improvement is still under progress (3, 6, 26).

Besides the simulation method, semiquantitative analyses are often used, consisting in measuring in EPR spectra a parameter that directly depends on the mobility of the spin label.

In the *fast motional regime of the spin label* ($10^{-11} < \tau_{local} < 10^{-9}$ s), the parameter of choice is the ratio of the peak-to-peak amplitude of the lateral lines, referred to as $h(M_I = \pm1)$, and the peak-to-peak amplitude of the central line, $h(M_I = 0)$ (Fig. 6.7).

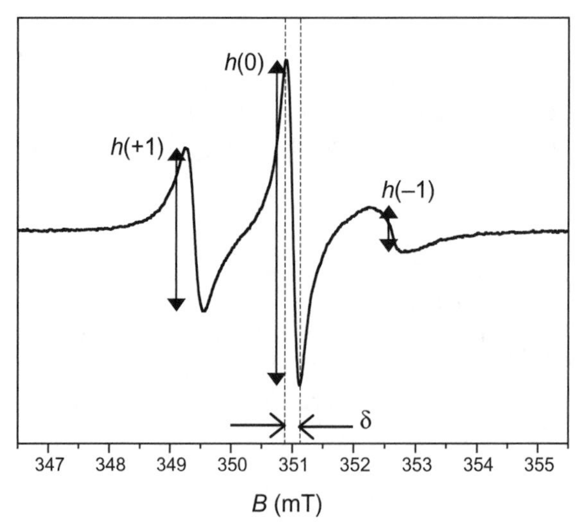

Figure 6.7 Illustration of the different parameters that can be measured on an EPR spectrum of a spin label. $h(+1)$, $h(0)$, and $h(-1)$ are the peak-to-peak amplitudes of the lines corresponding to M_I = +1, 0, and −1, respectively. δ is the central line width expressed in mT.

For an isotropic movement of the spin label, the width of the three lines centered at $B_{M_I} = \dfrac{h\upsilon - AM_I}{g\beta}$ is given by $\Delta B = W + A + BM_I + CM_I^2$, where W is the residual width, and A, B, and C are coefficients proportional to the rotational correlation time and depend on the principal values of the $\tilde{g}$ and $\tilde{A}$ magnetic tensors (32). It has been shown that B < 0, A and C > 0, and that A < C ≈ |B|. For a given rotational correlation time, this leads to the relations:

$$\Delta B_{M_I=0} \approx \Delta B_{M_I=1} \approx W + A$$

$$\Delta B_{M_I=0} < \Delta B_{M_I=-1} = W + A - B + C \approx W + A + 2C.$$

Since the peak-to-peak amplitude of a line is proportional to the reciprocal of its width, we obtain $h(-1) < h(1) \approx h(0)$ (see Fig. 6.8A), and as a consequence, the ratio $h(-1)/h(0)$ is a better indicator of the spin label mobility compared with $h(+1)/h(0)$ (64).

For a *fast anisotropic motion* of the spin label, the description of the mobility can be done by considering a rapid movement of the spin label around the y magnetic axis, characterized by the rotational correlation time $\tau_{//}$, and a slower movement around an axis normal to the previous one, characterized by the rotational correlation time $\tau_{\perp}$ (Fig. 6.8B) (32). As long as $\tau_{//} \leq 1\,\text{ns}$ and $\tau_{\perp} \leq 4\,\text{ns}$, an analytical expression of the line widths will be obtained, which depends on the two correlation times and on the principal values of the g and A tensors (32,56). This expression leads to the relation: $\Delta B_{M_I=-1} > \Delta B_{M_I=1} > \Delta B_{M_I=0}$, and thus $h(-1) < h(+1) < h(0)$. In order to assess whether, in this case, the $h(+1)/h(0)$ ratio was a better indicator of the spin label mobility as compared with the $h(-1)/h(0)$ ratio, the variation of these ratios as a function of typical rotational correlation times encountered in our experiments was measured from a set of simulated spectra. As shown in Figure 6.8C, the $h(+1)/h(0)$ ratio is a more suited and sensitive parameter than the $h(-1)/h(0)$ ratio to infer information about the radical mobility, where variations in the $h(-1)/h(0)$ ratios are approximately four times smaller than variations in the $h(+1)/h(0)$ ratios for a given variation of $\tau_{\perp}$ (60). Note that for isotropic movements of the label, the behavior of both ratios is reversed: $h(-1)/h(0)$ varies significantly, whereas $h(+1)/h(0)$ is almost stable (64).

Another parameter that is sensitive to label motion is the so-called scaled mobility that has been introduced by Hubbell and coworkers (37). It is based on the measurement of the peak-to-peak width of the central line δ (see Fig. 6.7) and is equal to:

$$M_S = \frac{\delta^{-1} - \delta_{slow}^{-1}}{\delta_{fast}^{-1} - \delta_{slow}^{-1}},$$

where δ_{slow} and δ_{fast} are the peak-to-peak widths of the central line for sites where the spin label has, respectively, the lowest and highest mobility. M_S is a normalized indicator ($0 \leq M_S \leq 1$) of the relative variations of the spin label

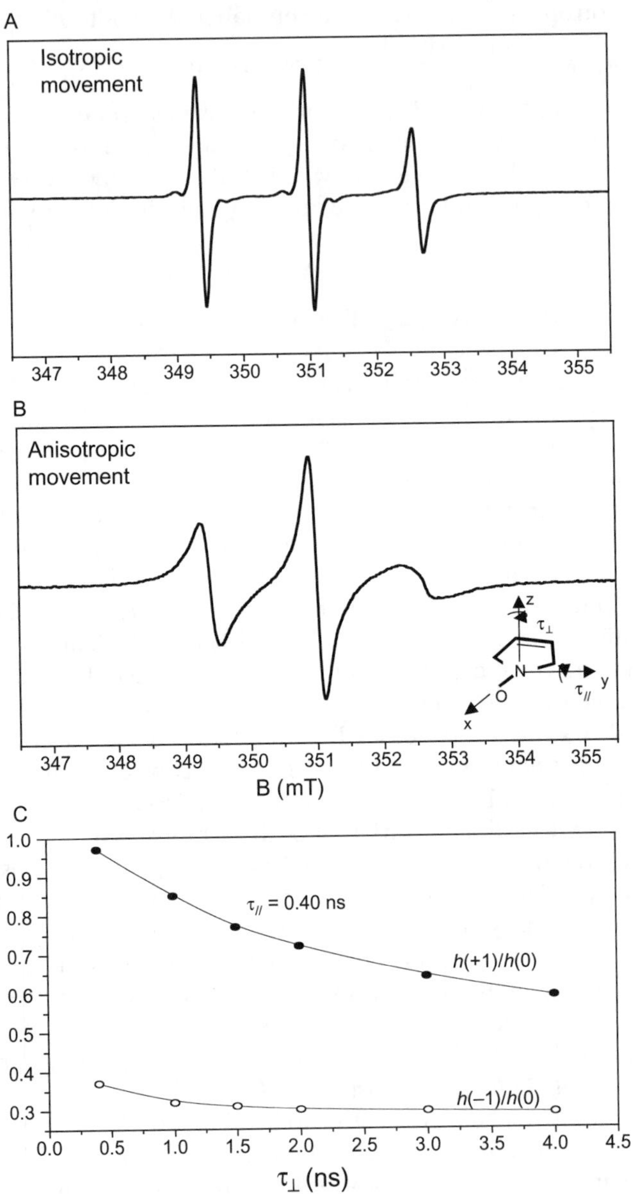

Figure 6.8 (A) EPR spectrum characteristic of an isotropic movement of the radical. (B) EPR spectrum characteristic of an anisotropic movement of the radical described by two rotational correlation times: $\tau_{//}$ corresponding to the rapid movement of the radical around the y magnetic axis and $\tau_{\perp}$ corresponding to the slow movement of the radical around the z magnetic axis. (C) Variation of the $h(+1)/h(0)$ ratio (•) and $h(-1)/h(0)$ ratio (o) versus $\tau_{\perp}$ as deduced from a set of simulated spectra with a constant value of $\tau_{//} = 0.4$ ns.

mobility for different sites that can be used for all mobility regimes. This parameter is well suited when the study involves a collection of spin labels grafted individually on a protein and has the advantage of being not restricted to a particular regime of mobility.

6.3.4 Mobility of MTSL Grafted onto a Protein

In 1996, McHaourab and coworkers established the basis for the interpretation of EPR line shape of spin labels grafted on protein by investigating the relationship between the mobility of the nitroxide side chain and the protein structure (57). McHaourab and coworkers designed, purified, and spin labeled 30 single cysteine mutants of T4 lysozyme, a model protein of well-known structure. The introduction of the spin label at various sites allowed the different structural elements of the protein to be probed. The first remarkable observation is that the mobility of the label depends on the protein site onto which it is grafted. In particular, it gradually increases following the order: buried sites < sites with tertiary contacts < solvent-accessible surfaces of α-helices < loops. A second important observation concerns the behavior of spin labels grafted on the solvent-exposed surface of α-helices. The authors showed, in fact, that the mobility of the nitroxide side chain (referred to as R_1, see Fig. 6.6A) is almost independent from the interactions with neighboring side chains, and rather reflects backbone fluctuations. McHaourab and coworkers proposed that the restricted mobility of the nitroxide side chain at these sites arises from the immobilization of the disulfide bond of the label through interaction with the protein. This interpretation was reinforced by the visual inspection of the crystallographic structure of T4 lysozyme spin labeled on exposed α-helical sites (51), which shows that the disulfide bond of the R_1 side chain is in interaction with the protein backbone and has a low thermal coefficient. Since, the importance of backbone fluctuations on EPR spectra has been studied in more detail using a collection of nitroxide MTSL reagents differing by a substituent on the nitroxide ring (16, 18). This work confirmed that the internal dynamic of the R_1 side chain may be considered as site independent and pointed that the site-specific differences inside chain motion can be assigned to dynamic contributions of the backbone, including both torsional oscillations around main chain bonds and rigid-body helical motions. This point is particularly relevant to the study of IDPs since the variations of their backbone flexibility can be inferred from their EPR spectra.

6.4 A CASE STUDY: THE INTRINSICALLY DISORDERED N-TAIL DOMAIN OF THE MEV NUCLEOPROTEIN

6.4.1 MeV Replicative Complex and Nucleoprotein Structural Organization

MeV is a negative-stranded RNA virus within the *Morbillivirus* genus of the *Paramyxoviridae* family. Its nonsegmented, single-stranded RNA genome is

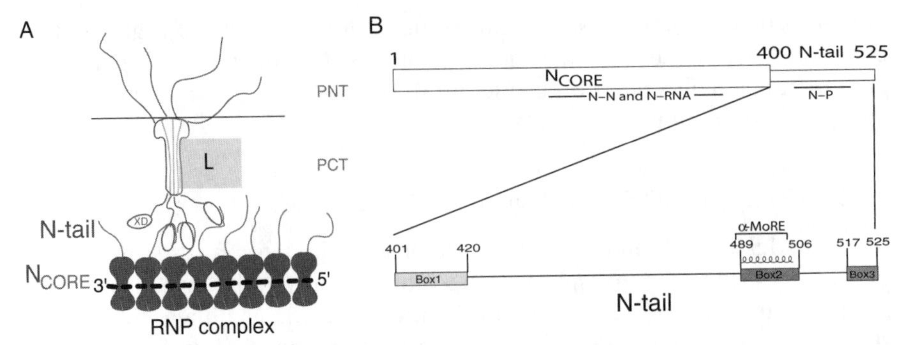

Figure 6.9 (A) Schematic representation of the RNP complex of MeV. The nucleoprotein is represented in dark gray. The disordered regions, including N-tail (aa 401–525) and PNT (aa 1–230), are represented by lines. The encapsidated RNA is shown as a black dotted line embedded in the middle of N. The tetrameric P (65) is shown bound to the nucleocapsid through three of its four C-terminal XD (aa 459–507) "arms," as in the model of Curran and Kolakofsky (20). The L protein is shown as a rectangle contacting P. (B) Organization of MeV N in a structured N_{CORE} region and in an intrinsically disordered N-tail domain. The location of N–N, N–P, and RNA binding sites is indicated. The three boxes of homology conserved in *Morbillivirus* N-tail, as well as the α-MoRE, are shown.

encapsidated by the nucleoprotein (N) to form a helical nucleocapsid. The viral RNA is tightly bound within the nucleocapsid and does not dissociate during RNA synthesis. Therefore, this ribonucleoproteic (RNP) complex, rather than naked RNA, is used as a template for both transcription and replication (Fig. 6.9A). These latter activities are carried out by the RNA-dependent RNA polymerase that is composed of the large (L) protein and of the phosphoprotein (P).

MeV nucleoprotein consists of two regions: an N-terminal moiety, N_{CORE} (aa 1–400), which contains all the regions necessary for self-assembly and RNA binding (43, 46), and a C-terminal domain, N-tail (aa 401–525) (Fig. 6.9B), which is responsible for binding to P (2, 46, 52, 54). Although N-tail is much less conserved in sequence than N_{CORE} throughout members of the *Morbillivirus* genus, it contains three regions of homology (namely, Box1, aa 401–420; Box2, aa 489–506; and Box3, aa 517–525) (Fig. 6.9B) (22). Using various hydrodynamic and spectroscopic approaches, we showed that N-tail is intrinsically unstructured (54).

IDPs are functional proteins that fulfill essential biological functions while lacking highly populated constant secondary and tertiary structure in isolation under physiological conditions (23, 25, 28, 75, 76, 79, 82, 84–86).

Although there are IDPs that carry out their function while remaining permanently disordered (e.g., entropic chains) (23), many of them undergo a disorder-to-order transition upon binding to their physiological partner(s), a process termed *induced folding* (24, 30, 81). The term induced folding is often associated with a notion of gain of regular secondary structure. However, coupled folding and binding can also occur without such gain (12, 24, 25, 29, 82). Moreover, IDPs tend to conserve their overall extended conformation even after binding to their targets (75). Thus, induced folding is characterized by a decrease in flexibility of the IDP due to selection by the partner of a particular conformer out of the otherwise numerous conformations that the free IDP can adopt in solution. Hence, the term induced folding designates all disorder-to-order transitions that IDPs undergo upon binding to their partner/ligand, regardless of whether they imply or not α or β transitions.

The disordered nature of N-tail confers to this N domain the ability to adapt to various partners and to form complexes that are critical for both transcription and replication. Indeed, thanks to its exposure at the surface of the viral nucleocapsid (33, 34, 43), N-tail establishes numerous interactions with various viral partners, including P (2, 52), the P–L complex (46, 54), and the matrix protein (19). Beyond viral partners, N-tail also interacts with various cellular proteins, including the heat shock protein Hsp72 (87, 88); the cell protein responsible for the nuclear export of N (67); the interferon regulator factor 3 (72), probably in complex with a yet unknown cellular cofactor (15); and, possibly, components of the cell cytoskeleton (21, 61). Moreover, N-tail within viral nucleocapsids released from infected cells also binds to the yet unidentified nucleoprotein receptor, NR, expressed at the surface of dendritic cells of lymphoid origin (both normal and tumoral) (50), and of T and B lymphocytes (49, 50). The N-tail–NR interaction triggers an arrest in the G_0/G_1 phase of cell cycle, leading to the immunosuppression that is a hallmark of MeV infections (49).

6.4.2 N-Tail Is a Premolten Globule That Undergoes α-Helical Folding upon Binding to P

Using a combination of various hydrodynamic and spectroscopic approaches (for a review, see Reference 66), we showed that N-tail belongs to the premolten globule subfamily within the family of IDPs (54). Premolten globules are mainly disordered proteins that nevertheless possess a certain degree of compactness in solution due to the presence of residual, transiently populated secondary and/or tertiary structure (79). As for the functional implications, it has been proposed that the residual intramolecular interactions that typify the premolten globule state may enable a more efficient start of the folding process induced by a partner (30, 31, 48, 74).

In agreement, we showed that N-tail undergoes α-helical folding upon binding to P (54). P plays a crucial role in transcription and replication as it allows recruitment of L onto the nucleocapsid template (Fig. 6.9A). P is a modular protein consisting of alternating structured and disordered regions (42, 44). Within P, we mapped the region responsible for binding to N-tail and for the α-helical-induced folding of this latter, to the C-terminal X domain

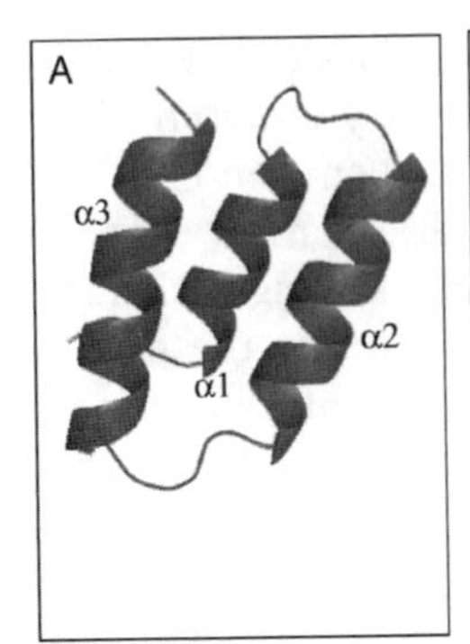

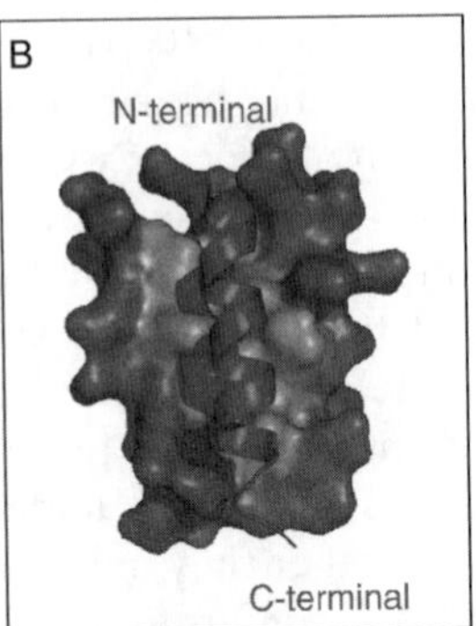

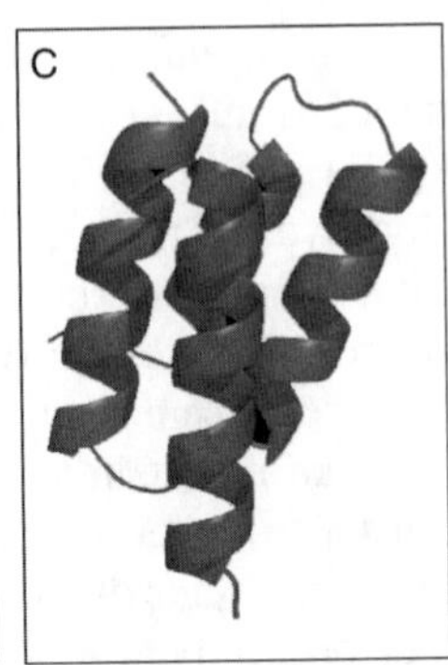

Figure 6.10 (A) Ribbon representation of the XD structure. (B) Model of the complex between XD and the α-MoRE of N-tail according to Reference 41. Hydrophobic residues are shown in orange. (C) Crystal structure of the chimeric construct between XD (blue) and the N-tail region encompassing residues 486–504 (red) (PDB code 1T6O) (45). The pictures were drawn using Pymol (DeLano, W.L., The PyMOL Molecular Graphics System [2002] DeLano Scientific, San Carlos, CA, USA; http://www.pymol.org). See color insert.

(XD, aa 459–507 of P) (41). We also solved the crystal structure of XD at 1.8 Å and showed that it consists of a triple α-helical bundle (Fig. 6.10A) (41).

Using a computational approach, a molecular recognition element (MoRE, aa 489–499) was predicted within a conserved region of N-tail (Box2, aa 489–506) (8) (see Fig. 6.9B). MoREs or molecular recognition features are relatively short protein segments (10–70 residues) within larger disordered regions with a propensity to bind to a partner, and thereby to undergo induced folding (59, 63, 83). MoREs can be divided into α-MoREs, β-MoREs, and I-MoREs, according to the nature of the structural transition occurring upon binding to the partner (i.e., α, β, and irregular, respectively) (59, 63, 83). In the case of N-tail, an α-MoRE is predicted (8), in agreement with secondary structure predictions that identify an α-helix within residues 488–504 as the sole secondary structure element of N-tail (see Fig. 6.9B) (54). The involvement of the predicted α-MoRE of N-tail in both binding to P and induced folding was experimentally confirmed (8, 41). We therefore proposed a model where the α-MoRE is embedded in a hydrophobic cleft of XD delimited by helices α2 and α3 to form a pseudo-four-helix arrangement occurring frequently in nature (Fig. 6.10B) (41). This structural model was successively validated by Kingston and coworkers who solved the crystal structure of a chimeric construct composed of XD and of the 488–504 region of N-tail (PDB code 1T6O) (Fig. 6.10C) (45).

Using small-angle X-ray scattering, we derived a low-resolution structural model of the complex between XD and N-tail (Fig. 6.11A). This model shows that the majority of N-tail (residues 401–488) remains disordered in the complex and does not establish contacts with XD, contrary to the 489–525

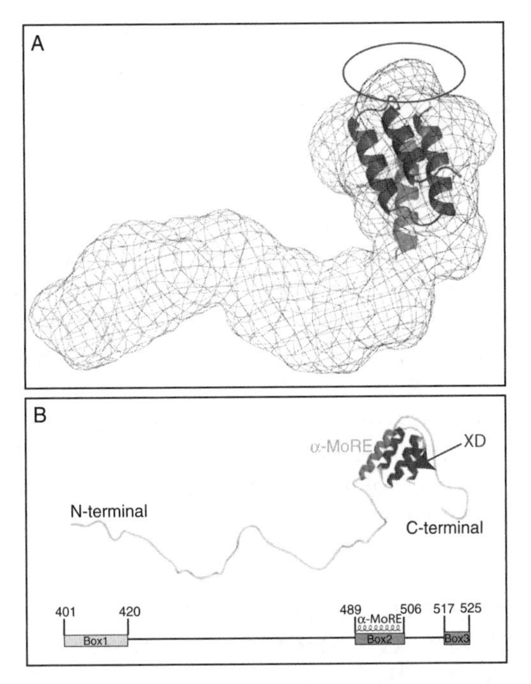

Figure 6.11 (A) Global shape of the N-tail–XD complex as derived by small-angle X-ray scattering studies (12). The red circle points to the lack of a protruding shape from the globular body of the model. The crystal structure of the chimera between XD (red) and the N-tail region encompassing residues 486–504 (blue) is shown. The picture was drawn using Pymol (DeLano, W.L., The PyMOL Molecular Graphics System). (B) Low-resolution model of the N-tail–XD complex showing that the 401–488 region of N-tail is disordered and exposed to the solvent, while the α-MoRE and the C-terminus of N-tail (Box3) are packed against XD. Data were taken from Reference 12. See color insert.

region of N-tail (Fig. 6.11B) (12). Additionally, the lack of a protruding shape from the globular, bulky part of the model indicates that the C-terminus of N-tail does not point toward the solvent and must rather be part of the globular region (Fig. 6.11A). This suggests that, beyond the α-MoRE, the C-terminus may be also involved in binding to XD (12).

Using surface plasmon resonance (SPR) and various spectroscopic approaches, we indeed showed that N-tail possesses an additional site (Box3, aa 517–525) involved in binding to XD (Fig. 6.9B). Specifically, SPR studies carried out with truncated N-tail proteins supported a role for Box3 in binding, where removal of either Box3 alone or Box2 *plus* Box3 results in a strong decrease (three orders of magnitude, K_D of 10 μM) in the equilibrium dissociation constant (12). When synthetic peptides mimicking Box1, Box2, and Box3 were used, we found that Box2 peptide displays an affinity toward XD (K_D of 20 nM) similar to that of N-tail (K_D of 80 nM), consistent with the role of Box2 as the primary binding site (S. Longhi and M. Oglesbee, unpublished data). Interestingly, however, Box3 peptide exhibits a significantly lower affinity for XD (K_D of approximately 1 mM) as compared with that of Box2 peptide (S. Longhi and M. Oglesbee, unpublished data). The discrepancy between the data obtained with N-tail truncated proteins and with peptides can be accounted for by assuming that Box3 would act only in the context of N-tail and not in isolation. Thus, according to this model, Box3 and Box2 would be functionally coupled in the binding of N-tail to XD. We can speculate that burying the hydrophobic side of the α-MoRE in the hydrophobic cleft formed by helices α2 and α3 of XD would provide the primary driving force in the N-tail–XD interaction and that Box3 would act as a "clamp." The combined interaction of Box2 and Box3 with XD would provide further stabilization of the complex as compared with a complex devoid of Box3.

Finally, heteronuclear NMR (^{1}H ^{15}N NMR) experiments showed that while Box2 undergoes α-helical folding upon binding to XD, Box3 does not gain any regular secondary structure (12) (for reviews, see References 9–11, 53).

6.5 INSIGHTS INTO THE MOLECULAR MECHANISM OF THE XD-INDUCED FOLDING OF N-TAIL BY SDSL EPR SPECTROSCOPY

EPR spectroscopy has been extensively used to study conformational changes within structured proteins (36), as well as folding and unfolding processes in the presence of denaturing agents, such as urea and guanidium chloride (47, 64). Moreover, it has been shown that MTSL is a well-suited spin probe to observe backbone fluctuations (see section 6.3.4). Hence, we reasoned that this technique could also be well suited to assess and investigate the induced folding that N-tail undergoes in the presence of XD.

To this endeavor, we targeted for spin labeling 14 different sites within N-tail, one falling in the N-terminal region (Box1, position 407), one located in the linker region between Box1 and Box2 (position 460), one immediately upstream of the α-MoRE (position 488), six within Box2 (positions 491, 493, 496, 499, 502, and 505), two within the linker region between Box2 and Box3 (positions 510 and 512), and three within Box3 (positions 517, 520, and 522) (Fig. 6.12A). We then monitored the variations in the mobility of the spin label upon addition of the secondary structure stabilizer 2,2,2-trifluoroethanol (TFE) or the physiological partner, XD (see also References 5 and 60).

6.5.1 Rational Design of N-Tail Cysteine-Substituted Variants

We designed 14 individual cysteine mutants, thus allowing introduction of the spin label at 14 different N-tail sites scattered through the whole sequence but mainly located in the Box2–Box3 regions (aa 488–525) (Fig. 6.12A), which are involved in the interaction with XD (9, 10, 12).

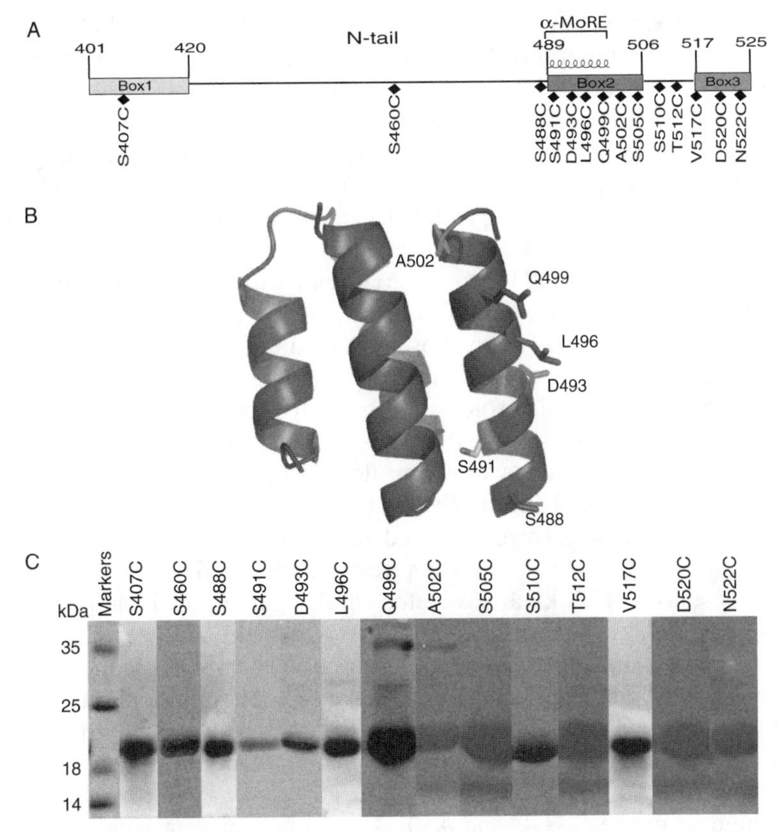

Figure 6.12 (A) Schematic representation of positions targeted for cysteine substitution and spin labeling (black diamonds). (B) Structure of the chimera (PDB code 1T6O) between XD (blue) and the N-tail region encompassing residues 486–504 (red) showing the orientation of the side chains of residues targeted for cysteine substitution and spin labeling. Note that the side chain of residue 491 points toward the surface of XD, while all the other side chains are solvent exposed. The picture was drawn using Pymol (DeLano, W.L., The PyMOL Molecular Graphics System). (C) Coomassie blue staining of a 12% SDS-PAGE loaded with N-tail variant proteins purified from the soluble fraction of *E. coli*. Data of panel C were taken from Reference 5. See color insert.

Whenever possible, serine to cysteine substitutions were done. In the case of Box1, Ser407 was targeted for site-directed mutagenesis, as this residue is not conserved within *Morbillivirus* members, contrary to Ser418 (22). Ser460 was chosen so as to introduce the spin label in the middle of the region connecting Box1 to Box2 (Fig. 6.12A). However, isosteric substitution was not the sole criterion we used to target the various positions.

In fact, in the case of Box2, the rationale for the choice of positions within Box2 was to target residues whose side chains point toward the solvent in the complex between XD and the α-MoRE of N-tail (PDB code: 1T6O) (45) so as to avoid steric hindrance. The only exception is represented by position 491, whose side chain is buried at the interface of interaction in the crystal structure of the chimera (Fig. 6.12B).

Contrary to Box2, no structural information is available for the 505–525 region in the presence of XD. Hence, the targeting criterion we used was to introduce the paramagnetic spin label in locations as scattered as possible while doing, whenever possible, isosteric side-chain substitutions. Hence, we targeted Ser510 and Thr512 for isosteric cysteine substitution within the region connecting Box2 to Box3. Finally, as no serine residue occurs within Box3, positions 517, 520, and 522 were chosen only on the basis of the former criterion (Fig. 6.12A).

All recombinant N-tail proteins were expressed in *Escherichia coli* and were recovered from the soluble fraction of bacterial lysates (data not shown). The N-tail mutated proteins were purified to homogeneity (>95%) in two steps: immobilized metal affinity chromatography and gel filtration. As shown in Figure 6.12C, all the N-tail mutated proteins migrate in sodium dodecyl sulfate polyacrylamide gel electrophoresis (SDS-PAGE) with an apparent molecular mass of 20 kDa (expected molecular mass is approximately 14.6 kDa). This abnormal migratory behavior has already been documented for N-tail, where mass spectrometry analysis and N-terminal sequencing gave the expected results (54). This anomalous electrophoretic mobility is rather due to a relatively high content of acidic residues, as frequently observed in IDPs (74). Likewise, the behavior of the mutated N-tail proteins can be accounted for by this sequence bias composition.

In the case of the Q499C and A502C variants, a protein band of approximately 36 kDa is also observed in SDS-PAGE (Fig. 6.12C). This band corresponds to a dimeric form of the protein, as judged from its disappearance upon reduction with a 1000-fold molar excess of dithiothreitol (DTT) (data not shown).

The number of different conformations of the N-tail mutated proteins is limited, as indicated by the sharpness of the peaks observed in gel filtration (data not shown). The Stokes radius (R_S) value of the N-tail mutated proteins is close to that observed for wt N-tail (data not shown). The corresponding hydrodynamic volume of N-tail proteins is consistent with the theoretical value expected for a premolten globule state (79), as already observed in the case of wt N-tail (54). Thus, these mutated proteins share similar hydrodynamic properties with wt N-tail, being all nonglobular while possessing a certain residual compactness typical of the premolten globule state (80).

The purified N-tail variants were successively spin labeled through a two-step procedure consisting of a DTT reduction, followed by covalent modification of the sulfydryl group by the MTSL nitroxide derivative as described in References 5 and 60. The concentration of spin labeled proteins was evaluated by double integration of the EPR signal recorded under nonsaturating conditions and comparison with that given by a 3-carboxy-proxyl standard sample. The labeling yields were estimated by calculating the ratio between the concentration of labeled proteins and the total protein concentration. Labeling yields ranged from 40% to 80%. Differences in labeling yields were not position dependent and were ascribed to more subtle differences in the experimental conditions rather than to a different extent of solvent exposure of the sulfydryl group, in agreement with the prevalently unfolded state of N-tail.

6.5.2 Analysis of Structural Propensities of N-Tail Variants

We addressed the question as to whether the cysteine substitution and the introduction of the covalently bound nitroxide radical could affect the overall secondary structure content of the various N-tail variants. Notably, secondary structure predictions point out no differences among the various N-tail variants, with an α-helix spanning residues 486–504 being predicted as the sole secondary structure element in all cases (data not shown). In order to directly assess the possible impact of the substitution and of the presence of the spin label on the N-tail structure, we recorded the far-UV circular dichroism (CD) spectra of spin labeled N-tail variants at neutral pH. The CD spectra of spin labeled N-tail proteins quite well superimpose onto that of wt N-tail and are all typical of unstructured proteins, as seen by their large negative ellipticity at 200 nm and moderate ellipticity at 185 nm (Fig. 6.13A). These data indicate that the cysteine substitution and the introduction of the spin label induce little, if any, structural perturbations. In agreement, most spin labeled N-tail variants possess an α-helical content similar to that of wt N-tail, with the only two exceptions being provided by the N522C and the S491C variants, which contain, respectively, a lower and higher α-helical content as compared with wt N-tail (Fig. 6.13B).

In order to further investigate the structural impact brought by the cysteine substitution and by the introduction of the spin label, we analyzed the structural propensities of the spin labeled variants in the presence of TFE. The solvent TFE mimics the hydrophobic environment experienced by proteins in protein–protein interactions and is therefore widely used as a probe to unveil disordered regions having a propensity to undergo an induced folding (35). We previously reported that the addition of increasing amounts of TFE to N-tail triggers a gain of α-helicity and that Box2/α-MoRE plays a major role in this α-helical transition (8, 12, 54). We thus recorded CD spectra of spin labeled N-tail proteins in the presence of 20% TFE (Fig. 6.13C) and calculated their α-helical content (Fig. 6.13D). All proteins show a gain of α-helicity upon addition of TFE (Fig. 6.13D), thus indicating that neither the cysteine substitution nor the presence of the spin label impair the ability of N-tail to undergo α-helical folding. Notably, the α-helical propensity of spin labeled S491C and D493C variants is even enhanced, as judged on the basis of their increased α-helical content at 20% TFE as compared with all the other spin labeled N-tail variants and with native N-tail (Fig. 6.13D). The more pronounced α-helical propensity of the spin labeled S491C and D493C variants, in which the spin label is grafted within the TFE-induced α-helix, may be ascribed to the formation of short-range interactions between the spin labels and the residues

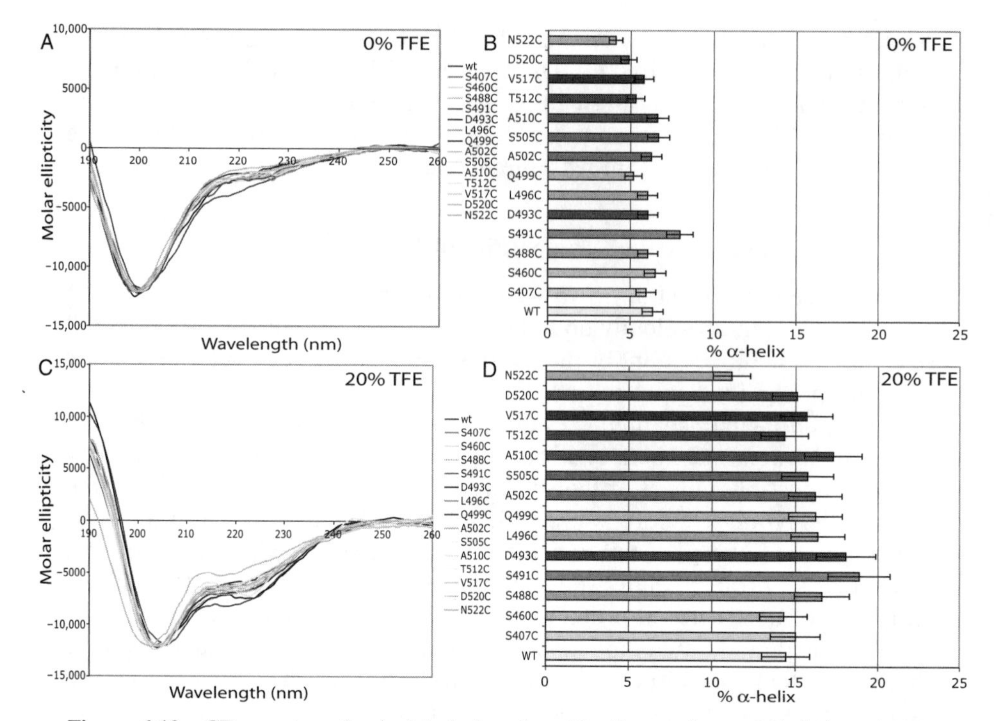

Figure 6.13 CD spectra of spin labeled and wt N-tail proteins at 0% (A) and 20% TFE (C). α-Helical content of spin labeled and wt N-tail proteins at 0% (B) and 20% (D) TFE. Each spectrum is the mean of three independent acquisitions. The error bar (10% of the value) corresponds to the experimentally determined standard deviation from three independent experiments. The α-helical content was derived from the ellipticity at 222 nm as described in Reference 62. Data were taken from Reference 5. See color insert.

building up the α-MoRE, thus leading to the stabilization of the α-helix. Conversely, the lower α-helical content of the spin labeled S522C variant at both 0% and 20% TFE may be accounted for by the reduced ability of this variant to establish long-range, stabilizing contacts with the neighboring TFE-induced α-helix.

6.5.3 EPR Data Analysis of Spin Labeled N-Tail Variants

We have already reported that the mobility of a spin label grafted within N-tail can be inferred from the $h(+1)/h(0)$ ratio (60) (see also section 6.3.3).

For all the spin labeled N-tail proteins, the EPR spectra are indicative of a high radical mobility, as judged on the basis of the $h(+1)/h(0)$ ratio (ranging from 0.83 to 0.98 ± 0.02) (Fig. 6.14, left panel). The EPR spectra exhibit a relatively narrow, single-component shape, with an outer line splitting of 3.22 (± 0.02) mT (Fig. 6.14, left panel). Despite the overall high mobility, slight

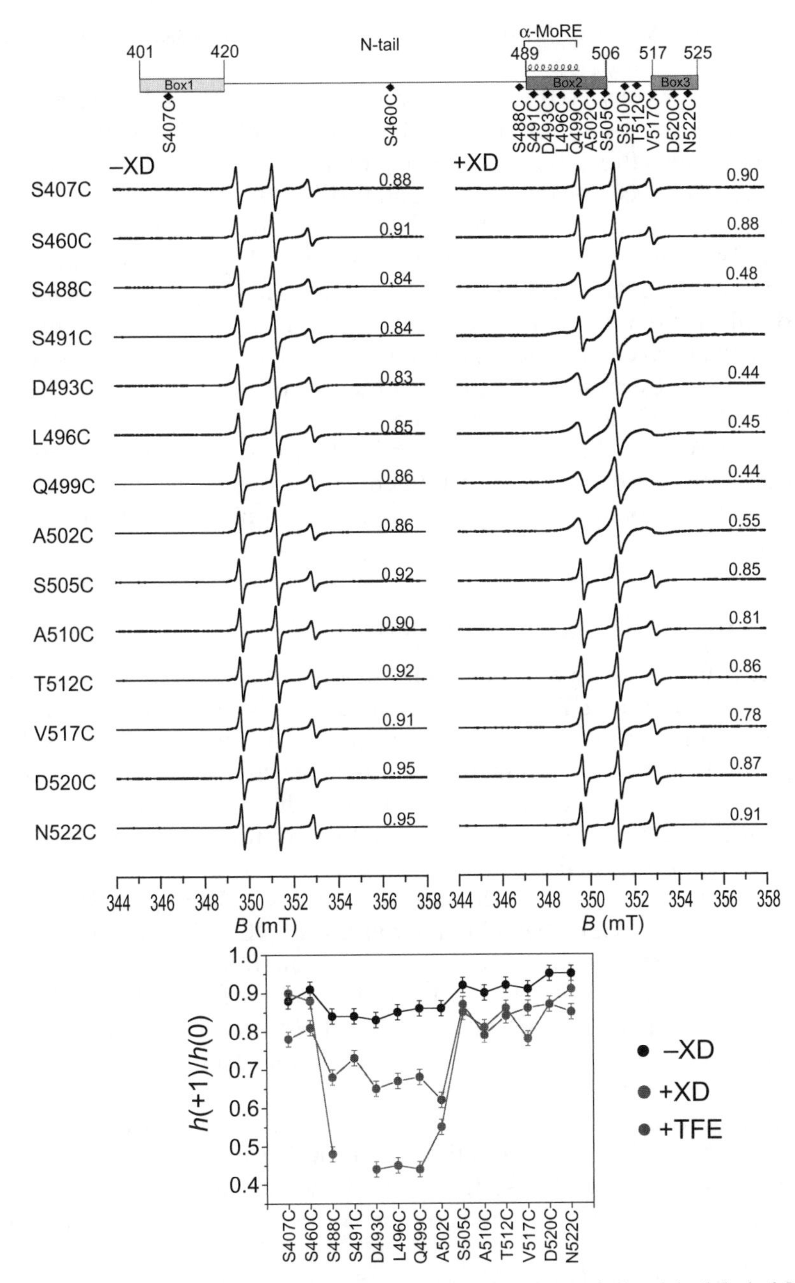

Figure 6.14 Normalized room temperature EPR spectra of the spin labeled N-tail proteins (100 μM) in the absence (left panel) or presence of a molar excess of XD (right panel). $h(+1)/h(0)$ ratios of the spin labeled N-tail proteins free and in the presence of either saturating amounts of XD or 20% TFE as a function of spin label position (bottom panel). Note that the $h(+1)/h(0)$ ratio of the spin labeled S491C variant was not indicated, as it is not a reliable indicator of the mobility of this spin label (see text). The schematic representation of the N-tail variants is shown on the top. Data were taken from Reference 5. For S407C, S488C, L496C, and V517C N-tail variants, see also Reference 60.

differences were observed in the different spin labeled N-tail proteins. A fast movement is observed for the nitroxide radical covalently bound to positions 407 (Box1), 460 (Box1–Box2 connecting region), 505 (end of Box2), 510 and 512 (Box2–Box3 connecting region), and 517, 520, and 522 (Box3), while a slower movement is observed for the nitroxide radicals covalently bound to positions 488, 491, 493, 496, 499, and 502 (Box2) (Fig. 6.14, left and bottom panels), as judged based on the $h(+1)/h(0)$ ratios. These differences in mobility could reflect differences in the content of transient secondary structure elements in the local environment experienced by each label. This is in agreement with previous observations pointing out that N-tail is not a (uniform) random coil but rather belongs to the premolten globule subfamily (54), with the Box2 region undergoing α-helical folding upon binding to XD (8,41,45). Accordingly, the flexibility of Box2 may be restrained by the occurrence of a fluctuating α-helix, which would be transiently populated even in the absence of the partner.

6.5.4 Effect of XD on N-Tail Flexibility

In order to investigate the folding events that N-tail undergoes in the presence of XD, we monitored EPR spectra of spin labeled N-tail variants in the presence of a molar excess of XD. For each variant, the molar excess of XD was progressively increased until saturation was achieved. Under saturating conditions, the extent of reduction of the mobility of the spin label is maximal, and further increasing the molar excess of XD does not result in any further variation in the radical mobility.

In all cases, addition of stoichiometric amounts of XD is not sufficient to lead to saturation, with a twofold molar excess of XD being required in most cases. Even larger XD molar excesses were required to achieve saturation in the case of spin labeled Q499C and A502C (fourfold), and of S491C (eightfold), thus suggesting that all the spin labeled N-tail variants have a reduced affinity toward XD as compared with wt N-tail, whose reported K_D for XD is 100 nM (12). In addition, the various spin labeled N-tail variants display significant differences in affinity as a function of the radical location, with the grafting of the spin label at position 491 resulting in the most pronounced reduction in the affinity (see section 6.5.6). However, we could not determine a K_D because of the rather low variations in the $h(+1)/h(0)$ ratio that were within the error bars.

As a negative control, we recorded EPR spectra of a mixture containing spin labeled L496C and an eightfold molar excess of an irrelevant protein (lysozyme). No variation in the radical mobility was observed (data not shown), thus indicating that the reductions in the radical mobility observed in the presence of XD are specific of complex formation.

The mobility of most spin labels is significantly reduced upon addition of XD, with the exception of spin labels grafted at positions 407 and 460, whose mobility is not affected even at XD molar excesses as high as eight (Fig. 6.14, right and bottom panel). The possibility that the lack of variations in the mobility of the spin label bound at these positions could be due to the inability of these mutated N-tail proteins to interact with XD, rather than to the lack of impact on the radical mobility, was checked and ruled out using equilibrium displacement experiments (5, 60). In particular, we showed that unlabeled S407C is able to interact with XD since it can displace the equilibrium between XD and the spin labeled L496C variant, thus leading to an increase in the mobility of the spin label grafted at this position (60). Furthermore, in the case of the S460C variant, we showed that the reduced, hydroxylamine-labeled, EPR-silent form of the variant is also able to displace the equilibrium between XD and the spin labeled L496C variant (5).

The most dramatic reductions in the radical mobility are observed for the N-tail variants whose spin labels are located within the α-helix observed in the crystal structure of the XD/α-MoRE chimera, namely, D493C, L496C, and Q499C, followed by S488C and A502C with variations in the $h(+1)/h(0)$ ratio of about 0.40 or 0.30, respectively (Fig. 6.14, bottom panel). In the case of S491C, addition of XD triggers a dramatic spectral transition: The EPR spectrum is in fact drastically different and reflects a highly restricted nitroxide radical mobility, with a broad outer line splitting (see section 6.5.6 and Fig. 6.14, right panel). Under these conditions, the $h(+1)/h(0)$ ratio is no longer a reliable parameter of the radical mobility. This particular case will be discussed in more detail in section 6.5.6.

However, the reduction in the mobility of the spin label grafted at positions 505 (i.e., at the end of Box2), 510 and 512 (region connecting Box2 to Box3), and 517 and 520 (Box3) is much less pronounced, with variations in $h(+1)/h(0)$ ratio of about 0.1 (Fig. 6.14, bottom panel). Position 522 is the least XD sensitive, with a borderline decrease in the $h(+1)/h(0)$ ratio (Fig. 6.14, bottom panel).

In order to further analyze the nature of the transition that the spin labeled N-tail proteins undergo in the presence of XD, we recorded EPR spectra of the N-tail variants in the presence of 30% sucrose. Under these conditions, the contribution of the entire protein mobility to the EPR spectral line shape is reduced due to an increase in the protein rotational correlation time by about a factor three (see section 6.3.2). Moreover, it has been shown that the addition of 10–40% sucrose does not affect the rotational mobility of the side chain relative to the protein at room temperature (57). The possibility that 30% sucrose might affect the overall secondary structure of the spin labeled N-tail proteins was checked and ruled out by using CD (data not shown). As expected, the mobility of all spin labels was significantly reduced in the presence of sucrose (Fig. 6.15) regardless of the position of the spin label. Upon addition of XD, the same pattern of variations in mobility is observed as in the absence of sucrose, with a dramatic impact of XD on the region spanning residues 488–502, modest impact on the region encompassing residues 505–520, borderline effect on position 522, and no significant impact on positions 407 and 460 (Fig. 6.15). The mobility of the nitroxide radicals of S488C,

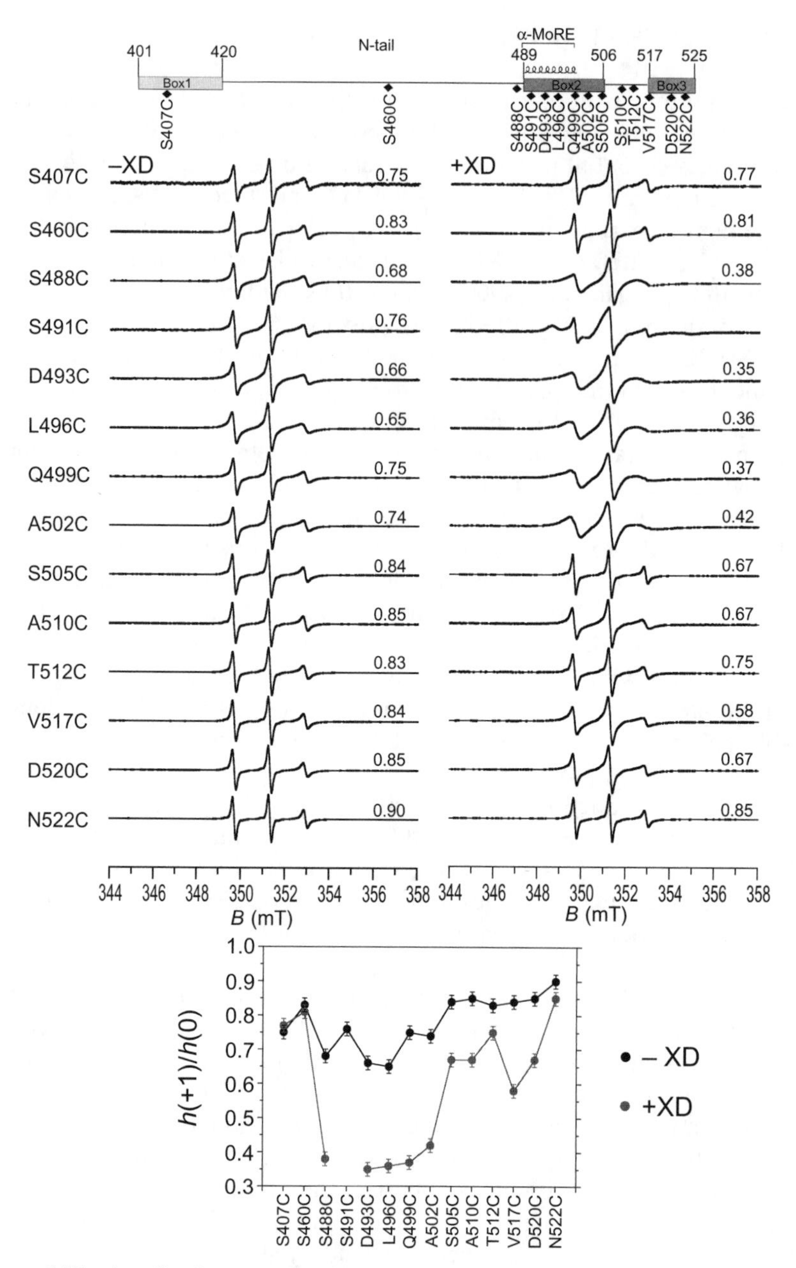

Figure 6.15 Amplitude-normalized room temperature EPR spectra of the spin labeled N-tail proteins (100 μM) in the presence of 30% sucrose and in the absence (left panel) or presence of a molar excess of XD (right panel). $h(+1)/h(0)$ ratios of the spin labeled N-tail proteins either free or in the presence of saturating amounts of XD as a function of spin label position (bottom panel). Note that the $h(+1)/h(0)$ ratio of the spin labeled S491C variant was not indicated, as it is not a reliable indicator of the mobility of this spin label (see text). The schematic representation of the N-tail variants is shown on the top. Data were taken from Reference 5. For S407C, S488C, L496C, and V517C N-tail variants, see also Reference 60.

D493C, L496C, Q499C, and A502C N-tail variants bound to XD is in the twilight zone between the high and intermediate regime of mobility (with $\tau_{//} < 1$ ns). Conversely, in the case of the S491C variant, a dramatic reduction in the radical mobility, approaching the so-called rigid-limit regime, is observed (see section 6.2.1.3 and Fig. 6.5).

Notably, in the presence of 30% sucrose and of a molar excess of XD, the spectra obtained for N-tail variants bearing the spin label grafted within the 488–502 region are reminiscent of those observed with spin labeled variants of proteins in which the radicals have been grafted at the surface of helices (17, 57). Taking also into account the fact that previous studies have shown that the XD-induced folding of N-tail implies the formation of an α-helix within the Box2 region (8, 12, 45), the EPR data herein presented are consistent with the formation of an α-helix in the 488–502 region induced by the presence of XD. In further support of this hypothesis, the drop in the radical mobility for the N-tail variants whose radical is grafted within the 488–502 region is much more pronounced than that of variants in which the spin label is bound to the downstream region.

Furthermore, in the presence of XD, as already observed in the absence of sucrose (Fig. 6.14, bottom panel), a slightly higher mobility is observed for the spin label at positions 488 and 502 as compared with positions 493, 496, and 499 (Fig. 6.15, right and bottom panel). This difference may reflect a more constrained motion of the spin labels in the central part of the α-MoRE. That the extremities of α-helices are more mobile than their central parts has already been well documented by SDSL EPR spectroscopy (57).

Altogether, these data indicate that the region spanning residues 488–520 is significantly affected by XD binding, while positions 407 (Box1) and 460 (middle of region connecting Box1 to Box2) are not, in agreement with previous spectroscopic and SPR data (12). In addition, the more pronounced drop in the radical mobility for the N-tail variants whose radical is grafted within Box2 as compared with that of variants in which the spin label is bound to the downstream region is consistent with previous data, indicating that Box3 does not undergo α-helical folding upon binding to XD (12). Notably, the mobility of the spin label grafted at position 522 is only weakly reduced in the presence of XD, indicating that this residue remains highly flexible within the complex.

6.5.5 Effect of TFE on N-Tail Flexibility

Experiments in the presence of sucrose showed that addition of XD does not trigger a transition to a folded state within the 505–522 region of N-tail. The slight drop in the mobility of the spin labels grafted within this region observed upon addition of XD could be ascribed either to a gain of rigidity arising from α-helical folding of the neighboring Box2 region or to the presence of XD, which might restrain the conformational space available to the radicals grafted within the C-terminal region of N-tail. In order to discriminate among these hypotheses, we monitored the gain of rigidity that N-tail undergoes in the presence of 20% TFE, a condition where the impact of the α-helical folding of Box2 upon the mobility of the radicals grafted downstream can be assessed and separated from the effect due to complex formation with XD. We recorded EPR spectra of spin labeled N-tail proteins in the presence of 20% TFE, a condition where all spin labeled N-tail variants gain α-helicity as shown by CD spectra (see Fig. 6.13). The addition of TFE triggers a decrease in the mobility of all spin labels (Fig. 6.14, bottom panel). The possibility that TFE could affect the mobility of the free radical in solution was checked and ruled out, by comparing the EPR spectra obtained with a 40-μM MTSL solution in the presence or absence of 40% TFE (data not shown). Hence, the variations in the radical mobility observed in the presence of TFE with the spin labeled N-tail proteins reflect changes in the protein environment in the proximity of the spin label. The most pronounced decrease in mobility is observed for the spin labels grafted in the 488–502 region, with mean variations in the $h(+1)/h(0)$ ratio of approximately 0.16 (±0.02) (Fig. 6.14, bottom panel). However, the mobility of radicals bound within the 505–522 region is only moderately affected by the addition of TFE, with the highest variation in the $h(+1)/h(0)$ being 0.1 (±0.02) (Fig. 6.14, bottom panel).

Notably, addition of 20% TFE triggers a reduction in the mobility of the spin labels grafted within the 505–522 region comparable to that observed with XD, while for the 488–502 region, the effect of XD is much more pronounced (Fig. 6.14, bottom panel). Taking also into account the fact that TFE does not promote α-helical folding within Box3 (12), these data suggest that the α-helical transition taking place within Box2 is responsible for the restrained motion of the downstream region rather than a direct interaction with XD. However, the reduction in the mobility of the spin labels grafted in the 488–502 region is more pronounced in the presence of XD than in the presence of 20% TFE (Fig. 6.14, bottom panel). This difference can likely be ascribed to the ability of XD to stabilize the formation of the α-MoRE, while the latter would be only transiently populated in the unbound form of N-tail in the presence of TFE.

Last, but not least, the higher impact of TFE on the mobility of the radicals grafted within the 488–502 region as compared with that of radicals grafted to other N-tail positions points out the relevance of studies making use of TFE. Indeed, the reliability of structural information derived by CD studies in the presence of TFE is still a matter of debate, with artifactual (i.e., nonnative) α-helical folding induced by this solvent being often evoked (13, 27, 55). However, the present results clearly indicate that the extent of TFE impact on radical mobility well reflects the inherent structural propensities of the various protein regions, with the 505–522 region, as well as positions 407 and 460, being "resistant" to undergoing α-helical folding. Altogether, these data show that, in the case of N-tail, TFE does not promote nonnative folding, thus pointing out the relevance of studies making use of TFE to infer information about protein structural propensities.

6.5.6 Effect of XD on the Mobility of the Spin Label Grafted on Position 491

As already mentioned, an eightfold molar excess of XD is required to achieve saturation in the case of the S491C N-tail variant, thus pointing out its reduced affinity toward XD with respect to all other variants (Fig. 6.16A). Notably, in the case of the S491C variant, saturation does not correspond to 100% of bound form. Indeed, the experimentally observed EPR spectrum is composed of two signals: one arising from the unbound spin labeled protein and one resulting from the spin labeled protein bound to XD. We estimated the percentage of complex as follows: (1) the spectrum of free spin labeled protein was subtracted to the composite spectrum so as to obtain the one-component

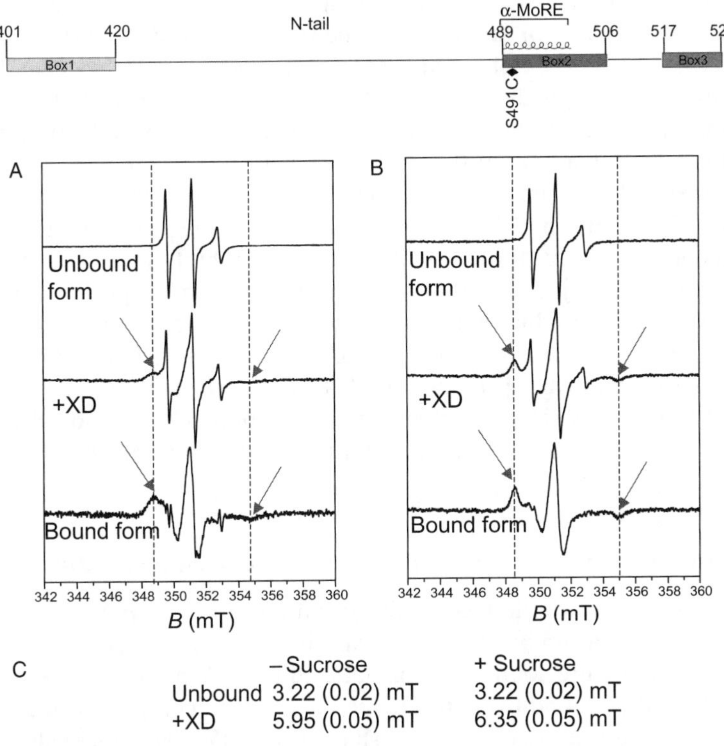

Figure 6.16 Normalized room temperature EPR spectra of the spin labeled S491C N-tail protein (100 μM) in the absence (A) or presence (B) of 30% sucrose either in the absence (top) or in the presence of saturating amounts of XD (middle). The EPR spectra shown at the bottom were obtained upon subtraction of the spectrum of the unbound form. The arrows point to the appearance of a new spectral shape with a highly restricted radical mobility. The dotted lines indicate the outer line splitting for the bound form. (C) Outer line splitting values in the various conditions given in mT. The schematic representation of the S491C N-tail variant is shown on the top. Data were taken from Reference 5.

spectrum corresponding to the spin labeled S491C/XD complex (the bound form of the labeled protein), (2) the integrated intensity of the resulting spectrum, I_{bound}, and of the composite spectrum, I_{total}, were calculated, and (3) the proportion of the bound form was calculated as the I_{bound}/I_{total} ratio. This allowed us to estimate the percentages of free and bound form to be 25 ± 5% and 75 ± 5%, respectively.

The inability of spin labeled S491C to bind to XD at 100% under saturating conditions suggests that two protein populations can occur in solution, one of which (25%) would be unable to bind to XD even at oversaturating XD concentrations. We can speculate that these two populations may reflect static disorder, that is, two alternate conformations of the side chain in position 491 (and hence of the spin label) with an occupancy of 0.75 and of 0.25. In the less populated one, the spin label would occur in a spatial position that would cause steric hindrance, thereby preventing formation of a complex with XD. Indeed, the side chain of the native serine residue at position 491 is buried at the interface in the complex with XD (PDB code 1T6O) (see Fig. 6.12B). Interestingly, although the grafting of the nitroxide radical at this position does not suppress complex formation, it causes a significant reduction in affinity and impairs the ability of the protein to yield 100% complex formation.

In agreement with the location of the radical buried at the N-tail/XD interface (see Fig. 6.12B), the EPR spectrum obtained in the presence of saturating amounts of XD is typical of a highly restricted mobility (τ_r in the ms range) of the spin label, as judged by the appearance of a novel spectral line shape with a broader outer line splitting (5.95 ± 0.05 mT) as compared with the unbound form (3.22 ± 0.02 mT) (Fig. 6.16A,C). As expected, this effect is even more pronounced in the presence of 30% sucrose (Fig. 6.16B,C), where the mobility of the radical is severely restricted and approaches the so-called rigid-limit regime obtained for a frozen solution of the free radical (see section 6.2.1.3).

6.5.7 Reversibility of Induced Folding and Functional Implications

In order to assess whether complex formation is a reversible process, we carried out equilibrium displacement experiments. We first recorded the EPR spectrum of a mixture containing spin labeled L496C and XD in the 1:2 molar ratio. Under these conditions, the percentage of the complex was estimated to be 100% as judged by the spectral shape (Fig. 6.17). We then evaluated the ability of wt N-tail to displace the equilibrium toward the unbound form of spin labeled L496C. To this endeavor, we recorded the EPR spectrum of a mixture containing spin labeled L496C, XD, and wt N-tail in the 1:2:1 ratio.

The spectrum obtained in these experimental conditions is composed of two signals: one arising from the unbound spin labeled protein and one resulting from the spin labeled protein bound to XD. Under these conditions, the $h(+1)/h(0)$ is 0.62 ± 0.02, a value consistent with a decrease in the amount of the spin labeled L496C/XD complex (Fig. 6.17). This decrease indicates that

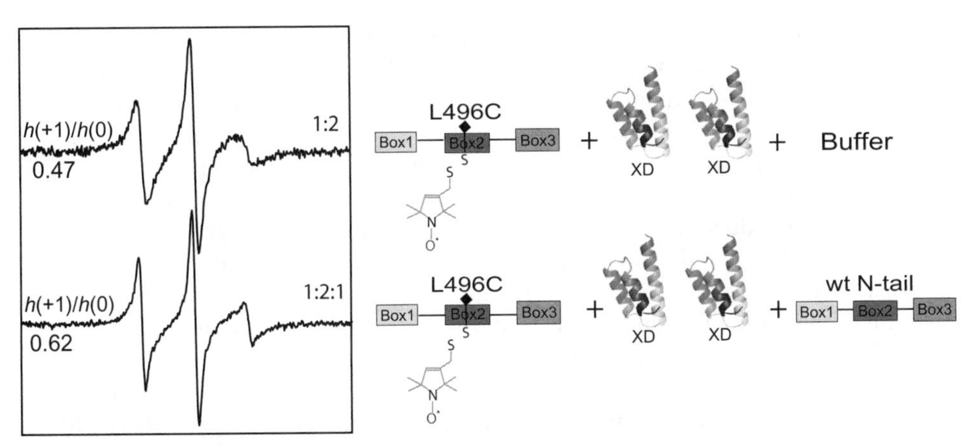

Figure 6.17 Dissociation of the spin labeled L496C–XD complex by wt N-tail (20 μM). Normalized room temperature EPR spectra of the various mixtures. The molar ratios and the $h(+1)/h(0)$ ratios are also indicated. The nitroxide radical is shown. Note that the spacing between N-tail boxes does not reflect actual sizes. Data were taken from Reference 60.

XD can dissociate from spin labeled L496C and can associate with unlabeled wt N-tail, thus pointing out that complex formation is a reversible process. As a negative control, we recorded an EPR spectrum of a mixture containing spin labeled L496C, XD, and lysozyme in 1:2:1 ratio. No variation in the percentage of complex was observed (data not shown), reflecting lysozyme's inability to interact with XD and to displace the equilibrium.

Therefore, by using equilibrium displacement experiments, we showed not only the reversibility of the N-tail–XD interaction, but also the reversibility of the induced folding of the α-MoRE upon dissociation of XD (60). Indeed, upon dissociation of XD, the spin labeled L496C protein exhibits a spectral signature corresponding to the unbound form, consistent with a loss of α-helicity (Fig. 6.17). Hence, N-tail adopts its original premolten globule conformation after dissociation of its partner.

This latter point is particularly relevant, taking into consideration that the contact between XD and N-tail within the replicative complex has to be dynamically made and broken to allow the polymerase to progress along the nucleocapsid template during both transcription and replication. Hence, the complex cannot be excessively stable for this transition to occur efficiently at a high rate.

In conclusion, using SDSL EPR spectroscopy, we mapped the N-tail region involved in α-helical folding in solution. We showed that different N-tail regions contribute to a different extent to the folding process induced by XD. We also showed that the interaction between N-tail and XD implies the stabilization of the helical conformation of the α-MoRE, which is otherwise only transiently populated in the unbound form. The occurrence of a transiently populated α-helix, even in the absence of the partner, suggests that the molecular mechanism governing the folding of N-tail induced by XD would rely at least partly on conformer selection (i.e., selection by the partner of a preexisting conformation) (77, 78) rather than on a "fly casting" mechanism (68), contrary to what has been reported for the pKID–KIX couple (71).

Stabilization of the helical conformation of the α-MoRE is also accompanied by a reduction in the mobility of the downstream region. The lower flexibility of the region downstream Box2 is not due to gain of α-helicity, nor can it be ascribed to a restrained motion brought by a direct interaction with XD. Rather, it likely arises from a gain of rigidity brought by α-helical folding of the neighboring Box2 region. In agreement with these data, titration studies using heteronuclear NMR suggest that XD does not establish direct interactions with Box3 (7), contrary to what has been previously proposed (12).

We tentatively propose that binding to XD might take place through a sequential mechanism that could involve binding and α-helical folding of Box2, followed by a conformational change of Box3, whose overall mobility is consequently reduced probably through tertiary contacts with the neighboring Box2 region (Fig. 6.18). That Box3 likely establishes contacts with Box2 is also supported by the SAXS-derived low-resolution model of the complex between N-tail and XD, which indicates that the C-terminus is not exposed to the solvent and is rather embedded in the globular, bulky part of the model accommodating XD and the α-MoRE (see Fig. 6.11 and Reference 12). Hence, Box3 would dynamically control the strength of the N-tail–XD interaction, by stabilizing the Box2–XD interaction. Modulation of XD/N-tail binding affinity could also be dictated by interactions between N-tail and cellular and/or viral

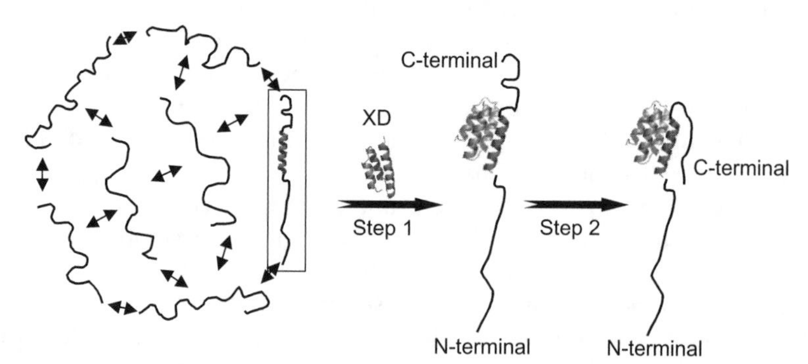

Figure 6.18 Schematic representation of the proposed molecular mechanism of the XD-induced folding of N-tail pointing out conformer selection by the partner and α-helical folding of the 488–502 region (helix) (step 1), followed by establishment of tertiary contacts between Box2 and the downstream region (step 2). Sequestration of the partially folded N-tail conformer by the partner will result in a shift in the equilibrium among N-tail conformers progressively leading to full complex formation with XD. Data were taken from Reference 5.

cofactors. Indeed, MeV transcription and replication are enhanced by the major heat shock protein (Hsp72), and this stimulation relies on interaction with N-tail (87, 88). Hsp72 and XD were shown to compete for binding to N-tail (87), suggesting that Hsp72 may serve as a processivity or transcription elongation factor that could act by modulating the strength of the interaction between the polymerase complex and the nucleocapsid template.

Finally, the large number of spin labeled N-tail variants that we generated and their analysis by EPR spectroscopy provided us with a panel of spectral signatures that have a high predictive value. Indeed, the EPR spectral shape reflects the specific environment in the proximity of the spin label and the extent of involvement of the radical in the interaction with a partner while providing information on possible α-helical transitions. As such, we will be able to infer information on the chemical environment and the implication of specific N-tail residues within complexes with other partners, such as the N-tail–Hsp72 complex for which no structural information is presently available.

ACKNOWLEDGMENTS

We wish to thank all the persons who contributed to the EPR spectroscopy studies herein described. In particular, we would like to thank Benjamin Morin, Jean-Marie Bourhis, and Elodie Liquière of the AFMB laboratory; Bruno Guigliarelli and Mireille Woudstra of the BIP laboratory; and Frédéric Carrière of the EIPL laboratory (UPR 9025-CNRS, Marseille, France). We also thank Janez Strancar for useful comments on the choice of N-tail positions targeted for spin labeling. We are also grateful to Michael Oglesbee and Denis Gerlier for stimulating discussions. The studies described in this chapter were partly carried out with the financial support of the Agence Nationale de la Recherche, specific program "Microbiologie et Immunologie," ANR-05-MIIM-035-02, "Structure and disorder of measles virus nucleoprotein: molecular partnership and functional impact," and of the National Institute of Neurological Disorders and Stroke, specific program "The cellular stress response in viral encephalitis," R01 NS031693-11A2.

REFERENCES

1. Altenbach, C., D. A. Greenhalgh, H. G. Khorana, and W. L. Hubbell. 1994. A collision gradient method to determine the immersion depth of nitroxides in lipid bilayers: application to spin-labeled mutants of bacteriorhodopsin. Proc Natl Acad Sci U S A **91**:1667–71.
2. Bankamp, B., S. M. Horikami, P. D. Thompson, M. Huber, M. Billeter, and S. A. Moyer. 1996. Domains of the measles virus N protein required for binding to P protein and self-assembly. Virology **216**:272–7.

3. Barnes, J. P., Z. Liang, H. S. McHaourab, J. H. Freed, and W. L. Hubbell. 1999. A multifrequency electron spin resonance study of T4 lysozyme dynamics. Biophys J **76**:3298–306.

4. Belle, V., A. Fournel, M. Woudstra, S. Ranaldi, F. Prieri, V. Thome, J. Currault, R. Verger, B. Guigliarelli, and F. Carriere. 2007. Probing the opening of the pancreatic lipase lid using site-directed spin labeling and EPR spectroscopy. Biochemistry **46**:2205–14.

5. Belle, V., S. Rouger, S. Costanzo, E. Liquiere, J. Strancar, B. Guigliarelli, A. Fournel, and S. Longhi. 2008. Mapping alpha-helical induced folding within the intrinsically disordered C-terminal domain of the measles virus nucleoprotein by site-directed spin-labeling EPR spectroscopy. Proteins **73**:973–88.

6. Bennati, M., G. J. Gerfen, G. V. Martinez, R. G. Griffin, D. J. Singel, and G. L. Millhauser. 1999. Nitroxide side-chain dynamics in a spin-labeled helix-forming peptide revealed by high-frequency (139.5-GHz) EPR spectroscopy. J Magn Reson **139**:281–6.

7. Bernard, C., S. Gely, J. M. Bourhis, X. Morelli, S. Longhi, and H. Darbon. 2009. Interaction between the C-terminal domains of N and P proteins of measles virus investigated by NMR. FEBS Lett **583**:1084–9.

8. Bourhis, J., K. Johansson, V. Receveur-Bréchot, C. J. Oldfield, A. K. Dunker, B. Canard, and S. Longhi. 2004. The C-terminal domain of measles virus nucleoprotein belongs to the class of intrinsically disordered proteins that fold upon binding to their physiological partner. Virus Res **99**:157–67.

9. Bourhis, J. M., B. Canard, and S. Longhi. 2005. Désordre structural au sein du complexe réplicatif du virus de la rougeole: implications fonctionnelles. Virologie **9**:367–83.

10. Bourhis, J. M., B. Canard, and S. Longhi. 2006. Structural disorder within the replicative complex of measles virus: functional implications. Virology **344**: 94–110.

11. Bourhis, J. M., and S. Longhi. 2007. Measles virus nucleoprotein: structural organization and functional role of the intrinsically disordered C-terminal domain, pp. 1–35. In S. Longhi (ed.), Measles Virus Nucleoprotein. Nova Publishers Inc., Hauppauge, NY.

12. Bourhis, J. M., V. Receveur-Bréchot, M. Oglesbee, X. Zhang, M. Buccellato, H. Darbon, B. Canard, S. Finet, and S. Longhi. 2005. The intrinsically disordered C-terminal domain of the measles virus nucleoprotein interacts with the C-terminal domain of the phosphoprotein via two distinct sites and remains predominantly unfolded. Protein Sci **14**:1975–92.

13. Buck, M. 1998. Trifluoroethanol and colleagues: cosolvents come of age. Recent studies with peptides and proteins. Q Rev Biophys **31**:297–355.

14. Budil, D. E., S. Lee, S. Saxena, and J. H. Freed. 1996. Non linear least-squares analysis of slow motion EPR spectra in one and two dimensions using a modified Levenberg-Marquardt algorithm. J Magn Reson A **120**:155–89.

15. Colombo, M., J. M. Bourhis, C. Chamontin, C. Soriano, S. Villet, S. Costanzo, M. Couturier, V. Belle, A. Fournel, H. Darbon, D. Gerlier, and S. Longhi. 2009. The interaction between the measles virus nucleoprotein and the Interferon Regulator Factor 3 relies on a specific cellular environment. Virol J **6**:59.

16. Columbus, L., and W. L. Hubbell. 2002. A new spin on protein dynamics. Trends Biochem Sci **27**:288–95.
17. Columbus, L., and W. L. Hubbell. 2004. Mapping backbone dynamics in solution with site-directed spin labeling: GCN4-58 bZip free and bound to DNA. Biochemistry **43**:7273–87.
18. Columbus, L., T. Kalai, J. Jeko, K. Hideg, and W. L. Hubbell. 2001. Molecular motion of spin labeled side chains in alpha-helices: analysis by variation of side chain structure. Biochemistry **40**:3828–46.
19. Iwasaki, M., M. Takeda, Y. Shirogane, Y. Nakatsu, T. Nakamura, and Y. Yanagi. 2009. The matrix protein of measles virus regulates viral RNA synthesis and assembly by interacting with the nucleocapsid protein. J Virol **83**:10374–83.
20. Curran, J., and D. Kolakofsky. 1999. Replication of paramyxoviruses. Adv Virus Res **54**:403–22.
21. De, B. P., and A. K. Banerjee. 1999. Involvement of actin microfilaments in the transcription/replication of human parainfluenza virus type 3: possible role of actin in other viruses. Microsc Res Tech **47**:114–23.
22. Diallo, A., T. Barrett, M. Barbron, G. Meyer, and P. C. Lefevre. 1994. Cloning of the nucleocapsid protein gene of peste-des-petits-ruminants virus: relationship to other morbilliviruses. J Gen Virol **75** (Pt 1):233–7.
23. Dunker, A. K., J. D. Lawson, C. J. Brown, R. M. Williams, P. Romero, J. S. Oh, C. J. Oldfield, A. M. Campen, C. M. Ratliff, K. W. Hipps, J. Ausio, M. S. Nisse n, R. Reeves, C. Kang, C. R. Kissinger, R. W. Bailey, M. D. Griswold, W. Chiu, E. C. Garner, and Z. Obradovic. 2001. Intrinsically disordered protein. J Mol Graph Model **19**:26–59.
24. Dyson, H. J., and P. E. Wright. 2002. Coupling of folding and binding for unstructured proteins. Curr Opin Struct Biol **12**:54–60.
25. Dyson, H. J., and P. E. Wright. 2005. Intrinsically unstructured proteins and their functions. Nat Rev Mol Cell Biol **6**:197–208.
26. Earle, K. A., D. E. Budil, and J. H. Freed. 1993. 250-GHZ EPR of nitroxides in the slow motional regime—models of rotational diffusion. J Phys Chem **97**:13289–97.
27. Fan, P., C. Bracken, and J. Baum. 1993. Structural characterization of monellin in the alcohol-denatured state by NMR: evidence for beta-sheet to alpha-helix conversion. Biochemistry **32**:1573–82.
28. Fink, A. L. 2005. Natively unfolded proteins. Curr Opin Struct Biol **15**:35–41.
29. Fletcher, C. M., A. M. McGuire, A. C. Gingras, H. Li, H. Matsuo, N. Sonenberg, and G. Wagner. 1998. 4E binding proteins inhibit the translation factor eIF4E without folded structure. Biochemistry **37**:9–15.
30. Fuxreiter, M., I. Simon, P. Friedrich, and P. Tompa. 2004. Preformed structural elements feature in partner recognition by intrinsically unstructured proteins. J Mol Biol **338**:1015–26.
31. Fuxreiter, M., P. Tompa, and I. Simon. 2007. Local structural disorder imparts plasticity on linear motifs. Bioinformatics **23**:950–6.
32. Goldman, A., G. V. Bruno, C. F. Polnaszek, and J. H. Freed. 1972. An ESR study of anisotropic rotational reorientation and slow tumbling in liquid and frozen media. J Chem Phys **56**:716–35.

33. Heggeness, M. H., A. Scheid, and P. W. Choppin. 1980. Conformation of the helical nucleocapsids of paramyxoviruses and vesicular stomatitis virus: reversible coiling and uncoiling induced by changes in salt concentration. Proc Natl Acad Sci U S A **77**:2631–5.
34. Heggeness, M. H., A. Scheid, and P. W. Choppin. 1981. The relationship of conformational changes in the Sendai virus nucleocapsid to proteolytic cleavage of the NP polypeptide. Virology **114**:555–62.
35. Hua, Q. X., W. H. Jia, B. P. Bullock, J. F. Habener, and M. A. Weiss. 1998. Transcriptional activator-coactivator recognition: nascent folding of a kinase-inducible transactivation domain predicts its structure on coactivator binding. Biochemistry **37**:5858–66.
36. Hubbell, W. L., C. Altenbach, C. M. Hubbell, and H. G. Khorana. 2003. Rhodopsin structure, dynamics, and activation: a perspective from crystallography, site-directed spin labeling, sulfhydryl reactivity, and disulfide cross-linking. Adv Protein Chem **63**:243–90.
37. Hubbell, W. L., D. S. Cafiso, and C. Altenbach. 2000. Identifying conformational changes with site-directed spin labeling. Nat Struct Biol **7**:735–9.
38. Hubbell, W. L., A. Gross, R. Langen, and M. A. Lietzow. 1998. Recent advances in site-directed spin labeling of proteins. Curr Opin Struct Biol **8**:649–56.
39. Hubbell, W. L., H. S. McHaourab, C. Altenbach, and M. A. Lietzow. 1996. Watching proteins move using site-directed spin labeling. Structure **4**:779–83.
40. Jeschke, G. 2002. Distance measurements in the nanometer range by pulse EPR. Chemphyschem **3**:927–32.
41. Johansson, K., J. M. Bourhis, V. Campanacci, C. Cambillau, B. Canard, and S. Longhi. 2003. Crystal structure of the measles virus phosphoprotein domain responsible for the induced folding of the C-terminal domain of the nucleoprotein. J Biol Chem **278**:44567–73.
42. Karlin, D., F. Ferron, B. Canard, and S. Longhi. 2003. Structural disorder and modular organization in Paramyxovirinae N and P. J Gen Virol **84**:3239–52.
43. Karlin, D., S. Longhi, and B. Canard. 2002. Substitution of two residues in the measles virus nucleoprotein results in an impaired self-association. Virology **302**:420–32.
44. Karlin, D., S. Longhi, V. Receveur, and B. Canard. 2002. The N-terminal domain of the phosphoprotein of morbilliviruses belongs to the natively unfolded class of proteins. Virology **296**:251–62.
45. Kingston, R. L., D. J. Hamel, L. S. Gay, F. W. Dahlquist, and B. W. Matthews. 2004. Structural basis for the attachment of a paramyxoviral polymerase to its template. Proc Natl Acad Sci U S A **101**:8301–6.
46. Kingston, R. L., A. B. Walter, and L. S. Gay. 2004. Characterization of nucleocapsid binding by the measles and the mumps virus phosphoprotein. J Virol **78**:8615–29.
47. Kreimer, D. I., R. Szosenfogel, D. Goldfarb, I. Silman, and L. Weiner. 1994. Two-state transition between molten globule and unfolded states of acetylcholinesterase as monitored by electron paramagnetic resonance spectroscopy. Proc Natl Acad Sci U S A **91**:12145–9.
48. Lacy, E. R., I. Filippov, W. S. Lewis, S. Otieno, L. Xiao, S. Weiss, L. Hengst, and R. W. Kriwacki. 2004. p27 binds cyclin-CDK complexes through a sequential

mechanism involving binding-induced protein folding. Nat Struct Mol Biol **11**:358–64.

49. Laine, D., J. Bourhis, S. Longhi, M. Flacher, L. Cassard, B. Canard, C. Sautès-Fridman, C. Rabourdin-Combe, and H. Valentin. 2005. Measles virus nucleoprotein induces cell proliferation arrest and apoptosis through NTAIL/NR and NCORE/FcgRIIB1 interactions, respectively. J Gen Virol **86**:1771–84.
50. Laine, D., M. Trescol-Biémont, S. Longhi, G. Libeau, J. Marie, P. Vidalain, O. Azocar, A. Diallo, B. Canard, C. Rabourdin-Combe, and H. Valentin. 2003. Measles virus nucleoprotein binds to a novel cell surface receptor distinct from FcgRII via its C-terminal domain: role in MV-induced immunosuppression. J Virol **77**:11332–46.
51. Langen, R., K. J. Oh, D. Cascio, and W. L. Hubbell. 2000. Crystal structures of spin labeled T4 lysozyme mutants: implications for the interpretation of EPR spectra in terms of structure. Biochemistry **39**:8396–405.
52. Liston, P., R. Batal, C. DiFlumeri, and D. J. Briedis. 1997. Protein interaction domains of the measles virus nucleocapsid protein (NP). Arch Virol **142**:305–21.
53. Longhi, S. 2009. Nucleocapsid structure and function. Curr Top Microbiol Immunol **329**:103–28.
54. Longhi, S., V. Receveur-Brechot, D. Karlin, K. Johansson, H. Darbon, D. Bhella, R. Yeo, S. Finet, and B. Canard. 2003. The C-terminal domain of the measles virus nucleoprotein is intrinsically disordered and folds upon binding to the C-terminal moiety of the phosphoprotein. J Biol Chem **278**:18638–48.
55. Luo, P., and R. L. Baldwin. 1997. Mechanism of helix induction by trifluoroethanol: a framework for extrapolating the helix-forming properties of peptides from trifluoroethanol/water mixtures back to water. Biochemistry **36**:8413–21.
56. Marsh, D., D. Kurad, and V. A. Livshits. 2002. High-field electron spin resonance of spin labels in membranes. Chem Phys Lipids **116**:93–114.
57. McHaourab, H. S., M. A. Lietzow, K. Hideg, and W. L. Hubbell. 1996. Motion of spin-labeled side chains in T4 lysozyme. Correlation with protein structure and dynamics. Biochemistry **35**:7692–704.
58. McHaourab, H. S., K. J. Oh, C. J. Fang, and W. L. Hubbell. 1997. Conformation of T4 lysozyme in solution. Hinge-bending motion and the substrate-induced conformational transition studied by site-directed spin labeling. Biochemistry **36**:307–16.
59. Mohan, A., C. J. Oldfield, P. Radivojac, V. Vacic, M. S. Cortese, A. K. Dunker, and V. N. Uversky. 2006. Analysis of molecular recognition features (MoRFs). J Mol Biol **362**:1043–59.
60. Morin, B., J. M. Bourhis, V. Belle, M. Woudstra, F. Carrière, B. Guigliarelli, A. Fournel, and S. Longhi. 2006. Assessing induced folding of an intrinsically disordered protein by site-directed spin-labeling EPR spectroscopy. J Phys Chem B **110**:20596–608.
61. Moyer, S. A., S. C. Baker, and S. M. Horikami. 1990. Host cell proteins required for measles virus reproduction. J Gen Virol **71**:775–83.
62. Myers, J. K., C. N. Pace, and J. M. Scholtz. 1997. Helix propensities are identical in proteins and peptides. Biochemistry **36**:10923–9.

63. Oldfield, C. J., Y. Cheng, M. S. Cortese, P. Romero, V. N. Uversky, and A. K. Dunker. 2005. Coupled folding and binding with alpha-helix-forming molecular recognition elements. Biochemistry **44**:12454–70.

64. Qu, K., J. L. Vaughn, A. Sienkiewicz, C. P. Scholes, and J. S. Fetrow. 1997. Kinetics and motional dynamics of spin-labeled yeast iso-1-cytochrome c: 1. Stopped-flow electron paramagnetic resonance as a probe for protein folding/unfolding of the C-terminal helix spin-labeled at cysteine 102. Biochemistry **36**:2884–97.

65. Rahaman, A., N. Srinivasan, N. Shamala, and M. S. Shaila. 2004. Phosphoprotein of the rinderpest virus forms a tetramer through a coiled coil region important for biological function. A structural insight. J Biol Chem **279**:23606–14.

66. Receveur-Bréchot, V., J. M. Bourhis, V. N. Uversky, B. Canard, and S. Longhi. 2006. Assessing protein disorder and induced folding. Proteins: Structure, Function and Bioinformatics **62**:24–45.

67. Sato, H., M. Masuda, R. Miura, M. Yoneda, and C. Kai. 2006. Morbillivirus nucleoprotein possesses a novel nuclear localization signal and a CRM1-independent nuclear export signal. Virology **352**:121–30.

68. Shoemaker, B. A., J. J. Portman, and P. G. Wolynes. 2000. Speeding molecular recognition by using the folding funnel: the fly-casting mechanism. Proc Natl Acad Sci U S A **97**:8868–73.

69. Steinhoff, H. J., N. Radzwill, W. Thevis, V. Lenz, D. Brandenburg, A. Antson, G. Dodson, and A. Wollmer. 1997. Determination of interspin distances between spin labels attached to insulin: comparison of electron paramagnetic resonance data with the X-ray structure. Biophys J **73**:3287–98.

70. Stoll, S., and A. Schweiger. 2006. EasySpin, a comprehensive software package for spectral simulation and analysis in EPR. J Magn Reson **178**:42–55.

71. Sugase, K., H. J. Dyson, and P. E. Wright. 2007. Mechanism of coupled folding and binding of an intrinsically disordered protein. Nature **447**:1021–5.

72. tenOever, B. R., M. J. Servant, N. Grandvaux, R. Lin, and J. Hiscott. 2002. Recognition of the measles virus nucleocapsid as a mechanism of IRF-3 activation. J Virol **76**:3659–69.

73. Timofeev, V. P., and V. I. Tsetlin. 1983. Analysis of mobility of protein side chains by spin-label technique. Biophys Struct Mech **10**:93–108.

74. Tompa, P. 2002. Intrinsically unstructured proteins. Trends Biochem Sci **27**:527–33.

75. Tompa, P. 2003. The functional benefits of disorder. J Mol Struct (Theochem) **666–7**:361–71.

76. Tompa, P. 2005. The interplay between structure and function in intrinsically unstructured proteins. FEBS Lett **579**:3346–54.

77. Tsai, C. D., B. Ma, S. Kumar, H. Wolfson, and R. Nussinov. 2001. Protein folding: binding of conformationally fluctuating building blocks via population selection. Crit Rev Biochem Mol Biol **36**:399–433.

78. Tsai, C. J., B. Ma, Y. Y. Sham, S. Kumar, and R. Nussinov. 2001. Structured disorder and conformational selection. Proteins **44**:418–27.

79. Uversky, V. N. 2002. Natively unfolded proteins: a point where biology waits for physics. Protein Sci **11**:739–56.

80. Uversky, V. N. 1993. Use of fast protein size-exclusion liquid chromatography to study the unfolding of proteins which denature through the molten globule. Biochemistry **32**:13288–98.
81. Uversky, V. N. 2002. What does it mean to be natively unfolded? Eur J Biochem **269**:2–12.
82. Uversky, V. N., C. J. Oldfield, and A. K. Dunker. 2005. Showing your ID: intrinsic disorder as an ID for recognition, regulation and cell signaling. J Mol Recognit **18**:343–84.
83. Vacic, V., C. J. Oldfield, A. Mohan, P. Radivojac, M. S. Cortese, V. N. Uversky, and A. K. Dunker. 2007. Characterization of molecular recognition features, MoRFs, and their binding partners. J Proteome Res **6**:2351–66.
84. Vucetic, S., H. Xie, L. M. Iakoucheva, C. J. Oldfield, A. K. Dunker, Z. Obradovic, and V. N. Uversky. 2007. Functional anthology of intrinsic disorder. 2. Cellular components, domains, technical terms, developmental processes, and coding sequence diversities correlated with long disordered regions. J Proteome Res **6**:1899–916.
85. Xie, H., S. Vucetic, L. M. Iakoucheva, C. J. Oldfield, A. K. Dunker, Z. Obradovic, and V. N. Uversky. 2007. Functional anthology of intrinsic disorder. 3. Ligands, post-translational modifications, and diseases associated with intrinsically disordered proteins. J Proteome Res **6**:1917–32.
86. Xie, H., S. Vucetic, L. M. Iakoucheva, C. J. Oldfield, A. K. Dunker, V. N. Uversky, and Z. Obradovic. 2007. Functional anthology of intrinsic disorder. 1. Biological processes and functions of proteins with long disordered regions. J Proteome Res **6**:1882–98.
87. Zhang, X., J. M. Bourhis, S. Longhi, T. Carsillo, M. Buccellato, B. Morin, B. Canard, and M. Oglesbee. 2005. Hsp72 recognizes a P binding motif in the measles virus N protein C-terminus. Virology **337**:162–74.
88. Zhang, X., C. Glendening, H. Linke, C. L. Parks, C. Brooks, S. A. Udem, and M. Oglesbee. 2002. Identification and characterization of a regulatory domain on the carboxyl terminus of the measles virus nucleocapsid protein. J Virol **76**:8737–46.

7

THE STRUCTURE OF UNFOLDED PEPTIDES AND PROTEINS EXPLORED BY VIBRATIONAL SPECTROSCOPY

Reinhard Schweitzer-Stenner, Thomas J. Measey, Andrew M. Hagarman, and Isabelle C. Dragomir

Department of Chemistry, Drexel University, Philadelphia, PA

ABSTRACT

The exploration of unfolded peptides and proteins is an emerging area of research in the field of protein biophysics and biochemistry. In this context, vibrational spectroscopy has become a valuable tool for exploring local structural preferences of amino acid residues in the unfolded state. After introducing the basic physical concepts, this article reviews the most recent utilization of UV resonance Raman, visible nonresonance Raman, vibrational circular dichroism, Raman optical activity, and, to a limited extent, electronic circular dichroism spectroscopy for the exploration of naturally unfolded peptides and proteins. With respect to peptides, this article puts an emphasis on recent work about alanine-based peptides, which, owing to the large abundance of alanine in proteins, have emerged as ideal model systems for attempts to understand structure and dynamics in the unfolded state.

Instrumental Analysis of Intrinsically Disordered Proteins: Assessing Structure and Conformation, Edited by Vladimir Uversky and Sonia Longhi

7.1 INTRODUCTION

At first glance the exploration of the structure of unfolded peptides or proteins seems to be a futile endeavor, because textbook wisdom suggests that they are structurally random (21). This means that each individual amino acid residue samples the entire sterically allowed fraction of the Ramachandran space with comparable probability (30, 85, 112). This sampling yields a statistical distribution of a large number of peptide/protein conformations (13). However, this view was challenged already 30 years ago by Tiffany and Krimm, who found that the so-called random coil states of poly-L-lysine (PLL) and poly-L-glutamic acid (PGA) give rise to a far-UV electronic circular dichroism (ECD) spectrum, which is very similar to that observed for *trans*-polyproline (113). This led the authors to conclude that local segments might predominantly adopt a polyproline II (PPII) conformation, which, in the ideal case, exhibits dihedral angles of $\phi = -78°$ and $\psi = 145°$ (19). After a controversial debate in the 1970s, further lack of experimental support led to the disappearance of the Tiffany–Krimm hypothesis, until it resurfaced in the 1990s due to the findings of Dukor and Keiderling (23), who provided very strong support for the hypothesis by means of vibrational circular dichroism (VCD) data for the same peptides investigated by Tiffany and Krimm. It took another 10 years before the relevance of the PPII conformation for the understanding of the unfolded state of peptides and proteins became a subject of an intense discussion (101). This particularly concerns alanine, whose propensity in the unfolded state has led to a very controversial debate (68, 69, 94, 99, 119, 125). While spectroscopic data indicate that different amino acids such as alanine and valine show rather different propensities (PPII and β-strand) (25, 26, 37), ECD and NMR data, as well as computational results, seem to suggest that nearly all amino acids except glycine have a rather high PPII propensity (98, 117).

In the context of peptide and protein research, vibrational spectroscopy is generally considered a low-resolution technique, which can solely be used to discriminate between helices, β-sheets, turns, and random coils (108). However, recent advances in the field have substantially improved the usability of vibrational spectroscopy for the structure analysis of peptides and proteins. Chiral techniques such as VCD or Raman optical activity (ROA) enable the discrimination between different helical and turn structures (10, 53). For small peptides, density functional theory (DFT) calculations, combined with infrared (IR) and VCD spectroscopy, can be utilized for discriminating between different helical, sheet, and turn structures (12, 61). UV resonance Raman spectroscopy has been developed to probe the distribution of ψ-angles from an analysis of the amide III band shape (3, 74). Finally, coherent femtosecond IR spectroscopy, as well as the combination of more conventional techniques such as IR, polarized visible Raman, and VCD, exploited the structural sensitivity of the amide I mode to obtain dihedral angles adopted by folded and unfolded peptides (123).

This review is aimed at providing an overview on how vibrational spectroscopy can be used for a detailed structural analysis of unfolded peptides. The relevance of peptide research for the understanding of intrinsically disordered proteins has been emphasized and outlined in a recent review (88). The present article is structured as follows: first, the reader will be familiarized with the vibrations of the peptide/protein backbone, which are most relevant for structure analysis (section 7.2). Second, the underlying physics of the discussed spectroscopies (IR, VCD, ROA, Raman) will be briefly described (section 7.3). Section 7.4, which is the core part of this article, describes results obtained from the application of these spectroscopies to unfolded peptides and proteins. This section includes a detailed description of work on peptides performed in our research group. An outlook delineating the relevance of the results performed with vibrational spectroscopies is given in section 7.5.

7.2 PEPTIDE AND PROTEIN VIBRATIONS

Articles describing the normal modes of peptide/protein vibrations have been published before (7, 58, 90). Here, we confine ourselves to the most prominent backbone vibrations and their use in different types of vibrational spectroscopies. For the purpose of illustration, Figure 7.1 depicts the polarized Raman spectra and the IR spectrum of neutral (zwitterionic) dialanine in H_2O and D_2O.

7.2.1 Amide I

Figure 7.2 exhibits the eigenvector (normal mode composition) of the amide I mode, which is the most prominent of the protein backbone modes, owing to the structural sensitivity of its wavenumber position (7, 90, 108). In the literature, amide I is generally described as a pure CO stretching mode (CO s). However, as depicted in Figure 7.2A, its eigenvector depicts substantial admixtures from NH in-plane bending (NH ip b) and CN stretching (CN s) vibrations. An admixture of CH bending (CH b) is also noteworthy. If water is used as a solvent, amide I mixes with bending modes of water molecules in the hydration shell (17, 102). This effect and the admixture of NH ip b are eliminated if D_2O, instead of H_2O, is used as a solvent (Fig. 7.2B). Thus, the amide I′ mode of deuterated peptide groups (NH is replaced by ND) is more dominated by CO s than amide I of naturally abundant peptides. The CO s contribution brings about a strong transition dipole moment that forms an angle of about 15° with the CO bond. Hence, it gives rise to a very intense band in the IR spectrum, which, for neutral dialanine in H_2O, is positioned at $1676\,cm^{-1}$ and downshifts to $1660\,cm^{-1}$ in D_2O. The respective (visible) Raman bands of amide I and I′ are of intermediate intensity and polarized. The intrinsic amide I wavenumber (in the absence of any intrapeptide vibrational coupling) is affected by the strength of hydrogen bonding between the solvent molecules and the peptide's CO and NH group, and depends on the solvent's

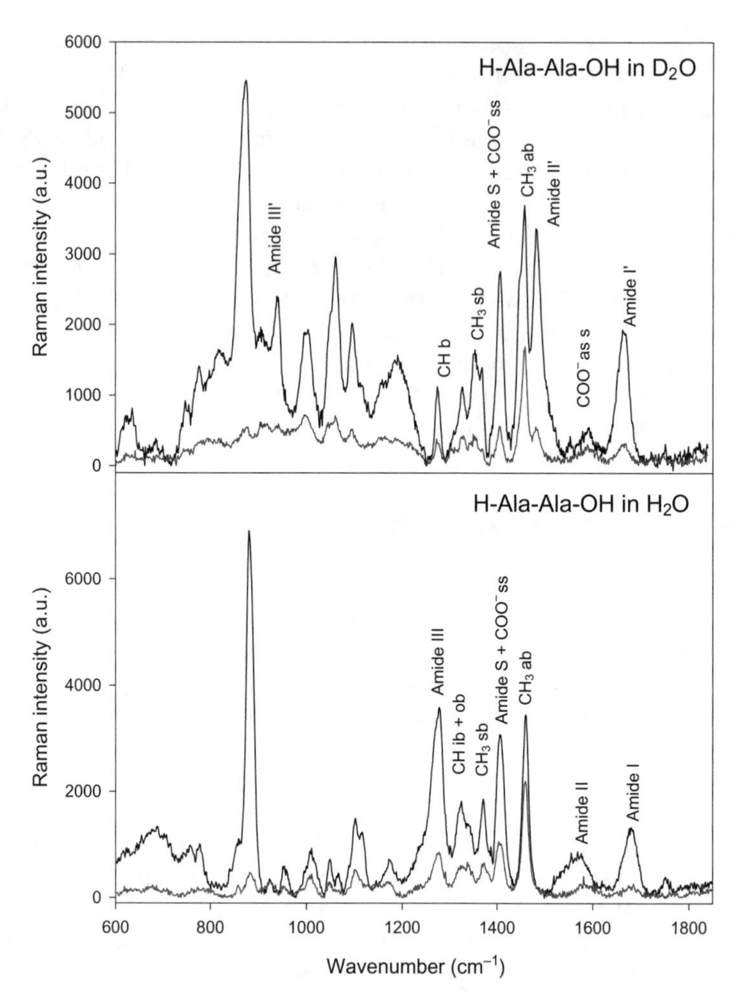

Figure 7.1 Parallel and perpendicular polarized Raman spectra of zwitterionic H-Ala-Ala-OH in (A) D_2O and (B) H_2O (black solid line: polarization parallel to the polarization of excitation; gray line: polarization perpendicular to the polarization of excitation). The assignments of the most prominent modes are indicated. a.u., arbitrary units; as s, antisymmetric stretch; ib, in-plane bending; ob, out-of-plane bending; ss, symmetric stretch.

dielectric properties (115, 121). The admixture of HOH b modes to the amide I eigenvector yields a broadening and a pronounced asymmetry of the band profile (102). A conformational dependence of the amide I frequency has been predicted based on DFT calculations (34, 38). Recently, Measey et al. reported that wavenumber, oscillator strength, and Raman depolarization ratios of amide I′ depend on the adjacent side chains (72). DFT calculations further suggest that the wavenumber depends on the dihedral angles of the connected residues (44).

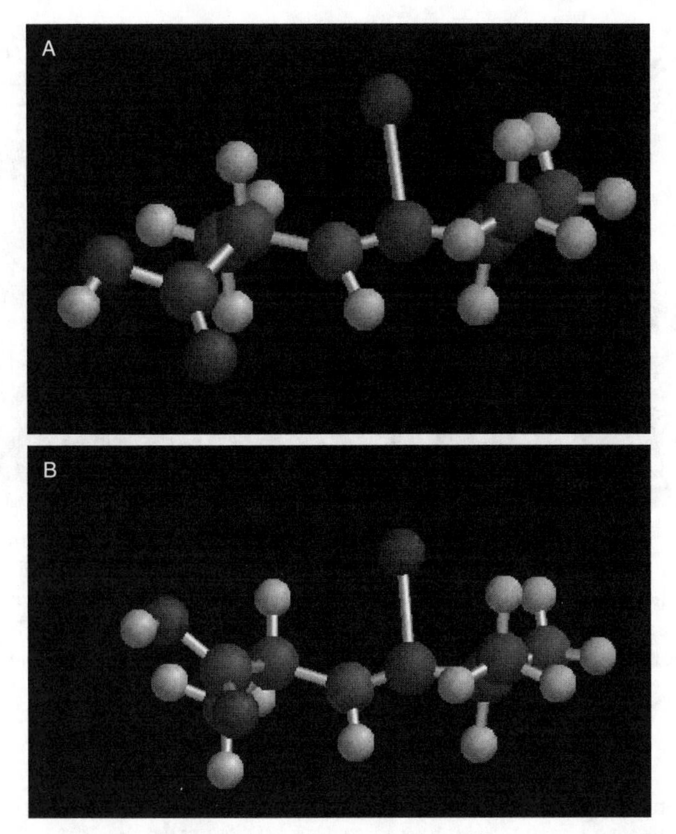

Figure 7.2 Snapshot of the amide I mode of (A) naturally abundant dialanine and (B) dialanine with deutrated amide groups *in vacuo*. The vibration was obtained by performing a normal mode calculation based on the semiempirical force field, AM1, which is part of the TITAN software from Schrödinger, Inc. (New York, NY). Figure 7.2A shows the displacement of CO, NH, and $C_\alpha H$. In Figure 7.2B, only the carbonyl group is significantly displaced.

The parameters discussed above all affect the intrinsic wavenumber of amide I. However, the structural sensitivity of this mode stems, to a major extent, from the fact that it is heavily delocalized in polypeptides (38–41, 58, 116). The delocalization arises from vibrational coupling between amide I modes, owing to through bond nearest neighbor and through space electrostatic interactions (115). There are two concepts to describe the interaction between amide I modes. The first concept is a classical one and based on a normal mode calculation for a given peptide and protein carried out by means of either an empirical force field (for proteins and large peptides) (58) or an *ab initio* calculated quantum chemical force field (12, 18, 38, 61–63). Figure 7.3 shows two amide I eigenvectors of trialanine in a PPII conformation obtained from a DFT calculation at the BL3LYP 6-31G** level of theory. The

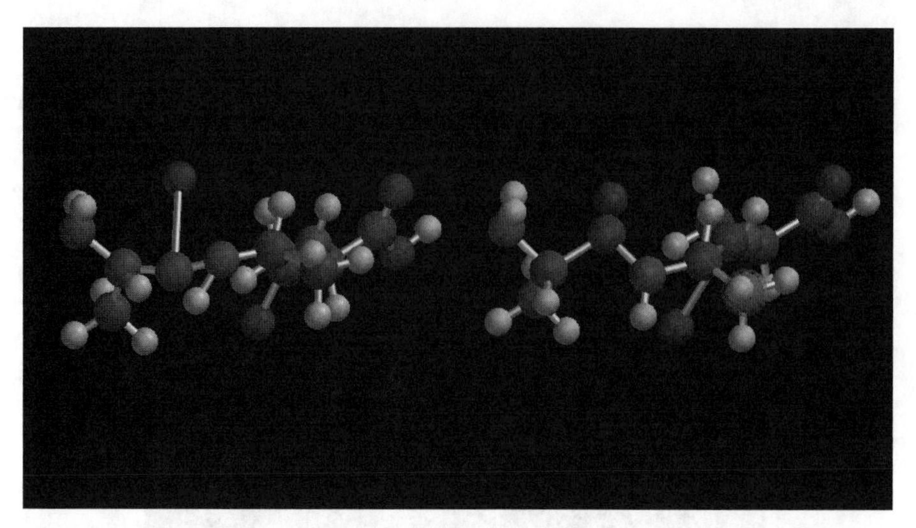

Figure 7.3 Snapshot of the amide I modes of trialanine with deuterated amide groups *in vacuo*. The vibrations were obtained by performing a normal mode calculation based on a force field obtained with a DFT calculation at a BL3LYP 6-31G** level of theory. The program is part of the TITAN software from Schrödinger, Inc. The in-phase combination is depicted in the right and the out-of-phase combination in the left figure.

calculation was performed for a structure that resulted from an optimization with the dihedral angles arrested to $(\phi, \psi) = (-68°, 150°)$. A normal mode calculation was then performed for the optimized conformation. The eigenvector assignable to the lower wavenumber mode of amide I displays an out-of-phase combination of the two interacting modes, with a larger amplitude for the C-terminal amide I mode, whereas the higher wavenumber mode is an in-phase combination, with a larger amplitude for the N-terminal. The mixing with the C-terminal CO s vibration is minimal.

The second concept invokes quantum mechanics and considers each amide I mode as a harmonic oscillator, which can be described by the Schrödinger equation:

$$\hat{H}_0 \prod_j |v_j\rangle = E_{v,j} \prod_j |v_j\rangle, \tag{7.1}$$

where $\hat{H}_0$ denotes the harmonic oscillator Hamiltonian and $|v_j\rangle$ is the wavefunction of the jth oscillator in the vth state in Dirac notation, and v represents the vibrational quantum number. Coupling between excited states of individual oscillators can be accounted for by the Hamiltonian (87, 88):

$$\hat{H}^{ex} = \sum_{\substack{i,j \\ i \neq j}} f_{ij}^{ex}(Q_i, Q_j) Q_i Q_j, \tag{7.2}$$

where f_{ij}^{ex} is the Q-dependent interaction potential between the ith and jth oscillator. The product of nuclear coordinates in Equation 7.2 can be expressed as:

$$Q_i Q_j = \frac{h}{2\pi c}\sqrt{\frac{1}{\tilde{\nu}_i \tilde{\nu}_j}}\left(b_i^+ + b_i^-\right)\left(b_j^+ + b_j^-\right), \tag{7.3}$$

where $\tilde{\nu}_i$ and $\tilde{\nu}_j$ are the intrinsic wavenumbers of the interacting oscillators, b_i^+, b_i^- and b_j^+, b_j^- denote the corresponding creation and annihilation operators, h is Planck's constant, and c is the velocity of light. Mixing between, for example, the excited vibrational states $|1,0\rangle$ and $|1,0\rangle$ of the two amide I modes of the aforementioned tripeptide is accounted for solely by the product terms, b_1^+, b_2^- and b_1^-, b_2^+. Since the interaction energy might be on the order of, or even larger than, the wavenumber difference between the interacting modes, the mixing between two excited quantum states can be substantial. As a consequence of this interaction, new delocalized eigenstates, so-called excitonic states, are created, which read as:

$$|\varphi_\alpha(1)\rangle = c_{\alpha\beta}|1_\beta\rangle, \tag{7.4}$$

where the use of Greek indexes is indicative of the Einstein convention for summations. The number 1 reflects the fact that only individual states associated with $\nu = 1$ are considered.

The quantum mechanical model assumes that the individual oscillators are independent of each other in the absence of coupling. One might question this assumption since it is easy to believe that adjacent amide I modes share the same internal coordinates. However, extended DFT calculations by Cho and associates on tripeptides have demonstrated the equivalence of the classical normal mode analysis and the excitonic approach (38). This is important, because the latter can be more easily used for the quantitative analysis of experimental amide I band profiles, as described in more detail in section 7.4.3 of this chapter.

Mode mixing yields asymmetric band profiles with different wavenumber positions in IR and Raman spectra. This particularly holds for both parallel and antiparallel β-sheets (7) but, to a lesser extent, also for all conformations assignable to the upper left quadrant of the Ramachadran plot. On the contrary, IR and Raman amide I bands nearly coincide for helical conformations (89, 115). Vibrational mixing (excitonic coupling) is the main cause for the structural sensitivity of amide I.

7.2.2 Amide II, III, and S

The spectrum of dialanine in water, shown in Figure 7.1, exhibits amide II and III at 1574 and 1277 cm^{-1}, respectively (the spectra were recorded in our laboratory). The corresponding normal modes have in common that they both

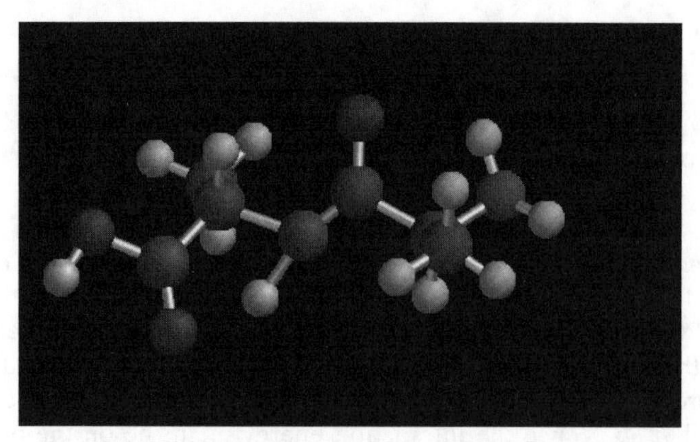

Figure 7.4 Snapshot of the amide II mode of dialanine *in vacuo*. The vibration was obtained by performing a normal mode calculation based on the semiempirical force field, AM1, which is part of the TITAN software from Schrödinger, Inc. The eigenvector exhibits the characteristic out-of-phase vibration of NH ib and CN s.

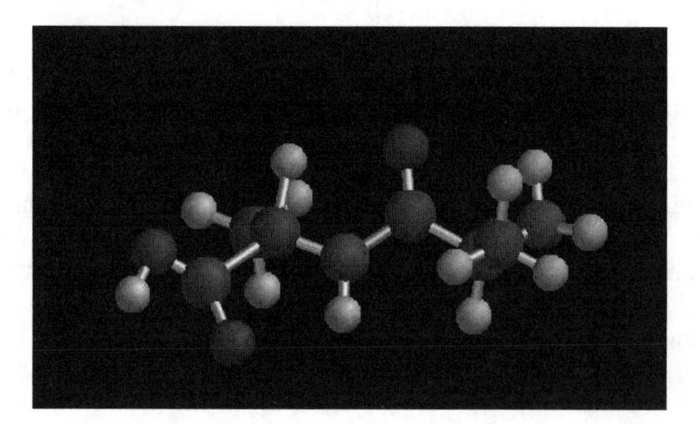

Figure 7.5 Snapshot of the amide III mode of dialanine *in vacuo*. The vibration was obtained by performing a normal mode calculation based on the semiempirical force field, AM1, which is part of the TITAN software from Schrödinger, Inc. The eigenvector exhibits the characteristic in-phase vibration of NH ib and CN s combined with an in-plane vibration of $C_{\alpha}H$.

contain substantial CN s and NH ip b (Figs. 7.4 and 7.5) (16). The combination is out of phase for amide II and in phase for amide III. The substitution of NH by ND in D_2O changes the normal mode composition dramatically. For dialanine, the amide II′ appears now at 1481 cm^{-1}. The respective normal mode pattern reads as a mixture of CN s and side-chain deformation vibrations. Amide III′ appears at a much lower wavenumber position (1004 cm^{-1}) than amide III, owing to the larger mass of deuterium.

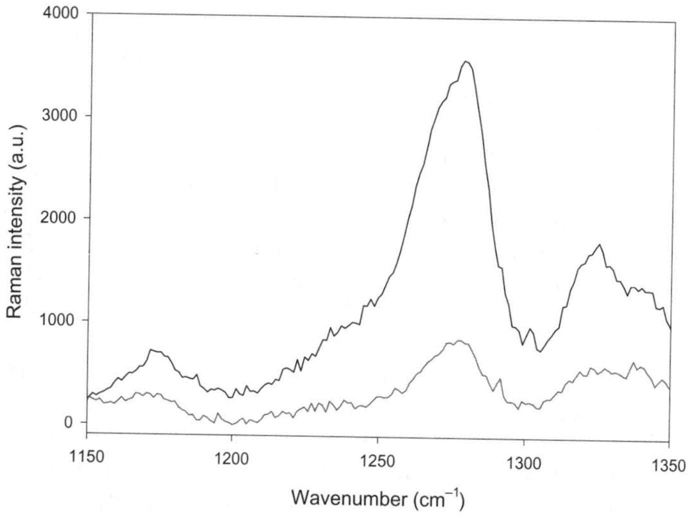

Figure 7.6 Zoom in of the amide III region of the polarized Raman spectra of zwitterionic H-Ala-Ala-OH in H_2O measured at neutral pH (black solid line: polarization parallel to the polarization of excitation; gray line: polarization perpendicular to the polarization of excitation).

Amide II is generally not very structure sensitive, but it is an ideal tool for probing the access of amide groups for H $\leftrightarrow$ D exchange in protein folding experiments (104). Amide III is a much more complex mode because of the interaction between NH ib (in-plane bending) and $C_\alpha H$ deformation modes. This gives rise to a multiplicity of the amide III band in both IR and Raman spectra. The assignment of individual components of the respective band profile is subject to a controversy that has been recently reviewed (87). Figure 7.6 exhibits three amide III bands in the polarized Raman spectra of dialanine, which were identified by means of a DFT-based normal mode calculation (92). In the present context, it is more important to mention that the wavenumber of one of its components has been shown to depend on the dihedral angle ψ owing to a mixture of NH ip b and the $C_\alpha H$ (3, 67, 92). In principle, this property makes amide III an important tool for secondary structure analysis, which has mostly been exploited by the Asher group using UV resonance Raman spectroscopy. Their work is briefly reviewed in section 7.3 of this chapter.

For polypeptides, it is presently unclear whether the vibrational states of amide II and amide III of different peptide groups are mixed, owing to the mechanism that yields a delocalization of amide I. The transition dipole moment for amide III is weak, so that only through bond, nearest neighbor coupling can be expected to be significant. Amide II, however, exhibits a significant transition dipole moment, so that one would expect nearest neighbor coupling, as well as non-nearest neighbor coupling. However, UV resonance Raman experiments carried out by Asher and coworkers indicate the contrary (75, 77). They found that for a series of short peptides and for the thermally unfolded state of a longer 21-residue polyalanine peptide dissolved in H_2O, D_2O, and 50%/50% H_2O/D_2O, the UV resonance Raman band profiles of amide II and amide III measured for the H_2O/D_2O mixture can be reproduced by adding 50% of the respective spectra obtained with the pure solutions. For the polyalanine peptide, they reported the same finding for amide I (75). These observations led them to conclude that vibrational mixing is generally absent for amides II and III, and also for amide I, if the peptide adopts a PPII conformation (78). With respect to amide I, the validity of this conclusion has recently been challenged by Measey and Schweitzer-Stenner (73), who combined polarized Raman and IR spectroscopy to show that the amide I modes of the octapeptide $(AAKA)_2$ in aqueous solution are coupled and that one can nevertheless reproduce the amide I band profile of the respective H_2O/D_2O mixture by the appropriate combination of profiles measured for the pure solvents. The above-mentioned DFT-based normal mode calculations for an extended PPII-like conformation of trialalaine yielded substantial vibrational mixing for both amides II and III. For amide II, this is demonstrated by the displacements visualized in Figure 7.7. The reason for the discrepancy between experiment and theory is currently investigated in our laboratory. The clarification of this issue is important for determining how these two bands can be used for structure analysis.

Spiro and coworkers identified another structure-sensitive backbone mode, which appears around $1400\,cm^{-1}$ (Fig. 7.1), which they term amide S. It is generally weak in IR spectra and of moderate intensity in visible Raman spectra, but it becomes resonance enhanced with UV excitation (50). Figure 7.8 illustrates this for diglycine in H_2O (103). The band is absent in the spectra of helical structures. The assignment of this band was initially the subject of a heavy debate (106, 122), but it eventually became clear that a $C_\alpha H$ bending mode contributes substantially to the eigenvector of this mode. Figure 7.9 exhibits the corresponding eigenvector calculated for dialanine. Based on the aforementioned UV resonance Raman spectra on peptides in a H_2O/D_2O mixture, Mix et al. proposed that this mode is localized and that its intensity can be used to estimate the number of residues in extended, nonhelical structures (77). This characterizes amide S as an important maker band for unfolded peptides and proteins.

7.3 BASIC THEORY OF VIBRATIONAL SPECTROSCOPY

The paragraphs below briefly describe the underlying physics of the spectroscopies discussed in this article, that is, IR, VCD, Raman, and ROA. More extended theoretical approaches based on the consideration of vibrational coupling between, for example, amide I vibrations of different peptide groups

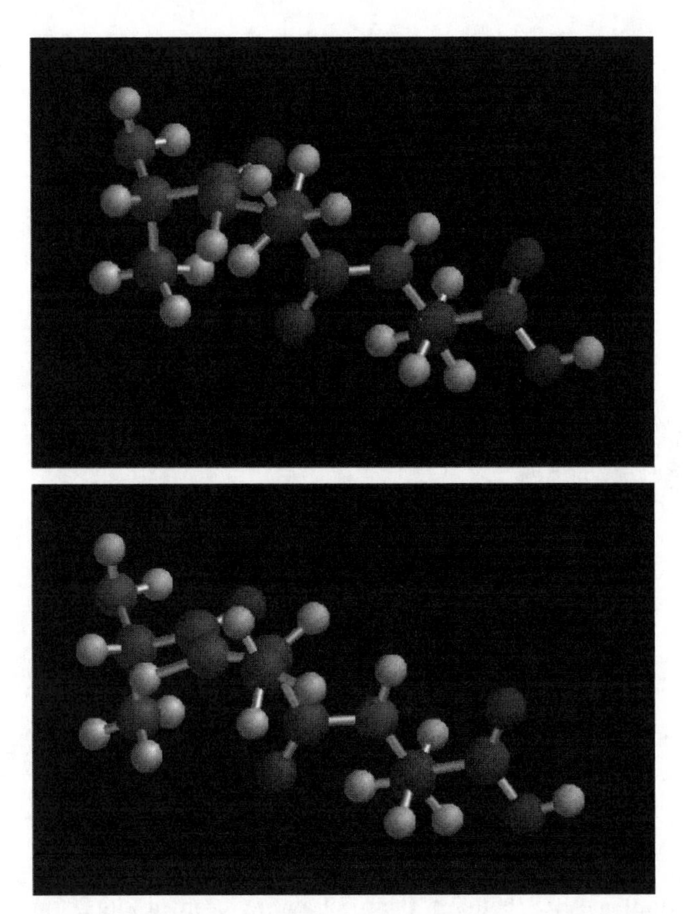

Figure 7.7 Two snapshots of one of the amide II modes of trialanine *in vacuo*. The vibrations were obtained by performing a normal mode calculation based on a force field obtained with a DFT calculation at a BL3LYP 6-31G** level of theory. The program is part of the TITAN software from Schrödinger, Inc. The snapshots reveal vibrational mixing with the larger amplitude residing on the N-terminal residue.

are discussed in the sections describing its utilization for the structure analysis of unfolded peptides and proteins.

7.3.1 IR Spectroscopy

The underlying physics of IR absorption is textbook knowledge for spectroscopists. Therefore, we confine ourselves on presenting those equations and definitions that are relevant for the subsequent sections of this chapter (e.g., section 7.4.3). The probability for the transition between two vibrational levels $|v''\rangle$ and $|v'\rangle$ of a harmonic oscillator is given by the expression:

$$P_{v'v''}(q) = \vec{\mu}'\langle v'|q|v''\rangle, \tag{7.5}$$

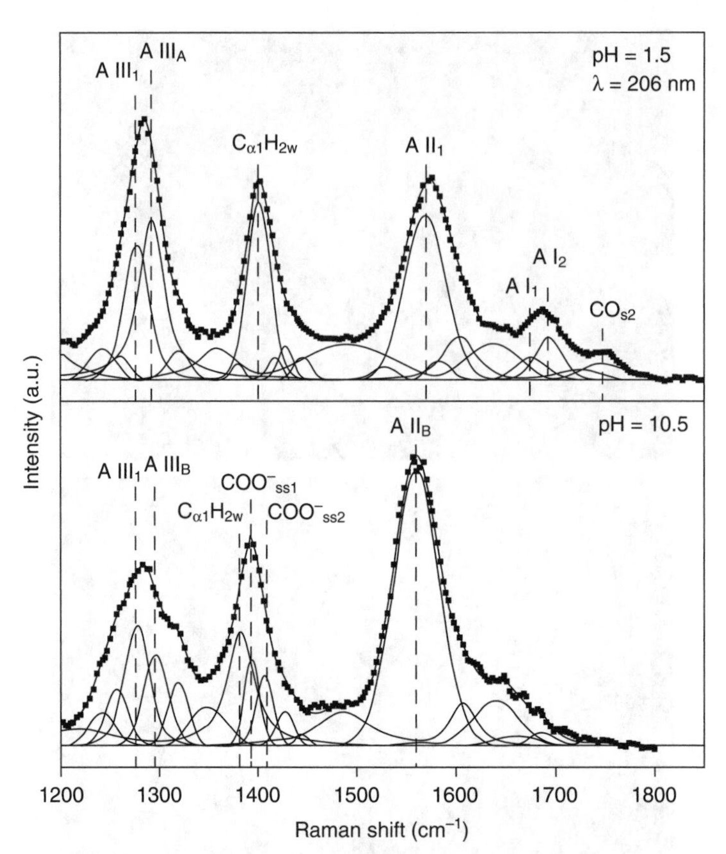

Figure 7.8 UV resonance Raman spectra of glycylglycine in aqueous buffer adjusted to pH 1.5 (top) and 10.5 (bottom). The spectra were recorded with 206 nm excitation. The most prominent Raman bands are indicated. The band profiles result from a self-consistent decomposition of the spectra. The spectra were taken from Reference 103 and modified.

where $\vec{\mu}' = \left\langle g \left| \frac{\partial \vec{\mu}}{\partial q} \right| g \right\rangle$ is the expectation value of the first derivative of the electronic dipole moment $\vec{\mu}$ with respect to the normal coordinate q of the considered mode in the electronic ground state $|g\rangle$. This transition dipole moment generally forms angles between 15 and 20° with the peptide carbonyl bond and points toward the CN bond (58, 115, 123). The second matrix element, $\langle v'|q|v''\rangle$, accounts for the transition between two vibrational states. For a harmonic oscillator, the vibrational states can be ascribed by Hermitian polynominals, and the selection rule is $v' = v'' \pm 1$. The absorptivity (e.g., the intensity of a given IR band in the spectrum) is proportional to the oscillator strength, $D_{v',v''}$, of the transition $v'' \rightarrow v'$:

$$D_{v'v''} = |P_{v'v''}|^2. \tag{7.6}$$

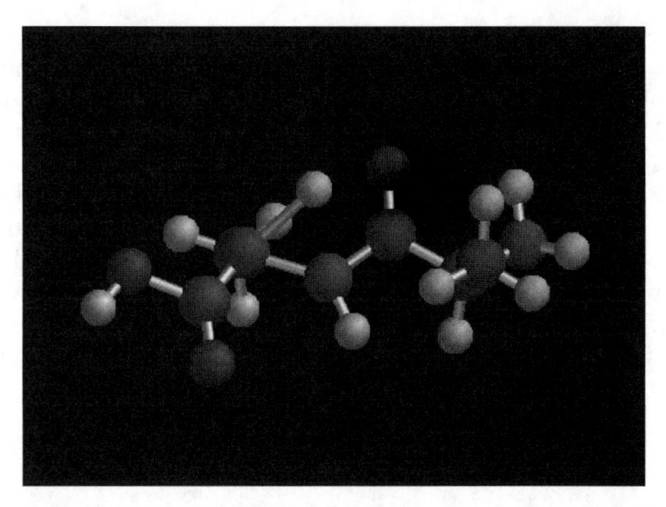

Figure 7.9 Snapshot of the amide S mode of dialanine *in vacuo*. The vibration was obtained by performing a normal mode calculation based on the semiempirical force field, AM1, which is part of the TITAN software from Schrödinger, Inc. The eigenvector exhibits the characteristic out-of-phase vibration of NH ib and $C_{\alpha}H$s.

The oscillator strength is now directly proportional to the integrated intensity of an absorption band. The absorptivity at a given wavenumber is written as:

$$\varepsilon(\tilde{\nu}) = c_{ir}\tilde{\nu}_0 P^2_{v'v''} \cdot f(\tilde{\nu} - \tilde{\nu}_0), \tag{7.7}$$

where $c_{ir} = 1.0869 \cdot 10^{38}\,M^{-1}\,esu^{-2}\,cm^{-2}$ (87, 95). The function $f(\tilde{\nu} - \tilde{\nu}_0)$ describes the normalized band profile at the wavenumber position, $\tilde{\nu}_0$, which is Voigtian in most cases but can sometimes be approximated by a Gaussian function (Chapter 8 and ref. 89). Hence, the oscillator strength of a $0 \rightarrow 1$ absorption can be inferred from the measured band profile by:

$$D_{10} = \frac{1}{c_{ir}\tilde{\nu}_0} \int_{-\infty}^{\infty} \varepsilon(\tilde{\nu})\, d\tilde{\nu}. \tag{7.8}$$

The amide I IR band is traditionally used to explore the secondary structure of proteins, as described in detail in earlier review articles (7, 58, 108) and also by Natalello and Doglia in chapter 8 of this book. Generally, this analysis is based on the assumption that the oscillator strength of amide I is independent of the neighboring residues and of the secondary structure. Measey et al., however, have shown that the amide I transition dipole moment of XA and AX dipeptides depend on the choice of residue X and can vary between $2.4 \cdot 10^{-19}$ and $3.8 \cdot 10^{-19}$ esu cm (72).

The harmonic approach does not fully describe the vibrational properties of peptide modes. Even the best normal mode calculations performed with very complex basis sets generally overestimate the wavenumbers of amide I modes (12, 44, 61–63). Gregurick and associates have provided computational evidence for anharmonic coupling between high-frequency amide I modes (36). Moreover, the fact that the wavenumbers of amides I, II, and III are all temperature dependent is indicative of anharmonic coupling with low-frequency modes associated with hydrogen bonding to solvent molecules (4). The anharmonicity of amide I has directly been probed by IR femtosecond spectroscopy (55). Generally, this can be accounted for by expanding the potential function of the Hamiltonian of a distinct vibrational mode q_i to fourth order:

$$V_i = V(q_{i0}) + \frac{1}{2}\frac{\partial^2 V}{\partial q_i} q_i^2 + \frac{1}{3!}\sum_j \left(\frac{\partial^3 V}{\partial q_i^2 \partial q_j} q_i^2 q_j \right) + \frac{1}{4!}\left(\frac{\partial^4 V}{\partial q_i^2 \partial q_j^2} q_i^2 q_j^2 \right) \quad (7.9)$$

where q_j denotes the vibrational mode to which q_i is anharmonically coupled. The first term on the right side of the expansion is the electronic energy for the equilibrium configuration of the molecule, the second term describes the harmonic part of the potential, the cubic term accounts for anharmonic coupling with low-frequency (wavenumber) modes, and the quartic term reflects anharmonic coupling with modes of similar energy (e.g., between amides I and II [36]).

7.3.2 VCD Spectroscopy

VCD spectroscopy involves the determination of the difference of IR absorption for right-handed and left-handed circular polarized light. In contrast to ECD (cf. chapter 10 in this book), VCD probes the local chirality of a molecule. The underlying physics of VCD spectroscopy has been described in detail by Nafie (79). For the current context, it is sufficient to outline the quantum mechanical description of the rotational strength associated with the $v'' \rightarrow v'$ transition of a vibrational mode, q, which is generally written as:

$$R_{v'v''} = \mathrm{Im}\langle g|\vec{\mu}'|g\rangle\langle g|\vec{m}'|g\rangle|\langle v'|q|v''\rangle|^2 \quad (7.10)$$

where $\vec{m}' = \dfrac{\partial m}{\partial q}$ is the magnetic transition dipole moment of the mode, q, in the electronic ground state. Nafie showed that it is convenient to write the two matrix elements of Equation 7.9 as tensors in the Cartesian space. For $0 \rightarrow 1$ transitions, Equation 7.9 is written as:

$$R_{01}^{a} = \frac{\hbar}{2}\sum_A P_{\alpha\beta}^{A} s_{A\alpha,\alpha} \sum_A M_{\gamma\beta}^{A'} s_{A\gamma,\alpha} \quad (7.11)$$

The atomic polar tensor:

$$P_{\alpha\beta}^{A} = \left(\frac{\partial \mu_\beta}{\partial R_{A\alpha}} \right)_{R=0}, \tag{7.12}$$

describes the derivative of the β-Cartesian component of the electronic dipole moment with respect to the nuclear displacement $R_{A\alpha}$ along the coordinate α. Following Nafie (79), we used the Einstein convention for the subscripts, that is, repeated Greek subscripts are to be summed over all three Cartesian coordinates. The corresponding atomic axial tensor of the magnetic moment reads as:

$$M_{\alpha\beta}^{A} = \left(\frac{\partial m_\beta}{\partial \dot{R}_{A\alpha}} \right)_{R=0}. \tag{7.13}$$

The vector connects Cartesian and normal coordinates:

$$s_{A\alpha,a} = \left(\frac{\partial R_{A\alpha}}{\partial q_a} \right) = \left(\frac{\partial \dot{R}_{A\alpha}}{\partial p_a} \right), \tag{7.14}$$

where p_a is the momentum associated with the normal mode q_A. If a vibration does not induce a magnetic dipole moment, or if it is perpendicularly oriented to the electronic dipole moment, the rotational strength is zero. In section 7.4.2, we show that rotational strength can be induced by excitonic coupling even in the absence of an intrinsic magnetic moment. The relationship between rotational strength and the measured VCD signal of a given mode is written as:

$$\Delta\varepsilon(\tilde{\nu}) = c_{rot} \tilde{\nu}_0 R_{v'v''} \cdot f(\tilde{\nu} - \tilde{\nu}_0), \tag{7.15}$$

where $c_{rot} = 4.348 \cdot 10^{38}$ ($M \cdot cm^2 \cdot esu^2$) (87). Δε can be positive or negative, thus reflecting the sign of $R_{v'v''}$. The rotational strength of a given $0 \rightarrow 1$ transition can be experimentally obtained from the integrated VCD profile:

$$R_{10} = \frac{1}{c_{rot} \tilde{\nu}_0} \int_{-\infty}^{\infty} \Delta\varepsilon(\tilde{\nu}) d\tilde{\nu}. \tag{7.16}$$

It should be mentioned that this formalism accounts solely for the rotational strength of fundamentals. For over and combination tones, higher order terms of the respective expansions of the electronic and magnetic dipole moment have to be considered.

7.3.3 Raman Spectroscopy

In first order, Raman scattering is caused by the electric field component, $\vec{E}$, of an electromagnetic field, which induces a dipole moment, $\vec{P}$:

$$\vec{P} = \hat{\alpha} \vec{E}, \tag{7.17}$$

where $\hat{\alpha}$ is the (second rank) polarizability tensor. For conventional, linear Raman scattering, the dependence of α on nuclear coordinates is accounted for by a Taylor series:

$$\hat{\alpha} = \hat{\alpha}(q_0) + \frac{\partial \hat{\alpha}}{\partial q}(q_0)q, \tag{7.18}$$

where q_0 denotes the equilibrium value of the normal coordinate q. The first term on the right side of the equation accounts for Rayleigh scattering, and the second one for Stokes and anti-Stokes scattering. In an arbitrary coordi nate system, the Raman tensor, $\hat{\alpha}' = \frac{\partial \hat{\alpha}}{\partial q}(q_0)q$, might contain on-diagonal and off-diagonal matrix elements, depending on the symmetry of the considered vibrational mode and the point group of the molecule. However, for nonresonance and preresonance Raman scattering, the tensor can be diagonalized by an orthogonal transformation to yield:

$$\hat{\alpha}'_d = \begin{pmatrix} \alpha'_{xx} & 0 & 0 \\ 0 & \alpha'_{yy} & 0 \\ 0 & 0 & \alpha'_{zz} \end{pmatrix}. \tag{7.19}$$

Pajcini et al. identified the principal axes of the Raman tensor (PARTs) of the main backbone modes of a diglycine peptide for visible (nonresonant) and UV (resonant) excitation (82). They are shown in Figure 7.10. The diagonalized Raman tensor can be decomposed into an isotropic (with the tensor element a) and anisotropic part (109):

$$\begin{pmatrix} \alpha_{xx} & 0 & 0 \\ 0 & \alpha_{yy} & 0 \\ 0 & 0 & \alpha_{zz} \end{pmatrix} = \begin{pmatrix} a & 0 & 0 \\ 0 & a & 0 \\ 0 & 0 & a \end{pmatrix} + \begin{pmatrix} \alpha_{xx} - a & 0 & 0 \\ 0 & a_{yy} - a & 0 \\ 0 & 0 & \alpha_{zz} - a \end{pmatrix}. \tag{7.20}$$

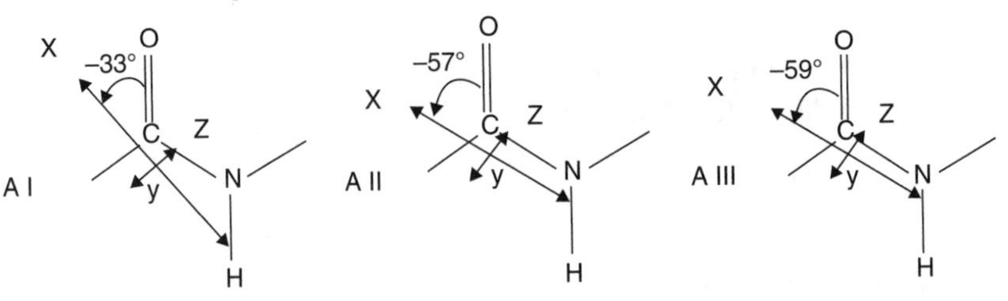

Figure 7.10 Representation of the orientation PARTs for the amide I, II, and III vibration of glycylglycine for visible excitation (514 nm). The angles are shown relative to the amide carbonyl bond. The lengths of the axes represent the (relative) Raman tensor values. Taken from Reference 82 and modified.

The differential Raman cross-section for a sample of nonoriented molecules can be obtained from invariants of the above tensor, which are independent of the choice of the molecular coordinate system:

$$\begin{aligned} I_{\perp} &= 3\xi\gamma'^2 \\ I_{\parallel} &= \xi\left(45\beta_s'^2 + 4\gamma'^2\right)' \end{aligned} \tag{7.21}$$

where $I_{\perp}$ and I_{II} are the intensities of Raman scattering measured perpendicular and parallel to the polarization of the exciting radiation, respectively. The tensor invariants are defined as follows:

$$\begin{aligned} \beta_s' &= \frac{1}{3}\alpha_{\alpha\alpha}' \\ \gamma'^2 &= \frac{1}{2}\left[3\alpha_{\alpha\beta}'\alpha_{\alpha\beta}' - \alpha_{\alpha\alpha}'\alpha_{\beta\beta}'\right] \end{aligned} \tag{7.22}$$

Note that the Einstein convention is again used for the subscripts. The constant, ξ, is written as:

$$\xi = \frac{\pi^2\tilde{\nu}_L(\tilde{\nu}_L - \tilde{\nu}_R)^3}{45\varepsilon_0^2}, \tag{7.23}$$

where $\tilde{\nu}_L$ and $\tilde{\nu}_R$ are the wavenumbers of the exciting radiation and the considered Raman-active vibration, respectively. ε_0 is the dielectricity constant *in vacuo*.

The theory described thus far accounts for an increase of Raman cross section with approximately the fourth power of the wavenumber of the exciting laser beam. This is generally termed nonresonance Raman scattering. If, however, the excitation wavenumber approaches the resonance energy of an optical transition, the Raman tensor itself exhibits a dispersion with respect to the excitation wavenumber. For peptides, this generally occurs if the molecule is excited in the intermediate (for aromatic side chains) and far-UV region (for amide modes). This is called resonance Raman scattering (66). While the theoretical treatment of resonance Raman scattering on aromatic side chains is complicated by interstate vibronic coupling, the resonance enhancement of amide modes, which is associated with the lowest $\pi \rightarrow \pi^*$ transitions of the peptide chromophore, can be accounted for by solely considering the Franck–Condon coupling (90). If one neglects multimode mixing and the possibility of degenerate excited states, the corresponding Raman tensor reads as:

$$\alpha_{\sigma'\rho'}' = \sum_{m_\sigma}\sum_{v_r^m}\left(\frac{\langle g|R_\sigma|m_\sigma\rangle\langle m_\sigma|R_\sigma|g\rangle}{\left(E_m + v_r^m\tilde{\nu}_r^m - \tilde{\nu}_L - i\Gamma_m\right)}\langle 1_r^g|v_r^m\rangle\langle v_r^m|0_r^g\rangle\right), \tag{7.24}$$

where $\langle g|R_\sigma|m_\sigma\rangle$, with $\sigma = x,y,z$, is the electronic dipole moment for the transition from the ground state into the excited state $|m_\sigma>$. The Franck–Condon overlap integrals $\langle 1_r^g | v_r^m \rangle \langle v_r^m | 0_r^g \rangle$ in the numerator account for transitions between the vibrational ground state of the electronic ground state and the manifold of vibrational states of the considered Raman-active vibration, q_r. Each of the vibronic eigenstates of $|m_\sigma>$ contribute a resonance position $E_m + v_r^m \tilde{\nu}_r^m$ to the energy denominator, where $\tilde{\nu}_r^m$ is the wavenumber of the Raman-active vibration in the excited state, $\tilde{\nu}_L$ is the wavenumber of the exciting radiation, and Γ_m is the damping factor associated with the lifetime of the excited state. It should be noted that Equation 7.24 applies only for homogeneous ensembles of molecules. Inhomogeneous broadening can be accounted for by convoluting the Lorentzian profile used for Equation 7.24 with a Gaussian profile.

It should be noted that the actual values for $\alpha_{\sigma'\rho'}$ depend on the choice of the molecular coordinate system. If, however, the contributions from transitions into a single electronic state dominate, it would be prudent to identify one of the coordinate axes with the polarization of the respective dipole transition. In this case $\sigma' = \rho' = \sigma$, so that the tensor has only one diagonal element, and as a consequence, the depolarization ratio $\rho = I_\perp = I_{//}$ is 1/3 (16, 17). If two electronic transitions with different polarization directions contribute, the tensor may exhibit diagonal, as well as nondiagonal elements, depending on the choice of the coordinate system (90).

When the wavenumber of the exciting laser beam approaches a resonance position, the Raman cross section can increase by orders of magnitudes. As a consequence, the bands of resonance Raman-active modes dominate the spectrum. Resonance Raman spectroscopy can therefore be used to selectively probe chromophores in very complex biological molecules, for example, metalloporphyrins in heme proteins or aromatic residues in all types of proteins (2). An excitation in the deep UV (~200 nm) causes a resonance enhancement predominantly of amide II and III owing to the vibronic coupling of these modes to the $\pi \rightarrow \pi^*$ transition between the π-HOMO (highest occupied molecular orbital) and LUMO (lowest unoccupied molecular orbital) of the peptide group (15). This transition is termed S1 in the literature (86). This enhancement is illustrated by the UV resonance Raman spectrum of diglycine in Figure 7.8 (103). Vibronic coupling means that the excited state is displaced along the normal coordinate of the considered Raman-active vibration, so that the overlap integrals in Equation 7.24 are both unequal to zero.

Interestingly, preresonance scattering can occur in far-off resonance. As shown by Schweitzer-Stenner et al., the dispersion of the Raman cross section of amide II and amide III in the visible region exceeds the $\tilde{\nu}_L^4$ dependence predicted by classical theory (Eq. 7.23) (92). The possibility of such preresonance enhancements has been predicted by Shorigyn (64), who described it by:

$$\alpha'_{\sigma'\rho'} = \sum_i \frac{K_{\rho\sigma}^i(r) E_i}{E_i^2 - \tilde{\nu}_L^2}, \quad (7.25)$$

where the sum runs over all vibronic contributions. $K^{i}_{\rho\sigma}$ contains the dipole moment of the *i-th* transition and the respective Franck–Condon integrals for the respective Raman mode.

7.3.4 ROA

In principle, ROA involves the measurement of the difference between the Raman scattering induced by left-handed and right-handed circular polarized light. However, it is a more complex spectroscopic tool, owing to the fact that four different setups can be employed with respect to the polarization properties of excitation and scattered light. They are schematically visualized in Figure 7.11 (79, 80). Among the setups depicted in this figure, incident circular polarization (ICP) and dual circular polarization (DCP_1) are the most prominent ones. The former involves the measurement of the difference between the intensities of unpolarized, scattered light obtained for right-handed and left-handed circular polarized excitation. In the DCP_1 mode, the difference between right-handed and left-handed polarized, scattered light obtained with excitation of the same respective polarization is determined. Generally, the theoretical treatment of ROA is very complex. Details can be inferred from several review articles and books by Barron and Nafie, who have both pioneered this field (8, 80). Here, for illustrative purposes, we confine ourselves on the nonresonant scattering. The ROA and Raman intensities for ICP can then be written as:

$$I_u^R - I_u^L = 8\frac{\xi}{c}\left[3\beta(G')^2 + \beta(A)^2\right], \quad I_u^R + I_u^L = \xi\left[45\beta_s^2 + 7\gamma^2\right] \tag{7.26}$$

where the indexes R, L, and u represent right-handed, left-handed, and unpolarized light, respectively. The superscripts denote the polarization state of the incident light, while the subscripts denote that of scattered light. c is the velocity of light *in vacuo*. Apparently, the second equation can be directly derived from Equation 7.21. The tensor invariants for the ROA intensity can be written as:

$$\beta(G')^2 = \frac{1}{2}\left(3\alpha'_{\alpha\beta}G'_{\alpha\beta} - \alpha'_{\alpha\alpha}G'_{\beta\beta}\right), \quad \beta(A)^2 = \pi c\tilde{\nu}_L \alpha'_{\alpha\beta}\varepsilon_{\alpha\gamma\delta}A_{\gamma,\delta\beta} \tag{7.27}$$

where the Einstein convention for the subscript is used again. $G'_{\alpha\beta}$ and $A_{\gamma,\delta\beta}$ are the components of the magnetic dipole and the electric quadrupole moment tensor, respectively. $\varepsilon_{\alpha\gamma\delta}$ is the Levi-Civitá tensor. It should be noted that the above formalism is valid only for a 180° backscattering geometry. As VCD involves the change of magnetic dipole moment along normal coordinates, ROA is related to the capability of the exciting electromagnetic radiation to induce a magnetic dipole moment, which changes along nuclear coordinates.

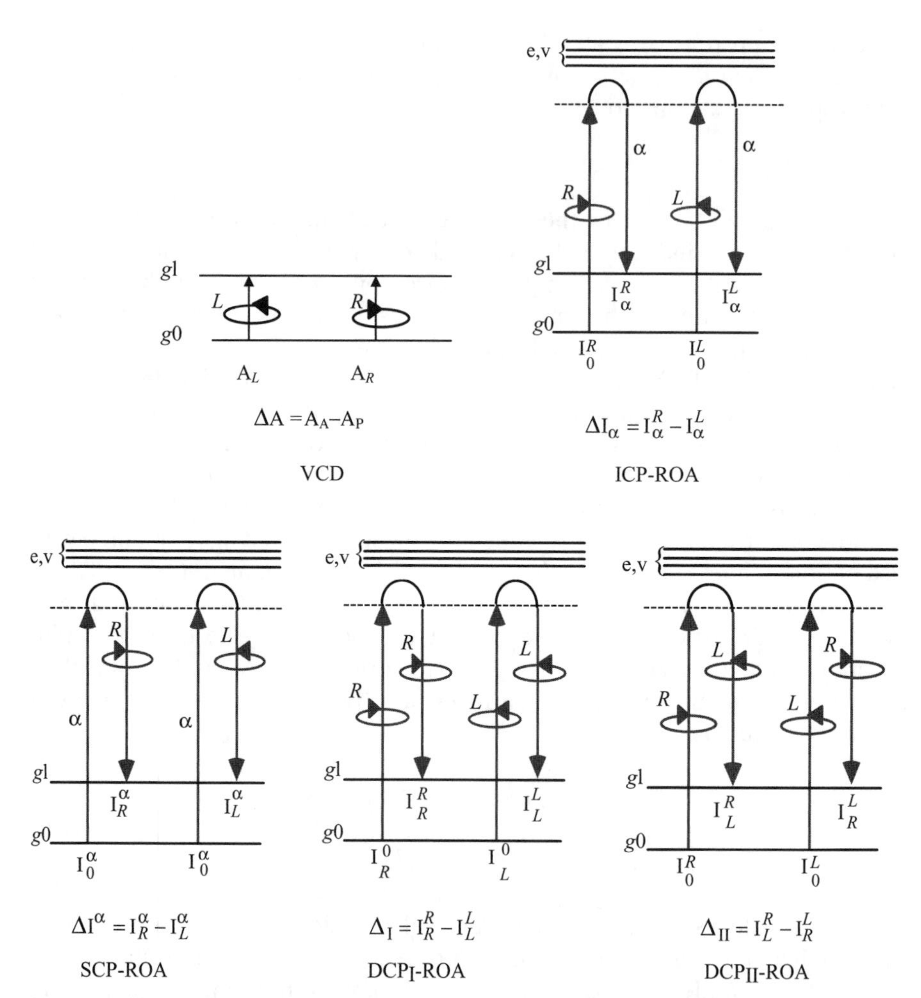

Figure 7.11 Diagrammatic representation illustrating the basic forms of circular polarized IR and ROA. Energy level diagram illustrating the definitions of VCD and four forms of ROA for a molecular transition from the zeroth vibrational level of the ground electronic state, g0, to the first vibrational level of the ground electronic state, g1. The figure was kindly provided by Dr. L. A. Nafie.

7.4 VIBRATIONAL SPECTROSCOPY ON UNFOLDED PEPTIDES AND PROTEINS

7.4.1 UV Resonance Raman Spectroscopy

The application of UV resonance Raman spectroscopy to biological systems, particularly proteins, has been pioneered by the research groups of Asher, Kitagawa, and Spiro. These applications have been documented by excellent review articles (1, 2, 57), which we do not intend to duplicate in this context.

Instead, we focus on what is relevant for the exploration of unfolded peptides and proteins.

Generally, the exploration of unfolded peptides by UV resonance Raman spectroscopy has begun only recently, but an earlier study by Song and Asher is noteworthy in this context (105). These authors used 218-nm excitation to obtain the respective Raman spectra of PGA and PLL. This excitation lies within the low-energy wing of backbone absorption associated with the S1 transition, which reflects transitions into the lower vibrational states of the excited electronic state (70). The secondary structure of the investigated peptides can be switched between helical, β-sheet, and what the authors called random coil by changes of pH and temperature. For reasons to be given below, we will, instead, use the term "disordered" rather than "random coil" throughout this chapter. Figure 7.12 shows the respective UV resonance Raman spectra of PLL as an example. Apparently, positions and intensities of all prominent amide bands in the spectrum of the disordered state differ significantly from what one observes for the helical state. The differences between the spectra of the β-sheet and the disordered state are not that pronounced but still significant. Compared with the spectra of these two secondary structures, amide III appears much weaker and at substantially higher wavenumbers in the spectrum of the helical structure. Concomitantly, the amide S band at $1400\,cm^{-1}$ has disappeared. It should be noted that ECD measurements as well as later vibrational spectroscopy investigations strongly suggest that the individual residues of PLL and PGA predominantly sample a PPII-like conformation in the disordered state (27, 107, 113), which would make them much more structurally confined than suggested by the canonical random coil model. Hence, one can interpret the spectrum in the upper panel of Figure 7.12 as a model UV resonance Raman spectrum of PPII.

A more systematic investigation of UV Raman spectra of unfolded peptides has been conducted by the Asher group over the last 7 years (3, 4, 74, 75, 78). Their research first focused on validating an earlier hypothesis of Lord, stipulating that the wavenumber of the amide III band of a distinct peptide group reflects mostly the ψ-angle of the adjacent residue on the C-terminal side (67). Asher et al. used UV resonance Raman spectroscopy to determine the amide III band position for several small peptides (containing one or two peptide groups) (3). In addition, they performed *ab initio* calculations for the model peptide, NH_2–$CHCH_3$–CONH–CH_3, *in vacuo* at a MP2 6-31G(d) level of theory. The ψ-angle of the alanine residue was constrained to different values, while the remaining coordinates of the peptide were subject to geometry optimization. Thus, they observed that the amide III wavenumber is linearly dependent on the distance between the amide and C_α proton, so that the relationship between ψ and the amide III wavenumber can be described by a periodic function.

The above results suggest that the amide III band can, in principle, be used as a very sensitive tool to probe the residue conformations of peptides and proteins. Unfortunately, its interpretation is somewhat complicated by the fact

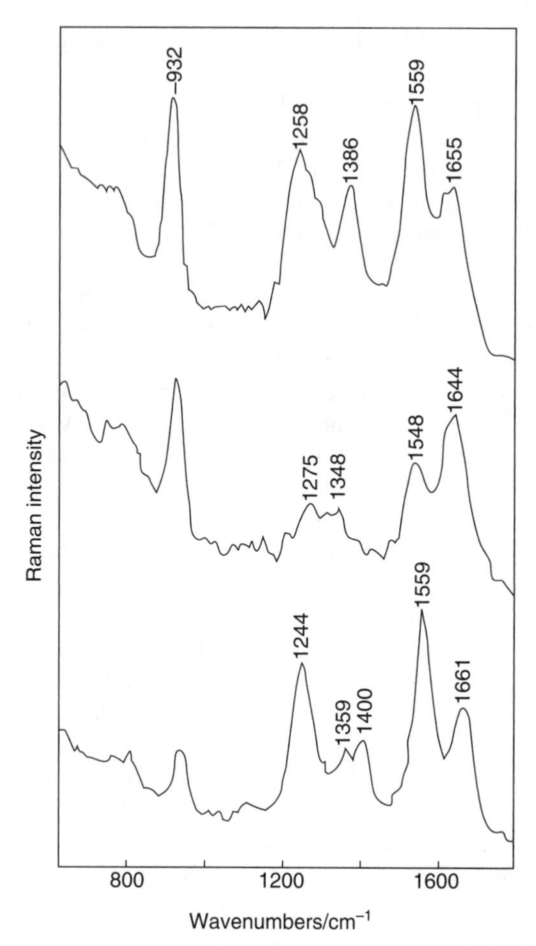

Figure 7.12 Resonance Raman spectrum of PLL in water taken with 218 nm excitation. Upper panel: spectrum of the disordered state; middle panel: spectrum of the helical conformation; lower panel: spectrum of the β-sheet conformation (modified from reference 105).

that there is more than one amide III-type vibration per peptide unit (81). Assignments are difficult and controversial, owing to a complex vibrational mixing pattern involving NH ib, CN s, $C_\alpha H$ ib, and $C_\alpha H$ ob vibrations. The controversy has been discussed in detail in earlier papers and reviews (92, 95). In the current context, however, it is important to note that assignments based on UV and visible Raman data agree in indicating that the amide III subband, which dominates the UV resonance Raman spectra of peptides, exhibits, indeed, the ψ-dependence proposed by Lord. Thus, with respect to this spectroscopic method, amide III is, in principle, a suitable tool for secondary structure exploration.

Asher et al. exploited the structure sensitivity of the main amide III band in their investigation of the unfolded state of two polyalanine peptides, namely, $Ac–X_2(A)_7O_2–NH_2$ (XAO, X and O denote diaminobutyric acid and ornithine, respectively) and $NH_2–A_5(AARA)_3A–OH$ (AP) (4). The XAO peptide has acquired some prominence after Shi et al. reported that its structure is predominantly PPII (99). Shi et al. reached this conclusion based on the peptide's CD spectrum and the $^3J_{C\alpha HNH}$ coupling constants of the alanine residues. AP is partially helical at room temperature. Asher et al. compared the UV Raman spectra of XAO and of thermally unfolded AP with a difference spectrum obtained from subtracting the UV Raman spectrum of A_3 from that of A_5. The similarity of XAO, AP, and difference spectrum led the authors to conclude that XAO forms a PPII helix and that AP unfolds into a PPII conformation rather than into a random distribution of conformations. Asher et al. based their conclusion on two assumptions, namely, that the difference spectrum reflects the spectral contributions from the three central alanine residues of A_5 and that they predominantly adopt a PPII helix. For reasons to be given in more detail below, the term PPII helix is inappropriate because it suggests a stable secondary structure. In the current context, we just mention in passing that several experimental and computational studies on XAO have provided compelling evidence that its structure is more disordered than suggested by Shi et al. (68, 69, 94, 119, 125).

In the above-cited paper, Asher et al. (4) proposed an interesting strategy for extracting conformational distributions along the ψ-coordinate. They modeled the ψ-dependence of amide III by a simple empirical function:

$$\tilde{\nu} = \tilde{\nu}_0 - 46.8\ \mathrm{cm}^{-1} \sin\left(\psi + 5.6^0\right), \tag{7.28}$$

and derived the Lorentzian bandwidth for the amide III band of each individual conformation from the respective band profile in the UV Raman spectrum of a GAL peptide crystal. Figure 7.13 shows the thus derived probability distributions of XAO, AP, and two of the central alanine residues of A_5, with respect to the coordinate, ψ. It is interesting to note that the distribution for XAO indicates the sampling of conformations with ψ-angles smaller than 100°, which is in agreement with a recent spectroscopic study of this peptide (94). While these plots do not allow discrimination between, and identification of, different conformers assignable to the upper left-hand quadrant of the Ramachandran plot, they are important if used in combination with other spectroscopic techniques owing to their potential to identify the sampling of helical conformations. It deserves to be mentioned that a more refined model for amide III appeared in the paper of Mikhonin et al., who elucidated the influence of hydrogen bonding on amide III in order to obtain the true conformational dependence of this mode (76).

Spiro and associates have recently combined UV resonance Raman and temperature jump in an effort to elucidate the conformational manifold of PLL. Their experiments focused on characterizing the α-helix ↔ β-sheet

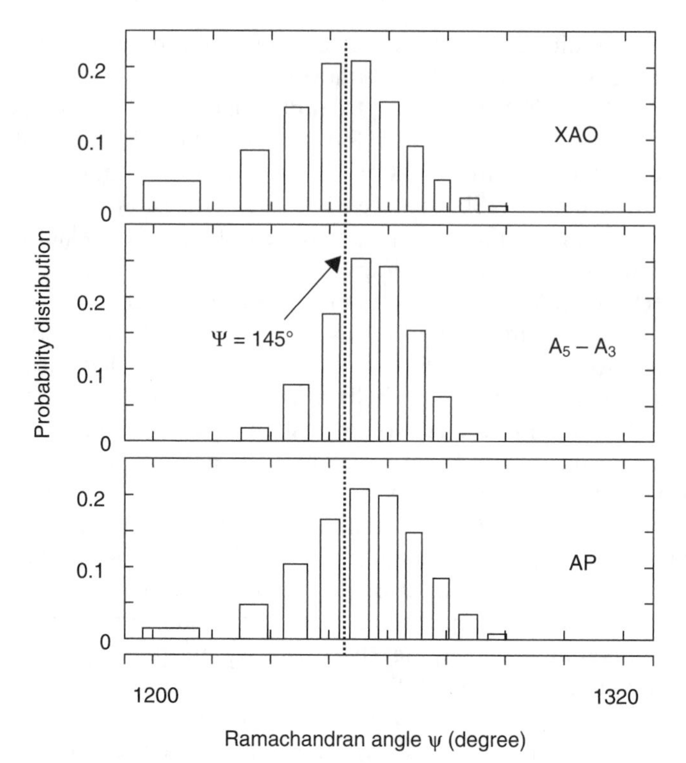

Figure 7.13 Probability distribution of the dihedral angle ψ for the indicated peptides as derived from a deconvolution of the amide III band profile. Taken from Reference 4 and modified.

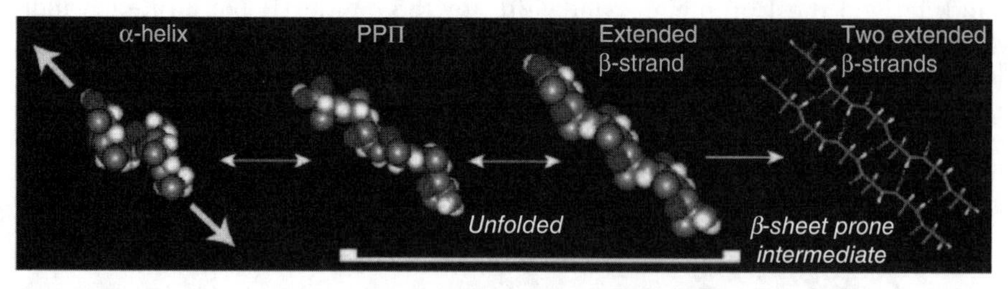

Figure 7.14 Schematic representation of a model assuming an equilibrium among α_R, PPII, and β-strand, and the penultimate step in β-sheet formation of hydrogen bonds between two β-strands.

transition of this peptide, which bears some relevance for the understanding of protein misfolding (49). They used the amide III band in T-jump UV Raman difference spectra to identify PPII as an intermediate of the α-helix ↔ β-sheet transition that converts into another intermediate, which is predominantly β-strand (Fig. 7.14). These results seem to underscore the notion that the random coil concept is insufficient for describing the unfolded state of peptides. In a more recent study, Balakrishnan et al. studied the unfolding of the helical peptide Ac–GSPEA3(KA_4)$_2$–CO–D–N–$CONH_2$ and identified PPII as one of the sampled conformations of the unfolded peptide (6).

This section does not discuss recent advances in deep UV resonance Raman spectroscopy pioneered by Lednev and coworkers, which are the subject of chapter 14. Taken together, far-UV resonance Raman has been developed into a valuable tool for exploring the conformations of folded as well as unfolded peptides. However, since UV resonance Raman mostly probes the average ψ-distributions, complementary methods are required for a comprehensive exploration of conformational manifolds.

7.4.2 Application of VCD and ROA Spectroscopy

The application of VCD spectroscopy to biological macromolecules has been pioneered by Nafie, Freedman, Diem, Keiderling, and their respective coworkers (10, 53, 79, 80). This paragraph focuses on their work on short peptides that do not form regular, canonical secondary structures. In the 1980s, the Nafie/Freedman group focused on measuring and analyzing the VCD spectra of amino acids and very small peptides (31). In this context, they focused on the VCD signals of CH b and stretching modes. As an example, Figure 7.15 exhibits the IR and VCD spectrum of neutral, zwitterionic trialanine in D_2O reported by Zuk et al. (127). The bands in the IR spectrum are all assignable to CH stretching modes. The highest wavenumber band at 2990 cm^{-1} correlates a comparatively small negative VCD band, whereas the band at 2940 cm^{-1} corresponds to a rather strong positive VCD signal. The authors interpreted the rotational strength as originating from a ring current in a loop formed by intrapeptide hydrogen bonding between CO and NH groups of the same residue and the covalent NC, CC, and CO bonds. The current is induced by the electronic flow associated with the CH vibration and creates a magnetic dipole moment perpendicular to the plane in which the current flows. Thus, the VCD of, for example, the CH s vibration of the central residue reflects its dihedral coordinates. Zuk et al. rationalized their data as indicating nearly a completely extended structure with dihedral angles close to 180°. As we will show below (section 7.4.3), this is unlikely to be the case. It is also unlikely that intrapeptide hydrogen bonding can compete with the high propensity of both the carbonyl and the amino groups for hydrogen bonding with water (or D_2O) molecules.

Trialanine has served as a model system for studying possible structural preferences of short peptides. Lee et al. measured the region between 1550 and 1750 cm^{-1} of the IR and VCD spectra of zwitterionic (neutral) and anionic (basic) trialanine in D_2O (65). For the former, the amide I′ band profile displays a doublet and an associated negative couplet as shown in Figure 7.16 (25). The respective spectrum of the alkaline species shows only one broad amide I′ band, and Lee et al. did not detect any VCD signal with their

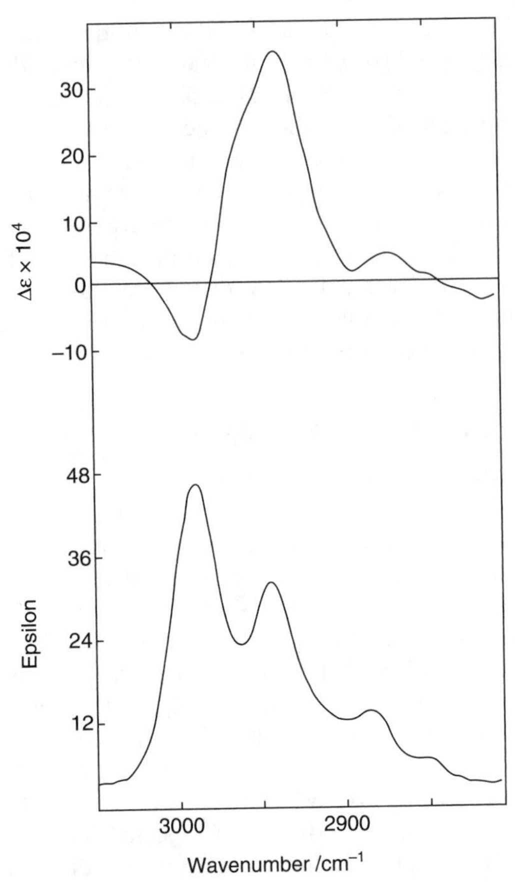

Figure 7.15 Absorption and VCD spectra of the CH stretching region of zwitterionic L-alanyl-L-alanyl-L-alanine in D_2O. Taken from Reference 127 and modified.

instrument (65). They used a coupling oscillator model to analyze their data, which considered a mixing of the excited vibrational states of the two amide I′ modes due to transition dipole coupling, as suggested earlier by Chabay and Holzwarth (45). This consideration yields the following expression for the excited state wavefunctions:

$$|\psi_+\rangle = \frac{1}{\sqrt{2}}(|1,0\rangle + |0,1\rangle),\qquad |\psi_-\rangle = \frac{1}{\sqrt{2}}(|1,0\rangle - |0,1\rangle) \tag{7.29}$$

where |1,0> and |0,1> denote the two excited states of the two uncoupled oscillators. The corresponding rotational strengths are written as:

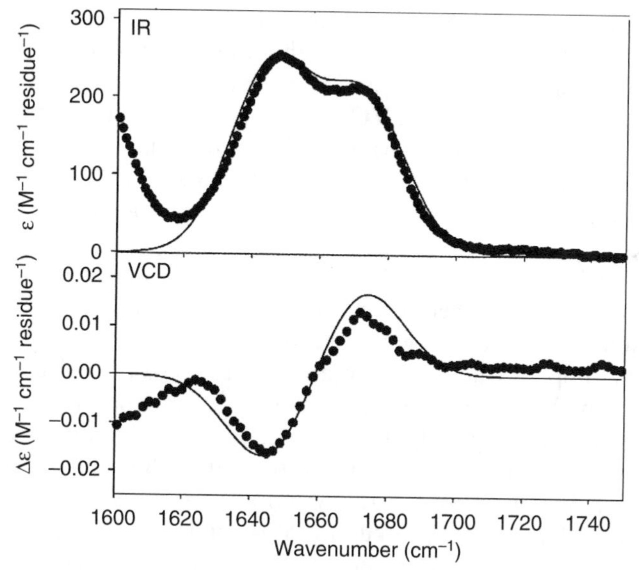

Figure 7.16 IR and VCD spectra of zwitterionic trialanine in D_2O between 1600 and 1650 cm^{-1}. The solid lines result from a simulation described in Reference 24, from where the figure has been taken and modified.

$$R^{\pm} = \mp(\pi\tilde{\nu}_0/2)\vec{T}_{12}\cdot\vec{\mu}_1'\times\vec{\mu}_2' q_1 q_2. \qquad (7.30)$$

Here, $\vec{T}_{12}$ is the distance vector between the center of masses of the interacting oscillators, $\vec{\mu}_1'$ and $\vec{\mu}_2'$ are the transition dipole moments of the respective peptide groups caused by the two interacting amide I modes, and $\tilde{\nu}_0$ is the central wavenumber position of the coupled transitions. The eigenenergies (wavenumbers) of the coupled modes are written as:

$$\tilde{\nu}_{\pm} = \tilde{\nu}_0 \pm V_{12}, \qquad (7.31)$$

where

$$V_{12} = \kappa\left[\frac{\vec{\mu}_1'\cdot\vec{\mu}_2'}{\left|\vec{T}_{12}\right|^3} - \frac{3\left(\vec{\mu}_1'\cdot\vec{T}_{12}\right)\left(\vec{\mu}_2'\cdot\vec{T}_{12}\right)}{\left|\vec{T}_{12}\right|^5}\right]q_1 q_2 \qquad (7.32)$$

and

$$\kappa = 9.047\cdot 10^{15}\ \text{esu}^{-1} \qquad (7.33)$$

describes transition dipole coupling. Lee et al. analyzed their data with this formalism and obtained two conformations with $(\phi, \psi) = (165°, -5°)$ and (120°,

–25°) (65). The latter is close to what the authors termed a left-handed bridge conformation. The authors argued that this conformation could be stabilized by Coulomb interactions between the charged termini, which would explain the absence of any couplet in the VCD spectrum of the alkaline state. However, we will show below (section 7.4.3) that (1) the spectrum of the anionic species does, in fact, show a negative couplet, and (2) the above theoretical approach incorrectly assumed that the uncoupled amide I modes are accidentally degenerate.

Even though the above results are questionable in view of more recent investigations to be discussed below, these results are still noteworthy because they were led by the assumption that short peptides can adopt stable structures rather than sampling the sterically allowed region of the Ramanchandran space, as assumed by the random and statistical coil models (30, 85). However, these papers did not explicitly discuss their respective results in this context, most likely because the proposed structures were believed to be stabilized by hydrogen bonding, whereas the really unfolded state is believed to be governed by the isolated pair hypothesis, that is, the independence of the conformation of a given residue from the conformations of its neighbors (30). The issue of conformational preference of individual residues was directly addressed by the VCD experiments that Dukor and Keiderling performed on PLL and PGAs (23). As was also rationalized by Tiffany and Krimm based on their far-UV electronic CD spectra (113), they found that the amide I mode of the ionized, unfolded state of these peptides exhibits a rather strong negative couplet like that observed for trialanine (Fig. 7.16), which, in turn, resembles that observed for the amide I band of poly-L-proline (23). Dukor and Keiderling concluded that (1) it is unlikely that a statistical ensemble that covers the entire sterically allowed region of the Ramachandran space can give rise to such a pronounced signal, and (2) the similarity of the observed couplet suggests that lysine and glutamic acid have a strong preference for PPII. The VCD couplet and, to a lesser extent, the IR band profiles are temperature dependent, in that the couplet is generally reduced at high temperatures. This could be interpreted as indicative of a more random conformation, again resembling observations obtained using ECD (113).

The result of Dukor and Keiderling suggests that VCD is a very suitable tool for probing the unfolded state of peptides and even proteins. More recently, this notion was underscored by computational studies conducted by the Keiderling group (12, 52, 53, 59–63). They obtained DFT-based force fields for small peptides, which were subsequently used to construct force fields for longer peptides. IR and VCD spectra were calculated by using the formalism outlined in section 7.3.4. Thus, vibrational mixing between, for example, amide I modes, which Lee et al. (65) used to describe the respective VCD couplet of trialanine, is automatically taken into consideration. Kubelka et al. calculated the IR and VCD spectra of several model peptides for different secondary structures including PPII and found that the observed strong, negative couplet is, indeed, diagnostic for this conformation. To

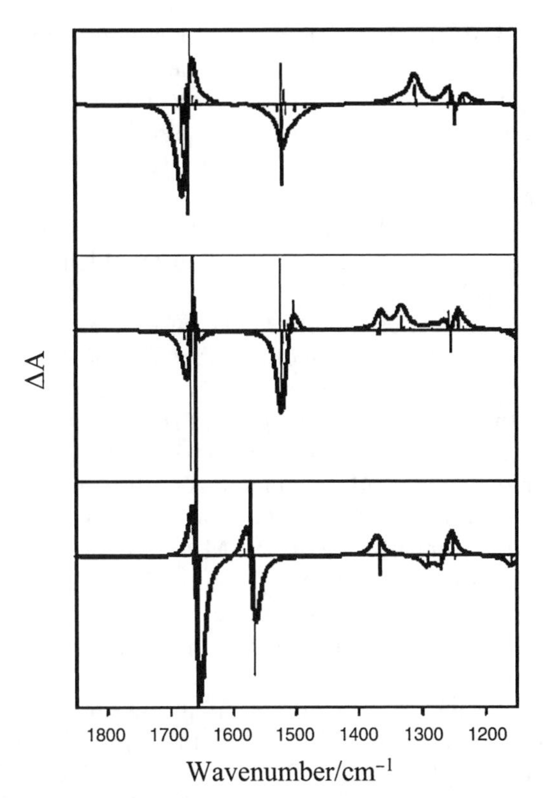

Figure 7.17 VCD spectra of Ac-A_{20}-NH-CH_3 in an α-helical (upper panel), 3_{10}-helical (middle panel), and PPII helical (lower panel) conformation calculated with force field, oscillator, and rotational strengths obtained from a fully DFT calculation. Vertical lines indicate positions and rotational strengths of delocalized vibrational modes. Taken from Reference 61 and modified.

illustrate this point, Figure 7.17 shows the computed VCD spectra of Ac-A_{20}-NH in an α-helical, 3_{10}-helical, and PPII conformation (59). This laboratory has also computed the VCD spectra of β-strand-like conformations for other model peptides and found the amide I signal to be comparatively weak (60).

While the work of Keiderling and coworkers confirmed the notion that longer homopeptides like PLL exhibit a preference for PPII, comparatively weak amide I signals obtained for, for example, trilysine, led them to conclude that short peptides are essentially conformationally random (52). Eker et al., however, showed that even trilysine adopts a PPII-like conformation (27). In section 7.4.3, we will present experiments on short peptides, which show that their conformation is generally not random.

VCD is still not a routine method for peptide and protein structure analysis, but it has become more common over the last 10 years, particularly due to the fact that excellent instrumentation is now commercially available. Even though this is also true for ROA (46, 71), it is still a less prominent method in spite of its enormous potential for a more detailed secondary structure analysis (10). In what follows, we briefly present some of the results Barron and coworkers obtained from their investigations of unfolded, misfolded, and naturally disordered peptides and proteins.

Figure 7.18 depicts the ICP-ROA spectra and the unpolarized Raman spectra of a series of polyalanine peptides of different length (71). The spectra were taken with visible excitation. McColl et al. identified a positive ROA band at ~1319 cm^{-1} as diagnostic of the PPII conformation, based on its appearance in the corresponding spectrum of unfolded PGA (71). This band appears in spectra of all polyalanines investigated. From its respective intensities, the authors inferred that A_4 contained a higher PPII fraction than A_2 and A_3. This has been confirmed by other spectroscopic investigations (37, 91) and also by some molecular dynamics (MD) simulations with a modified Amber force field (33). Overall, these results suggest that alanine has a very high individual PPII propensity. A recent very detailed NMR study of Graf et al. confirmed the high PPII propensity of alanine (35), but it also suggests that it does not increase with the number of alanine residues in a polypeptide chain. All these studies agree in suggesting that unfolded alanine peptides deviate from statistical coil behavior.

The Barron group has applied their technique also to the study of unfolded proteins. Figure 7.19 shows the ROA and Raman spectra of human lysozyme and hen lysozyme at pH 5.4 (T = 20°C) and 2.0 (reduced, T = 57°C) (11). These correspond to the native and thermally unfolded state, respectively. The ROA spectrum of the folded state displays two positive bands at 1305 and 1345 cm^{-1}, which are diagnostic of helices in a hydrophobic and aqueous environment, reflecting the substantial helical fraction of the protein. The spectra of the unfolded molecules display a strong and broad negative couplet, with peaks at 1244/1306 and 1244/1298 cm^{-1} for human and hen lysozyme, respectively, indicative of a very high β-strand fraction. The positive peak at 1318 cm^{-1} reflects a substantial PPII fraction and appears only in the spectrum of human lysozyme. It is interesting to note that only human lysozyme can aggregate to form amyloid fibrils upon unfolding. This led Blanch et al. to speculate that PPII is a necessary precursor for amyloid formation, which is consistent with the observation that amyloid β-peptides, which form the fibrils involved in Alzheimer's disease, contain a substantial PPII fraction (28, 47). It should be noted that the PPII band is also absent in the spectrum of reduced bovine ribonuclease A. These results seem to suggest that PPII is not always abundant in unfolded proteins and that the latter can differ in terms of the conformational manifold sampled by the individual amino acid residues.

Smythe et al. measured the Raman and ROA spectra of bovine β-casein, human γ-synuclein, and human tau 46 P301L, which all have in common that they are naturally disordered (110). These proteins are also involved in amy-

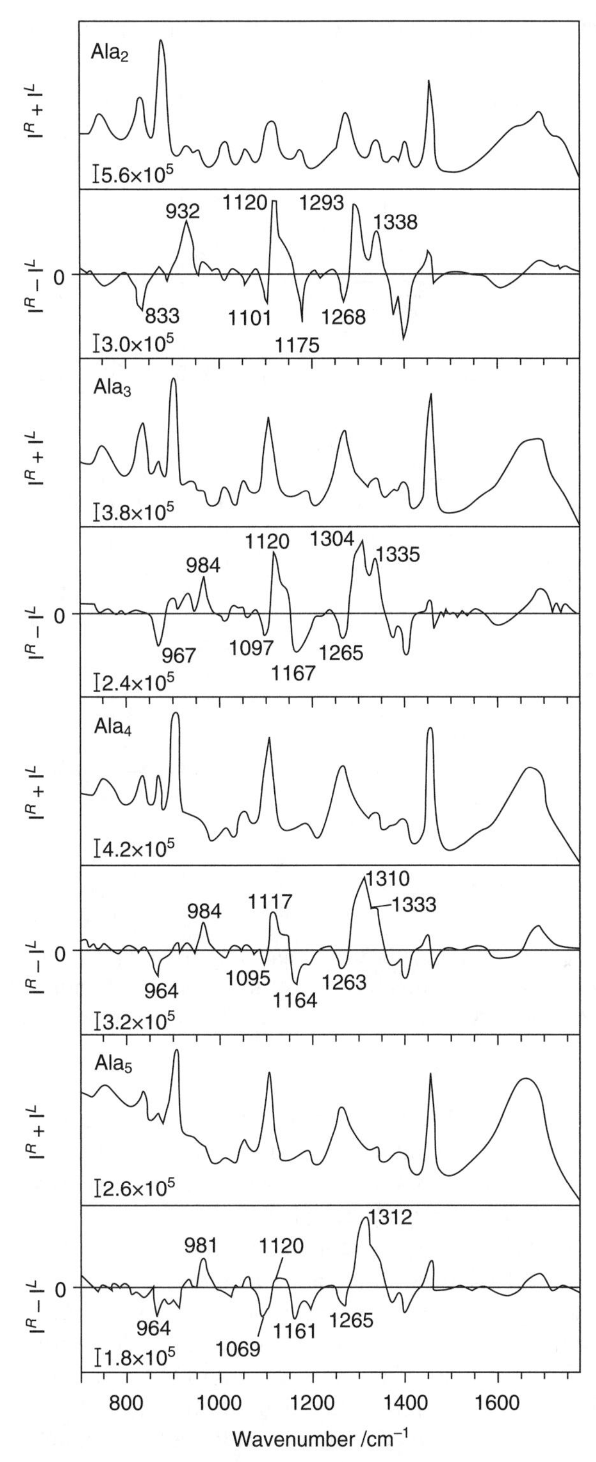

Figure 7.18 Raman and ROA spectra of cationic A_2 to A_5 taken with 514 nm excitation. The most important bands are indicated. Taken from Reference 71 and modified.

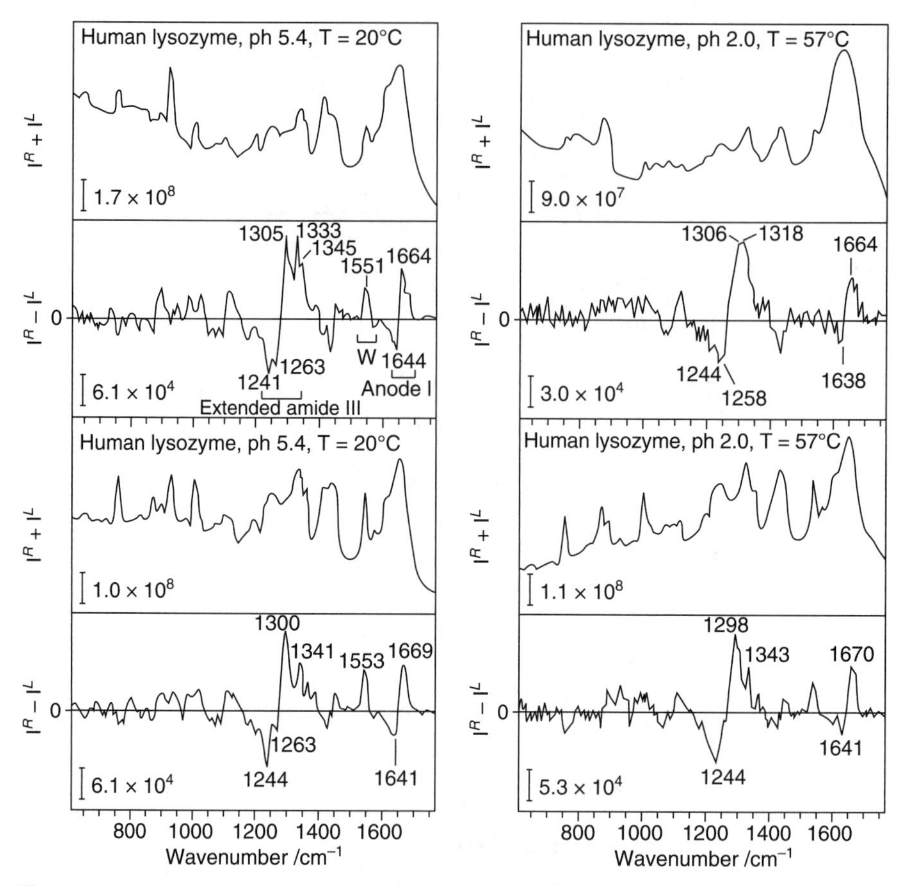

Figure 7.19 Raman and ROA spectra of native lysozyme (top pair) and native hen lysozyme (bottom pair), measured at pH 5.4, and room temperature (left column) as well as at pH 2.0 and 57°C (right column). Taken from Reference 11 and modified.

loid-forming processes. Interestingly, the corresponding ROA spectra displayed in Figure 7.20 all exhibit the PPII signal at ~1320 cm^{-1}. These findings support the hypothesis of Blanch et al. concerning the relevance of PPII for the formation of amyloid aggregates (11).

Among all vibrational spectroscopies that have been used thus far for the investigation of unfolded peptides and proteins, ROA has made the most headway in the direction of analyzing the unfolded states of naturally folded proteins, as well as the structure of naturally unfolded proteins, which has become an important field of protein research (114). Barron and coworkers have recently developed mathematical methods to analyze complex ROA spectra of folded and unfolded proteins, which allow a much more detailed secondary structure analysis than the conventional application of, for example, IR spectroscopy (9).

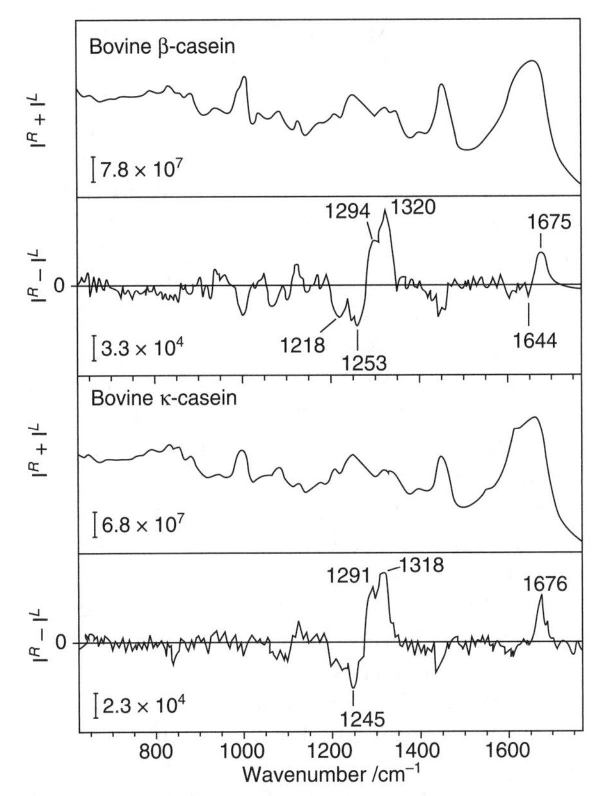

Figure 7.20 The Raman and ROA spectra of bovine β-casein (top pair) and κ-casein (bottom pair) measured at room temperature (»20°C) and at pH 7.0. Taken from Reference 110 and modified.

7.4.3 The Use of Amide I for the Structure Analysis of Unfolded Peptides

The utilization of the amide I band for structure analysis stems from the fact that the corresponding modes of a polypeptide chain are vibrationally coupled. This leads to a delocalization of excited vibrational states. The basic theory has been briefly outlined in section 7.2.1. We mentioned this coupling also when we discussed the VCD profile of trialanine reported by Diem and coworkers (section 7.4.2) (65). There are different approaches to describe the vibrational states of amide I, which have recently been reviewed (87).

An ideal probe to explore such excitonic states is two-dimensional IR spectroscopy. This technique has been pioneered by Hochstrasser, Hamm, and associates (56, 123). In the context of this article, the work of Hamm and coworkers on small peptides are particularly relevant. Since we have reviewed this technique recently (87), we now confine ourselves on a brief description. A narrow laser pulse that only partially overlaps with the amide I band profile in the IR spectrum and a rather broad probe pulse, which covers the entire band region, are used in the experiment. The pump wavenumber is scanned for different pump frequencies. The experiment is generally conducted with different polarizations. The off-diagonal peaks, which can be identified in the difference spectra obtained by subtracting the spectrum taken with parallel polarization from that measured with perpendicular polarization, result from the excitonic coupling between the amide I modes of cationic trialanine in D_2O. A quantitative analysis of the spectrum revealed a coupling constant of $6\,cm^{-1}$ and an angle of 106° between the transition dipole moments of amide I modes. Woutersen and Hamm used the *ab initio* calculations of Torii and Tasumi (115), who had calculated the excitonic coupling constant as a function of ϕ and ψ, for identifying a representative conformation, namely, PPII with $(\phi, \psi) = (-60°, 140°)$. A later analysis based on DFT as well as MD calculations yielded a mixture of 80% PPII and 20% α-helix for trialanine (124).

The results of Woutersen and Hamm are in line with what has been inferred from ROA-experiments (cf. section 7.4.2) and from NMR/CD experiments on alanine-based peptides (99, 101), namely, that alanine has a preference for PPII rather than for a statistical coil distribution. In what follows, we will now review some experiments and theoretical investigations carried out in our own group, which address this issue by investigating the amide I′ profiles of polyalanine peptides with IR, VCD, and polarized Raman spectroscopy. Our work about other (unfolded) peptides has been summarized in a recent book article (88), and it is therefore not discussed in the current context.

In the presence of excitonic coupling, the Raman tensor and transition dipole moment of a given excitonic state can be written as linear combinations of Raman tensors and dipole moments of individual amide I modes:

$$\begin{aligned}\hat{\alpha}'_\beta &= c_{\beta\gamma}\hat{\alpha}'_\gamma \\ \vec{\mu}'_\beta &= c_{\beta\gamma}\vec{\mu}'_\gamma\end{aligned}. \quad (7.34)$$

The subscript γ denotes the number of the peptide linkages (number starts from the N-terminal), whereas β reflects the excitonic state. The coefficients are defined by the eigenvectors described by Equation 7.4. However, to actually utilize Equation 7.34, Raman tensors and transition dipole moments have to be transformed into a common coordinate system. Figure 7.21 exhibits the coordinate systems used for describing the amide I band profiles of trialanine. The x-axes coincide with the NC_α bonds of the respective residues, the y-axes lie in the peptide plane and point into the direction of the peptide carbonyl bond, and the z-axes (not shown in Fig. 7.21) are perpendicular to the peptide plane and follow the three-finger rule. If the peptide has N oscillators, the Raman tensor and the transition dipole moment of the *i-th* peptide group have to be subjected to the operation:

$$\begin{aligned}\hat{\alpha}'(S_N)_i &= \prod_{i=1}^{N-1}(\Gamma^T_{ii+1})\hat{\alpha}'(S_i)_i\prod_{i=1}^{N-1}\Gamma_{ii+1} \\ \vec{\mu}'(S_N)_i &= \prod_{i=1}^{N-1}(\Gamma_{ii+1})\vec{\mu}'_i\end{aligned}, \quad (7.35)$$

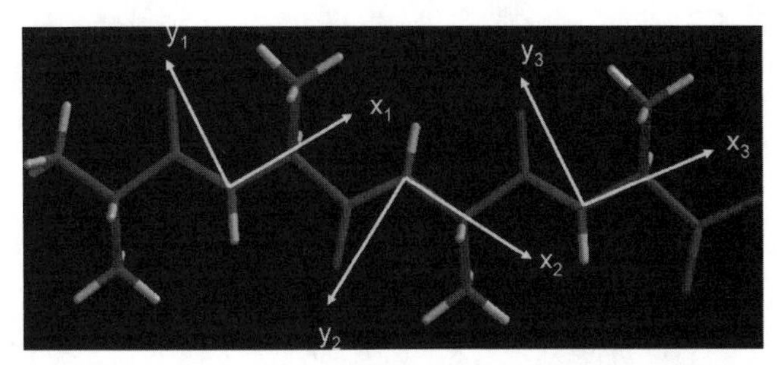

Figure 7.21 Planar structure of tetraalanine (ϕ = 180°, ψ = 180°). The coordinate systems S1(x1, y1, z1), S2(x2, y2, z2), and S3(x3, y3, z3) were used to express the Raman tensors of the individual, uncoupled amide I modes and their transition dipole moments (the z-component for S2 has been omitted for the sake of clarity). The structure was obtained by using the program TITAN from Schrödinger, Inc. See color insert.

with

$$\Gamma_{ii+1} = R(\phi'_{ii+1})R(\xi_{ii+1})R(\psi_{ii+1})R(\omega_{ii+1}), \tag{7.36}$$

which transforms them into the coordinate system S_N at the C-terminal. R is a rotational matrix that rotates a chosen coordinate system about the following angles: $\phi' = \phi - \pi$, where ϕ is one of dihedral coordinates, the $NC_\alpha C$ angle ξ, the dihedral angle ψ, and the $C_\alpha CN$ angle ω. These transformed tensors and dipole moments can then be used in Equation 7.34 for the description of the excitonic state. Thus, the orientational dependence of IR absorption and Raman cross section are accounted for.

Equation 7.21 suggests that one can use polarized Raman scattering to discriminate between isotropic and anisotropic scattering:

$$\begin{aligned} I_{iso} &= I_{II} - \frac{4}{3} I_{\perp} \\ I_{aniso} &= I_{\perp} \end{aligned}. \tag{7.37}$$

The VCD expression for the excitonic states is rather complex and is written as:

$$R_\beta = \mathrm{Im}\left[c_{\beta\gamma}\vec{\mu}'_\lambda\left[c_{\beta\gamma}\vec{m}'_\lambda - \frac{i\pi}{2}\left(\tilde{\nu}_{\gamma\delta}\vec{T}_{\gamma\delta} \times \left(c_{\beta\gamma}\vec{\mu}'_\gamma - c_{\beta\delta}\vec{\mu}'_\delta\right)\right)\right]\right], \tag{7.38}$$

where the subscripts β and γ denote the numbering of the involved peptide groups to which the Einstein convention for summation applies and $\vec{m}_\lambda$ is the intrinsic magnetic transition moments of these peptide groups. Special versions of this equation have been derived earlier by Diem and coworkers (65). It should be mentioned that Equation 7.38 differs from the simpler formalism

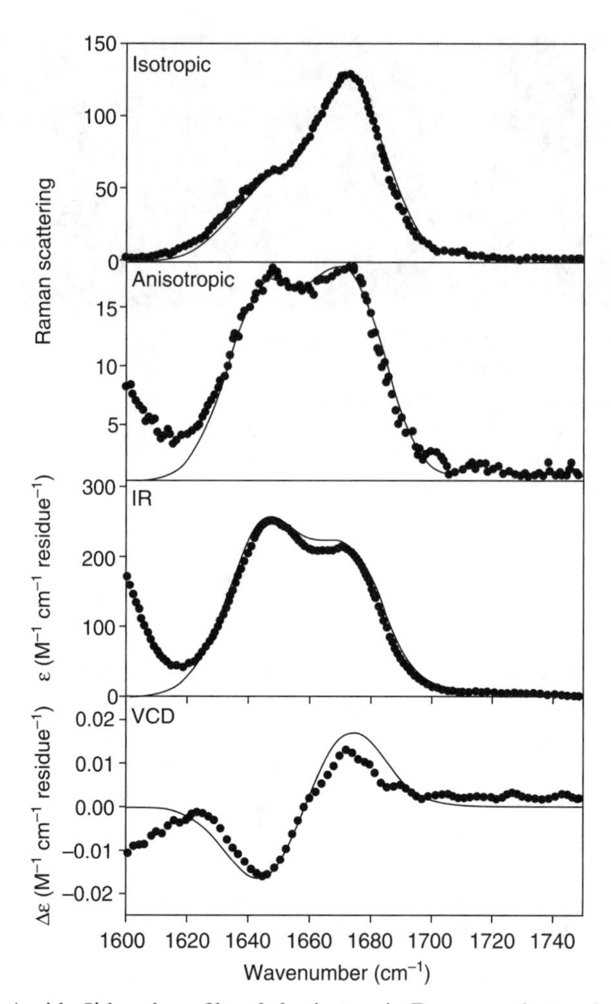

Figure 7.22 Amide I′ band profile of the isotropic Raman, anisotropic Raman, IR, and VCD spectrum of AAA measured at neutral pH 6 (taken from Reference 96). The solid line results from a simulation based on a three-state (per residue) model, encompassing PPII, β, and helix-like conformations as described in the text. The figure was reprinted from Reference 96 with permission from the American Chemical Society.

of Equation 7.30 because it is not based on the following two assumptions, namely, that (1) the interacting oscillators have the same frequency, and (2) any intrinsic magnetic transition dipole moment is absent.

The above formalism was first used by Eker et al. to determine an effective structure of all three protonation states of trialanine in water (D_2O) (25). The respective amide I′ band profiles in the IR, isotropic, and anisotropic Raman spectra of the zwitterionic peptide show two bands whose intensity ratio is significantly different in the three spectra (Fig. 7.22). The higher wavenumber

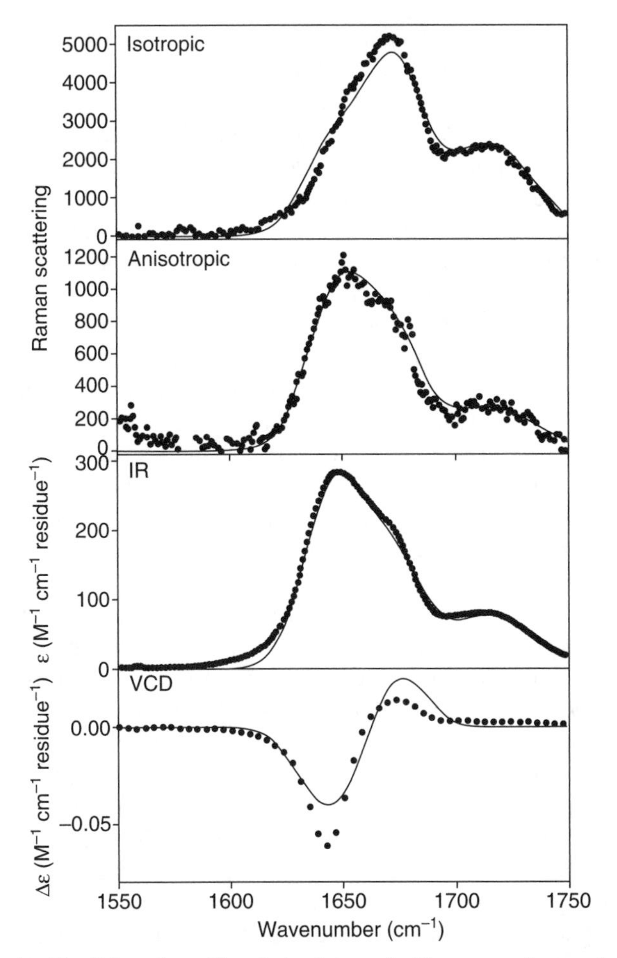

Figure 7.23 Amide I′ band profile of the isotropic Raman, anisotropic Raman, IR, and VCD spectrum of AAAA measured at pD = 1. The solid line results from a simulation based on a three-state (per residue) model, encompassing PPII, β, and helix-like conformations as described in the text. The band of the C-terminal carbonyl stretching modes has been modeled for the sake of completeness by fitting a Gaussian profile to the experimental data with wavenumber and half-width as free parameters. The figure was taken from Reference 92 with permission from ACS.

band dominates the profile in the isotropic Raman spectrum, whereas the lower wavenumber band is slightly more intense in the IR spectrum. Both bands have comparable intensity in the anisotropic Raman spectrum. The VCD spectrum displays a rather strong negative couplet, which is normally indicative of a predominance of PPII. The spectrum of cationic tetraalanine measured by Schweitzer-Stenner et al. (also in D_2O) is shown in Figure 7.23 (91). The different intensities of the three amide I′ bands yield asymmetric band profiles with a clear noncoincidence between the peak positions in the

IR and isotropic Raman spectrum. The peak of the anisotropic spectrum is closer to the respective IR than to the isotropic Raman band position. The couplet in the VCD spectrum is clearly negatively biased. The respective amide I′ band profile of AAKA (alanyl-alanine-lysine-alanine) and the octapeptide $(AAKA)_2$ reported by Schweitzer-Stenner et al. (96), and Measey and Schweitzer-Stenner (73), respectively, look very similar. The noncoincidence between IR and isotropic Raman have been theoretically shown to reflect extended structures associated with the upper left quadrant of the Ramachdran plot; the VCD suggest again a PPII-like conformation.

Two levels of analyses have been performed on the above spectra. Eker et al. used the intensity ratios of the two amide I′ bands of trialanine to derive a possible conformation, for which the VCD spectrum was subsequently calculated. The conformation yielding the best agreement was then reported as the representative structure. For zwitterionic trialanine, the obtained dihedral angles were $(\phi, \psi) = (-120^\circ, 164^\circ)$. Values obtained for the other two protonation states are similar, which rules out that the terminal protonation has a major impact on the conformation of the central residue that is probed by the amide I′ band profile. The obtained conformation lies between PPII and a more extended β-strand conformation. Eker et al. showed (25) that their data could also be explained by assuming a nearly 50:50 mixture of PPII (with the coordinates reported by Woutersen and Hamm [123]) and a β-strand conformation with $(\phi, \psi) = (-165^\circ, 50^\circ)$, which was found to be consistent with the temperature dependence of the ECD spectrum of cationic trialanine (29). Eker et al. used the same strategy to analyze the amide I band profile of trivaline (VVV) and found it to be locked in a β-strand conformation (25, 29). The difference between trialanine and trivaline is clearly reflected by their respective ECD spectra shown in Figures 7.24 and 7.25. Trialanine displays the negatively biased couplet typical for PPII, whereas the respective signal of trivaline is very weak and diagnostic of a more extended β-strand conformation.

Schweitzer-Stenner et al. used the method of Eker et al. to analyze the amide I′ profiles of cationic tetraalanine (92). Thus, they obtained a representative structure with $(\phi, \psi)_1 = (-70^\circ, 155^\circ)$ and $(\phi, \psi)_2 = (-80^\circ, 145^\circ)$. Apparently, this is close to the canonical PPII structure, and the results therefore led to the conclusion that the PPII propensity of alanine is larger in tetraalanine than in trialanine. The aforementioned ROA spectra of these peptides in Figure 7.18 are also consistent with this notion (71). A length dependence of the propensity of alanine also seem to be consistent with the respective ECD spectra of AA, AAA, and AAAA in Figure 7.25, which also indicate that the conformational distributions in these peptides might be different.

The ECD spectra in Figure 7.25 deserve some further comments. We have recently discovered that the use of some cuvettes with very short path lengths in submillimeter regime leads to erroneous results below 220 nm. We have recently purchased cuvettes from International Crystal Laboratories (Garfield, NJ) that do not show this artifact. The ECD spectra of cationic AA, AAA, and AAAA in Figure 7.25 were taken with these new cuvettes. They are still

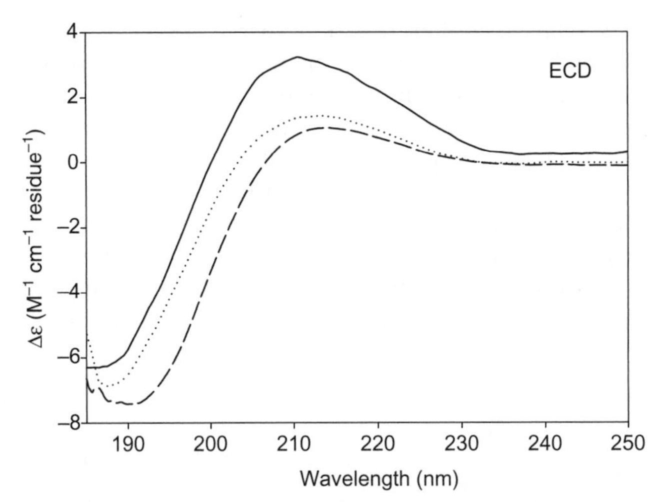

Figure 7.24 Far-UV ECD spectra of AA (solid line), AAA (dotted line), and AAAA (dashed line) measured at room temperature (20°C) in D_2O at pD = 1.0. Units of $\Delta\varepsilon$ in $M^{-1}\,cm^{-1}\,residue^{-1}$.

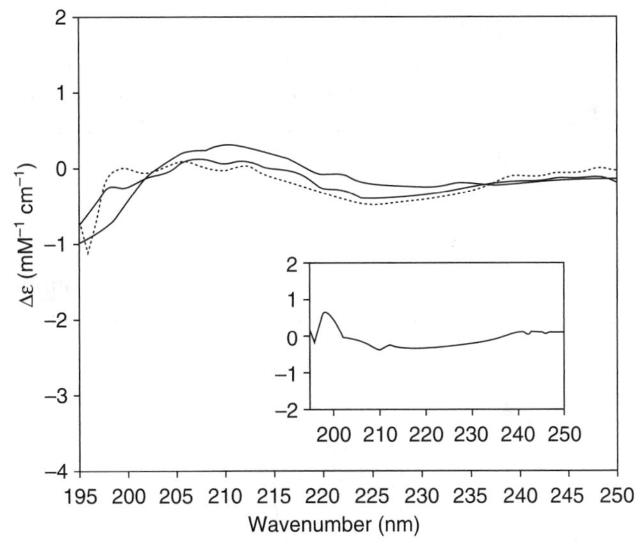

Figure 7.25 Temperature-dependent CD spectra of VVV in D2O at pD = 1.0 (solid line: 20°C; medium dash: 40°C; dotted line: 65°C). Inset: CD difference spectrum for VVV between 65 and 20°C. The figures were taken from Reference 29 with permission from ACS.

qualitatively similar to earlier reported spectra (37), in that they show a minimum below 200 nm and a less pronounced maximum between 210 and 220 nm. However, all minima appear blueshifted with respect to the earlier detected positions, the spectrum of AA now appears much less symmetric than reported by Hagarman et al. (37), and the difference between AAA and AAAA is less pronounced than reported in Reference 91.

The second type of analysis of the above-discussed amide I′ band profiles has been developed more recently and is based on an explicit consideration of different conformations sampled by the residues probed by the excitonic states of amide I. The total intensity of an amide I′ band profile at a given wavenumber is written in terms of a statistical ensemble:

$$I(\Omega) = \frac{\sum_{i=1}^{N} \sum_{j_1, j_2, \ldots, j_N = 1}^{n_{c1}, n_{c2}, \ldots, n_{cN}} \left(I_{ij_1, j_2, \ldots, j_N}(\Omega) e^{-\left(\sum_{k=1}^{N} G_{jk}\right) / RT} \right)}{Z}, \tag{7.39}$$

where N is the number of amide I oscillators and G_{jk} is the Gibbs energy of the *k-th* residue with conformation *j*. The subscript *i* labels the excitonic states of the amide I′ oscillators. $I_{ij_1, j_2, \ldots, j_N}$ is thus the intensity profile of the *i-th* excitonic state associated with the configuration $(j_1, j_2, \ldots, j_N)$ of N nonterminal residues of the peptide. The conformation of the C-terminal residue does not affect the amide I′ band profiles. R is the gas constant, T the absolute temperature, Z the partition sum, and $n_{c1}, n_{c2}, \ldots, n_{cN}$ are the numbers of conformations considered for the respective residues.

In addition to the amide I profiles, $^3J_{C\alpha HNH}$ coupling constants were also measured and incorporated into the analysis. The dependence of the coupling constants on the dihedral angle ϕ was described by the Karplus equation as modified by Vuister and Bax (120):

$$^3J_{C\alpha HNH} = A\cos^2(\phi - \pi/3) + B\cos(\phi - \pi/3) + C, \tag{7.40}$$

with A = 6.52 Hz, B = −1.76 Hz, and C = 1.6 Hz. The $^3J_{C\alpha HnH}$ coupling constants were used to restrict the choice of possible conformational mixtures by calculating:

$$^3J_k = \frac{\sum_{jk}^{n_{ck}} \chi_{jk}\, {}^3J_{jk}}{\sum_{jk}^{n_{ck}} {}^3J_{jk}}, \tag{7.41}$$

where ^3J_k is the experimental coupling constant obtained for the *k-th* residue, χ_{jk} is the mole fraction of the *j-th* conformation of the *k-th* residue, and $^3J_{jk}$ is the respective coupling constant.

TABLE 7.1 Mole Fractions of Residue Conformations Obtained from an Analysis of the Amide I′B and Profiles of Tetrapeptides and Tripeptides (96)

	AAAA	AAKA	AAA
PPII	0.66, 0.73	0.63, 0.36	0.52
β	0.14, 0.17	0.07, 0.34	0.23
Helical (right handed)	0.2, 0.1	0.3, 0.3	0.25

We simulated the amide I′ band profiles of A_3, A_4, and A_2KA by allowing three conformations per central residue (one for A_3, and two for A_4 and A_2KA), namely, PPII ($j = 1$, $(\phi, \psi) = (-65°, 150°)$, β-strand ($j = 2$, $(\phi, \psi) = (-125°, 115°)$, and helix ($j = 3$, $(\phi, \psi) = (-65°, -30°)$). The conformations are representative in that they correspond to the maxima of coil library distributions obtained by Avbelj and Baldwin (5). This model led to the profiles illustrated by the solid lines in Figures 7.22 and 7.23. The coordinates of the PPII conformation had to be slightly modified for A_4 to obtain an optimal agreement with the experimental data. The thus obtained molar fractions of the probed residues are listed in Table 7.1. The results confirm the notion that (1) alanine has a clear propensity for PPII, and that (2) this propensity is higher in A_4 than in A_3. This observation can be understood in terms of an optimal hydration of the peptide backbone of A_4 if the molecule adopts a PPII-like conformation. The role of water in the stabilization of PPII has been emphasized by several computational and experimental studies (32). Our analysis also reveal the presence of a minor fraction of helical structures, in agreement with theoretical predictions and distributions in coil libraries (5, 97). Recently, Graf et al. performed a very thorough NMR study on polyalanines A_n, with n varying between 3 and 7 (35). The experimental work involved the determination of various coupling constants to assess the distribution of both ϕ and ψ values. The results suggest an even higher PPII propensity for alanine ($\chi_{PPII} = 0.9$), which is independent of the number of residues. The remaining 10% comprises mostly of helix-like fractions. This study provides additional strong support for the notion that alanine has a high propensity for PPII.

What do the above-discussed results mean for our understanding of the unfolded state? The pure random coil model suggests that in the unfolded state, alanine and other amino acid side chains (with the exception of proline) sample the entire sterically allowed region of the Ramachandran space with comparable probability (85). The statistical coil model proposed by Flory and Scheraga is a somewhat more realistic modification of the random coil model (30, 111), in that it takes into account the fact that not all accessible conformations are energetically equivalent. We use the term statistical coil, then, if three canonical troughs of the Ramachandran plot, which are associated with right-handed helices, PPII, and β-strand, are sampled in a comparable way, with some admixtures of less frequently populated conformations like left-handed helical or turn-like conformations. As a matter of fact, this description seems to apply to lysine in AAKA, for which Schweitzer-Stenner et al. observed a nearly identical sampling of the three canonical conformations (96). A somewhat cruder analysis of the spectra of $(AAKA)_2$ yields a similar result. This is surprising in view of the fact that homopeptides like K_3, K_7, and even PLL have been shown to be predominantly PPII (27, 107, 113). Hence, we can conclude that some residues (like lysine) in a distinct context (in this case, alanine) exhibit a statistical coil behavior, while others (like alanine and valine) deviate from this model by preferring one of the canonical conformations (PPII or β). We would like to reiterate in this context that by mentioning conformations like PPII and β, we think in terms of distributions that cluster around a certain coordinate in the Ramachandran space rather than a definitive (ϕ, ψ) pair.

The high PPII propensity of alanine has been questioned by Scheraga and coworkers (68, 69, 119). They focused on the aforementioned XAO peptide, which had been introduced by Shi et al. as a classical example of a peptide whose alanine residues adopt predominantly PPII conformations (99). This had led to the erroneous notion that the peptide can actually form a real stable PPII helix as it is found, for example, in collagen, even though that was not the intention of Shi et al. Vila et al., based on MD simulations and their own analysis of the CD spectra of Shi et al., came to the conclusion that the PPII fraction, per alanine residue, in XAO does not exceed 30% and that the peptide itself exhibits a statistical coil structure (119). Zagrovic et al. performed small-angel X-ray scattering measurements on this peptide and obtained a radius of gyration of 7.4 Å. This value is, by far, too short for a peptide with its residues either in PPII or β conformations (125). Makowska et al. used the $^3J_{C\alpha HNH}$ coupling constants of the peptides' residues to constrain the conformational search carried out by means of MD Amber force field-based simulations and found that, particularly, the alanine residues located at the termini of the heptaline stretch, and their respective neighbors X and O, can sample various types of turn structures (69). They calculated the expectation value for the radius of gyration for their ensemble of conformations and found it to be close to the experimental value. In a subsequent paper, Makowska et al. reported calorimetric measurements (68), which rule out that the molecule undergoes a PPII → β transition as suggested by Shi et al. (99). They argued that PPII is a local conformation and that the temperature dependence of CD spectra and $^3J_{C\alpha HNH}$ values reflects a conformational redistribution rather than a process like the helix ↔ coil transition.

A direct investigation of the structural manifold sampled by the amino acid residues of XAO has been carried out by Schweitzer-Stenner and Measey (94). Figure 7.26 shows the amide I′ profile of the IR, isotropic Raman, anisotropic Raman, and VCD spectra of this peptide. The solid line therein results from a simulation that was built on an ensemble of conformations representing distributions predicted by MD simulations carried out by Makowska et al. and restricted by the reported $^3J_{C\alpha HNH}$ coupling constants (69). They calculated an

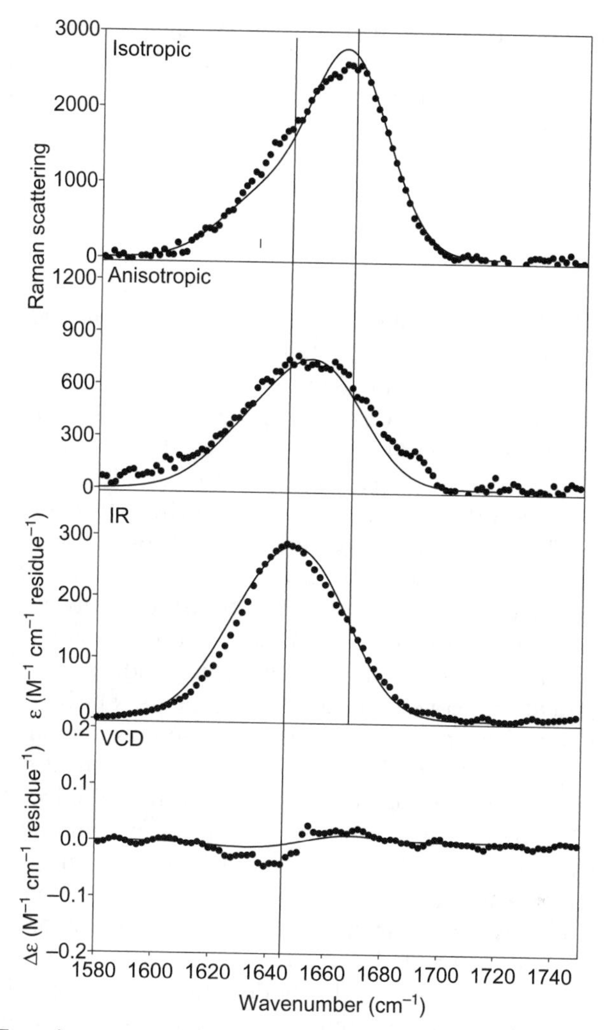

Figure 7.26 Experimental isotropic and anisotropic Raman, FTR, and VCD of XAO, pD = 2.2 (heavy black dots), and the spectra simulated using the conformations listed in Table 7.1, as described in the text. The figures were taken from Reference 91 with permission.

average end-to-end distance of 19.1 Å, which is consistent with the radius of gyration reported by Zagrovic et al. (125). Schweitzer-Stenner and Measey (91) obtained a higher PPII propensity than Makowska et al. (68) for the alanine residues that are not involved in turn structures. In this regard, their result parallels the finding on shorter polyalanine peptides. However, these authors explicitly agree with the notion that XAO does not form a PPII-like helix, as suggested by Asher et al. (4). Instead, it is more likely that PPII is a local conformation, for which alanine residues have a certain propensity. It should be mentioned, however, that short "all PPII" segments encompassing three or four residues can temporarily form and that the respective molar fractions are not negligible.

7.5 SIGNIFICANCE OF PPII

The investigations described above have in common that the respective results indicate that the PPII conformation plays a pivotal role in understanding the unfolded state of peptides and proteins. This point has been particularly emphasized in recent review articles by Kallenbach and his collaborators (100, 101). In what follows, we reflect some aspects of this issue in the context of the ongoing discussion about how to understand the unfolded state.

First of all, it should be reiterated that unfolded peptides cannot be described as adopting a stable PPII state comparable to helical, turn, or β-sheet states. The term "state" refers to the entire molecule and corresponds to a minimum of the total Gibbs energy (20). For alanine, the existence of a PPII-like state can be ruled out based on the fact that the transitions between the PPII and β-strand are noncooperative (14, 91). This notion might not strictly apply to polypeptides with a high diversity of amino acid residues, as indicated by the study of Gräslund and coworkers on *A*β fragments, which revealed that, at least, segments of these peptides can adopt something similar to a PPII-like state. Generally, however, the PPII conformation is the most predominant of a set of conformations that alanine can adopt in the unfolded state. The persistence length corresponds to segments of four to five amino acids (48, 117), which implies that local order can temporarily exist in the unfolded state. Changes of temperature cause a redistribution of conformations, that is, a shift of the Gibbs energy minimum along an ordered parameter coordinate, as pointed out by Makowska et al. (68, 69).

The newest results for the model peptide XAO indicate that the complexity of the conformational sampling increases with increasing chain length, so that if a peptide exceeds a certain length, it might, in fact, adopt a more compact rather than an extended structure (69, 94, 125). A similar observation has been made for another alanine-based peptide (51, 118). Schweitzer-Stenner and Measey (91) speculated that sequences like XXA and AOO, which contain alanine and charged peptides, might be much more conformationally flexible than, for example, a pure alanine sequence. Investigations in our laboratory are under way to address this issue.

The forces that stabilize the PPII conformation are still a matter of debate between computational chemists. The very thorough DFT calculations on the alanine dipeptide in explicit water showed that an energy minimum in the PPII trough of the Ramachandran plot requires four water molecules that are hydrogen bonded to the carbonyl and amide group of the peptide linkages (42). The relevance of water was also emphasized by the MD simulations on polyalanines (32, 33, 54). Pappu and Rose argued that water stabilizes PPII by minimizing the electrostatic interactions between peptide linkages (83). Eker et al. showed that Ac-A_2-COOH resembles the behavior of tri-alanine in water, while it predominantly samples the β-strand in dimethyl sulfoxide (24).

A somewhat different approach has been suggested by Hinderacker and Raines (43). The authors argued that even for proline, the well-known PPII conformation of the *trans*-isomer is not only stabilized by steric forces but also, additionally, by an $n \rightarrow \pi^*$ transition between adjacent peptide carbonyls. If valid, this argument would also apply to nonproline residues, as pointed out by Shi et al. (100). The difficulty with this interpretation is that the transition energy of an $n \rightarrow \pi^*$ transition is very high. This has recently been demonstrated by Dragomir et al., who identified weak absorption bands in the far-UV spectra of zwitterionic dipeptides, which are assignable to $n \rightarrow \pi^*$ and $\pi \rightarrow \pi^*$ transitions from the carboxylate to the peptide group (22, 82). These charge transfer spectra were obtained by subtracting the absorption spectra of the cationic species from those of the zwitterionic species. The result for a series of AX peptides is shown in Figure 7.27.

The paper by Dragomir et al. combined with a ECD/NMR study on AX peptides by Hagarman et al. (37) suggest that even though different amino acid residues at the C-terminal can still be differentiated with respect to their conformational preference, they seem to be conformationally more flexible than residues at more central positions. Dragomir et al. found that the $n \rightarrow \pi^*$ band is inhomogeneously broadened with more β-strand conformations appearing at the low-energy side of the band. The greater heterogeneity of C-terminal residues has recently been confirmed by MD simulations for Ac-AibAA-OMe and Ac-AAibA-OMe peptides in water (93).

Finally, we would like to discuss the relevance of PPII for other amino acid residues. The "single conformation analysis" of AXA tripeptides by Eker et al. (26) and the "two-conformation" analysis of AX dipeptides by Hagarman et al. (37) both indicate that besides alanine and proline, leucine and charged residues prefer PPII over more extended conformations, while valine, isoleucine, phenylalanine, and threonine show a propensity for β-strand. While Shi et al. agree with this assessment in quantitative terms, their analysis of ${}^3J_{C\alpha HNH}$ coupling constants of Ac-GGXGG-NH_2 suggest that nearly all nonproline residues can sample PPII to a significant extent (98). The determination of a "context free" propensity of amino acid residues in water is of utmost importance for modeling the conformational manifolds of unfolded peptides and even proteins. Corresponding investigations on G × G are currently under way in our laboratory.

Finally, we like to emphasize that the propensity of a distinct residue is likely to be context dependent. Above, we mentioned the different behavior of lysine in AAKA and ionized K_3 (27, 96). The propensity of lysine for PPII and valine for β-strand (25) might depend on whether they are flanked by lysine or valine, respectively. Such a context dependence has been theo-

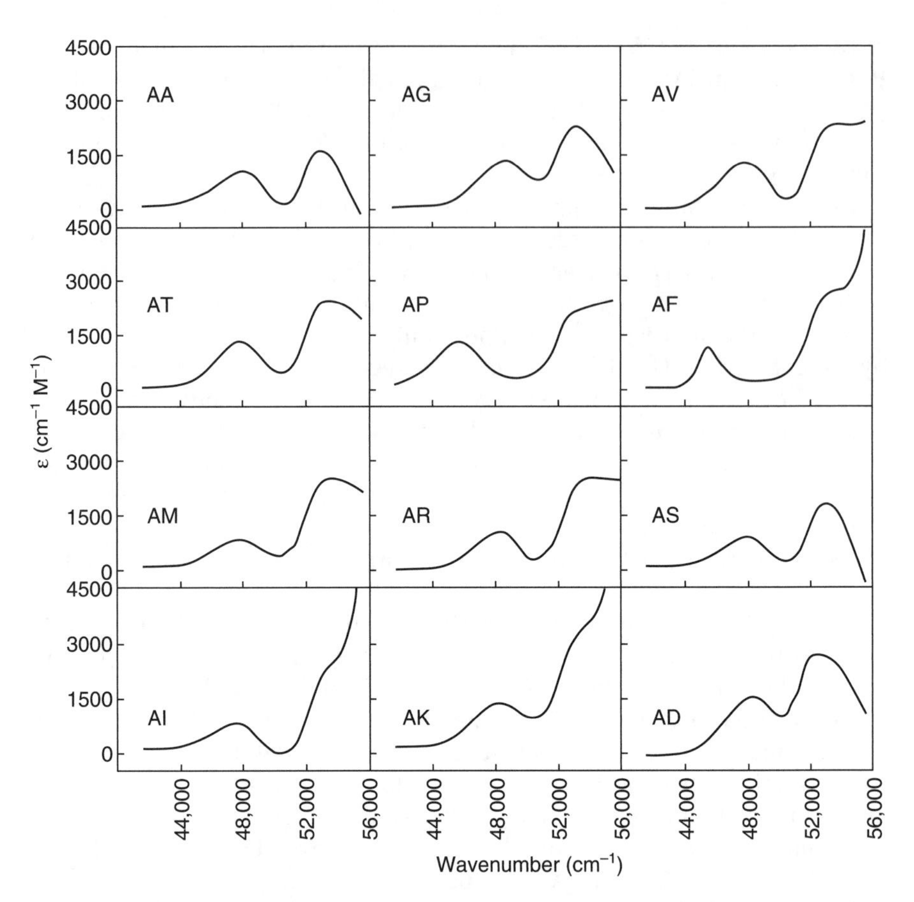

Figure 7.27 Absorption difference spectra of AX dipeptides obtained by subtracting the spectrum of the indicated dipeptide observed at pH 1 from that measured at pH 7. Taken from Reference 22 with permission.

retically predicted (117, 126) and, if existing, would indicate a violation of Flory's isolated pair hypothesis (30). A violation of this principle has also been predicted by Pappu et al., who showed that if a distinct residue adopts a right-handed helical conformation, the probability for the adjacent residue to adopt a β-strand conformation is reduced (84). If this is, indeed, the case, one would expect that if a given residue is locked in the helical conformation, the propensity of the neighbor should be shifted either toward PPII or even right-handed helical. This is, indeed, what we have recently obtained for Ac-AibAA-OMe. The additional methylation of the α-carbon forces Aib either into a right-handed or left-handed helical conformation. We found that the next alanine residue differs from the normal alanine behavior in that it samples predominantly the helical trough of the Ramachandran plot (93).

REFERENCES

1. Asher, S. A. 2001. UV-resonance Raman spectroscopy, pp. 557–71. In J. M. Chalmers and P. Griffiths (eds.), Handbook of Vibrational Spectroscopy. John Wiley & Sons, New York.
2. Asher, S. A. 1988. UV resonance Raman studies of molecular structure and dynamics, applications for physical and biophysical chemistry. Ann Rev Phys Chem **39**:537–88.
3. Asher, S. A., A. Ianoul, G. Mix, M. N. Boyden, A. Karnoup, M. Diem, and R. Schweitzer-Stenner. 2001. Dihedral ψ angle dependence of the amide III vibration: a unique sensitive UV resonance Raman secondary structural probe. J Am Chem Soc **123**:11775–81.
4. Asher, S. A., A. V. Mikhonin, and S. Bykov. 2004. UV Raman demonstrates that α-helical polyalanine peptides melt to polyproline II conformations. J Am Chem Soc **126**:8433–40.
5. Avbelj, F., and R. L. Baldwin. 2003. Role of backbone solvation and electrostatics in generating preferred peptide backbone conformations: distributions of phi. Proc Natl Acad Sci U S A **100**:5742–7.
6. Balakrishnan, G., Y. Hu, G. M. Bender, Z. Getahun, W. F. De Grado, and T. G. Spiro. 2007. Enthalpic and entropic stages in α-helical peptide unfolding, from laser T-Jump/UV Raman spectroscopy. J Am Chem Soc **129**:12801–8.
7. Bandekar, J. 1992. Amide modes and protein conformation. Biochim Biophys Acta **1120**:123–43.
8. Barron, L. D. 2004. Molecular Light Scattering and Optical Activity, 2nd Edition. Cambridge, UK: Cambridge University Press.
9. Barron, L. D., E. W. Blanch, and L. Hecht. 2002. Unfolded proteins studied by Raman optical activity. Adv Protein Chem **62**:51–90.
10. Barron, L. D., L. Hecht, E. W. Blanch, and A. F. Bell. 2000. Solution structure and dynamics of biomolecules from Raman optical activity. Prog Biophys Mol Biol **73**:1–49.
11. Blanch, E. W., L. A. Morozowa-Roche, D. A. E. Cochran, A. J. Doig, L. Hecht, and L. D. Barron. 2000. Is polyproline II helix the killer conformation? A Raman optical activity study of the amyloidogenic prefibrillar intermediate of human lysozyme. J Mol Biol **301**:553–63.
12. Bour, P., and T. A. Keiderling. 2005. Vibrational spectral simulation for peptides of mixed secondary structure: method comparisons with the Trpzip model hairpin. J Phys Chem B **123**:23687–97.
13. Brant, D. A., and P. J. J. Flory. 1965. The configuration of random polypeptide chains. II. Theory. J Am Chem Soc **87**:2791–800.
14 Chen, K., Z. Liu, and N. R. Kallenbach. 2004. The polyproline II conformation in short alanine peptides is non cooperative. Proc Natl Acad Sci U S A **101**: 15352–7.
15. Chen, X. G., S. A. Asher, R. Schweitzer-Stenner, N. G. Mirkin, and S. Krimm. 1995. UV Raman determination of the ππ* excited state geometry of N-methylacetamide: vibrational enhancement pattern. J Am Chem Soc **117**: 2884–95.

16. Chen, X. G., R. Schweitzer-Stenner, S. A. Asher, N. G. Mirkin, and S. Krimm. 1995. Vibrational assignments of trans-N-methylacetamide and some of its deuterated isotopomers from band decomposition of IR, visible, and resonance Raman spectra. J Phys Chem **99**:3074–83.
17. Chen, X. G., R. Schweitzer-Stenner, N. G. Mirkin, S. Krimm, and S. A. Asher. 1994. N-methylacetamide and its hydrogen-bonded water molecules are vibrationally coupled. J Am Chem Soc **116**:11141–2.
18. Choi, J.-H., J.-S. Kim, and M. Cho. 2005. Amide I vibrational circular dichroism of polypeptides: generalized fragmentation approximation method. J Chem Phys **122**:174903-1–11.
19. Cowan, P. M., and S. Mc Gavin. 1955. Structure of poly-L-proline. Nature **176**: 501–3.
20. Dill, K. A., and S. B. Bromber. 2003. Molecular Driving Force. Garland Science, New York.
21. Dill, K. A., and D. Shortle. 1991. Denatured state of proteins. Annu Rev Biochem **60**:795–825.
22. Dragomir, I., T. J. Measey, A. M. Hagarman, and R. Schweitzer-Stenner. 2006. Environment-controlled interchromophore charge transfer transitions in dipeptides probed by UV absorption and electronic circular dichroism spectroscopy. J Phys Chem B **110**:13235–41.
23. Dukor, R., and T. A. Keiderling. 1991. Reassessment of the random coil conformation: vibrational CD study of proline oligopeptides and related polypeptides. Biopolymers **31**:179–61.
24. Eker, F., X. Cao, L. Nafie, K. Griebenow, and R. Schweitzer-Stenner. 2003. The structure of alanine based tripeptides in water and dimethyl sulfoxide probed by vibrational spectroscopy. J Phys Chem B **107**:358–65.
25. Eker, F., X. Cao, L. Nafie, and R. Schweitzer-Stenner. 2002. Tripeptides adopt stable structures in water. A combined polarized visible Raman, FTIR and VCD spectroscopy study. J Am Chem Soc **124**:14330–41.
26. Eker, F., K. Griebenow, X. Cao, L. Nafie, and R. Schweitzer-Stenner. 2004. Preferred peptide backbone conformations in the unfolded state revealed by the structure analysis of alanine-based (AXA) tripeptides in aqueous solution. Proc Natl Acad Sci U S A **101**:10054–9.
27. Eker, F., K. Griebenow, X. Cao, L. Nafie, and R. Schweitzer-Stenner. 2004. Tripeptides with ionizable side chains adopt a perturbed polyproline II structure in water. Biochemistry **43**:613–21.
28. Eker, F., K. Griebenow, and R. Schweitzer-Stenner. 2004. Aβ1-28 fragment of the amyloid peptide predominantly adopts a polyproline II conformation in an acidic solution. Biochemistry **43**:6893–8.
29. Eker, F., K. Griebenow, and R. Schweitzer-Stenner. 2003. Stable conformations of tripeptides in aqueous solution studied by UV circular dichroism spectroscopy. J Am Chem Soc **125**:8178–85.
30. Flory, P. J. 1969. Statistical Mechanics of Chain Molecules. John Wiley & Sons, New York.
31. Freedman, T. B., A. C. Chernovitz, W. M. Zuk, M. G. Paterlini, and L. A. Nafie. 1988. Vibrational circular dichroism in the methine bending modes of amino acids and dipeptides. J Am Chem Soc **110**:6970–4.

32. Garcia, A. E. 2004. Characterization of non-alpha conformations in Ala peptides. Polymer **120**:885–90.
33. Gnanakaran, S., and A. E. Garcia. 2003. Validation of an all-atom protein force field: from dipeptides to larger peptides. J Phys Chem B **107**:12555–7.
34. Gorbunov, R. D., P. H. Nguyen, M. Kobus, and G. Stock. 2007. Quantum-classical description of the amide I vibrational spectrum of trialanine. J Chem Phys **126**:054509.
35. Graf, J., P. H. Nguyen, G. Stock, and H. Schwalbe. 2007. Structure and dynamics of the homologous series of alanine peptides: a joint molecular dynamics/NMR study. J Am Chem Soc **129**:1179–89.
36. Gregurick, S. K., G. M. Chaban, and R. B. Gerber. 2002. Ab initio and improved empirical potentials for the calculation of the anharmonic vibrational states and intramolecular mode coupling of N-methylacetamide. J Phys Chem B **106**: 8696–707.
37. Hagarman, A., F. Eker, T. Measey, K. Griebenow, and R. Schweitzer-Stenner. 2006 Conformational analysis of XA and AX dipeptides in water by electronic circular spectroscopy. J Phys Chem B **110**:6979–86.
38. Ham, S., S. Cha, J.-H. Choi, and M. Cho. 2003. Amide I modes of tripeptides: Hessian matrix reconstruction and isotope effects. J Chem Phys **119**:1452–61.
39. Ham, S., and M. Cho. 2003. Amide I modes in the N-methylacetamide dimer and glycine dipeptide analog: diagonal force constants. J Chem Phys **118**:6915–22.
40. Ham, S., S. Hahn, C. Lee, T.-K. Kim, K. Kwak, and M. Cho. 2004. Amide I modes of a-helical polypeptide in liquid water: conformational fluctuation, phase correlation, and linear and non-linear vibrational spectra. J Phys Chem B **108**:9333–45.
41. Hamm, P., M. Lim, W. F. DeGrado, and R. Hochstrasser. 1999. Stimulated photon echos from amide I vibrations. J Phys Chem B **103**:10049–53.
42. Han, W.-G., K. J. Jakanen, M. Elstner, and S. Suhai. 1998. Theoretical study of aqueous N-acetyl-L-alanine N-methylamide: structures and Raman, VCD, and ROA spectra. J Phys Chem B **102**:2587–602.
43. Hinderaker, M. P., and R. T. Raines. 2003. An electronic effect on protein structure. Protein Sci **12**:1188–94.
44. Ho, J.-H., S. Ham, and M. Cho. 2003. Local amide I frequencies and coupling constants in polypeptides. J Phys Chem B **107**:9132–8.
45. Holzwarth, G., and I. Chabay. 1972. Optical activity of vibrational transitions: a coupled oscillator model. J Chem Phys **57**:1632–8.
46. Hug, W., and G. Hangartner. 1999. A novel high-throughput Raman spectrometer for polarization difference measurements. J Raman Spectrosc **30**:841.
47. Jarvet, J., P. Damberg, J. Danielson, I. Johansson, L. E. Erikson, and A. Gräslund. 2003. A left-handed 3_1 helical conformation in the Alzheimer Aβ(12–28) peptide. FEBS Lett **555**:371–4.
48. Jha, A. K., A. Colubri, M. H. Zaman, S. Koide, T. R. Sosnick, and K. F. Freed. 2005. Helix, sheet and polyproline II frequencies and strong nearest neighbor effects in a restricted coil library. Biochemistry **44**:9691–702.
49. JiJi, R. D., G. Balakrishnan, and T. G. Spiro. 2006. Intermediacy of poly(L-proline) II and -strand conformations in poly(L-lysine) β-sheet formation probed by temperature-jump/UV resonance Raman spectroscopy. Biochemistry **45**:34.

50. Jordan, T., and T. G. Spiro. 1994. Enhancement of C_a H vibrations in the resonance Raman spectra of peptides. J Raman Spectrosc **25**:537–43.
51. Jun, S., J. S. Becker, M. Yonkunas, R. Coalson, and S. Saxena. 2006. Unfolding of alanine-based peptides using electron spin resonance distance measurements. Biochemistry **45**:11666–73.
52. Keiderling, T. A. 1996. Vibrational circular dichroism: application to conformational analysis of biomolecules, p. 555. In G. D. Fasman (ed.), Circular Dichroism and the Conformational Analysis of Biomolecules. Plenum Press, New York.
53. Keiderling, T. A., and Q. Xu. 2002. Unfolded proteins studied with IR and VCD spectra. Adv Protein Chem **62**:111–61.
54. Kentsis, A., M. Mezei, T. Gindin, and R. Osman. 2004. Unfolded state of polyalanine is a segmented polyproline II helix. Proteins **55**:493–501.
55. Kim, S., and R. M. Hochstrasser. 2007. The 2D IR responses of amide and carbonyl modes in water cannot be described by Gaussian frequency fluctuations. J Phys Chem B **111**:9697–701.
56. Kim, Y. S., J. Wang, and R. M. Hochstrasser. 2005. Two-dimensional infrared spectroscopy of the alanine dipeptide in aqueous solution. J Phys Chem B **109**: 7511–21.
57. Kitagawa, T., and S. Hirota. 2002. Raman spectroscopy of proteins, pp. 3427–46. In J. M. G. Chalmers and P. R. Griffiths (eds.), Handbook of Vibrational Spectroscopy. John Wiley & Sons, New York.
58. Krimm, S., and J. Bandekar. 1986. Vibrational spectroscopy of peptides and proteins. Adv Protein Chem **38**:181.
59. Kubelka, J., T. A. Huang, and T. A. Keiderling. 2005. Solvent effects on IR and VCD spectra of helical peptides: DFT-based static spectral simulations with explicit water. J Phys Chem B **109**:8231–43.
60. Kubelka, J., and T. A. Keiderling. 2001. The anomalous infrared amide I intensity distribution in 13C isotopically labeled peptide β-sheets comes from extended, multiple-stranded structures. An ab initio study. J Am Chem Soc **123**:6142–50.
61. Kubelka, J., and T. A. Keiderling. 2001. Differentiation of β-sheet-forming structures: ab initio-based simulations of IR absorption and vibrational CD for model peptide and protein β-sheets. J Am Chem Soc **123**:12048–58.
62. Kubelka, J., J. Kim, P. Bour, and T. A. Keiderling. 2007. Contribution of transition dipole coupling to amide coupling in IR spectra of peptide secondary structures. Vibr Spectrosc **42**:63–73.
63. Kubelka, J., R. A. G. D. Silva, and T. A. Keiderling. 2002. Discrimination between peptide 3_{10}- and α-helices. Theoretical analysis of the impact of methyl substitution on experimental spectra. J Am Chem Soc **124**:5325–32.
64. Kushinskii, L. L., and P. P. Shorygin. 1961. Raman scattering from light near and far from resonance. Opt Spectrosc **11**:80–9.
65. Lee, O., G. M. Roberts, and M. Diem. 1989. IR vibrational CD in alanyl tripeptide: indication of a stable solution conformer. Biopolymers **28**:1759–70.
66. Long, D. A. 2002. The Raman Effect. A Unified Treatment of the Theory of Raman Scattering by Molecules. John Wiley & Sons, New York.
67. Lord, R. 1977. Strategy and tactics in the Raman spectroscopy of biomolecules. App Spectrosc **31**:187–94.

68. Makowska, J., S. Rodziewicz, S. Baginska, M. Makowski, J. A. Vila, A. Liwo, L. Chmurzyński, and H. A. Scheraga. 2007. Further evidence for the absence of polyproline II stretch in the XAO peptide. Biophys J **92**:2904–17.

69. Makowska, J., S. Rodziewicz-Motowidlo, K. Baginska, J. A. Vila, A. Liwo, L. Chmurzynski, and H. A. Scheraga. 2006. Polyproline II conformation is one of many local conformational states and is not an overall conformation of unfolded peptides and proteins. Proc Natl Acad Sci U S A **103**:1744–9.

70. Mayne, L. C., L. D. Ziegler, and B. Hudson. 1985. Ultraviolet resonance Raman studies of N-methylacetamide. J Phys Chem **89**:3395.

71. McColl, I. H., E. W. Blanch, L. Hecht, N. R. Kallenbach, and L. D. Barron. 2004. Vibrational Raman optical activity characterization of poly(L-proline II) helix in alanine oligopeptides. J Am Chem Soc **126**:5076–7.

72. Measey, T., A. Hagarman, F. Eker, K. Griebenow, and R. Schweitzer-Stenner. 2005. Side chain dependence of intensity and wavenumber position of amide I' in IR and visible Raman spectra of XA and AX dipeptides. J Phys Chem B **109**: 8195–205.

73. Measey, T., and R. Schweitzer-Stenner. 2006. The conformations adopted by the octamer peptide $(AAKA)_2$ in aqueous solution probed by FTIR and polarized Raman spectroscopy. J Raman Spectrosc **37**:248–54.

74. Mikhonin, A. V., Z. Ahmed, A. Ianoul, and S. A. Asher. 2004. Assignments and conformational dependencies of the amide III peptide backbone UV resonance Raman bands. J Phys Chem B **108**:19020–8.

75. Mikhonin, A. V., and S. A. Asher. 2005. Uncoupled peptide bond vibrations in a a-helical and polyproline II conformation of polyalanine peptides. J Phys Chem B **109**:3047–52.

76. Mikhonin, A. V., S. V. Bykov, N. S. Myshakina, and S. A. Asher. 2006. Peptide secondary structure folding reaction coordinate: correlation between UV Raman amide III frequency, Ramachandran angle ψ, and hydrogen bonding. J Phys Chem B **110**:1928.

77. Mix, G., R. Schweitzer-Stenner, and S. A. Asher. 2000. Uncoupled adjacent amide vibrations in small peptides. J Am Chem Soc **112**:9028.

78. Myshakina, N. S., and S. A. Asher. 2007. Peptide bond vibrational coupling. J Phys Chem B **111**:4271–9.

79. Nafie, L. A. 1997. Infrared and Raman vibrational optical activity. Annu Rev Phys Chem **48**:357.

80. Nafie, L. A., R. K. Dukor, and T. B. Freedman. 2002. Dichroism and optical activity in optical spectroscopy, pp. 731–44. In J. Chalmers and P. Griffiths (eds.), Handbook of Vibrational Spectroscopy. John Wiley & Sons, New York.

81. Oboodi, M. R., C. Alva, and M. Diem. 1984. Solution-phase Raman studies of alanyl dipeptides and various isotopomers: a reevaluation of the amide III vibrational assignment. J Phys Chem **88**:501.

82. Pajcini, V., X. G. Chen, R. W. Bormett, S. J. Geib, P. Li, S. A. Asher, and E. G. Lidiak. 1996. Glycylglycine $\pi \rightarrow \pi^*$ and charge transfer transition moment orientations: near resonance Raman single crystal measurements. J Am Chem Soc **118**: 9716–26.

83. Pappu, R. V., and G. D. Rose. 2002. A simple model for polyproline II structure in unfolded states of alanine-based peptides. Protein Sci **11**:2437–55.

84. Pappu, R. V., R. Srinivasan, and G. D. Rose. 2000. The Flory isolated-pair hypothesis is not valid for polypeptide chains: implications for protein folding. Proc Natl Acad Sci U S A **97**:12565–70.

85. Ramachandran, G. N., C. Ramachandran, and V. Sasisekharan. 1963. Stereochemistry of polypeptide chain configurations. J Mol Biol **7**:95–9.

86. Robin, M. B. 1975. Higher Excited States of Polyatomic Molecules, vol. II. Academic Press, New York.

87. Schweitzer-Stenner, R. 2006. Advances in vibrational spectroscopy as a sensitive probe of peptide and protein structure. A critical review. Vib Spectrosc **42**:98–117.

88. Schweitzer-Stenner, R. 2008. Conformational analysis of unfolded peptides by vibrational spectroscopy, pp. 101–42. In T. A. Creamer (ed.), Unfolded Proteins. From Denatured States to Intrinsically Disordered. Novalis Press, New York.

89. Schweitzer-Stenner, R. 2004. Secondary structure analysis of polypeptides based on an excitonic coupling model to describe the band profile of amide I of IR, Raman and vibrational circular dichroism spectra. J Phys Chem B **108**: 16965–75.

90. Schweitzer-Stenner, R. 2001. Visible and UV-resonance Raman spectroscopy on model peptides. J Raman Spectrosc **32**:711.

91. Schweitzer-Stenner, R., F. Eker, K. Griebenow, X. Cao, and L. Nafie. 2004. The conformation of tetraalanine in water determined by polarized Raman, FTIR and VCD spectroscopy. J Am Chem Soc **126**:2768–76.

92. Schweitzer-Stenner, R., F. Eker, Q. Huang, K. Griebenow, P. A. Mroz, and P. M. Kozlowski. 2002. Structure analysis of dipeptides in water by exploring and utilizing the structural sensitivity of amide III by polarized visible Raman, FTIR-spectroscopy and DFT based normal coordinate analysis. J Phys Chem B **106**:4294–304.

93. Schweitzer-Stenner, R., W. Gonzales, J. T. Bourne, J. A. Feng, and G. A. Marshall. 2007. Conformational manifold of α-aminoisobutyric acid (Aib) containing alanine-based tripeptides in aqueous solution explored by vibrational spectroscopy, electronic circular dichroism spectroscopy, and molecular dynamics simulations. J Am Chem Soc **129**:13095–109.

94. Schweitzer-Stenner, R., and T. Measey. 2007. The alanine-rich XAO peptide adopts a heterogeneous population, including turn-like and PPII conformations. Proc Natl Acad Sci U S A **104**:6649–54.

95. Schweitzer-Stenner, R. 2008. Conformational analysis of unfolded peptides by vibrational spectroscopy. In T. Creamer (ed.), Unfolded Proteins: From Denatured States to Intrinsically Disordered. Novalis Press, pp. 101–42.

96. Schweitzer-Stenner, R., T. Measey, L. Kakalis, F. Jordan, S. Pizzanelli, C. Forte, and K. Griebenow. 2007. Conformations of alanine-based peptides in water probed by FTIR, Raman, vibrational circular dichroism, electronic circular dichroism, and NMR spectroscopy. Biochemistry **46**:1587–96.

97. Serrano, L. 1995. Comparison between the Φ-distribution of the amino acids in the protein data base and NMR data indicates that amino acids have various Φ propensities in the random coil conformation. J Mol Biol **254**:322–33.

98. Shi, Z., K. Chen, Z. Liu, A. Ng, W. C. Bracken, and N. R. Kallenbach. 2005. Polyproline II propensities from GGXGG peptides reveal an anticorrelation with β-sheet scales. Proc Natl Acad Sci U S A **102**:17964–8.
99. Shi, Z., C. A. Olson, G. A. Rose, R. L. Baldwin, and N. R. Kallenbach. 2002. Polyproline II structure in a sequence of seven alanine residues. Proc Natl Acad Sci U S A **99**:9190–5.
100. Shi, Z., K. Shen, Z. Liu, and N. R. Kallenbach. 2006. Conformation in the backbone in unfolded proteins. Chem Rev **106**:1877–97.
101. Shi, Z., R. W. Woody, and N. R. Kallenbach. 2002. Is polyproline II a major backbone conformation in unfolded proteins? Adv Protein Chem **62**:163–240.
102. Sieler, G., and R. Schweitzer-Stenner. 1997. The amide I mode of peptides in aqueous solution involves vibrational coupling between the peptide group and water molecules of the hydration shell. Am Chem Soc **119**:1720–6.
103. Sieler, G., R. Schweitzer-Stenner, J. S. W. Holtz, V. Pajcini, and S. A. Asher. 1999. Different conformers and protonation states of dipeptides probed by polarized Raman, UV-resonance Raman, and FTIR spectroscopy. J Phys Chem B **103**: 372–84.
104. Sola, R., and K. Griebenow. 2006. Influence of modulated structural dynamics on the kinetics of a-chymotrypsin catalysis. Insights through chemical glycosylation, molecular dynamics and domain motion analysis. FEBS J **273**:5303–19.
105. Song, S., and S. A. Asher. 1989. UV resonance Raman studies of peptide conformation in poly(L-lysine), poly(L-glutamic acid), and model complexes: the basis for protein secondary structure determinations. J Am Chem Soc **111**:4295–305.
106. Song, S., S. A. Asher, S. Krimm, and J. Bandekar. 1988. UV resonance Raman studies of peptide conformation in poly(L-lysine), poly(L-glutamic acid), and model complexes: the basis for protein secondary structure determinations. J Am Chem Soc **110**:8547.
107. Stapley, B. J., and T. P. Creamer. 1999. Lysine peptides revisited. Protein Sci **8**:587.
108. Surewicz, W. K., and H. H. Mantsch. 1988. New insight into protein secondary structure from resolution-enhanced infrared spectra. Biochim Biophys Acta **952**:115–30.
109. Suschtschinskij, M. M. 1974. Ramanspektren von Molekülen und Kristallen. Heyden & Son, Rheine, Germany.
110. Syme, C. D., E. W. Blanch, C. Holt, J. Ross, M. Goedert, L. Hecht, and L. D. Barron. 2002. A Raman optical activity study of rheomorphism in caseins, synucleins and tau. New insight into the structure and behaviour of natively unfolded proteins. Eur J Biochem **269**:148–56.
111. Tanaka, S., and H. A. Scheraga. 1976. Statistical mechanical treatment of protein conformation. II. A three-state model for specific-sequence copolymers of amino acids. Macromolecules **9**:150–67.
112. Tanford, C. 1968. Protein denaturation. Adv Protein Chem **23**:121–282.
113. Tiffany, M. L., and S. Krimm. 1968. New chain conformations of poly(glutamic acid) and polylysine. Biopolymers **6**:1767–70.
114. Tompa, P. 2002. Intrinsically unstructured proteins. Trends Biochem Sci **27**:527–33.

115. Torii, H., and M. Tasumi. 1998. Ab initio molecular orbital study of the amide I vibrational interactions between the peptide groups in di- and tripeptides and considerations on the conformation of the extended helix. J Raman Spectrosc **29**:81–6.

116. Torii, H., and M. Tasumi. 1992. Model calculations on the amide-I infrared bands of globular proteins. J Chem Phys **96**:3379–87.

117. Tran, H. T., X. Wang, and R. V. Pappu. 2005. Reconciling observations of sequence-specific conformational propensities with the generic polymeric behavior of denatured proteins. Biochemistry **44**:11369–80.

118. Tucker, M. J., R. Oyola, and F. Gai. 2005. Conformational distribution of a 14-residue peptide in solution: a fluorescence resonance energy transfer study. J Phys Chem B **109**:4788–95.

119. Vila, J. A., H. A. Baldoni, D. R. Ripoli, A. Gosh, and H. Scheraga. 2004. Polyproline II helix conformation in a proline-rich environment: a theoretical study. Biophys J **86**:731–42.

120. Vuister, G. W., and A. Bax. 1993. Quantitative J correlation: a new approach for measuring homonuclear three-bond J (HNCHα) coupling. J Am Chem Soc **115**:7772–7.

121. Wang, Y., R. Purello, S. Georgiou, and T. G. Spiro. 1991. UVRR spectroscopy of the peptide bond. 2. Carbonyl H-bond effects on the ground- and excited-state structures of N-methylacetamide. J Am Chem Soc **113**:6368–77.

122. Wang, Y., R. Purello, T. Jordan, and T. G. Spiro. 1991. UVRR spectroscopy of the peptide bond. 1. Amide S, a nonhelical structure marker, is a $C_{\alpha}H$ bending mode. J Am Chem Soc **113**:6359–6368.

123. Woutersen, S., and P. Hamm. 2000. Structure determination of trialanine in water using polarized sensitive two-dimensional vibrational spectroscopy. J Phys Chem B **104**:11316–20.

124. Woutersen, S., R. Pfister, P. Hamm, Y. Mu, D. S. Kosov, and G. Stock. 2002. Peptide conformational heterogeneity revealed from nonlinear vibrational spectroscopy and molecular-dynamics simulations. J Chem Phys **117**:6833–40.

125. Zagrovic, B., J. Lipfert, E. J. Sorin, I. S. Millett, W. F. van Gunsteren, S. Doniach, and V. S. Pande. 2005. Unusual compactness of a polyproline II structure. Proc Natl Acad Sci U S A **102**:11698–703.

126. Zaman, M. H., M.-Y. Shen, R. S. Berry, K. F. Freed, and T. R. Sosnick. 2003. Investigations into sequence and conformational dependence of backbone entropy, inter-basin dynamics and the Flory isolated-pair hypothesis for peptides. J Mol Biol **331**:693–711.

127. Zuk, W. M., T. B. Freedman, and L. A. Nafie. 1989. Vibrational CD studies of the solution conformation of simple alanyl-peptides as a function of pH. Biopolymers **28**:2025.

8

INTRINSICALLY DISORDERED PROTEINS AND INDUCED FOLDING STUDIED BY FOURIER TRANSFORM INFRARED SPECTROSCOPY

Antonino Natalello and Silvia Maria Doglia

Department of Biotechnology and Biosciences,
University of Milano-Bicocca, Milan, Italy

ABSTRACT

Infrared spectroscopy has been proved to be a powerful tool to study protein conformation and dynamics and, therefore, to characterize the structural properties of intrinsically disordered proteins (IDPs) and their induced folding in different environmental conditions. In this chapter, we present a general survey of the standard experimental methods to obtain the infrared absorption spectrum of a protein. The procedures required to identify the protein absorption components in the amide I region (1700–1600 cm^{-1}) and to assign them to the secondary structures are discussed, together with the data analysis that enable to evaluate the secondary structure content. Interestingly, this spectroscopy allows to examine proteins in different environmental conditions, both in solution and in solid form as protein films. We illustrate the potential of infrared spectroscopy on selected studies of IDP-induced folding by different effectors, such as DNA, partner proteins, and osmolytes. IDPs undergoing amyloid aggregation, as α-synuclein and a prion peptide, are also reported.

Instrumental Analysis of Intrinsically Disordered Proteins: Assessing Structure and Conformation, Edited by Vladimir Uversky and Sonia Longhi

8.1 INTRODUCTION

Intrinsically disordered proteins (IDPs) do have a biological activity, but in physiological conditions lack a well-defined three-dimensional structure. According to their secondary and tertiary structures, IDPs include random coil-like, premolten globules, and molten globule-like proteins (52). The majority of IDPs undergo a conformational transition from their disordered native state into a structured one, induced by the interaction with different physiological partners. It is therefore desirable to be able to characterize the structural properties of IDPs and their induced folding in different environmental conditions. To this goal, among several optical techniques (48, 52), infrared spectroscopy has been proved to be a powerful tool to obtain information on protein secondary structure and aggregation through the analysis of the protein absorption spectrum in the mid-infrared region (wavelengths of 2.5–10 μm, corresponding to wavenumbers of 4000–1000 cm^{-1}). In particular, the amide I absorption band, due to the stretching vibration of the C=O peptide bond from 1600 to 1700 cm^{-1}, is sensitive to the C=O backbone environment, therefore enabling to identify the secondary structural elements of the protein (4, 5, 7, 8, 16, 19, 22, 58, 60, 61, 64).

In recent years, Fourier transform infrared (FT-IR) spectroscopy has been successfully applied to IDPs for the characterization of their native state and induced folding. Indeed, as this spectroscopy allows to examine also highly scattering samples, proteins can be studied in different environmental conditions, both in solution and in solid form as protein films. This capability is very useful, for instance, for the study of IDPs undergoing amyloid aggregation, as in the case of α-synuclein and prion peptides.

This chapter will give a general survey of the standard experimental methods to obtain the absorption spectrum of a protein, as well as of the data analyses required to evaluate protein secondary structure, conformational changes, and aggregation. We will illustrate this approach by reporting selected examples of FT-IR studies on IDPs. Advanced FT-IR techniques, such as vibrational circular dichroism, were already discussed in chapter 7 by Reinhard Schweitzer-Stenner.

8.2 FT-IR SPECTROSCOPY FOR PROTEIN STUDIES

FT-IR spectroscopy is an established method to investigate protein secondary structure, stability, and aggregation through the analysis of the protein absorption spectrum. In the following pages, we describe how the infrared spectrum of proteins can be measured by different sampling techniques in solution and in form of protein film. The spectral analysis required to identify and assign the amide I components will be illustrated on a natively folded protein, taken as an example, before extending the method to IDPs.

8.2.1 How to Measure the FT-IR Absorption Spectrum of Proteins in Transmission and in Attenuated Total Reflection

8.2.1.1 Transmission Measurements The absorption spectra of protein in water solution can be measured in spite of the high water absorption in the amide I region, using highly performing FT-IR spectrometers that, thanks to their good signal-to-noise ratio and baseline stability, allow to perform the subtraction of the solvent spectrum (22, 50, 73). Typically, protein solutions at a concentration ranging from 1 to 15 mg/mL are required for measurements in water, while concentrations can decrease down to 0.3 mg/mL when measurements are performed in D_2O (7). However, the total protein amount is about 10–100 µg, since only 10–20 µL are required for a single measurement.

To illustrate the method, we report here the spectrum of *Candida rugosa* Lipase 1 (CRL1) (41) in water solution at 10 mg/mL concentration. The absorption spectrum was measured in transmission using a demontable cell (Wilmad, Buena, NJ, USA) with two BaF_2 windows separated by a 15-µm spacer, requiring a sample volume of about 10 µL. The absorption spectra of CRL1 solution and that of water (Fig. 8.1A) are almost identical and domi-

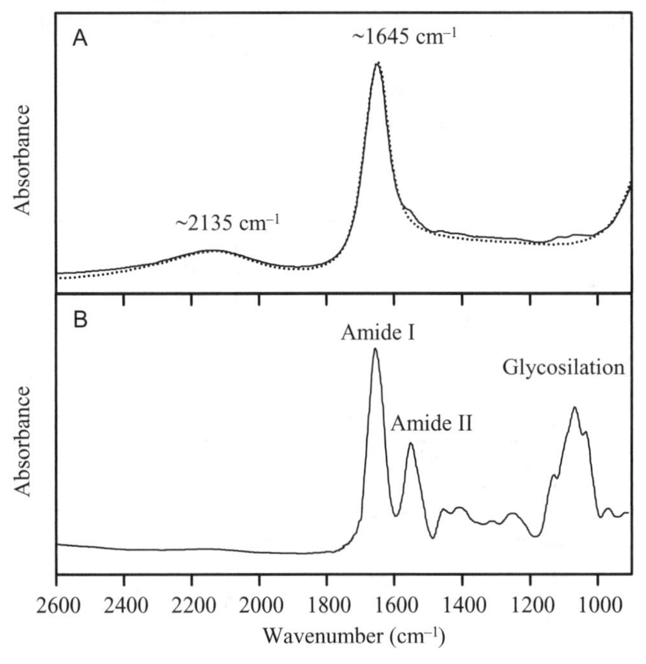

Figure 8.1 (A) FT-IR absorption spectra of CRL1 water solution (continuous line) and of water (dotted line) in the 900–2600 cm^{-1} range. (B) Absorption spectrum of CRL1 obtained after subtraction of the water spectrum from that of the protein solution. FT-IR spectra were measured at a resolution of 2 cm^{-1} by a FTS-40A spectrometer (Bio-Rad, Digilab Division, Cambridge, MA, USA) with a deuterated triglycine sulphate detector.

nated by the bending vibration of water around 1645 cm^{-1}. To obtain the infrared response of the protein, it is therefore necessary to perform the subtraction of the water spectrum. This can be iteratively repeated until a flat baseline is obtained in the 1800–2300 cm^{-1} region, which contains the water combination band at ~2135 cm^{-1}, but no protein absorption (22, 50, 73).

To obtain a high-quality protein spectrum after water subtraction, few crucial experimental requirements are necessary. The temperature of the sample and of the subtracted water buffer need to be carefully controlled since the infrared absorption of water displays strong temperature dependence. Furthermore, an efficient purging (by nitrogen or dry air) of the spectrometer is required to reduce the water vapor absorption. Under this condition, the residual vapor could be corrected by subtracting the vapor spectrum from that of the protein.

Figure 8.1B reports the CRL1 FT-IR spectrum, after water subtraction, containing the two principal protein absorption bands amide I and amide II, occurring respectively in the range 1700–1600 cm^{-1} and 1600–1500 cm^{-1}.

Figure 8.2 presents the spectrum of CRL1 in D_2O solution at the concentration of 10 mg/mL, where it is possible to recognize, already, the protein absorption before solvent subtraction since the D_2O spectrum displays only a broad absorption band in the amide I and amide II region. The protein spectrum, after D_2O subtraction, is reported in the inset of Figure 8.2. Actually, thanks to the low absorption of D_2O in the amide I region, a good spectrum can be also obtained for protein at a concentration of about 0.5 mg/mL, using an optical path of about 50–100 μm.

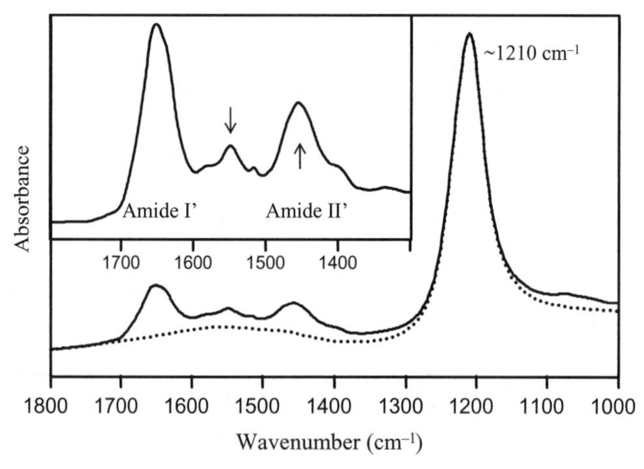

Figure 8.2 FT-IR absorption spectra of CRL1 D_2O solution (continuous line) and of D_2O (dotted line) in the 1000–1800 cm^{-1} range. The band peaked at 1210 cm^{-1} corresponds to the bending absorption of D_2O. Inset: FT-IR spectrum of CRL1 obtained after subtraction of the D_2O spectrum from that of the protein solution. The arrows point to the H/D exchange effect on the amide II (in H_2O before exchange) and amide II′ (in D_2O after exchange) bands.

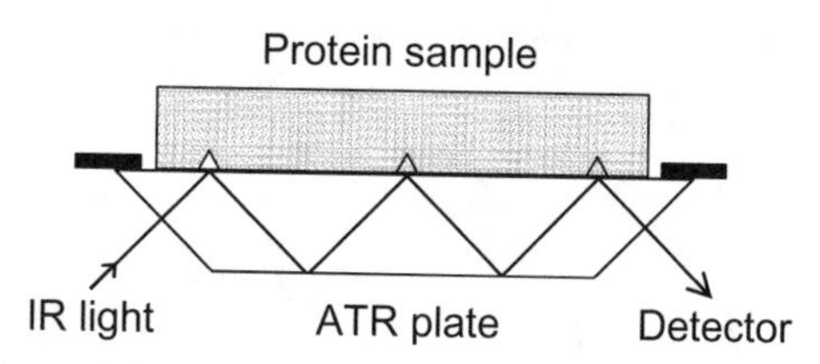

Figure 8.3 Scheme of a horizontal ATR system.

8.2.1.2 Attenuated Total Reflection (ATR) Measurements Protein solutions in water as well as in D_2O can be also measured in ATR, a technique where the sample is placed on an ATR element having a refractive index higher than that of the sample (Fig. 8.3) (19, 64).

The infrared beam reaches the ATR-plate/sample-interface at an angle larger than the limit angle of total reflection. Under this condition, the beam is totally reflected by the interface and penetrates, as evanescent wave, into the sample where it can be absorbed. The optical path of the penetrating beam, which is of the order of the infrared wavelength, depends on the wavelength, on the incident angle, as well as on the refractive indices of the sample and of the ATR element. After one or several reflections, the attenuated beam is collected on the detector. The measured ATR spectrum should be, then, corrected for the dependence of the beam penetration depth upon the infrared wavelength (19). This technique is very useful to keep constant—and limited to a few microns—the optical path of the sample, enabling to easily study proteins in water solution (thanks to the sampling easiness). Also, it allows to measure proteins in the form of films—films obtained by evaporation of diluted protein solutions on the ATR plate—using only a few micrograms of protein. An interesting advantage of ATR measurements is related to the possibility of investigating the interaction of proteins with surface and synthetic membranes, and their consequent conformational changes (19, 46, 64, 65).

8.2.2 Spectral Analysis by Resolution Enhancement Procedures

The amide I band is a broad envelop due to the overlapping of the peptide bond absorption in the different secondary structures of the protein, as can be seen for CRL1 in Figure 8.4 (41). To disclose these components, two resolution enhancement procedures can be used: the second derivative analysis of the spectra and the Fourier self-deconvolution (FSD) method (5, 59, 61).

8.2.2.1 Second Derivative Analysis of the Spectra This is a mathematical operation that can be applied to the measured spectrum after proper smoothing. This procedure enables to identify the overlapping components in the spectrum as negative bands in the second derivative (Fig. 8.4). Indeed, in this way, an enhanced resolution of the spectrum is obtained since the component

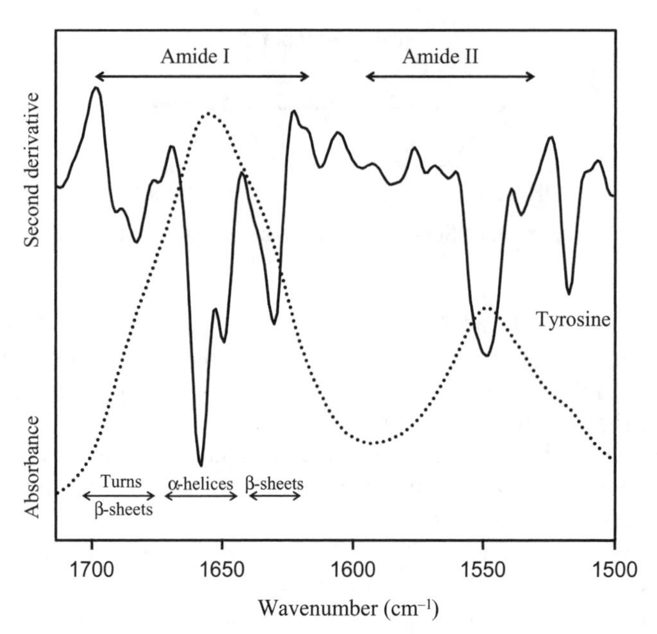

Figure 8.4 Absorption spectrum of CRL1 in water (dotted line), as in Figure 8.1, and its second derivative in the amide I and amide II regions (continuous line). The tyrosine band around 1515 cm^{-1} is indicated.

band half-width in the second derivative is reduced to about a factor 2.7, while the band positions are the same as those in the spectrum (61). The intensity of the negative band components is proportional to the original intensity, but inversely proportional to the square of the half-width in the measured spectrum (61). It should be noted that the inverse proportionality of the band intensity in the second derivative spectrum upon the square of the original band half-width greatly enhances the relative contribution of sharp lines in the spectrum, such as those of noise and vapor. This underlines the need of a high signal-to-noise ratio in the spectrum and of an accurate purging of the spectrometer in order to avoid residual vapor absorption.

A correct quantitative estimate of the components cannot be, in principle, obtained directly from the second derivative spectrum for the above considerations. However, a successful evaluation has been performed in a few cases (13, 14). Furthermore, relative conformational changes induced by different factors, such as temperature, environment, and mutations, can be accurately monitored through the variations in intensity and peak position of the second derivative spectrum (81).

8.2.2.2 Fourier Self-Deconvolution To resolve the different components of the protein amide I band, the FSD method has been proposed (5, 29, 58, 61). The basis of this deconvolution is that a narrowing of the bandwidth in the measured spectrum can be obtained by decreasing the decay rate of the corresponding component in the Fourier time domain. To this goal, a Fourier reverse transformation of the measured spectrum is performed, followed by a multiplication of an exponentially increasing weighting function and an appropriate apodization.

In this deconvolution, two parameters are required: the half-width at half-height (σ) of the unresolved bands in the spectrum and the efficiency factor (K), which determines the extent of narrowing. For σ, the same value is assumed for each unresolved components, even if they have different widths. For the amide I band deconvolution, typical values of σ range from 6.5 to 15 cm^{-1} (27, 61), with the last value being used to resolve random coil secondary structures. Considering the choice for K, its maximum value should be evaluated from the signal-to-noise ratio of the spectrum as $K_{max} = \log(S/N)$, in general limited to values lower than 3 (61). The choice of these parameters is then crucial to avoid artifacts on the deconvoluted spectrum. The goodness of the parameters can be estimated by comparing the number of components and peak positions in the FSD spectrum to those in the second derivative.

8.2.3 Principal Infrared Absorption Bands of Proteins

The principal absorption bands that have been widely used for protein characterization are the so-called amide I (1700–1600 cm^{-1}), amide II (1600–1500 cm^{-1}), and amide III (1400–1200 cm^{-1}), which are due to the vibrational modes of the protein backbone.

The absorption of amino acid side chains, occurring in a wide spectral range, has been recently discussed in detail in several reviews (6, 8, 77) and will not be examined here.

8.2.3.1 Amide I (1700–1600 cm^{-1}). This band, mainly due to the stretching vibration of the C=O peptide bond, is very sensitive to the C=O environment and, therefore, to the secondary structure of the protein. This band is the most used for secondary structure analysis.

8.2.3.2 Amide II (1600–1500 cm^{-1}) This band is due to the contribution of several backbone modes—NH in-plain bending and CN stretching, with small contributions from C=O bending, CC and NC stretching vibrations. It is also sensitive to the protein secondary structure, but its analysis is complicated by the overlapping of the different vibrational modes. Nevertheless, it has been successfully used to study the flexibility of protein structures (19) since it is very sensitive to the hydrogen/deuterium (H/D) exchange. Indeed, as a consequence of the H/D exchange, the amide II band decreases in intensity and shifts to a new deuterated amide II′ band, around 1490–1460 cm^{-1} (Fig. 8.2). The kinetics of the exchange can be evaluated by the time course of the amide II and amide II′ intensity changes (15, 19).

8.2.3.3 Amide III (1400–1200 cm^{-1}) This band is due to contribution of several vibrations—the in-phase combination of NH bending and CN stretching, with small contributions from C=O bending and CC stretching modes. It is also sensitive to the backbone structure, but for its complexity and low intensity is not extensively used. However, due to the reduced water absorption in this spectral region, the analysis of the amide III band may be informative for secondary structure analysis (9, 10, 72).

8.2.4 Amide I Band Assignment to Protein Secondary Structures

A central issue for protein secondary structure analysis is the assignment of the distinct amide I components that have been identified by the resolution enhancement procedures described above. Thanks to experimental FT-IR investigations and computational studies on model compounds, polypeptides and proteins of known three-dimensional structure (see for a review (8, 61)), it has been possible to assign the distinct amide I components to the protein-specific secondary structures according to their peak position (4, 5, 8, 16, 18, 61). However, the absorption of the secondary structure elements can partially overlap in the spectrum, making it difficult—and sometimes arbitrary—to assign the band components. For this reason, it is crucial to validate the assignment with other FT-IR experiments, such as H/D exchange, thermal and chemical protein denaturation. In addition, the FT-IR band assignment can be further assessed by complementary techniques, as for instance circular dichroism (CD) (48). In this way, reliable assignments of the amide I components to α-helix and random coil structures can be made. We report in Table 8.1 the peak positions of the different secondary structures in the amide I region (4, 5, 8, 16, 18, 61). Peak positions of intermolecular β-sheets were taken from References 1, 2, 27, 41–44, 56, 71, and 80.

8.2.4.1 α-Helices The absorption of α-helices occurs in water in the region 1660–1648 cm^{-1} that downshifts of a few wavenumbers in D_2O. The band position depends on the α-helix length, flexibility, and hydration. In particular, higher wavenumbers characterize short and flexible helices, while lower wavenumbers are associated to long and rigid structures (5). For example (65), phospholipase A2 free in solution displays a single α-helix absorption around 1650 cm^{-1}; however, when bound to lipid bilayers, an additional α-helix component is observed at 1658 cm^{-1}, reflecting a more flexible α-helix upon binding. This increased flexibility has been also demonstrated by H/D exchange experiments (65), a powerful technique to evaluate the flexibility of secondary structures.

Furthermore, α-helices forming H-bonds with the solvent display an important shift toward low wavenumbers (1640–1630 cm^{-1} in D_2O), reflecting the weakening of the C=O bond (76). These hydrated α-helices have been observed in a designed globular protein α_3D (76) and in the equilibrium and

Table 8.1 Protein Amide I Band Assignment

Secondary Structure	Band Position (cm^{-1})			
	H_2O		D_2O	
	Average	Extremes	Average	Extremes
Turn	1672	1686–1662	1671	1691–1653
α-Helix	1654	1660–1648	1652	1660–1642
Random coil	1654	1657–1642	1645	1654–1639
β-Sheet	1633	1640–1623	1630	1638–1615
intramolecular	1686	1695–1674	1679	1694–1672
β-Sheet	1625	1630–1620	1620	1630–1611
intermolecular*	1695	1698–1692	1686	1690–1680

Average band positions and spectral ranges taken from Goormaghtigh et al. (1994) (18) and from Arrondo et al. (1993) (5).
*Taken from References 1, 2, 4, 27, 41–44, 56, 71, and 80.

kinetic intermediate states of apomyoglobin (45) and of single-chain monellin (31). Interestingly, an α-helix absorption around 1632–1637 cm^{-1} in D_2O was also observed in alanine-based peptides as a consequence of their hydration and peculiar structure (21, 39, 47, 63, 79). Upon thermal unfolding, these α-helices undergo a transition into random coil structures, which absorb around 1642 cm^{-1} (39).

We should add that, as it will be further discussed, the absorption of α-helices overlaps to that of the random coil structures. To discriminate between them, it is possible to compare the protein spectrum measured in H_2O and D_2O since the H/D exchange is different for the two secondary structures.

8.2.4.2 Intramolecular β-Sheets Intramolecular β-sheets display two absorption bands of different intensity. For antiparallel structures, a low-frequency band occurs in water around 1633 cm^{-1} and a high-frequency band of lower intensity around 1686 cm^{-1}; both bands are downshifted in D_2O (see Table 8.1). For parallel β-sheets, the low-frequency band is usually upshifted a few wavenumbers, while the high-frequency band—which in some native proteins is of very low intensity—is expected to be downshifted (8). In fact, from computational studies (see Reference 8 for a review), a reduced splitting between the two components is expected for parallel β-sheet. However, it is difficult to discriminate between parallel and antiparallel β-sheets (8, 32), as pointed out by several experimental works (30, 62), since a similar infrared response has been observed for the two structures in native proteins. Indeed, it should be noted that the band positions and intensities of β-sheet structures depend critically on the geometry of the structure, such as the strand twist angle (80), the number of strands per sheet (80), and the H-bond strength (67).

During protein unfolding, the low-frequency band of β-sheet structures was often found to shift toward higher wavenumbers due to the weakening of the H-bonded structures. For instance, during the β-sheet to α-helix transition of β-lactoglobulin, an upshift from 1632 cm^{-1} (native β-sheet) to 1637 cm^{-1} was observed for the β intermediate state, indicating the presence of looser β-sheet structures (28).

8.2.4.3 Intermolecular β-Sheets Intermolecular β-sheet structures in protein assemblies display a similar absorption of native intramolecular β-sheets, but shifted in peak positions (see Table 8.1). In amyloid and thermal aggregates, as well as in bacterial inclusion bodies, the low-frequency band was found to be in the range 1630–1620 cm^{-1} in water (1630–1611 cm^{-1} in D_2O), while the high-frequency band was observed between 1698–1692 cm^{-1} in water (1690–1680 cm^{-1} in D_2O).

Also in this case, it is difficult to discriminate between parallel and antiparallel β-sheet in aggregates using the infrared spectra. However, as suggested in literature for several aggregates, practically, it is possible to assign intermolecular parallel β-sheet structures when only the β-sheet band at low frequency is observed (12, 43, 49, 78).

A reliable assignment is not easy if the two β-sheet components are both present in the spectrum, as for instance in the case of Aβ1-40 fibrils. Indeed, solid-state NMR indicates a parallel orientation of β-sheets in Aβ1-40 fibrils (3), while their FT-IR spectrum displays the two β-sheet components (17).

8.2.4.4 Random Coil The absorption of the C=O group in random coil structures is due to the contributions of the peptide bonds in different environments, therefore leading to a broad band centered around 1654 cm^{-1} (Table 8.1) in water. Unfortunately, this band is superimposed to the α-helix secondary structure. Upon H/D exchange in D_2O, the band downshifts to 1645 cm^{-1} and can be, therefore, discriminated from that of the α-helix structure in the protein spectrum. However, it should be noted that in this region can be also found the absorption of open loops, which is around 1643 cm^{-1}, both in H_2O and in D_2O. The comparison between spectra in water and heavy water can, then, be useful to evaluate the contribution of random coils, α-helix, and loops (53, 54, 55).

The large bandwidth of random coil structures makes it difficult to detect them from the amide I second derivative spectrum. As we noted above, the intensity of the negative peaks in the second derivative spectrum is inversely proportional to the bandwidth of the corresponding component in the measured spectrum. Therefore, the sharp contributions of the other secondary structures might dominate the derivative spectrum, hiding the contribution of the broad band of random coil structures. To study these structures in IDPs, the FSD analysis is often used, as it will be shown here in few examples.

8.2.4.5 Other Secondary Structures The absorption of other secondary structures occurs in the wide amide I range (4, 8, 18, 61, 64). Among them, β-turns can be found from 1686 to 1660 cm^{-1} in water and from 1691 to 1653 cm^{-1} in D_2O (18); also the 3_{10} helix structures occur in the range from 1670 to 1660 cm^{-1} in water and from 1645 to 1634 cm^{-1} in D_2O (18).

8.2.5 Protein Secondary Structure Evaluation

8.2.5.1 Relative Secondary Structure Changes Changes in protein secondary structures induced by different environmental conditions and interactions, as in the case of the induced folding in IDPs, can be easily studied through the FT-IR absorption spectra and their second derivative or FSD analysis. Indeed, the variation of protein secondary structures can be determined by monitoring the intensity changes of the corresponding absorption bands in the resolution-enhanced spectra.

In this way, even if no information on the secondary structure percentage is obtained, folding, unfolding, and aggregation can be monitored directly from the resolution-enhanced spectra.

The thermal denaturation of CRL1 in D_2O (41) is presented to illustrate this point. In Figure 8.5, the unfolding of the secondary structures and the concomitant formation of thermal aggregates are monitored by the second derivative spectra in the amide I region. In the figure, two α-helix components are resolved in the CRL1 spectra in D_2O (41).

8.2.5.2 Quantitative Evaluation of Secondary Structure For a quantitative evaluation of the fraction of the different secondary structural elements of a protein, a further analysis of the spectra is required. The most common procedure is to perform a curve fitting of the amide I band as a linear combination of a set of Lorentian or Gaussian functions, each assigned to a given secondary structural element. This fitting can be performed on the measured absorption spectrum (4, 16) as well as on the FSD (18).

In the curve fitting of the measured spectra, the parameters of the different band components—heights, widths and positions—are adjusted iteratively in order to find a set of best fitting functions. The fractional area of each Lorentian/Gaussian function—over the component total area—represents the percentage of the corresponding secondary structure. Considering the complexity of this fitting due to the large number of parameters, attention should be given to the choice of their initial values. The number of components and the initial values of their band positions can be taken from second derivative and FSD spectra. More problematic is the choice of the initial values of band heights and widths. To this goal, several approaches have been proposed (4, 16, 18, 41, 75). For instance, in Vila et al. (75), the initial values of band height are set for the main components at 90% of those of the measured spectrum, and at 70% for the other components. Then, a first fitting of the measured

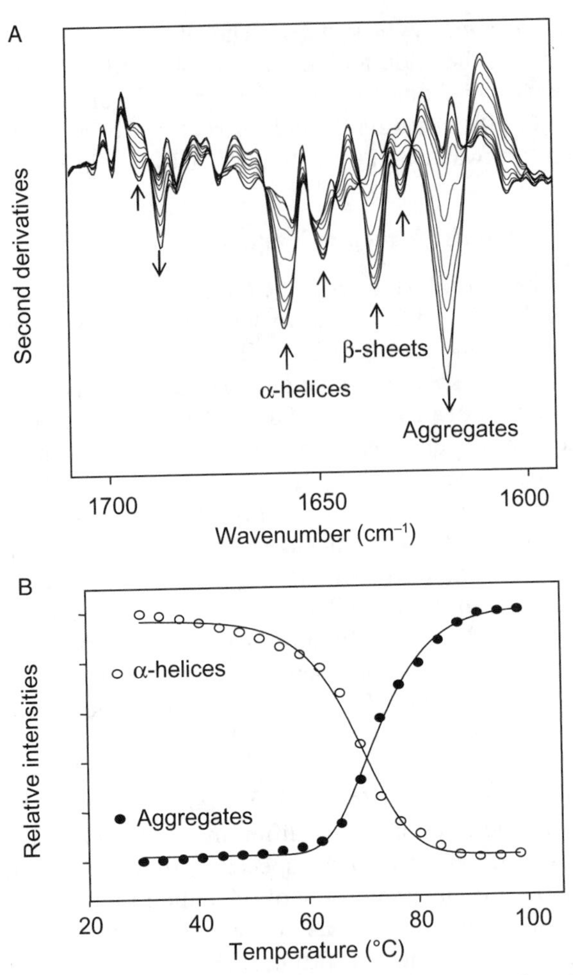

Figure 8.5 (A) Second derivative spectra of CRL1 in D_2O as a function of temperature in the range 20–100°C. The arrows point to increasing temperature. Spectra were measured as in Figure 8.1, but at 1 cm^{-1} resolution. (B) Temperature dependence of α-helix and aggregate components taken from (A).

spectrum is performed by fixing the band positions and by letting free to adjust iteratively height and width parameters. The resulting values can be taken as initial parameters for a second curve fit where they are all free to adjust to the final best fitting values.

Approaches based on pattern recognition can also be used for secondary structure prediction, similar to what is done in CD spectroscopy. These methods, including factor analysis and partial least square analysis, require a calibration set of infrared spectra from proteins of known X-ray structure ([23] and references therein). Neural network approach has been also successfully used (25).

8.3 IDPs AND INDUCED FOLDING BY FT-IR SPECTROSCOPY

In this section, we present selected examples about IDP structural properties studied by FT-IR spectroscopy.

8.3.1 Induced Folding of Histone H1° Protein by DNA Interaction

The histone H1° protein, which has been found to play a role in the stabilization of nucleosome and chromatine structure, is made by a central globular domain and by two disordered amino-terminal and carboxy-terminal tails. These disordered regions undergo a structural transition into an ordered state upon interaction with DNA. This process has been studied by FT-IR spectroscopy, examining first a short carboxy-terminal peptide (CH1) from residue 99 to 121 (75), the entire carboxy-terminal domain (53), as well as two peptides within the amino terminal domain (74).

The absorption spectrum of CH1 in D_2O, measured in transmission, is reported in the amide I region in Figure 8.6A, where also the FSD spectrum and the curve fitting of the measured data are presented (75). As shown by the curve fitting results, a broad band around 1643 cm^{-1} characterizes the spectrum and can be assigned to random coil structures. In addition, a component around 1662 cm^{-1} is observed and assigned to turn structures. The relative weights of these components are, respectively, 38% and 25% of the total secondary structures, indicating that CH1 is mainly in a disordered state. In the presence of 90% trifluoroethanol (TFE), an induced folding is observed as shown in Figure 8.6B. Under this condition, the spectrum of CH1 in D_2O is dominated by two main components around 1654 and 1670 cm^{-1}, respectively assigned to α-helix and turn structures. The relative weight of these two components is 51% for the α-helices and 20% for turns.

When the CH1 interacts with mouse DNA and with polynucleotides poly[dA-dT]•poly[dA-dT] and poly[dA]•poly[dT], the peptide undergoes a structural transition from a disordered to a folded state (Fig. 8.7). In particular, the interaction with different DNAs induces the formation of α-helical structures, with peak positions around 1647 and 1657 cm^{-1}, and of turns at 1670 cm^{-1}, each component having a relative weight of about 20% of the total secondary structure. Interestingly, the two α-helical components at 1647 and 1657 cm^{-1} might reflect a different environment for these structures when the protein interacts with DNA, which can induce a helix distortion (75).

This study has been extended to the entire carboxyl-terminal domain (53), which was also found to be mainly in a disordered state. Again, the interaction with DNA induced a folding into α-helix, turns, and also in β-sheet structures, which were observed to be highly stable against thermal treatments. Interestingly, the DNA-induced folding has been found to depend on the presence of salt. At low NaCl concentration (10 mM), a residual random coil structure is retained, which disappeared at physiological concentration (140 mM NaCl).

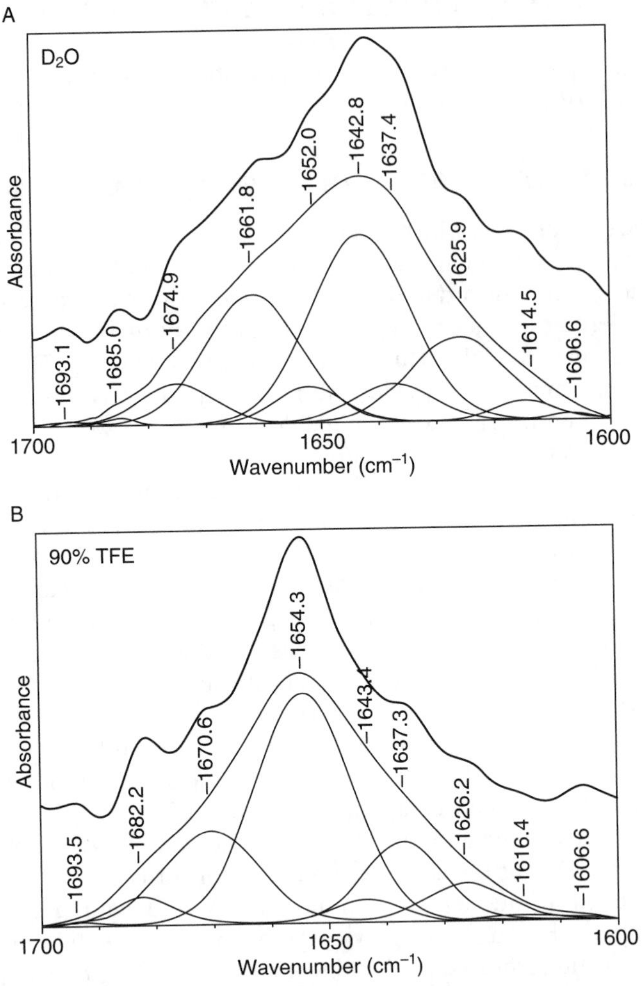

Figure 8.6 Amide I absorption spectrum and curve fitting of the CH1 peptide in D_2O solution. The spectrum of CH1 (4.6 mg/mL) in D_2O (A) and in presence of 90% TFE (B).The curve fitted spectra as a linear combination of Gaussian components (dashed line) are completely superimposed to the measured spectra (continuous line). On the top of (A) and (B) are reported the Fourier self-deconvolution of the measured spectra used to identify the component peak positions. The FT-IR spectra were measured at $2\,cm^{-1}$ resolution by a Nicolet Magna II 550 spectrometer with a mercury cadmium telluride detector, using a CaF_2 transmission cell with $50\,\mu m$ optical path. Taken from Vila et al. (75).

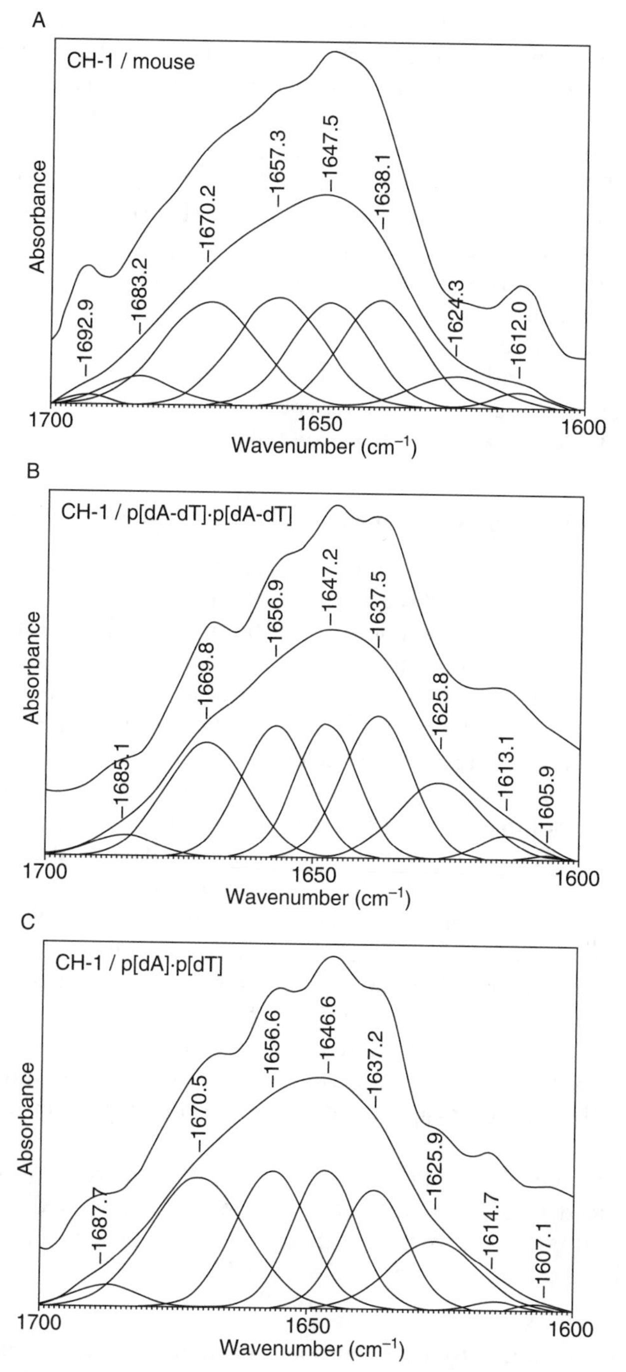

Figure 8.7 Induced folding of CH1 by DNA interaction. Amide I absorption spectra of CH-1 in presence of DNA, at peptide/DNA ratio = 0.7 (w/w). The measured spectra (obtained as in Fig. 8.6) were corrected for the DNA absorption by subtraction of the DNA spectrum, collected under the same conditions. Taken from Vila et al. (75).

We should note that the authors performed the FT-IR measurements both in D_2O as well as in water to confirm the assignment of the random coil and of the induced secondary structures. Indeed, while stable secondary structures downshift only a few wavenumbers in D_2O, random coil structures undergo a major shift due to their efficient H/D exchange.

Recently, FT-IR spectroscopy has been also successfully employed to characterize conformational changes induced on the H1 carboxyl-terminal domain by crowding agents (fycoll 70 and PEG 6000) (54) and by phosphorylation (55). Interestingly, in this second case, phosphorylation does not induce any structural change of the disordered domain free in solution, but only in the presence of the bound DNA.

Furthermore, the induced folding of two N-terminal disordered peptides of H1° has been also studied by FT-IR spectroscopy as complementary techniques to CD and to NMR spectroscopy (74).

8.3.2 Induced Folding of α-Synuclein in Water–Alchool Mixtures

α-Synuclein is a small IDP that plays a crucial role in neurodegenerative disorders, including Parkinson's disease. FT-IR spectroscopy has been extensively used as complementary technique to other biophysical methods to study conformational properties and aggregation of this IDP (26, 37, 38, 40, 51, 68, 69).

In particular, the induced folding of the protein in methanol (MeOH) and hexafluoro-2-propanol (Hfip) has been investigated by ATR spectroscopy (40). The FT-IR spectrum of α-synuclein in water at pH 7.5 (Fig. 8.8A) is dominated by the large absorption of the random coil structures around $1649\,cm^{-1}$. The addition of 10% MeOH (Fig. 8.8B) induces a partial folding of the protein mainly into β-sheet structures as indicated by the band around $1635\,cm^{-1}$. Interestingly, at higher MeOH concentration (40%), a new component grows around $1624\,cm^{-1}$ and it can be assigned to intermolecular β-sheet structures in oligomers (Fig. 8.8C). A different folding intermediate is induced by 20% Hfip, as shown in Fig. 8.8D, where an important α-helical structure is indicated by the presence of the $1653\,cm^{-1}$ component, in addition to β-sheet components around 1637 and $1625\,cm^{-1}$. Furthermore, the FT-IR spectrum of α-synuclein fibrils formed in water (Fig. 8.8E) displays a high-intensity band at $1629\,cm^{-1}$ that can be assigned to a β-sheet protein–protein interaction in the aggregates.

All these results highlight the potential of FT-IR spectroscopy to characterize the induced folding of α-synuclein under different conditions. Indeed, different band components in the amide I region allow to identify distinct partially structured intermediates of α-synuclein.

8.3.3 Induced Folding by Osmolytes and Partner Protein Interaction

In the following examples, FT-IR spectroscopy has been applied to characterize the conformational transitions of the intrinsically disordered domain AF1

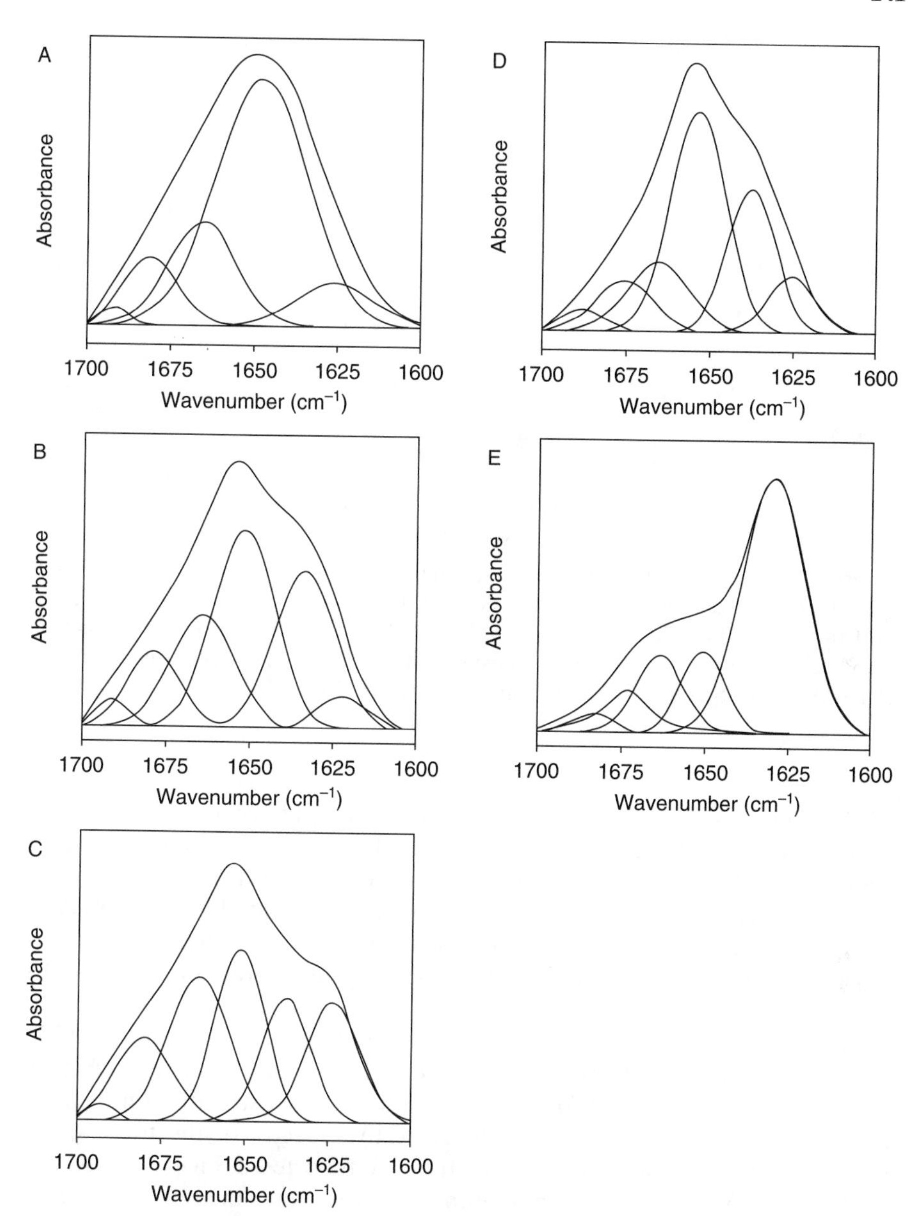

Figure 8.8 Secondary structures induced on α-synuclein. ATR/FT-IR absorption spectra of the protein in form of hydrated film and curve fitting analysis: (A) IDP at pH 7.4; (B), (C), and (D) partially folded intermediates induced respectively by 10% MeOH, 40% MeOH, and 20% Hfip; (E) α-synuclein fibrils obtained in aqueous buffer. The ATR/FT-IR spectra, at a resolution of 1 cm^{-1}, were collected by a Thermo-Nicolet Nexus 670 spectrometer equipped with MCT detector, using a germanium internal reflectance element. Taken from Munishkina et al. (40).

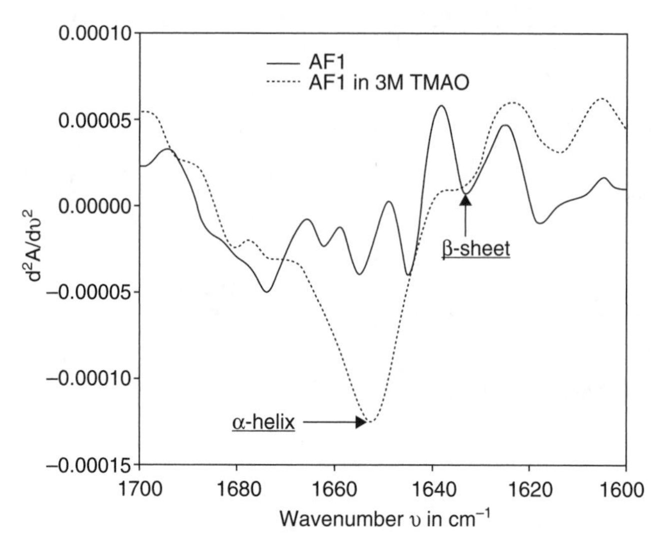

Figure 8.9 Induced folding of AF1 domain by osmolytes. Second derivative of FT-IR spectra of the androgen receptor AF1 domain (0.2 mM), free in water solution (continuous line), and in presence of 3 MTMAO (dashed line). The absorption spectra at 4 cm^{-1} resolution were measured in transmission by a BOMEM MB Series spectrometer with a DTGS detector. Taken from Kumar et al. (33).

of the androgen receptor (33). The second derivative spectrum of AF1 in water, reported in Figure 8.9, displays a broad envelop in the amide I region, where a few low-intensity components can be also observed, suggesting an overall disordered structure. The addition of 3M trimethylamine-N-oxide (TMAO) induces the rising of an intense component at 1656 cm^{-1}, indicating that a conformational transition from the natively disordered state into α-helix structures takes place.

Among the physiological partners of the androgen receptor, the RAP74 protein is a subunit of the transcription factor TFIIF complex. The C-terminal domain of this protein was found to induce an ordered structure on the AF1 as shown in Figure 8.10. The second derivative spectra of the two partner proteins, free in solution and their mixture, are reported in Figure 8.10A. The AF1 spectrum indicates an unordered structure, while the spectrum of its partner is characterized by an α-helix component around 1656 cm^{-1}. The mixture spectrum is again dominated by a strong component at 1656 cm^{-1}, which, however, cannot be obtained by the simple addition of the two partner spectra (Fig. 8.10B). This result indicates that in agreement with what is observed in the presence of TMAO, a conformational transition into α-helical structures is induced by the interaction with RAP74 protein (33).

Furthermore, the induced folding of the AF1 domain of the glucocorticoid receptor has been studied by FT-IR spectroscopy (34). Also in this case, the

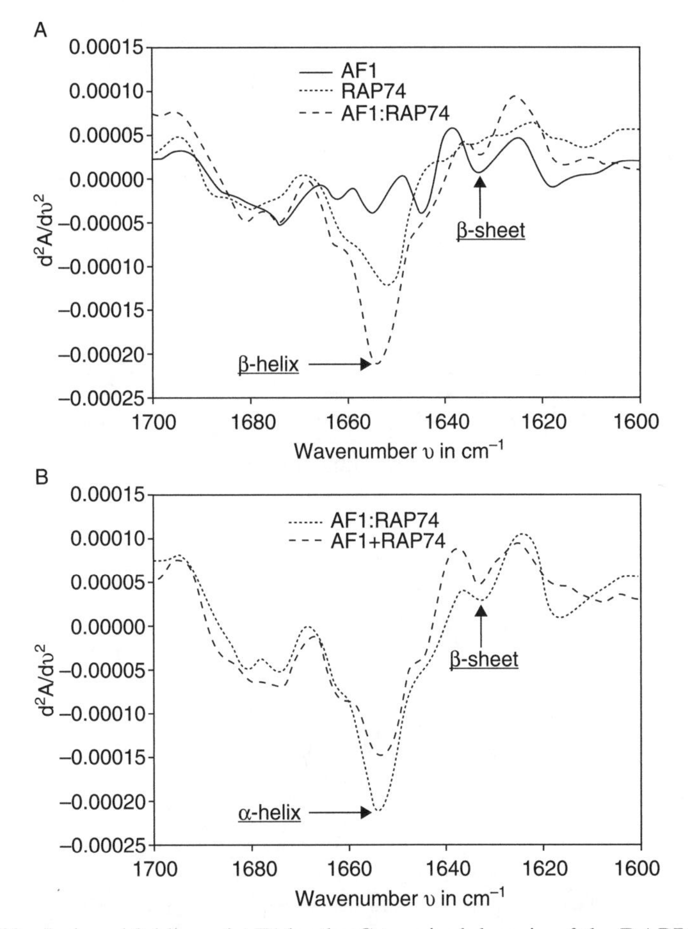

Figure 8.10 Induced folding of AF1 by the C-terminal domain of the RAP74 protein. (A) Second derivative spectra of AF1 (continuous line), of RAP74 (dashed line), and of the two protein mixture (dashed-dotted line). (B) Algebric sum of the spectra of the two proteins free in solution (dashed) and spectrum of the two-protein mixture (dashed-dotted line). Taken from Kumar et al. (33).

disordered AF1 domain acquires an α-helical structure when interacting with the TATA box-binding protein, a part of the general transcription machinery (34).

8.3.4 Induced Folding by Dehydration

Late embryogenesis abundant (LEA) proteins are IDPs that play an important role in desiccation tolerance in plants and seeds. Recently, it has been discovered that a LEA protein is also expressed during dehydration in

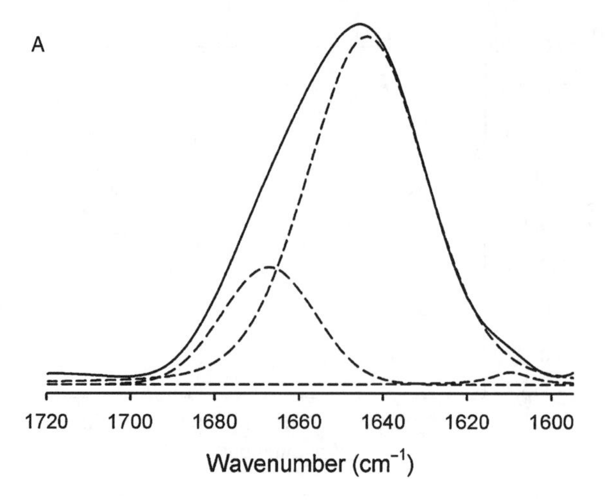

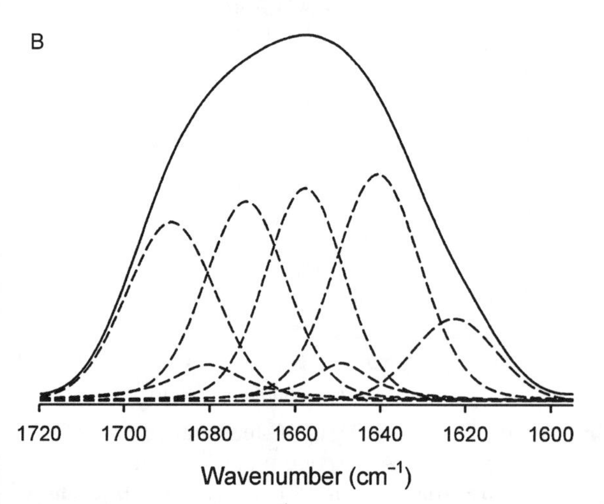

Figure 8.11 Induced folding by dehydration of AavLEA1 protein. FT-IR spectra in D_2O solution (A) and after drying (B). In (A), 15 mg/mL of protein solution was measured in a transmission cell with 50 μm optical path. In (B), the protein solution—as in (A)—was deposed on a CaF_2 infrared support and dried under vacuum for 30 min. The spectrum of the protein film was then measured. Both (A) and (B) report the results of the curve fitting into Gaussian/Lorentzian components. The spectra were measured in transmission with a Bruker Equinox 55 FT-IR spectrometer at 2 cm^{-1} resolution. Taken from Goyal et al. (20).

anhydrobiotic animals (20). Among them, the nematode protein Aav-LEA1 has been found to undergo a transition into a more folded α-helical structure after drying. The absorption spectra of Aav-LEA1, in solution and dried, are reported in Figure 8.11, together with the decomposition of the amide I band into secondary structure components. The spectrum in solution is dominated by a 1644 cm^{-1} band due to random coil structures. Upon drying, which is a natural stress on anhydrobiotes, the protein spectrum changes drastically, displaying major components at 1641, 1658, and 1672 cm^{-1}. This peculiar pattern has been suggested to result from α-helix elements arranged as coiled coils (24). This FT-IR result, confirmed by other complementary studies, clearly indicates that a transition into ordered α-helices is induced by dehydration.

A similar transition has been observed also for mitochondrial LEA protein expressed in plant seeds, where the protein bound to liposomes was found to play a protective role against dehydration (66).

8.3.5 Conformational Changes during Aggregation of Disordered Proteins and Peptides

A number of IDPs and disordered peptides that are involved in amyloid diseases (11, 70) undergo a transition from random coil to cross-β-sheet structures in fibrils. This transition has been characterized by FT-IR spectroscopy for several proteins, such as α-synuclein (40, 51, 69), prion peptides (27, 35, 43), and Aβ peptides (17, 49).

Here we report the FT-IR study of the aggregation kinetics of the human prion peptide fragment spanning residues 82–146 (PrP82-146) (43). In Figure 8.12A, the absorption spectrum of the peptide immediately after solubilization in heavy water is characterized by a broad band around 1645 cm^{-1}, reflecting its random coil structure, in addition to a 1673 cm^{-1} band, due to a residual trifluoroacetic acid. Upon incubation at 37°C, a fibril formation has been observed by means of electron and atomic force microscopies after 1 week. The spectrum of fibrils is also reported in Fig. 8.12A, where the peak at 1626 cm^{-1} indicates the formation of intermolecular β-sheet structures. The second derivative spectra taken during the time course of aggregation are also reported in Figure 8.12B,C. In the early stages of aggregation up to 6–7 days, the formation of oligomers can be detected by the presence of two bands of low intensity at 1623 and 1690 cm^{-1} that can be assigned to intermolecular β-sheet structures. In the enlarged Figure 8.12C, the presence of an isosbestic point around 1627 cm^{-1} indicates that the disordered peptides were converting into β-sheet aggregates. After a lag time of about 3–7 days, the 1626 cm^{-1} fibril band suddenly appears and increases, reaching a plateau after 10–11 days (Fig. 8.12D). The rising of the 1626 cm^{-1} band was not accompanied by the growth of the second β-sheet component at higher wavenumbers, indicating that a parallel β-sheet structure characterizes PrP86-142 fibrils.

These results indicate therefore that PrP82-146 undergoes a transition from random coil to intermolecular β-sheets in oligomers and to intermolecular parallel β-sheet structures in fibrils (43).

Interestingly, other aggregation pathways of PrP82-146 have been characterized, leading to end products with a different infrared response. In

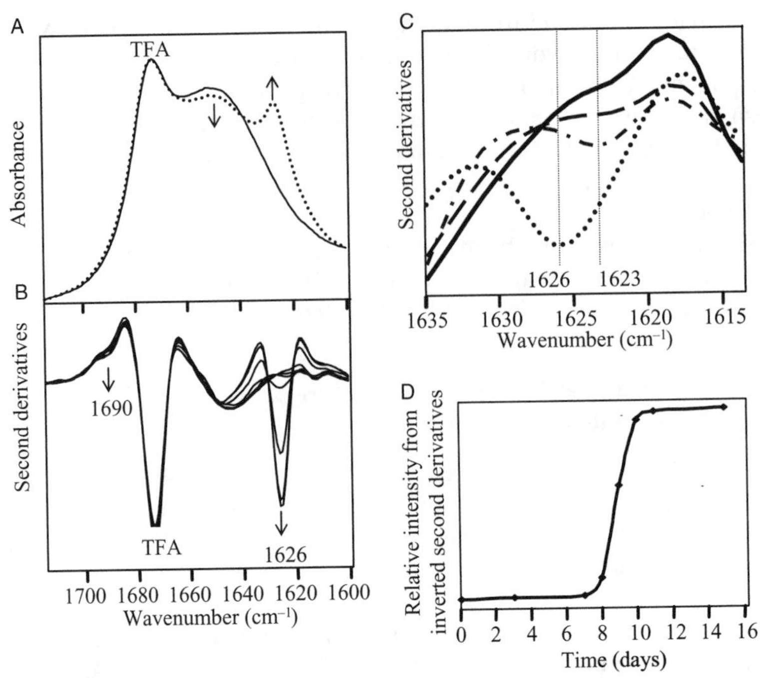

Figure 8.12 Kinetics of fibril formation of PrP82-146 in D_2O. (A) FT-IR absorption spectra of the peptide immediately after sample preparation (continuous line) and 11 days of incubation at 37°C (dotted line). (B) FT-IR second derivative spectra of the peptide immediately after sample preparation and 3, 7–11 days of incubation at 37°C. Arrows point to increasing time. (C) FT-IR second derivative spectra at time 0 (continuous line) and 3 (dashed line), 7 (dashed–dotted line), and 8 (dotted line) days taken from (B) and reported on enlarged scale. (D) Kinetics of aggregation monitored by the fibril β-sheet band at 1626 cm^{-1}.

particular, thermal treatment induced the formation of aggregates with infrared absorption at 1622 and 1690 cm^{-1} due to β-sheet structures. Random aggregates (36) induced by dehydration of the peptide solution displayed several secondary structure components at 1659, 1633–1623, and 1698 cm^{-1}. While the two last components are due to β-sheets, the 1659 cm^{-1} component can be assigned to α-helical structures. Indeed, it has been shown by solid-state NMR that similar prion peptides in random aggregates form α-helical structures (36).

All these results demonstrated that PrP82-146 has a high plasticity as indicated by its capability to undergo different aggregation pathways, with end products characterized by distinct FT-IR responses. Thanks to the specific signature of β-sheet intermolecular structures, FT-IR spectroscopy has been successfully used to characterize protein aggregates (27, 56, 80), even allowing prion strain discrimination (27, 57).

ACKNOWLEDGMENTS

We are grateful to Prof. José Luis Arrondo for the critical reading of the manuscript and helpful suggestions. We thank Carla Smeraldi for kindly revising the language. S. M. D. acknowledges the financial support of the grant Fondo di Ateneo per la Ricerca of the University of Milano-Bicocca. A. N. acknowledges a postdoctoral fellowship of the University of Milano-Bicocca.

REFERENCES

1. Ami, D., A. Natalello, P. Gatti-Lafranconi, M. Lotti, and S. M. Doglia. 2005. Kinetics of inclusion body formation studied in intact cells by FT-IR spectroscopy. FEBS Lett **579**:3433–6.
2. Ami, D., A. Natalello, G. Taylor, G. Tonon, and S. M. Doglia. 2006. Structural analysis of protein inclusion bodies by Fourier transform infrared microspectroscopy. Biochim Biophys Acta **1764**:793–9.
3. Antzutkin, O. N., J. J. Balbach, R. D. Leapman, N. W. Rizzo, J. Reed, and R. Tycko. 2000. Multiple quantum solid-state NMR indicates a parallel, not antiparallel, organization of ß-sheets in Alzheimer's ß-amyloid fibrils. Proc Natl Acad Sci U S A **97**:13045–50.
4. Arrondo, J. L. R., and F. M. Goni. 1999. Structure and dynamics of membrane proteins as studied by infrared spectroscopy. Prog Biophys Mol Biol **72**: 367–405.
5. Arrondo, J. L. R., A. Muga, J. Castresana, and F. M. Goni. 1993. Quantitative studies of the structure of proteins in solution by Fourier-transform infrared spectroscopy. Prog Biophys Mol Biol **59**:23–56.
6. Barth, A. 2000. The infrared absorption of amino acid side chains. Prog Biophys Mol Biol **74**:141–73.
7. Barth, A. 2007. Infrared spectroscopy of proteins. Biochim Biophys Acta **1767**:1073–101.
8. Barth, A., and C. Zscherp. 2002. What vibrations tell us about proteins. Q Rev Biophys **35**:369–430.
9. Cai, S. W., and B. R. Singh. 1999. Identification of beta-turn and random coil amide III infrared bands for secondary structure estimation of proteins. Biophys Chem **80**:7–20.
10. Cai, S. W., and B. R. Singh. 2004. A distinct utility of the amide III infrared band for secondary structure estimation of aqueous protein solutions using partial least squares methods. Biochemistry **43**:2541–9.
11. Chiti, F., and C. M. Dobson. 2006. Protein misfolding, functional amyloid, and human disease. Annu Rev Biochem **75**:333–66.
12. Chitnumsub, P., W. R. Fiori, H. A. Lashuel, H. Diaz, and J. W. Kelly. 1999. The nucleation of monomeric parallel β-sheet-like structures and their self-assembly in aqueous solution. Bioorg Med Chem **7**:39–59.

13. Dong, A., P. Huang, and W. S. Caughey. 1990. Protein secondary structures in water from 2Nd-derivative amide-I infrared-spectra. Biochemistry **29**:3303–8.
14. Dong, A., B. Kendrick, L. kreilgard, J. Matsuura, M. C. Manning, and J. F. Carpenter. 1997. Spectroscopic study of secondary structure and thermal denaturation of recombinant human factor XIII in aqueous solution. Arch Biochem Biophys **347**:213–20.
15. Dong, A., R. M. Hyslop, and D. L. Pringle. 1996. Differences in conformational dynamics of ribonucleases A and S as observed by infrared spectroscopy and hydrogen-deuterium exchange. Arch Biochem Biophys **333**:275–81.
16. Echabe, I., and J. L. R. Arrondo. 1997. Structural analysis of proteins by infrared spectroscopy, pp. 9–23. In G. Pifat-Mrzljak (ed.), Supramolecular Structure and Function 5. Balaban Publishers, Rehovot, Israel.
17. Fraser, P. E., J. T. Nguyen, H. Inouye, W. K. Surewicz, D. J. Selkoe, M. B. Podlisny, and D. A. Kirschner. 1992. Fibril formation by primate, rodent, and Dutch-hemorrhagic analogs of Alzheimer amyloid ß-protein. Biochemistry **31**: 10716–23.
18. Goormaghtigh, E., V. Cabiaux, and J. M. Ruysschaert. 1994. Determination of soluble and membrane protein structure by Fourier transform infrared spectroscopy. III. Secondary structures. Subcell Biochem **23**:405–50.
19. Goormaghtigh, E., V. Raussens, and J. M. Ruysschaert. 1999. Attenuated total reflection infrared spectroscopy of proteins and lipids in biological membranes. Biochim Biophys Acta **1422**:105–85.
20. Goyal, K., L. Tisi, A. Basran, J. Browne, A. Burnell, J. Zurdo, and A. Tunnacliffe. 2003. Transition from natively unfolded to folded state induced by desiccation in an anhydrobiotic nematode protein. J Biol Chem **278**:12977–84.
21. Graff, D. K., B. PastranaRios, S. Y. Venyaminov, and F. G. Prendergast. 1997. The effects of chain length and thermal denaturation on helix-forming peptides: a mode-specific analysis using 2D FT-IR. J Am Chem Soc **119**:11282–94.
22. Haris, P. I., and D. Chapman. 1994. Analysis of polypeptide and protein structures using Fourier transform infrared spectroscopy. Methods Mol Biol **22**:183–202.
23. Haris, P. I., and F. Severcan. 1999. FTIR spectroscopic characterization of protein structure in aqueous and non-aqueous media. J Mol Catal B Enzym **7**:207–21.
24. Heimburg, T., J. Schunemann, K. Weber, and N. Geisler. 1999. FTIR-spectroscopy of multistranded coiled coil proteins. Biochemistry **38**:12727–34.
25. Hering, J. A., P. R. Innocent, and P. I. Haris. 2002. Automatic amide I frequency selection for rapid quantification of protein secondary structure from Fourier transform infrared spectra of proteins. Proteomics **2**:839–49.
26. Hokenson, M. J., V. N. Uversky, J. Goers, G. Yamin, L. A. Munishkina, and A. L. Fink. 2004. Role of individual methionines in the fibrillation of methionine-oxidized α-synuclein. Biochemistry **43**:4621–33.
27. Jones, E. M., and W. K. Surewicz. 2005. Fibril conformation as the basis of species-and strain-dependent seeding specificity of mammalian prion amyloids. Cell **121**:63–72.
28. Kauffmann, E., N. C. Darnton, R. H. Austin, C. Batt, and K. Gerwert. 2001. Lifetimes of intermediates in the β-sheet to α-helix transition of β-lactoglobulin by using a diffusional IR mixer. Proc Natl Acad Sci U S A **98**:6646–9.

29. Kauppinen, J. K., D. J. Moffatt, H. H. Mantsch, and D. G. Cameron. 1981. Fourier self-deconvolution—a method for resolving intrinsically overlapped bands. Appl Spectrosc **35**:271–6.
30. Khurana, R., and A. L. Fink. 2000. Do parallel β-helix proteins have a unique Fourier transform infrared spectrum? Biophys J **78**:994–1000.
31. Kimura, T., A. Maeda, S. Nishiguchi, K. Ishimori, I. Morishima, T. Konno, Y. Goto, and S. Takahashi. 2008. Dehydration of main-chain amides in the final folding step of single-chain monellin revealed by time-resolved infrared spectroscopy. Proc Natl Acad Sci U S A **105**:13391–6.
32. Kubelka, J., and T. A. Keiderling. 2001. Differentiation of beta-sheet-forming structures: ab initio-based simulations of IR absorption and vibrational CD for model peptide and protein β-sheets. J Am Chem Soc **123**:12048–58.
33. Kumar, R., R. Betney, J. Q. Li, E. B. Thompson, and I. J. Mcewan. 2004. Induced α-helix structure in AF1 of the androgen receptor upon binding transcription factor TFIIF. Biochemistry **43**:3008–13.
34. Kumar, R., D. E. Volk, J. Q. Li, J. C. Lee, D. G. Gorenstein, and E. B. Thompson. 2004. TATA box binding protein induces structure in the recombinant glucocorticoid receptor AF1 domain. Proc Natl Acad Sci U S A **101**:16425–30.
35. Kundu, B., N. R. Maiti, E. M. Jones, K. A. Surewicz, D. L. Vanik, and W. K. Surewicz. 2003. Nucleation-dependent conformational conversion of the Y145Stop variant of human prion protein: structural clues for prion propagation. Proc Natl Acad Sci U S A **100**:12069–74.
36. Laws, D. D., H. L. Bitter, K. Liu, H. L. Ball, K. Kaneko, H. Wille, F. E. Cohen, S. B. Prusiner, A. Pines, and D. E. Wemmer. 2001. Solid-state NMR studies of the secondary structure of a mutant prion protein fragment of 55 residues that induces neurodegeneration. Proc Natl Acad Sci U S A **98**:11686–90.
37. Li, J., V. N. Uversky, and A. L. Fink. 2001. Effect of familial Parkinson's disease point mutations A30P and A53T on the structural properties, aggregation, and fibrillation of human α-synuclein. Biochemistry **40**:11604–13.
38. Li, J., V. N. Uversky, and A. L. Fink. 2002. Conformational behavior of human α-synuclein is modulated by familial Parkinson's disease point mutations A30P and A53T. Neurotoxicology **23**:553–67.
39. Martinez, G., and G. Millhauser. 1995. Ftir spectroscopy of alanine-based peptides: assignment of the amide i' modes for random coil and helix. J Struct Biol **114**:23–7.
40. Munishkina, L. A., C. Phelan, V. N. Uversky, and A. L. Fink. 2003. Conformational behavior and aggregation of α-synuclein in organic solvents: modeling the effects of membranes. Biochemistry **42**:2720–30.
41. Natalello, A., D. Ami, S. Brocca, M. Lotti, and S. M. Doglia. 2005. Secondary structure, conformational stability and glycosylation of a recombinant Candida rugosa lipase studied by Fourier-transform infrared spectroscopy. Biochem J **385**:511–7.
42. Natalello, A., S. M. Doglia, J. Carey, and R. Grandori. 2007. Role of flavin mononucleotide in the thermostability and oligomerization of Escherichia coli stress-defense protein WrbA. Biochemistry **46**:543–53.
43. Natalello, A., V. V. Prokorov, F. Tagliavini, M. Morbin, G. Forloni, M. Beeg, C. Manzoni, L. Colombo, M. Gobbi, M. Salmona, and S. M. Doglia. 2008.

Conformational plasticity of the Gerstmann-Sträussler-Scheinker disease peptide as indicated by its multiple aggregation pathways. J Mol Biol **381**:1349–61.

44. Natalello, A., R. Santarella, S. M. Doglia, and A. de Marco. 2008. Physical and chemical perturbations induce the formation of protein aggregates with different structural features. Protein Expr Purif **58**:356–61.
45. Nishiguchi, S., Y. Goto, and S. Takahashi. 2007. Solvation and desolvation dynamics in apomyoglobin folding monitored by time-resolved infrared spectroscopy. J Mol Biol **373**:491–502.
46. Noinville, S., M. Revault, M. H. Baron, A. Tiss, S. Yapoudjian, M. Ivanova, and R. Verger. 2002. Conformational changes and orientation of Humicola lanuginosa lipase on a solid hydrophobic surface: an in situ interface Fourier transform infrared-attenuated total reflection study. Biophys J **82**:2709–19.
47. Pastrana-Rios, B. 2001. Mechanism of unfolding of a model helical peptide. Biochemistry **40**:9074–81.
48. Pelton, J. T., and L. R. Mclean. 2000. Spectroscopic methods for analysis of protein secondary structure. Anal Biochem **277**:167–76.
49. Peralvarez-Marin, A., A. Barth, and A. Graslund. 2008. Time-resolved infrared spectroscopy of pH-Induced aggregation of the Alzheimer $A\beta_{1-28}$ peptide. J Mol Biol **379**:589–96.
50. Rahmelow, K., and W. Hubner. 1997. Infrared spectroscopy in aqueous solution: difficulties and accuracy of water subtraction. Appl Spectrosc **51**:160–70.
51. Ramakrishnan, M., P. H. Jensen, and D. Marsh. 2006. Association of α-synuclein and mutants with lipid membranes: spin-label ESR and polarized IR. Biochemistry **45**:3386–95.
52. Receveur-Brechot, V., J. M. Bourhis, V. N. Uversky, B. Canard, and S. Longhi. 2006. Assessing protein disorder and induced folding. Proteins **62**:24–45.
53. Roque, A., I. Iloro, I. Ponte, J. L. R. Arrondo, and P. Suau. 2005. DNA-induced secondary structure of the carboxyl-terminal domain of histone H1. J Biol Chem **280**:32141–7.
54. Roque, A., I. Ponte, and P. Suau. 2007. Macromolecular crowding induces a molten globule state in the C-terminal domain of histone H1. Biophys J **93**:2170–7.
55. Roque, A., I. Ponte, J. L. Arrondo, and P. Suau. 2008. Phosphorylation of the carboxy-terminal domain of histone H1: effects on secondary structure and DNA condensation. Nucleic Acids Res **36**:4719–26.
56. Seshadri, S., R. Khurana, and A. L. Fink. 1999. Fourier transform infrared spectroscopy in analysis of protein deposits. Methods Enzymol **309**:559–76.
57. Spassov, S., M. Beekes, and D. Naumann. 2006. Structural differences between TSEs strains investigated by FT-IR spectroscopy. Biochim Biophys Acta **1760**:1138–49.
58. Surewicz, W. K., H. H. Mantsch, and D. Chapman. 1993. Determination of protein secondary structure by Fourier-transform infrared-spectroscopy—a critical-assessment. Biochemistry **32**:389–94.
59. Surewicz, W. K., and H. H. Mantsch. 1988. New insight into protein secondary structure from resolution-enhanced infrared spectra. Biochim Biophys Acta **952**:115–30.

60. Susi, H. 1972. Infrared spectroscopy-conformation. Methods Enzymol **26**: 455–472.
61. Susi, H., and D. M. Byler. 1986. Resolution-enhanced Fourier transform infrared spectroscopy of enzymes. Methods Enzymol **130**:290–311.
62. Susi, H., and D. M. Byler. 1987. Fourier-transform infrared study of proteins with parallel β-chains. Arch Biochem Biophys **258**:465–9.
63. Takekiyo, T., A. Shimizu, M. Kato, and Y. Taniguchi. 2005. Pressure-tuning FT-IR spectroscopic study on the helix-coil transition of Ala-rich oligopeptide in aqueous solution. Biochim Biophys Acta **1750**:1–4.
64. Tamm, L. K., and S. A. Tatulian. 1997. Infrared spectroscopy of proteins and peptides in lipid bilayers. Q Rev Biophys **30**:365–429.
65. Tatulian, S. A., R. L. Biltonen, and L. K. Tamm. 1997. Structural changes in a secretory phospholipase A2 induced by membrane binding: a clue to interfacial activation? J Mol Biol **268**:809–15.
66. Tolleter, D., M. Jaquinod, C. Mangavel, C. Passirani, P. Saulnier, S. Manon, E. Teyssier, N. Payet, M. H. Avelange-Macherel, and D. Macherel. 2007. Structure and function of a mitochondrial late embryogenesis abundant protein are revealed by desiccation. Plant Cell **19**:1580–9.
67. Torii, H., T. Tatsumi, and M. Tasumi. 1998. Effects of hydration on the structure, vibrational wavenumbers, vibrational force field and resonance Raman intensities of N-methylacetamide. J Raman Spectrosc **29**:537–46.
68. Uversky, V. N., J. Li, and A. L. Fink. 2001. Evidence for a partially folded intermediate in α-synuclein fibril formation. J Biol Chem **276**:10737–44.
69. Uversky, V. N., J. Li, P. Souillac, I. S. Millett, S. Doniach, R. Jakes, M. Goedert, and A. L. Fink. 2002. Biophysical properties of the synucleins and their propensities to fibrillate: inhibition of α-synuclein assembly by beta- and gamma-synucleins. J Biol Chem **277**:11970–8.
70. Uversky, V. N., C. J. Oldfield, and A. K. Dunker. 2008. Intrinsically disordered proteins in human diseases: introducing the D2 concept. Annu Rev Biophys **37**:215–46.
71. van de Weert, M., P. I. Haris, W. E. Hennink, and D. J. A. Crommelin. 2001. Fourier transform infrared spectrometric analysis of protein conformation: effect of sampling method and stress factors. Anal Biochem **297**:160–9.
72. Vecchio, G., F. Zambianchi, P. Zacchetti, F. Secundo, and G. Carrea. 1999. Fourier-transform infrared spectroscopy study of dehydrated lipases from Candida antarctica B and Pseudomonas cepacia. Biotechnol Bioeng **64**:545–51.
73. Venyaminov, S. Y., and F. G. Prendergast. 1997. Water (H_2O and D_2O) molar absorptivity in the 1000–4000 cm-1 range and quantitative infrared spectroscopy of aqueous solutions. Anal Biochem **248**:234–45.
74. Vila, R., I. Ponte, M. Collado, J. L. R. Arrondo, M. A. Jimenez, M. Rico, and P. Suau. 2001. DNA-induced alpha-helical structure in the NH2-terminal domain of histone H1. J Biol Chem **276**:46429–35.
75. Vila, R., I. Ponte, M. Collado, J. L. R. Arrondo, and P. Suau. 2001. Induction of secondary structure in a COOH-terminal peptide of histone H1 by interaction with the DNA—an infrared spectroscopy study. J Biol Chem **276**:30898–903.

76. Walsh, S. T. R., R. P. Cheng, W. W. Wright, D. O. V. Alonso, V. Daggett, J. M. Vanderkooi, and W. F. DeGrado. 2003. The hydration of amides in helices; a comprehensive picture from molecular dynamics, IR, and NMR. Protein Sci **12**:520–31.

77. Wolpert, M., and P. Hellwig. 2006. Infrared spectra and molar absorption coefficients of the 20 alpha amino acids in aqueous solutions in the spectral range from 1800 to 500 cm(−1). Spectrochim Acta A Mol Biomol Spectrosc **64**:987–1001.

78. Yamada, N., K. Ariga, M. Naito, K. Matsubara, and E. Koyama. 1998. Regulation of β-sheet structures within amyloid-like β-sheet assemblage from tripeptide derivatives. J Am Chem Soc **120**:12192–9.

79. Yoder, G., P. Pancoska, and T. A. Keiderling. 1997. Characterization of alanine-rich peptides, Ac-(AAKAA)(n)-GY-NH2 (n = 1 − 4), using vibrational circular dichroism and Fourier transform infrared. Conformational determination and thermal unfolding. Biochemistry **36**:15123–33.

80. Zandomeneghi, G., M. R. H. Krebs, M. G. McCammon, and M. Fandrich. 2004. FTIR reveals structural differences between native β-sheet proteins and amyloid fibrils. Protein Sci **13**:3314–21.

81. Zhang, J., and Y. B. Yan. 2005. Probing conformational changes of proteins by quantitative second-derivative infrared spectroscopy. Anal Biochem **340**:89–98.

9

GENETICALLY ENGINEERED POLYPEPTIDES AS A MODEL OF INTRINSICALLY DISORDERED FIBRILLOGENIC PROTEINS: DEEP UV RESONANCE RAMAN SPECTROSCOPIC STUDY

NATALYA I. TOPILINA, VITALI SIKIRZHYTSKI, SEIICHIRO HIGASHIYA, VLADIMIR V. ERMOLENKOV, JOHN T. WELCH, AND IGOR K. LEDNEV

Department of Chemistry, University at Albany, State University of New York, Albany, NY

ABSTRACT

Protein misfolding plays an important role in many neurodegenerative diseases associated with proteinaceous aggregates having an extended cross-β-sheet structure. Intrinsically disordered proteins (IDPs) can also be involved in the amyloid diseases. Genetic engineering facilitates examination of folding and fibrillation mechanisms by probing the influence of the primary polypeptide sequence on kinetic and equilibrium properties of the protein. This chapter focuses on the use of genetic engineering in the study of the mechanism of fibrillation of large biopolymers that are excellent models for intrinsically disordered proteins. The folding mechanism was probed using deep UV resonance Raman (DUVRR) spectroscopy, a novel method for acquisition of quantitative information on the peptide backbone conformation of large

Instrumental Analysis of Intrinsically Disordered Proteins: Assessing Structure and Conformation, Edited by Vladimir Uversky and Sonia Longhi

fibrillar aggregates. The first two examples of cross-β-core Raman signatures, those obtained for a genetically engineered polypeptide and lysozyme, are indicative of a highly ordered, crystalline-like structure for the antiparallel β-sheet. We believe that the selective design of the polypeptide sequence by genetic engineering is a great tool for studies of the relationship between the sequence and the cross-β-sheet structure. The data from these studies may be especially useful in computational modeling of the polypeptide sequence–fibrillar structure relationship.

9.1 INTRODUCTION

Protein misfolding and the fate of misfolded proteins are focal points of modern biology and molecular medicine. Protein misfolding, a quite common event in nature, is often the basis for human disease (63, 82). The majority of misfolding diseases, referred as amyloid diseases, includes various neurodegenerative disorders and systemic amyloidoises (32, 45, 46). These diseases are associated with formation of proteinaceous aggregates principally composed of a protein or peptide specific to a particular disease (32, 45, 46, 52, 82, 104). Despite structural differences among the native folded states, all amyloidogenic proteins and peptides have been shown to form surprisingly similar aggregates, most often highly ordered nonbranched fibrils typically 2–5 nm wide (46, 172). One of the unique features of the fibrils is the folding of proteins to form an extended cross-β-sheet structure regardless of the native protein fold (32, 45, 46). Recently, the formation of fibrillar aggregates with morphological, structural, and tinctorial attributes of amyloidogenic fibrils as well as the aforementioned extended intermolecular β-sheet structure, was discovered on aggregation of several proteins and peptides unrelated to protein deposition diseases (1, 3, 33, 65, 108, 133, 143, 173, 188, 200). From these findings, it has been suggested that amyloid-like folding with extended cross-β-sheet structure is not restricted to specific amino acid sequences (21, 47, 81, 86, 143, 156), but is the result of an alternate folding path facilitated and stabilized by intermolecular interactions (47).

Since folded, partially unfolded and intrinsically disordered proteins can be involved in the amyloid diseases, it is likely that the initial steps in protein deposition are different for each protein. Initiation of amyloid formation by ordered proteins always requires partial destabilization and unfolding (17, 47, 57, 87, 97, 146, 188, 190, 191, 217). On the contrary, for intrinsically unfolded proteins, some stabilization and formation of local, partially folded conformations is important (187, 188). In every case, the process of protein aggregation (106) requires a protein to pass through an ensemble of intermediates (or misfolded states) for specific intermolecular interactions (50, 175).

The majority of human amyloid diseases including Alzheimer's, Parkinson's, Huntington's, and Creutzfeld–Jacob diseases involve deposition of natively unfolded proteins (185, 189). These proteins do not have rigid tertiary structures under native conditions and exist as dynamic ensembles of various conformations (56, 186, 189). From multiple analyses (142, 186), IDPs have been found to lack order-promoting amino acids, both hydrophobic (Val, Ile, Leu) and aromatic (Trp, Tyr, Phe), that can initiate protein folding via hydrophobic collapse and lead to the construction of the hydrophobic core of the folded protein. On the other hand, these polypeptides are enriched in disorder-promoting polar and charged amino acid residues (Ala, Arg, Gly, Gln, Ser, Pro, Glu, and Lys (49, 142, 148, 203)). The specific combination of reduced hydrophobicity and large net charge results in IDP chain flexibility as well as high protein solubility.

The process by which essentially flexible and unfolded proteins develop intermolecular bonds and aggregate into ordered fibrils is a question of both fundamental and medical importance. Two main aspects could be considered in respect to this question: the formation of folding intermediates capable for specific intermolecular interactions during the initiation stage and the decrease of overall solubility of ID proteins to form stable aggregates. Inherently, IDPs are extremely soluble to inhibit the formation of stable secondary structure within the protein as well as to prevent protein aggregation in the native state at a global level. However, IDPs are known not only to interact with various molecules including proteins, nucleic acids, membranes, and small molecules but also to commonly undergo induced folding on complexation. These interactions stimulate the changes in the protein local environment and affect the balance of forces governing folding and aggregation. Other factors that affect protein folding and solubility, such as increases in hydrophobicity, decreases in net charge or specific environmental changes, can also increase the amyloidogenic properties of IDP. Environmental factors such as temperature, pH, ionic strength, the presence of cosolvents, small polyanionic molecules as well as protein sequence mutations influence IDP amyloidogenesis (185, 189).

9.1.1 Why Genetic Engineering

Chimeric proteins and de novo designed peptide models can afford valuable insights that are not easily accessible (221) by simplifying the process of folding to test only specific aspects of the overall very complicated process. These molecules are especially useful in identification of folding-initiation sites given that the misfolding of natural proteins is an unfavorable process under physiological conditions. Investigations of misfolding normally require nonnative conditions such as high concentrations, lowered pH, or increased ionic strength of the medium (19, 48, 51, 117, 174). The relevance of this approach to fibril formation that occurs under native conditions is clearly suspect.

The use of de novo designed polypeptides can however facilitate an atomic-level understanding of how intermediates compatible with the formation of amyloid assemblies are populated. Model compounds with materials such as those described below are not only simple by virtue of their highly repetitive nature but also enable dissection of various contributions to the folding and fibrillation processes. Based on the information reviewed above, it was determined that β-sheet formation by a polypeptide designed from first principles could be illustrative (15, 16, 53). The use of model systems to explore the forces that control β-sheet formation has been stymied for many years by the perception that small β-sheet domains would necessarily aggregate (154). Recently, subsequent to descriptions of two-stranded antiparallel β-sheets ("β hairpin") or three-stranded antiparallel β-sheets (60, 94) by short peptides (9–16 amino acids), we developed methods for the construction of much larger β-sheet-forming peptides (100, 199). The investigations employing these peptides were extremely fruitful in probes of the correlation of local amino acid sequence and folding in the absence of tertiary interactions (153).

Genetic engineering affords excellent opportunities to verify the mechanism of folding and fibrillation by tuning selectively the primary polypeptide sequence and testing the influence of those changes on kinetic and equilibrium properties. As an example, our recent study (199) has demonstrated that a de novo, genetically engineered 687-residue peptide with 32-amino acid repeats utilizing carbamylated lysines, GH6((GA)3GY(GA)3GE(GA)3GH(GA)3G K)21GAH6 (YEHK21), self-assembles into a well-defined antiparallel β-sheet structure (see Fig. 9.5). After folding to a single "native" conformation initially, YEHK21 completely and reversibly denatures on heating. With these properties, YEHK21 is an excellent model for large β-sheet proteins, which are known to form well-organized fibrillar aggregates.

9.1.2 Why Deep UV Resonance Raman Spectroscopy

Amyloid fibrils are noncrystalline and insoluble, and thus are not amenable to conventional X-ray crystallography and solution NMR, the classical tools of structural biology (32). Advances in solid-state NMR (SSNMR; 4, 13, 14, 23, 27, 75, 83, 84, 112, 139, 141, 164, 169, 184, 194, 202) and the recent successes in growing microcrystals (52, 128, 129, 151) of small peptide fragments that have characteristics of amyloid fibrils and yet appropriate for X-ray studies have contributed the most to our knowledge about the cross-β-sheet structure (32). SSNMR, however, requires site-specific ^{13}C and/or ^{15}N labels while the application of atomic-resolution X-ray crystallography to the full-length protein fibrils is yet to be demonstrated. Low-resolution techniques, such as transmission electron microscopy (TEM), scanning probe microscopy (SPM) (32), and wide-angle X-ray scattering (171) have been developed and routinely used in fibril studies. TEM and SPM probe the topology and morphology of fibrillar aggregates, while wide-angle X-ray scattering allows for estimating intersheet and interstrand spacing in cross-β structures, which provides constraints for structure refinement by high-resolution methods. Recent EPR studies coupled with site-directed spin labeling have testified to the in-register parallel β-sheet structure of the fibrillar core of several disease-related proteins (36). This approach, however, is insensitive to subtle differences in the Ψ and Φ torsional angles determining the conformation of the polypeptide backbone.

Raman spectroscopy has been proven to be an efficient technique for characterizing highly scattering gelatinous and solid samples. Raman spectroscopy, in general, and resonance Raman spectroscopy, in particular, have been widely used for structural characterization of biological systems (2, 5–7, 26, 34, 35, 38, 64, 66, 67, 72, 78, 85, 105, 111, 124, 126, 132, 137, 147, 150, 152, 155, 170, 183, 192, 207–209). We have recently demonstrated that DUVRR spectroscopy is a powerful tool for protein structural characterization at all stages of fibrillation (157, 160, 211, 214). In particular, this method is capable of (1) detecting structural intermediates at early stages of fibrillation and determining their sequential order using 2D correlation analysis (157, 160) and (2) characterizing the cross-β-core structure of amyloid fibrils prepared from entire proteins by the hydrogen/deuterium (H/D) exchange combined with DUVRR spectroscopy and advanced statistical analysis (213). DUVRR spectroscopy is a novel method to acquire quantitative information on the peptide backbone conformation in large fibrillar aggregates. This method does not require isotope labeling and, consequently, opens the opportunity for comparative characterization of β-sheet structure in fibrils prepared from entire proteins and short peptides as well as in model segment microcrystals suitable for X-ray crystallography. We also envision using this methodology for structural comparison of fibrils prepared *in vivo* and *in vitro* in the future. In addition, DUVRR spectroscopic method complements X-ray crystallography and SSNMR by opening the opportunity for the real-time kinetic study of amyloid fibril formation including the process of peptide aggregation, protofibril formation, and fibril maturation. The kinetic study could be accomplished by DUVRR spectroscopy using much smaller amounts of material in comparison with SSNMR requirements.

9.2 ULTRAVIOLET RAMAN SPECTROSCOPY FOR CHARACTERIZING PROTEINS IN GENERAL AND IDPS AND FIBRILLOGENIC PROTEINS IN PARTICULAR

Numerous applications of ultraviolet resonance Raman spectroscopy in various areas of science and technology have been documented in recent years including microelectronics, analytical chemistry, and biology (5–7, 34, 35, 38, 64, 67, 72, 78, 85, 105, 111, 124, 126, 132, 137, 147, 150, 152, 155, 170, 192, 207–209, 220). Thin surface layers of semiconductor materials (127) have been characterized by UV Raman microscopy. A great potential of UV resonance Raman spectroscopic detector coupled with capillary electrophoresis has been demonstrated for the identification of nucleotides and other analytes (44). Being coupled with chemometrics, UV resonance Raman spectroscopy could be used for rapid detection and identification of microorganisms (110, 207). Several comprehensive manuscripts describing various applications of UV Raman spectroscopy for studying biological systems have been published over the years (5–7, 26, 64, 66, 98, 170, 183).

The resonance enhancement not only decreases the required sample amount, but also allows for probing specific parts of biomolecules. For example, Raman spectra of heme proteins generated by excitation with visible light are dominated by the contribution from heme vibrational modes and provide information about the heme (iron spin and oxidation state, coordination, planarity) as well as on its interactions with the close neighborhood (90, 136). A vibrational signature of aromatic amino acid residues, tryptophan and tyrosine, have been shown to be responsive to contacts between secondary structure elements and exposure to water and, consequently, are used for characterizing the tertiary structure of proteins (28, 29, 76, 134, 182). UV resonance Raman spectroscopy with excitation from 229 to 280 nm has been utilized for obtaining tryptophan and tyrosine vibrational signatures (28–30, 74, 76). A 229-nm excitation has been also used to probe histidine ligation via resonance enhancement of imidazole ring modes (181, 208). The important breakthrough in the biological application of Raman spectroscopy has started with utilizing DUVRR spectroscopy with excitation around and below ~210 nm (37, 58, 78, 98, 114, 145, 168, 201). Deep UV excitation resonantly enhances Raman scattering from the amide chromophore, a building block of a polypeptide backbone (5, 42, 183). The amide chromophore Raman signature is sensitive to Ψ and Φ dihedral angles and provides direct quantitative information about the secondary structure of proteins (10–12, 31, 37, 73, 80, 122).

9.2.1 Polypeptide Backbone Conformation

The amide chromophore exhibits two $\pi\pi^*$ absorption peaks in the far-ultraviolet and vacuum region at 190 nm ($\varepsilon = 5600$) and 165 nm, respectively (42). Although aromatic amino acids have 6- to 10-fold larger molar absorption coefficient in the deep UV region, the amide chromophore makes a major contribution to the absorption spectrum of proteins because the relative concentration of amide groups is much higher than that of aromatic amino acid residues. Similarly, the resonance Raman scattering from amide chromophores makes the major contribution to Raman spectra of proteins at deep UV excitation (28, 30, 31, 37, 73, 98, 218). The Raman excitation profiles obtained for N-methylacetamide, the simplest model of the polypeptide backbone, have indicated that all amide bands are resonantly enhanced by the allowed 190-nm transition (8). In contrast, the allowed 165-nm transition causes destructive interference, eliminating amide I (Am I) band at 184-nm excitation (8). The amide chromophore Raman signature is extraordinarily sensitive to the polypeptide backbone conformation and provides direct quantitative information about the secondary structure of proteins (11, 12, 31, 37, 73). A high sensitivity of deep UVRR spectra to the protein secondary structure is based on extremely complex nature of amide modes involving coupling of various vibrations. Am I mode consists of carbonyl C=O stretching, with smaller contribution from C–N stretching and N–H bending (26, 31, 198). Both

Am I and amide II (Am II) bands involve significant C–N stretching, N–H bending, and C–C stretching. The C_α–H bending vibrational mode involves C_α–H symmetric bending and C–C_α stretching (31, 198).

Amide III (Am III) and (C)C_α–H bending vibrational modes are the most sensitive Raman bands to the amide backbone conformation. Asher et al. (9) have recently demonstrated that this sensitivity results from the coupling of amide N–H motion to (C)C_α–H motion that depends on a Ramachandran dihedral Ψ angle. The vibrations are strongly mixed at $\Psi \sim 120°$, corresponding to both random coil and β-sheet conformations, but are almost completely uncoupled at $\Psi \sim -60°$ for α-helix conformation. In addition to a strong dependence on the Ψ dihedral angle, the Am III vibrational frequency shows a modest dependence on the Φ dihedral angle (80, 123). Steric constraints allow a very limited region of Φ angles to be populated. Most recently, Asher et al. (10) developed a new method for estimating the Ramachandran Ψ-angular distributions from an Am III deep UV Raman band shape.

The application of DUVRR spectroscopy for structural characterization of IDPs has not yet been exploited. Although there is every evidence that DUVRR spectroscopy should provide important structural information by evaluating the distribution of the Ramachandran Ψ angle and the extent of intramolecular hydrogen bonding. Deep UV Raman spectroscopy has been recently used to evaluate the conformation of disordered polypeptides and proteins (10, 100). Specifically, DUVRR spectroscopic studies confirmed the occurrence of polyproline II (PPII) conformations in large and small peptides (10, 100). A de novo 687-amino acid residue polypeptide with a regular 32-amino acid repeat sequence, (GA)3GY(GA)3GE(GA)3GH(GA)3GK, forms large β-sheet assemblages which exhibit remarkable folding properties and, as well, form fibrillar structures (100). The polypeptide assumes a fully folded antiparallel β-sheet/turn structure at room temperature, and yet is completely and reversibly denatured at 125°C adopting a predominant PPII conformation. A 21-amino acid peptide, predominantly composed of alanines, has been shown to exist below room temperature as a mixture of α-helix and PPII helix (10). Both steady-state and nanosecond time-resolved T-jump experiments (101–103) showed no significantly populated intermediates when the alanine polypeptide underwent a thermal melting transition between the α-helix and a PPII conformation (10).

A semiempirical method developed by Asher and colleagues (121, 122) allows for quantitative evaluation of the Ramachandran Ψ angle distribution from DUVRR spectra and, thus, characterization of the polypeptide conformation. A specific relationship has been derived for a specific number and a type of amide hydrogen bonds at different temperatures. The application of the developed approach is limited to the systems in which the state of hydrogen bonding is known, that normally is the case if a known secondary structural motif is studied (121). Additionally, the state of hydrogen bonding can be elucidated based on the position of Am I vibrational mode probed with DUVRR spectroscopy or IR absorption.

9.2.2 Aromatic Amino Acids as UV Raman Probes of Protein Tertiary Structure

Aromatic amino acids are often utilized as natural probes of protein tertiary structure. Various spectroscopic techniques have been used to assess the local environment of aromatic amino acid side chains and the extent of hydrogen bonding. Specifically, steady-state (177) and time-resolved (107, 125, 215) fluorescence, UV absorption (193), and circular dichroism (62) spectroscopic methods have used mainly tryptophan and tyrosine for protein structural characterization. Phenylalanine has a very low fluorescence quantum yield that limits its use for fluorescence studies (42). The major limitation in applying fluorescence, absorption and circular dichroism spectroscopic methods arises from the difficulty in separating the spectral contributions from different aromatic amino acids. UV resonance Raman spectra of tryptophan and tyrosine are comprised of narrow bands that could be used to differentiate the contributions of different residues (29, 58, 71, 144). Near UV resonance Raman spectroscopic methods have utilized mainly tryptophan and tyrosine because of their superior absorbance over phenylalanine at wavelengths longer than ~230 nm, where commercial Raman instruments operate (28, 29, 71, 134). Nonresonance Raman spectroscopy with excitation in the visible light range has utilized all three aromatic amino acid residues, tryptophan, tyrosine, and phenylalanine, for structural characterization of biological systems, but large protein concentration is required (134, 176). The strong ν_{12} phenylalanine Raman band has been reported to be sensitive to the local environment (71, 212). The 229-nm Raman cross section of this band doubles when the ethylene glycol content in water increases to 80%. Ethylene glycol is known to be used for modeling the local environment of aromatic amino acids in hydrophobic parts of proteins. The variations in the ν_{12} band Raman intensity of phenylalanine have been attributed to the environmental influences on the benzene ring π electron clouds since phenylalanine could not be involved in specific hydrogen bonding or other polar interactions. The red shift of L_a absorption band of phenylalanine in hydrophobic environment has been suggested to result in the increase in the ν_{12} band Raman cross section (71). We have recently observed a similar increase in the ν_{12} Raman cross section of phenylalanine exposed to the hydrophobic environment on 195-nm excitation when this vibrational mode was strongly enhanced due to the resonance with the $B_{a,b}$ electronic transition (211). The potential of phenylalanine as a deep UV Raman probe of protein conformations is well documented (98, 99, 212).

9.2.3 DUVRR Spectroscopy for Studying the Kinetic Mechanism of Fibrillation

DUVRR spectroscopy has been used for structural characterization of hen eggwhite lysozyme (HEWL) at all stages of fibril formation (157, 211–214). The evolution of the protein secondary structure as well as the local environment of phenylalanine, a new natural deep ultraviolet Raman marker, was documented. Concentration-independent irreversible helix melting was quantitatively characterized as the first step of the fibrillation. The native HEWL, composed initially of 32% helix, transforms monoexponentially to an unfolded intermediate with 6% helix with a characteristic time of 29h (211, 212). The phenylalanine residues in lysozyme fibrils are accessible to solvent in contrast to those in the native protein. The local environment of phenylalanine residues changes concomitantly with the secondary structure transformation. A similarity of unfolding kinetics found for the secondary and tertiary structures of lysozyme using DUVRR and tryptophan fluorescence spectroscopy leads to a hypothesis that the unfolding might be an all-or-none transition (214). Chemometric analysis, including abstract factor analysis, target factor analysis, evolving factor analysis, multivariate curve resolution—alternating least square (ALS), and genetic algorithm were employed to verify the presence of only two principal components contributing to the DUVRR and fluorescence spectra of soluble fraction of lysozyme (excluding insoluble fibrils) during the fibrillation process (158, 162, 163, 214). However, a definite conclusion on the number of conformers cannot be made based solely on the above spectroscopic data even if the chemometric analysis strongly suggested the existence of two principal components. Therefore, electrospray ionization mass spectrometry (ESI-MS) was also utilized to address the hypothesis (214). The protein ion charge state distribution (CSD) envelopes of ESI mass spectra provide information on protein conformational changes. By using chemometric analysis, the CSD envelopes of the incubated lysozyme were well fitted with two principal components. Based on the spectroscopic/spectrometric data along with chemometric analysis, the partial unfolding of lysozyme during *in vitro* fibrillation was characterized quantitatively and proved to be an all-or-none transition. The combination of ESI-MS, Raman and fluorescence spectroscopies with advanced statistical analysis was demonstrated to be a powerful methodology for studying protein structural transformations (214).

The early stages of HEWL fibrillation have been quantitatively characterized by two-dimensional correlation deep UV resonance Raman spectroscopy (2D-DUVRR) in terms of the sequential order of events and their characteristic times (157, 160). Figure 9.1 shows DUVRR spectra of the supernatant part of lysozyme solution incubated at pH 2.0 and 65°C for various times.

These spectra exhibit pronounced amide bands which report on the protein secondary structure (100). Am I mode consists of carbonyl C=O stretching, with a small contribution from C–N stretching and N–H bending. Am II and Am III bands involve significant C–N stretching, N–H bending, and C–C stretching. The C_α–H bending vibrational mode involves C_α–H symmetric bending and C–C_α stretching. Tyr and Phe designate Raman peaks originating from the side-chain groups of Tyrosine and Phenylalanine amino acid residues, respectively. As seen in Figure 9.1, the C_α–H bending band intensity increased with the incubation time, indicating the melting of α-helix and the formation of β-sheet and random coil. The development of new β-sheet was also evident from the apparent sharpening of Am I band. The pronounced

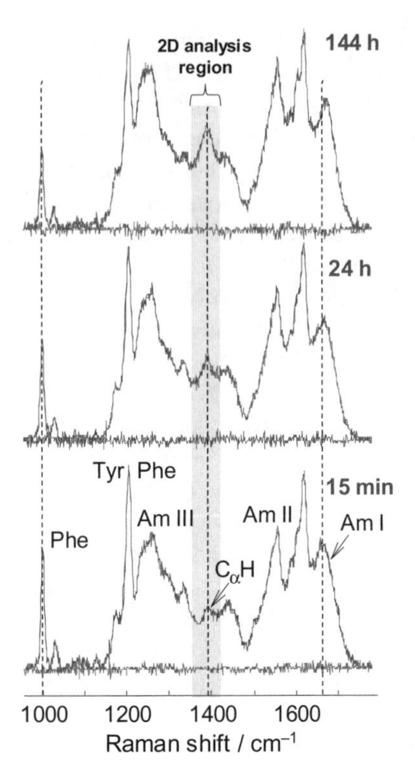

Figure 9.1 Experimental (blue), modeled (red), and difference (green) DUVRR spectra of lysozyme incubated for various times. The experimental Raman spectra were fitted with three pure component spectra, that is, the spectra of "nucleus" β-sheet and partially unfolded intermediate calculated by the independent component analysis, and the experimental spectrum of native lysozyme. A mixed soft-hard modeling approach provided the refined DUVRR spectra of β-sheet and partially unfolded intermediate, kinetic profiles for all three species, and the characteristic times for each step of lysozyme transformation (157). The highlighted C_α–H bending region of the spectra has been used for the initial 2D correlation analysis (157). Adopted from Shashilov et al. (157) with permission from the American Chemical Society. See color insert.

decrease in the 1000-cm^{-1} phenylalanine band intensity during the incubation corresponds to melting of lysozyme tertiary structure. The crucial mechanism-relevant information about the sequential order of lysozyme unfolding and β-sheet formation events, however, cannot be gleaned from the visual inspection of DUVRR spectra. The C_α–H bending DUVRR region contributed by β-sheet (~1396 cm^{-1}) and random coil (~1387 cm^{-1}) conformations is especially sensitive to secondary structural transformations of proteins. In particular, C_α–H bending band is strong in deep UVRR spectra of random coil and β-sheet, where the adjacent C_α–H and N–H bending vibrations are coupled, and

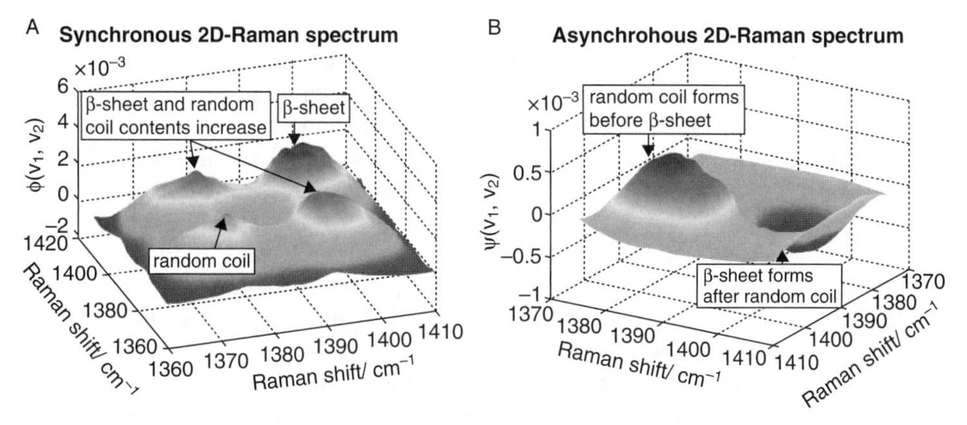

Figure 9.2 Synchronous (A) and asynchronous (B) 2D correlation spectra of the CαH bending region of lysozyme DUVRR spectra. (A) Two auto-peaks at 1387/1387 and 1396/1396 cm^{-1} correspond to random coil and β-sheet, respectively, and two positive cross-peaks at 1387/1396 cm^{-1} and 1396/1387 cm^{-1} show positive correlation between β-sheet and random coil changes. (B) Two opposite-sign areas at 1387/1396 cm^{-1} and 1396/1387 cm^{-1} indicate asynchronous formation of random coil and β-sheet; valley at 1396/1387 cm^{-1} shows that β-sheet appeared after random coil. See color insert.

weak for peptides in the α-helical form. Accordingly, 2D correlation analysis of the C_α–H bending region was performed in order to distinguish the highly correlated processes of random coil and β-sheet formation. Synchronous $\Phi(\nu_1,\nu_2)$ and asynchronous $\Psi(\nu_1,\nu_2)$ 2D-Raman spectra were calculated following Noda's approach (131):

$$\Phi(\nu_1,\nu_2)+i\Psi(\nu_1,\nu_2)=\frac{1}{\pi(T_{\max}-T_{\min})}\int_0^\infty \tilde{Y}_1(\omega)\cdot\tilde{Y}_2^*(\omega)\,d\omega \qquad (9.1)$$

where $\tilde{Y}_1(\omega)$ and $\tilde{Y}_2^*(\omega)$ were derived from experimental spectral intensities $\tilde{Y}_1(\nu,t)$ for all wavenumbers ν and incubation times t.

Two positive peaks at 1396 and 1387 cm^{-1} on the synchronous 2D correlation spectrum $\Phi(\nu_1,\nu_2)$ in Figure 9.2(A) represented a codirectional change (increase) in β-sheet and random coil fractions as a function of time. The asynchronous 2D-Raman correlation map is shown in Figure 9.2(B). The peak and the valley centered at 1385 and 1400 cm^{-1}, respectively, illustrated (1) that changes in random coil and β-sheet spectral regions occurred asynchronously and (2) that the formation of β-sheet was delayed with respect to the formation of random coil. This result unambiguously indicated that the appearance of β-sheet and random coil at the early stages of lysozyme fibrillation was incompletely correlated.

These data were used to distinguish between two alternative mechanisms of β-sheet formation (157). In a parallel process mechanism, random coil and β-sheet are produced directly from the native protein and should be completely correlated. In a step-by-step mechanism, β-sheet develops from the partially unfolded intermediate. In the latter case, the formation of β-sheet and the partially unfolded intermediate could correlate, but only partially. Consequently, the step-by-step mechanism proposed by Dobson and coworkers (18) was in complete agreement with our analysis above. Following the proposed mechanism, the newly formed β-sheet in the solution part of the incubated samples was provisionally assigned to the fibrillation nucleus. To further support this assignment, the supernatant of a lysozyme sample incubated for 48 h was used for seeding the fibrillation of fresh lysozyme. The seeding was successful and the fibrillation lag-phase was eliminated. The rigorous quantitative characterization of the step-by-step nucleation mechanism calls for the evaluation of all reaction-specific characteristic times provided that the evolution profiles and DUVRR spectra of all reacting species are known (157). These studies showed that DUVRR spectroscopy combined with 2D correlation spectroscopy, independent component analysis and advanced ALS modeling (159, 214) is a powerful tool for the quantitative characterization of protein structural rearrangements and could be used for various protein folding problems.

9.2.4 DUVRR Spectroscopy for Structural Characterization of the Fibril Core

As indicated above, our current knowledge about the detail structure of fibril core is mainly based on the application of SSNMR (4, 13, 14, 23, 27, 75, 83, 84, 112, 139, 141, 164, 169, 184, 194, 202) and X-ray crystallography (52, 128, 129, 151) for fibrils and microcrystals, respectively, prepared from small peptide fragments that have characteristics of amyloid fibrils. We have recently developed a novel method based on the combination of H/D exchange and DUVRR spectroscopy for structural characterization of the fibril core (213). This method does not require isotope labeling and, consequently, opens the opportunity for comparative characterization of β-sheet structure in fibrils prepared from entire proteins and short peptides as well as in model segment microcrystals suitable for X-ray crystallography. We also envision using this methodology for structural comparison of fibrils prepared *in vivo* and *in vitro* in the future. In addition, the proposed method complements X-ray crystallography and SSNMR by opening the opportunity for the real-time kinetic study of amyloid fibril formation including the process of peptide aggregation, protofibril formation, and fibril maturation. The kinetic study could be accomplished by DUVRR spectroscopy using much smaller amounts of material in comparison with SSNMR requirements.

The main idea of the method is quite straightforward. The exchange between hydrogen and deuterium can occur in the areas of fibrils where water

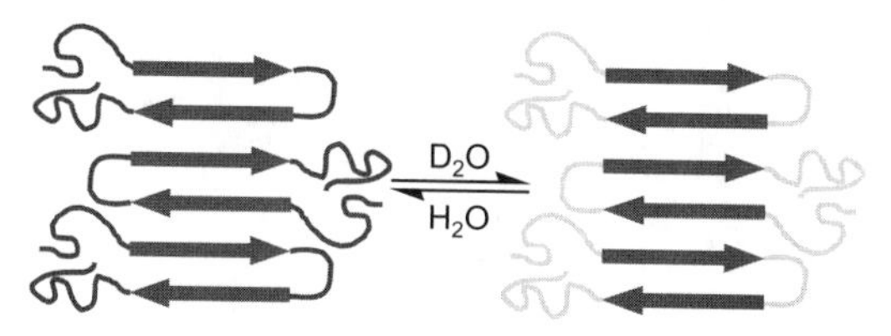

Figure 9.3 H/D exchange takes place in the unordered parts of fibrils leaving the fibril core protonated.

penetrates. When mature fibrils prepared in water are transferred into D_2O, the exchange occurs in the unordered parts of fibrils while the core stays protonated (Fig. 9.3).

H/D exchange is a valuable tool for characterizing protein structure, solvation, and water exposure when combined with NMR (70), mass spectrometry (88, 89), and vibrational spectroscopy (41). In an amino acid residue, the main-chain NH group and O, N, S bound protons exchange easily whereas carbon-bound hydrogens do not. In the protein hydrophobic core or strongly hydrogen-bonded secondary structures, the H/D exchange rates are strongly reduced due to shielding of exchangeable sites (41). It is well established (88, 89) that N–H protons in unordered fragments of amyloid fibrils should exchange readily whereas those hidden from water in the cross-β structure will remain protonated. As shown by *Mikhonin* and *Asher* (120), H/D exchange causes a downshift of the Am II DUVRR band from ~1555 cm^{-1} to ~1450 cm^{-1} (Am II′) and the virtual disappearance of the Am III band in an unordered protein. We hypothesized and then proved experimentally (213) that the H/D exchange-DUVRR spectroscopic method resolves the spectroscopic signature of the cross-β-core from that of water-accessible moieties, including unordered structures and β-turns. We have gradually changed the composition of the solvent from 100% H_2O to 100% D_2O and acquired Raman spectra for various solvent compositions. Then, we have applied advanced statistical analysis (161) to retrieve Raman spectroscopic signatures of the protonated fibril core (Fig. 9.4). It has been discovered that the cross-β-core of lysozyme fibrils is highly ordered. No inhomogeneous broadening of Raman bands due to various amino acid residues was found. This is in contrast to the Raman spectra of globular protein β-sheets, which exhibit broader Raman peaks than those of homopolypeptides (77, 122). We utilized Asher's approach (121) to estimate the Ramachandran Ψ dihedral angle of the amide group from the band frequency of the Am III vibrational mode:

$$\nu^{\beta}_{\mathrm{AmIII}}(\Psi) = 1244\left(\mathrm{cm}^{-1}\right) - 54\left(\mathrm{cm}^{-1}\right)\cdot\sin\left(\Psi + 26^{\circ}\right) \qquad (9.2)$$

Using the above equation, a Ψ angle of 133° was estimated for the cross-β-core structure. This value falls within the range of 129–133° of the

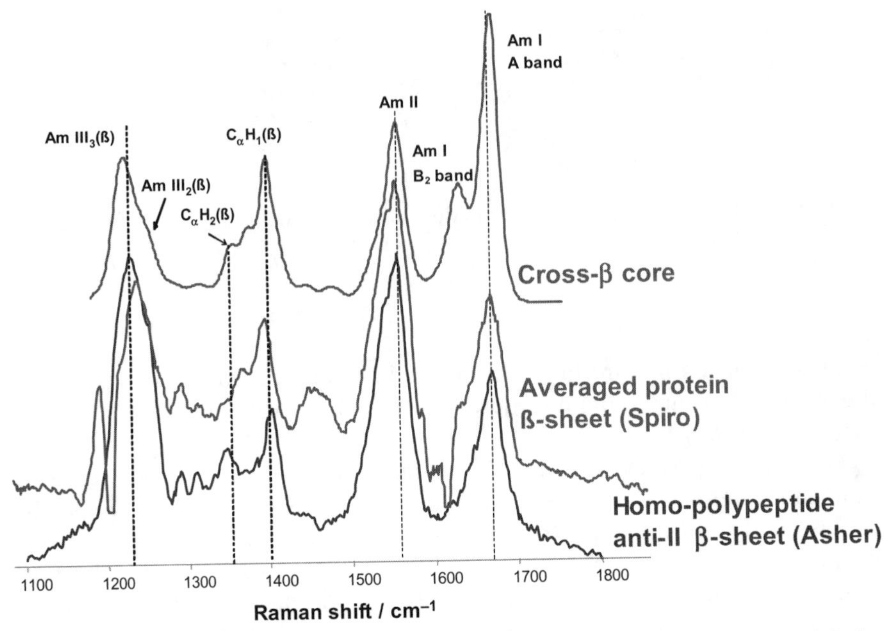

Figure 9.4 Pure DUVRR spectra of cross-β-core, the β-sheet structure of globular protein, and homopolypeptides.

antiparallel β-sheet. Consequently, the resolved Raman signature indicated that the antiparallel β-sheet is the dominant secondary structural conformation of the core. This is certainly a very promising but, at the same time, very tentative conclusion. The ability of the proposed method to differentiate parallel and antiparallel β-sheet would be very valuable for the field. Currently, a specific labeling is required to determine the type β-sheet in the fibril core by SSNMR, EPR, or fluorescence spectroscopy (4, 43, 96, 141, 180).

9.3 GENETIC ENGINEERING OF POLYPEPTIDES EXHIBITING FIBRILLOGENIC PROPERTIES

9.3.1 Polypeptide Design and Construction

Libraries of antiparallel β-sheet-forming repetitive polypeptide blocks have previously been prepared in the course of investigations of elastin-based polymers. Several groups have also prepared high molecular weight polypeptides which can adopt a β-sheet conformation (20, 22, 61, 68, 69, 93, 140, 167, 197, 200, 204, 210, 216). Two principal approaches have been developed for creation of β-sheet polypeptide libraries. In the first approach, a polypeptide homopolymer with an appropriate repeating unit was constructed by a single

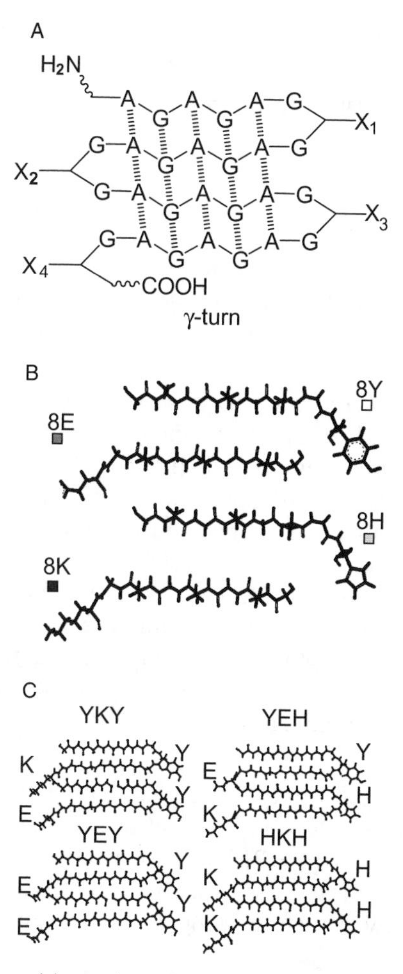

Figure 9.5 Design of repetitive unit architecture. (A) General design of polypeptide repetitive unit. (B) Peptide fragments that are coded by smallest DNA building blocks (strand + turn). (C) Representative repetitive units YKYE, YEHK, YEYE, and HKHK.

concatenation of the repetitive unit. An alternate approach, based on the formation of polypeptide libraries where the peptide would assume a β-sheet structure, relied upon a binary patterning of β-strand elements, that is, alternation of hydrophobic and hydrophilic amino acids. This assembly did however consistently employ a specified amino acid sequence for turn introduction at the desired site (20, 22, 68, 69, 197, 200, 210). In our work, each β-sheet-forming polypeptide block was derived from the assembly of DNA building blocks that encoded the elementary constructs for a single strand and turn (Fig. 9.5A,B). This strategy afforded precise control of the polypeptide sequence while permitting flexible variation of the polymer architecture. The constituents of any particular turn or the position of a strand within the repetitive polypeptide were easily designated.

9.3.2 Polypeptide Repeat Units

GA repeats were utilized for the β-strand forming motif with pendant amino acids X at the turns (Fig. 9.5A) after work by Tirrell (24, 93, 138, 140, 167, 196, 216). The (GA)3GX repetitive polypeptides form antiparallel β-sheets by regular adjacent-reentry chain folding through γ-turns.

9.3.2.1 Polypeptide Libraries To enhance the solubility of the polypeptides, various functional groups were incorporated at the turn sites with a resultant introduction of amphiphilicity. Amphiphilic polypeptide design includes the introduction of hydrophilic and hydrophobic edges to the β-sheet to enhance water solubility. Starting with the four building blocks (8Y, 8E, 8H, 8K where each coding block consists of a β-strand forming AGAGAG sequence and a turn sequence GX, Fig. 9.5B), the heterogeneous unit YEHK can be assembled. Negatively charged glutamate and protonated lysine, introduced to generate a hydrophilic edge, can form stabilizing salt bridges. The aromatic tyrosine and histidine residues were postulated to form an intramolecular π-stacked hydrophobic array.

9.3.2.2 β-Strand Length Polypeptides consisting of seven repeats have also been prepared with 32, 40, or 48 repetitive units, as in (GA)nGY(GA)nGE(GA)nGH(GA)nGK (n = 3, 4, or 5, respectively, 32YEHK7, 40YEHK7, or 48YEHK7, Fig. 9.6). The cross-β-sheet core structure of each of the polypeptides, differing only in β-strand length, was similar experimentally (166).

9.3.2.3 Other Variations The system of β-strand subunits described above (9.3.2.2) enables examination of the impact of the systematic displacement of the turn residues, for example, the creation of YKYE where histidinyl turn residues of the YEHK repeat are replaced with tyrosines. Deletion of the lysine residues from YKYE forms the simple distrand repeat 16YE which results in an unbalanced negative charge in the resultant polypeptides. By analogy, the HK peptide presents histidine residues along one edge and array of protonated lysines on the other (Fig. 9.5C). These systems involve a finite but large set of weakly interacting β-strands in a single molecule to enable study of the role of intramolecular/intermolecular nucleation in the absence of tertiary contacts and importance of intramolecular and intermolecular β-sheet propagation.

As part of a long-term strategy for the study of protein folding that focuses on aggregation and amyloid formation, a family of large, fibril-forming polypeptides has been prepared and a structure relationship between

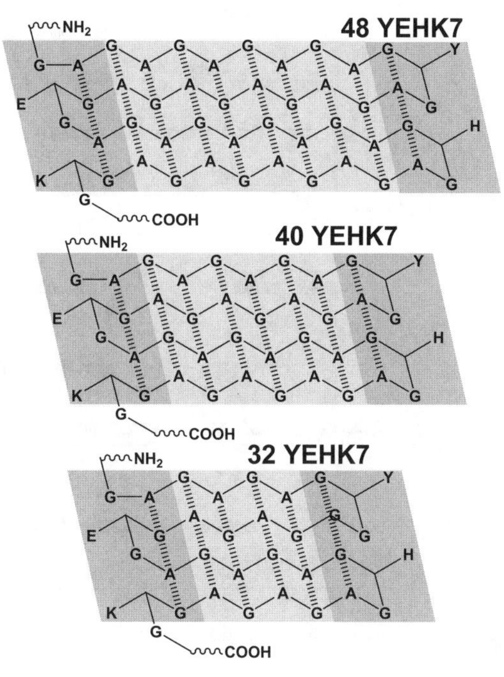

Figure 9.6 Hypothetical scheme of YEHK fibril cross-β-core (166).

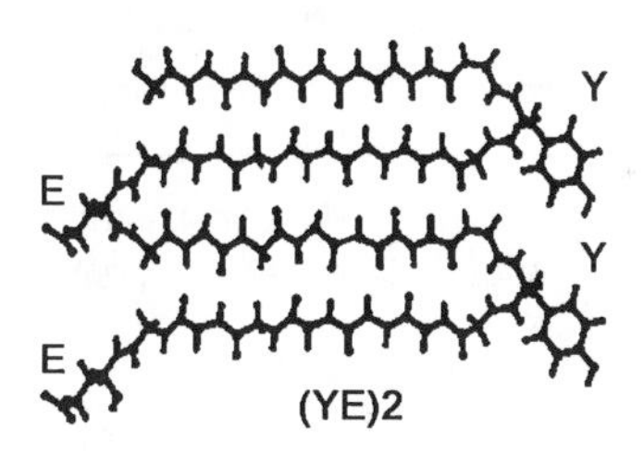

Figure 9.7 Schematic representation of a hypothetical conformation of a de novo YE2 polypeptide folded into the cross-β-sheet structure.

polypeptide sequence and folding/aggregation properties has been identified. For example, the YE8 polypeptide illustrates all properties of a typical amyloid fibril-forming protein (see section 9.5). Aggregation and folding kinetics of GH6((GA)3GY(GA)3GE)8GAH6 (YE8), which consists of 16 repeats of a very simple six-amino acid strand, GAGAGA, similar to the β-strand motif found in Bombyx mori silk and in other de novo designed polypeptides (95, 179, 195) as well as the appropriate turn residues has been investigated.

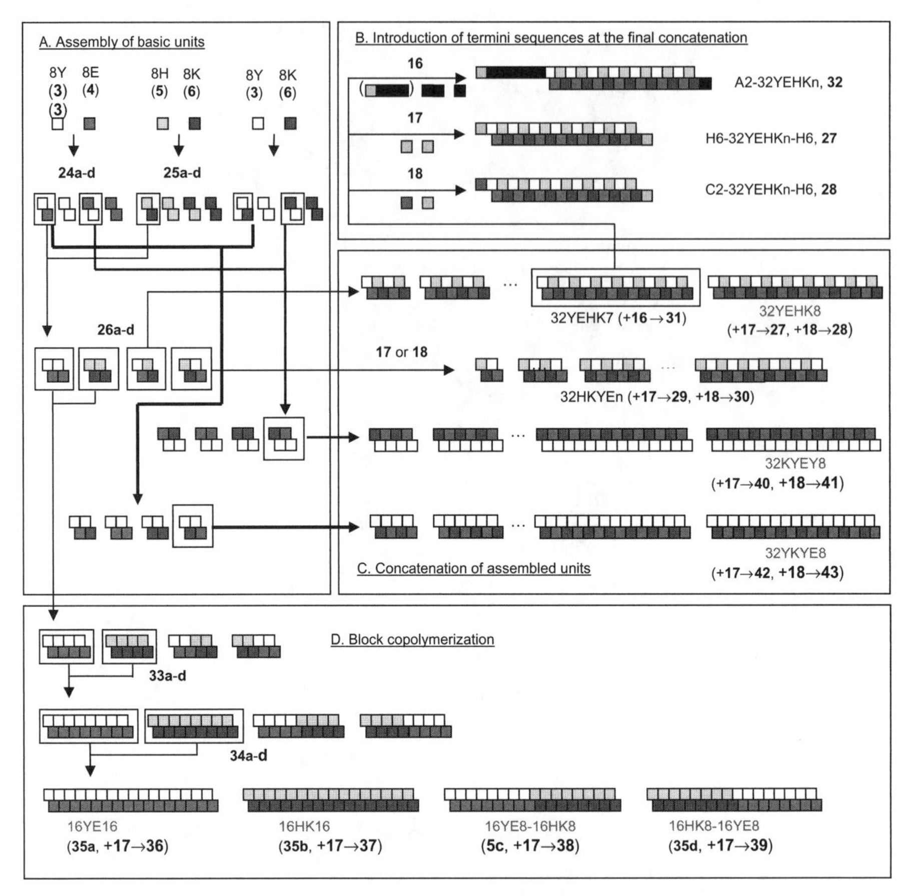

Figure 9.8 Schematic representation of the construction of the polypeptide libraries. Adopted from Biomacromolecules 2007, **8**, 1487–97, page 1492. See color insert.

9.3.3 Library Construction

An overview of the process of library construction is illustrated in Figure 9.8.

9.3.3.1 Building Block Assembly and Oligomerization Oligonucleotide units can be assembled based on unidirectional, head-to-tail oligomerization (25, 39, 55, 59, 115) with adaptive DNA sequences for incorporation at the appropriate cloning sites (118). Each of the minimal DNA coding units can be concatenated by minimal coding units. Utilization of the adaptive DNA sequences allows cloning at virtually any restriction site with common, conventional, commercially available cloning vectors.

9.3.3.2 The Adaptive DNA for Cloning Adaptive DNA containing the recognition sites for type IIs restriction endonucleases can be introduced at

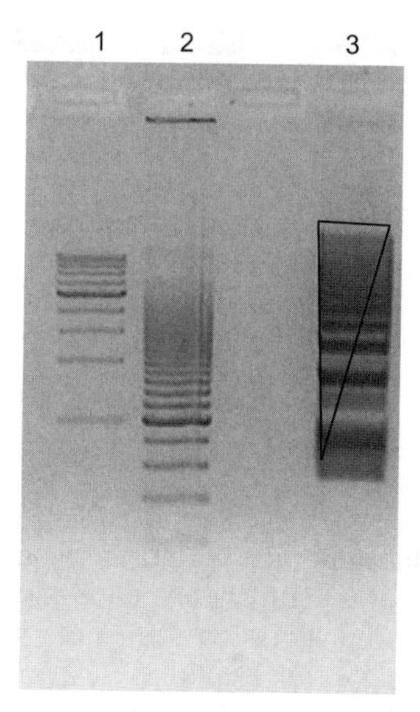

Figure 9.9 Oligomerization with adaptive sequences, agarose gel electrophoresis, and enrichment of longer adapted oligomers. Lane 1: 1 kb molecular marker ladder, lane 2: high 200 bp molecular marker ladder, lane 3: C2-32YEHKn-H6 oligomers. Enclosed triangular area was purified and used for cloning.

both the 3′ and 5′ termini to facilitate recovery of assembled DNA units and thereby enable recursive, seamless oligomerization and block copolymerization (54, 61, 116, 118, 135, 219). Self-ligation of the adaptive sequences during the concatenation is suppressed by selective phosphorylation of 5′-hydroxyl groups of synthetic adaptive DNA strands. Oligomerized DNA is purified by agarose gel electrophoresis so the bands containing the desired adapted oligomers can be enriched (Fig. 9.9) and cloned into recipient vectors.

9.3.3.3 Adaptive DNA for Expression Adaptive DNA for expression can be utilized in the last concatenation step to terminate the oligomerization and to enable insertion of the construct into the desired expression vector (93, 140, 167, 216). Simultaneously, the appropriate fusion amino acid sequences at both the head and the tail of the repetitive and block-copolymerized polypeptides can be introduced thereby enabling rapid shotgun library construction with various terminal amino acids (Fig. 9.8B). Resulting polypeptides have the repetitive sequence fused with a 4.9-kDa domain that can be easily removed to enable recovery of the purified repetitive polypeptide sequence.

9.3.3.4 Oligomerization In contrast with the previously described methods (119, 205, 206), the sequential oligomerization of the DNA product recovered from the first concatenation step can be effected in the presence of the appropriate adaptive DNA sequence prior to insertion into an expression vector

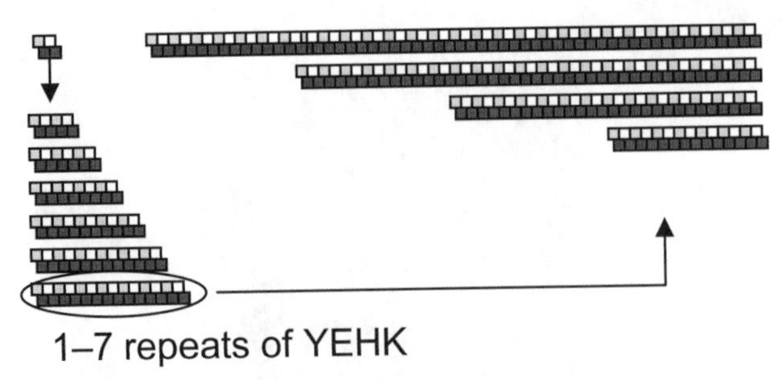

Figure 9.10 Schematic representation of the extent of oligomerization at different steps of concatenation.

(Figs. 9.8C and 9.10). This approach permits a second level of concatenation. In a single step, a library of repetitive polypeptides can be formed bearing the appropriate C- and N-terminal fusion sequences where there is a gradual increase in molecular weight, for example, $1x$, $2x$, $3x$, $4x$, $5x$ (where x is the selected concatemer from previous oligomerization). This approach affords a diversity of oligomers not accessible by arithmetic synthesis based on ligation of two blocks, for example, $1x$, $2x$, $4x$, $8x$, $16x$ series (119, 205). Block copolymers prepared via this stepwise concatenation strategy have distinct domains comprised of the simple motifs (Fig. 9.8D).

9.3.3.5 Amphiphilic Character Repetitive and block-copolymerized polypeptides 27, 28, 32 (YEHKn), 29, 30 (HKYEn), 36 (16YEn), 37 (16HKn), 38 ([YE8HK8]n), 39 ([HK8YE8]n), 40, 41 (KYEYn), and 42, 43 (YKYEn) were constructed with various N- and C-terminal sequences by combination of 8Y, 8E, 8H, and 8K coding DNA units. The results of expression of the repetitive block copolymer 39a-g (H6-[HK8YE8]n-H6) and repetitive polypeptides 27a-h (H6-YEHKn-H6) and 29a-d (H6-HKYEn-H6) are shown in Figure 9.11, lanes 1–7 and Figure 9.12. Some of the shorter peptides, such as 27a, 28a-b, 29a, and 30a-b, were not detected by sodium dodecyl sulfate polyacrylamide gel electrophoresis (SDS-PAGE) or Western blot (27a, lane 1 and 29a, lane 9, Fig. 9.12).

9.4 CHARACTERIZATION OF OLIGOPEPTIDES

9.4.1 Peptide Isolation and Functionalization

On purification, the desired β-sheet polypeptide (H6YEHK21H6) can be isolated. The negatively charged glutamic acid (E) and positively charged lysine (K) residues were introduced at adjacent β-turn sites to facilitate antiparallel β-sheet assembly by salt bridge formation (Fig. 9.5). It was found however that when the crude cell lysate was denatured with a urea solution prior to affinity chromatography, the lysines were carbamylated to homocitrulline residues (109). Denaturation of the lysate with guanidinium

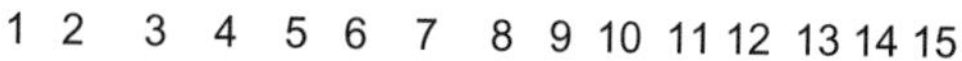

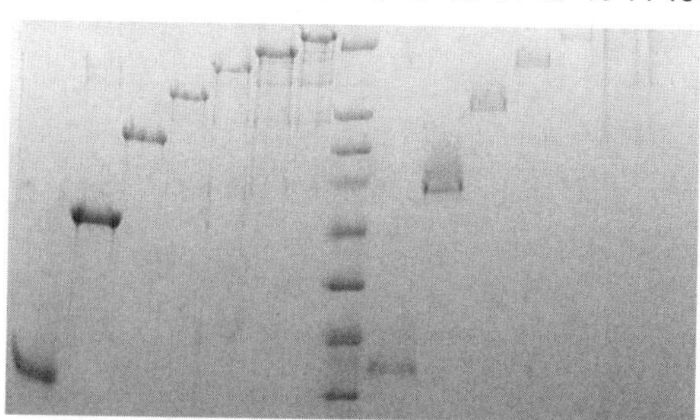

Figure 9.11 An 8% SDS-PAGE of intact and carbamylated repetitive polypeptides H6-(HK8YE8)n-H6 (n = 1–7, 39a-g). Lanes 1–7; lanes 8: molecular weight marker (from the top, 205, 116, 97, 84, 66, 55, 45, 36 kD); lanes 9–15: carbamylated 39a"-g," respectively.

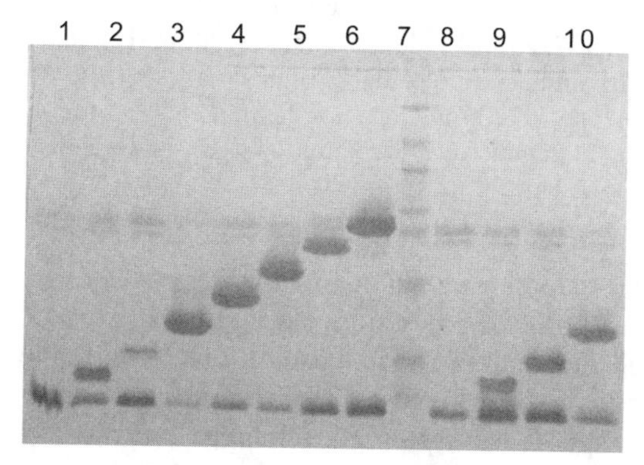

Figure 9.12 A 15% SDS-PAGE of purified short repetitive polypeptides H6-YEHKn-H6 (27a–h) and H6-HKYEn-H6 (29a–d). Lanes 1–8: 27a–h; lane 9: molecular weight marker (from the top, 66, 45, 36, 29, 24, 20, 14.2 kD); lanes 10–13: 29a–d, respectively.

hydrochloride solutions obviated this conversion. Confirmation of this effect can be seen by gel electrophoretic analysis (Fig. 9.13). Polypeptide constructs containing homocitrulline residues exhibited a more pronounced propensity to undergo intermolecular aggregation to form fibrils than a material prepared where the conversion was avoided. Therefore, our studies have focused on these homocitrulline-containing peptides.

9.4.2 Mobility of Polypeptides on Gel Electrophoresis

The polypeptides constructed by the described method are comprised of the same small peptide building blocks yet differ in the architecture of repetitive

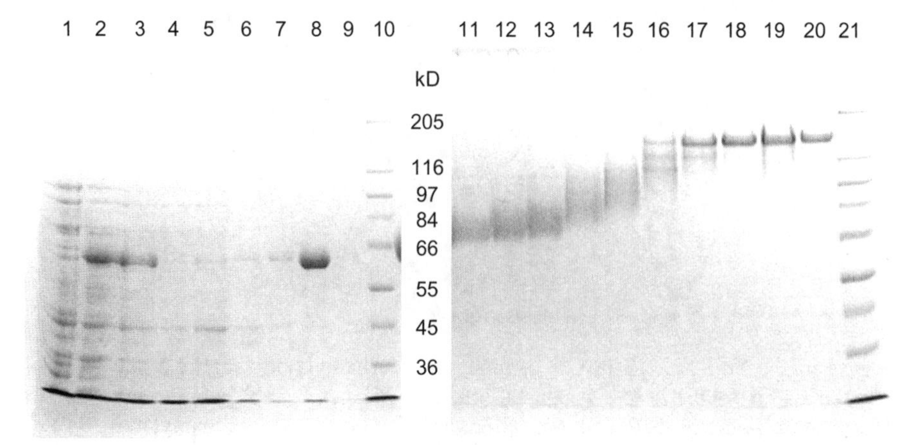

Figure 9.13 An 8% SDS-PAGE gel electrophoretic analysis of induction, purification, and modification of H6YEHK21H6. Lane 1, before induction; lane 2, 4-h induction; lane 3, whole cell lysate; lane 4, flow through 1; lane 5, flow through 2; lane 6, 10mM imidazole; lane 7, 20mM imidazole; lane 8, 300mM imidazole; lane 9, 500mM imidazole; lanes 10 and 21, molecular weight marker (SIGMAMARKER Wide Molecular Range); lane 11, 300mM fraction heated at 98°C for 1min; lane 12, 2min; lane 13, 3min; lane 14, 4min; lane 15, 5min; lane 16, 7min; lane 17, 10min; lane 18, 20min; lane 19, 60min; lane 20, 180min.

units. The degree of polymerization or block copolymerization of the polypeptides varies as do the fusion sequences employed at the C- and N-termini. Posttranslational chemical transformations allow quantitative modifications of all selected residues in the polymers with a consequent variation of the polymer properties. Together, these systematic changes enable investigation of the influence of the polypeptide structure on the physical properties of the polypeptide. As an example, gel electrophoresis under the denaturing conditions demonstrated structure–physical property correlations. Under denaturing conditions, these polypeptides can be viewed as unstructured biopolymers with incorporation of charged amino acids at precise position in the molecule.

Though SDS binds the polypeptides in a molecular weight-specific ratio and therefore enables polypeptide separation by the molecular weight, side-chain modifications, such as glycosylation or phosphorylation, significantly affect gel mobility (149). In contrast to native polypeptides, the mobility of carbamylated compounds was sensitive to the repetitive unit and terminal sequences. The HKYE repetitive polypeptides had greater gel mobility than the corresponding YEHK-derived polypeptide. Interestingly, polypeptide libraries with H6 sequences at both termini had a slightly higher mobility than those with a combination of C2 and H6 termini.

The polypeptides in their modified and unmodified forms represent different types of charged polymers (Fig. 9.14): *negatively charged polypeptides*:

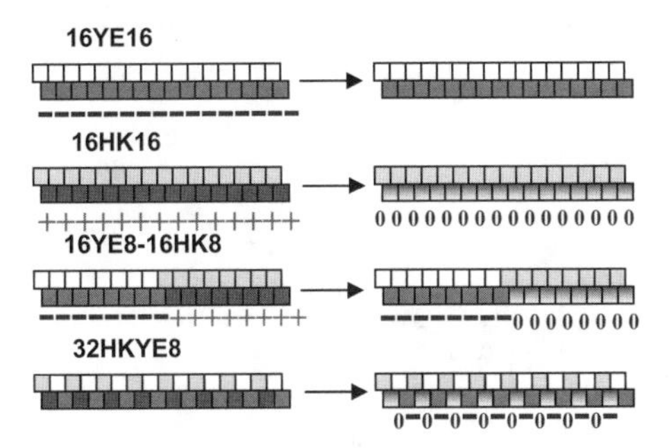

Figure 9.14 The schematic representation of the polypeptide charge distribution before and after posttranslational chemical modification. See color insert.

type 1a—possesses a uniform distribution of charge, type 1b—with cluster of negative charge; *positively charged polypeptides*: type 2; *neutral polypeptides*: type 3a—where there is a uniform distribution of positive and negative charge, type 3b—with clusters of negative and positive charge, and type 3c—composed of neutral amino acids.

9.4.3 Polypeptides with High Molecular Weights

The tendencies observed for low molecular weight polypeptide, such as the influence of the repetitive building block sequence, posttranslational chemical modification, or block copolymerization, could be extended over the full range of polypeptide molecular weights (Figs. 9.15 and 9.16). The uniformly negatively charged polypeptides H6-YEn-H6 (**36**) and carbamylated C2-YEHKn-H6 (**28′**) (type 1a) demonstrate the lowest gel mobility, while positively charged H6-HKn-H6 (**37**) (type 2) polypeptides and neutral polypeptides with uniformly distributed charge H6-YEHKn-H6 (**27**), H6-HKYEn-H6 (**29**), H6-KYEYn-H6 (**40**), and H6-YKYE-K6 (**42**) (type 3a) are more mobile. The neutral block copolymers such as H6-(YE8HK8)n-H6 (**38**) and H6-(HK8YE8)n-H6 (**39**) (type 3b), where negative and positive charges lie in separate domains, have an intermediate gel mobility that only decreases slightly on carbamylation.

9.5 KINETICS OF FIBRILLATION OF A GENETICALLY ENGINEERED POLYPEPTIDE

YE8 polypeptide consisting of eight repeats of the 16-amino acid monomer (see above) has been found to adopt unordered conformation at neutral pH and form amyloid-type fibrils after incubation at pH 3.5 (178). The atomic force microscopy (AFM; Fig. 9.17A) and TEM images (Fig. 9.17B) of the YE8 aggregates clearly illustrate the presence of fibrillar aggregates. Thus, this

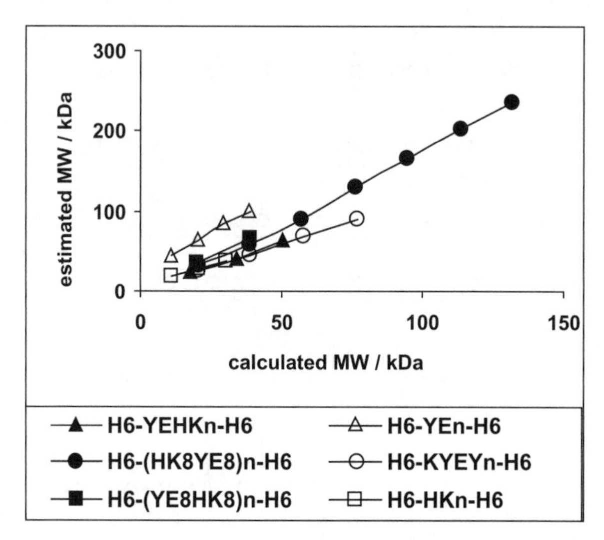

Figure 9.15 Calculated and estimated molecular weight from the mobility of the polypeptide from libraries with different architectures of repetitive unit but the same N- and C-terminal groups. Filled triangles: H6-YEHKn-H6 (27); triangles: H6-YEn-H6 (36); squares: H6-HKn-H6 (37); filled circles: H6-(HK8YE8)n-H6 (39); circles: H6-KYEYn-H6 (40); filled squares: H6-(YE8HK8)n-H6 (38).

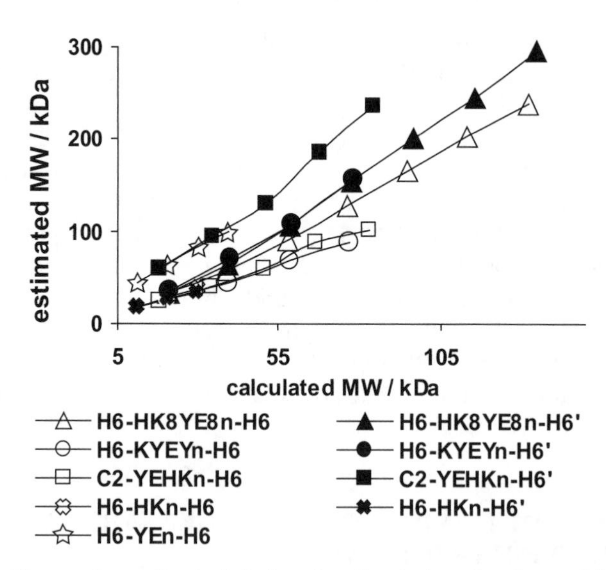

Figure 9.16 Comparison of calculated and estimated molecular weight of selected repetitive polypeptides in their modified and unmodified state Empty: intact polypeptides; filled: carbamylated polypeptides; squares: C2-YEHKn-H6 (28, 28′); stars: H6-YEn-H6 (36); pluses: H6-HKn-H6 (37, 37′); triangles: H6-(HK8YE8)n-H6 (39, 39′); circles: H6-KYEYn-H6 (40, 40′).

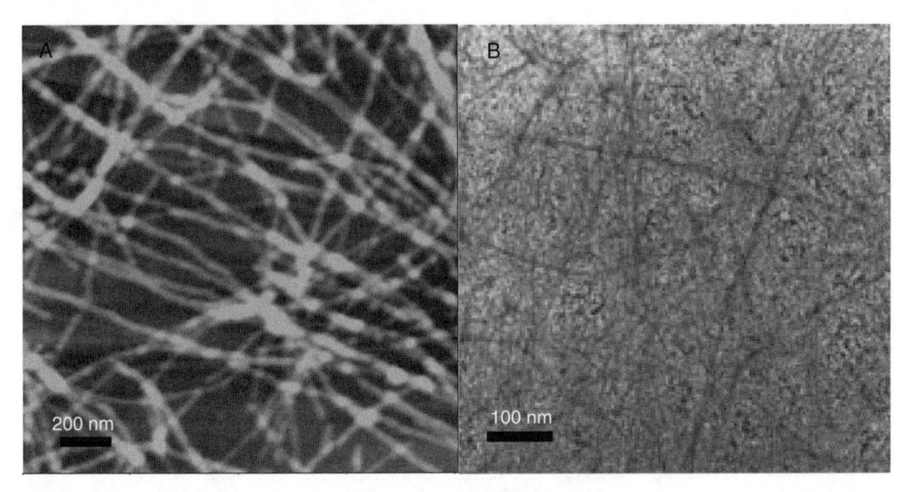

Figure 9.17 YE8 polypeptide aggregation incubated at pH 3.5. (A) TM AFM topographs on HOPG and (B) TEM micrograph on carbon coated Cu grid.

genetically engineered biopolymer suits perfectly to model an intrinsically disordered fibrillogenic protein. YE8 has been designed such that the GAGAGA repeats form the 16 antiparallel β-strands of the β-sheet structure with differentiated turn groups along the edges (Fig. 9.7): one edge with arrays of tyrosine residues (Y) and the other with charged glutamic acids (E). Variation of pH results in protonation or deprotonation of the glutamic acid and histidine residues and thereby enables management of the overall polypeptide charge. Consequently, pH variation leads to control of solubility, folding, and aggregation.

YE8 solutions with concentrations varied from 44 to 9 μM and pH 3.5 have been incubated at room temperature and probed frequently by DUVRR and CD spectroscopies (178). Both spectroscopic techniques indicated that in the initial solutions, the predominant conformation of YE8 was random coil, although, the spectroscopic signature of a PPII conformation was evident in the DUVRR spectra (98, 100). Figure 9.18 shows DUVRR spectra of a pH 3.5 solution of YE8 incubated for various times at room temperature. The spectra are dominated by contributions from the amide chromophore and tyrosine. The changes in YE8 DUVRR spectra due to incubation at pH 3.5, the appearance of a narrow and intense Am I band in particular, clearly indicate the formation of a β-sheet. The plot of the Am I intensity as a function of incubation time had a sigmoidal shape showing a lag phase of about 7 days and reached saturation after about 50 days. The amide part of the 48-day Raman spectrum, in particular the Am I band, had a shape similar to that of DUVRR spectrum of the YEHK21 fibrils, predominantly that of a β-sheet (100). This indicates that β-sheet should be a predominant conformation of YE8 polypeptide incubated at pH 3.5 for 48 days. This conclusion was

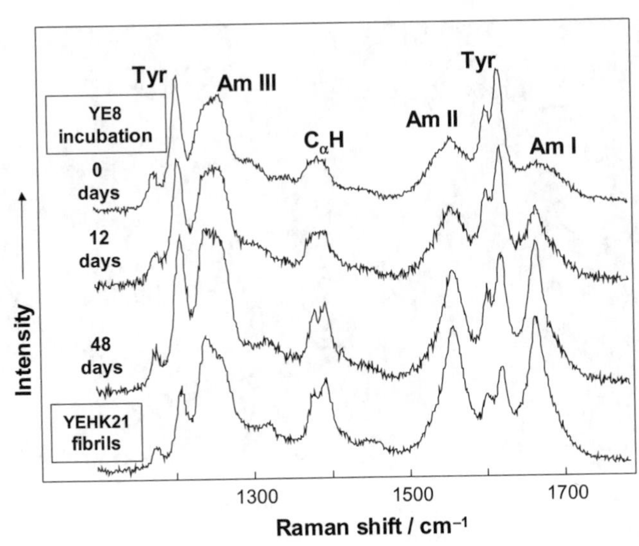

Figure 9.18 A 197-nm excited Raman spectra of YE8 polypeptide (44 μM) at various stages of incubation at pH 3.5 and room temperature, and fibrillar aggregates formed by a de novo genetically engineered 687-residue peptide YEHK21 in a predominant β-sheet conformation (100). Adopted from Topilina et al. (178) with permission from Biopolymers journal.

supported qualitatively by CD spectroscopic analysis, although the exact assignment of the folded YE8 CD spectrum was complicated by its "unusual" shape, as described previously for the YEHK21 peptide (100), and because of increased light scattering over the course of incubation. AFM and TEM confirmed the formation of amyloid-like fibrilar aggregates (178). The AFM (Fig. 9.17A) and TEM images (Fig. 9.17B) of the YE8 aggregates clearly illustrate the presence of fibrillar aggregates.

It was also reported (178) that the concentration of polypeptide influenced β-sheet folding of YE8 at pH 3.5 in a manner consistent with known fibrillation processes. The folding reaction slowed dramatically at lower concentrations of the polypeptide. Figure 9.19 shows the change in Am I intensity with the incubation time for various YE8 concentrations. In addition, a substantial increase in lag time was obvious when the polypeptide concentration decreased. At 9 μM of YE8, the lag time exceeded 30 days, clearly suggesting that if there was any intramolecular assistance to β-sheet folding of YE8 then the characteristic time of this process should be over 30 days. This also suggested that the β-sheet folding of YE8 at concentrations of 25 μM and higher was completely aggregation driven. Moreover, all the formed β-sheet stayed within aggregated species as evident from the following experiments. On centrifugation of 44 μM YE8 samples incubated at pH 3.5 for various times, the resultant

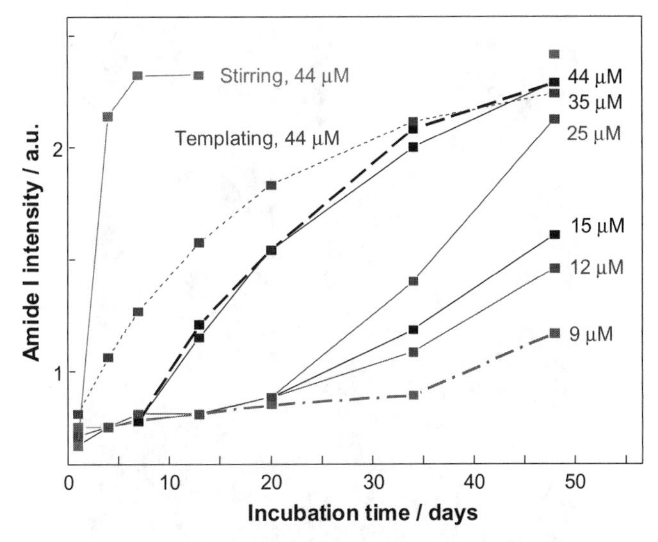

Figure 9.19 Kinetics of β-sheet formation during the incubation of YE8 polypeptide at pH 3.5 and room temperature. The increase in YE8 concentration as well as templating and constant stirring accelerated the β-sheet folding of the de novo polypeptide in a similar way to the acceleration of fibrillation reported for many global proteins. Adopted from Topilina et al. (178) with permission from Biopolymers journal.

gelatinous portion and supernatant were characterized separately. The gelatinous portions exhibited Raman spectra with the β-sheet vibrational signature while the supernatant spectra were indicative of an unordered conformation.

Templates and agitation are known to accelerate the fibrillation of global proteins *in vitro* (130). Both these effects have been tested for β-sheet formation by YE8. No lag phase was evident for YE8 fibrillation under templated conditions (Fig. 9.19). Except for the lag phase, the shape of the kinetic curves and the characteristic times of fibrillation were similar for the 44-μM YE8 samples incubated with and without templating. Dramatic acceleration of β-sheet formation was found on agitation of a 44-μM solution of YE8 with a magnetic stirrer. The changes in the DUVRR spectrum of YE8 were complete after 5 days of continuous stirring (Fig. 9.19). The shape of the final DUVRR spectrum, the intensity of the Am I band in particular, was the same for the stirred sample and those incubated without stirring. Consequently, a de novo YE8 polypeptide exhibits all the major properties of a fibrillogenic protein and provides an excellent opportunity for detailed study of the fibrillation mechanism. At neutral pH, YE8 is soluble in disordered form that makes the polypeptide an excellent model of an intrinsically disordered fibrillogenic protein.

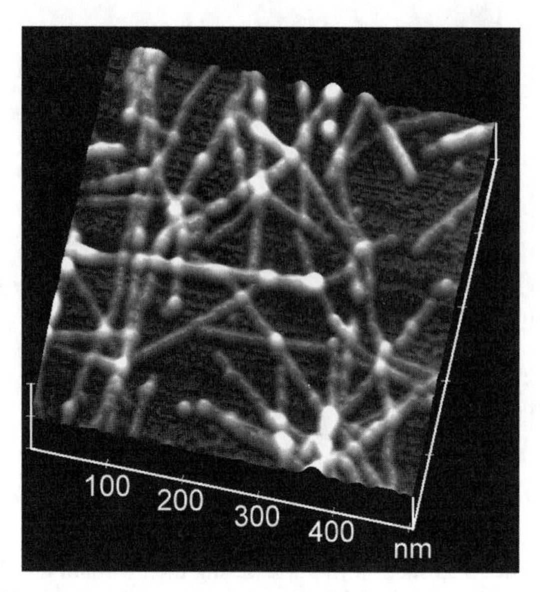

Figure 9.20 TM-AFM topograph of H6YEHK21H6 fibrils on HOPG.

9.6 FIBRIL MORPHOLOGY

9.6.1 Characterization of H6YEHK21H6 by AFM

Controlled self-assembly of the polypeptide on a defined surface has also been reported. Purified H6YEHK21H6 dissolved in doubly distilled water was deposited on highly oriented pyrolytic graphite (HOPG) or polycrystalline Ni surfaces. Deposition from the liquid phase was followed by rinsing with doubly deionized water and drying under a stream of N_2. The topography of the deposited polypeptide was examined utilizing tapping-mode atomic force microscopy (TM-AFM; Fig. 9.20).

The most notable feature in Figure 9.20 is the presence of highly linear fibrillar H6YEHK21H6 structures. These structures are stable under ambient conditions and exhibit no conformational change after extended storage times. The fibril lengths range from tens to thousands of nanometers. The H6YEHK21H6 fibril width is uniform with an average value of 15 ± 2 nm as determined prior to deconvolution.

Figure 9.21 displays a probability density plot of fibril thickness derived from an analysis of 79 fibril samplings derived from multiple deposition experiments. Observed fibril thicknesses as determined by TM-AFM did not exhibit a continuum of thicknesses as might be expected from lamellar stacking (140). Fibril thickness varies in discrete increments of approximately 0.8 ± 0.2 nm. Figure 9.21 clearly shows a substantially higher probability for fibrils with a thickness of 1.7 ± 0.2 nm, a thickness not inconsistent with a loose bilayer or "ribbon" configuration (1; Fig. 9.22). The observed fibril thickness increments

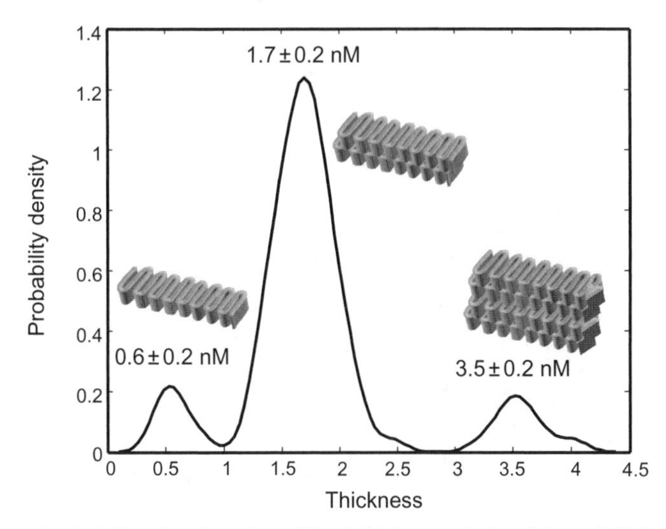

Figure 9.21 Probability density plot of fibril thickness derived from AFM topographic data of H6YEHK21H6 fibrils. A total of 79 samplings were employed to generate the distribution profile.

are consistent with thickness estimates from the computationally derived H6YEHK21H6 polypeptide β-sheet structure.

Apparently, the intrasheet interactions and the disposition of the hexahistidinyl tracts at the H6YEHK21H6 termini strongly favor a bilayer configuration. Nonetheless, turn formation and the implication of turn motif on amphiphilicity is a complex relationship. Differing turn motifs can be assumed by β-hairpins as reflected in the number of amino acids per turn and the number of hydrogen bonds formed between the distal strands (40, 165) with a consequence that the faces of β-sheet may be differentiated by the orientation of the methyl groups of constituent strands toward a single face of the sheet. It has been previously shown that closely related constructs poly([AG]3EG), poly([AG]3YG) and poly([AG]3KG) form amphiphilic β-sheet structures via the intermediacy of γ-turns to redirect the β-strand subunits (140). If a γ-turn-containing structure is assumed for H6YEHK21H6 as shown in Figure 9.22, the amphiphilic character of the resultant β-sheet would be consistent with the observed bilayer formation. The nonpolar alanine methyl-bearing faces would be shielded from the solvent, while the glycyl-derived moieties would be exposed to the aqueous environment.

9.6.2 Characterization of H6YEHK21H6 by TEM

The proposed structure of H6YEHK21H6 fibrils and the morphology of those fibrils were investigated by TEM. H6YEHK21H6 was deposited on carbon-coated support grids utilizing deposition protocols similar to those for HOPG

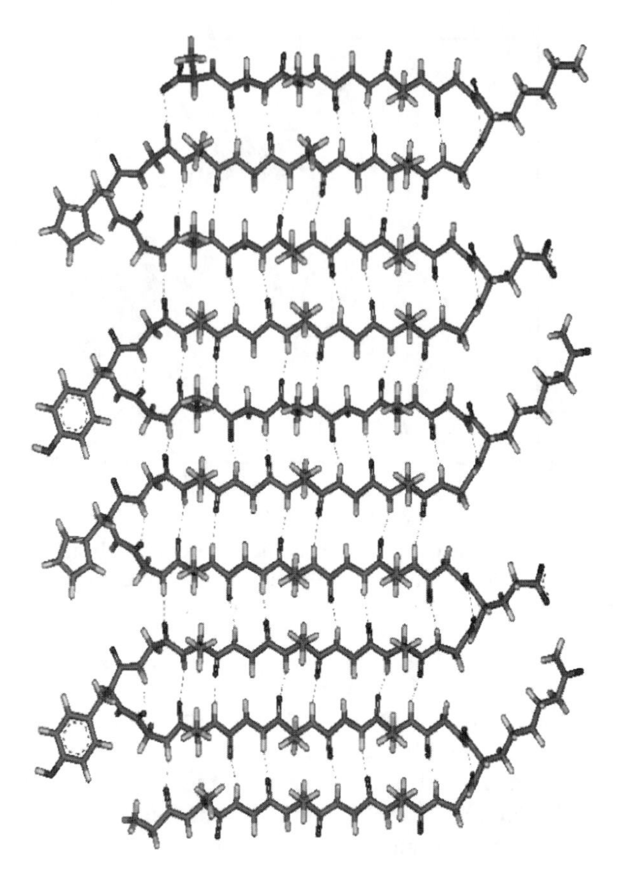

Figure 9.22 YEHK repeat units with γ-turns forming antiparallel β-sheet structure. Width (turn-strand-turn) of computed model, 3.4–3.8 nm; thickness, 0.72 nm.

substrates. A uranyl acetate stain was utilized on dried samples to enhance edge contrast of the fibril assemblies in the TEM. Figure 9.23 displays resulting TEM micrographs. The highly linear assemblies in the upper image of Figure 9.23 agree qualitatively with the TM-AFM topographs of Figure 9.20. The lower image of Figure 9.23 displays a higher magnification micrograph of the fibril assemblies. On analysis of the TEM data, the average width of the fibrillar structures was found to be about 6.5 ± 1 nm as shown in Figure 9.24. As expected, TM-AFM data yield a much larger average fibril width of 15 ± 2 nm due to tip-convolution effects. The inclusion of tip convolution-induced feature broadening of 8 ± 1 nm led to very good agreement between the TEM and TM-AFM fibril-width measurements.

An average measured fibril width of 6.5 ± 1 nm was predominant for all TEM data studied. This is in approximate agreement with the width of a structure comprised of two β-sheets side by side where the hydrophobic turn groups (Tyr and His) of the two sheets pack closely (see Figure 9.22).

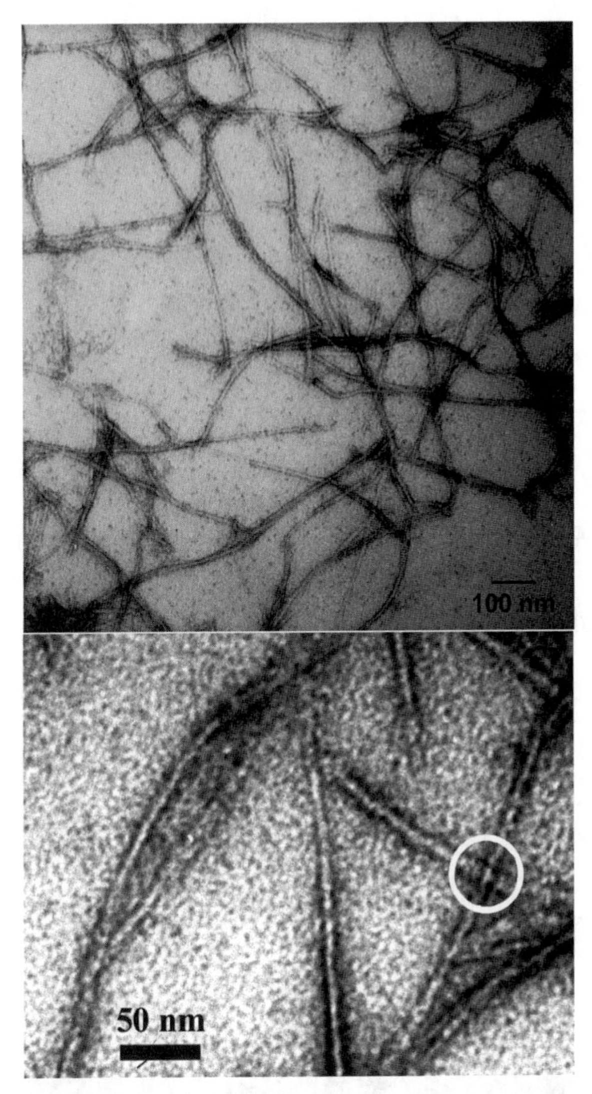

Figure 9.23 TEM micrographs of H6YEHK21H6 fibrils on carbon-coated Cu grids.

9.6.3 Fibril Formation

A higher-resolution TM-AFM micrograph of crossed fibrils is shown in Figure 9.25. Note the apparent increase in the fibril thickness at the intersections of the fibrils. A similar crossing phenomenon is evidenced in the lower image of Figure 9.23 (circle) where the distinct edges of two overlapping fibrils are seen via TEM. Both these sets of image data imply that the fibrils form in solution prior to deposition on the substrate. This contrasts with the surface-templated

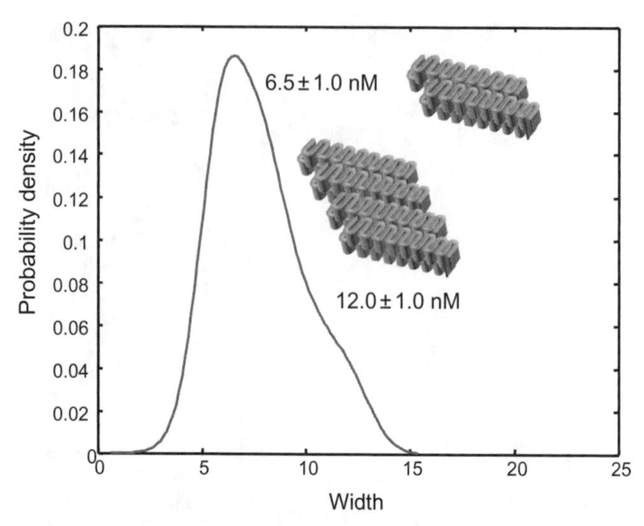

Figure 9.24 Probability density plot of fibril width derived from TEM of H6YEHK21H6 fibrils. A total of 161 samplings were employed to generate the distribution profile.

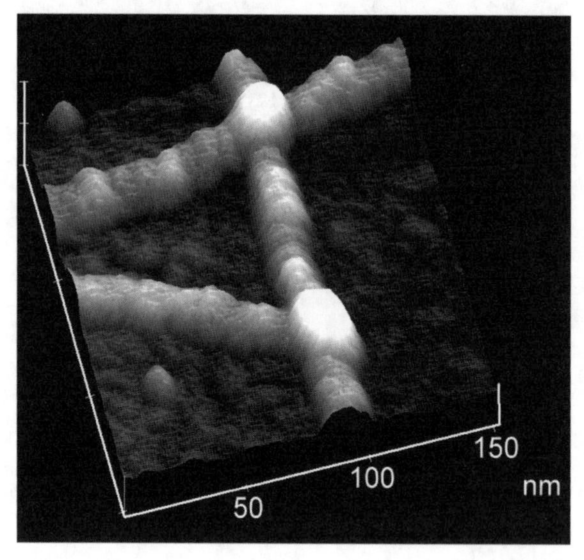

Figure 9.25 AFM topography of assembled domains of H6YEHK21H6 on HOPG.

assembly of low molecular weight polypeptide β-sheets on HOPG that has been previously reported (22, 91, 92). The surface disposition of the fibrillar structures presented herein more closely resembles observations by Marini et al. (113), although the polypeptide β-sheets reported in that work consisted of only eight residues, nearly two orders of magnitude fewer than the number of amino acids in the polypeptides here.

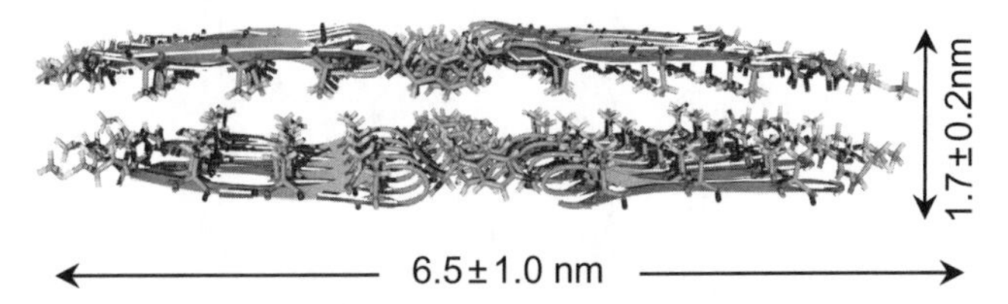

Figure 9.26 In a tentative structural model for the most common fibrils observed, both the dimensions of those fibrils and amphiphilic nature of proposed antiparallel β-sheet can be accommodated by the interaction of four polypeptide molecules.

It is important to note that β-sheets frequently exhibit a right-handed twist. An increase in the number of strands with a concomitant increase in the number of H-bonds should improve sheet stability, however the twist induces dissymmetry in the interaction of side chains from neighboring strands (94). Cooperative β-sheet formation has been reported in the formation of helical ribbons based upon the self-assembly of the octapeptide FKFEFKFE (79) where the KE and F domains lead to self-assembly of the individual molecular strands into a ribbon with an extended β-sheet geometry. Over longer times, the ribbons tend to form super helices around the initial strand (113). It has also been reported that low molecular weight peptides assemble into elongated β-sheet-containing tapes, which then dimerize to form helical ribbons and subsequently aggregate to form fibrils by forming twisted lamellae (1). In this work, helical assemblies were not commonly observed for H6YEHK21H6 in contrast to either of the above materials or the products described in Reference 113.

In Figure 9.26, depicted is a possible assemblage of four H6YEHK21H6 molecules, which can accommodate the observed dimensions of the fibrils that apparently form in solution. The amphiphilic character induced by the involvement of the γ-turn in reversing strand direction is consistent with the facile solution formation of bilayer thick fibrils. By design, the β-sheet-forming H6YEHK21H6 molecule has relatively hydrophobic and hydrophilic edges dominated by the turn groups. As a consequence, the more hydrophobic surfaces and turn groups of the molecules are shielded from the aqueous environment in the model. Further experiments to validate this hypothesis are under way.

9.7 GENETIC ENGINEERING FOR FIBRIL CORE STRUCTURE ELUCIDATION

The topology of amyloid-like fibrils prepared from three de novo designed polypeptides consisting of seven repeats of 32-, 40-, or 48-amino acid repeats,

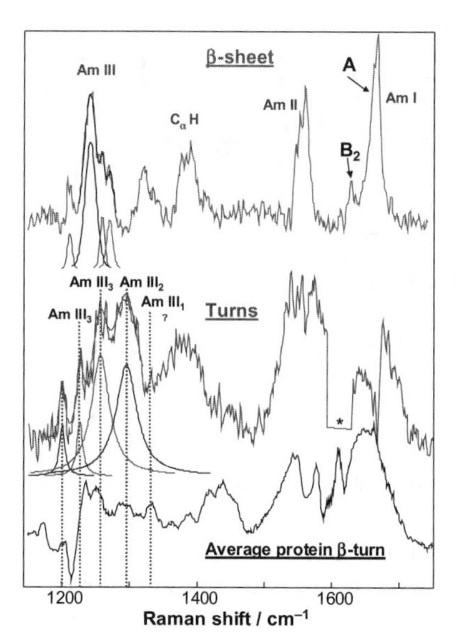

Figure 9.27 DUVRR spectra of YEHK fibril β-sheet and turns, and the average spectrum of native protein β-turns adopted from Reference 77. * = Spectral range of strong contribution of aromatic amino acids. Adopted from Sikirzhytski et al. (166) with permission from the American Chemical Society. See color insert.

$(GA)_nGY(GA)_nGE(GA)_nGH(GA)_nGK$ ($n = 3, 4$, or 5, respectively, 32YEHK7, 40YEHK7, or 48YEHK7) was consistent with the cross-β-core structure shown in Figure 9.6. DUVRR spectra measured for the three types of fibrils showed noticeable differences. At the same time, consistent with the postulated structural similarity of fibrils derived from the three peptides, the three differential Raman spectra (all possible combinations of subtractions between the Raman spectra of individual fibrillar samples) were identical to each other within the experimental error. These differential spectra represent the Raman signature for the β-strand portion of the fibrils (Fig. 9.27). The pure variable method was utilized for determining the second component (Fig. 9.27, "turns") of the two-element set, which was assigned as the Raman signature of turns. An excellent fit of all three spectra with the two components confirmed the structural similarity of the three fibrillar types, and hence, the validity of the approach used.

The obtained Raman signatures of the fibril core were used for evaluating the peptide backbone conformation based on Asher's approach described above (121). The dominant 1241–cm^{-1} peak in the Am III region of "β-sheet" spectrum (Figure 9.27) corresponds to Ramachandran angle of $\Psi = 150°$ as would be associated with an antiparallel β-sheet structure. The Am II frequency (1556 cm^{-1}) corresponds with the value assigned to the β-sheet structure of globular proteins (77) and homopolypeptides (122) although the peak is much narrower in the case of the YEHK fibrils. The Am I region of the "β-sheet" spectrum is comprised of two vibrational modes, A (1665 cm^{-1}) and B_2 (1629 cm^{-1}) (213). The observed shift of these frequencies in the case of YEHK fibrils, with respect to those found in the study of lysozyme (213), was tentatively attributed to stronger hydrogen bonding associated with the YEHK fibrils in comparison to lysozyme fibrils. The reported Ramachandran Ψ angular distribution determined from the β-sheet spectrum of YEHK fibrils (166) was much narrower than that attributed to a "pure" β-sheet prepared from poly-L-lysine-poly-L-glutamic acid (122). The Am II and Am III peaks are very narrow as well, 18 and 14 cm^{-1}, respectively, suggesting that the YEHK fibril β-sheet is highly ordered. These findings are consistent with our report above on lysozyme fibrils (213) where no inhomogeneous broadening was found for the cross-β-core Raman signature.

The calculated YEHK "turn" spectrum (Fig. 9.27) has many similarities with the "average protein β-turn" Raman spectrum reported by Spiro (77). The Am III region of the YEHK "turn" spectrum is well fitted by three peaks. Two of these features (1228 and 1256 cm^{-1}) correspond well with the Am III_3 peaks of β-turn type I (1223 and 1260 cm^{-1}) or type VIII (1226 and 1260 cm^{-1}) according to the Mikhonin and Asher assignments (121). The third peak at 1294 cm^{-1} was tentatively assigned to an Am III_2 vibration (120). There is no known definitive DUVRR spectra assignment of Am III_2 and Am III_1. However, since Ramachandran angles ($\Psi = -30°$, in particular) of $i + 1$ amide bond of a β-turn are close to those of an α-helix, similar Raman bands could be expected for both conformations. Indeed, the DUVRR spectrum (120) of an α-helix has three Am III lines at 1261 cm^{-1} (Am III_3), 1306 cm^{-1} (Am III_2), and 1337 cm^{-1} (weak, Am III_1), two of which are consistent with 1256 and 1294 cm^{-1} bands in the YEHK "turn" spectrum. A weak Am III_1 band might be obscured by the broad C_αH-bending band of the YEHK turns. A 3D modeling of the cross-β-core based on the estimated Ψ Ramachandran angles for turns and β-sheet was consistent with the allowed bond distance-angle space of interacting polypeptide chains.

9.8 CONCLUDING REMARKS AND PROSPECTIVES

Protein misfolding plays an important role in many neurodegenerative diseases and, consequently, is a focus of modern biology and molecular medicine. These diseases are characterized by the formation of proteinaceous aggregates principally composed of a protein or peptide specific to the particular disease. Despite structural differences between the native folded states, all amyloidogenic proteins and peptides form surprisingly similar aggregates. A unique feature of the fibrils is folding to form an extended cross-β-sheet structure regardless of the native protein tertiary structure. It has been suggested that amyloid-like folding with extended cross-β-sheet structure is not restricted to specific amino acid sequences, but results from intermolecular interactions that facilitate an alternate folding path. IDPs can be involved in the amyloid diseases as well. Genetic engineering affords excellent opportunities to test the influence of primary polypeptide sequence on folding and fibrillation mechanisms. This chapter focuses on utilization of genetic engineering in the study of the fibrillation of large biopolymers that are excellent models for IDPs. Specifically, β-sheet folding of high molecular weight synthetic fibrillogenic polypeptides was investigated by establishment of the primary sequence–property relationship governing β-sheet nucleation and the role of aggregation on formation of β-sheet nucleus. The mechanism of folding was probed using DUVRR spectroscopy combined with advanced statistical analysis including 2D correlation spectroscopy. The first two examples of cross-β-core Raman signatures, those obtained for YEHK polypeptide (166) and lysozyme (213), indicate the very ordered, crystalline-like structure of the antiparallel β-sheet. No inhomogeneous broadening by diverse amino acids is evident, yet Ramachandran ψ angle differs in the YEHK and lysozyme fibrillar β-sheets. To the best of our knowledge, this is the first direct evidence that the structure of the fibril cross-β-core is sequence-dependent. We believe that the directed variation of the polypeptide sequence by genetic engineering is a great tool for studies of the relationship between the sequence and the cross-β-sheet structure. These data may be especially useful in computational modeling of the polypeptide sequence-fibrillar structure relationship.

ACKNOWLEDGMENT

This material is based upon work supported by the National Science Foundation under CHE-0809525.

REFERENCES

1. Aggeli, A., I. A. Nyrkova, M. Bell, R. Harding, L. Carrick, T. C. B. McLeish, A. N. Semenov, and N. Boden. 2001. Hierarchical self-assembly of chiral rod-like molecules as a model for peptide β sheet tapes, ribbons, fibrils, and fibers. Proc Natl Acad Sci U S A **98**:11857–62.
2. Ahmed, Z., I. A. Beta, A. V. Mikhonin, and S. A. Asher. 2005. UV-resonance Raman thermal unfolding study of Trp-cage shows that it is not a simple two-state miniprotein. J Am Chem Soc **127**:10943–50.
3. Altman, M., P. Lee, A. Rich, and S. Zhang. 2000. Conformational behavior of ionic self-complementary peptides. Protein Sci **9**:1095–105.
4. Antzutkin, O. N., J. J. Balbach, R. D. Leapman, N. W. Rizzo, J. Reed, and R. Tycko. 2000. Multiple quantum solid-state NMR indicates a parallel, not antiparallel, organization of β-sheets in Alzheimer's β-amyloid fibrils. Proc Natl Acad Sci U S A **97**:13045–50.
5. Asher, S. A. 2001. Ultraviolet Raman spectrometry, pp. 557–71. In J. M. Chalmers and P. R. Griffiths (eds.), Handbook of Vibrational Spectroscopy, vol. 1. John Wiley & Sons, Ltd., New York.

6. Asher, S. A. 1993. UV resonance Raman spectroscopy for analytical, physical, and biophysical chemistry. Part 1. Anal Chem **65**:59A–66A.
7. Asher, S. A. 1993. UV resonance Raman spectroscopy for analytical, physical, and biophysical chemistry. Part 2. Anal Chem **65**:201A–10A.
8. Asher, S. A., Z. Chi, and P. Li. 1998. Resonance Raman examination of the two lowest Amide $\pi\pi$ excited states. J Raman Spectrosc **29**:927–31.
9. Asher, S. A., A. Ianoul, G. Mix, M. N. Boyden, A. Karnoup, M. Diem, and R. Schweitzer-Stenner. 2001. Dihedral psi angle dependence of the amide III vibration: a uniquely sensitive UV resonance Raman secondary structural probe. J Am Chem Soc **123**:11775–81.
10. Asher, S. A., A. V. Mikhonin, and S. Bykov. 2004. UV Raman demonstrates that α-helical polyalanine peptides melt to polyproline II conformations. J Am Chem Soc **126**:8433–40.
11. Asher, S. A., C. H. Munro, and Z. Chi. 1997. UV lasers revolutionize Raman spectroscopy. Laser Focus World **33**:99–109.
12. Austin, J. C., K. R. Rodgers, and T. G. Spiro. 1993. Protein structure from ultraviolet resonance Raman spectroscopy. Methods Enzymol **226**:374–96.
13. Balbach, J. J., Y. Ishii, O. N. Antzutkin, R. D. Leapman, N. W. Rizzo, F. Dyda, J. Reed, and R. Tycko. 2000. Amyloid fibril formation by $A\beta_{16-22}$, a seven-residue fragment of the Alzheimer's β-amyloid peptide, and structural characterization by solid state NMR. Biochemistry **39**:13748–59.
14. Balbach, J. J., A. T. Petkova, N. A. Oyler, O. N. Antzutkin, D. J. Gordon, S. C. Meredith, and R. Tycko. 2002. Supramolecular structure in full-length Alzheimer's β-amyloid fibrils: evidence for a parallel β-sheet organization from solid-state nuclear magnetic resonance. Biophys J **83**:1205–16.
15. Baltzer, L. 1999. Functionalization and properties of designed folded polypeptides. Topics Curr Chem **202**:39–76.
16. Baltzer, L., H. Nilsson, and J. Nilsson. 2001. De novo design of proteins-what are the rules? Chem Rev (Washington, DC) **101**:3153–63.
17. Bellotti, V., P. Mangione, and M. Stoppini. 1999. Biological activity and pathological implications of misfolded proteins. Cell Mol Life Sci **55**:977–91.
18. Booth, D. R., M. Sunde, V. Bellotti, C. V. Robinson, W. L. Hutchinson, P. E. Fraser, P. N. Hawkins, C. M. Dobson, S. E. Radford, C. C. F. Blake, and M. B. Pepys. 1997. Instability, unfolding and aggregation of human lysozyme variants underlying amyloid fibrillogenesis. Nature (London) **385**:787–93.
19. Bouchard, M., J. Zurdo, E. J. Nettleton, C. M. Dobson, and C. V. Robinson. 2000. Formation of insulin amyloid fibrils followed by FTIR simultaneously with CD and electron microscopy. Protein Sci **9**:1960–7.
20. Bradley, L. H., P. P. Thumfort, and M. H. Hecht. 2006. De novo proteins from binary-patterned combinatorial libraries. Methods Mol Biol (Totowa, NJ) **340**:53–69.
21. Brooks, C. L., III. 2002. Protein and peptide folding explored with molecular simulations. Acc Chem Res **35**:447–54.
22. Brown, C. L., I. A. Aksay, D. A. Saville, and M. H. Hecht. 2002. Template-directed assembly of a de novo designed protein. J Am Chem Soc **124**:6846–8.

23. Bu, Z., Y. Shi, D. J. E. Callaway, and R. Tycko. 2007. Molecular alignment within β-sheets in $A\beta_{14-23}$ fibrils: solid-state NMR experiments and theoretical predictions. Biophys J **92**:594–602.
24. Cantor, E. J., E. D. T. Atkins, S. J. Cooper, M. J. Fournier, T. L. Mason, and D. A. Tirrell. 1997. Effects of amino acid side-chain volume on chain packing in genetically engineered periodic polypeptides. J Biochem (Tokyo) **122**:217–25.
25. Cappello, J., J. Crissman, M. Dorman, M. Mikolajczak, G. Textor, M. Marquet, and F. Ferrari. 1990. Genetic engineering of structural protein polymers. Biotechnol Prog **6**:198–202.
26. Carey, P. R. 1982. Molecular Biology: Biochemical Applications of Raman and Resonance Raman Spectroscopies. Academic Press, New York.
27. Chan, J. C. C., N. A. Oyler, W.-M. Yau, and R. Tycko. 2005. Parallel β-sheets and polar zippers in amyloid fibrils formed by residues 10-39 of the yeast prion protein ure2p. Biochemistry **44**:10669–80.
28. Chi, Z., and S. A. Asher. 1999. Ultraviolet resonance Raman examination of horse apomyoglobin acid unfolding intermediates. Biochemistry **38**:8196–203.
29. Chi, Z., and S. A. Asher. 1998. UV Raman determination of the environment and solvent exposure of Tyr and Trp residues. J Phys Chem B **102**:9595–602.
30. Chi, Z., and S. A. Asher. 1998. UV resonance Raman determination of protein acid denaturation: selective unfolding of helical segments of horse myoglobin. Biochemistry **37**:2865–72.
31. Chi, Z., X. G. Chen, J. S. Holtz, and S. A. Asher. 1998. UV resonance Raman-selective amide vibrational enhancement: quantitative methodology for determining protein secondary structure. Biochemistry **37**:2854–64.
32. Chiti, F., and C. M. Dobson. 2006. Protein misfolding, functional amyloid, and human disease. Annu Rev Biochem **75**:333–66.
33. Chiti, F., P. Webster, N. Taddei, A. Clark, M. Stefani, G. Ramponi, and C. M. Dobson. 1999. Designing conditions for in vitro formation of amyloid protofilaments and fibrils. Proc Natl Acad Sci U S A **96**:3590–4.
34. Clarkson, J., D. N. Batchelder, and D. A. Smith. 2001. UV resonance Raman study of streptavidin binding of biotin and 2-iminobiotin: comparison with avidin. Biopolymers **62**:307–14.
35. Clarkson, J., and D. A. Smith. 2001. UV Raman evidence of a tyrosine in apo-human serum transferrin with a low pK(a) that is elevated upon binding of sulphate. FEBS Lett **503**:30–4.
36. Cobb, N. J., F. D. Sonnichsen, H. McHaourab, and W. K. Surewicz. 2007. Molecular architecture of human prion protein amyloid: a parallel, in-register β-structure. Proc Natl Acad Sci U S A **104**:18946–51.
37. Copeland, R. A., and T. G. Spiro. 1987. Secondary structure determination in proteins from deep (192-223-nm) ultraviolet Raman spectroscopy. Biochemistry **26**:2134–9.
38. Couling, V. W., P. Fischer, D. Klenerman, and W. Huber. 1998. Ultraviolet resonance Raman study of drug binding in dihydrofolate reductase, gyrase, and catechol O-methyltransferase. Biophys J **75**:1097–106.
39. Creel, H. S., M. J. Fournier, T. L. Mason, and D. A. Tirrell. 1991. Genetically directed syntheses of new polymeric materials: efficient expression of a

monodisperse copolypeptide containing fourteen tandemly repeated—(AlaGly)4ProGluGly—elements. Macromolecules **24**:1213–4.

40. de Alba, E., M. A. Jimenez, and M. Rico. 1997. Turn residue sequence determines β-hairpin conformation in designed peptides. J Am Chem Soc **119**:175–83.
41. DeFlores, L. P., and A. Tokmakoff. 2006. Water penetration into protein secondary structure revealed by hydrogen-deuterium exchange two-dimensional infrared spectroscopy. J Am Chem Soc **128**:16520–1.
42. Demchenko, A. P. 1986. Ultraviolet Spectroscopy of Proteins. Springer-Verlag, Berlin.
43. Deng, W., A. Cao, and L. Lai. 2008. Distinguishing the cross-β spine arrangements in amyloid fibrils using FRET analysis. Protein Sci **17**(6):1102–5.
44. Dijkstra, R. J., E. V. Efremov, F. Ariese, U. A. Brinkman, and C. Gooijer. 2003. Capillary electrophoresis coupled on-line with ultraviolet resonance Raman spectroscopy. Anal Chem **75**:5697–702.
45. Dobson, C. M. 2005. An overview of protein misfolding diseases. Protein Folding Handbook **5**:1093–113.
46. Dobson, C. M. 2006. Protein aggregation and its consequences for human disease. Protein Pept Lett **13**:219–27.
47. Dobson, C. M. 1999. Protein misfolding, evolution and disease. Trends Biochem Sci **24**:329–32.
48. Dumoulin, M., A. M. Last, A. Desmyter, K. Decanniere, D. Canet, G. Larsson, A. Spencer, D. B. Archer, J. Sasse, S. Muyldermans, L. Wyns, C. Redfield, A. Matagne, C. V. Robinson, and C. M. Dobson. 2003. A camelid antibody fragment inhibits the formation of amyloid fibrils by human lysozyme. Nature (London) **424**:783–8.
49. Dunker, A. K., J. D. Lawson, C. J. Brown, R. M. Williams, P. Romero, J. S. Oh, C. J. Oldfield, A. M. Campen, C. M. Ratliff, K. W. Hipps, J. Ausio, M. S. Nissen, R. Reeves, C. Kang, C. R. Kissinger, R. W. Bailey, M. D. Griswold, W. Chiu, E. C. Garner, and Z. Obradovic. 2001. Intrinsically disordered protein. J Mol Graph Model **19**:26–59.
50. Dyson, H. J., P. E. Wright, and H. A. Scheraga. 2006. The role of hydrophobic interactions in initiation and propagation of protein folding. Proc Natl Acad Sci U S A **103**:13057–61.
51. Eakin, C. M., J. D. Knight, C. J. Morgan, M. A. Gelfand, and A. D. Miranker. 2002. Formation of a copper specific binding site in non-native states of β-2-microglobulin. Biochemistry **41**:10646–56.
52. Eisenberg, D., R. Nelson, M. R. Sawaya, M. Balbirnie, S. Sambashivan, M. I. Ivanova, A. O. Madsen, and C. Riekel. 2006. The structural biology of protein aggregation diseases: fundamental questions and some answers. Acc Chem Res **39**:568–75.
53. Fabiola, F., and V. Pattabhi. 2002. The structure based design of peptide motifs. PINSA-A: Proc Indian Natl Sci Acad, Part A: Phys Sci **68**:251–65.
54. Ferrari, F. A., and J. Cappello. 1997. Biosynthesis of protein polymers. Protein-Based Mater:37–60.
55. Ferrari, F. A., C. Richardson, J. Chambers, S. Causey, T. J. Pollock, J. Cappello, and J. W. Crissman. 19950607. 2000. Peptides comprising repetitive units of amino

acids and DNA sequences encoding the same for production of fibers for use in prosthetics. Application: US patent 95-4820856018030.

56. Fink, A. L. 2005. Natively unfolded proteins. Curr Opin Struct Biol **15**:35–41.

57. Fink, A. L. 1998. Protein aggregation: folding aggregates, inclusion bodies and amyloid. Fold Des **3**:R9–23.

58. Fodor, S. P. A., R. P. Rava, T. R. Hays, and T. G. Spiro. 1985. Ultraviolet resonance Raman spectroscopy of the nucleotides with 266-, 240-, 218-, and 200-nm pulsed laser excitation. J Am Chem Soc **107**:1520–9.

59. Fukushima, Y. 1998. Genetically engineered syntheses of tandem repetitive polypeptides consisting of glycine-rich sequence of spider dragline silk. Biopolymers **45**:269–79.

60. Gellman, S. H. 1998. Minimal model systems for β sheet secondary structure in proteins. Curr Opin Chem Biol **2**:717–25.

61. Goeden-Wood, N. L., V. P. Conticello, S. J. Muller, and J. D. Keasling. 2002. Improved assembly of multimeric genes for the biosynthetic production of protein polymers. Biomacromolecules **3**:874–9.

62. Goldbeck, R. A., R. M. Esquerra, and D. S. Kliger. 2002. Hydrogen bonding to Trp b37 is the first step in a compound pathway for hemoglobin allostery. J Am Chem Soc **124**:7646–7.

63. Gregersen, N., P. Bross, S. Vang, and J. H. Christensen. 2006. Protein misfolding and human disease. Annu Rev Genomics Hum Genet **7**:103–4.

64. Gremlich, H.-U., and B. Yan (eds.). 2001. Infrared and Raman Spectroscopy of Biological Materials, vol. 24. Dekker, New York.

65. Guijarro, J. I., C. J. Morton, K. W. Plaxco, I. D. Campbell, and C. M. Dobson. 1998. Folding kinetics of the SH3 domain of PI3 kinase by real-time NMR combined with optical spectroscopy. J Mol Biol **276**:657–67.

66. Harada, I., and H. Takeuchi. 1986. Raman and ultraviolet resonance Raman spectra of proteins and related compounds, p. 547. In R. J. H. Clark and R. E. Hester (eds.), Advances in Spectroscopy, vol. **13**: Spectrosc Biol Syst. John Wiley & Sons, Chichester, UK.

67. Hashimoto, S., M. Sasaki, H. Takeuchi, R. Needleman, and J. K. Lanyi. 2002. Changes in hydrogen bonding and environment of tryptophan residues on helix F of bacteriorhodopsin during the photocycle: a time-resolved ultraviolet resonance Raman study. Biochemistry **41**:6495–503.

68. Hecht, M. H. 1994. De novo design of β-sheet proteins. Proc Natl Acad Sci U S A **91**:8729–30.

69. Hecht, M. H., A. Das, A. Go, L. H. Bradley, and Y. Wei. 2004. De novo proteins from designed combinatorial libraries. Protein Sci **13**:1711–23.

70. Henkels, C. H., and T. G. Oas. 2006. Ligation-state hydrogen exchange: coupled binding and folding equilibria in ribonuclease p protein. J Am Chem Soc **128**:7772–81.

71. Hildebrandt, P. G., R. A. Copeland, T. G. Spiro, J. Otlewski, M. Laskowski, Jr., and F. G. Prendergast. 1988. Tyrosine hydrogen-bonding and environmental effects in proteins probed by ultraviolet resonance Raman spectroscopy. Biochemistry **27**:5426–33.

72. Holt, A., G. Alton, C. H. Scaman, G. R. Loppnow, A. Szpacenko, I. Svendsen, and M. M. Palcic. 1998. Identification of the quinone cofactor in mammalian semicarbazide-sensitive amine oxidase. Biochemistry **37**:4946–57.
73. Holtz, J. S. W., R. W. Bormett, Z. Chi, N. Cho, X. G. Chen, V. Pajcini, S. A. Asher, L. Spinelli, P. Owen, and M. Arrigoni. 1996. Applications of a new 206 5-nm continuous-wave laser source: UV Raman determination of protein secondary structure and CVD diamond material properties. Appl Spectrosc **50**:1459–68.
74. Holtz, J. S. W., J. H. Holtz, Z. Chi, and S. A. Asher. 1999. Proteins—ultraviolet Raman examination of the environmental dependence of bombolitin I and bombolitin III secondary structure. Biophys J **76**:3227–34.
75. Hou, L., and M. G. Zagorski. 2004. Sorting out the driving forces for parallel and antiparallel alignment in the Ab peptide fibril structure. Biophys J **86**:1–2.
76. Hu, X., and T. G. Spiro. 1997. Tyrosine and tryptophan structure markers in hemoglobin ultraviolet resonance Raman spectra: mode assignments via subunit-specific isotope labeling of recombinant protein. Biochemistry **36**:15701–12.
77. Huang, C.-Y., G. Balakrishnan, and T. G. Spiro. 2006. Protein Secondary Structure from Deep UV Resonance Raman Spectroscopy. J Raman Spectrosc **37**:277–82.
78. Hudson, B., and L. Mayne. 1986. Ultraviolet resonance Raman spectroscopy of biopolymers. Methods Enzymol **130**:331–50.
79. Hwang, W., D. M. Marini, R. D. Kamm, and S. Zhang. 2003. Supramolecular structure of helical ribbons self-assembled from a β-sheet peptide. J Chem Phys **118**:389–97.
80. Ianoul, A., M. N. Boyden, and S. A. Asher. 2001. Dependence of the peptide amide III vibration on the phi dihedral angle. J Am Chem Soc **123**:7433–4.
81. Iguchi, K. 2002. Statistical mechanical foundation for the two-state transition in protein folding of small globular proteins. Int J Mod Phys B **16**:1807–39.
82. Jahn, T. R., and S. E. Radford. 2005. The yin and yang of protein folding. FEBS J **272**:5962–70.
83. Jaroniec, C. P., C. E. MacPhee, N. S. Astrof, C. M. Dobson, and R. G. Griffin. 2002. Molecular conformation of a peptide fragment of transthyretin in an amyloid fibril. Proc Natl Acad Sci U S A **99**:16748–53.
84. Jaroniec, C. P., C. E. MacPhee, V. S. Bajaj, M. T. McMahon, C. M. Dobson, and R. G. Griffin. 2004. High-resolution molecular structure of a peptide in an amyloid fibril determined by magic angle spinning NMR spectroscopy. Proc Natl Acad Sci U S A **101**:711–6.
85. Juszczak, L. J., C. Fablet, V. Baudin-Creuza, S. Lesecq-Le Gall, R. E. Hirsch, R. L. Nagel, J. M. Friedman, and J. Pagnier. 2003. Conformational changes in hemoglobin S (βE6V) imposed by mutation of the βGlu_7-beta Lys_{132} salt bridge and detected by UV resonance Raman spectroscopy. J Biol Chem **278**: 7257–63.
86. Kammerer, R. A., D. Kostrewa, J. Zurdo, A. Detken, C. Garcia-Echeverria, J. D. Green, S. A. Mueller, B. H. Meier, F. K. Winkler, C. M. Dobson, and M. O. Steinmetz. 2004. Exploring amyloid formation by a de novo design. Proc Natl Acad Sci U S A **101**:4435–40.
87. Kelly, J. W. 1998. The alternative conformations of amyloidogenic proteins and their multi-step assembly pathways. Curr Opin Struct Biol **8**:101–6.

88. Kheterpal, I., K. D. Cook, and R. Wetzel. 2006. Hydrogen/deuterium exchange mass spectrometry analysis of protein aggregates. Methods Enzymol **413**: 140–66.

89. Kheterpal, I., and R. Wetzel. 2006. Hydrogen/deuterium exchange mass spectrometry—a window into amyloid structure. Acc Chem Res **39**:584–93.

90. Kincaid, J. R. 2000. Resonance Raman spectra of heme proteins and model compounds. In K. M. Kadish, K. M. Smith, and R. Guilard (eds.), Theoretical and Physical Characterization, The Porphyrin Handbook, vol. 7. Academic Press, New York, pp. 225–91.

91. Kogan, M. J., I. Dalcol, P. Gorostiza, C. Lopez-Iglesias, M. Pons, F. Sanz, D. Ludevid, and E. Giralt. 2001. Self-assembly of the amphipathic helix (VHLPPP)8. A mechanism for zein protein body formation. J Mol Biol **312**:907–13.

92. Kowalewski, T., and D. M. Holtzman. 1999. In situ atomic force microscopy study of Alzheimer's β-amyloid peptide on different substrates: new insights into mechanism of β-sheet formation. Proc Natl Acad Sci U S A **96**:3688–93.

93. Krejchi, M. T., S. J. Cooper, Y. Deguchi, E. D. T. Atkins, M. J. Fournier, T. L. Mason, and D. A. Tirrell. 1997. Crystal structures of chain-folded antiparallel β-sheet assemblies from sequence-designed periodic polypeptides. Macromolecules **30**:5012–24.

94. Lacroix, E., T. Kortemme, M. L. De la Paz, and L. Serrano. 1999. The design of linear peptides that fold as monomeric β-sheet structures. Curr Opin Struct Biol **9**:487–93.

95. Langer, R., and D. A. Tirrell. 2004. Designing materials for biology and medicine. Nature (London) **428**:487–92.

96. Lansbury, P. T., P. R. Costa, J. M. Griffiths, E. J. Simon, M. Auger, K. J. Halverson, D. A. Kocisko, Z. S. Hendsch, T. T. Ashburn, R. G. S. Spencer, B. Tidor, and R. G. Griffin. 1995. Structural model for the β-amyloid fibril based on interstrand alignment of an antiparallel-sheet comprising a C-terminal peptide. Nat Struct Mol Biol **2**:990–8.

97. Lansbury, P. T., Jr. 1999. Evolution of amyloid: what normal protein folding may tell us about fibrillogenesis and disease. Proc Natl Acad Sci U S A **96**:3342–4.

98. Lednev, I. K. 2006. Vibrational spectroscopy: biological applications of ultraviolet Raman spectroscopy. In V. N. Uversky and E. A. Permyakov (eds.), Protein Structures, Methods in Protein Structures and Stability Analysis. Nova Science Publishers, Inc., New York.

99. Lednev, I. K., V. V. Ermolenkov, W. He, and M. Xu. 2005. Deep-UV Raman spectrometer tunable between 193 and 205 nm for structural characterization of proteins. Anal Bioanal Chem **381**:431–7.

100. Lednev, I. K., V. V. Ermolenkov, S. Higashiya, L. A. Popova, N. I. Topilina, and J. T. Welch. 2006. Reversible thermal denaturation of a 60-kDa genetically engineered β-sheet polypeptide. Biophys J **91**:3805–18.

101. Lednev, I. K., A. S. Karnoup, M. C. Sparrow, and S. A. Asher. 1999. α-helix peptide folding and unfolding activation barriers: A nanosecond UV resonance Raman study. J Am Chem Soc **121**:8074–86.

102. Lednev, I. K., A. S. Karnoup, M. C. Sparrow, and S. A. Asher. 1999. Nanosecond UV resonance Raman examination of initial steps in α-helix secondary structure evolution. J Am Chem Soc **121**:4076–7.

103. Lednev, I. K., A. S. Karnoup, M. C. Sparrow, and S. A. Asher. 2001. Transient UV Raman spectroscopy finds no crossing barrier between the peptide α-helix and fully random coil conformation. J Am Chem Soc **123**:2388–92.

104. Lin, J.-C., and H.-L. Liu. 2006. Protein conformational diseases: from mechanisms to drug designs. Curr Drug Discov Technol **3**:145–53.

105. Lin, S. W., G. G. Kochendoerfer, K. S. Carroll, D. Wang, R. A. Mathies, and T. P. Sakmar. 1998. Mechanisms of spectral tuning in blue cone visual pigments. Visible and Raman spectroscopy of blue-shifted rhodopsin mutants. J Biol Chem **273**:24583–91.

106. Lindorff-Larsen, K., P. Rogen, E. Paci, M. Vendruscolo, and C. M. Dobson. 2005. Protein folding and the organization of the protein topology universe. Trends Biochem Sci **30**:13–9.

107. Lipman, E. A., B. Schuler, O. Bakajin, and W. A. Eaton. 2003. Single-molecule measurement of protein folding kinetics. Science (Washington, DC) **301**:1233–5.

108. Litvinovich, S. V., S. A. Brew, S. Aota, S. K. Akiyama, C. Haudenschild, and K. C. Ingham. 1998. Formation of amyloid-like fibrils by self-association of a partially unfolded fibronectin type III module. J Mol Biol **280**:245–58.

109. Liu, Z., and J. Pawliszyn. 2004. Capillary isoelectric focusing with laser-induced fluorescence whole column imaging detection as a tool to monitor reactions of proteins. J Proteome Res **3**:567–71.

110. Lopez-Diez, E. C., and R. Goodacre. 2004. Characterization of microorganisms using UV resonance Raman spectroscopy and chemometrics. Anal Chem **76**:585–91.

111. Maiti, N. C., T. Tomita, T. Kitagawa, K. Okamoto, and T. Nishino. 2003. Resonance Raman studies on xanthine oxidase: observation of Mo(VI)-ligand vibrations. J Biol Inorg Chem **8**:327–33.

112. Makin, O. S., and L. C. Serpell. 2005. Structures for amyloid fibrils. FEBS J **272**:5950–61.

113. Marini, D. M., W. Hwang, D. A. Lauffenburger, S. Zhang, and R. D. Kamm. 2002. Left-handed helical ribbon intermediates in the self-assembly of a β-sheet peptide. Nano Lett **2**:295–9.

114. Mayne, L. C., L. D. Ziegler, and B. Hudson. 1985. Ultraviolet resonance Raman studies of N-methylacetamide. J Phys Chem **89**:3395–8.

115. McGrath, K. P., D. A. Tirrell, M. Kawai, T. L. Mason, and M. J. Fournier. 1990. Chemical and biosynthetic approaches to the production of novel polypeptide materials. Biotechnol Prog **6**:188–92.

116. McMillan, R. A., T. A. T. Lee, and V. P. Conticello. 1999. Rapid assembly of synthetic genes encoding protein polymers. Macromolecules **32**:3643–8.

117. McParland, V. J., N. M. Kad, A. P. Kalverda, A. Brown, P. Kirwin-Jones, M. G. Hunter, M. Sunde, and S. E. Radford. 2000. Partially unfolded states of b2-microglobulin and amyloid formation in vitro. Biochemistry **39**:8735–46.

118. McPherson, D. T., J. Xu, and D. W. Urry. 1996. Product purification by reversible phase transition following *Escherichia coli* expression of genes encoding up to 251 repeats of the elastomeric pentapeptide GVGVP. Protein Expr Purif **7**:51–7.

119. Meyer, D. E., and A. Chilkoti. 2002. Genetically encoded synthesis of protein-based polymers with precisely specified molecular weight and sequence by recursive directional ligation: examples from the elastin-like polypeptide system. Biomacromolecules **3**:357–67.

120. Mikhonin, A. V., and S. A. Asher. 2005. Uncoupled peptide bond vibrations in α-helical and polyproline II conformations of polyalanine peptides. J Phys Chem **109**:3047–52.

121. Mikhonin, A. V., S. V. Bykov, N. S. Myshakina, and S. A. Asher. 2006. Peptide secondary structure folding reaction coordinate: correlation between UV Raman amide III frequency, Psi Ramachandran angle, and hydrogen bonding. J Phys Chem B **110**:1928–43.

122. Mikhonin, A. V., N. S. Myshakina, S. V. Bykov, and S. A. Asher. 2005. UV resonance Raman determination of polyproline II, extended 2.5_1-helix, and β-sheet Psi angle energy landscape in poly-L-lysine and poly-L-glutamic acid. J Am Chem Soc **127**:7712–20.

123. Mirkin, N. G., and S. Krimm. 2002. Amide III mode, dependence in peptides: a vibrational frequency map. J Phys Chem A **106**:3391–4.

124. Mukerji, I., and A. P. Williams. 2002. UV resonance Raman and circular dichroism studies of a DNA duplex containing an A(3)T(3) tract: evidence for a premelting transition and three-centered H-bonds. Biochemistry **41**:69–77.

125. Munoz, V., P. A. Thompson, J. Hofrichter, and W. A. Eaton. 1997. Folding dynamics and mechanism of β-hairpin formation. Nature **390**:196–9.

126. Nagatomo, S., M. Nagai, N. Shibayama, and T. Kitagawa. 2002. Differences in changes of the α1-β2 subunit contacts between ligand binding to the α and β subunits of hemoglobin A: UV resonance Raman analysis using Ni-Fe hybrid hemoglobin. Biochemistry **41**:10010–20.

127. Nakashima, S.-I., H. Okumura, T. Yamamoto, and R. Shimidzu. 2004. Deep-ultraviolet Raman microspectroscopy: Characterization of wide-gap semiconductors. Appl Spectrosc **58**:224–9.

128. Nelson, R., and D. Eisenberg. 2006. Recent atomic models of amyloid fibril structure. Curr Opin Struct Biol **16**:260–5.

129. Nelson, R., M. R. Sawaya, M. Balbirnie, A. O. Madsen, C. Riekel, R. Grothe, and D. Eisenberg. 2005. Structure of the cross-β spine of amyloid-like fibrils. Nature (London) **435**:773–8.

130. Nielsen, L., R. Khurana, A. Coats, S. Frokjaer, J. Brange, S. Vyas, V. N. Uversky, and A. L. Fink. 2001. Effect of environmental factors on the kinetics of insulin fibril formation: elucidation of the molecular mechanism. Biochemistry **40**:6036–46.

131. Noda, I., and Y. Ozaki. 2004. Two-dimensional correlation spectroscopy : applications in vibrational and optical spectroscopy. John Wiley & Sons, Hoboken, NJ.

132. Okada, A., T. Miura, and H. Takeuchi. 2003. Zinc- and pH-dependent conformational transition in a putative interdomain linker region of the influenza virus matrix protein M1. Biochemistry **42**:1978–84.

133. Otzen, D. E., O. Kristensen, and M. Oliveberg. 2000. Designed protein tetramer zipped together with a hydrophobic Alzheimer homology: a structural clue to amyloid assembly. Proc Natl Acad Sci U S A **97**:9907–12.

134. Overman, S. A., and G. J. Thomas, Jr. 1995. Raman spectroscopy of the filamentous virus Ff (fd, f1, M13): structural interpretation for coat protein aromatics. Biochemistry **34**:5440–51.

135. Padgett, K. A., and J. A. Sorge. 1996. Creating seamless junctions independent of restriction sites in PCR cloning. Gene **168**:31–5.

136. Palaniappan, V., and D. F. Bocian. 1994. Acid-induced transformations of myoglobin. Characterization of a new equilibrium heme-pocket intermediate. Biochemistry **33**:14264–74.

137. Pan, D., Z. Ganim, J. E. Kim, M. A. Verhoeven, J. Lugtenburg, and R. A. Mathies. 2002. Time-resolved resonance Raman analysis of chromophore structural changes in the formation and decay of rhodopsin's BSI intermediate. J Am Chem Soc **124**:4857–64.

138. Panitch, A., K. Matsuki, E. J. Cantor, S. J. Cooper, E. D. T. Atkins, M. J. Fournier, T. L. Mason, and D. A. Tirrell. 1997. Poly(L-alanylglycine): multigram-scale biosynthesis, crystallization, and structural analysis of chain-folded lamellae. Macromolecules **30**:42–9.

139. Paravastu, A. K., A. T. Petkova, and R. Tycko. 2006. Polymorphic fibril formation by residues 10-40 of the Alzheimer's β-amyloid peptide. Biophys J **90**:4618–29.

140. Parkhe, A. D., S. J. Cooper, E. D. T. Atkins, M. J. Fournier, T. L. Mason, and D. A. Tirrell. 1998. Effect of local sequence inversions on the crystalline antiparallel β-sheet lamellar structures of periodic polypeptides: implications for chain-folding. Int J Biol Macromol **23**:251–8.

141. Petkova, A. T., Y. Ishii, J. J. Balbach, O. N. Antzutkin, R. D. Leapman, F. Delaglio, and R. Tycko. 2002. A structural model for Alzheimer's β-amyloid fibrils based on experimental constraints from solid state NMR. Proc Natl Acad Sci U S A **99**:16742–7.

142. Radivojac, P., L. M. Iakoucheva, C. J. Oldfield, Z. Obradovic, V. N. Uversky, and A. K. Dunker. 2007. Intrinsic disorder and functional proteomics. Biophys J **92**:1439–56.

143. Ramirez-Alvarado, M., J. S. Merkel, and L. Regan. 2000. A systematic exploration of the influence of the protein stability on amyloid fibril formation in vitro. Proc Natl Acad Sci U S A **97**:8979–84.

144. Rava, R. P., and T. G. Spiro. 1984. Selective enhancement of tyrosine and tryptophan resonance Raman spectra via ultraviolet laser excitation. J Am Chem Soc **106**:4062–4.

145. Rava, R. P., and T. G. Spiro. 1985. Ultraviolet resonance Raman spectra of insulin and α-lactalbumin with 218- and 200-nm laser excitation. Biochemistry **24**:1861–5.

146. Rochet, J.-C., and P. T. Lansbury, Jr. 2000. Amyloid fibrillogenesis: themes and variations. Curr Opin Struct Biol **10**:60–8.

147. Rodriguez-Casado, A., and G. J. Thomas, Jr. 2003. Structural roles of subunit cysteines in the folding and assembly of the DNA packaging machine (portal) of bacteriophage p22. Biochemistry **42**:3437–45.

148. Romero, P., Z. Obradovic, X. Li, E. C. Garner, C. J. Brown, and A. K. Dunker. 2001. Sequence complexity of disordered protein. Protein Struct Funct Genet **42**:38–48.

149. Sambrook, J. F., D. W. Russell, and Editors. 2000. Molecular Cloning: A Laboratory Manual, third edition. Cold Spring Harbor Laboratory Press, Cold Spring Harbor, NY.

150. Samuni, U., D. Dantsker, I. Khan, A. J. Friedman, E. Peterson, and J. M. Friedman. 2002. Spectroscopically and kinetically distinct conformational populations of sol-gel-encapsulated carbonmonoxy myoglobin. A comparison with hemoglobin. J Biol Chem **277**:25783–90.

151. Sawaya, M. R., S. Sambashivan, R. Nelson, M. I. Ivanova, S. A. Sievers, M. I. Apostol, M. J. Thompson, M. Balbirnie, J. J. W. Wiltzius, H. T. McFarlane, A. O. Madsen, C. Riekel, and D. Eisenberg. 2007. Atomic structures of amyloid cross-β spines reveal varied steric zippers. Nature **447**:453–7.

152. Schulze, H. G., L. S. Greek, C. J. Barbosa, M. W. Blades, B. B. Gorzalka, and R. F. Turner. 1999. Measurement of some small-molecule and peptide neurotransmitters in-vitro using a fiber-optic probe with pulsed ultraviolet resonance Raman spectroscopy. J Neurosci Methods **92**:15–24.

153. Searle, M. S. 2005. Design and stability of peptide β-sheets. Protein Folding Handbook **1**:314–42.

154. Searle, M. S. 2001. Peptide models of protein β-sheets: design, folding and insights into stabilising weak interactions. J Chem Soc Perkin Trans 1 **2**:1011–20.

155. Serban, D., S. F. Arcineigas, C. E. Vorgias, and G. J. Thomas, Jr. 2003. Structure and dynamics of the DNA-binding protein HU of B. stearothermophilus investigated by Raman and ultraviolet-resonance Raman spectroscopy. Protein Sci **12**:861–70.

156. Serrano, L. 2000. The relationship between sequence and structure in elementary folding units. Adv Protein Chem **53**:49–85.

157. Shashilov, V., M. Xu, V. V. Ermolenkov, L. Fredriksen, and I. K. Lednev. 2007. Probing a fibrillation nucleus directly by deep ultraviolet Raman spectroscopy. J Am Chem Soc **129**:6972–3.

158. Shashilov, V. A. 2007. Development of Mathematical Methods for Quantitative Resonance Raman Spectroscopy. PhD University at Albany, State University of New York, Albany.

159. Shashilov, V. A., V. V. Ermolenkov, and I. K. Lednev. 2006. Multiple bicyclic diamide-lutetium complexes in solution: chemometric analysis of deep-UV Raman spectroscopic data. Inorg Chem **45**:3606–12.

160. Shashilov, V. A., and I. K. Lednev. 2008. 2D correlation deep UV resonance Raman spectroscopy of early events of lysozyme fibrillation: kinetic mechanism and potential interpretation pitfalls. J Am Chem Soc **130**:309–17.

161. Shashilov, V. A., and I. K. Lednev. 2007. Bayesian extraction of deep UV resonance Raman signature of fibrillar cross-b sheet core based on H-D exchange data. AIP Conf Proc **954**:450–7.

162. Shashilov, V. A., and I. K. Lednev. 2006. Novel methods for latent variable analysis of Raman spectra. Abstracts of Papers, 231st ACS National Meeting, Atlanta, GA, March 26–30, 2006:ANYL-134.

163. Shashilov, V. A., M. Xu, V. V. Ermolenkov, and I. K. Lednev. 2006. Latent variable analysis of Raman spectra for structural characterization of proteins. J Quant Spectrosc Radiat Transf **102**:46–61.

164. Shewmaker, F., R. B. Wickner, and R. Tycko. 2006. Amyloid of the prion domain of Sup35p has an in-register parallel β-sheet structure. Proc Natl Acad Sci U S A **103**:19754–9.

165. Sibanda, B. L., T. L. Blundell, and J. M. Thornton. 1989. Conformation of β-hairpins in protein structures. A systematic classification with applications to modeling by homology, electron density fitting and protein engineering. J Mol Biol **206**:759–77.

166. Sikirzhytski, V., N. I. Topilina, S. Higashiya, J. T. Welch, and I. K. Lednev. 2008. Genetic engineering combined with deep UV resonance Raman spectroscopy for structural characterization of amyloid-like fibrils. J Am Chem Soc **130**:5852–3.

167. Smeenk, J. M., M. B. J. Otten, J. Thies, D. A. Tirrell, H. G. Stunnenberg, and J. C. M. van Hest. 2005. Controlled assembly of macromolecular β-sheet fibrils. Angew Chem Int Ed Engl **44**:1968–71.

168. Song, S., and S. A. Asher. 1989. UV resonance Raman studies of peptide conformation in poly(L-lysine), poly(L-glutamic acid), and model complexes: the basis for protein secondary structure determinations. J Am Chem Soc **111**:4295–305.

169. Spencer, R. G., K. J. Halverson, M. Auger, A. E. McDermott, R. G. Griffin, and P. T. Lansbury, Jr. 1991. An unusual peptide conformation may precipitate amyloid formation in Alzheimer's disease: application of solid-state NMR to the determination of protein secondary structure. Biochemistry **30**:10382–7.

170. Spiro, T. G. (ed.). 1987. Biological Applications of Raman Spectroscopy. Vol. 1: Raman Spectra and the Conformations of Biological Macromolecules, vol. 1. John Wiley & Sons, New York.

171. Squires, A. M., G. L. Devlin, S. L. Gras, A. K. Tickler, C. E. MacPhee, and C. M. Dobson. 2006. X-ray scattering study of the effect of hydration on the cross-β structure of amyloid fibrils. J Am Chem Soc **128**:11738–9.

172. Stefani, M. 2004. Protein misfolding and aggregation: new examples in medicine and biology of the dark side of the protein world. Biochim Biophys Acta **1739**:5–25.

173. Stefani, M., and C. M. Dobson. 2003. Protein aggregation and aggregate toxicity: new insights into protein folding, misfolding diseases and biological evolution. J Mol Med (Heidelberg, Germany) **81**:678–99.

174. Swietnicki, W., M. Morillas, S. G. Chen, P. Gambetti, and W. K. Surewicz. 2000. Aggregation and fibrillization of the recombinant human prion protein huPrP90-231. Biochemistry **39**:424–31.

175. Thirumalai, D., and D. K. Klimov. 2007. Intermediates and transition states in protein folding. Methods Mol Biol (Totowa, NJ) **350**:277–303.

176. Thomas, G. J., Jr., B. Prescott, and L. A. Day. 1983. Studies of virus structure by laser Raman spectroscopy. Part XI. Structure similarity, difference and variability in the filamentous viruses fd, If1, IKe, Pf1 and Xf. Investigation by laser Raman spectroscopy. J Mol Biol **165**:321–56.

177. Thompson, P. A., V. Munoz, G. S. Jas, E. R. Henry, W. A. Eaton, and J. Hofrichter. 2000. The Helix-coil kinetics of a heteropeptide. J Phys Chem B **104**:378–89.

178. Topilina, N. I., V. V. Ermolenkov, S. Higashiya, J. T. Welch, and I. K. Lednev. 2007. β-Sheet folding of 11-kDa fibrillogenic polypeptide is completely aggregation driven. Biopolymers **86**:261–4.

179. Topilina, N. I., S. Higashiya, N. Rana, V. V. Ermolenkov, C. Kossow, A. Carlsen, S. C. Ngo, C. C. Wells, E. T. Eisenbraun, K. A. Dunn, I. K. Lednev, R. E. Geer, A. E. Kaloyeros, and J. T. Welch. 2006. Bilayer fibril formation by genetically engineered polypeptides: preparation and characterization. Biomacromolecules **7**:1104–11.

180. Torok, M., S. Milton, R. Kayed, P. Wu, T. McIntire, C. G. Glabe, and R. Langen. 2002. Structural and dynamic features of Alzheimer's Aβ peptide in amyloid fibrils studied by site-directed spin labeling. J Biol Chem **277**:40810–15.

181. Toyama, A., Y. Takahashi, and H. Takeuchi. 2004. Catalytic and structural role of a metal-free histidine residue in bovine Cu-Zn superoxide dismutase. Biochemistry **43**:4670–9.

182. Tsuboi, M., S. A. Overman, K. Nakamura, A. Rodriguez-Casado, and G. J. Thomas, Jr. 2003. Orientation and interactions of an essential tryptophan (trp-38) in the capsid subunit of pf3 filamentous virus. Biophys J **84**:1969–76.

183. Tu, A. T. 1986. Peptide backbone conformation and microenvironment of protein side chains, pp. 47–175. In R. J. H. Clark and R. E. Hester (eds.), Spectroscopy of Biological Systems, vol. 13. John Wiley & Sons, Chichester, UK.

184. Tycko, R. 2006. Molecular structure of amyloid fibrils: insights from solid-state NMR. Q Rev Biophys **39**(1):1–55.

185. Uversky, V. N. 2008. Amyloidogenesis of natively unfolded proteins. Curr Alzheimer Res **5**:260–87.

186. Uversky, V. N. 2002. Natively unfolded proteins: a point where biology waits for physics. Protein Sci **11**:739–56.

187. Uversky, V. N. 2003. Protein folding revisited. A polypeptide chain at the folding—misfolding—nonfolding cross-roads: which way to go? Cell Mol Life Sci **60**:1852–71.

188. Uversky, V. N., and A. L. Fink. 2004. Conformational constraints for amyloid fibrillation: the importance of being unfolded. Biochim Biophys Acta **1698**:131–53.

189. Uversky, V. N., C. J. Oldfield, and A. K. Dunker. 2008. Intrinsically disordered proteins in human diseases: introducing the D2 concept. Annu Rev Biophys Biomol Struct **37**:215–46.

190. Uversky, V. N., A. Talapatra, J. R. Gillespie, and A. L. Fink. 1999. Protein deposits as the molecular basis of amyloidosis. Part I. Systemic amyloidoses. Med Sci Monit **5**:1001–12.

191. Uversky, V. N., A. Talapatra, J. R. Gillespie, and A. L. Fink. 1999. Protein deposits as the molecular basis of amyloidosis. Part II. Localized amyloidosis and neurodegenerative disorders. Med Sci Monit **5**:1238–54.

192. Vaillancourt, F. H., C. J. Barbosa, T. G. Spiro, J. T. Bolin, M. W. Blades, R. F. Turner, and L. D. Eltis. 2002. Definitive evidence for monoanionic binding of 2,3-dihydroxybiphenyl to 2,3-dihydroxybiphenyl 1,2-dioxygenase from UV resonance Raman spectroscopy, UV/Vis absorption spectroscopy, and crystallography. J Am Chem Soc **124**:2485–96.

193. Van Dael, H., P. Haezebrouck, L. Morozova, C. Arico-Muendel, and C. M. Dobson. 1993. Partially folded states of equine lysozyme. Structural characterization and significance for protein folding. Biochemistry **32**:11886–94.

194. van der Wel, P. C., J. R. Lewandowski, and R. G. Griffin. 2007. Solid-state NMR study of amyloid nanocrystals and fibrils formed by the peptide GNNQQNY from yeast prion protein Sup35p. J Am Chem Soc **129**:5117–30.

195. van Hest, J. C. M., and D. A. Tirrell. 2001. Protein-based materials, toward a new level of structural control. Chem Commun (Camb) **2001**(19):1897–904.

196. Wang, J., A. D. Parkhe, D. A. Tirrell, and L. K. Thompson. 1996. Crystalline aggregates of the repetitive polypeptide {(AlaGly)3GluGly(GlyAla)3GluGly}10: Structure and dynamics probed by 13C magic angle spinning nuclear magnetic resonance spectroscopy. Macromolecules **29**:1548–53.

197. Wang, W., and M. H. Hecht. 2002. Rationally designed mutations convert de novo amyloid-like fibrils into monomeric β-sheet proteins. Proc Natl Acad Sci U S A **99**:2760–5.

198. Wang, Y., R. Purrello, S. Georgiou, and T. G. Spiro. 1991. UVRR spectroscopy of the peptide bond. 2. Carbonyl H-bond effects on the ground- and excited-state structures of N-methylacetamide. J Am Chem Soc **113**:6368–77.

199. Welch, J. T., S. Higashiya, S. C. Ngo, N. Rana, V. Ermolenkov, C. A. Kossow, A. Carlsen, K. S. Bousman, C. C. Wells, E. T. Eisenbraun, I. Lednev, R. E. Geer, and A. E. Kaloyeros. 2006. Novel nanofibril formation motif by genetically engineered polypeptides. Biomacromolecules **7**:1104–11.

200. West, M. W., W. Wang, J. Patterson, J. D. Mancias, J. R. Beasley, and M. H. Hecht. 1999. De novo amyloid proteins from designed combinatorial libraries. Proc Natl Acad Sci U S A **96**:11211–6.

201. Whittaker, M. M., V. L. DeVito, S. A. Asher, and J. W. Whittaker. 1989. Resonance Raman evidence for tyrosine involvement in the radical site of galactose oxidase. J Biol Chem **264**:7104–6.

202. Wickner, R. B., F. Dyda, and R. Tycko. 2008. Amyloid of Rnq1p, the basis of the (PIN+) prion, has a parallel in-register β-sheet structure. Proc Natl Acad Sci U S A **105**:2403–8.

203. Williams, R. M., Z. Obradovi, V. Mathura, W. Braun, E. C. Garner, J. Young, S. Takayama, C. J. Brown, and A. K. Dunker. 2001. The protein non-folding problem: amino acid determinants of intrinsic order and disorder. Pacific Symposium on Biocomputing 2001, Mauna Lani, HI, January 3–7, **2001**:89–100.

204. Winkler, S., D. Wilson, and D. L. Kaplan. 2000. Controlling β-sheet assembly in genetically engineered silk by enzymatic phosphorylation/dephosphorylation. Biochemistry **39**:12739–46.

205. Won, J.-I., and A. E. Barron. 2002. A new cloning method for the preparation of long repetitive polypeptides without a sequence requirement. Macromolecules **35**:8281–7.

206. Wright, E. R., and V. P. Conticello. 2002. Self-assembly of block copolymers derived from elastin-mimetic polypeptide sequences. Adv Drug Deliv Rev **54**:1057–73.

207. Wu, Q., T. Hamilton, W. H. Nelson, S. Elliott, J. F. Sperry, and M. Wu. 2001. UV Raman spectral intensities of E. coli and other bacteria excited at 228.9, 244.0, and 248.2 nm. Anal Chem **73**:3432–40.

208. Wu, Q., F. Li, W. Wang, M. H. Hecht, and T. G. Spiro. 2002. UV Raman monitoring of histidine protonation and H-(2)H exchange in plastocyanin. J Inorg Biochem **88**:381–7.

209. Wu, Q., W. H. Nelson, J. M. Treubig, Jr., P. R. Brown, P. Hargraves, M. Kirs, M. Feld, R. Desari, R. Manoharan, and E. B. Hanlon. 2000. UV resonance Raman detection and quantitation of domoic acid in phytoplankton. Anal Chem **72**:1666–71.

210. Xu, G., W. Wang, J. T. Groves, and M. H. Hecht. 2001. Self-assembled monolayers from a designed combinatorial library of de novo β-sheet proteins. Proc Natl Acad Sci U S A **98**:3652–7.

211. Xu, M., V. V. Ermolenkov, W. He, V. N. Uversky, L. Fredriksen, and I. K. Lednev. 2005. Lysozyme fibrillation: deep UV Raman spectroscopic characterization of protein structural transformation. Biopolymers **79**:58–61.

212. Xu, M., V. V. Ermolenkov, V. N. Uversky, and I. K. Lednev. 2008. Hen egg white lysozyme fibrillation: a deep-UV resonance Raman spectroscopic study. J Biophotonics **1**:215–29.

213. Xu, M., V. Shashilov, and I. K. Lednev. 2007. Probing the cross-β core structure of amyloid fibrils by hydrogen-deuterium exchange deep ultraviolet resonance Raman spectroscopy. J Am Chem Soc **129**:11002–3.

214. Xu, M., V. A. Shashilov, V. V. Ermolenkov, L. Fredriksen, D. Zagorevski, and I. K. Lednev. 2007. The first step of hen egg white lysozyme fibrillation, irreversible partial unfolding, is a two-state transition. Protein Sci **16**:815–32.

215. Yang, W. Y., and M. Gruebele. 2004. Detection-dependent kinetics as a probe of folding landscape microstructure. J Am Chem Soc **126**:7758–9.

216. Yoshikawa, E., M. J. Fournier, T. L. Mason, and D. A. Tirrell. 1994. Genetically engineered fluoropolymers. Synthesis of repetitive polypeptides containing p-fluorophenylalanine residues. Macromolecules **27**:5471–5.

217. Zerovnik, E. 2002. Amyloid-fibril formation. Proposed mechanisms and relevance to conformational disease. Eur J Biochem **269**:3362-71.

218. Zhao, X., R. Chen, C. Tengroth, and T. G. Spiro. 1999. Solid-state tunable kHz ultraviolet laser for Raman applications. Appl Spectrosc **53**:1200–5.

219. Zhou, Y., S. Wu, and V. P. Conticello. 2001. Genetically directed synthesis and spectroscopic analysis of a protein polymer derived from a flagelliform silk sequence. Biomacromolecules **2**:111–25.

220. Ziegler, L. D., and B. Hudson. 1981. Resonance Raman scattering of benzene and benzene-d6 with 212.8 nm excitation. J Chem Phys **74**:982–92.

221. Zurdo, J. 2005. Polypeptide models to understand misfolding and amyloidogenesis and their relevance in protein design and therapeutics. Protein Pept Lett **12**:171–87.

10

CIRCULAR DICHROISM OF INTRINSICALLY DISORDERED PROTEINS

Robert W. Woody

Department of Biochemistry and Molecular Biology, Colorado State University, Fort Collins, CO

ABSTRACT

Circular dichroism (CD) in the far ultraviolet region is one of the most widely used methods for characterizing the secondary structure of proteins. Unordered polypeptides have a characteristic far-UV CD spectrum. Proteins that are disordered throughout their sequence are readily recognized by CD and CD has been used to identify and characterize a large number of IDPs. A common type of IDP in which a protein has folded domains interspersed with extensive unordered regions is more difficult to diagnose. Limited proteolysis can locate the boundaries between compact and unordered portions of the protein. Expression of the individual domains identified by proteolysis, followed by CD analysis can be used to elucidate the structure of such proteins.

10.1 INTRODUCTION

Circular dichroism (CD) in the far ultraviolet region is a widely used method for determining and monitoring the secondary structure of proteins (29, 32, 46). α-Helices, β-sheets, β-turns, and unordered polypeptides give rise to characteristic CD spectra that can easily be recognized when one of these

Instrumental Analysis of Intrinsically Disordered Proteins: Assessing Structure and Conformation, Edited by Vladimir Uversky and Sonia Longhi

conformations is dominant. Typical proteins contain most or all of these conformational types, but methods have been developed to analyze the CD spectrum of a protein and determine the fraction of residues in each conformation (5, 8, 28, 41, 47, 49). IDPs that lack any significant organized secondary structure give the characteristic CD spectrum of an unordered polypeptide, with a strong negative band near 200 nm and either a weak negative shoulder or a weak positive maximum near 220 nm. Such IDPs are readily recognized by CD. It is more challenging to recognize and characterize IDPs that contain folded domains intermingled with extensive unfolded regions. Progress has recently been made with some specific IDPs by defining the domains through limited proteolysis, followed by cloning and expression of individual domains that can be analyzed by CD and other methods.

In this chapter, the CD of unfolded proteins will be described, starting with model polypeptides and proceeding to globular proteins unfolded by various means. The CD of IDPs will then be reviewed, beginning with full-length IDPs and concluding with IDPs that have both ordered and unordered regions.

This chapter considers only CD in the UV region, that is, electronic CD. CD in the infrared, vibrational CD (VCD), and the related phenomenon of Raman optical activity are discussed in chapter 11 by Reinhard Schweitzer-Stenner.

10.2 THE CD OF UNFOLDED PROTEINS

10.2.1 Model Systems

The most widely used models for unfolded proteins are the homopolypeptides poly(Glu) and poly(Lys) at neutral pH (6, 26). The repulsions among the charged side chains of these polypeptides preclude formation of α-helix, which forms at low and high pH, respectively. The CD spectrum of poly(Lys) at room temperature and pH 7 is shown in Figure 10.1. The spectrum has a strong negative band at 198 nm and a weaker positive band at 217 nm. The spectrum of poly(Glu) is nearly identical.

Tiffany and Krimm (51, 52, 54) challenged the conventional view that poly(Glu) and poly(Lys) in neutral aqueous solution are fully unordered polypeptides. They noted that the CD spectra of poly(Glu) and poly(Lys) bear a striking resemblance to that of poly(Pro) II, the left-handed, threefold helical form that poly(Pro) adopts in water and other polar solvents. Figure 10.2 shows the spectra of poly(Glu) and poly(Pro), which have nearly identical shapes but differ by a factor of ~2 in intensity and a wavelength shift of ca. 10 nm. Tiffany and Krimm interpreted this strong spectral resemblance as evidence that poly(Glu) and poly(Lys) are not completely unordered at room temperature but have short regions of poly(Pro) II (P_{II}) helix, interspersed with unordered regions. Tiffany and Krimm's proposal was met with considerable skepticism at the time, but a large body of evidence now supports their interpretation (44, 63).

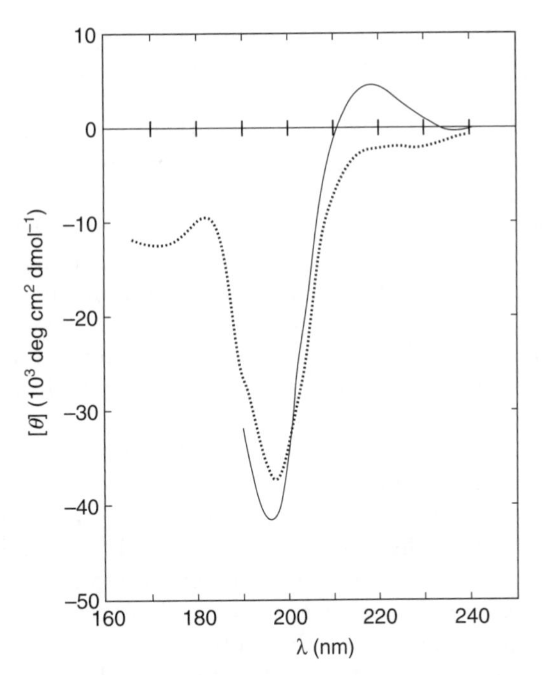

Figure 10.1 CD spectra of unordered polypeptides. Poly(Lys) in water at pH 5.7 (———) (23); poly(Pro-Lys-Leu-Lys-Leu) in salt-free water (·····) (6).

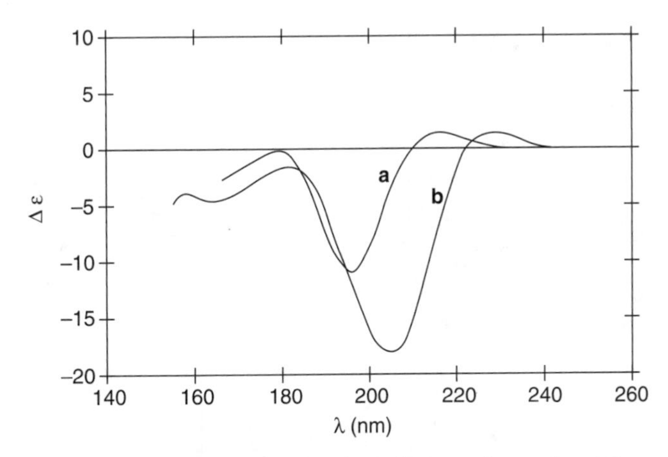

Figure 10.2 CD spectra of the poly (Pro) II (P_{II}) conformation. The spectrum of poly(Pro) in water (27) (B) is prototypical, with a strong negative band at 206 nm and a weak positive band at 225 nm. The spectrum of poly(Glu) in water (30) (A) is dominated by the P_{II} conformation (54, 63), as indicated by the CD spectrum, which resembles that of poly (Pro). The blue shift of ~10 nm results from the difference between secondary and tertiary amides. (From Reference 63, with permission.)

Figure 10.1 shows the CD spectrum of poly(Pro-Lys-Leu-Lys-Leu) [poly(PKLKL)] in salt-free water at pH 7 (6), conditions under which this copolymer is unordered. The spectrum of poly(PKLKL) has a negative band at 197 nm that is comparable in intensity to that of poly(Glu) and poly(Lys), but it has a weak negative shoulder at longer wavelengths instead of a positive band.

The CD spectrum of poly(Lys) has been measured over a nearly 200-K range by Drake et al. (12) (Figure 10.3). As the temperature decreases toward −100°C, the long-wavelength positive band increases in intensity by more than fourfold relative to that at room temperature. The 200-nm negative band increases about twofold in amplitude. Above room temperature, the positive long-wavelength band becomes a negative shoulder and ultimately a weak negative maximum at ~80°C.

The CD spectra of poly(Lys) (12) (Fig. 10.3) also show a sharp isodichroic point at 203 nm over the entire temperature range. This implies that poly(Lys) behaves as a two-state system over this broad range. These two states can be readily interpreted with the Tiffany and Krimm (54) model. The low-temperature form is the P_{II} helix and the high-temperature form is a truly unordered form in which the polypeptide chain samples all of accessible Ramachandran space. This interpretation implies that poly(Lys) and poly(Glu) have significant P_{II} helix at room temperature (~50% based on the amplitude of the 200-nm negative band), whereas poly(PKLKL) has less P_{II} helix and is closer to a

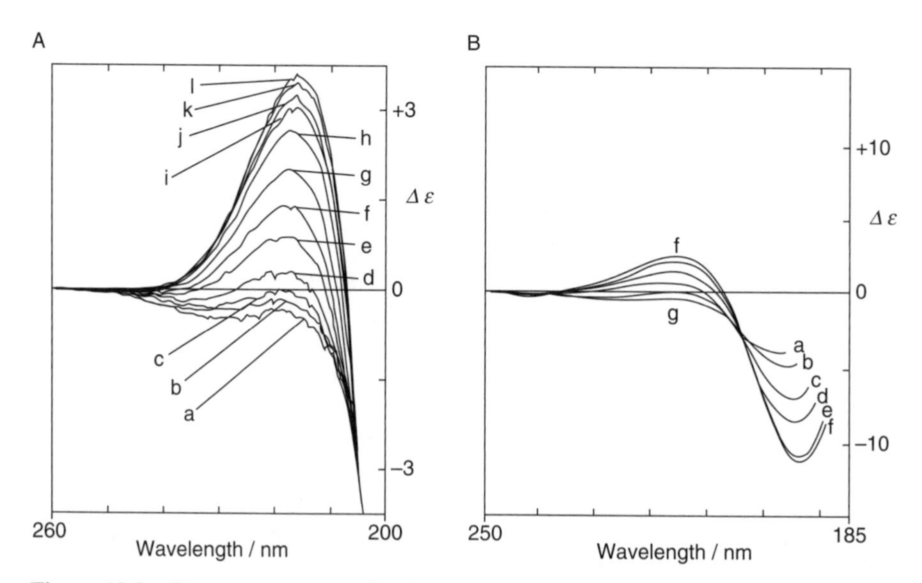

Figure 10.3 CD spectra of poly(Lys) in a 2:1 1,2-ethanediol-water mixture, pH 7.6 (12). (A) 260–200 nm region: (a) 82, (b) 69, (c) 60, (d) 45, (e) 21, (f) 2, (g) −26, (h) −55, (i) −75, (j) −85, (k) −95, (l) −105° C. (B) 250–185 nm region: (a) 84, (b) 60, (c) 22, (d) −28, (e) −8, (f) −102°C. (From Reference 12, with permission.)

fully unordered polypeptide. It is probable that poly(PKLKL) at lower temperatures would exhibit a positive CD band near 220 nm.

The CD spectrum of the neutral homopolymer poly(N^5-ω-hydroxyethyl-Gln) (36) at 15°C shows a strong negative band at 195–196 nm, a weak positive band at 217 nm, and a weaker negative band at 233 nm. At 75°C, the short-wavelength negative band is somewhat less intense and the long-wavelength CD is entirely negative with a maximum near 230 nm. In this polypeptide, the P_{II} helix appears to be intermediate in stability between that in poly(Lys) and that in poly(PKLKL).

Tiffany and Krimm (53) showed that the classical denaturants urea and guanidinium chloride (GuCl) increase the magnitude of the long-wavelength positive band of poly(Pro), poly(Glu), and poly(Lys). (The band near 200 nm is inaccessible in the denaturant solutions because of strong absorption by urea or GuCl.) The increase in the 226-nm band of poly(Pro) is interpreted as reflecting a stabilization of the P_{II} conformation by urea or GuCl. The corresponding increases for poly(Glu) and poly(Lys) are attributed to enhancement of the P_{II} conformation in these polypeptides by the denaturants.

Park et al. (38) have used a 17-residue Ala-rich peptide with a Pro at the central position, AcYEAAAKEAP_9AKEAAAKANH_2 (P9), as a model for unfolded polypeptides in helix-coil transition studies. The CD spectrum of this peptide at 0°C has a strong negative band at ~196 nm and a broad, weak positive band at long wavelengths with $[\theta]_{222}$ = 120 deg cm^2 dmol^{-1}. As shown in Figure 10.4, the spectrum of this peptide fits into a family of spectra that range

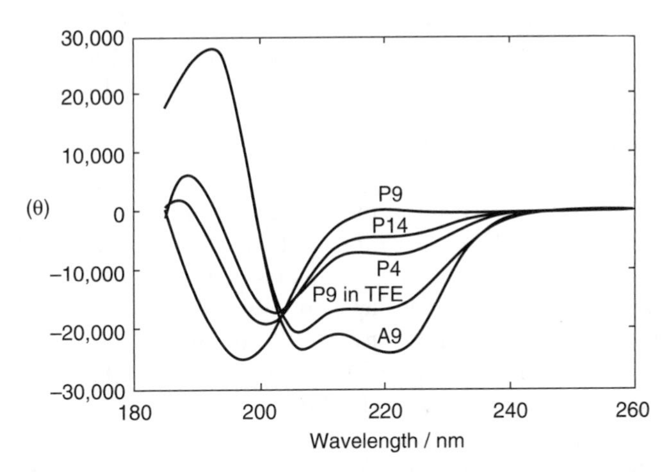

Figure 10.4 CD spectra of peptides based on the peptide AcYEAAAKEAAAKEAAAKANH_2 (38). Peptide A9 is the parent peptide, with Ala at the central position, 9. In P9 the central Ala is replaced by Pro. In P4 and P14, the Ala residues at positions 4 and 14, respectively, are replaced by Pro. The spectra were measured in 10 mM NaCl and 1 mM phosphate buffer, pH 7.0, at 0°C, except for that designated "P9 in TFE," which used 82% by volume trifluorethanol. (From Reference 38, with permission.)

from highly α-helical (peptide A9 with Ala at position 9) to the unordered spectrum of the peptide P9. The spectra exhibit an approximate isodichroic point at 203 nm. Urea and GuCl were found to increase the long-wavelength positive CD to values of ~3500 deg cm^2 dmol^{-1} at 0°C and 6–8 M denaturant. Park et al. measured the temperature dependence of the CD at 222 nm for the P9 peptide in 8 M GuCl and found that it fits a two-state transition curve. The data of Drake et al. (12) for poly(Lys) in water and in a 2:1 ethanediol-water mixture fall on the same curve. The fit to the thermal transition curve gave $[\theta]_{222}$ values of +9580 and −5560 deg cm^2 dmol^{-1}, respectively, for the P_{II} and the fully unordered conformations.

Bienkiewicz et al. (4) used the data of Park et al. (38) to estimate the fraction of P_{II} conformation in a nominally unordered polypeptide. Bienkiewicz et al. also estimated limiting values of $[\theta]_{200}$ for the P_{II} and truly unordered forms of −45700 and −9100 deg cm^2 dmol^{-1}, respectively. The P_{II} contents derived from the CD at these two wavelengths must be considered as rough estimates because the limiting values will depend on the solvent, temperature, and amino acid composition, especially Pro content. When both measures can be used, a comparison of the results from the two wavelengths provides a useful test of the reliability. In most cases, only the $[\theta]_{222}$-based value will be accessible for urea- or GuCl-denatured proteins, but Matsuo et al. (35) were able to make measurements through the 200-nm band of urea-denatured proteins.

Toumadje et al. (55) proposed a scale for estimating P_{II} content in nominally unfolded polypeptides that uses $[\theta]_{200}$ values for poly(Pro) for 100% P_{II}, and for the peptide $(TSDSR)_3$ at 80°C for 100% truly unordered. However, the assumption that $(TSDSR)_3$ is fully disordered at 80°C appears questionable as the $\Delta[\theta]_{200}$ between 60 and 80°C is comparable to that from 40 to 60°. The value of $[\theta]_{200}$ = −14000 deg cm^2 dmol^{-1}, according to the method of Bienkiewicz et al. (4), indicates 13% residual P_{II} conformation.

How should the fraction of P_{II} helix be interpreted? It has been argued (44) that the positive 220-nm band requires three or more successive residues in the P_{II} conformation, based upon the studies of Dukor and Keiderling (13), who observed this band for $(Pro)_3$ but not for $(Pro)_2$. However, Dukor and Keiderling studied *unblocked* $(Pro)_n$ oligomers, that is, the terminal groups were free amino and carboxyl groups, not amides. Thus, for $n = 2$, for which they observed a single negative band at 210 nm, there is a *single* amide chromophore. Helbeque and Loucheux-Lefebvre (25) observed a positive band at 221 nm for cationic $Gly(Pro)_2OH$, which has two amide groups. It appears that the minimum requirement for observing a positive long-wavelength band is a *single* P_{II} residue flanked by two peptide groups. The P_{II} residue holds the two adjacent amide groups in the proper relative geometry to give the characteristic 220-nm positive and 200-nm negative bands. In a polypeptide chain, except for the terminal residues, each C_α is flanked by two peptide groups, so each residue in the P_{II} conformation can be expected to give the characteristic P_{II} CD signal. This is consistent with the results of Sreerama and Woody (48), who found the characteristic P_{II} CD spectrum in globular proteins even when single P_{II} residues were counted. Thus, f_{PII} reflects the fraction of residues in the P_{II} conformation. The truly unordered form that is observed at higher temperatures is an ensemble of conformers distributed over the Ramachandran map, with residues in the α-, β-, and P_{II}-regions. Theoretical and experimental studies of di- and tripeptides (15–18, 22, 43, 44) show that the P_{II} conformation is at or near the global free energy minimum for several types of residues, so isolated P_{II} residues or pairs of P_{II} residues will be significant components of this ensemble, perhaps 30% or more, as well as individual residues or pairs of residues in the α- and β-regions. However, in the unordered ensemble, the P_{II} CD signature is obscured by the contributions of the other conformers. Thus, the f_{PII} evaluated from CD spectra should be regarded as a lower limit to the actual P_{II} content.

10.2.2 Unfolded Proteins

Several studies have compared the CD of proteins unfolded by urea or GuCl (U), heat (H), cold (C), and acid (A) (11, 31, 35, 37, 40). (The notation U, H, etc. was suggested by Matsuo et al. [35]. The native form is denoted N, a widely used convention.) The spectra for various unfolded forms of five globular proteins were reported by Privalov et al. (40). H- and A-forms of proteins have a strong negative CD band near 200 nm, characteristic of unfolded proteins. (The H-form of apomyoglobin [apoMb] has a CD spectrum that differs significantly from those of H-forms of the other four globular proteins studied by Privalov et al. [40]. H-form apoMb appears to form a β-sheet, probably through aggregation.) For each protein, $[\theta]_{222}$ becomes increasingly negative in the order U < A < H. The reverse is true for $[\theta]_{200}$, although we can only compare the A- and H-forms at the 200-nm maximum. According to the method of Bienkiewicz et al. (4, 44), urea-unfolded proteins have 30% or more P_{II} whereas heat-unfolded proteins have less than 10% P_{II}. Acid-unfolded proteins are intermediate, with 10–20% P_{II}. The larger variability of P_{II} content for acid-unfolded proteins and the sizeable spread of values estimated from $[\theta]_{200}$ and $[\theta]_{222}$ may result from residual structure. The two representatives of cold-denatured proteins have P_{II} contents that are comparable to those of acid-denatured proteins.

Matsuo et al. (35) reported CD spectra (Fig. 10.5) for denatured metmyoglobin (metMb), staphylococcal nuclease (SN), and thioredoxin (trx) measured to ~175 nm using a synchrotron light source. Based upon the CD at 220 and 200 nm (4), the P_{II} content of unfolded metMb decreases in the order U > A > H. (The C-form of metMb appears to be incompletely unfolded, as judged by the substantial positive CD band near 190 nm.) For SN, the order is U > A > H > C. In the case of trx, P_{II} content decreases as U > C > H. Here it is the A-form that appears to be incompletely unfolded, with the spectrum of the putative A-form showing α-helix-like features, both in the long-wavelength region and near 190 nm. In all cases, the estimates based upon 220- and 200-nm CD follow the same order as those given by Matsuo et al., which were

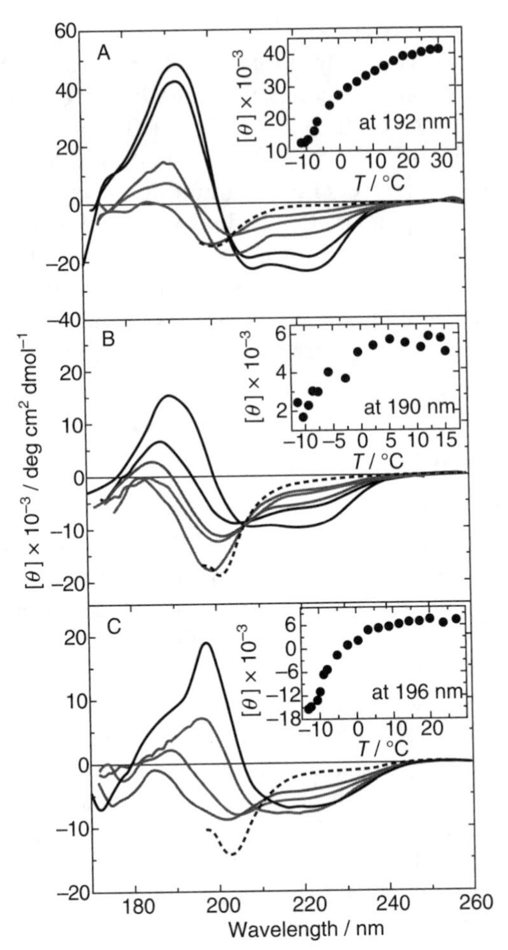

Figure 10.5 Vacuum-UV CD spectra of metmyoglobin (A), staphylococcal nuclease (B), and thioredoxin (C) in various conformational states: native (solid black line), heat-denatured (red line), cold-denatured (blue line), acid-denatured (green line), partially acid-denatured (brown line), and urea-denatured (dotted black line) (35). The inset shows the temperature dependence of the molar ellipticity at the indicated wavelength for each protein. (From Reference 35, with permission.) See color insert.

obtained from a more detailed analysis using a set of globular reference proteins. An interesting aspect of the work of Matsuo et al. is that they were able to measure the CD of the U-form through the 200-nm negative maximum.

Intrinsically disordered proteins (IDPs) have CD spectra that exhibit the characteristic features of unfolded polypeptides—a strong negative band near 200 nm and a weak negative CD at 220 nm. Uversky (56) has compiled CD data for over 100 IDPs and represented these data in a useful plot of $[\theta]_{222}$ versus $[\theta]_{200}$, as shown in Figure 10.6. This plot shows that IDPs fall into two

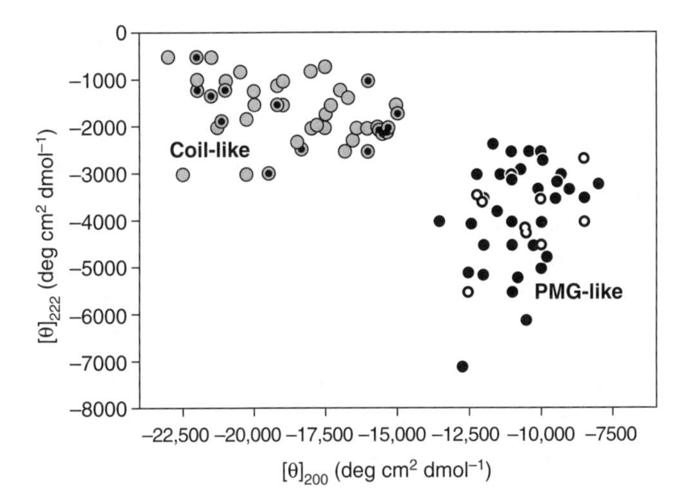

Figure 10.6 Far-UV CD spectra of IDPs represented by a plot of $[\theta]_{222}$ versus $[\theta]_{200}$ distinguish between coil-like (gray circles) and premolten globule-like (black circles) classes (56). IDPs that have been classified as PMG-like or coil-like on the basis of hydrodynamic measurements are marked by white-dotted and black-dotted symbols, respectively. (From Reference 56, with permission.)

classes. One class, with $[\theta]_{200} = -19{,}000 \pm 2800\,\mathrm{deg\,cm^2\,dmol^{-1}}$ and $[\theta]_{222} = -1700 \pm 700\,\mathrm{deg\,cm^2\,dmol^{-1}}$, is called coil-like by Uversky. The other, with $[\theta]_{200} = -10{,}700 \pm 1300\,\mathrm{deg\,cm^2\,dmol^{-1}}$ and $[\theta]_{222} = -3900 \pm 1100\,\mathrm{deg\,cm^2\,dmol^{-1}}$, is designated PMG-like, that is, like a premolten globule (7, 59, 60). The difference in these two classes of IDPs lies in their P_{II} content. Uversky's coil-like class contains ~30% P_{II} whereas the PMG-like class has ~10% P_{II}. Thus, the coil-like IDPs have substantial fractions of P_{II} helix, whereas the PMG-like IDPs have smaller fractions. In both cases, the non-P_{II} residues are distributed over the Ramachandran map in an unordered ensemble, although small regions of helix or β-strand may be present.

Uversky (57) has discussed the temperature dependence of the CD of IDPs. At low temperatures (<25°C), the CD spectra are characteristic of Uversky's coil-like class (56), and they change to the PMG-like class in a smooth, noncooperative transition as the temperature increases. This transition was interpreted as "temperature-induced formation of secondary structure" (57) but, in the absence of independent evidence for secondary structure formation (58), it should be viewed as temperature-induced melting of P_{II} helix. Although higher temperature leads to a conversion of P_{II} to α- and β-conformation, the α- and β-conformation thus formed generally consists of isolated residues or very short stretches of α- or β-residues and not coherent α-helices or β-strands.

The approximate secondary structure content of unfolded proteins, including IDPs, can be obtained by analyzing their CD spectra. Current methods of analysis use the CD spectra of proteins of known structure and adapt the analysis to the specific protein being analyzed (24, 34, 46). However, the use of only globular proteins in such an analysis gives questionable results. When a set of unfolded proteins was analyzed using a reference set containing only folded, globular proteins, the average secondary structure content was 10% α-helix, 20% β-sheet, 25% β-turns, and 45% unordered (45). Similar results were obtained by Matsuo et al. (35) for three proteins unfolded by various methods and using data extending into the vacuum UV. A substantial content of organized secondary structure and an unordered content of less than 50% seems improbable for a denatured protein and indicates that the results should not be interpreted in the same way as those for a folded protein. The α-helical and β-sheet residues indicated in the analysis are generally not part of recognizable helices or strands but are isolated residues or short segments that make up the ensemble of conformers in the unordered conformation.

Sreerama et al. (45) expanded the set of globular reference proteins to include five denatured proteins. The construction of this reference set required the specification of the secondary structure contents of the denatured proteins. It was assumed that they were 90% unfolded and that each of the other five structural categories was 2%. With the expanded reference set, the analysis gave more reasonable results for the denatured proteins: 6% α-helix, 11% β-sheet, 8% β-turns, and 75% unordered. This expanded reference set including denatured proteins is recommended for unfolded proteins or for unfolding transitions,

10.3 CHARACTERIZING IDPS BY CD

10.3.1 Recognizing Whole Proteins as IDPs

Many native proteins are now known to be disordered over their entire length (56). These proteins are readily recognized by their CD spectra, which have the characteristic strong negative band near 200 nm, as described in the previous section. The CD spectrum can be analyzed using the methods described previously (45–47).

There is a class of compact, globular proteins that can easily be mistaken for IDPs on the basis of their CD spectra (33, 64). The serine proteases, soybean trypsin inhibitor, and wheat-germ agglutinin are classified as β-rich proteins based upon their crystal structures, but their CD spectra exhibit strong negative bands near 200 nm, where model β-sheets and other β-rich proteins, for example, concanavalin A and β-lactoglobulin, have positive CD. Wu et al. (64) designated the β-rich proteins with negative CD bands near 200 nm as β-II proteins and those with positive CD near 200 nm as β-I proteins. Wu et al. also proposed two criteria by which β-II proteins could be distinguished from unordered proteins.

1. Unordered proteins show a smooth temperature variation of CD at any given wavelength, whereas β-II proteins exhibit a cooperative transition, manifested as a sigmoidal variation in CD.
2. The near-UV CD spectrum of a β-II proteins typically exhibits bands associated with the aromatic side chains having a well-defined location and conformation. By contrast, the near-UV CD of an unordered protein is generally weak and featureless because the aromatic side chains have no well-defined geometry.

Sreerama and Woody (50) analyzed the X-ray structures of eight β-I and eight β-II proteins, focusing on the relative amounts of β-strand and P_{II} conformations. The β-I proteins generally have more β-strand and less P_{II} than the β-II proteins, but there is some overlap in these fractions between the two classes. The ratio f_{PII}/f_{β}, however, is a useful criterion. For all of the β-II proteins, this ratio is >0.40 and for all β-I proteins it is <0.40.

Early CD studies of IDPs were reviewed by Shi et al. (44). Selected examples of more recent work are discussed below.

α-Synuclein is a 140-residue protein that is a major component of Lewy bodies and Lewy neurites that are characteristic of Parkinson's disease and of a form of dementia clinically similar to Alzheimer's disease. It is an IDP as evidenced by its CD spectrum, Stokes radius, and small-angle X-ray scattering (SAXS) (58, 62). Uversky et al. (58) investigated the effects of decreasing pH and increasing temperature on the conformation of α-synuclein and on its ability to form amyloid fibrils. Figure 10.7 shows the effects of decreasing pH on the CD spectrum and on the fluorescence of added 1-anilinonapohthalene-8-sulfonate (ANS). The CD spectra change with decreasing pH in a manner consistent with melting of P_{II} helix to the unordered conformation.

However, the large increase in ANS fluorescence, which shows the same pH dependence as the CD (Figure 10.7B), argues that another process must be occurring (58). ANS fluorescence is enhanced by a hydrophobic environment, and the 5–6-fold enhancement of ANS fluorescence observed in the inset to Figure 10.7A indicates that the lower pH must lead to the formation of hydrophobic binding pockets in α-synuclein, which is not consistent with simple melting of P_{II} helix. Uversky et al. also monitored the effects of lower pH by infrared absorption in the amide I band. They observed that a shoulder developed at ~1630 cm^{-1} when the pH was lowered from 7.5 to 3. An amide I band in the 1630 cm^{-1} region is characteristic of β-strands. Based upon these data, Uversky et al. proposed that a partially folded intermediate with enhanced β-strand content relative to the form that prevails at neutral pH is formed at lower pH. The low-pH form is still predominantly unfolded as indicated by Kratky (21) plots of SAXS data, but local β-strand/sheet structure may provide hydrophobic pockets that can bind ANS. Uversky et al. further proposed that these local β-strand/sheet regions can form the nucleus for fibril formation. As shown in Figure 10.7B, the lag time for fibril formation decreases with decreasing pH, which is in parallel with the CD and fluorescence changes.

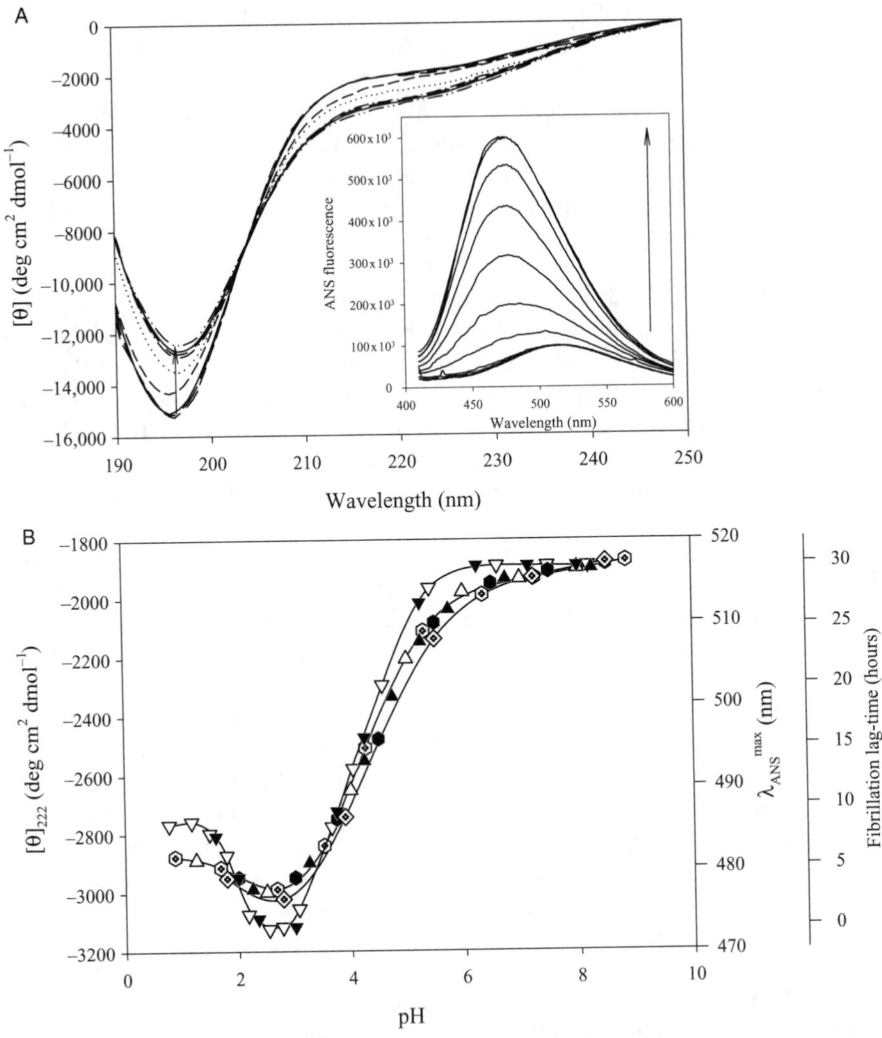

Figure 10.7 Effect of pH on the structural properties of α-synuclein (42). (A) Far-UV CD spectra as a function of pH. pH values ranged from 8.9 to 0.9 in the direction shown by the arrow. The inset represents ANS fluorescence spectra measured at pH values ranging from 8.2 to 2.5, in the direction shown by the arrow. (B) Comparison of the effect of pH on far-UV CD (△ and ▲) and ANS fluorescence (▽ and ▼) spectra. The results of the initial titration (decrease in pH) and reverse (increase in pH) experiments are presented as *open* and *closed* symbols, respectively. The cell path length was 0.1 and 10 mm for far-UV CD and fluorescence measurements, respectively. Measurements were carried out at 20°C. Protein concentration was 0.1 (*circles*), 1.0 (*squares*), and 0.01 (*triangles*) mg/mL. Data for the pH effect on the lag time of α-synuclein fibrillation (diamonds and hexagons) are also shown for comparison. (Figure provided by V. N. Uversky based upon Reference 58.)

Increasing temperature leads to changes in the far-UV CD that are similar to those induced by low pH. They also lead to decreased lag time for fibril formation as with the pH-induced changes and were interpreted as evidence for a thermally induced partial folding, driven by enhanced hydrophobic interactions.

10.3.2 Characterizing Disordered Regions in Proteins

Proteins that have extensive disordered regions interspersed with ordered domains are much more numerous than those that are entirely disordered (14). Characterizing these regions by CD is a challenging problem. One valuable approach is to define the structured regions by resistance to proteolytic digestion (20), to clone and express the portions of the protein between resistant regions, and to characterize these putative disordered regions. Another is to introduce mutations that destabilize or stabilize specific types of secondary structure.

p53 is a critically important protein that regulates cell proliferation, apoptosis, and tumor suppression. The 393-residue protein consists of three major domains and several subdomains: an N-terminal domain (NTD), residues 1–93, which consists of a transcription activation subdomain (TAD) and a Pro-rich subdomain; a DNA-binding domain (DBD) from residues 102 to 292; and a C-terminal domain (CTD), residues 293–393, containing the tetramerization domain (TD) and the C-terminal regulatory domain (CTRD). Bell et al. (2) studied the full-length protein (p53); the DBD (p53core); the NTD-DBD fragment, lacking the CTD (p53C312); and the DBD-CTD fragment, lacking the NTD (N93p53). CD spectra were analyzed by a neural network method (5) and by CDSStr (28). Each fragment was more than 50% unordered, with the DBD being the least unordered (53%) and having the most ordered structure (8% α and 38% β). The secondary structures of the NTD and CTD were calculated by subtracting the secondary structure contents of the DBD from those of the appropriate truncation construct. The NTD was 11% α, 10% β, and 79% unordered, whereas the CTD was 26% α, 9% β, and 65% unordered.

The conclusion that the NTD of p53 is largely, if not completely, disordered was corroborated in a subsequent study (10) of the NTD alone. The CD spectrum of the NTD is typical for an unfolded polypeptide chain, with a negative maximum at 200 nm and a weak negative shoulder near 220 nm. The ellipticity values at 200 and 222 nm place the NTD in the "coil-like" category of Uversky (56). The CD at 222 nm shows a smooth increase in magnitude with increasing temperature and is almost linear from 0 to 60°C. The family of spectra at various temperatures shows an isoelliptic point near 210 nm, which is consistent with a two-state transition between P_{II} helix and an unordered state. The longer wavelength of the isoelliptic point relative to that for poly(Lys) (12) is attributable to the relatively high Pro content of the NTD. The near-UV CD spectrum was very weak, with a simple, broad negative band at ~290 nm. Other lines of evidence for IDP character in NTD include Trp fluorescence spectra indicative of full solvent exposure and the limited dispersion of ^{1}H and ^{15}N NMR spectra.

The cyclin-dependent kinase (Cdk) inhibitor p27^{Kip1} is an IDP, as evidenced by its CD spectrum, with a strong negative band at 200 nm and a weak negative shoulder near 220 nm (3). Residues 22–97 of p27^{Kip1} constitute a domain that binds to and inhibits the cyclin A-Cdk2 complex. The crystal structure of this ternary complex has been reported (42), and in this complex, the inhibitory domain adopts a well-defined conformation with an α-helix, a short 3_{10}-helix, a β-hairpin, and a β-strand.

The isolated inhibitory domain is largely unordered as evidenced by CD, fluorescence, and NMR (3, 19). However, the CD spectrum shows evidence of nascent helical structure, especially at 5°C (3): the negative maximum is at 204 nm, rather than at 200 nm or lower; the negative shoulder near 220 nm is more pronounced than that observed for fully unordered polypeptides; the CD is positive below ~194 nm. Estimates of the helix content give 10–16% (8, 47), which are significantly higher than the values of 2–6% for the full-length p27^{Kip1}. The CD changes upon increasing temperature to 30°C show that the helix is of low stability. Mutations to Pro of residues between 38 and 59 in the sequence, which form an α-helix in the ternary complex, lead to CD changes characteristic of decreased helix content. Such changes are not observed when residues in the short region that forms a 3_{10}-helix in the ternary complex are mutated. By contrast, simultaneous replacement of three non-contiguous residues in the 38–59 region by Ala, which stabilizes α-helix, led to an increase in helix content to 17–22% at 5°C. These observations support the hypothesis that the nascent helix in the isolated inhibitor domain resides in the sequence that becomes α-helical when the domain forms the ternary complex.

Csizmok et al. (9) have investigated the major proteolytic fragments of calpastatin (CSD1) and a microtubule-associated protein 2 (MAP2c). Using brief digestions at very low protease:protein ratios (<~1:1000), only one or two sites were cleaved for each protein by a given protease. MAP2c was cleaved into approximately equal fragments, each of which gave similar CD spectra characteristic of unordered proteins. The sum of the spectra for the two fragments was nearly identical with the spectrum of the intact MAP2c, indicating a lack of interactions between the two halves in the full-length protein. By contrast, the N- and C-terminal halves of CSD1 have significantly different CD spectra. Although both display a single negative band, the C-terminal half has its maximum at 205 nm, suggesting that it has more secondary structure than the N-terminal portion, with its maximum at 200 nm. Moreover, the sum of the spectra of the two halves is significantly more negative than the spectrum of the intact CSD1 at wavelengths above 205 nm. The mean residue ellipticity of the CD of the C-terminal half is independent of concentration, demonstrating that the more structured character of this fragment is intrinsic and does not result from association. Therefore, interactions between the two halves of CSD1 suppress secondary structure formation in the C-terminal half.

MeCP2 is a nuclear protein that has at least one ordered domain and extensive disordered regions, as predicted by FoldIndex (39). Adams et al. (1) have studied MeCP2 with a variety of biophysical techniques, including CD, and with proteolytic digestion, to elucidate the domain structure. The CD spectrum of the full-length protein was analyzed using a set of reference proteins that included five denatured proteins (45). The analysis gave 4% α, 21% β, 13% turns, and 59% unordered, consistent with a protein that not only has extensive disorder but also one or more ordered domains.

Protease digestion experiments yielded several fragments, two of which closely coincided with previously recognized domains: MBD, the methylcytosine-binding domain, and TRD, the transcriptional repression domain. The CD spectra of these domains are shown in Figure 10.8. The spectrum of MBD was consistent with a substantially folded domain and gave 9% α, 31% β, 20% turn, and 40% unordered structure on CDPro analysis using the expanded reference set containing denatured proteins (45). The structure of the MBD has been determined by NMR (61) and the CD data are in agreement with the NMR structure. The spectrum of the TRD is that of an unordered protein and this is corroborated by CDPro (47) analysis: 3% α, 7% β, 5% turn, and 85% unordered. The largely unordered character of this domain appears to conflict with its resistance to trypsin, given that it contains 24 potential tryptic cleavages sites. FoldIndex (39) predicts short folded regions interspersed with unfolded regions throughout the sequence of MeCP2. Adams et al. suggest that the tertiary structure of MeCP2 is dominated by β-strands connected by disordered regions, and this is a reasonable interpretation of the structure of

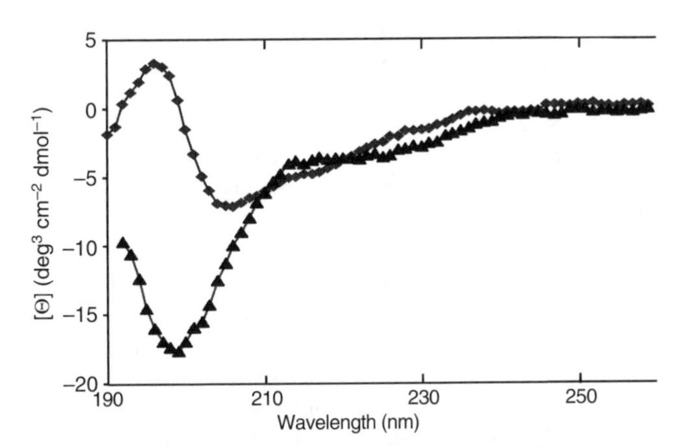

Figure 10.8 CD spectra of isolated MBD (◆◆◆◆) and TRD (▲▲▲▲) fragments from MeCP2. (Figure provided by K. Hite, V. H. Adams and J. C. Hansen based upon Reference 1.)

the TRD domain as well. It is also possible that the TRD is analogous to β-II proteins(64), with β-strand, P_{II}-, and unordered conformation, in which the P_{II}-conformation dominates the CD spectrum. Thermal melting, near-UV CD, and sedimentation velocity experiments on the TRD domain will be necessary to test this possibility.

REFERENCES

1. Adams, V. H., S. J. McBryant, P. A. Wade, C. L. Woodcock, and J. C. Hansen. 2007. Intrinsic disorder and autonomous domain function in the multifunctional nuclear protein, MeCP2. J Biol Chem **282**:15057–64.
2. Bell, S., C. Klein, L. Muller, S. Hansen, and J. Buchner. 2002. p53 Contains large unstructured regions in its native state. J Mol Biol **322**:917–27.
3. Bienkiewicz, E. A., J. N. Adkins, and K. J. Lumb. 2002. Functional consequences of preorganized helical structure in the intrinsically disordered cell-cycle inhibitor $p27^{Kip1}$. Biochemistry **41**:752–9.
4. Bienkiewicz, E. A., A.-Y. M. Woody, and R. W. Woody. 2000. Conformation of the RNA polymerase II C-terminal domain: circular dichroism of long and short fragments. J Mol Biol **297**:119–33.
5. Bohm, G., R. Muhr, and R. Jaenicke. 1992. Quantitative analysis of protein far UV circular-dichroism spectra by neural networks. Protein Eng **5**:191–5.
6. Brahms, S., J. Brahms, G. Spach, and A. Brack. 1977. Identification of β, β-turns and unordered conformations in polypeptide chains by vacuum ultraviolet circular dichroism. Proc Natl Acad Sci U S A **74**:3208–12.
7. Chaffotte, A. F., J. I. Guijarro, Y. Guillou, M. Delepierre, and M. E. Goldberg. 1997. The "pre-molten globule", a new intermediate in protein folding. J Protein Chem **16**:433–9.
8. Chen, Y. H., J. T. Yang, and H. M. Martinez. 1972. Determination of secondary structures of proteins by circular dichroism and optical rotatory dispersion. Biochemistry **11**:4120–31.
9. Csizmok, V., M. Bokor, P. Banki, T. Klement, K. F. Medzihradszky, P. Friedrich, K. A. Tompa, and P. Tompa. 2005. Primary contact sites in intrinsically unstructured proteins: the case of calpastatin and microtubule-associated protein 2. Biochemistry **44**:3955–64.
10. Dawson, R., L. Muller, A. Dehner, C. Klein, H. Kessler, and J. Buchner. 2003. The N-terminal domain of p53 is natively unfolded. J Mol Biol **332**:1131–41.
11. Dolgikh, D. A., L. V. Abaturov, I. A. Bolotina, E. V. Brazhnikov, V. E. Bychkova, R. I. Gilmanshin, Y. O. Lebedev, G. V. Semisotnov, E. I. Tiktopulo, and O. B. Ptitsyn. 1985. Compact state of a protein molecule with pronounced small-scale mobility—bovine α-lactalbumin. Eur Biophys J **13**:109–21.
12. Drake, A. F., G. Siligardi, and W. A. Gibbons. 1988. Reassessment of the electronic circular dichroism criteria for random coil conformations of poly(L-lysine) and the implications for protein folding and denaturation studies. Biophys Chem **31**:143–6.

13. Dukor, R. K., and T. A. Keiderling. 1991. Reassessment of the random coil conformation: vibrational CD study of proline oligopeptides and related polypeptides. Biopolymers **31**:1747–61.
14. Dunker, A. K., J. D. Lawson, C. J. Brown, R. M. Williams, P. Romero, J. S. Oh, C. J. Oldfield, A. M. Campen, C. R. Ratliff, K. W. Hipps, J. Ausio, M. S. Nissen, R. Reeves, C. H. Kang, C. R. Kissinger, R. W. Bailey, M. D. Griswold, M. Chiu, E. C. Garner, and Z. Obradovic. 2001. Intrinsically disordered protein. J Mol Graph Model **19**:26–59.
15. Eker, F., X. L. Cao, L. Nafie, and R. Schweitzer-Stenner. 2002. Tripeptides adopt stable structures in water. A combined polarized visible Raman, FTIR, and VCD spectroscopy study. J Am Chem Soc **124**:14330–41.
16. Eker, F., K. Griebenow, X. L. Cao, L. A. Nafie, and R. Schweitzer-Stenner. 2004. Preferred peptide backbone conformations in the unfolded state revealed by the structure analysis of alanine-based (AXA) tripeptides in aqueous solution. Proc Natl Acad Sci U S A **101**:10054–9.
17. Eker, F., K. Griebenow, X. L. Cao, L. A. Nafie, and R. Schweitzer-Stennert. 2004. Tripeptides with ionizable side chains adopt a perturbed polyproline II structure in water. Biochemistry **43**:613–21.
18. Eker, F., K. Griebenow, and R. Schweitzer-Stenner. 2003. Stable conformations of tripeptides in aqueous solution studied by UV circular dichroism spectroscopy. J Am Chem Soc **125**:8178–85.
19. Flaugh, S. L., and K. J. Lumb. 2001. Effects of macromolecular crowding on the intrinsically disordered proteins c-Fos and $p27^{Kip1}$. Biomacromolecules **2**:538–40.
20. Fontana, A., P. P. de Laureto, B. Spolaore, E. Frare, P. Picotti, and M. Zambonin. 2004. Probing protein structure by limited proteolysis. Acta Biochim Pol **51**:299–321.
21. Glatter, O., and O. Kratky. 1982. Small Angle X-ray Scattering. Academic Press, London.
22. Graf, J., P. H. Nguyen, G. Stock, and H. Schwalbe. 2007. Structure and dynamics of the homologous series of alanine peptides: a joint molecular dynamics/NMR study. J Am Chem Soc **129**:1179–89.
23. Greenfield, N., and G. D. Fasman. 1969. Computed circular dichroism spectra for evaluation of protein conformation. Biochemistry **8**:4108–16.
24. Greenfield, N. J. 1996. Methods to estimate the conformation of proteins and polypeptides from circular dichroism data. Anal Biochem **235**:1–10.
25. Helbecque, N., and M. H. Loucheux-Lefebvre. 1982. Critical chain-length for polyproline-II structure formation in H-Gly-$(Pro)_n$-OH—circular dichroism approach. Int J Pept Protein Res **19**:94–101.
26. Holzwarth, G., and P. Doty. 1965. The ultraviolet circular dichroism of polypeptides. J Am Chem Soc **87**:218–28.
27. Jenness, D. D., C. Sprecher, and W. C. Johnson, Jr. 1976. Circular dichroism of collagen, gelatin, and poly(proline)II in vacuum ultraviolet. Biopolymers **15**:513–21.
28. Johnson, W. C. 1999. Analyzing protein circular dichroism spectra for accurate secondary structures. Proteins Struct Funct Genet **35**:307–12.

29. Johnson, W. C., Jr. 1990. Protein secondary structure and circular dichroism—a practical guide. Proteins Struct Funct Genet **7**:205–14.
30. Johnson, W. C., Jr., and I. Tinoco, Jr. 1972. Circular dichroism of polypeptide solutions in vacuum ultraviolet. J Am Chem Soc **94**:4389–90.
31. Katou, H., M. Hoshino, H. Kamikubo, C. A. Batt, and Y. Goto. 2001. Native-like beta-hairpin retained in the cold-denatured state of bovine beta-lactoglobulin. J Mol Biol **310**:471–84.
32. Kelly, S. M., T. J. Jess, and N. C. Price. 2005. How to study proteins by circular dichroism. Biochim Biophys Acta **1751**:119–39.
33. Manavalan, P., and W. C. Johnson, Jr. 1983. Sensitivity of circular dichroism to protein tertiary structure class. Nature **305**:831–2.
34. Manavalan, P., and W. C. Johnson, Jr. 1987. Variable selection method improves the prediction of protein secondary structure from circular dichroism spectra. Anal Biochem **167**:76–85.
35. Matsuo, K., Y. Sakurada, R. Yonehara, M. Kataoka, and K. Gekko. 2007. Secondary structure analysis of denatured proteins by vacuum-ultraviolet circular dichroism spectroscopy. Biophys J **92**:4088–96.
36. Mattice, W. L., J.-T. Lo, and L. Mandelkern. 1972. A comparison of the effects of salt and temperature on charged and uncharged polypeptides in water. Macromolecules **5**:729–34.
37. Nolting, B., R. Golbik, A. S. Soler-Gonzalez, and A. R. Fersht. 1997. Circular dichroism of denatured barstar suggests residual structure. Biochemistry **36**:9899–905.
38. Park, S.-H., W. Shalongo, and E. Stellwagen. 1997. The role of PII conformations in the calculation of peptide fractional helix content. Protein Sci **6**:1694–700.
39. Prilusky, J., C. E. Felder, T. Zeev-Ben-Mordehai, E. H. Rydberg, O. Man, J. S. Beckmann, I. Silman, and J. L. Sussman. 2005. FoldIndex©: a simple tool to predict whether a given protein sequence is intrinsically unfolded. Bioinformatics **21**: 3435–8.
40. Privalov, P. L., E. I. Tiktopulo, S. Y. Venyaminov, Y. V. Griko, G. I. Makhatadze, and N. N. Khechinashvili. 1989. Heat capacity and conformation of proteins in the denatured state. J Mol Biol **205**:737–50.
41. Provencher, S. W., and J. Glockner. 1981. Estimation of globular protein secondary structure from circular dichroism. Biochemistry **20**:33–7.
42. Russo, A. A., P. D. Jeffrey, A. K. Patten, J. Massague, and N. P. Pavletich. 1996. Crystal structure of the $p27^{Kip1}$ cyclin-dependent-kinase inhibitor bound to the cyclin A Cdk2 complex. Nature **382**:325–31.
43. Schweitzer-Stenner, R., F. Eker, A. Perez, K. Griebenow, X. L. Cao, and L. A. Nafie. 2003. The structure of tri-proline in water probed by polarized Raman, Fourier transform infrared, vibrational circular dichroism, and electric ultraviolet circular dichroism spectroscopy. Biopolymers **71**:558–68.
44. Shi, Z., R. W. Woody, and N. R. Kallenbach. 2002. Is polyproline II a major backbone conformation in unfolded proteins? Adv Protein Chem **62**:163–240.
45. Sreerama, N., S. Y. Venyaminov, and R. W. Woody. 2000. Estimation of protein secondary structure from circular dichroism spectra: inclusion of denatured proteins with native proteins in the analysis. Anal Biochem **287**:243–51.

46. Sreerama, N., and R. W. Woody. 2004. Computation and analysis of protein circular dichroism spectra. Methods Enzymol **383**:318–51.
47. Sreerama, N., and R. W. Woody. 2000. Estimation of protein secondary structure from circular dichroism spectra: comparison of CONTIN, SELCON, and CDSSTR methods with an expanded reference set. Anal Biochem **287**:252–60.
48. Sreerama, N., and R. W. Woody. 1994. Poly(Pro)II helices in globular proteins: identification and circular dichroic analysis. Biochemistry **33**:10022–5.
49. Sreerama, N., and R. W. Woody. 1993. A self-consistent method for the analysis of protein secondary structure from circular dichroism. Anal Biochem **209**:32–44.
50. Sreerama, N., and R. W. Woody. 2003. Structural composition of β-I and β–II proteins. Protein Sci **12**:384–8.
51. Tiffany, M. L., and S. Krimm. 1969. Circular dichroism of random polypeptide chain. Biopolymers **8**:347–59.
52. Tiffany, M. L., and S. Krimm. 1972. Effect of temperature on circular dichroism spectra of polypeptides in extended state. Biopolymers **11**:2309–16.
53. Tiffany, M. L., and S. Krimm. 1973. Extended conformation of polypeptides and proteins in urea and guanidine hydrochloride. Biopolymers **12**:575–87.
54. Tiffany, M. L., and S. Krimm. 1968. New chain conformations of poly(glutamic acid) and polylysine. Biopolymers **6**:1379–82.
55. Toumadje, A., and W. C. Johnson, Jr. 1995. Systemin has the characterisics of a poly(L-proline) II type helix. J Am Chem Soc **117**:7023–4.
56. Uversky, V. N. 2002. Natively unfolded proteins: a point where biology waits for physics. Protein Sci **11**:739–56.
57. Uversky, V. N. 2002. What does it mean to be natively unfolded? Eur J Biochem **269**:2–12.
58. Uversky, V. N., J. Li, and A. L. Fink. 2001. Evidence for a partially folded intermediate in α-synuclein fibril formation. J. Biol. Chem. **276**:10737–44.
59. Uversky, V. N., and O. B. Ptitsyn. 1996. Further evidence on the equilibrium "premolten globule state": four-state guanidinium chloride-induced unfolding of carbonic anhydrase B at low temperature. J Mol Biol **255**:215–28.
60. Uversky, V. N., and O. B. Ptitsyn. 1994. Partly folded state, a new equilibrium state of protein molecules—four-state guanidinium chloride-induced unfolding of β-lactamase at low temperature. Biochemistry **33**:2782–91.
61. Wakefield, R. I. D., B. O. Smith, X. S. Nan, A. Free, A. Soteriou, D. Uhrin, A. P. Bird, and P. N. Barlow. 1999. The solution structure of the domain from MeCP2 that binds to methylated DNA. J Mol Biol **291**:1055–65.
62. Weinreb, P. H., W. G. Zhen, A. W. Poon, K. A. Conway, and P. T. Lansbury. 1996. NACP, a protein implicated in Alzheimer's disease and learning, is natively unfolded. Biochemistry **35**:13709–15.
63. Woody, R. W. 1992. Circular dichroism and conformation of unordered polypeptides. Adv Biophys Chem **2**:37–79.
64. Wu, J., J. T. Yang, and C. S. C. Wu. 1992. β-II conformation of all-β proteins can be distinguished from unordered form by circular-dichroism. Anal Biochem **200**:359–64.

11

FLUORESCENCE SPECTROSCOPY OF INTRINSICALLY DISORDERED PROTEINS

EUGENE A. PERMYAKOV[1] AND VLADIMIR N. UVERSKY[1,2]

[1]*Institute for Biological Instrumentation of the Russian Academy of Sciences, Pushchino, Moscow, Russia*

[2]*Institute for Intrinsically Disordered Protein Research, Center for Computational Biology and Bioinformatics, and Department of Biochemistry and Molecular Biology, Indiana University School of Medicine, Indianapolis, IN*

ABSTRACT

Fluorescence spectroscopy can be successfully used in studies of intrinsically disordered proteins (IDPs). IDPs are usually characterized by surface location of tryptophan residues with redshifted tryptophan fluorescence spectra with maxima at 340–353 nm. Such tryptophans are readily accessible to external fluorescence quenchers. Interactions of these proteins with another proteins and peptides usually transfer tryptophan residues to a more hydrophobic or more rigid environment, which results in a blueshift of fluorescence maximum position. These spectral effects can be used for evaluation of interaction parameters. Spectral probes and labels are also widely used for the investigation of natively disordered proteins. ANS and bis-ANS have a tendency to bind to IDPs, but not to the tightly packed ordered proteins.

Instrumental Analysis of Intrinsically Disordered Proteins: Assessing Structure and Conformation, Edited by Vladimir Uversky and Sonia Longhi

11.1 INTRODUCTION

Luminescence is an emission of photons by electron-excited states of molecules. There are two types of luminescence—fluorescence and phosphorescence. Fluorescence is emitted due to radiation transitions between singlet states of a molecule, while phosphorescence arises due to radiation transitions between triplet and singlet states. In this review, we will deal mostly with fluorescence.

Parameters of fluorescence reflect properties of excited states of fluorophores, characteristics of electron transitions in fluorophores, and interactions of fluorophores with their environment (reviewed in References 17, 23, and 24). Just the latter circumstance allows using aromatic amino acid residues and various dyes as fluorescent reporter groups in studies of proteins. The high sensitivity of fluorescence to changes in environment of fluorophores is very useful in studies of changes in structural and physicochemical properties of proteins.

Fluorescence spectroscopy is one of the most popular and widespread methods in biophysics and biochemistry. Due to relative cheapness of spectrofluorimeters (especially for steady-state measurements), these instruments became usual for many biophysical and biochemical laboratories. The method is widely used for studies of various proteins, including intrinsically disordered proteins (IDPs).

11.2 FLUORESCENCE PARAMETERS

Fluorescence is characterized by several major parameters: fluorescence quantum yield, fluorescence intensity at a fixed wavelength, fluorescence lifetime, position of fluorescence spectrum maximum, fluorescence spectrum shape, and fluorescence anisotropy.

Fluorescence quantum yield is a ratio of the number of photons emitted from an excited state to the number of photons absorbed during the transitions from the ground to the excited state by the same molecule per time unit. Fluorescence quantum yield is proportional to the area under the fluorescence spectrum. This parameter reflects the effectiveness of radiationless deactivation of excited states of the molecule. The effectiveness of radiationless processes in organic molecules depends upon their environment (availability of quenching groups, their type, and degree of their mobilities).

Fluorescence intensity at a fixed wavelength is a parameter, which is proportional to the fluorescence quantum yield.

By definition, *fluorescence lifetime* is the time required by a population of N excited fluorophores to decrease exponentially to N/e by losing excitation energy through fluorescence and other deactivation pathways. Fluorescence lifetime is inversely proportional to the effectiveness of radiationless processes in a molecule.

11.2.1 Position of Fluorescence Spectrum Maximum

This parameter is used mostly in the case of structureless emission spectra. For fluorophores possessing dipole moments in both ground and excited states, it reflects their interactions with the surrounding molecular ensemble and connected with mobility of their polar environment.

According to the most widespread classification, all types of intermolecular interactions are divided into universal and specific ones. Universal interactions between molecules occur in all cases without exception and depend upon their physicochemical properties. After configuration and space averaging, they characterize the collective effects of the environment on the properties of a given molecule. Specific interactions are individual, quasi-chemical, exchange interactions. These interactions are characterized by high selectivity to molecular properties and result in formation of rather strong links between molecules.

The potential of the universal van der Waals interaction of two molecules is a sum of potentials of orientation, induction, and dispersion interactions. In the condensed medium, each molecule is affected not by a single neighboring molecule, but by an ensemble of molecules. The simplest assumption of the potential of collective interactions is the assumption as to the additivity of the total energy of interactions, the main contribution being given by the molecules of the first coordination sphere.

Let us consider the mechanisms of influence of the universal interactions in the liquid phase on the parameters of the emission bands of organic chromophores. Let us assume that the emitting molecule (chromophore) possesses a dipole moment both in the ground and excited states. Since the lifetime of the molecule in the ground state is infinitely long, surrounding molecules around the unexcited chromophore are oriented in such a way that the energy of the chromophore-surrounding polar molecules system is minimal. Absorption of a light quantum causes a transition of the molecule to the excited state. However, due to a redistribution of electron density, this results in a change of the value and orientation of the dipole moment of the molecule. According to the Frank–Condon principle, the electron transition occurs very rapidly so that the configuration of the surrounding dipoles around the chromophore does not significantly change during the electron transition. The energy level occupied by the molecule in this state is a nonequilibrium one; it is called Frank–Condon level. After the electron transition, the surrounding polar molecules start to reorient around the excited chromophore to reach a new configuration, which would be the most energetically favorable for the new electron density distribution. If their orientational relaxation time τ_r is much longer than the lifetime of the excited state of the chromophore τ_f, the relaxation has no time to occur and the emission will proceed from the practically nonequilibrium Frank–Condon level. The emission spectrum will have a maximum with shorter wavelength position in this case.

In the opposite case, when $\tau_r << \tau_f$, there is enough time for relaxation to be completed, and during the excitation lifetime, the surrounding polar molecules reorient around the chromophore to the most favorable configuration. The emission will occur from the equilibrium excited level corresponding to the most favorable orientation of the surrounding molecules around the excited molecule. The emission spectrum will have maximum with a longer wavelength position in this case.

In the case when τ_f is comparable with τ_r, the relaxation will be incomplete and the emission will occur from some intermediate level located between the levels corresponding to the situation when $\tau_r << \tau_f$ and $\tau_r >> \tau_f$. The position of the emission spectrum maximum will be intermediate.

As a rule, it is very difficult to take into account specific interactions leading to formation of stoichiometric complexes of the solute molecule with the surrounding molecules. Their energy is often higher than that of the universal interactions, and sometimes a change in the electron configuration of the chromophore molecule is so significant that it can be considered as an entirely different molecule. The most studied specific interactions are now probably hydrogen bonds and charge-transfer complexes.

Fluorescence spectrum shape depends upon the intensity of interactions between fluorophore and its environment. If the interaction is weak, fluorescence spectra of some substances demonstrate vibrational structure; if the interaction is strong, fluorescence spectra become smooth.

Fluorescence anisotropy is defined as the ratio of the difference between the emission intensity parallel to the polarization of the electric vector of the exciting light ($I_{\|}$) and that perpendicular to that vector ($I_{\perp}$) divided by the total intensity (I_T):

$$A = (I_{\|} - I_{\perp})/(I_{\|} + 2I_{\perp}) \tag{11.1}$$

The anisotropy of emission (A) is related to the correlation time of the fluorophore (τ_c) through the Perrin equation:

$$A_o/A - 1 = \tau/\tau_c \tag{11.2}$$

where A_o is the limiting anisotropy of the probe, which depends on the angle between the absorption and emission transition dipoles, and τ is the fluorescence lifetime.

The measurements of fluorescence anisotropy can be used to obtain hydrodynamic information concerning macromolecules and macromolecular complexes. Since the tryptophan residues of proteins and covalently attached dye molecules nearly always exhibit local rotational motion in addition to depolarization through global Brownian tumbling of the macromolecule, it is possible to measure the time-resolved anisotropy as well as the steady-state anisotropy. Motions of probes on macromolecules are quite complex and have been the subject of numerous reports. The basic idea is that a fluorophore excited by polarized light will also emit polarized light. However, if a molecule is moving, it will tend to decrease the polarization of the light by radiating at a different direction from the incident light. The polarization decreasing effect is greatest with fluorophores freely tumbling in solution and decreases with decreased rates of tumbling. Upon interaction of two protein molecules, a complex is formed which will tumble more slowly (thus, increasing the polarization of the emitted light and reducing the polarization decreasing effect).

11.2.2 Fluorescence Resonance Energy Transfer

This is the radiationless transfer of energy from an excited donor fluorophore to a suitable acceptor fluorophore, which is a physical process that depends on spectral overlap and proper dipole alignment of the two fluorophores. The transfer of excitation energy between the donor and the acceptor originates only upon the fulfillment of several conditions: (1) the absorption (excitation) spectrum of the acceptor overlaps with the emission spectrum of the donor; (2) spatial proximity of the donor and the acceptor (usually up to several tens of angstroms); (3) a sufficiently high emission quantum yield of the donor; and (4) a favorable spatial orientation of the donor and acceptor. The fluorescence resonance energy transfer (FRET) can be used as a molecular ruler to measure distances between the donor and acceptor. According to Förster, the efficiency of energy transfer, E, from the excited donor, D, to the nonexcited acceptor, A, located from the D at a distance R_{DA} is determined by the equation:

$$E = 1/\left[1 + (R_{DA}/R_0)^6\right] \tag{11.3}$$

where R_0 is the characteristic donor-acceptor distance, the so-called Förster distance, which has a characteristic value for any given donor-acceptor pair.

11.3 INTRINSIC PROTEIN FLUORESCENCE

Most proteins in water solution possess intrinsic fluorescence in the ultraviolet region of spectrum and this emission can be used for structural and physicochemical studies of proteins (reviewed in References 23 and 24). Aromatic amino acid residues of tryptophan, tyrosine, and phenylalanine are major fluorescent groups in proteins. Their fluorescence parameters reflect properties of the excited states of the fluorophores, characteristics of electron transitions in the fluorophores, and interactions with their microenvironment. Just the last circumstance allows using of aromatic amino acids as natural reporter groups in proteins.

Tryptophan fluorescence (300–450 nm) provides the richest information about proteins. Most proteins, however, possess multiple tryptophan residues and the total protein emission yields only average information on the protein structure. To extract and evaluate the contribution of each reporter, and thereby track the conformational changes occurring in different parts of the macromolecule, is a difficult challenge.

Since indole group of tryptophan is characterized by a relatively large dipole moment both in the ground and excited states, it extensively interacts with polar and charged groups in its environment, which is reflected in changes of tryptophan fluorescence parameters. When a tryptophan residue is located in a rigid hydrophobic environment inside a protein molecule, its fluorescence spectrum displays distinct vibrational structure with extremely short wavelength position of the main maximum at 308 nm (azurin as an example). Buried tryptophans with some polar groups in their environment are characterized also by slightly structured spectra with maximum at 316–325 nm (L-asparaginase as an example). Occurrence of more mobile polar groups in the environment of buried tryptophans shifts its fluorescence maximum up to 330–335 nm (actin, chimotrypsin). The emission spectrum is structureless in this case. Tryptophan residues located at the protein surface in contact with bound water molecules are characterized by emission spectra with maxima at 340–345 nm. Tryptophans in the environment of freely relaxing water molecules (totally unfolded proteins) have fluorescence spectra with maxima at 350–353 nm. Thus, the position of fluorescence spectrum maximum reflects relaxation properties of the chromophore polar environment, which are connected to the location of tryptophan residues in protein molecule.

Fluorescence maximum position is used to elucidate location of tryptophan residues in proteins and to characterize properties of their environment. Using extremely high sensitivity of luminescence to changes in the fluorophore microenvironment, one can study structural and physicochemical changes in protein molecules, including functionally significant changes. Practical absence of inertia of the method permits kinetic monitoring of these changes.

IDPs are usually characterized by surface location of tryptophan residues, which is reflected in redshifted tryptophan fluorescence spectra with maxima at 340–353 nm. Sometimes the fluorescence spectrum of an IDP coincides with the emission spectrum of free tryptophan in water. Interactions of these proteins with another proteins and peptides usually transfer tryptophan residues to a more hydrophobic or more rigid environment, which results in a blueshift of fluorescence maximum position.

Figure 11.1 represents tryptophan fluorescence spectra of C-terminal domain of chicken gizzard caldesmon (CD136). Free CaD136 does not have globular structure, has low secondary structure content, is essentially noncompact as it follows from the results of size exclusion chromatography, and is characterized by the absence of distinct heat absorption peaks, that is, it belongs to the family of natively unfolded (or intrinsically unstructured) proteins (26). It is clearly seen that the spectrum of free CaD136 has maximum at about 350.2 nm; that is, it is close to that of tryptophan in water (353 nm). This indicates that the CaD136 tryptophan residues are almost totally exposed to water.

Chicken gizzard caldesmon contains five tryptophan residues per protein molecule; three of them, Trp674, Trp707, and Trp737, are located within the CaD136 domain. On the other hand, calmodulin, ubiquitous calcium sensor

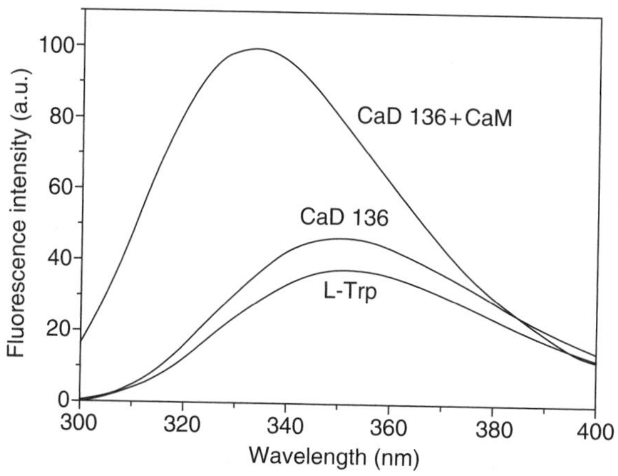

Figure 11.1 Tryptophan fluorescence spectra of CaD136 in the free- and calmodulin-bound states at 20°C (pH 7.6; 10 mM HEPES-KOH buffer, 5 mM $CaCl_2$). Fluorescence was excited at 296.7 nm. Spectrum of L-Trp in water solution under the same absorption at 296.7 nm is shown for comparison.

protein, does not have any Trp residues. This allows the use of changes in the CaD136 intrinsic tryptophan fluorescence as fast and easy method to detect caldesmon–calmodulin complex formation. Figure 11.1 compares the tryptophan fluorescence spectra (excitation at 296.7 nm) of CaD136 measured in the absence or presence of calmodulin. Notably, excess of Ca^{2+} (5 mM $CaCl_2$) was used to saturate calmodulin by calcium, which is a necessary condition of effective binding of calmodulin by caldesmon. It can be seen that calmodulin binding to CaD136 led to the considerable (1.9-fold) increase in the fluorescence quantum yield and a pronounced (17 nm) blueshift of the CaD136 fluorescence spectrum, reflecting the transfer of tryptophans into a less mobile and/or less polar environment. This most likely reflects some calmodulin-induced compaction of a polypeptide chain, at least in the vicinity of tryptophans, or these changes in fluorescence are due to the insertion of Trp into a hydrophobic binding pocket on the calmodulin molecule. This is a very typical effect for interaction of IDPs with other proteins.

These observations have been used to evaluate the equilibrium association parameters of CaD136-calmodulin complex (26) (see Fig. 11.2). It should be noted that the characteristic bend of the calmodulin-titration curve in Figure 11.2, corresponding to saturation of CaD136 by calmodulin, takes place around calmodulin to CaD136 ratio about 1, which is an evidence of the binding of a single calmodulin molecule per CaD136 molecule. The experimental data for the changes in fluorescence intensity were fitted by a theoretical curve computed according to the one-site binding scheme. The best fit was achieved with dissociation constant $K_d = (1.4 \pm 0.2)\,\mu M$.

Here is one more practical example. Baskakov et al. (6) studied recombinant human glucocorticoid receptor fragments consisting of residues 1–500

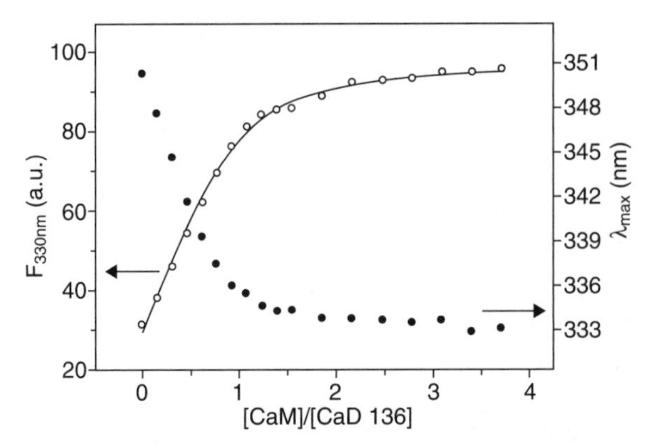

Figure 11.2 Spectrofluorimetric titration of caldesmon fragment CaD136 (9.5 μM) by calmodulin at 20°C (pH 7.6; 10 mM HEPES-KOH buffer, 5 mM $CaCl_2$). Fluorescence was excited at 296.7 nm. ○, tryptophan fluorescence intensity at 330 nm, F_{330nm}, arbitrary units; •, fluorescence spectrum maximum position, λ_{max}, nm. Points are experimental and the curve is theoretically fitted to the experimental points according to the single site binding scheme.

and residues 77–262. Far-UV circular dichroism (CD) and fluorescence spectroscopy showed that both fragments were largely disordered in aqueous solution (fluorescence maximum position at about 340 nm suggests an external location of tryptophans). Trimethylamine N-oxide induced a condensed structure in the large fragment, indicated by the substantial enhancement in intrinsic fluorescence and by a blueshift of fluorescence maximum by more than 10 nm.

Some proteins have no tryptophan residues giving the most informative spectral data, but it is possible to introduce tryptophan residues in protein sequence artificially. For example, Bourhis et al. (7) have designed an intrinsically disordered C-terminal domain, N_{TAIL} (amino acids 401–525), of the measles virus nucleoprotein bearing a tyrosine to tryptophan substitution at position 518. Introduction of a tryptophan residue in Box 3 allowed binding events to be monitored by fluorescence spectroscopy, while maximizing the conservative nature of the substitution.

Fluorescence quantum yield and fluorescence lifetime of tryptophans in proteins depend upon quenching properties of surrounding groups and their mobilities. These parameters can be used to monitor interactions of IDPs with other proteins. At the same time, changes in tryptophan residue location can both increase and decrease fluorescence quantum yield and fluorescence lifetime. The effect depends upon specific features of tryptophan environment.

Tyrosine fluorescence spectrum is a structure-less emission band with maximum at 303 nm. The maximum position of tyrosine fluorescence spectrum practically does not depend upon properties of the tyrosine residues environment and the only informative parameters in this case are fluorescence quantum yield and fluorescence lifetime.

Phenylalanine residues are characterized by fluorescence spectra with distinct vibrational structure and main maximum at 282 nm. As in the case of tyrosine, the only informative parameters of phenylalanine fluorescence are fluorescence quantum yield and fluorescence lifetime.

Information about the accessibility of protein chromophores to solvent (and thus on relative compactness of a protein molecule) can be obtained from the analysis of dynamic quenching of intrinsic fluorescence by small molecules. The method of selective fluorescence quenching by external quenchers is used to determine location of chromophores in protein molecules. The method is based on the use of quenchers, which deactivate excited states of the chromophores during direct collisions. Such quenchers quench fluorescence of surface chromophores more effectively than the emission of buried chromophores. The collisional quenching is described by Stern–Volmer equation:

$$I_0/I = (1 + K_{SV}[Q])e^{V[Q]} \tag{11.4}$$

where I_0 and I are the fluorescence intensities in the absence and presence of quencher, K_{SV} is the dynamic quenching constant (Stern–Volmer constant), V is a static quenching constant, and [Q] is the quencher concentration.

If a solution contains two types of fluorescent chromophores (buried and surface located), which give different contributions to the total emission spectrum, a_1 and a_2, and which are characterized by different Stern–Volmer constants, K_1 and K_1, the Stern–Volmer equation looks like that (without static quenching):

$$I = a_1 I_0/(1 + K_1[Q]) + a_2 I_0/(1 + K_2[Q]) \tag{11.5}$$

Commonly used quenchers for tryptophan fluorescence are Cs^+, I^-, CSN^-, NO_3^- ions, and acrylamide. IDPs usually contain exposed to solvent aromatic amino acid residues which are characterized by high values of Stern–Volmer constants. In distinction from the deuterium exchange, this method can be used to evaluate the amplitude and timescale of dynamic processes by using quenchers of different size, polarity, and charge.

In the work of Muro-Pastor et al. (21), the fluorescence quenching method was used to examine whether the tertiary structure around Trp42 in the protein IF7 is absent at pH 6.5. They used iodide as an extrinsic fluorescence quencher. It turned out that the Stern–Volmer constant remained essentially unchanged (3.6–4.4 M^{-1}) in both the absence and the presence of any chemical denaturant, suggesting that the aromatic rings were solvent-exposed.

UreG from *Bacillus pasteurii* was found to behave as an intrinsically unstructured dimeric protein (22). The features of the emission band of the unique tryptophan residue (W192) located on the C-terminal helix of this protein, as well as the rate of bimolecular quenching by potassium iodide, indicated that in the native state, W192 is protected from the aqueous polar solvent, while upon addition of denaturant, a conformational change occurs that causes solvent exposure of the indole side chain. This structural change, mainly affecting the C-terminal helix, is associated with the release of static quenching, as shown by resolution of the decay-associated spectra.

Acrylamide, a neutral quencher, is one of the most widely used quenchers of intrinsic protein fluorescence. Usually the globular protein interior is inaccessible to the acrylamide molecule. Thus, acrylamide actively quenches only the intrinsic fluorescence of solvent-exposed tryptophan residues. Acrylamide quenching was shown to decrease by two orders of magnitude as unstructured polypeptide chains transitioned to globular structure (9, 10). More importantly, the degree of shielding of tryptophan residues by the intramolecular environment of the molten globule state was shown to be close to that determined for the native globular proteins, whereas the accessibility of tryptophans to acrylamide in the premolten globule state was closer to that in the unfolded polypeptide chain (35). One of the shortcomings of acrylamide is its ability to bind to some proteins and even penetrate to their interior.

Fluorescence lifetimes of tryptophans in proteins lie usually in the nanosecond region, which complicates their measurement. Rather sophisticated time-domain or frequency-domain kinetic methods are used to measure such short lifetimes. Moreover, interpretation of fluorescence lifetime data is a separate problem. One can measure several exponential fluorescence decay components for proteins with two to three tryptophans and identify them with emission of separate tryptophans. Through measuring fluorescence decay curves at different wavelengths, it is possible to obtain decay-associated spectra and to try to monitor the emission from separate tryptophans. However, fluorescence decay curves even of proteins with single tryptophan residues often are not described by a single exponent. This fact can be interpreted as a reflection of principal nonexponentiality of tryptophan fluorescence decay in proteins or as the existence of several real exponential components in the decay curve arising from the emission of several protein conformers with different tryptophan environments. For example, fluorescence decay curve of the single tryptophan in whiting parvalbumin is adequately approximated only by two exponents (25). The data were interpreted in terms of two parvalbumin conformers with different tryptophan environments.

Conformational heterogeneity of proteins in water solution as a cause of the nonexponential fluorescence decay was mentioned by many authors (1, 14, 20). Within the frames of this interpretation, it is assumed that the most short-living fluorescence decay components arise in conformational states with tryptophans interacting with nearest functional groups causing fluorescence quenching due to the excited state electron transfer. Since these quenching interactions compete with fluorescence decay, they require movements of tryptophan and/or quenching groups within the decay time scale (from picoseconds to nanoseconds). Therefore, in this interpretation, the heterogeneity of the fluorescence decay is explained by existence of conformational substates with different dynamic properties (30). Dipole relaxation also seems to contribute to the fluorescence decay heterogeneity.

Site-directed mutagenesis is widely used in intrinsic fluorescence studies. This method facilitates these studies by targeting individual tryptophan residues. For example, the replacement of a tryptophan residue in a protein with two to three tryptophans per molecule by another nonfluorescent amino acid may help to decompose the total protein fluorescence spectrum into elementary contributions from separate tryptophans. A major concern, however, in such studies relates to the structural similarity of a natural protein and its designed analogs. Indeed, some residues within the protein structure might be forced to occupy alternate, "incorrect" conformations. As a consequence, a point mutation can result in conformational changes distorting the native protein conformation. The fluorescence of the remaining reporter groups could be affected by these induced changes, which, in any case, need to be evaluated.

Site-specific information on the protein conformation can be obtained by biosynthetic incorporation of an unnatural amino acid, 5-fluorotryptophan (5FW), into recombinant protein. Such study was carried out on α-synuclein, the main protein component of fibrillar deposits found in Parkinson's disease, which is intrinsically disordered *in vitro* by Winkler et al. (41). Using fluorescence and ^{19}F NMR spectroscopy, they have characterized three proteins with 5FW at positions 4, 39, and 94. Steady-state emission spectra (maxima at 353 nm; quantum yields approximately 0.2) indicate that all three indole side chains are exposed to aqueous medium. Virtually identical single-exponential excited-state decays ($\tau \approx 3.4$ ns) were observed in all three cases.

Fluorescence polarization or anisotropy is used to study mobility of IDPs and their association with other proteins. The degree of fluorescence depolarization depends upon the following factors that characterize the structural state of the protein molecules: (1) mobility of the chromophores and (2) energy transfer between similar chromophores (17). More importantly, the relaxation times of tryptophan residues determined from polarized luminescence data reflect the compactness of the polypeptide chain.

11.4 LUMINESCENT PROBES AND LABELS

11.4.1 Fluorescent Labels

Spectral probes and labels are widely used for the investigation and determination of proteins (34). Traditional luminescence probes include fluorescent derivatizing reagents, fluorescent probes, and chemiluminescence probes which continue to develop. Proteins have at least two functional groups where a derivatization may take place: the amino group and the carboxyl group. Whereas the carboxylic group at the C-terminus is less active and it must first be activated itself before derivatization, it is rarely used in protein-labeling procedures. On the contrary, amino groups at the N-terminus are easily derivatized. The fluorescent labels reacting with primary amino groups include fluorescein-5-isothiocyanate, *o*-phthaldialdehyde, naphthalene-2,3-dialdehyde/cyanide, 5-fluroylquinoline-3-carboxaldehyde, 6-aminoquinolyl N-hydroxysuccinimidyl carbamate, fluorenylmethyloxycarbonyl chloride, 3-(4-carboxybenzoyl)-2-quinolinecarboxaldehyde, and others. Almost all the derivatizing regents mentioned above react with primary or with both primary and secondary amines. Only 4-(N,N-dimethylamino-sulfonyl)-7-fluoro-2,1,3-benzoxadiazole (DBD-F) can react with secondary amines.

Introducing a fluorophore to thiol group is also a very common method to form a highly fluorescent product. Ammonium 7-fluoro-2,1,3-benzoxadiazole-4-sulfonate can be used for derivatization of SH groups in proteins.

The derivatizing reagents mentioned above interact with proteins through covalent labeling. The labeling substantially improves the detectability and the sensitivity of protein determination. But there are some problems arising from inefficient chemistry and multiple derivatives for trace analysis of protein. Moreover, photophysics and photochemistry of most of these fluorescent labels were not studied in detail; therefore, interpretation of spectral data obtained with these labels is complicated.

Fluorescence energy transfer is often used for studies of interactions of IDPs with other proteins. For example, Anderluh et al. (2) have studied binding of the colicin N translocation domain to its periplasmic receptor TolA by FRET using fluorescent probes attached to engineered cysteine residues. They measured resonance energy transfer efficiency between the tryptophan residues and IAEDANS 5-({2-[(iodoacetyl)amino]ethyl}amino)naphthalene-1-sulfonic acid. The domain exhibits a random coil far-UV CD spectrum. However, FRET revealed that guanidinium hydrochloride denaturation caused increases in all measured intramolecular distances showing that, although natively unfolded, the domain is not extended. Furthermore, NMR reported a compact hydrodynamic radius of 18 Å. Nevertheless, the FRET-derived distances changed upon binding to TolA indicating a significant structural rearrangement (2).

11.4.2 Fluorescent Dyes

The dyes serving as noncovalent probes of proteins are almost all anionic dyes. These dyes can bind to the positively charged amino acid residues of proteins; therefore, pH is an important parameter for their use. Upon binding to proteins, the fluorescence intensity of the dyes may be enhanced or quenched. The enhancement of dyes' fluorescence mainly comes from a change in the microenvironment in which those dyes exist. Very often, these probe reagents are nonfluorescent in water, but highly fluorescent in apolar media. These dyes can bind to the hydrophobic regions of a protein through noncovalent binding and their fluorescence yields are enhanced greatly. Typical probes of this type are naphthalene derivatives, Sypro dyes, and Nile red (reviewed by Sun et al. 34).

The aromatic chromophore 1-anilino-8-naphthalene sulfonate (ANS) is feebly fluorescent in water, but its spectrum is blueshifted and its intensity is dramatically increased in nonpolar solvents or when it binds to nonpolar sites of proteins (33). ANS is a widely used fluorescence probe, yet, despite this popularity, significant questions remain concerning its binding selectivity for hydrophobic "patches" as well as the precise origin of its enhanced quantum yield when bound to such regions (16). Kirk et al. (15) found that in proteins where the quantum yield of ANS fluorescence is appreciable, as in organic solvents, the preferred conformation of ANS is often with the phenyl ring nearly (65–85°) orthogonal to the naphthalene. The major consequence of this geometry is water molecules exclusion from the critical zone of ANS, from where the largest amount of solvent dipolar relaxation originates. This, in turn, leads to a depression of the rate of electron transfer to the surroundings, together with other effects, which results in pronounced increase of fluorescence quantum yield and a blueshift of fluorescence spectrum maximum.

ANS and bis-ANS have a tendency to bind to more unfolded protein structures. Therefore, they readily bind to IDPs, but not to the majority of tightly packed ordered proteins (except to ordered proteins with large hydrophobic pockets surrounding active sites). The fluorescence decay of free ANS is monoexponential, but the formation of complexes between ANS and proteins makes fluorescence decay multiexponential.

An interest to this probe reached its highest point when it was shown that there is a predominant interaction of ANS with the equilibrium and with kinetic intermediates accumulating during the folding of globular proteins in comparison with the folded and completely unfolded proteins or with coil-like, α-helical, or β-structural hydrophilic homopolypeptides (12, 27, 29, 31, 32). The term "large affinity to ANS" became one of the general structural characteristics of a protein molecule in the molten globule state. This "preferable interaction" of ANS with the molten globule was shown to be accompanied by interaction of ANS with a protein molecule that is reflected in a pronounced blueshift of maximal fluorescence, a significant increase of the probe fluorescence intensity, and a change in the fluorescence lifetime. Some of these properties (a blueshift of maximal fluorescence and an increase in fluorescence intensity) were used for the visualization of the formation of the molten globule state in the course of protein folding (32).

The ability of ANS to interact with folded proteins possessing solvent-accessible hydrophobic regions and with highly flexible partially folded conformations brought important question on how to discriminate these dye-protein two complexes. Three approaches developed to answer this important question are described below.

11.4.2.1 ANS Fluorescence Lifetimes The ANS interaction with proteins (both folded and partially folded) is accompanied by the characteristic changes in the fluorescence lifetime of the probe (39). The fluorescence decay of free ANS in aqueous or organic solvents is well described by the monoexponential law, whereas formation of complexes of this probe with proteins results in a more complicated dependence (39). Analysis of the ANS fluorescence lifetimes in a number of proteins revealed that at least two types of ANS-protein interactions might exist. At interaction of the first type (characterized by fluorescence lifetime of about 1–5 ns), the probe molecules are bound to the surface hydrophobic clusters of the protein molecule and are in a relatively good contact with the solvent. At interaction of the second type (characterized by fluorescence lifetime of about 10–17 ns), the probe molecules are embedded into the protein molecule and are poorly accessible to the solvent and external quencher (39).

The changes in the long lifetime component correlate well with the overall conformational changes of the protein molecule observed upon its denaturation and unfolding (38, 39). In fact, the values of the longest lifetime component of fluorescence decay were measured for complexes of ANS with six different proteins: three of them, β-lactamase, lysozyme, and β-lactoglobulin, can bind this dye in the native state, whereas the others, bovine carbonic anhydrase, β-lactamase, β-lactoglobulin, human and bovine α-lactalbumins, acquire large affinity to 8-ANS after the transformation to the molten globule state. It has been shown that the ANS interaction with the native proteins is characterized by the shorter fluorescence decay time compared with the ANS–molten globule complexes ($\tau_N \sim 10$–11 ns and $\tau_{MG} \sim 15$–17 ns, respectively) (38, 39). This sensitivity of ANS molecules to be inserted either into the folded protein or into the molten globule ($\tau_N \leq 12$ ns as compared with $\tau_{MG} \geq 15$ ns) was explained by taking into account the capability of the dye to self-associate (38). The penetration of the dye into the rigid hydrophobic pocket(s) of native proteins is not necessarily accompanied by its dissociation due to the considerable steric limitations, while the liquid-like core of the molten globules cannot prevent such dissociation. As a result, ANS molecules, being embedded into the native proteins, exist as self-oligomers and exhibit fluorescence with a shorter lifetime. Embedding into the molten globule proteins leads to the dissociation of ANS oligomers and, as a consequence, to the increase in characteristic times of fluorescence decay (38). Therefore, the interaction of ANS with both molten globules and premolten globules of different proteins results in fluorescence lifetimes characteristic of the second type, with molten globules reacting more strongly than premolten globules. In other words, there are some "magic numbers"—the values of 8-ANS fluorescence lifetime $\tau_N \leq 12$ ns and $\tau_{MG} \geq 15$ ns—that show which conformational state (the folded or the molten globule) of a protein molecule one deals with (38).

11.4.2.2 Urea Titration of ANS Fluorescence This method is based on the important observation that ordered proteins unfold cooperatively, whereas unfolding of molten globular forms typically is much less cooperative (37). It has been found that urea titration of ANS fluorescence therefore could be used to distinguish between the ANS binding to the hydrophobic pocket of an ordered protein as compared with the binding of ANS to a molten globular

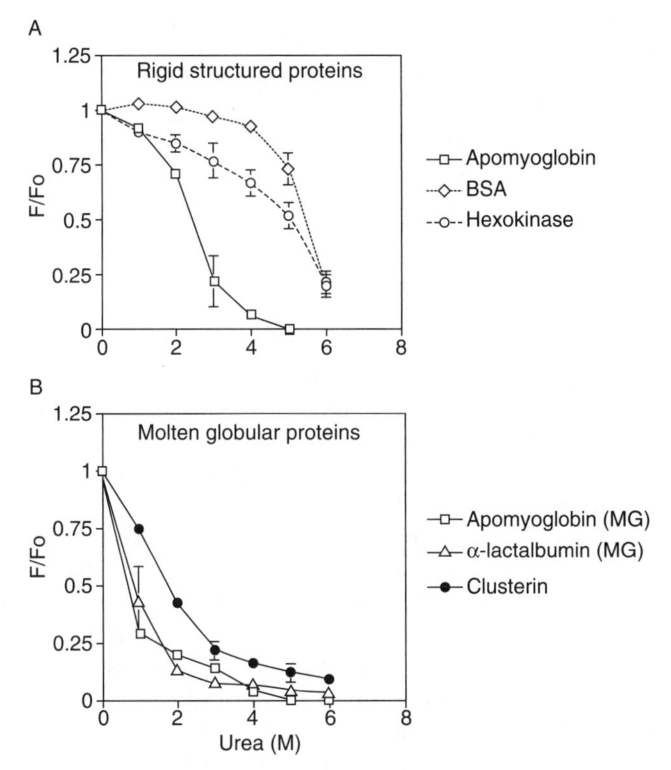

Figure 11.3 Urea titration curves of protein-ANS complexes. Rigid structured proteins (A) and molten globule proteins and clusterin (B) at a concentration of 1 μM were incubated with 50 μM ANS and titrated with urea to observe the change in the fluorescence (F/F_0) of ANS versus increasing urea. As the concentration of the denaturant increases, the proteins unfold with a concomitant decrease in the F/F_0 of ANS. The characteristic shape of these curves reflect the differences between proteins with rigid-structured binding sites and the more unstructured molten globular sites which are more suitable for ANS binding.

form. Results of this analysis are presented in Figure 11.3, which shows that the urea-induced unfolding of such ordered proteins as BSA, apomyoglobin, and hexokinase is characterized by typical sigmoidal curves, whereas unfolding of molten globular forms of apomyoglobin and α-lactalbumin is much less cooperative.

11.4.2.3 Comparison of the Stern–Volmer Quenching by Acrylamide and by Trichloroethanol (TCE) Folded and molten globular ANS binders might be potentially discriminated by comparing the Stern–Volmer quenching curves for a polar quencher, acrylamide, with the quenching curves for a nonpolar quencher, TCE. The essence of this method is based on the following: If the hydrophobic groups surrounding a fluorophore (e.g., ANS) are rigidly packed, then both acrylamide and TCE are both excluded and so both show little

TABLE 11.1 Fluorescence Emission Maxima and Intensities for fd Forms and Tryptamine

Protein/Sample	K_{sv} (acrylamide)	K_{sv} (TCE)	λ_{max} (nm)	References
fd	3.1	3.0	343	(3)
fd I-forms	3.6	27	342	(28)
fd S-forms	2.6	100	341	(28)
Tryptamine	33	22	355	(11)
Tryptamine/SDS	10.5	42	338	(11)

quenching. On the other hand, if the hydrophobic groups surrounding the fluorophore are loosely packed and dynamic, then the hydrophilic quencher, acrylamide, is still excluded and so continues to show little quenching. However, the hydrophobic quencher, TCE, actually partitions into the hydrophobic region surrounding the fluorophore. This leads to quenching that is much stronger than if the fluorophore were completely exposed on the protein surface. These concepts were proven in a model system in which a tryptamine fluorophore was complexed with SDS micelles (11). This approach was applied for the characterization of three different forms of fd phage (28). The fluorescence emission maxima and intensities for the tryptophans in all three forms are nearly identical. Furthermore, there is very little difference in the acrylamide quenching (Table 11.1), suggesting that the indole rings are in tightly packed environments. On the other hand, for the two contracted forms (I- and S-forms), quenching by TCE is enormously stronger than the quenching by acrylamide. Moreover, the data show that the quenching by TCE is even stronger than the quenching for a naked indole ring in water. How can this be? As shown in Table 11.1, the data for tryptamine in SDS show a similar behavior. For SDS, the original interpretation was that the internal, dynamic micelle could actually dissolve the TCE, so its local concentration around the fluorophore is higher than in the surrounding solution. Thus, these data suggest a similar interpretation for the contracted forms of fd phage: the residues surrounding the tryptophan are likely dynamic similar to the inside of an SDS micelle, thus leading to high local accumulation of TCE and very high quenching values (28).

11.4.3 Examples of IDP Analysis with Fluorescence Dyes

UreG is an essential protein for the *in vivo* activation of urease. The exposure of protein hydrophobic sites, monitored using the fluorescent probe bis-ANS, indicated that the native dimeric state of BpUreG is disordered even though it maintains a significant amount of tertiary structure (22). ANS fluorescence also indicated that, upon addition of a small amount of GuHCl, a transition to a molten globule state occurs, followed by formation of a premolten globule state at a higher denaturant concentration. The hydrodynamic parameters obtained by time-resolved fluorescence anisotropy at maximal denaturant concentrations (3 M GdmHCl) confirmed the existence of a disordered but stable dimeric protein core.

As it has been mentioned above, spectrofluorimetric urea titration experiments with the use of ANS fluorescence could distinguish the binding of ANS to the hydrophobic pocket of an ordered protein from the binding of ANS to a molten globule (including an IDP). This approach was utilized in analysis of clusterin (5) and the N-terminal transactivation domain of the human androgen receptor (18).

Clusterin, also known as sulfated glycoprotein-2, TRPM-2, GP-80, SP 40,40, and ApoJ, has been found in many tissues including prostate, brain, kidney, liver, and in plasma in many species including rat, human, ram, and bovine (5). Clusterin was proposed to have a wide range of biological functions including cell–cell interactions, sperm maturation, complement inhibition, and lipid transport. Furthermore, it was shown to have chaperone-like activity, preventing the precipitation of denatured proteins *in vitro* (13). Finally, clusterin is able to interact specifically with a wide range of biological ligands including proteins such as complement components, peptides such as amyloid β_{1-40}, and lipids such as those found in high-density lipoproteins (8, 40). Clusterin is associated with cellular injury, lipid transport, apoptosis, and it may be involved in the clearance of cellular debris caused by cell injury or death. A model was proposed where clusterin acted as a "biological detergent," binding to hydrophobic complexes and denatured proteins to aid in their clearance from ducts or lumen during tissue remodeling (4). To do so, clusterin was hypothesized to have a flexible or dynamic binding site or sites to allow numerous associations to take place. This hypothesis was tested by several methods, including prediction of disorder from amino acid sequence, limited protease digestion, far-UV CD, and ANS fluorescence (5). The comparison of the response of the clusterin-ANS complex to the increasing urea concentrations with those of proteins with structured binding pockets and molten globular forms of proteins revealed that clusterin likely contains a molten globule-like domain in its native state (5).

The androgen receptor is a ligand-activated transcription factor that mediates the actions of the steroid hormones testosterone and dihydrotestosterone. Its N-terminal domain (NTD) is intrinsically disordered and structurally flexible and participates in multiple protein–protein interactions (19). Using a set of computational and experimental approaches, it has been shown that this domain, being intrinsically disordered, exists in a collapsed disordered conformation, distinct from extended disordered (random coil) and a stable globular fold (5). Particularly, the interaction of purified domain and the hydrophobic fluorescence probe ANS was investigated. Incubation of the transactivation domain with ANS resulted in a significant increase in fluorescence intensity and a blueshift for the maximum emission to 465 nm (18). Furthermore, the urea-induced unfolding of this domain was compared with that of the structured protein BSA and α-lactalbumin in the molten globule conformation.

Similar to the data reported in Figure 11.3, BSA was shown to unfold cooperatively, whereas the unfolding of the molten globular α-lactalbumin was noncooperative. The transactivation domain of the androgen receptor was similarly sensitive to urea and unfolded in a noncooperative manner. Based on these data, it has been concluded that the domain in a native state is able to bind ANS and exhibited ANS binding characteristics similar to those of a well-characterized molten globule state protein (18).

An ability to gain some ordered structure at extreme pH values is a characteristic property of extended IDPs (see chapter 20 for more details). pH-Induced folding of a typical IDP, α-synuclein, was analyzed using a variety of biophysical techniques, including changes in the ANS fluorescence (36). Figure 11.4 shows that in the case of α-synuclein, a decrease in pH led to a noticeable blueshift of the ANS fluorescence maximum (from ~515 to ~475 nm), reflecting the pH-induced transformation from the natively unfolded state to the partially folded partially compact conformation. The transition from the natively unfolded to a partially folded conformation took place between pH 5.5 and 3.0, and was completely reversible (Fig. 11.4, open and solid symbols).

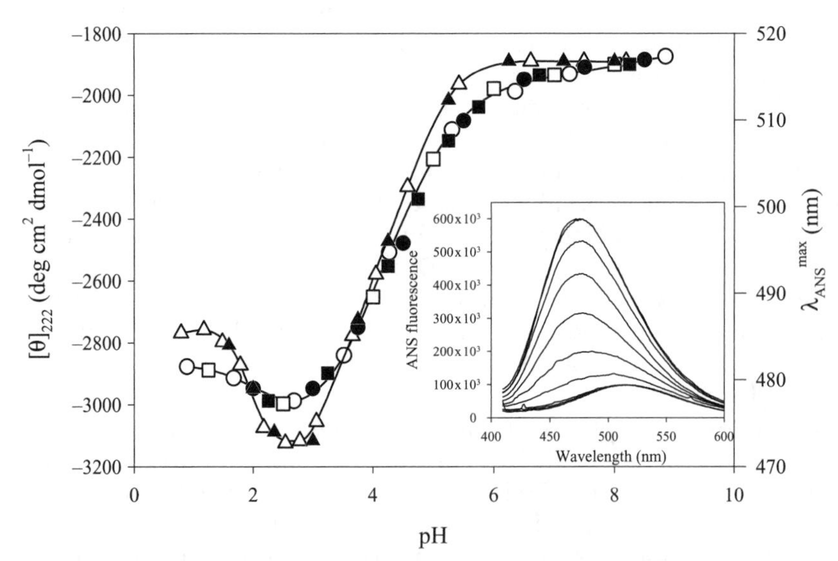

Figure 11.4 Effect of pH on the structural properties of α-synuclein. Comparison of the effect of pH on far-UV circular dichroism (circles and squares) and ANS fluorescence spectra (triangles). The results of the initial titration (decrease in pH) and reverse experiments (increase in pH) are presented as open and solid symbols respectively. For far-UV CD and fluorescence measurements, the cell path lengths were 0.1 and 10 mm, respectively. Measurements were carried out at 20°C. Protein concentration was 0.1 (circles) 1.0 (squares) and 0.01 mg/mL (triangles). The inset represents ANS fluorescence spectra measured in the presence of α-synuclein under at pHs: 8.2, 7.5, 6.6, 5.4, 4.6, 4.0, 3.7, 3.1, 2.8, and 2.5 (in order of increasing intensity).

Furthermore, Figure 11.4 clearly shows that the pH-induced structural transitions observed by ANS fluorescence and far-UV CD ellipticity occurred simultaneously in a rather cooperative manner. This means that the α-synuclein protonation resulted in the transformation of this natively unfolded protein into a conformation with a significant amount of ordered secondary structure and with increased affinity to ANS. The position of the transition indicated that the protonation of one or more carboxylates was responsible for the detected structural changes (36).

11.5 CONCLUDING REMARKS

The data provided in this short review clearly show that luminescent spectroscopy can be successfully used for studies of IDPs. The best results can be obtained when this method is used in combination with other physical and chemical methods.

ACKNOWLEDGMENTS

This work was partially supported by the Program of the Russian Academy of Sciences "Molecular and Cellular Biology" (E. A. P and V. N. U.), and by grants R01 LM007688-01A1 (V. N. U.) and GM071714-01A2 (V. N. U.) from the National Institutes of Health. We gratefully acknowledge the support of the IUPUI Signature Centers Initiative.

REFERENCES

1. Alcala, J. R., E. Gratton, and F. G. Prendergast. 1987. Interpretation of fluorescence decays in proteins using continuous lifetime distributions. Biophys J **51**:925–36.
2. Anderluh, G., Q. Hong, R. Boetzel, C. MacDonald, G. R. Moore, R. Virden, and J. H. Lakey. 2003. Concerted folding and binding of a flexible colicin domain to its periplasmic receptor TolA. J Biol Chem **278**:21860–8.
3. Arnold, G. E., L. A. Day, and A. K. Dunker. 1992. Tryptophan contributions to the unusual circular dichroism of fd bacteriophage. Biochemistry **31**:7948–56.
4. Bailey, R., and M. D. Griswold. 1999. Clusterin in the male reproductive system: localization and possible function. Mol Cell Endocrinol **151**:17–23.
5. Bailey, R. W., A. K. Dunker, C. J. Brown, E. C. Garner, and M. D. Griswold. 2001. Clusterin, a binding protein with a molten globule-like region. Biochemistry **40**:11828–40.
6. Baskakov, I. V., R. Kumar, G. Srinivasan, Y. S. Ji, D. W. Bolen, and E. B. Thompson. 1999. Trimethylamine N-oxide-induced cooperative folding of an intrinsically unfolded transcription-activating fragment of human glucocorticoid receptor. J Biol Chem **274**:10693–6.

7. Bourhis, J. M., V. Receveur-Brechot, M. Oglesbee, X. Zhang, M. Buccellato, H. Darbon, B. Canard, S. Finet, and S. Longhi. 2005. The intrinsically disordered C-terminal domain of the measles virus nucleoprotein interacts with the C-terminal domain of the phosphoprotein via two distinct sites and remains predominantly unfolded. Protein Sci **14**:1975–92.
8. Calero, M., T. Tokuda, A. Rostagno, A. Kumar, B. Zlokovic, B. Frangione, and J. Ghiso. 1999. Functional and structural properties of lipid-associated apolipoprotein J (clusterin). Biochem J **344** (Pt 2):375–83.
9. Eftink, M. R., and C. A. Ghiron. 1975. Dynamics of a protein matrix revealed by fluorescence quenching. Proc Natl Acad Sci U S A **72**:3290–4.
10. Eftink, M. R., and C. A. Ghiron. 1977. Exposure of tryptophanyl residues and protein dynamics. Biochemistry **16**:5546–51.
11. Eftink, M. R., and C. A. Ghiron. 1976. Fluorescence quenching of indole and model micelle systems. J Phys Chem **80**:486–93.
12. Goto, Y., and A. L. Fink. 1989. Conformational states of beta-lactamase: molten-globule states at acidic and alkaline pH with high salt. Biochemistry **28**:945–52.
13. Humphreys, D. T., J. A. Carver, S. B. Easterbrook-Smith, and M. R. Wilson. 1999. Clusterin has chaperone-like activity similar to that of small heat shock proteins. J Biol Chem **274**:6875–81.
14. Hutnik, C. M., and A. G. Szabo. 1989. Confirmation that multiexponential fluorescence decay behavior of holoazurin originates from conformational heterogeneity. Biochemistry **28**:3923–34.
15. Kirk, W., E. Kurian, and W. Wessels. 2007. Photophysics of ANS. V. Decay modes of ANS in proteins: the IFABP-ANS complex. Biophys Chem **125**:50–8.
16. Klimtchuk, E., S. Venyaminov, E. Kurian, W. Wessels, W. Kirk, and F. G. Prendergast. 2007. Photophysics of ANS. I. Protein-ANS complexes: intestinal fatty acid binding protein and single-trp mutants. Biophys Chem **125**:1–12.
17. Lackowicz, J. 1999. Principles of Fluorescence Spectroscopy. Kluwer Academic/ Plenum Publishers, New York.
18. Lavery, D. N., and I. J. McEwan. 2008. Structural characterization of the native NH2-terminal transactivation domain of the human androgen receptor: a collapsed disordered conformation underlies structural plasticity and protein-induced folding. Biochemistry **47**:3360–9.
19. Lavery, D. N., and I. J. McEwan. 2005. Structure and function of steroid receptor AF1 transactivation domains: induction of active conformations. Biochem J **391**:449–64.
20. Millar, D. P. 1996. Time-resolved fluorescence spectroscopy. Curr Opin Struct Biol **6**:637–42.
21. Muro-Pastor, M. I., F. N. Barrera, J. C. Reyes, F. J. Florencio, and J. L. Neira. 2003. The inactivating factor of glutamine synthetase, IF7, is a "natively unfolded" protein. Protein Sci **12**:1443–54.
22. Neyroz, P., B. Zambelli, and S. Ciurli. 2006. Intrinsically disordered structure of Bacillus pasteurii UreG as revealed by steady-state and time-resolved fluorescence spectroscopy. Biochemistry **45**:8918–30.
23. Permyakov, E. A. 1993. Luminescent Spectroscopy of Proteins. CRC Press, Boca Raton, Ann Arbor, London, Tokyo.

24. Permyakov, E. A. 2003. The Method of Intrinsic Protein Luminescence. Nauka, Moscow.
25. Permyakov, E. A., A. V. Ostrovsky, E. A. Burstein, P. G. Pleshanov, and C. Gerday. 1985. Parvalbumin conformers revealed by steady-state and time-resolved fluorescence spectroscopy. Arch Biochem Biophys **240**:781–91.
26. Permyakov, S. E., I. S. Millett, S. Doniach, E. A. Permyakov, and V. N. Uversky. 2003. Natively unfolded C-terminal domain of caldesmon remains substantially unstructured after the effective binding to calmodulin. Proteins **53**:855–62.
27. Ptitsyn, O. B., R. H. Pain, G. V. Semisotnov, E. Zerovnik, and O. I. Razgulyaev. 1990. Evidence for a molten globule state as a general intermediate in protein folding. FEBS Lett **262**:20–4.
28. Roberts, L. M., and A. K. Dunker. 1993. Structural changes accompanying chloroform-induced contraction of the filamentous phage fd. Biochemistry **32**:10479–88.
29. Rodionova, N. A., G. V. Semisotnov, V. P. Kutyshenko, V. N. Uverskii, and I. A. Bolotina. 1989. [Staged equilibrium of carbonic anhydrase unfolding in strong denaturants]. Mol Biol (Mosk) **23**:683–92.
30. Royer, C. A. 1993. Understanding fluorescence decay in proteins. Biophys J **65**:9–10.
31. Semisotnov, G. V., N. A. Rodionova, V. P. Kutyshenko, B. Ebert, J. Blanck, and O. B. Ptitsyn. 1987. Sequential mechanism of refolding of carbonic anhydrase B. FEBS Lett **224**:9–13.
32. Semisotnov, G. V., N. A. Rodionova, O. I. Razgulyaev, V. N. Uversky, A. F. Gripas, and R. I. Gilmanshin. 1991. Study of the "molten globule" intermediate state in protein folding by a hydrophobic fluorescent probe. Biopolymers **31**:119–28.
33. Stryer, L. 1965. The interaction of a naphthalene dye with apomyoglobin and apohemoglobin. A fluorescent probe of non-polar binding sites. J Mol Biol **13**:482–95.
34. Sun, C., J. Yang, L. Li, X. Wu, Y. Liu, and S. Liu. 2004. Advances in the study of luminescence probes for proteins. J Chromatogr B Analyt Technol Biomed Life Sci **803**:173–90.
35. Uversky, V. N., A. S. Karnoup, D. J. Segel, S. Seshadri, S. Doniach, and A. L. Fink. 1998. Anion-induced folding of Staphylococcal nuclease: characterization of multiple equilibrium partially folded intermediates. J Mol Biol **278**:879–94.
36. Uversky, V. N., J. Li, and A. L. Fink. 2001. Evidence for a partially folded intermediate in alpha-synuclein fibril formation. J Biol Chem **276**:10737–44.
37. Uversky, V. N., and O. B. Ptitsyn. 1996. All-or-none solvent-induced transitions between native, molten globule and unfolded states in globular proteins. Fold Des **1**:117–22.
38. Uversky, V. N., S. Winter, and G. Lober. 1998. Self-association of 8-anilino-1-naphthalene-sulfonate molecules: spectroscopic characterization and application to the investigation of protein folding. Biochim Biophys Acta **1388**:133–42.
39. Uversky, V. N., S. Winter, and G. Lober. 1996. Use of fluorescence decay times of 8-ANS-protein complexes to study the conformational transitions in proteins which unfold through the molten globule state. Biophys Chem **60**:79–88.

40. Wilson, M. R., and S. B. Easterbrook-Smith. 2000. Clusterin is a secreted mammalian chaperone. Trends Biochem Sci **25**:95–8.
41. Winkler, G. R., S. B. Harkins, J. C. Lee, and H. B. Gray. 2006. Alpha-synuclein structures probed by 5-fluorotryptophan fluorescence and 19F NMR spectroscopy. J Phys Chem B **110**:7058–61.

12

HYDRATION OF INTRINSICALLY DISORDERED PROTEINS FROM WIDE-LINE NMR

KÁLMÁN TOMPA,[1] MONIKA BOKOR,[1] AND PETER TOMPA[2]

[1]*Research Institute for Solid State Physics and Optics, Hungarian Academy of Sciences, Budapest, Hungary*

[2]*Institute of Enzymology, Biological Research Center, Hungarian Academy of Sciences, Budapest, Hungary*

ABSTRACT

The principal aim of our work is to characterize structural and dynamical properties of interfacial water at the protein surface by wide-line NMR spectroscopy and nuclear relaxation time measurements for the identification and characterization of intrinsically disordered proteins (IDPs) and to make a distinction between IDPs and globular proteins. Our approach is to explore the structure↔interface relations of IDPs and globular proteins. In this chapter, we provide a detailed description of the theoretical background and practice of this approach, followed by the description of its implementation on two proteins, the IDP early responsive to dehydration 10 (ERD10), and globular bovine serum albumin (BSA).

The main results are the direct determination of the number of hydration water molecules, the elements of hydration water dynamics (activation energy and correlation times), and the differences in dynamics as seen by the different time windows provided by the different types of relaxation rates (R_1, $R_{1\rho}$ and R_2).

We show by these two examples that IDPs are distinguished from globular proteins by their more extended interfacial region (hydration), their stronger

Instrumental Analysis of Intrinsically Disordered Proteins: Assessing Structure and Conformation, Edited by Vladimir Uversky and Sonia Longhi

(spin diffusion- and chemical-) interactions between protein and bound water at low temperatures and their higher relaxation rates and activation energies at high temperatures. The different values for the slow-motion characteristics for bovine serum albumin BSA and for ERD10 at intermediate temperatures are of potential further significance.

12.1 INTRODUCTION

This report points out the scarcity of literary data on the interfacial water in aqueous solutions of intrinsically disordered proteins (IDPs), and outlines an approach to take the first steps in this field. Although high-resolution NMR techniques (such as the homo- and hetero-nuclear Overhauser methods) are widely and successfully applied in the field of protein research, we show that the more traditional approach of pulsed wide-line NMR spectrum and relaxation time measurements can also be of significant utility in addressing basic issues of IDP research. The results and the still-remaining open questions of the experimental and of the molecular dynamics simulation (MDS) investigations on the structural and the dynamical properties of globular protein-water interfacial region have been described and documented in many papers (see for instance in References 3, 17, 26, and 36 and the references therein). B. Halle (17) summarized the present state of our knowledge on the hydration of globular proteins, with the main diagnostic conclusions that "the progress is slow and erratic, and the results given by different experimental methods are contradictory and more or less model-dependent in the interpretation in spite of that the water in biological systems has been studied for well over a century." He provided the definition of hydration, structure of interface, the interrelation of structure and dynamics, and proposed the method of nuclear magnetic relaxation dispersion (NMRD) as a therapy to get a better approach (i.e., longitudinal relaxation rate measurements as a function of resonance frequency, carried out almost exclusively at room temperature). Bizzarri and Cannistraro gave the present cross-section of the MDS field (3). They have called attention to the fact that the analysis of the number of different water molecules engaged in H-bonds provides information about the dynamics of H-bond network at the protein-solvent interface.

In these studies, it was found that the temperature trend in the number of water molecules mentioned exhibits a critical transition at about –73 to –53°C (see, e.g., Reference 2). Protein-water systems have two transitions between –93°C and the temperature of denaturation (see, e.g., Reference 22). In hydrated proteins, at about –53°C, the rate of the atomic motional amplitude increase with temperature suddenly becomes enhanced, signaling the onset of more "liquid-like" motion. This "dynamical transition" of proteins may be triggered by coupling of the protein with the hydration water through hydrogen bonding since hydration water shows a dynamic transition at a similar temperature.

To come back to a more basic problem, the water structure and dynamics, and their modeling left a few unanswered questions to the present workers of science (10). On the other hand, Uversky has called the attention to that "Natively unfolded proteins: A point where biology waits for physics" (45). We suggest a direct approach for the model-free characterization of IDPs. Our motivation is only partly influenced by the above pessimistic conclusions because the IDP class was not investigated before us and we proposed and applied the cooling and step-by-step reheating and partly the vacuum drying technologies together with the multiexperimental wide-line NMR approach. These technologies and approach open a not yet used window to the field. We demonstrate the approach by applying it to an IDP, which functions in dehydration stress as a late embryogenesis abundant (LEA) protein. LEA proteins comprise several groups, most of which fall into the group of IDPs (13, 42, 46). They lack conventional secondary structure, and attempts to crystallize purified LEA proteins for X-ray crystallography have reportedly been unsuccessful (e.g., References 14 and 23). Research into LEA proteins has been ongoing for more than 20 years. Although there is a strong association of LEA proteins with abiotic stress tolerance, particularly dehydration and cold stresses, their molecular function is largely obscure, and despite much effort for molecular characterizations, they present more puzzles than answers (see the review in Reference 44). LEA proteins can have several possible functions. These include roles as antioxidants and as membrane and protein stabilizers during water stress, either by direct interaction or by acting as molecular shields. They might also serve as "space fillers" to prevent cellular collapse at low water activities. This multifunctional capacity of the LEA proteins is probably attributable in part to their structural plasticity, as they are largely lacking in secondary structure in the fully hydrated state, but can become more folded during water stress and/or through association with membrane surfaces. Our recent work (43), using ^{1}H-NMR and differential scanning calorimetry measurements, suggests that LEA early responsive to dehydration 10 (ERD10) has a large capacity for binding solute ions, that is, to function as an ion sink. Using wide-line NMR relaxation technique, we showed that ERD10 also binds significantly more water than globular proteins (4). The globular proteins were represented by serum albumin, the most plentiful protein in blood plasma, which is the carrier of fatty acids in the blood that can carry seven fatty acid molecules. These bind in deep crevices in the protein, burying their carbon-rich tails safely away from the surrounding water. Bovine serum albumin (BSA) also binds many other water-insoluble molecules. We use it here as a control to contrast the behavior of an IDP with a well-folded, globular protein.

12.2 EXPERIMENTAL DETAILS AND APPLIED FORMALISM

Considering the scarcity of literary observations, experimental investigation of the hydration properties of proteins, especially IDPs, appears to be as actual a task as ever. A complex experimental approach has been applied by us (9, 41, 43). The interface-oriented experimental approach consists of a cooling then step-by-step reheating (freeze-thawing) technology and multiple-experimental measurements at every temperature. The freeze-thawing technique can be complemented with vacuum drying for powder samples, whereas the phrase "multiple-experimental" means the application of pulsed wide-line NMR spectroscopy and relaxation time measurements, which connect events on different timescales. In this respect, wide-line NMR consists of several different pulse-excitation techniques. The term wide-line and the experiments are not limited to "solid state", neither are they to "solid samples." As responses to the excitation, we measure nuclear magnetization, free induction decay signal (i.e., the inverse Fourier-transformed NMR spectrum), echoes (5, 16, 24, 30), and all the nuclear relaxation rates in the laboratory and rotating coordinate systems. In a heterogeneous sample (i.e., in a multi-fraction spin system), the response contains the global information referring to the total spin system, or to only one of the fractions of the whole sample (e.g., to one differently "bound" part of molecules). In our case, the distinction or classification of H_2O fractions is based on the differences in the mobility of water molecules (hydrogen nuclei). Mobility or its lack can be detected on the basis of the "motional narrowing" phenomenon in NMR spectroscopy (see for instance in Reference 37, p. 213 and Reference 25, p. 424). The reduction of sample heterogeneity and the consequent NMR response complexity can be observed by changing the temperature (or by vacuum drying) and/or by the proper selection of radio-frequency pulse excitation (41), respectively. The low-temperature (below 0°C) mobile-water fraction can be identified as corresponding to interfacial water (or in other words as the hydration of solutes), and its size, dynamics, and evolution can be investigated as a function of temperature.

The water molecular motion starts in the temperature range of about −73 to −53°C in all of our investigated protein-water systems (see References 4, 9, 39, and 43 and unpublished results). The understanding of the correlation between the measured and calculated quantities (3, 17, 26, 36) is the major challenge we would like to live up to.

In the present subsection, we give a list of different pulsed-NMR excitations and the proper response functions. These are the free induction decay signal (FID [t], i.e., the inversely Fourier-transformed spectrum [ω]), the echo signals (t), where t and ω in parentheses denote the time and frequency domains, and the spin-spin and spin-lattice relaxation rates R_i. All the quantities are measured as a function of temperature. The characteristics of the experimental technique in question are also given and the literary data on the proteins investigated are presented. We refer to the possible conclusions coming from the responses.

12.2.1 The Applied Pulse Sequences, the NMR Responses, and the Types of Contributing Nuclei

The following are the applied pulse sequences, the NMR responses, and the types of contributing nuclei:

1. Preparation pulse, $P_{x'}$ ⇒ response: FID signal, which is the inversely Fourier-transformed NMR spectrum. All the protons in the sample contribute to the FID signal. The amplitude of this response is $M_o = n_o B_o / T$ at time zero, where B_0 is the magnetic induction, T is the absolute temperature, and n_0 refers to the number of nuclear spins (protons) in the sample. Otherwise, all the following responses are proportional to the number (n_0) of the contributing resonant spins.
2. $P_{x'} - \tau - P_{x'}$ or $2P_{x'} - \tau - P_{x'}$ pulse sequences give ⇒ nonequilibrium FID (for short τ) and two methods for the longitudinal (spin-lattice) relaxation rate (R_1) measurements. ⇒ R_1 connects the spins with the "lattice." The sample is considered here as a thermal reservoir and is denoted by the term lattice. This experiment serves as the basis of the generally used NMR dispersion $R_1(\omega)$ method (NMRD) that gives the correlation time τ_q and its temperature dependence, if $R(\omega,T)$ was measured.
3. Preparation pulses, ($P_{x'} - \tau - P_{x'}$ or $P_{x'} - \tau - P_{y'}$) and $P_{x'} - \tau - 2P_{x'}$ or $P_{x'} - \tau - 2P_{y'}$ ⇒ response: (Hahn-) Carr-Purcell-echo (H-CP) (4, 16).
4. The extended version of the H-CP pulse sequence contains a series of pulses, as $P_{x'} - \tau - 2P_{y'} - 2\tau - 2P_{y'} - 2\tau - 2P_{y'} - 2\tau - \ldots$, ⇒ response: Carr-Purcell-Meiboom-Gill (CPMG) (5, 24) echo-train. Only those protons contribute to the response signal for which the average dipolar-field is zero because of the "motional narrowing," that is, only the mobile protons are detected. CPMG echo-train enables the measurements of transversal (spin-spin) relaxation rate R_2.
5. Preparation pulse $P_{x'} - \tau - P_{y'}$ ⇒ response: solid-echo (30). Only those protons contribute to a solid-echo signal for which the average dipolar-field is nonzero because the "motional narrowing" is not effective, that is, only the "protons in a rigid landscape" are detected.
6. $P_{x'} - 0 -$ (*long spin-lock pulse*)$_{y'}$ ⇒ response: nonequilibrium FID and a method for the rotating-frame longitudinal relaxation rate ($R_{1\rho}$) measurements.

τ is the time unit between the pulses in the given excitation (preparation) pulse sequences, x' and y' are axes of the rotating reference system.

12.2.2 Nuclear Relaxation Rates

The nuclear relaxation rates can be interpreted by using the density matrix formalism or the conventional perturbation theory (for the details see chapter 5 in Reference 37). Both methods are based on the fluctuating local magnetic field at the position of resonant nuclei (local field approximation) and give a simple and clear physical picture for the source of transversal and longitudinal relaxation in single- and multi-fraction spin systems in the state of motional narrowing. The theoretical background (the formalism) and the physical interpretation we use here are given in References 25, 28, 37 and in the references cited therein.

In accord, the formalism (we have used) is as follows.

12.2.2.1 Single-Phase (Single-Fraction) Spin System

12.2.2.1.1 Nuclear Relaxation Rate Expressions The longitudinal and transversal relaxation rates (R_1 and R_2, respectively) for a homogeneous sample (single-spin system) can be written as ((25) and references cited therein)

$$R_1 = \gamma_I^2 \sigma_1^2 \tau_1 / (1 + \omega_0^2 \tau_1^2), \tag{12.1a}$$

$$R_2 = \gamma_I^2 (\sigma_0^2 \tau_0 + \sigma_1^2 \tau_1 / 2(1 + \omega_0^2 \tau_1^2)), \tag{12.1b}$$

$$R_2 - R_1/2 = \gamma_I^2 \sigma_0^2 \tau_0, \tag{12.1c}$$

where γ_I is the gyromagnetic ratio of the proton. The equations are derived for a statistical sequence of rectangular pulses of local fluctuation of magnetic induction $B_q(t) = \pm a_n$ with a mean jump time τ_q (Poisson process) and a statistical amplitude distribution of $\langle a_n \rangle = 0$ and $\langle a_n^2 \rangle = \sigma_q^2 = \langle B_{loc}^2 \rangle / 3$.

Equation (1c) provides a simple opportunity for the experimental determination of the secular contribution (see the definition in Reference 37) to transversal relaxation rate $R_2' = R_2 - R_1/2$. The quantity is independent of ω_0 and the temperature dependence comes from that of τ_0. The longitudinal relaxation rate in the rotating frame ($R_{1\rho}$) can be described as (37)

$$R_{1\rho} = \gamma_I^2 (\sigma_0^2 \tau_0 / (1 + 4\omega_1^2 \tau_0^2) + \sigma_1^2 \tau_1 / 2(1 + \omega_0^2 \tau_1^2)), \tag{12.2}$$

where $\omega_1 = \gamma B_1$, that is the resonance frequency in the rotating magnetic induction B_1. The mean jump time τ_q follows the Arrhenius law as

$$\tau_q = \tau_{q0} \exp(E/RT), \tag{12.3}$$

where E is the activation energy of the molecular motion and R is the molar gas constant. The longitudinal relaxation rate has an extreme at $\omega_0 \tau_q = 1$. The interaction strength σ_q^2 can be determined from the maximum R_1 value. The ratios R_2/R_1 and $R_{1\rho}/R_1$ at the extreme of R_1 give substantial information on the homogeneity of the spin system. There are two important deviations, first, if a distribution of correlation time τ_q exists (25) and second, if two substantially different correlation times τ_{q1} and τ_{q2} describe the otherwise homogeneous spin system (26). The consequences of the first one are that the R_1 versus T curve around the maximum is not so sharp and the distance between R_1 and R_2 on the R_i versus T map is increased with respect to τ_q without distribution. In the second case, there are two terms independent of ω_0 in the R_2 and $R_{1\rho}$

expressions with two different σ_{qi}^2 and τ_{qi} values. The consequences of both deviations will be seen in section 12.3.

12.2.2.1.2 Decay of Magnetization The relaxation functions (longitudinal and transversal, $i = 1$, 1ρ and 2, respectively) for a single-fraction spin system are of the form

$$M_i(t) = M_0 \exp(-R_i t), \qquad (12.4)$$

where t is the time and M_0 is the equilibrium nuclear magnetization. M_0 is proportional to the number of contributing nuclear spins and the exponent R_i is the relaxation rate. R_2 represents a narrower time window than R_1 for the low-frequency motions.

12.2.2.1.3 Inhomogeneous Magnetic Fields, Translational Diffusion At the end of this subsection, let us recall the variant of the transversal relaxation function, which is valid in the case of inhomogeneous magnetic fields and for those samples where the translational diffusion of resonant nuclei is not negligible. The following formula (37) is valid in the case of a uniaxial magnetic induction gradient $\partial B/\partial z$.

$$M_2(n2\tau) = M_0 \exp[-F((\partial B/\partial z), D, \tau^3)]\exp[-\{n2\tau\}R_2], \qquad (12.5)$$

where $t = n2\tau$ is the time measured between the exciting pulses in the CPMG train in 2τ units and the first exponent is the F function of field gradient and the D diffusion constant characteristic to translational diffusion. We reproduced the formula here in a short form, because we should like to use it only for the demonstration of the existence or absence of the translational diffusion.

12.2.2.2 Two-Phase (Two-Fraction) Spin Systems The relaxation functions for a two-phase system (phases a and b) at intermediate exchange rates between the phases are formally similar to that given in Equation (12.6) but they are constituted of probabilities and relaxation rates having different complicated form and meaning (for the details see Eqs. (2.1)–(2.5) in Reference 28). The formalism in the case of *slow exchange*, where $\tau_{a,b} >> 1/R_{ia,b}$ (i.e., for two-exponential relaxation) is however as simple as follows:

$$M_i(t) = p_a \exp(-R_{ia}t) + p_b \exp(-R_{ib}t), \qquad (12.6)$$

where $p_a + p_b = 1$ are the abundance probabilities of the fractions and $R_{ia,b}$ denotes the relaxation rates.

The single exponential character is also valid in the case of *slow exchange*, if $R_{ib}/R_{ia} << 1$, $\tau_a/R_{ia} << 1$, and $R_{ia} \rightarrow R'_{ia}$ and we finally get the following equation.

$$R'_{iav} = p_a/[p_b(1/R_{ia} + \tau_a)] + R_{ib} \tag{12.7}$$

The relaxation function for a two-phase system in the case of *fast exchange*, where $\tau_{a,b} << 1/R_{ia,b}$ (single exponential relaxation) is as follows:

$$M_i(t) = M_0 \exp(-R_{iav}t) \tag{12.8}$$

12.2.2.3 More than Two Fractions, Intermediate Exchange Rate For more than two fractions of *intermediate exchange rates*, no general formula that would correspond to the modified form of Equation (12.6) exists (for the details consult the references given in Reference 28). However, the limiting cases Equations (12.6–12.8) can be applied after a simple generalization (adding terms with an index $j = 1, \ldots, n$, where $a = 1, b = 2$ in Eq. [12.6]). The relaxation function for the case of slow exchange (multi-exponential relaxation) is consequently as follows:

$$M_i(t) = \sum_{j=1}^{n} p_{ij} \exp(-R_{ij}t) \tag{12.9}$$

And finally, for the case of *fast exchange* (single exponential relaxation), the measured relaxation rate R_i^M is as follows:

$$R_i^M = \sum_{j=1}^{n} p_j R_{ij} \tag{12.10}$$

12.2.3 Changing Numbers of Spin Fractions

The consequence of *varying the temperature* (fast cooling and reheating) is the change in the number of contributing spin fractions (as far as their motional states are concerned) and in the number of relaxation channels. One can follow the changes by the direct or reversed application of the formalism that was proposed to vacuum dehydration (31–33).

12.2.4 Hydrogen Fractions—NMR Responses

To the generally used sources of information of NMR (e.g., spectrum, moments, shifts, relaxation rates) a new one was added (not used in this field up to now), namely the amplitude of the NMR response function. The amplitude of any NMR response is directly proportional to the equilibrium-state nuclear magnetization M_0 and so, to the number of contributing nuclei, which, in turn, is proportional to the mass of the sample.

The FID signal can be separated into distinct fractions for a multicomponent system. Its tail is generated by the mobile protons in the sample (Fig. 12.1). The extrapolated amplitude of the tail to zero time corresponds to the magnetization induced by these nuclei.

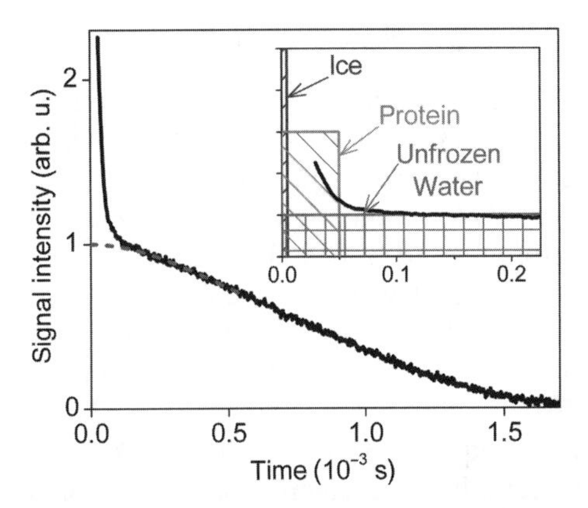

Figure 12.1 Illustration of the method applied to measure the fraction of the unfrozen water component x_{unfrozen}. The slowly decaying part of the time-domain FID signal was extrapolated to $t = 0$ by applying Lorentzian approximation (dashed red line). The extrapolated signal intensity was then compared with the signal intensity measured above 0°C when the whole sample is in liquid state, to get the x_{unfrozen} value. The inset shows the typical spread in time of FID signals produced by ice protons, protein protons, and unfrozen water protons. See color insert.

$$M_{i0} = n_i B_0 / T, \tag{12.11}$$

where n_i refers to the number of spins (protons) in the ith fraction. Equation (12.11) gives the direct measure of each proton fraction and the primary NMR signal is used for this purpose without any further processing. Spectrum integration is avoided this way, which is burdened by several inherent sources of error.

The response functions FID signal and CPMG echo-train are basically different (37). "Motional averaging" (motion: translational diffusion or rotation) or in other words, an inhomogeneous local field is necessary for creating a CPMG echo while the FID signal contains information on all ^{1}H nuclei of the sample irrespective of their motional state. An immobile (rigid) proton system, for which the dipolar-dipolar interaction is not zero, does not give CPMG echoes (37). These phenomena provide the theoretical background for the separation of mobile and immobile hydrogen atoms. Both methods were used in the determination of the mobile hydrogen components.

In principle, FID could give the spectrum for the all hydrogen atoms in the sample, but the dead time of the spectrometer and the unknown mathematical form of the shape of the spectrum puts serious limitations on the measurements. The uncertainty is greater in the case of powder samples than for aqueous solutions. The solid-echo excitation helps in the case of powder samples because it reconstructs the entire FID at time 2τ for solids (see section 12.2.1 number 5) and overcomes the dead-time problem. The second derivative of the solid-echo signal taken at its maximum gives the second central moment of the rigid proton system (immobile hydrogen atoms) (30). This rigid proton system can be identified as the protein backbone in our cases.

12.2.5 Hydrogen Fractions—Nomenclature

As far as the term "bound water" is concerned, the nomenclature proposed by Cooke and Kuntz (8) is used: water molecules which are in the vicinity of, and interact strongly with, macromolecular surfaces and have detectably different properties from those of the medium are referred to as "bound water." The remaining fraction of water is called "bulk water." By the term "unfrozen water," the actual fraction of water molecules in a mobile state at a given temperature is denoted. The unfrozen water term can therefore refer to a phase composed of either bound water molecules only or bound plus bulk water molecules also.

In our approach, the aqueous solution samples are frozen in order to separate the various water phases present. The phases of ice protons, organic protons, and bound water protons are clearly separated in the FID signal by virtue of large differences in the transversal relaxation rate (Fig. 12.1). Ice protons yield a signal fraction characteristic of solids with a typical decay rate of the order of $10^5\,s^{-1}$. This signal is buried in the dead time of the spectrometer. Organic protons and/or irrotationally bound water protons (25) also yield a solid-like signal fraction with a one order of magnitude smaller, but still large decay rate. The proton NMR signal of unfrozen water has a much smaller time-domain decay rate, typically $2000\,s^{-1}$. This enables specific recording of the FID signal that belongs to bound water molecules (Fig. 12.1). The zero-time extrapolated peak amplitude of the CPMG train gives the fraction of protons that belong to bound (mobile) water molecules directly.

12.2.6 NMR—Experimental Details

The temperature was controlled by an open-cycle Oxford cryostat with an uncertainty better than ±1°C. ^{1}H NMR measurements and data acquisition were accomplished by a Bruker SXP 4-100 NMR pulse spectrometer at frequencies of 44.1 and 82.6 MHz with a stability of better than $\pm 10^{-6}$, the dead time of the spectrometer is 10 μ. The data points in the figures are based on spectra recorded by averaging signals to reach a signal-to-noise ratio better than 50. We varied the number of averaged NMR signals to achieve the desired signal quantity for each samples and unfrozen water quantities. We controlled the sensitivity of the NMR spectroscope by measuring the length of the $P_{x'}$-pulse during measurements (41) to obtain reliable $M_i(n_i)$ values. The extrapolation to zero time was done by fitting a stretched exponential function (Fig. 12.1).

12.2.7 Samples

Double-distilled water was measured to obtain calibration data and parameters for the temperature correction. Tris(hydroxymethyl)-aminoethane (Tris) solution contained 50 mM Tris (Sigma) and 1 mM ethylenediaminetetraacetic acid (EDTA, Sigma) at pH 7.0. Buffer solution contained 150 mM NaCl, 50 mM Tris, and 1 mM EDTA at pH 7.0. Samples of 100 μL were used in closed teflon capsules. The aqueous protein solutions were prepared by dissolving the proper amounts of BSA (Sigma) or ERD10 (prepared as described previously [4]) in the above buffer solution. For determining the amount of water bound per unit protein, and having noted that the measurement of IDPs is error-prone due to their unusual reactions with colorimetric dyes, we directly measured the amount of protein dissolved by determining the mass of samples lyophilized from distilled water. This measure provided the absolute concentration of the protein, which could be directly used for calculating the absolute average concentration of its constituent amino acids.

For each solution composition, we carried out the NMR measurements on three to five samples prepared independently. The data obtained were reproducible within the given statistical errors.

12.2.7.1 BSA Powder Sample: NMR Characterization The room-temperature time-domain FID and echo signals for BSA powder comprise a fast and a slow component. The fast component comes from the immobile part (protein) of the sample and the long tail (slow component) is the response of the mobile water (mobile protons) in the sample. The ratio of magnetizations M_{s0}/M_{f0} gives the measure of water content in the sample (s: slow component, f: fast component). The water origin of the slow signal component is proved by the temperature dependence of signal parameters spectrum width and intensity (Fig. 12.2). The Fourier transform of the time-domain long tail corresponds to a narrow line in the frequency domain (*fwhm* = 0.7 kHz = $0.16 10^{-4}$T) at +43°C. The absence of this narrow line at −69°C is characteristic of the stopping of proton motion (immobile water molecules). The broad spectrum (fast

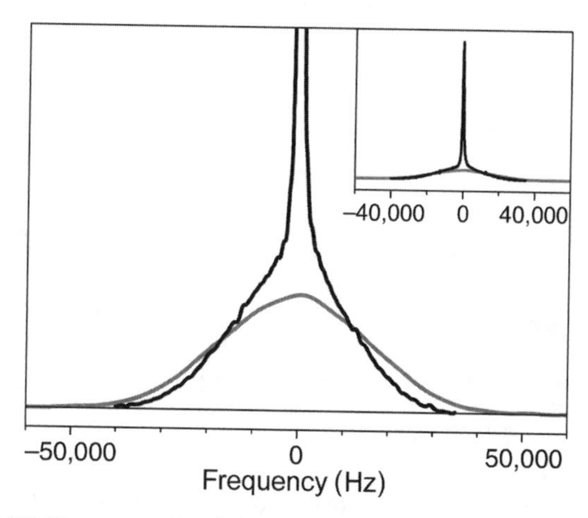

Figure 12.2 ^{1}H NMR spectra for BSA powder at +43°C (black line) and at −69°C (gray line).

component in the time domain) allows the determination of the second moment for protein protons. The estimated values are $1.04\,10^4\,\text{kHz}^2$ ($14.5\,10^{-8}\text{T}^2$) at −69°C and they are $5.9\,10^3\,\text{kHz}^2$ ($8.2\,10^{-8}\text{T}^2$) at +43°C. The value of the second moment refers to the rotation around a symmetry axis of a high-symmetry molecular group (e.g., methyl groups) in the low-temperature solid phase (15). The protons in mobile water give a substantially narrower line *fwhm* < 1 kHz and a second moment of <0.18 kHz² ($2.5\,10^{-12}\text{T}^2$). The water protons in a rigid (static) water molecule give a line of *fwhm* ~ 40 kHz and a second moment of $\sim 1.1\,10^4\,\text{kHz}^2$ ($16\,10^{-8}\text{T}^2$).

Finally, the water content of the powder sample is approximately 9.3 wt% ($h = 0.10$).

12.3 RESULTS AND DISCUSSION (EXAMPLES)

12.3.1 Unfrozen Water: NMR Intensity versus Temperature

NMR intensity versus temperature curves were measured for ERD10, BSA, and the buffer solutions (Fig. 12.3). These data were presented and interpreted previously (4). The main conclusion was that NMR enabled the direct measurement of the number of mobile (bound or unfrozen) water molecules in a wide temperature range below 0°C (Eq. 12.11). Figure 12.3 gives a direct guide for the classification of temperature subregions (*a* to *e*) and to get the fraction masses for the renormalization algorithm in the relaxation time data (see section 12.2.2.3.). There are no steps (except at the low-temperature freezing/melting) and there is no hysteresis on the curves of protein solutions, which

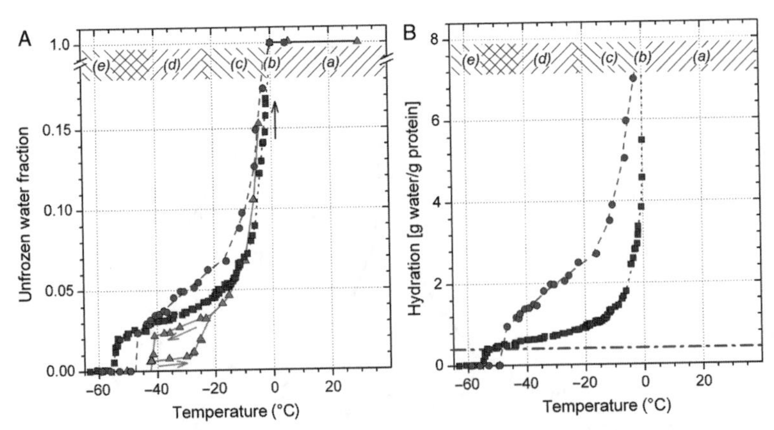

Figure 12.3 (A) Unfrozen water fraction as a function of temperature for buffer (green circles), BSA-buffer (blue squares), and ERD10-buffer (red triangles) solutions as measured by extrapolated FID and CPMG techniques. (B) The same data for BSA– and ERD10–buffer solutions, renormalized to hydration expressed as g water per g protein. The hydration level 0.4 is marked by the dash-dotted line. See color insert.

suggests that the cooling–heating process is fully reversible (except at around 0°C). The zero amplitude in region (*e*) does not mean that there is no visible NMR signal in that range, but rather that all the molecular motions of water are absent. The slope of the low-temperature part of the curves in region (*d*) (i.e., the temperature dependence of the number of contributing protons) reveals the heterogeneity of the mobile states of the bound water molecules. This slope is more pronounced (approximately 2–3 times higher) and, consequently, the heterogeneity of the mobile states of the bound water molecules is higher for the intrinsically disordered ERD10 than for the globular BSA. There are no zero-slope regions on these curves, so new water (proton spin) fractions of different motional characteristics are added as temperature changes. The existence of a "single-spin fraction" in these cases can be stated by the measurements of the other NMR characteristics, for example, longitudinal relaxation rate. The results also establish that there is no cooperative phase transition of the protein structure in the investigated temperature range, otherwise step-like changes would have been seen on these smooth curves.

The single low-temperature step at –43°C for buffer solution, at –54°C for BSA solution, and at –48°C for ERD10 solution (Fig. 12.3) suggests that all the molecular motions of water start suddenly at these temperatures, which is a direct experimental evidence of the change in the motional state of hydration water molecules from a solid-like to a fluid-like state. At low temperatures, proteins exist in a "glassy state" (18), which is a solid-like structure without conformational flexibility. The onset of liquid-like motion takes place at ~ –53°C for hydrated globular proteins (e.g., myoglobin and crystalline ribonuclease A) (11, 12, 27, 34). This "dynamical" transition of proteins is believed to be triggered by their strong coupling with the hydration water through a network of hydrogen bonds. Chen et al. (6, 7, 20) suggested that the dynamic transitions are not intrinsic properties of the biomacromolecules themselves but are imposed by the hydration water on their surfaces. Moreover, we observed the same transition for the buffer solution and its temperature falls within the interval suggested for biomacromolecules. NMR intensity values for the buffer solution are not analyzed here because it was done rather extensively in Reference 43. It was found that both the thermal hysteresis in the NMR intensity curve and the small endothermic peak in the DSC trace, which is characteristic of the buffer solution, disappeared in the protein–buffer solution at a given protein concentration. The limits and character of disappearance are markedly different for the globular BSA and the disordered ERD10.

The above dynamic crossover of hydration water deserves special attention as our NMR intensity data recorded at around its temperature provide direct experimental values of the minimum amount of solvent water needed to restore the conformational flexibility of proteins essential for their functions. The amounts of unfrozen water fractions measured just above the temperature where molecular motions of water start are $h = 0.020 \pm 0.002 = 2.5 \pm 0.2$ (mol H_2O)/(mol amino acid) = 0.39 ± 0.04 (g H_2O)/(g protein) for BSA solution and $h = 0.024 \pm 0.002 = 5.9 \pm 0.5$ (mol H_2O)/(mol amino acid) = 0.94 ± 0.08 (g H_2O)/(g protein) for ERD10 solution. The value of h measured for BSA solution agrees well with $h \approx 0.3$–0.4 g water per g protein (19) generally accepted for globular proteins to be sufficient to cover most of the protein surface with a single layer of water molecules, thought to be necessary and sufficient for protein activity (21, 29, 35). The IDP ERD10 has a remarkably higher starting h value, probably greater as a consequence of its larger solvent-accessible surface area than that of globular proteins. To our knowledge, this is the first measurement of fragile-to-dynamic crossover of hydration water for an IDP. The higher temperature for the start of proton/water molecular motion in ERD10 than in BSA suggests a higher viscosity (more dense landscape) on the surface of IDP.

12.3.2 NMR Spectrum Details

Examples of measured FID signals and the parameters of fitted stretched exponential curves ($a \cdot \exp(t \cdot R_{2\mathrm{eff}})^c$) are presented in Figures 12.4A–D at about −35 and −21°C. The lower-temperature examples were selected to coincide with the extreme of the fitted R_1 versus T curve. The room-temperature results

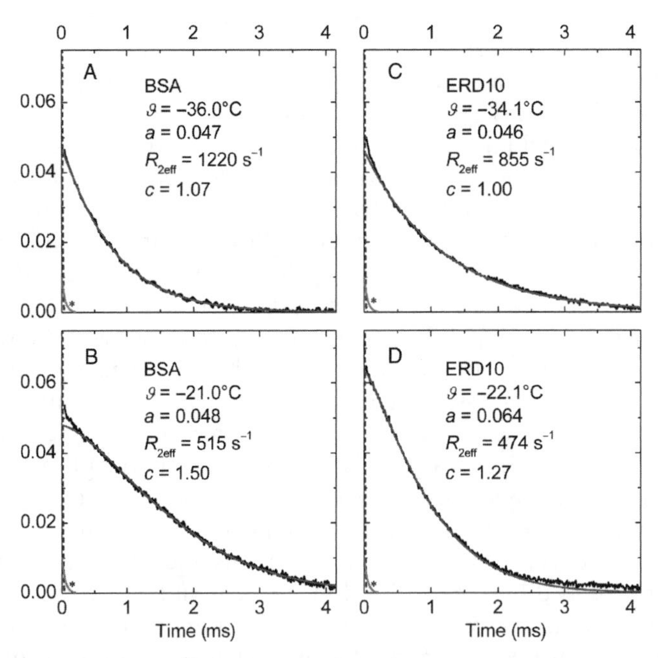

Figure 12.4 FID signals (thin line) of protons in BSA-buffer (A and B) and ERD10-buffer solutions (C and D) at two characteristic temperatures. Typical FID signals for protons of ice (dashed heavy line) and of a protein (heavy line marked with a star) are given for comparison. The FID signal of protons in the mobile water fraction of the solutions is approximated by stretched exponential function $a\ \exp(-tR_{2\mathrm{eff}})^c$ (heavy lines) and the fraction of unfrozen water is obtained as a values (see text). See color insert.

are not given here because the long tails in the time domain are strongly influenced by the inhomogeneous external magnetic fields and so they do not reveal the real biophysical picture. The transversal relaxation rate is substantially smaller at room temperature than in the lower temperature region (Fig. 12.4) as evaluated by Equation 12.4 ($i = 2$). The contribution of inhomogeneous, external magnetic fields is relatively small at sub-zero temperatures and qualitative conclusions can be drawn from these results as follows. All the samples (BSA and ERD10 solutions, and buffer [not shown here]) can be characterized by a single exponential function. The c exponents change from 1.00 at the lowest temperatures to ≈2 at just below 0°C (from a Lorentzian to a nearly Gaussian line in the frequency domain). The Gaussian character comes from the inhomogeneity of the magnetic fields, and the Lorentzian signal is characteristic of a spin system in which motional narrowing takes place. It is of practical importance that the Lorentzian character was found at lower temperatures and the near Gaussian at higher ones. Moreover, the smallest transversal relaxation rate, that is, the highest proton mobility exists in the buffer solution, the next one is in the ERD10 solution, and the lowest values were found in the BSA solution. The inhomogeneous field contribution and the real transversal relaxation rate can be separated as usual by $R_{2\text{eff}} = R_2 + R_2'$, where $R_{2\text{eff}}$ means the measured value (these are given in Fig. 12.3) and the quantity without upper index is the real transversal relaxation rate as it is measured by the CPMG sequence. We have to state that the inhomogeneities do not influence the extrapolated amplitude, that is, the FID amplitude measurements.

12.3.3 CPMG Train and Translational Diffusion

The CPMG results together with the measurement characteristics and fitting parameters of Equation 12.4 are given in Figure 12.5 for ERD10 solutions (for buffer and for BSA, the results are qualitatively similar but quantitatively different). Each curve is based on approximately 2000–10,000 measured echo amplitudes obtained by averaging few hundreds of echo trains, so the scatter of points on the figures practically represents the standard error. Equation 12.4 was used to get transversal relaxation rates and for the demonstration of existence or absence of translational diffusion. The CPMG amplitude train reveals systematic τ dependence above 0°C that proves the existence of translational diffusion of water molecules at these temperatures (Fig. 12.5B). The τ dependence (the translational diffusion) is absent below –4°C, as demonstrated in Figure 12.5A, which suggests that the motional narrowing of the hydration-shell protons comes from reorientation of bound water molecules. Two real transversal relaxation rate components can be found for ERD10 solutions and only one for BSA solutions. The magnitude of the transversal relaxation rates shows similar trends for both samples as was mentioned in the interpretation of FID signals in connection with the proton mobility.

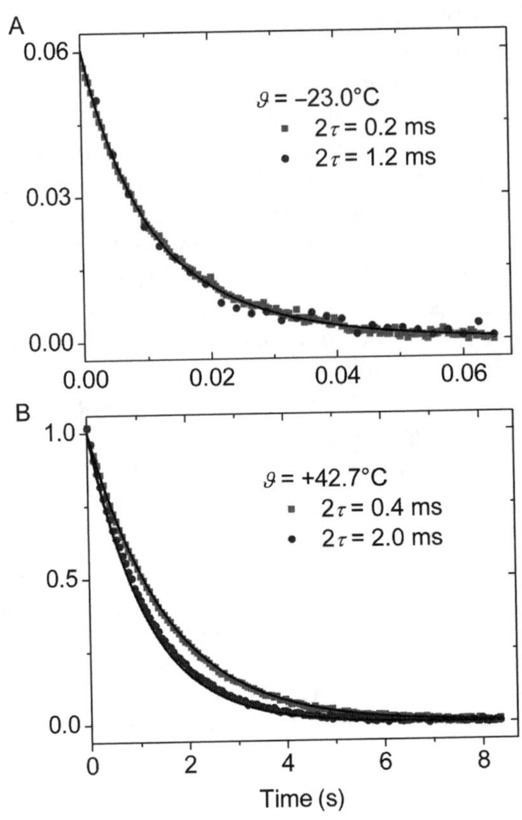

Figure 12.5 CPMG echo trains for ERD10–buffer solutions at characteristic temperatures of (A) −23.0°C and (B) +42.7°C. Pulse delay times of $2\tau = 0.2, 0.4, 0.6, 1.2$ and 2.0 ms were applied. Lines are fits of Equation (12.5).

The magnitude of the magnetic field gradient can be estimated from the combination of FID and CPMG transversal relaxation rate measurements for BSA solutions at about −23°C where both methods give a single-rate value. The measured quantities ($R_{2\text{eff}} = 515\,\text{s}^{-1}$ and $R_2 = 369\,\text{s}^{-1}$) were substituted into equation $R_{2\text{eff}} = R_2 + R_2'$ and the result $R_2 \approx 146\,\text{s}^{-1}$ was obtained. This value gives an inhomogeneous line broadening of approximately 170 Hz, which corresponds to a 2 ppm magnetic field gradient inside the sample. Only its existence and magnitude are important in this context. Its existence reveals the reason of τ dependence of CPMG results at ambient temperatures (of course, it is not the reason of the translational diffusion). The results demonstrate the existence of the translational diffusion term only (Fig. 12.5B) when using Equation 12.4. For more exact measurements, the application of pulsed-field gradient technique (38) is necessary. The real transversal relaxation rates (free from the inhomogeneous contribution) show trends similar to those mentioned in connection with FID signals. However, at low temperatures, the narrower time window for ERD10 gives two transversal relaxation rates and BSA has only one as a consequence of a different structure. At higher temperatures, the measurements on both samples can be fitted by a single value because of the wider time window.

12.3.4 Relaxation Rates and Hydrogen Dynamics

Longitudinal (both laboratory and rotating frame) and transversal relaxation rates were measured as a function of temperature for BSA (Fig. 12.6) and ERD10 (Fig. 12.7) solutions (4). The data corrected according to the renormalization (*gray lines*; see section 12.2.2.3) and the fitted curves using Equation 12.1 (*black lines*) are also given for R_1. We did the same correction for R_2 data but it resulted in differences much smaller than the errors of experimental values. The borderlines on the figures separate the temperature intervals from

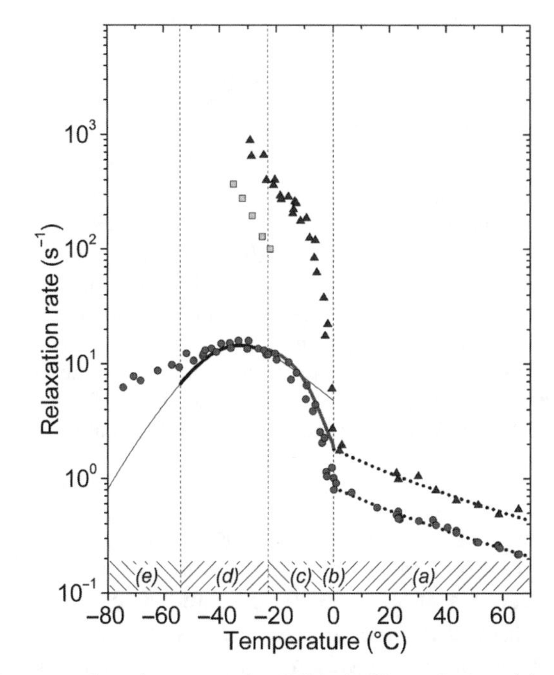

Figure 12.6 Proton relaxation rates for BSA–buffer solution (circles: longitudinal relaxation rate R_1; squares: rotating-frame longitudinal relaxation rate $R_{1\rho}$; triangles: transversal relaxation rate R_2). Letters (*a*) through (*d*) denote temperature regions within which the contribution of water phases to the NMR signal show characteristic differences (see text). Dotted black lines in region (*a*) correspond to Equation 12.1 fitted to measured R_1 and R_2 data (parameters see in Table 12.2). Gray line in region (*c*) gives the corrected longitudinal relaxation rates as calculated by Equation 12.10 (see also section 12.2.3). Heavy, solid black line in region (*d*) is a fit of Equation (12.1) to measured R_1 data (parameters see in Table 12.2). Thin black lines are extrapolations of this fit.

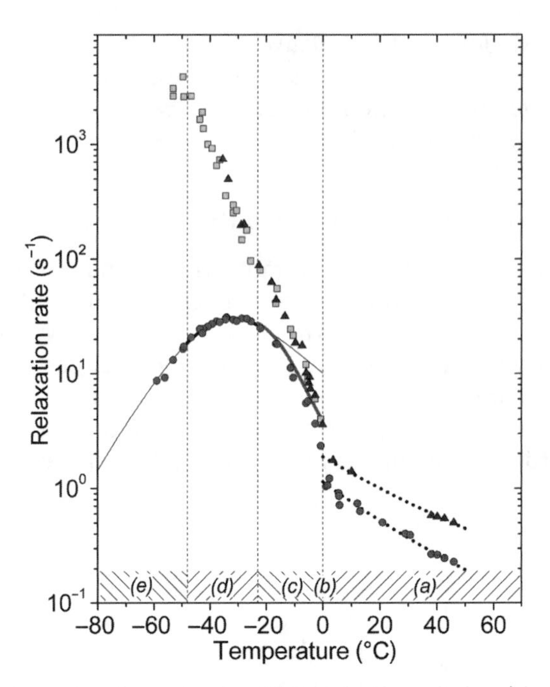

Figure 12.7 Proton relaxation rates for ERD10–buffer solution (circles: longitudinal relaxation rate R_1; squares: rotating-frame longitudinal relaxation rate $R_{1\rho}$; triangles: transversal relaxation rate R_2). Letters (*a*) through (*d*) denote temperature regions within which the contribution of water phases to the NMR signal show characteristic differences (see text). Dotted black lines in region (*a*) correspond to Equation 12.1 fitted to measured R_1 and R_2 data (parameters see in Table 12.2). Gray line in region (*c*) gives the corrected longitudinal relaxation rates as calculated by Equation 12.12. Heavy, solid black line in region (*d*) is a fit of Equation 12.1 to measured R_1 data (parameters see in Table 12.2). Thin black lines are extrapolations of this fit.

(*a*) to (*e*) as shown in Figure 12.3. The experimental data were collected for 3–5 independently prepared samples, and the experimental points in the figures are the averages of 5000–50,000 detected spectra.

In region (*e*), using the inversion-recovery sequence for the R_1 measurements, the NMR response (sampled near to the second pulse) is coming from the mobile protein protons; no mobile water proton contribution exists at these temperatures. There is a substantial difference in the transitions from regions (*d*) to (*e*) for BSA and ERD10, namely a continuation of the fitted R_1 versus T curve for ERD10 and a deviation for BSA. Probably, the continuation of the curve proves a much stronger spin-diffusion connection (more effective protein-water proton spin exchange) in the case of ERD10 than in BSA. In this transition temperature range (from *d* to *e*), the protons in protein and hydration shell behave as a single-spin system for ERD10 and not for BSA solutions.

The R_1 vervus T curve in region (*d*) can be described well by Equation 12.1a. The spin fraction corresponds to the first hydration shell of the protein–buffer complex. The reorientation of bound water molecules can be characterized by three quantities: activation energy (E), correlation time (τ_{q0}), and the strength of the interaction between the proton spins σ_q^2. These quantities are summarized in Table 12.1 for both proteins. The quantities $(R_2/R_1)_{R_{1\max}}$ and $(R_{1\rho}/R_1)_{R_{1\max}}$, that is, the ratios at the maximum of the longitudinal relaxation rates are 12 and 9 for ERD10 and 82 and 29 for BSA (Table 12.2). The ratio $(R_2/R_1)_{R_{1\max}}$ has to be 3/2 for a single-spin system and since $\omega_1\tau_q$ is very nearly zero, the statement is also valid for $(R_{1\rho}/R_1)_{R_{1\max}}$—the two ratios are equal (see for instance ERD10, Fig. 12.7). If we suppose another process with a much

TABLE 12.1 High-Temperature Relaxation Rates, Activation Energies (Above 0°C, Region [*a*]) and Low-Temperature Dynamic Parameters Calculated from Proton Relaxation Rates (Below –23°C, Region [*d*]) for BSA and ERD10 Solutions

Temperature Region	Parameter	BSA Solution	ERD10 Solution
Room temperature	R_1 [s^{-1}]	0.49 ± 0.02	0.51 ± 0.03
	R_2 [s^{-1}]	0.99 ± 0.10	0.97 ± 0.10
Region (*a*) R_1	E [kJ/mol]	15.7 ± 0.4	24.9 ± 0.6
	E [kJ/mol]	17.3 ± 0.4	21.9 ± 0.9
Region (*a*) R_2'	τ_{00} [10^{-13} s]	24 ± 15	5 ± 5
	E [kJ/mol]	17 ± 2	20 ± 2
Region (*d*) R_1	σ_q^2 [10^{-8} T^2]	10.5 ± 0.2	22.0 ± 0.8
	τ_{q0} [10^{-16} s]	8 ± 2	6 ± 5
	E [kJ/mol]	29.3 ± 0.5	30.2 ± 1.6

TABLE 12.2 Correlation Times $\tau_{q1,2}$ and Interaction Strengths $\sigma_{q1,2}^2$ Obtained by the Analysis of Relaxation Rate Values at the Temperature of the Longitudinal Relaxation Rate Maximum

	BSA	ERD10
$\omega_0/2\pi$ [MHz]	82.557	82.570
$T(R_1 = R_{1\max})$ [°C]	–33.0 ± 0.1	–32.0 ± 0.1
$R_{1\max}$ [s^{-1}]	10.5 ± 0.5	30.3 ± 1.5
$R_{1\rho}$ [s^{-1}]	302 ± 45	281 ± 42
R_2 [s^{-1}]	856 ± 86	375 ± 37
$R_{1\rho}/R_{1\max}$	29 ± 6	9 ± 2
$R_2/R_{1\max}$	82 ± 16	12 ± 2
$\tau_{q1} = 1/\omega_0$ [10^{-9}s]	1.9278 ± 0.0004	1.9275 ± 0.0004
σ_{q1}^2 [10^{-8} T^2]	10.6 ± 0.5	22 ± 1
τ_{q2} [10^{-6}s]	0.31 ± 0.03	1.0 ± 0.1
σ_{q2}^2 [10^{-8} T^2]	4.2 ± 0.7	0.44 ± 0.07

longer residence time (long τ_{q2}), then both R_2 and $R_{1\rho}$ contain additional terms (19) described by the parameters σ_{q2}^2 and τ_{q2} (Table 12.2) and give a formal explanation for our measured differences. Similar relaxation rate studies of the bound water content in human fish and bird eye lenses (31–33) showed the same behavior and the above calculation yielded parameter values of the same magnitude as for BSA now. This is an expectable agreement since bird lenses contain about 30% protein and 70% water. The other possible reason of the deviations is that the distribution of correlation times is insufficient to be applied for the whole region (*c*) because it would give a smaller slope in the high-temperature side of the R_1 versus T curve, in contrast to the measurements. The smaller slope in the low-temperature side followed in region (*e*) for BSA solution cannot be explained by a simple distribution of activation energies. It is remarkable that R_2 and $R_{1\rho}$ are not equal at the maximum of the longitudinal relaxation rate for the BSA solution, but these quantities are nearly equal for the ERD10 solution.

R_2 versus T and $R_{1\rho}$ versus T curves in region (*d*) seem to be exponential, but give the activation energies of 58 ± 1 kJ/mol for BSA solution and 75 ± 1 kJ/mol for ERD10 solution, which are approximately twice as high as the values obtained from R_1 versus T curves. The resolution of the apparent contradiction is that these relaxation times cannot be described by a single Arrhenius function because they contain two temperature-dependent terms, as it was shown in the former paragraph.

In region (*c*), no conclusions are drawn from these results because of the multivariable character existing in this range. Anyhow, the corrected curves are between the measured points and the fitted curve calculated by Equation 12.1a showing that the different spin fractions can be characterized by different longitudinal relaxation times and that a simple relaxation process described by higher activation energy cannot be used.

The room temperature R_1 and R_2 values are given in Table 12.1 together with the activation energies in region (*a*) derived from the measured longitudinal (R_1) and the total (R_2) and secular (R_2', Eq. [12.1c]) transversal relaxation rate data. They all give the same results within experimental error. R_1, R_2, and R_2' show a simple temperature dependence described by Arrhenius function Equation 12.3, which is acceptable if $\omega\tau_q$ is much smaller than one in Equation 12.1. The temperature dependence of the curves gives the activation energy for a complex molecular motion, namely the mixture of translation and reorientation fast enough to mimic the behavior of a single-spin system. The ratio of R_2/R_1, however, equals 2 within a good approximation for both proteins and it is not 3/2 as it would be for a single-spin system of one correlation time. The invitation of a slow motion component into R_2 and R_2', which enhances the transversal relaxation rates, offers the solution again as it was the situation in region (*d*). An interpretation based on Equation 12.8 can be excluded because the measured relaxation rate is proportional to the correlation time in that case, which is not true for our results. The magnitudes of R_1 and R_2 (Table 12.1) are nearly the same for both proteins at room temperature. The concentration dependence of relaxation rates (11, 12, 27, 34) leads to a difference of approximately 100% in the intrinsic relaxation rates of water in the ERD10 and the BSA solutions when corrected with the actual protein concentrations of 25 mg ERD10 and 50 mg BSA per milliliter of solution. This means that ERD10 solution has a relaxation rate twice as fast as that of BSA solution at the same protein concentration. The greater relaxation rate is due to longer correlation time, that is, water molecules spend longer time at the protein-water interface of ERD10 than that of BSA. This is a substantial difference measurable at ambient temperature which offers an opportunity of rapid and simple identification for the folded or unfolded nature of proteins not yet characterized.

The novel approach is based on the simultaneous measurements of different NMR characteristics in a wide temperature range. It gives a more detailed picture on the protein-water interface compared with that obtained form the measurement and interpretation of a single quantity at ambient temperature (e.g., NMRD measurements). The main results are the quantitative determination of the number of hydration water molecules, the elements of hydration water dynamics (activation energy and correlation times), and the differences in dynamics as seen by the different time windows provided by the different types of relaxation rates (R_1, $R_{1\rho}$, and R_2).

12.3.5 Conclusion: Main Differences between Globular Proteins and IDPs

Understanding the functioning of IDPs demands a thorough characterization of their surface properties with particular reference to their water interface. The main conclusions of our studies are the larger size of their hydration shell, the stronger spin diffusion interaction between protein and bound water protons in the low-temperature ranges (*d*) and (*e*), and the higher relaxation rates and activation energies in the high-temperature range (*a*). The different values for the slow-motion characteristics for BSA and for ERD10 in region (*d*) are of potential further significance.

All these specific features add up to characterize the unique surface properties of IDPs. Due to their unfolded and largely exposed nature and their frequent functioning in molecular recognition (1, 18), such a detailed description of the structure and dynamics of their water interface will advance our understanding of the atomic-level mechanism of their function.

The extension of our studies to further globular proteins and IDPs will be necessary and is planned in order to ascertain that the aforementioned differences are characteristic of the interfaces of the two groups of proteins. Molecular dynamics calculations will help deepen our understanding of the measured physical parameters. Building blocks of proteins (amino acids) and lager architectural elements (e.g., tripeptides) will be investigated by the novel approach to get the final conclusions, the main interface differences between the two classes of proteins. The extension of the novel approach to the other NMR methods mentioned (NMRD and NOE) would also be promising. A further globular and ID protein pair has been investigated presently to extend our database: the small and well-known globular ubiquitin (40) and a Drosophila protein of chromatin decondensation, which was newly identified as IDP (39).

ACKNOWLEDGMENTS

This research was supported by grants OTKA K60694 from the Hungarian Scientific Research Fund, ETT 245/2006 from the Hungarian Ministry of Health, and the International Senior Research Fellowship ISRF 067595 from the Wellcome Trust.

REFERENCES

1. Angell, C. A. 1995. Formation of glasses from liquids and biopolymers. Science **267**:1924–35.
2. Arcangeli, C., A. R. Bizzarri, and S. Cannistraro. 1998. Role of interfacial water in the molecular dynamics-simulated dynamical transition of plastocyanin. Chem Phys Lett **291**:7–14.
3. Bizzarri, A. N., and S. Cannistraro. 2002. Molecular dynamics of water at the protein-solvent interface. J Phys Chem B **106**:6617–33.
4. Bokor, M., V. Csizmók, D. Kovács, P. Bánki, P. Friedrich, P. Tompa, and K. Tompa. 2005. NMR relaxation studies on the hydrate layer of intrinsically unstructured proteins. Biophys J **88**:2030–7.
5. Carr, H. Y., and E. M. Purcell. 1954. Effects of diffusion on free precession in nuclear magnetic resonance experiments. Phys Rev **94**:630–8.
6. Chen, S.-H., L. Lip, X. Chu, Y. Zhang, E. Fratini, P. Baglioni, A. Faraone, and E. Mamontov. 2006. Experimental evidence of fragile-to-strong dynamic crossover in DNA hydration water. J Chem Phys **125**:171103.
7. Chen, S.-H., L. Lip, E. Fratini, P. Baglioni, A. Faraone, and E. Mamontov. 2006. Observation of fragile-to-strong dynamic crossover in protein hydration water. Proc Natl Acad Sci U S A **103**:9012–6.
8. Cooke, R., and I. D. Kuntz. 1974. The properties of water in biological systems. Annu Rev Biophys Bioeng **3**:95–126.
9. Csizmók, V., M. Bokor, P. Bánki, É. Klement, K. F. Medzihradszky, P. Friedrich, K. Tompa, and P. Tompa. 2005. Primary contact sites in intrinsically unstructured proteins: the case of calpastatin and microtubule-associated protein 2. Biochemistry **44**:3955–64.
10. Dill, K. A., T. M. Truskett, V. Vlachy, and B. Hribar-Lee. 2005. Modelling water, the hydrophobic effect and ion solvatation. Annu Rev Biophys Biomol Struct **34**:173–99.
11. Doster, W., S. Cusack, and W. Petry. 1990. Dynamic instability of liquidlike motions in a globular protein observed by inelastic neutron scattering. Phys Rev Lett **65**:1080–3.

12. Doster, W., S. Cusack, and W. Petry. 1989. Dynamical transition of myoglobin revealed by inelastic neutron scattering. Nature **337**:754–6.
13. Dunker, A. K., J. D. Lawson, C. J. Brown, R. M. Williams, P. Romero, J. S. Oh, C. J. Oldfield, A. M. Campen, C. M. Ratliff, K. W. Hipps, J. Ausio, M. S. Nissen, R. Reeves, C. Kang, C. R. Kissinger, R. W. Bailey, M. D. Griswold, W. Chip, E. C. Garner, and Z. Obradovic. 2001. Intrinsically disordered protein. J Mol Graphics Model **19**:26–59.
14. Goyal, K., L. Tisi, A. Basran, J. Browne, A. Burnell, J. Zurdo, and A. Tunnacliffe. 2003. Transition from natively unfolded to folded state induced by desiccation in an anhydrobiotic nematode protein. J Biol Chem **278**:12977–84.
15. Grüner, G., and K. Tompa. 1968. Molekuláris mozgások vizsgálata szilárdtestekben NMR módszerrel (Study of molecular motions in solids by the NMR method). Kémiai Közlemények **30**:315–56.
16. Hahn, E. L. 1950. Spin echoes. Phys Rev **80**:580–94.
17. Halle, B. 2004. Protein hydration dynamics in solutions. Phil Trans R Soc Lond B **359**:1207–24.
18. Iben, I. E. T., D. Braunstein, W. Doster, H. Frauenfelder, M. K. Hong, J. B. Johnson, S. Luck, P. Ormos, A. Schulte, P. J. Steinbach, A. H. Xie, and R. D. Young. 1989. Glassy behavior of a protein. Phys Rev Lett **62**:1916–19.
19. Knispel, R. R., R. T. Thomson, and M. M. Pintar. 1974. Dispersion of proton spin-lattice relaxation in tissues. J Magn Reson **14**:44–51.
20. Kumar, P., Z. Yan, L. Xu, M. G. Mazza, S. V. Buldyrev, S.-H. Chen, S. Sastry, and H. E. Stanley. 2006. Glass transition in biomolecules and the liquid-liquid critical point of water. Phys Rev Lett **97**:177802.
21. Kuntz, I. D. J., and W. Kauzmann. 1974. Hydration of proteins and polypeptides. Adv Protein Chem **28**:239–345.
22. Mallamace, F., S.-H. Chen, M. Broccio, C. Corsaro, V. Crupi, D. Majolino, V. Venuti, P. Baglioni, E. Fratini, C. Vannucci, and H. E. Stanley. 2007. Role of the solvent in the dynamical transitions of proteins: the case of the lysozyme-water system. J Chem Phys **125**:045104.
23. McCubbin, W. D., C. M. Kay, and B. G. Lane. 1985. Hydrodynamic and optical properties of the wheat germ Em protein. Can J Biochem Cell Biol **63**:803–11.
24. Meiboom, S., and D. Gill. 1958. Modified spin-echo method for measuring nuclear relaxation times. Rev Sci Instrum **29**:688–91.
25. Noack, F. 1971. Nuclear magnetic relaxation spectroscopy, pp. 83–144. *In* P. Diehl, E. Fluck, and F. Kosfeld (eds.), NMR Basic Principles and Progress, vol. 3. Springer Verlag, Berlin.
26. Pal, S. K., J. Preon, B. Bagchi, and A. H. Zewall. 2002. Biological water: femtosecond dynamics of macromolecular hydration. J Phys Chem B **106**:12376–95.
27. Parak, F., and E. W. Knapp. 1984. A consistent picture of protein dynamics. Proc Natl Acad Sci U S A **81**:7088–92.
28. Pfeifer, H. 1972. Nuclear magnetic resonance and relaxation of molecules absorbed on solids, pp. 53–153. In P. Diehl, E. Fluck, and F. Kosfeld (eds.), NMR Basic Principles and Progress, vol. 7. Springer Verlag, Berlin.
29. Poole, P. L., and J. L. Finney. 1983. Sequential hydration of a dry globular protein. Biopolymers **22**:255–60.

30. Powles, J. G., and P. Mansfield. 1962. Double-pulse nuclear-resonance transients in solids. Phys Lett **2**:58–9.

31. Rácz, P., K. Tompa, and I. Pócsik. 1979. The state of water in normal and senile cataractous lenses studied by nuclear magnetic resonance. Exp Eye Res **28**:129–35.

32. Rácz, P., K. Tompa, and I. Pócsik. 1979. The state of water in normal human, bird and fish eye lenses. Exp Eye Res **29**:601–8.

33. Rácz, P., K. Tompa, I. Pócsik, and P. Bánki. 1983. Water fractions in normal and senile cataractous eye lenses studied by NMR. Exp Eye Res **36**:663–9.

34. Rasmussen, B. F., A. M. Stock, D. Ringe, and G. A. Petsko. 1992. Crystalline ribonuclease A loses function below the dynamical transition at 220 K. Nature **357**:423–4.

35. Rupley, J. A., and G. Careri. 1991. Protein hydration and function. Adv Protein Chem **41**:37–172.

36. Schröder, C., T. Rudas, S. Boresch, and O. Steinhauser. 2006. Simulation studies of protein-water interface. 1. Properties at the molecular resolution. J Chem Phys **124**:234907.

37. Slichter, C. P. 1990. Principles of Magnetic Resonance, third enlarged and updated edition, vol. 1. Springer Verlag, Berlin.

38. Stejskal, E. O., and J. E. Tanner. 1965. Spin diffusion measurements: Spin echoes in the presence of a time-dependent field gradient. J Chem Phys **42**:288–92.

39. Szőllősi, E., M. Bokor, A. Bodor, A. Perczel, É. Klement, K. F. Medzihradszky, K. Tompa, and P. Tompa. 2007. Intrinsic structural disorder of DF31, a Drosophila protein of chromatin decondensation and remodeling activities. J Proteome Res **7**:2291–9.

40. Tompa, K., P. Bánki, M. Bokor, P. Kamasa, G. Lasanda, E. Szőllősi, and P. Tompa. 2009. Interfacial water at protein surfaces: wide-line NMR and DSC characterization of the hydration of ubiquitin. Biophys J **96**:2789–98.

41. Tompa, K., P. Bánki, M. Bokor, G. Lasanda, and J. Vasáros. 2003. Diffusible and residual hydrogen in an amorphous Ni(Cu)-Zn-H alloy. J Alloys Compd **350**:52–5.

42. Tompa, P. 2002. Intrinsically unstructured proteins. Trends Biochem Sci **27**:527–33.

43. Tompa, P., P. Bánki, M. Bokor, P. Kamasa, D. Kovács, G. Lasanda, and K. Tompa. 2006. Protein-water and protein–buffer interactions in the aqueous solution of an intrinsically unstructured plant dehydrin: NMR intensity and DSC aspects. Biophys J **91**:2243–9.

44. Tunnacliffe, A., and M. J. Wise. 2007. The continuing conundrum of the LEA proteins. Naturwissenschaften **94**:791–812.

45. Uversky, V. N. 2002. Natively unfolded proteins: a point where biology waits for physics. Protein Sci **11**:739–56.

46. Uversky, V. N., J. R. Gillespie, and A. L. Fink. 2000. Why are "natively unfolded" proteins unstructured under physiologic conditions? Proteins **41**:415–27.

PART III

SINGLE-MOLECULE TECHNIQUES

13

SINGLE-MOLECULE SPECTROSCOPY OF UNFOLDED PROTEINS

Benjamin Schuler
Biochemisches Institut, Universität Zürich, Zürich, Switzerland

ABSTRACT

Single-molecule spectroscopy has developed into a versatile method to probe distances, distance distributions, and dynamics of unfolded proteins. Single-molecule Förster resonance energy transfer has been used most extensively to study long-range intramolecular distance distributions and dynamics of unfolded proteins from timescales of nanoseconds to seconds. The methods developed in this context will also be helpful for studying the behavior of intrinsically disordered proteins.

13.1 INTRODUCTION

The folding of a protein is the first part of its function: it has to find its native three-dimensional structure—encoded in the amino acid sequence—to be able to act as an enzyme, a signal transducer, a membrane channel, or a stabilizing element of a cell, to name but a few of the many roles that proteins can take. Contrary to widespread belief, folding is not a unique, singular event in the life of a protein. Many proteins are marginally stable and will fold and unfold many times during their functional life. As described in the other chapters of

Instrumental Analysis of Intrinsically Disordered Proteins: Assessing Structure and Conformation, Edited by Vladimir Uversky and Sonia Longhi

this book, in many cases proteins even fold only in the presence of stabilizing ligands or binding partners, illustrating the close coupling of folding and function. Correspondingly, the unfolded or denatured state is of central relevance to understanding the protein-folding reaction, especially in the context of intrinsically disordered proteins (IDPs) (24, 102).

13.2 SINGLE-MOLECULE SPECTROSCOPY

In the past decade, single-molecule methods have increasingly been employed to study protein folding. The two main methods used are force-probe techniques and Förster resonance energy transfer (FRET). Experiments using atomic force microscopy and laser tweezers have provided a lot of previously inaccessible information on the mechanical stability and folding of proteins, and the behavior of unfolded polypeptides under force, and the reader is referred to a number of recent reviews on this topic (7, 11, 27, 28, 111). Here, we focus on the investigation of protein folding and especially unfolded proteins using single-molecule FRET (Fig. 13.1). First demonstrated about 10 years ago (35), single-molecule FRET has since become an important approach to study intramolecular distances and dynamics in biomolecules (106), including protein folding (37, 62, 76, 86–88).

13.2.1 FRET

The first quantitative test of Förster's theory (29, 103) and the crucial experiment that put FRET on the map of biochemistry was published by Stryer and Haugland in 1967 (97). They attached dansyl and naphthyl groups to the

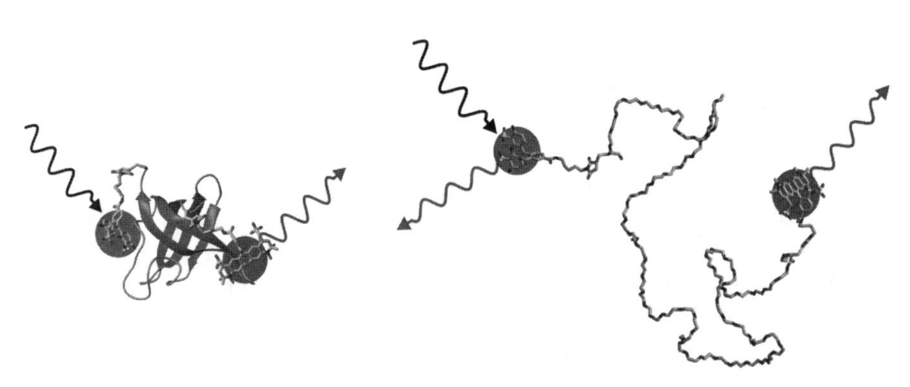

Figure 13.1 Schematic structures of folded and unfolded protein labeled with donor (Alexa 488) and acceptor (Alexa 594) dyes. (A) Folded CspTm, a five-stranded, 66-residue β-barrel protein (PDB-code 1G6P) (49), (B) unfolded CspTm. A laser excites the donor dye, which can transfer excitation energy to the acceptor dye.

termini of polyproline peptides and measured the transfer efficiency between them as a function of the length of the peptide. As predicted for the dipole–dipole coupling between the donor and the acceptor dye (29), they found the transfer efficiency E to depend on the inverse sixth power of the interchromophore distance r, in agreement with Theodor Förster's famous equation

$$E = \frac{R_0^6}{R_0^6 + r^6} \tag{13.1}$$

where R_0 is the Förster radius, which is the characteristic distance that results in a transfer efficiency of 50%. R_0 is calculated in Förster's theory according to

$$R_0^6 = \frac{9000(\ln 10)\kappa^2 Q_D J}{128\pi^5 n^4 N_A} \tag{13.2}$$

where J is the overlap integral between the donor emission and the acceptor absorption spectra, Q_D is the donor's fluorescence quantum yield, κ^2 is the orientation factor depending on the relative orientation of the chromophores, n the refractive index of the medium between the dyes, and N_A is Avogadro's constant (29, 103). The idea of such a "spectroscopic ruler" (97) has had a huge impact on the investigation of biomolecular structure and dynamics on distances in the range of about 1–10 nm (36, 53, 92, 96). More recently, renewed interest has come from the realization that FRET can be used for obtaining distance information in experiments on single biomolecules (21, 35) including proteins (39, 61, 111).

Experimentally, transfer efficiencies can be determined in a variety of ways (103), but for single-molecule FRET, two approaches are particularly useful. One is the measurement of the fluorescence intensities from both the donor and the acceptor chromophores, and the calculation of the transfer efficiency according to

$$E = \frac{n_A}{n_A + n_D} \tag{13.3}$$

where n_A and n_D are the numbers of photons detected from the acceptor and donor chromophore, respectively, corrected for the quantum yields of the dyes, the efficiencies of the detection system in the corresponding wavelength ranges, and related effects (23, 85). A second approach, which can be combined with the first (108), is the determination of the fluorescence lifetime of the donor in the presence (τ_{DA}) and absence (τ_D) of the acceptor, yielding the transfer efficiency as

$$E = 1 - \frac{\tau_{DA}}{\tau_D} \tag{13.4}$$

Especially for unfolded proteins, we are usually dealing with broad distributions of distances. Correspondingly, we have to analyze the transfer efficiencies in terms of distance distributions (4, 42, 89, 90). I would like to stress that

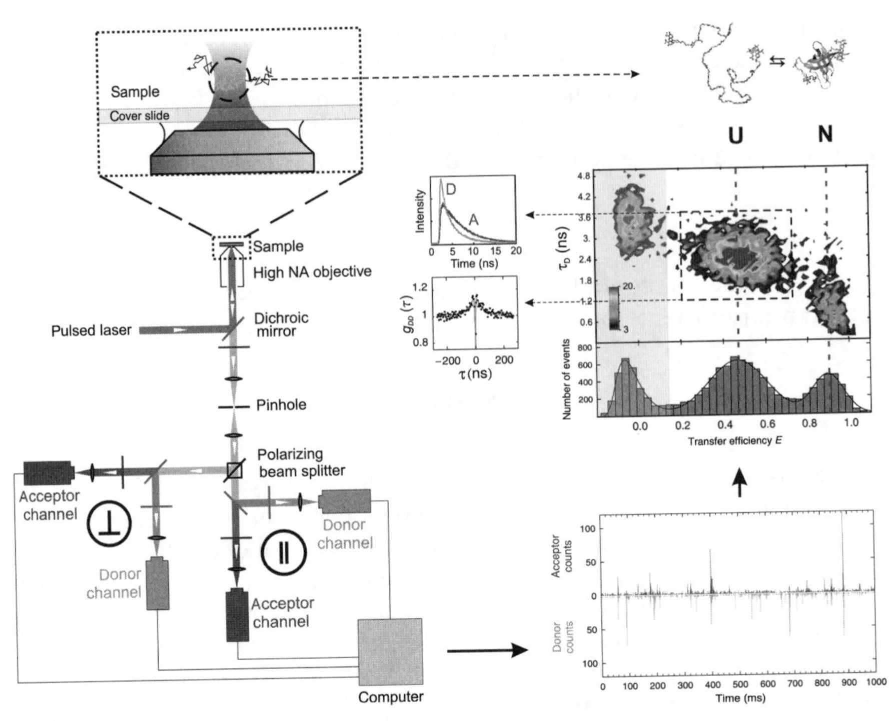

Figure 13.2 Overview of instrumentation and data reduction in confocal single-molecule spectroscopy. The scheme on the left illustrates the main components of a four-channel confocal single-molecule instrument that collects fluorescence photons separated by polarization and wavelength and records their individual arrival times. (A) Sample of a trajectory of detected photons recorded from molecules freely diffusing in solution (in this example, Csp*Tm* in 1.5 M GdmCl [42, 89]), where every burst corresponds to an individual molecule traversing the diffraction-limited confocal volume (see upper left of the scheme). (B) Transfer efficiency histogram calculated from individual bursts, resulting in subpopulations that can be assigned to the folded and unfolded protein and molecules without active acceptor at $E \approx 0$ (shaded in gray). (C) Two-dimensional histogram of donor fluorescence lifetime τ_D versus transfer efficiency E. (D) Subpopulation-specific time-correlated single photon-counting histograms from donor and acceptor photons from all bursts assigned to unfolded molecules that can be used to extract distance distributions (42, 54, 60). (E) Subpopulation-specific donor intensity correlation function, in this case, reporting on the nanosecond reconfiguration dynamics of the unfolded protein (71).

the transfer efficiency distributions observed in single-molecule FRET experiments (Figs. 13.2 and 13.3) always contain a significant contribution from shot noise (the variation in count rates about fixed means due to the discrete nature of the signals) (23, 89) and can therefore not directly be translated into distance distributions. Only the excess width beyond the shot noise value can be assigned to heterogeneity or slow dynamics (1, 30, 77), if suitable controls are available.

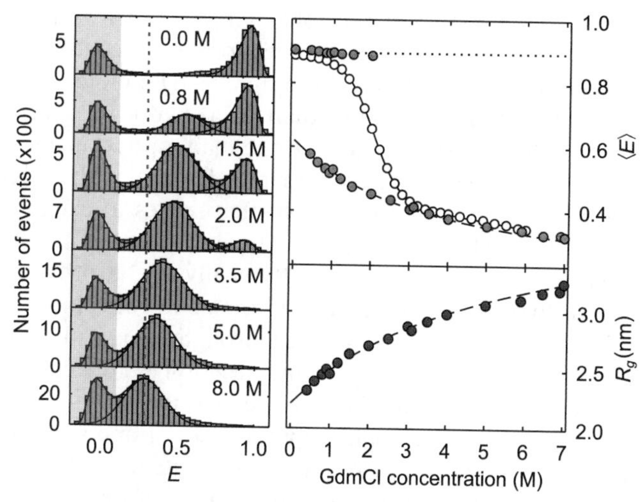

Figure 13.3 Left panels: transfer efficiency histograms of Csp*Tm* labeled with Alexa 488 and 594 at the chain termini at several GdmCl concentrations. The peak at high E corresponds to folded protein, the peak at lower E to unfolded protein. The signal at $E \approx 0$ originates from molecules with inactive acceptor. Upper right: mean transfer efficiencies $\langle E \rangle$ of folded (top, filled circles, corrected for refractive index change) and unfolded (bottom, filled circles) subpopulations from single-molecule measurements compared with an ensemble unfolding transition of the same sample (open circles) with a two-state fit (black line). Lower right: radius of gyration calculated from the $\langle E \rangle$ of the unfolded chain assuming the distance distribution of a Gaussian chain, showing the collapse of the unfolded state. The dashed line is an empirical fit to the equivalent of a binding model. Data taken from References 42, 71, 89.

An important aspect in this context is the relevant averaging regime (90). For instance, the intramolecular dynamics of a completely unfolded protein typically occur on a timescale much faster than the duration of a fluorescence burst (see 2.2) (71), such that there is complete averaging over the distance distribution during the observation time of a millisecond. As a result, no broadening of the transfer efficiency distributions is expected from this process. But chain dynamics are slow compared with the fluorescence lifetime of the donor, and therefore, the distance distribution can be obtained directly from fluorescence lifetime distributions (71) using procedure long established for ensemble experiments (32).

13.2.2 Instrumentation

The instrumentation is based on the optics, detectors and electronics developed for optical single-molecule spectroscopy (6, 65) and fluorescence correlation spectroscopy (FCS) (26, 38). Experimental setups for single-molecule FRET (34) typically involve either confocal excitation and detection using a pulsed or continuous wave laser and avalanche photodiodes (APDs), or wide-field microscopy with two-dimensional detectors such as a CCD camera, often in combination with total internal reflection fluorescence (2). Wide-field imaging allows the collection of data from many single molecules in parallel, albeit at much lower time resolution than in a confocal experiment using APDs. Figure 13.2 shows a schematic with the main optical elements for confocal epifluorescence detection. A laser beam is focused with a high aperture objective to a diffraction-limited focal spot that serves to excite the labeled molecules. In the simplest experiment, the sample molecules are freely diffusing in solution at very low concentrations (typically 10–100 pM), ensuring that the probability of two molecules residing in the confocal volume at the same time is negligible. When a molecule diffuses through the laser beam, the donor dye is excited, fluorescence from donor and acceptor is collected through the objective and gets focused onto the pinhole, a small aperture serving as a spatial filter. A dichroic mirror finally separates donor and acceptor emission into the corresponding detectors, from where the data are collected with multichannel scalers or suitable counting cards. The setup can be extended to sorting photons by additional colors, for example, if more than two chromophores are used (17, 43), or by both color and polarization (108). The advantage of observing freely diffusing molecules is that perturbations from surface interactions can largely be excluded, but the observation time is limited by the diffusion times of the molecules through the confocal volume. Typically, every molecule is observed for no more than a few milliseconds. Alternatively, the molecules can be immobilized on the surface and then observed for a more extended period of time, typically a few seconds, until one of the chromophores undergoes photodestruction. The complications in this case are interactions with the surface that can easily perturb the sensitive equilibrium of protein folding. The details of single-molecule instrumentation can be found in several recent reviews (6, 34, 61). An important development for the wide application of single-molecule methods to the study of biomolecules is the recent availability of comprehensive commercial instrumentation (104).

13.2.3 Protein Labeling

To our misfortune, protein chemistry has not made it easy for us to investigate polypeptides in single-molecule experiments (with the exception of the family of fluorescent proteins (84, 95)). As of today, even tryptophan, the natural amino acid with the highest fluorescence quantum yield (~13%), is not suitable for single-molecule detection (unless the molecule contains a very large number of tryptophan residues [56]) due to the low photostability of the indole ring. Labeling with extrinsic fluorophores is thus unavoidable, and complicated by the need for suitable reactive groups for site-specific attachment. For FRET, two (or more) chromophores are needed, and their specific placement on the protein ideally requires groups with orthogonal chemistries. For simple systems, such as short peptides, sequences can be designed to introduce only single copies of residues with suitable reactive side chains (89, 90). In chemical solid phase peptide synthesis, protection groups and the incorporation of non-natural amino acids can be used to increase specificity, but for longer chains, chemical synthesis becomes inefficient and shorter chains have to be ligated (20) to obtain the desired product (22).

Considering the maturity and versatility of heterologous recombinant protein expression, the production of proteins of virtually any size and sequence in microorganisms is the method of choice to obtain very pure material in sufficiently large amounts for preparative purposes. However, the number of functional groups that can be used for specific labeling is very limited. Sufficiently specific reactivity is only provided by the sulfhydryl groups of cysteine residues, the amino groups of lysine side chains, and the free α-amino group of the N-terminal amino acid. However, except for small peptides, the statistical, and therefore, often multiple occurrence of cysteine and especially lysine residues in one polypeptide prevents the specific attachment of exactly one label to a protein. For some applications, such as *in vivo* imaging, the degree of labeling is only of secondary importance, but for FRET, specificity is strictly required.

Currently, the most common approach is to rely exclusively on cysteine derivatization. Increased specificity can be achieved by removing unwanted natural cysteines by site-directed mutagenesis or introducing cysteines with different reactivity due to different molecular environments within the protein (81). Labeling is usually combined with multiple chromatography steps to purify the desired adducts. Alternative methods (46) are native chemical ligation of recombinantly expressed and individually labeled protein fragments or intein-mediated protein splicing (19), the specific reaction with thioester derivatives of dyes (91), puromycin-based labeling using *in vitro* translation (109), or introduction of nonnatural amino acids (18). Most of the latter methods are not yet used routinely and must be considered under development. But at least a wide variety of suitable organic dyes with various functional groups for protein labeling have become commercially available. Examples of particularly popular chromophores for single-molecule FRET are the cyanine dyes (67) or the Alexa Fluor series (79). Semiconductor quantum dots (14, 70) are promising candidates due to their extreme photostability, but they are not yet available with single functional groups; so far they can only be used as donors because of their broad absorption spectra; and they are themselves of the size of a small protein.

But even for the smaller organic dyes, interactions with the protein can interfere both with the photophysics of the chromophores and the stability of the protein. This needs to be taken into account both for the design of the labeled variants and the control experiments. Due to the substantial size of the fluorophores, they can only be positioned on the solvent-exposed surface of the protein if the folded structure is to be conserved. Even then, the use of hydrophobic dyes can lead to aggregation of the protein, or interactions with the protein surface can cause a serious reduction in fluorescence quantum yield, a problem that has been minimized by the introduction of charged groups in many of the popular dyes (67, 79). Important control experiments are equilibrium or time-resolved fluorescence anisotropy measurements (22, 40, 89, 90), which are sensitive to the rotational flexibility of the dyes and can therefore provide indications for undesirable interactions with the protein surface. It is also essential to ensure by direct comparison with unmodified protein that labeling has not substantially altered the protein's stability or folding mechanism (22, 89).

13.3 SINGLE-MOLECULE SPECTROSCOPY OF THE UNFOLDED STATE

Förster radii in the range of 5 nm for the dye pairs currently available for single-molecule FRET allow the measurement of long-range intramolecular distances and dynamics, and thus, make the method ideal for investigating denatured proteins, many aspects of which have been difficult to study in detail with ensemble methods because of their large structural heterogeneity. A main advantage that has been exploited in single-molecule experiments is the separation of folded and unfolded subpopulations, thus giving access to the properties of unfolded proteins even at low concentrations of denaturant (22, 89), where the majority of molecules in the ensemble are folded. Under these conditions, which are of course physiologically most relevant, the ensemble-averaged signal would be dominated by the native state. Aspects that have received particular attention recently are the overall structural properties of the unfolded state, especially its denaturant-dependent collapse, and the dynamics of the unfolded chain.

13.3.1 Unfolded State Collapse and Structure

Since the first unequivocal identification of unfolded state collapse in the small cold shock protein Csp*Tm* (55, 58, 89) (Figs. 13.1 and 13.3), single-molecule FRET has revealed a very similar behavior in a number of other proteins, including chymotrypsin inhibitor 2 (CI2) (54), for which previous experiments had already been suspected to indicate collapse (22), acyl-CoA binding protein (ACBP) (54), RNase HI (51, 52), protein L (60, 94), the B domain of protein A (44), the immunity protein Im9 (100), and the prion-determining domain of Sup35 (68). The growing number of examples showing collapse and the agreement with the behavior found even for unstructured peptides (10, 66) indicate that chain compaction with decreasing denaturant concentration is a generic phenomenon of polypeptides (78), in spite of conflicting evidence from small-angle X-ray scattering experiments for some cases (e.g., protein L [80]).

Even though the data are still too sparse for a systematic analysis of collapse in terms of, for example, sequence composition or native state structure, several interesting aspects of unfolded state collapse have been addressed in recent experiments. Laurence et al. (54) introduced subpopulation-specific fluorescence lifetime analysis to investigate structural distributions upon collapse. From their observation that the variance of the transfer rates contributing to the decays was greater than expected from simulations of simple polymer models, the authors inferred the presence of transient residual structure in the unfolded state of both CI2 and ACBP. For ACBP, this observation might be related to recent evidence for the population of an intermediate state (99). Similarly, McCarney et al. (59) found evidence for site-specific deviations from Gaussian-chain behavior for the FynSH3 domain in the GdmCl and trifluoroethanol-unfolded states.

To map the collapse of Csp*Tm* in more detail, Hoffmann et al. (42) investigated variants of the protein with the labels positioned such that different segments of the chain could be probed. Surprisingly, a combined analysis of transfer efficiency histograms and subpopulation-specific fluorescence lifetimes was in agreement with intramolecular distance distributions of a Gaussian chain for all variants, even at low denaturant concentrations, where the chain is compact. This indicates that any residual structure can affect only short segments relative to the contour length of the probed segments and is probably highly dynamic. Kinetic synchrotron radiation circular dichroism experiments in fact provided evidence for the presence of some β-structure in the compact unfolded state (42), and kinetic ensemble FRET experiments probing a short segment that forms a β-strand in the folded state indicate the local formation of extended structure even in the compact unfolded state of a closely related Csp (57).

Two recent studies investigated the collapse of protein L. Merchant et al. (60) compared it directly with CspTm and complemented their experiments with simulations. The mean FRET efficiencies for unfolded protein L and Csp*Tm* are identical at high denaturant concentration (60), consistent with the finding from small X-ray scattering measurements that the size of unfolded proteins depends only on the length of the polypeptide and is independent of sequence (48). However, they differ at lower denaturant concentration when the folded state is also present, with protein L being more compact, possibly as a result of its more hydrophobic sequence. It would be extremely difficult to reliably detect such differences in equilibrium small-angle X-ray scattering (SAXS) experiments where both folded and unfolded molecules contribute to the observed scattering (64). Sherman and Haran (94) employed an analytical polymer model to describe the change in size of unfolded protein L as a coil-globule transition and interpreted the results in terms of the solvation properties of the unfolded chain.

13.3.2 Unfolded State Dynamics

While considerable information about the overall dimensions and residual structure of unfolded proteins has been obtained especially from methods such as SAXS (63) and NMR (25), their often rapid dynamics in the microsecond range and below have only recently started to emerge. The importance of these timescales has become particularly obvious through the identification of proteins that fold in a few microseconds and faster (50). In this regime, the free energy barrier to folding is assumed to be extremely low or even absent, and diffusive chain dynamics become the dominant factor in folding kinetics. An ideal way to probe the dynamics of the heterogeneous ensemble of non-native protein conformations is single-molecule spectroscopy: the absence of averaging over many molecules allows spontaneous intramolecular distance fluctuations to be observed at equilibrium, without the need for perturbations to synchronize the ensemble.

Chattopadhyay et al. used FCS to study the dynamics of the denatured state of the intestinal fatty acid-binding protein (15, 16). Two identical chromophores were positioned 48 residues apart in the sequence such that they could self-quench in the denatured state, resulting in fluorescence intensity fluctuations on the timescale of contact formation. The corresponding component in the correlation function was shown to increase as the protein was denatured, and at 3 M GdmCl, the relaxation time extracted was 1.6 μs. The authors noted that this time is significantly longer than predicted from the dynamics of Gaussian chain, suggesting that conformational mobility of the protein is retarded compared with a random coil, but a quantitative analysis is complicated by the absence of information about the dimensions of the denatured protein.

A similar technique was used by Neuweiler et al. to study the dynamics of a 20-amino acid polypeptide, the Trp-cage (74). Here, the quencher was a tryptophan (residue 6), while the extrinsic chromophore was MR121 attached to a nearby lysine (residue 8). The fluorescence correlation curve of the polypeptide exhibits a temperature-dependent relaxation component on the microsecond timescale, which was assigned to the folding/unfolding dynamics. This approach was recently extended to the study of unstructured glycine-serine peptides and the dependence of their dynamics on solvent viscosity, chain length, and temperature (75). Another interesting example for quenching of extrinsic fluorescent dyes by aromatic amino acids was shown for the natively unfolded prion-determining domain of Sup35 (68), where dynamics on timescales between 20 and 300 ns were detected and assigned to quenching of Alexa 488 by the tyrosine residues in the unfolded chain.

Dynamics in a very broad range of timescales were investigated in fluorescence trajectories of individual immobilized RNase HI molecules labeled with a FRET pair (51). Cross-correlation analysis of donor and acceptor emission exhibited an anticorrelated component with a time constant of 20 μs, which was interpreted as the polypeptide reconfiguration time in the unfolded state. Surprisingly, transitions of the transfer efficiency within the range assigned to the unfolded state occurred with characteristic times of about 2 s. The authors concluded that such slow transitions must be due to high free energy barriers originating from substantial structure in the denatured state. Such pronounced features of the energy landscape in the unfolded state have not been observed for other proteins, and their structural basis remains to be understood.

The most comprehensive analysis of unfolded state dynamics to date is the determination of the reconfiguration time in the unfolded state of Csp*Tm* (71). The authors employed the fluctuations in the intensity of donor and acceptor fluorophores undergoing FRET when the distance between them fluctuates (71) (Fig. 13.2). Subpopulation-specific correlation methods with picosecond time resolution (73) were combined with the information on the unfolded state dimensions (42). Based on a model that describes chain dynamics as a diffusive process on a one-dimensional free energy surface (31, 71, 105), very rapid reconfiguration times could be extracted, increasing from about 20 to 60 ns upon collapse of the chain with decreasing denaturant concentration. The resulting intramolecular diffusion coefficients are a key parameter in Kramers-like descriptions of protein folding (9). The dependence of internal friction on chain dimensions provides an indication for the potential variability of intramolecular dynamics as the protein progresses toward the native state (5, 13). Recent experiments (72) indicate the absence of significant distance fluctuations of unfolded Csp*Tm* on slower timescales, suggesting that elementary chain dynamics dominate the dynamics in this protein.

13.4 FUTURE DIRECTIONS

Whereas single-molecule fluorescence experiments have already contributed substantially to our understanding of unfolded state behavior, method development will still be required for many aspects, for example, on the way to time-resolved observations of individual folding events. Important contributions are expected to come from new protein-labeling techniques and fluorophore development, instrumentation, data acquisition, data analysis, and theory. Protein-labeling techniques were recently reviewed comprehensively (46); I will thus summarize recent developments in some of the other areas.

13.4.1 Instrument Development

Because of the large number of observables in single-molecule fluorescence experiments and the continuous improvements in light sources and detection technologies, the development of single-molecule instrumentation and the corresponding data analysis is still a very active area of research. Some of the prominent recent developments are the use of several detectors that allow the measurement of multiple parameters simultaneously (108), including fluorescence intensities, lifetimes, polarization, and the parameters derived from them, such as transfer efficiencies (or related measures of distance), anisotropies, and intensity correlations. An important next step will be the use of more than one acceptor dye simultaneously (18, 43), which can provide information about multiple distances, but its application will be complicated by the difficulties in labeling proteins with three dyes in specific locations. Alternating excitation of donor and acceptor (45, 69) allows molecules to be eliminated spectroscopically if they are incorrectly labeled or contain a photoinactivated chromophore. This procedure not only relaxes the requirements for specific labeling, but also extends the range of measurable transfer efficiencies close to zero, where otherwise an overlap with the signal from "donor-only" molecules interferes with data analysis. Another important extension of single-molecule equipment are ways of observing individual molecules under nonequilibrium conditions. Continuous-flow mixing methods based on microfabricated devices (55) or nanopipettes (107) have already allowed time-resolved measurements of protein-folding kinetics and promise to become generally available tools that will also be valuable for studying the dynamics of protein–protein interactions.

13.4.2 Extending the Observation Time

The observation of individual molecules for times longer than the typical millisecond diffusion times through the confocal volume requires surface immobilization or related approaches. Experiments of this type may be the most promising to ultimately time-resolve the actual folding process of a protein, but the low conformational stability of many proteins demands particular care in minimizing surface interactions that might interfere with the folding process. Several methods are currently in use. Encapsulation in surface-immobilized lipid vesicles (8) has allowed first observations of longer fluorescence trajectories of both two-state (82) and kinetically more heterogeneous larger proteins (83). Covalent immobilization on optimized polyethylene glycol-coated surfaces (33) has also been used successful (51), and the embedding in silica gels has been applied (3, 12). A complementary method that has recently been developed and does not require immobilization involves the observation of individual molecules for several seconds as they flow down a capillary (47). A new development that may solve the problem of photobleaching is the rapid thermal cycling between ambient and cryo-conditions (112): a small volume in a film of frozen liquid is heated locally by infrared radiation, allowing a molecule to evolve in the liquid phase, followed by observation after rapid freezing when the infrared laser is turned off.

From the preceding discussion, it is clear that single-molecule fluorescence studies have already contributed to a better understanding of protein folding, particularly on the structure and dynamics of the unfolded state. A prospect of single-molecule FRET is the possibility of investigating protein-folding mechanisms in more complex environments or in nonnative states such as aggregates (41). Application include the influence of cellular factors, ranging from the ribosome and molecular chaperones (40, 93, 98, 101, 110), protein translocation and membrane protein folding, to protein unfolding and degradation by proteases. Even though the number of single-molecule experiments on the behavior of IDPs is still very limited, it is obvious that this will change soon.

ACKNOWLEDGMENTS

Our work has been supported by the Schweizerische Nationalfonds, the National Center for Competence in Research in Structural Biology, the European Research Council, the Human Frontier Science Program, the VolkswagenStiftung, and the Deutsche Forschungsgemeinschaft.

REFERENCES

1. Antonik, M., S. Felekyan, A. Gaiduk, and C. A. M. Seidel. 2006. Separating structural heterogeneities from stochastic variations in fluorescence resonance energy transfer distributions via photon distribution analysis. J Phys Chem B **110**:6970–8.
2. Axelrod, D., T. P. Burghardt, and N. L. Thompson. 1984. Total internal-reflection fluorescence. Annu Rev Biophys Bioeng **13**:247–68.
3. Baldini, G., F. Cannone, and G. Chirico. 2005. Pre-unfolding resonant oscillations of single green fluorescent protein molecules. Science **309**:1096–100.
4. Best, R., K. Merchant, I. V. Gopich, B. Schuler, A. Bax, and W. A. Eaton. 2007. Effect of flexibility and cis residues in single molecule FRET studies of polyproline. Proc Natl Acad.Sci U S A **104**:18964–9.
5. Best, R. B., and G. Hummer. 2006. Diffusive model of protein folding dynamics with Kramers turnover in rate. Phys Rev Lett **96**:228104.
6. Böhmer, M., and J. Enderlein. 2003. Fluorescence spectroscopy of single molecules under ambient conditions: methodology and technology. Chem Phys Chem **4**:793–808.
7. Borgia, A., P. M. Williams, and J. Clarke. 2008. Single-molecule studies of protein folding. Annu Rev Biochem **77**:101–25.
8. Boukobza, E., A. Sonnenfeld, and G. Haran. 2001. Immobilization in surface-tethered lipid vesicles as a new tool for single biomolecule spectroscopy. J Phys Chem B **105**:12165–70.
9. Bryngelson, J. D., J. N. Onuchic, N. D. Socci, and P. G. Wolynes. 1995. Funnels, pathways, and the energy landscape of protein folding: a synthesis. Proteins **21**:167–95.
10. Buscaglia, M., L. J. Lapidus, W. A. Eaton, and J. Hofrichter. 2006. Effects of denaturants on the dynamics of loop formation in polypeptides. Biophys J **91**:276–88.
11. Bustamante, C., Y. R. Chemla, N. R. Forde, and D. Izhaky. 2004. Mechanical processes in biochemistry. Annu Rev Biochem **73**:705–48.
12. Cannone, F., S. Bologna, B. Campanini, A. Diaspro, S. Bettati, A. Mozzarelli, and G. Chirico. 2005. Tracking unfolding and refolding of single GFPmut2 molecules. Biophysl J **89**:2033–45.
13. Chahine, J., R. J. Oliveira, V. B. P. Leite, and J. Wang. 2007. Configuration-dependent diffusion can shift the kinetic transition state and barrier height of protein folding. Proc Natl Acad Sci U S A **104**:14646–51.

14. Chan, W. C., D. J. Maxwell, X. Gao, R. E. Bailey, M. Han, and S. Nie. 2002. Luminescent quantum dots for multiplexed biological detection and imaging. Curr Opin Biotechnol **13**:40–6.
15. Chattopadhyay, K., E. L. Elson, and C. Frieden. 2005. The kinetics of conformational fluctuations in an unfolded protein measured by fluorescence methods. Proc Natl Acad Sci U S A **102**:2385–9.
16. Chattopadhyay, K., S. Saffarian, E. L. Elson, and C. Frieden. 2005. Measuring unfolding of proteins in the presence of denaturant using fluorescence correlation spectroscopy. Biophys J **88**:1413–22.
17. Clamme, J.-P., and A. A. Deniz. 2005. Three-color single-molecule fluorescence resonance energy transfer. Chem Phys Chem **6**:74–7.
18. Cropp, T. A., and P. G. Schultz. 2004. An expanding genetic code. Trends Genet **20**:625–30.
19. David, R., M. P. Richter, and A. G. Beck-Sickinger. 2004. Expressed protein ligation. Method and applications. Eur J Biochem **271**:663–77.
20. Dawson, P. E., and S. B. Kent. 2000. Synthesis of native proteins by chemical ligation. Annu Rev Biochem **69**:923–60.
21. Deniz, A. A., M. Dahan, J. R. Grunwell, T. J. Ha, A. E. Faulhaber, D. S. Chemla, S. Weiss, and P. G. Schultz. 1999. Single-pair fluorescence resonance energy transfer on freely diffusing molecules: observation of Forster distance dependence and subpopulations. Proc Natl Acad Sci U S A **96**:3670–5.
22. Deniz, A. A., T. A. Laurence, G. S. Beligere, M. Dahan, A. B. Martin, D. S. Chemla, P. E. Dawson, P. G. Schultz, and S. Weiss. 2000. Single-molecule protein folding: diffusion fluorescence resonance energy transfer studies of the denaturation of chymotrypsin inhibitor 2. Proc Natl Acad Sci U S A **97**:5179–84.
23. Deniz, A. A., T. A. Laurence, M. Dahan, D. S. Chemla, P. G. Schultz, and S. Weiss. 2001. Ratiometric single-molecule studies of freely diffusing biomolecules. Annu Rev Phys Chem **52**:233–53.
24. Dyson, H. J., and P. E. Wright. 2005. Intrinsically unstructured proteins and their functions. Nat Rev Mol Cell Biol **6**:197–208.
25. Dyson, H. J., and P. E. Wright. 2004. Unfolded proteins and protein folding studied by NMR. Chem Rev **104**:3607–22.
26. Eigen, M., and R. Rigler. 1994. Sorting single molecules: application to diagnostics and evolutionary biotechnology. Proc Natl Acad Sci U S A **91**:5740–7.
27. Fisher, T. E., A. F. Oberhauser, M. Carrion-Vazquez, P. E. Marszalek, and J. M. Fernandez. 1999. The study of protein mechanics with the atomic force microscope. Trends Biochem Sci **24**:379–84.
28. Forman, J. R., and J. Clarke. 2007. Mechanical unfolding of proteins: insights into biology, structure and folding. Curr Opin Struct Biol **17**:58–66.
29. Förster, T. 1948. Zwischenmolekulare Energiewanderung und Fluoreszenz. Annalen der Physik **6**:55–75.
30. Gopich, I. V., and A. Szabo. 2005. Theory of photon statistics in single-molecule Förster resonance energy transfer. J Chem Phys **122**:1–18.
31. Gopich, I. V., and A. Szabo. 2006. Theory of the statistics of kinetic transitions with application to single-molecule enzyme catalysis. J Chem Phys **124**: 154712.

32. Grinvald, A., E. Haas, and I. Z. Steinber. 1972. Evaluation of distribution of distances between energy donors and acceptors by fluorescence decay. Proc Natl Acad Sci U S A **69**:2273–2277.

33. Groll, J., E. V. Amirgoulova, T. Ameringer, C. D. Heyes, C. Rocker, G. U. Nienhaus, and M. Moller. 2004. Biofunctionalized, ultrathin coatings of cross-linked star-shaped poly(ethylene oxide) allow reversible folding of immobilized proteins. J Am Chem Soc **126**:4234–9.

34. Ha, T. 2001. Single-molecule fluorescence resonance energy transfer. Methods **25**:78–86.

35. Ha, T., T. Enderle, D. F. Ogletree, D. S. Chemla, P. R. Selvin, and S. Weiss. 1996. Probing the interaction between two single molecules: fluorescence resonance energy transfer between a single donor and a single acceptor. Proc Natl Acad Sci U S A **93**:6264–8.

36. Haas, E., E. Katchalskikatzir, and I. Z. Steinberg. 1978. Brownian-motion of ends of oligopeptide chains in solution as estimated by energy-transfer between chain ends. Biopolymers **17**:11–31.

37. Haran, G. 2003. Single-molecule fluorescence spectroscopy of biomolecular folding. J Phys Condens Matter **15**:R1291–317.

38. Hess, S. T., S. Huang, A. A. Heikal, and W. W. Webb. 2002. Biological and chemical applications of fluorescence correlation spectroscopy: a review. Biochemistry **41**:697–705.

39. Heyduk, T. 2002. Measuring protein conformational changes by FRET/LRET. Curr Opin Biotechnol **13**:292–6.

40. Hillger, F., D. Hänni, D. Nettels, S. Geister, M. Grandin, M. Textor, and B. Schuler. 2008. Probing protein-chaperone interactions with single molecule fluorescence spectroscopy. Angew Chem Int Ed Engl **47**:6184–8.

41. Hillger, F., D. Nettels, S. Dorsch, and B. Schuler. 2007. Detection and analysis of protein aggregation with confocal single molecule fluorescence spectroscopy. J Fluoresc **17**:759–65.

42. Hoffmann, A., A. Kane, D. Nettels, D. E. Hertzog, P. Baumgartel, J. Lengefeld, G. Reichardt, D. A. Horsley, R. Seckler, O. Bakajin, and B. Schuler. 2007. Mapping protein collapse with single-molecule fluorescence and kinetic synchrotron radiation circular dichroism spectroscopy. Proc Natl Acad Sci U S A **104**:105–10.

43. Hohng, S., C. Joo, and T. Ha. 2004. Single-molecule three-color FRET. Biophys J **87**:1328–37.

44. Huang, F., S. Sato, T. D. Sharpe, L. M. Ying, and A. R. Fersht. 2007. Distinguishing between cooperative and unimodal downhill protein folding. Proc Natl Acad Sci U S A **104**:123–7.

45. Kapanidis, A. N., N. K. Lee, T. A. Laurence, S. Doose, E. Margeat, and S. Weiss. 2004. Fluorescence-aided molecule sorting: analysis of structure and interactions by alternating-laser excitation of single molecules. Proc Natl Acad Sci U S A **101**:8936–41.

46. Kapanidis, A. N., and S. Weiss. 2002. Fluorescent probes and bioconjugation chemistries for single-molecule fluorescence analysis of biomolecules. J Chem Phys **117**:10953–64.

47. Kinoshita, M., K. Kamagata, A. Maeda, Y. Goto, T. Komatsuzaki, and S. Takahashi. 2007. Development of a technique for the investigation of folding dynamics of single proteins for extended time periods. Proc Natl Acad Sci U S A **104**:10453–58.
48. Kohn, J. E., I. S. Millett, J. Jacob, B. Zagrovic, T. M. Dillon, N. Cingel, R. S. Dothager, S. Seifert, P. Thiyagarajan, T. R. Sosnick, M. Z. Hasan, V. S. Pande, I. Ruczinski, S. Doniach, and K. W. Plaxco. 2004. Random-coil behavior and the dimensions of chemically unfolded proteins. Proc Natl Acad Sci U S A **101**:12491–6.
49. Kremer, W., B. Schuler, S. Harrieder, M. Geyer, W. Gronwald, C. Welker, R. Jaenicke, and H. R. Kalbitzer. 2001. Solution NMR structure of the cold-shock protein from the hyperthermophilic bacterium Thermotoga maritima. Eur J Biochem **268**:2527–39.
50. Kubelka, J., J. Hofrichter, and W. A. Eaton. 2004. The protein folding "speed limit". Curr Opin Struct Biol **14**:76–88.
51. Kuzmenkina, E. V., C. D. Heyes, and G. U. Nienhaus. 2005. Single-molecule Forster resonance energy transfer study of protein dynamics under denaturing conditions. Proc Natl Acad Sci U S A **102**:15471–6.
52. Kuzmenkina, E. V., C. D. Heyes, and G. U. Nienhaus. 2006. Single-molecule FRET study of denaturant induced unfolding of RNase H. J Mol Biol **357**:313–24.
53. Lakowicz, J. R. 1999. Principles of Fluorescence Spectroscopy, second edition. Kluwer Academic/Plenum Publishers, New York.
54. Laurence, T. A., X. X. Kong, M. Jager, and S. Weiss. 2005. Probing structural heterogeneities and fluctuations of nucleic acids and denatured proteins. Proc Natl Acad Sci U S A **102**:17348–53.
55. Lipman, E. A., B. Schuler, O. Bakajin, and W. A. Eaton. 2003. Single-molecule measurement of protein folding kinetics. Science **301**:1233–5.
56. Lippitz, M., W. Erker, H. Decker, K. E. van Holde, and T. Basche. 2002. Two-photon excitation microscopy of tryptophan-containing proteins. Proc Natl Acad Sci U S A **99**:2772–7.
57. Magg, C., J. Kubelka, G. Holtermann, E. Haas, and F. X. Schmid. 2006. Specificity of the initial collapse in the folding of the cold shock protein. J Mol Biol **360**:1067–80.
58. Magg, C., and F. X. Schmid. 2004. Rapid collapse precedes the fast two-state folding of the cold shock protein. J Mol Biol **335**:1309–23.
59. McCarney, E. R., J. H. Werner, S. L. Bernstein, I. Ruczinski, D. E. Makarov, P. M. Goodwin, and K. W. Plaxco. 2005. Site-specific dimensions across a highly denatured protein; a single molecule study. J Mol Biol **352**:672–82.
60. Merchant, K. A., R. B. Best, J. M. Louis, I. V. Gopich, and W. A. Eaton. 2007. Characterizing the unfolded states of proteins using single-molecule FRET spectroscopy and molecular simulations. Proc Natl Acad Sci U S A **104**: 1528–33.
61. Michalet, X., A. N. Kapanidis, T. Laurence, F. Pinaud, S. Doose, M. Pflughoefft, and S. Weiss. 2003. The power and prospects of fluorescence microscopies and spectroscopies. Annu Rev Biophys Biomol Struct **32**:161–82.

62. Michalet, X., S. Weiss, and M. Jäger. 2006. Single-molecule fluorescence studies of protein folding and conformational dynamics. Chem Rev **106**:1785–813.
63. Millett, I. S., S. Doniach, and K. W. Plaxco. 2002. Toward a taxonomy of the denatured state: small angle scattering studies of unfolded proteins. Adv Protein Chem **62**:241–62.
64. Millet, I. S., L. E. Townsley, F. Chiti, S. Doniach, and K. W. Plaxco. 2002. Equilibrium collapse and the kinetic "foldability" of proteins. Biochemistry **41**:321–5.
65. Moerner, W. E. 2002. A dozen years of single-molecule spectroscopy in physics, chemistry, and biophysics. J Phys Chem B **106**:910–27.
66. Möglich, A., K. Joder, and T. Kiefhaber. 2006. End-to-end distance distributions and intrachain diffusion constants in unfolded polypeptide chains indicate intramolecular hydrogen bond formation. Proc Natl Acad Sci U S A **103**:12394–9.
67. Mujumdar, R. B., L. A. Ernst, S. R. Mujumdar, C. J. Lewis, and A. S. Waggoner. 1993. Cyanine dye labeling reagents: sulfoindocyanine succinimidyl esters. Bioconjug Chem **4**:105–11.
68. Mukhopadhyay, S., R. Krishnan, E. A. Lemke, S. Lindquist, and A. A. Deniz. 2007. A natively unfolded yeast prion monomer adopts an ensemble of collapsed and rapidly fluctuating structures. Proc Natl Acad Sci U S A **104**:2649–54.
69. Müller, B. K., E. Zaychikov, C. Bräuchle, and D. C. Lamb. 2005. Pulsed interleaved excitation. Biophys J **89**:3508–22.
70. Murphy, C. J. 2002. Optical sensing with quantum dots. Anal Chem **74**: 520A–6A.
71. Nettels, D., I. V. Gopich, A. Hoffmann, and B. Schuler. 2007. Ultrafast dynamics of protein collapse from single-molecule photon statistics. Proc Natl Acad Sci U S A **104**:2655–60.
72. Nettels, D., A. Hoffmann, and B. Schuler. 2008. Unfolded protein and peptide dynamics investigated with single-molecule FRET and correlation spectroscopy from picoseconds to seconds. J Phys Chem B **112**:6137–46.
73. Nettels, D., and B. Schuler. 2007. Subpopulation-resolved photon statistics of single-molecule energy transfer dynamics. IEEE J Sel Top Quant Electron **13**:990–5.
74. Neuweiler, H., S. Doose, and M. Sauer. 2005. A microscopic view of miniprotein folding: enhanced folding efficiency through formation of an intermediate. Proc Natl Acad Sci U S A **102**:16650–5.
75. Neuweiler, H., M. Lollmann, S. Doose, and M. Sauer. 2007. Dynamics of unfolded polypeptide chains in crowded environment studied by fluorescence correlation spectroscopy. J Mol Biol **365**:856–69.
76. Nienhaus, G. U. 2006. Exploring protein structure and dynamics under denaturing conditions by single-molecule FRET analysis. Macromol Biosci **6**:907–22.
77. Nir, E., X. Michalet, K. M. Hamadani, T. A. Laurence, D. Neuhauser, Y. Kovchegov, and S. Weiss. 2006. Shot-noise limited single-molecule FRET histograms: comparison between theory and experiments. J Phys Chem B **110**:22103–24.
78. O'Brien, E. P., G. Ziv, G. Haran, B. R. Brooks, and D. Thirumalai. 2008. Effects of denaturants and osmolytes on proteins are accurately predicted by the molecular transfer model. Proc Natl Acad Sci U S A **105**:13403–8.

79. Panchuk-Voloshina, N., R. P. Haugland, J. Bishop-Stewart, M. K. Bhalgat, P. J. Millard, F. Mao, and W. Y. Leung. 1999. Alexa dyes, a series of new fluorescent dyes that yield exceptionally bright, photostable conjugates. J Histochem Cytochem **47**:1179–88.

80. Plaxco, K. W., I. S. Millett, D. J. Segel, S. Doniach, and D. Baker. 1999. Chain collapse can occur concomitantly with the rate-limiting step in protein folding. Nat Struct Biol **6**:554–6.

81. Ratner, V., E. Kahana, M. Eichler, and E. Haas. 2002. A general strategy for site-specific double labeling of globular proteins for kinetic FRET studies. Bioconjug Chem **13**:1163–70.

82. Rhoades, E., M. Cohen, B. Schuler, and G. Haran. 2004. Two-state folding observed in individual protein molecules. J Am Chem Soc **126**:14686–7.

83. Rhoades, E., E. Gussakovsky, and G. Haran. 2003. Watching proteins fold one molecule at a time. Proc Natl Acad Sci U S A **100**:3197–202.

84. Sako, Y., and T. Uyemura. 2002. Total internal reflection fluorescence microscopy for single-molecule imaging in living cells. Cell Struct Funct **27**:357–65.

85. Schuler, B. 2007. Application of single molecule Forster resonance energy transfer to protein folding. Methods Mol Biol **350**:115–38.

86. Schuler, B. 2006. Application of single molecule förster resonance energy transfer to protein folding, pp. 115–38. In Y. Bai and R. Nussinov (eds.), Protein Folding Protocols, vol. **366**. Humana, Totowa, NJ.

87. Schuler, B. 2005. Single-molecule fluorescence spectroscopy of protein folding. Chemphyschem **6**:1206–20.

88. Schuler, B., and W. A. Eaton. 2008. Protein folding studied by single-molecule FRET. Curr Opin Struct Biol **18**:16–26.

89. Schuler, B., E. A. Lipman, and W. A. Eaton. 2002. Probing the free-energy surface for protein folding with single-molecule fluorescence spectroscopy. Nature **419**:743–7.

90. Schuler, B., E. A. Lipman, P. J. Steinbach, M. Kumke, and W. A. Eaton. 2005. Polyproline and the "spectroscopic ruler" revisited with single molecule fluorescence. Proc Natl Acad Sci U S A **102**:2754–9.

91. Schuler, B., and L. K. Pannell. 2002. Specific labeling of polypeptides at amino-terminal cysteine residues using Cy5-benzyl thioester. Bioconjug Chem **13**:1039–43.

92. Selvin, P. R. 2000. The renaissance of fluorescence resonance energy transfer. Nature Struct Biol **7**:730–4.

93. Sharma, S., K. Chakraborty, B. K. Muller, N. Astola, Y. C. Tang, D. C. Lamb, M. Hayer-Hartl, and F. U. Hartl. 2008. Monitoring protein conformation along the pathway of chaperonin-assisted folding. Cell **133**:142–53.

94. Sherman, E., and G. Haran. 2006. Coil-globule transition in the denatured state of a small protein. Proc Natl Acad Sci U S A **103**:11539–43.

95. Shimomura, O. 2005. The discovery of aequorin and green fluorescent protein. J Microsc **217**:1–15.

96. Stryer, L. 1978. Fluorescence energy transfer as a spectroscopic ruler. Annu Rev Biochem **47**:819–46.

97. Stryer, L., and R. P. Haugland. 1967. Energy transfer: a spectroscopic ruler. Proc Natl Acad Sci U S A **58**:719–26.
98. Taguchi, H., T. Ueno, H. Tadakuma, M. Yoshida, and T. Funatsu. 2001. Single-molecule observation of protein-protein interactions in the chaperonin system. Nat Biotechnol **19**:861–5.
99. Teilum, K., F. M. Poulsen, and M. Akke. 2006. The inverted chevron plot measured by NMR relaxation reveals a native-like unfolding intermediate in acyl-CoA binding protein. Proc Natl Acad Sci U S A **103**:6877–82.
100. Tezuka-Kawakami, T., C. Gell, D. J. Brockwell, S. E. Radford, and D. A. Smith. 2006. Urea-induced unfolding of the immunity protein Im9 monitored by spFRET. Biophys J **91**:L42–4.
101. Ueno, T., H. Taguchi, H. Tadakuma, M. Yoshida, and T. Funatsu. 2004. GroEL mediates protein folding with a two successive timer mechanism. Mol. Cell **14**:423–34.
102. Uversky, V. N., C. J. Oldfield, and A. K. Dunker. 2005. Showing your ID: intrinsic disorder as an ID for recognition, regulation and cell signaling. J Mol Recognit **18**:343–84.
103. Van Der Meer, B. W., G. Coker, III, and S.-Y. Simon Chen 1994. Resonance Energy Transfer: Theory and Data. VCH Publishers, Inc., New York, Weinheim, Cambridge.
104. Wahl, M., F. Koberling, M. Patting, H. Rahn, and R. Erdmann. 2004. Time-resolved confocal fluorescence imaging and spectrocopy system with single molecule sensitivity and sub-micrometer resolution. Curr Pharm Biotechnol **5**:299–308.
105. Wang, Z. S., and D. E. Makarov. 2003. Nanosecond dynamics of single polypeptide molecules revealed by photoemission statistics of fluorescence resonance energy transfer: a theoretical study. J Phys Chem B **107**:5617–22.
106. Weiss, S. 1999. Fluorescence spectroscopy of single biomolecules. Science **283**:1676–83.
107. White, S. S., S. Balasubramanian, D. Klenerman, and L. M. Ying. 2006. A simple nanomixer for single-molecule kinetics measurements. Angew Chem Int Ed Engl **45**:7540–3.
108. Widengren, J., V. Kudryavtsev, M. Antonik, S. Berger, M. Gerken, and C. A. M. Seidel. 2006. Single-molecule detection and identification of multiple species by multiparameter fluorescence detection. Anal Chem **78**:2039–50.
109. Yamaguchi, J., N. Nemoto, T. Sasaki, A. Tokumasu, Y. Mimori-Kiyosue, T. Yagi, and T. Funatsu. 2001. Rapid functional analysis of protein-protein interactions by fluorescent C-terminal labeling and single-molecule imaging. FEBS Lett **502**:79–83.
110. Yamasaki, R., M. Hoshino, T. Wazawa, Y. Ishii, T. Yanagida, Y. Kawata, T. Higurashi, K. Sakai, J. Nagai, and Y. Goto. 1999. Single molecular observation of the interaction of GroEL with substrate proteins. J Mol Biol **292**:965–72.
111. Zhuang, X., and M. Rief. 2003. Single-molecule folding. Curr Opin Struct Biol **13**:88–97.
112. Zondervan, R., F. Kulzer, H. van der Meer, J. A. Disselhorst, and M. Orrit. 2006. Laser-driven microsecond temperature cycles analyzed by fluorescence polarization microscopy. Biophys J **90**:2958–69.

14

MONITORING THE CONFORMATIONAL EQUILIBRIA OF MONOMERIC INTRINSICALLY DISORDERED PROTEINS BY SINGLE-MOLECULE FORCE SPECTROSCOPY

MASSIMO SANDAL, MARCO BRUCALE, AND BRUNO SAMORÌ
Department of Biochemistry "G. Moruzzi", Bologna, Italy

ABSTRACT

In this chapter, we describe an experimental approach based on the single-molecule force spectroscopy (SMFS) technique that can be employed to gain insights on the conformational equilibria of monomeric intrinsically disordered proteins (IDPs). We report how this approach was utilized for characterizing the conformational diversity of α-synuclein, a Parkinson-involved IDP. We also offer a brief introduction to the SMFS technique, a description of the various SMFS experimental approaches that were applied to protein folding studies, and a summary of the available SMFS studies on IDPs or disordered segments. Finally, we propose plausible SMFS research perspectives that could be helpful in gaining further knowledge on the fascinating subject of IDPs.

Instrumental Analysis of Intrinsically Disordered Proteins: Assessing Structure and Conformation, Edited by Vladimir Uversky and Sonia Longhi

14.1 INTRODUCTION

A large variety of physical techniques have been brought into play to unveil the fundamental rules that make proteins fold into different structures, and with different dynamics (21). Theoretical studies have indicated how energy landscapes with a single dominant basin and an overall funnel topography (see Fig. 14.1) can efficiently drive a protein toward its native structure by progressively organizing the ensemble of the partially folded structures assumed on the way (68).

The funnel is roughened by local frustrations along the chain (e.g., steric hindrances, nonnative contacts) that lead to local barriers that are a few times larger than the thermal fluctuations (2). During folding, this ruggedness of the funnel dictates the kinetics of the process by trapping the folding molecules. Folding/refolding processes are thus expected to take place within complex energy landscapes with many possible intermediates. Within such a complex funneled multidimensional energy landscape, different protein molecules, in spite of having the same sequence, can follow markedly different trajectories during their folding and also in their thermal fluctuations after having reached their "native" structure. In fact, one molecule can be driven into one funnel trap, while a different molecule can visit another one, and so on.

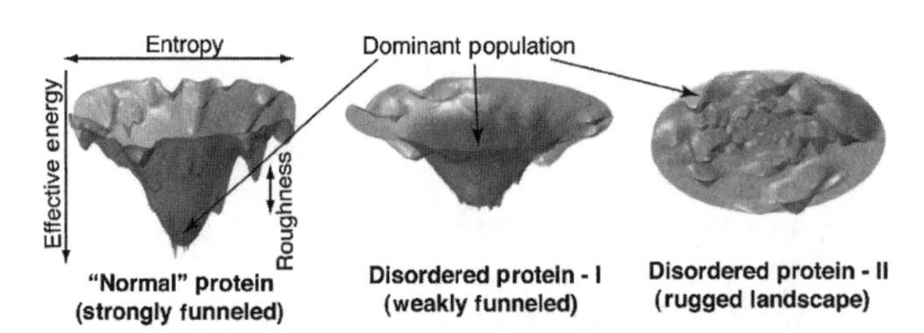

Figure 14.1 Schematic representation of the interplay between protein's energy landscape and its folding. Left panel: energy landscapes of globular proteins are represented as deep, steep funnels. Their native state is thus an energy minimum both locally and globally, maximizing thus its kinetic accessibility. In this type of diagram, it is common practice to partition the global free energy into chain conformational entropy (horizontal axes) and all the remaining part (vertical axis, "effective energy"), including entropic contributions due to solvent interactions. Center panel: example of energy landscape of a mostly disordered protein, capable of assuming discrete conformations. While the overall shape of the energy landscape is still reminiscent of a funnel, its walls are less steep and its bottom is less deep than in the previous case. Other local minima are present in its corrugated walls, allowing the protein to populate many other "disordered" conformations in addition to that having the highest thermodynamic stability. Contact with specific targets could generate favorable interactions, deepening the funnel and increasing folding drive energy. Right panel: mostly flat, highly frustrated energy landscape. Search for a specific conformation among those thermodynamically allowed is extremely inefficient. The ensemble of highly diverse populated conformations labels the protein as "disordered." Taken from Reference 70. See color insert.

Through such a multiplicity of conformational paths, peculiar structures could be assumed or particular motions could be made even by only a few molecules of the ensemble. Those structures might be selected or those motions might be rectified to make a specific biological function possible, and the same function would be inaccessible for all the other molecules at that same moment. On this basis, examining protein conformational equilibrium and folding kinetics at a single-molecule level has become a necessity, and it is currently considered a great challenge in experimental biology.

14.1.1 A Single Molecule Approach to the Study of IDPs

The capability of resolving the properties of individual protein molecules and quantify subpopulations is particularly crucial in the case of the intrinsically disordered proteins (IDP). In this case, the sequence does not encode a funneled energy landscape but one with a set of shallow minima. Different structures would correspond to any of these minima (see Fig. 14.1, central panel). The conformational landscape of an IDP would be thus characterized not by a stable native structure, but by a set of marginally stable interconverting conformations, whose equilibrium is driven by the depths and profiles of their energy minima and by the effect of the environment upon them.

The potential multiple biological functions of these proteins rely on this multiminima energy landscape, that is, on their capability of acquiring different structures (88). The natural selection of proteins able to fulfill more than one function has made possible to increase the complexity of the metabolic networks without increasing the number of underlying proteins (88), and therefore, without increasing the dimension of the genome. This is a very risky game because many of these IDPs display a tendency to aggregate. Most neurodegenerative diseases find their origin in this tendency (27, 53). The seeding of the aggregation process of an IDP is expected to be triggered by the propensity to aggregate of one or more than one of its conformations. A conformation with a high content of β-sheet is a strong candidate as an aggregation precursor, taking into account that the transitions of an IDP from its monomeric state to fibrils or oligomeric aggregates is a process of acquiring a β-structure (90).

The capability of observing directly and quantify conformers with high aggregation propensity becomes crucial for our understanding of the aggregation process. To this end, the role of single-molecule techniques can be fundamental because they can circumvent two main limitations of the traditional in-bulk methodologies in studying IDPs and their aggregation processes.

The first limitation is connected to the limited capability to single out the monomeric state of the investigated IDPs. In the experimental conditions, the monomeric state of the protein might easily occur to be accompanied by soluble oligomers when the latter form quickly in solution at the concentrations required by the sample preparations. Single-molecule methods operating at low concentrations or with immobilized molecules allow the properties of

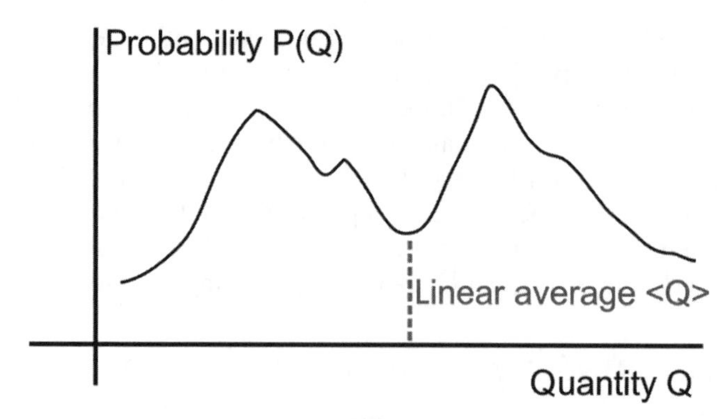

Figure 14.2 The ensemble averaging problem: the average of an observable quantity Q can give little information on the overall distribution of Q (inspired by Reference 97).

the monomeric species to be monitored without any interference by the already started aggregation process.

The second limitation of the traditional in-bulk methodologies is connected to their time and ensemble averaging. They can observe only average properties of 10^{14}–10^{17} molecules at a time. Even in the case in which no oligomers are present in the sample, the intrinsic ensemble averaging of in-bulk methodologies might prevent to single out the aggregation-prone structures because they are most likely poorly populated in physiological conditions (see Fig. 14.2). Commonly, the recording of a specific signal that can be attributed to them was made possible by shifting the conformational equilibria toward these structures through additions to the buffer solution of chemicals, such as methanol or trifluoroethanol, and metal cations. On the other hand, these additions were moving the experimental conditions far away from physiological ones.

Two single-molecule approaches have been reported so far in studies of the conformational properties of amyloidogenic IDPs involved in neurodegenerative diseases. The first relies on single-molecule fluorescence resonance energy transfer (SM-FRET). The technique reports changes in the mean distance between two residues (steady-state FRET) or intramolecular distance distributions (time-resolved FRET) making thus possible to study the conformational equilibria in timescales shorter than a few milliseconds. The group of Ashok Deniz was the first to apply SM-FRET to an amyloidogenic IDP, the yeast prion Sup35 (61), evidencing fast and complex fluctuations on the nanosecond timescale.

The other approach was proposed by us. It relied instead on the AFM-based single-molecule force spectroscopy (SMFS). AFM-based SMFS is particularly sensitive to the formation of secondary structures, and probes timescales from milliseconds to seconds, much longer than those of SM-FRET. We applied this approach so far to α-synuclein (10, 78). The SMFS methodology is increasingly used to study the folding of structured proteins and its capability to unveil fundamental details of the process is more and more evidenced (14, 33, 64, 77, 96). The methodology was first applied to the study of disordered proteins by J. Fernandez and coworkers to the study of the unstructured PEVK segment of titin, in the context of disclosing the basis of its mechanical role in muscles (see section section 14.3.1). In this chapter, we will remind the essential concepts necessary to understand the application of AFM-based SMFS to the study of disordered proteins and peptides, and we will show how the technique has brought unique insights to the problem of the conformational equilibrium of IDPs.

14.2 BASIC ELEMENTS OF THE SMFS METHODOLOGY FOR PROTEIN FOLDING STUDIES

SMFS is a general term that includes different techniques of single-molecule nanomanipulation. Distinguishing on the basis of the nanomanipulator used, the two most used SMFS techniques are optical tweezers-based SMFS (OT-SMFS) and atomic force microscope-based SMFS (AFM-SMFS). The molecules are pulled clamping either the pulling speed or the applied force. The different methodologies developed so far for studies of protein nanomechanics have been reviewed recently (14, 77, 96).

In this chapter, we will talk about the constant speed mode of the AFM-SMFS. An AFM tip is basically a dynamometer that works on the nanoscale. Originally developed for imaging, AFMs make it possible to approach the tip to a substrate upon which proteins have been previously deposited, to retract until a desired distance, and then to retract the tip: this cycle is usually repeated thousands of times in sequence. The cycles operate blindly: whenever a molecular bridge is formed between the two moving surfaces, either because the tip has picked up one end of a molecule or because a bond between two molecules (e.g., one attached on the tip and one attached on the surface) was formed; the force acting on this bridge is reported *versus* the tip displacement. This plot, that is basically a report of the mechanical response of the single molecule to the applied force, is usually called a *force curve*.

The plot is composed of two curves: the first is called the approaching curve, that is recorded when the tip is approaching the surface; the second is the retracting curve, that is recorded when the tip is retracting from the surface. Ideally, if nothing has been captured by the tip, the two curves superimpose perfectly, apart from noise. Each of the two curves features two different regions: (1) the noncontact region, which is the horizontal section where the tip is not touching the surface and where signals appear whenever we have picked up a molecule; and (2) the vertical contact region (see Fig. 14.3).

The point located at the boundary between these two regions is the contact point. The contact point is of paramount importance as a reference point since its location determines where the tip has touched the surface on the force

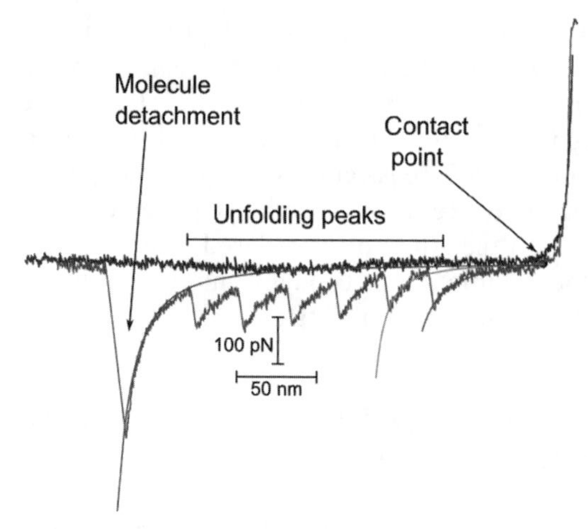

Figure 14.3 Example of a force curve from the unfolding of a multimodular protein, as obtained by the authors. Six unfolding peaks coming from individual I27 folded modules, plus the final detachment peaks, are clearly visible. A WLC fit is superimposed to the two first unfolding peaks and the detachment peak (see text).

curve. The contact point location is often marred by aspecific peaks due to aspecific interactions between the tip and the surface.

14.2.1 Interpreting the Force Curve

When a molecule has been picked up by the tip, force curves usually display a sawtooth shape composed of a rising part followed by a sudden drop. Modeling the system as a simple flexible chain kept together by point-like interactions (see Fig. 14.4), the rising part corresponds to the stretching of the flexible chain that is not kept enclosed by an interaction (free chain). The drop corresponds to the breaking of the interaction enclosing the chain loop. The height of the drop is a measure of the rupture force of the interaction. If the interaction was sustaining the bridge, the mechanical bridge will be broken. If it was an intramolecular interaction, the length of the chain enclosed by the interaction (hidden loop) adds to the free chain. The bridge between the tip and the surface will break through continued pulling.

The rising portion of each peak has a characteristic, curved profile that is usually dominated by the entropic elasticity contribution. When a stretching force is applied to a polymer, its conformational space is shrunk: less and less conformations become available, until at full elongation there would be ideally only one conformation available (a straight, rigid chain). Given that a large majority of conformations are virtually at the same energy level, entropy dominates the process, generating an opposing force. It is like when one wants

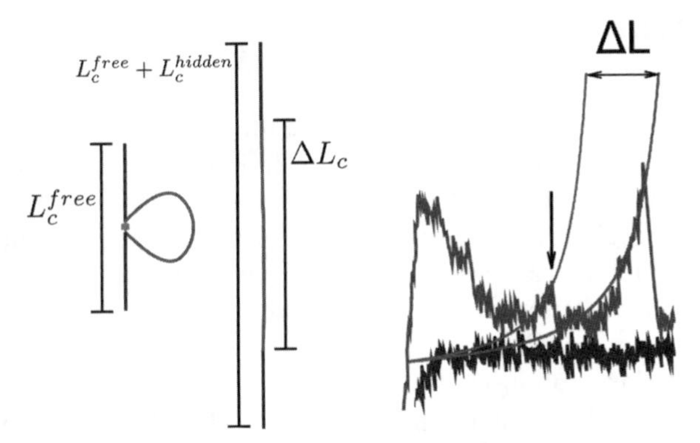

Figure 14.4 Left: model of a polymer chain with a loop (red) enclosed by an interaction. The contour length available to mechanical stretching is L_c^{free}. Center: the same chain stretched after the breaking of the interaction: the contour length is now given by $L_c^{free} + L_c^{hidden}$. Right: a force curve that follows the previous model, with relevant WLC fits (red) superimposed on the force peaks. The difference ΔL between the asymptotes of the fits gives the length L_c^{hidden} enclosed by the interaction.

to hold a certain volume of gas under a piston: a squeezing force must be loaded against the pressure of the gas.

$$F(x) = \frac{k_b T}{p}\left[\frac{x}{L} + \frac{1}{4\left(1-\frac{x}{L}\right)^2} - \frac{1}{4}\right] \tag{14.1}$$

The force (F) versus extension (x) profile of entropic elasticity can be modeled by the worm-like chain model (WLC) of entropic elasticity, a statistical model for polymer chain conformation that models the chain as a flexible, continuous string (see Eq. 14.1). The parameters describing the chain in the model are L, which is the contour length, and p, which is the persistence length of the chain molecule. The latter is defined as the distance at which the curvature of the chain in a single point can influence the curvature of a neighboring point, and it is a measure of chain elasticity: the more the persistence length is high, the more the molecule is rigid. The WLC model describes very well the entropic elasticity of molecules as long as $L \gg p$.

The stretching of the molecule may, in addition, induce an increase of its contour length through force-induced conformational transitions which extend along the chain. Such transitions are revealed by kinks, plateaux, or other deviations from the entropic elasticity profile. In this case, the elasticity has acquired an enthalpic component.

14.2.2 The Polyprotein Strategy

The protein under investigation is commonly inserted in a multimodular construct, which contains independently folded modules whose mechanical properties are well known. These modules act as linkers for pulling the protein and as internal gauges of the point at which the construct has been picked up by the tip at each single-molecule pulling event. This tool is also employed—and has been developed—to investigate folded globular proteins. This has been called the "polyprotein strategy" (14). There are three essential reasons for this strategy to be used.

First, a globular protein is often a short object, with lateral dimensions of a few nanometers. This means that when the AFM begins to unravel it, we are still well into the "blind window" possibly masked by the aspecific interactions between the tip and the surface. Distinguishing reliably the protein signal from aspecific signals in these conditions can be next to impossible. Second, we do not have any control on where the protein is picked up: the tip curvature radius is often about 10 times larger than the protein itself; when the tip squashes down the protein, it can be physisorbed to the tip and to the surface from any point. So, in this case, we do not know what portion of the protein we are stretching and we do not even know the direction of the force vector with respect to the protein, which is of fundamental importance for data interpretation (see section 14.2.4). Third, a globular protein does not give any control or gauge of having stretched only one protein molecule from the very end.

A multimodular protein construct instead makes it possible to circumvent these problems. First, such a protein can be long enough so that the unfolding of first modules will happen well away from the tip–surface aspecific interactions (e.g., a fully folded protein made of 6 Ig-like modules in tandem is more than 20 nm long). Second, in this case, we are sure that if our multimodular protein is made of modules connected by their N- and C-termini, the force vector pulling the protein is parallel to the N—C direction, allowing us to make considerations on the actual unfolding pathway of the protein. The point at which the construct has been picked up by the tip is indicated by the number of peaks corresponding to the catastrophic unfolding of the linker domains.

14.2.3 Designing and Building Polyproteins

The simplest polyprotein design is the homomeric polyprotein, that is, a protein construct made only of identical repeats of the same module $(M)_n$ (see Fig. 14.5A). Unfortunately, not all proteins are suitable for such a design. The signal of the protein may be particularly complex, and if weak signals add all together at the beginning of the force curve, interpretation can become difficult. Aggregation-prone proteins (e.g., amyloid-forming proteins) could prove difficult to express when assembled together in a polyprotein.

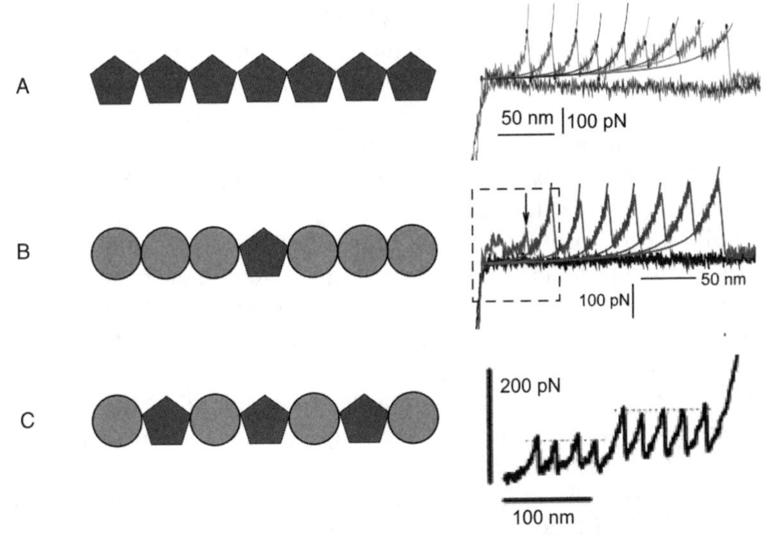

Figure 14.5 Polyprotein strategies for SMFS and corresponding experimental curves. (A) Left: diagram of an homomeric polyprotein, where the analyte is replicated several times in tandem. Right: example of a force curve coming from the stretching of a poly-GB1 polyprotein. (B) Left: diagram of a polyprotein following the sandwich pattern. The analyte (red pentagon) is flanked by marker modules (blue circles) at left and right. If at least four signals from the markers appear, the analyte will surely have been stretched. Right: example of force curve from a sandwich polyprotein (I27)3-synuclein-(I27)3, from the author. (C) Left: diagram of alternate pattern polyprotein. Bottom: example of force curve from alternate pattern. Notice that, as expected, the order of peaks reflects the hierarchy of mechanical stability of different modules and not their actual order in the construct. Experimental curves in (A) and (B) from the authors, (C) from (63).

Furthermore, not all modules are guaranteed to fold independently in a polyprotein construct. Finally, if the signal of the protein is not unambiguous, we might lose the structural gauges described above.

For these reasons, heteromeric polyproteins are today the strategy of choice for the SMFS study of IDPs. In this strategy, two different kinds of protein modules are inserted in the construct: here we call them the markers and the analyte. The marker should be a well-characterized protein module, proved to independently fold robustly in a multimodular construct, that yields a simple, strong unfolding peak that is immediately characterizable. Usually, the Ig-like titin modules like I27 and I28 satisfy correctly all these requisites. The analyte is the protein under investigation.

The "sandwich" pattern M_nAM_n, (M = marker, A = analyte) (see Fig. 14.5B) is usually the choice because it has the main advantage that we know that a single unit of the analyte has been stretched whenever we record $n + 1$ or more peaks coming from the unfolding of the markers.

Another possibility is the alternate $(MA)_n$ polyprotein (see Fig. 14.5C). An alternate polyprotein allows for amplified data acquisition, but might increase the difficulty of interpretation of signals if weak signals sum up in a complex fashion at the beginning of the stretching. Choice between the approaches thus depends on the nature of the analyte; the sandwich polyprotein is usually the safest alternative, mostly because with this configuration, the signal of each single analyte is more safely identified.

For building polyproteins, two strategies have been developed. The most used strategy relies on the genetic engineering of multimers by means of DNA concatamers (15). This is the most used current strategy. This strategy is the more flexible because it allows for specific modules to be inserted at a desired position (see above). However, only N—C terminal orientation is possible.

A possible alternative is a solid-state strategy, where the already expressed protein modules are assembled, chemically joining monomers via disulfide bonds (106). It requires the protein to be easily crystallized. Residues on the protein surface that are in close proximity with other monomers in the unit cell are modified with cysteines. Solid-state oxidation in the crystal leads to assembly of monomers into disulfide-linked strings. The same strategy can be employed without crystallization (by oxidizing the proteins in fluid and engineering-exposed cysteines in desired positions), but possibly losing directionality (23). This technique permits orientations alternative to the N—C, and it can be faster than the molecular biology-based one, but it is less flexible, allowing currently the building of homomeric polyproteins only. An analogous strategy has been developed for the building of protein constructs for optical tweezers (17).

14.2.4 Structural and Mechanical Parameters from SMFS Force Curves

Force spectroscopy can basically yield three raw variables related to the behavior of a protein molecule under tension: the increase in contour length ΔL_c, the unfolding force F_u, and the persistence length p. By fitting with the WLC model the rising part of each peak in the force curve, we can obtain the length ΔL_c of the hidden loops being exposed by the unfolding of each domain (see section 14.2.2 and Fig. 14.4). However, we have also seen that, formally, each peak indicates the release of an interaction that needs to be broken to proceed along the unraveling path of the polymer. Whenever unfolding intermediates are met along, the unfolding pathway "humps" (54) or smaller peaks (31) will appear superimposed on the entropic elasticity profile. The ΔL_c relative to the intermediates can, in most cases, be measured, thus uncovering further information on the actual interactions involved in the mechanical unfolding pathway.

The WLC fit also yields the persistence length p. A change in p usually indicates a change in the structure of the polymer: p for polypeptide chains is usually in the range between 0.4 and 0.8 nm (74), close to the length of one or a few amino acid units. The analysis of p can give also otherwise inaccessible information on the conformational equilibria of the polymer (see section 14.3.1).

Protein mechanical unfolding is a thermally driven probabilistic process: actually, the force never directly breaks a bond. Instead, force tilts the protein energy landscape until the barrier to unfolding becomes of the order of kT or smaller. Therefore, strictly speaking, we cannot talk about an "unfolding force" F_u, but only of a most probable unfolding force. The constant speed mechanical unfolding is not an equilibrium process but a kinetic process (96). This is the reason why the actual force distribution depends on the pulling speed (more precisely on the loading rate, that is, dF/dT). This latter parameter must be specified for any given unfolding force.

The distribution of the unfolding forces measured for a given module should contain information on the distribution of thermal energy available to each module prior to unfolding. This distribution is not Gaussian, but it has a characteristic shape skewed toward the lower forces (16, 26). This shape contains also information on the kinetic parameters that describe the mechanical unfolding energy landscape: more precisely, in a single-barrier, single-well approximation, K_{off} (interaction lifetime at zero force) and X_b (barrier position along the reaction coordinate X).

However, in practice, experimental errors can make it difficult to extract reliably those parameters from force distributions alone. The method of choice to obtain the kinetic parameters is based on the dependence of the unfolding force on the pulling speed (12, 25, 26, 96)

The difference area between the approaching and retracting curves gives the energy irreversibly dissipated by the system during the unfolding process. Exploiting the thermodynamic Jarzynski's equality, approaches have been attempted to obtain free unfolding energy information from the work dissipated in unfolding curves (34, 36, 103).

14.2.5 The Mechanical Resistance Depends on the Protein Secondary Structure

Table 14.1 evidences how β-sheet folds are, on average, more mechanically resistant than α-helical folds, with mixed α-β folds somewhere in the middle.

No known α-helical protein unfolds at forces above 60 pN in standard AFM-SMFS conditions, while β-sheet domains almost always unfold at forces above 60 pN. However, one could wonder if this datum is significative, given that only a few α-helical folds have been studied so far, while most studies concentrated on β-sheet folds like immunoglobulin-like (Ig-like) or fibronectin-like (Fn-like) folds.

To answer this question, J. I. Sulkowska and M. Cieplak employed coarse-grained simulations based on Go-like models to investigate the mechanical stability of 7510 protein structures gathered from the Protein Data Bank (87). Of these structures, 3813 have been assigned a class in the CATH structural classification scheme (71), allowing to easily correlate the structural class with

Table 14.1 Mechanical Resistance of β-Sheet, α-Helical, and Mixed α-β Folds.

Secondary Structure	Typical Force Range (Data from Reference 14 Unless Otherwise Specified)
Unstructured	<20 pN (N2B, PEVK, elastin-like peptides, random-coil conformation of α-synuclein [from Reference 79])
All α	<20 pN (calmodulin) to 60 pN (lysozyme); can give irregular peaks (spectrin) or *plateaux* (myosin)
α-β	~30 (DHFR) to 70–90 pN (barnase; DHFR with ligands)
Mostly or completely β	Usually 60 pN (C2A domain of synaptotagmin) to 300 pN (I28 domain of titin); most folds between 80 and 200 pN ; depending on force vector orientation can get as low as <20 pN (E2lip3)

the mechanical stability. The force distributions of the different structural classes, although broad, were consistent with the previous experimental data: with a few exceptions, α-helical folds are not expected to sustain strong forces, while the distribution of forces for α/β and β-sheet folds shows large tails at high unfolding forces. In particular, the set of strongest proteins (top 1.8% of the sample) is dominated by β-sandwiches (60%) and α-β rolls (30%), with Ig-like and ubiquitin-like topologies being by far the most represented. SMFS can therefore successfully monitor the presence of substantial β-structure in a protein fold at the single-molecule level, being this strongly correlated with high unfolding forces.

Finally, we have to remark that there is no full overlap between chemical and mechanical stability of a given protein fold. Thermodynamical stability ΔG or melting temperature $T_m = \Delta G/\Delta S$ do not correlate with mechanical stability (7). This is because the pathways of chemical and mechanical unfolding are not the same (28), and as such, are likely to experience alternative unfolding barriers.

14.2.6 The Protein Mechanical Resistance Depends on the Pulling Geometry

Mechanical stability is not a property of the protein fold itself, but of the protein fold and its orientation with respect to the pulling force vector.

In practice, the most important factor in deciding the mechanical resistance of a given fold/vector orientation couple is the geometry of the hydrogen bond network with respect to the orientation of the force vector. Two different general classes of geometries can be distinguished: *unzipping* geometries and *shearing* geometries. In unzipping geometries, two different protein strands are connected by hydrogen bonds parallel to the force direction. In this case, with each moment, force is loaded on one or a very few bonds. Each single bond thus rapidly breaks, leaving the force acting on the next one (see Fig. 14.6). The net force required to unfold the protein is thus never much larger

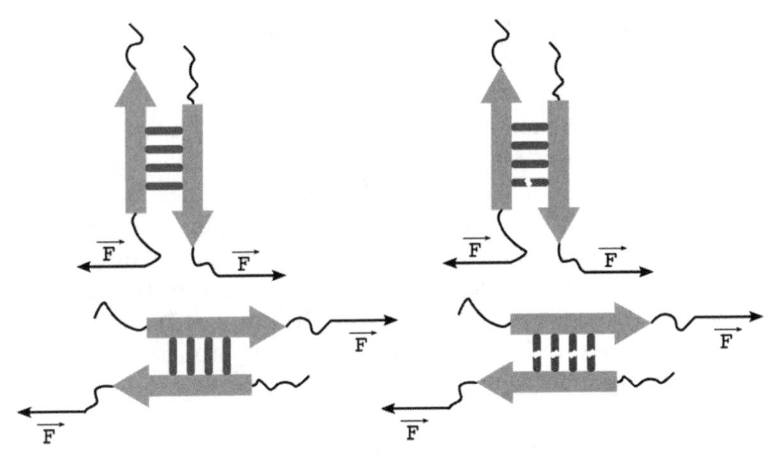

Figure 14.6 Unzipping and shearing geometries of hydrogen bond networks with respect to force. Top left: two β-strands (orange) connected in a zipping geometry with respect to the force vector F. The hydrogen bonds (red) are parallel to the force vector. Top right: under the stretching force, hydrogen bonds break one at a time (blue). Bottom left: two β-strands connected in a shearing geometry with respect to the force vector F. The hydrogen bonds are perpendicular to the force vector. Bottom right: to detach the two strands, all the hydrogen bonds must break at the same time (blue).

than that required to break a single hydrogen bond. Unzipping geometries have the lowest mechanical resistance, sometimes even below the AFM detection threshold.

In shearing geometries instead bonds are perpendicular to force direction. Thus, many bonds are put into tension at the same time. The contemporary rupture of all of them is much less probable, of course, than the breaking of a single bond, and thus much larger forces are needed to make it happen. Shearing geometries display unfolding forces usually above 60–100 pN. This kind of geometry is practically displayed only by β-sheet-rich proteins: α-helices in α-helix bundles, in fact, tend to peel from each other in a zipping-like fashion (as dramatically demonstrated by the case of the double helix coiled-coil of myosin [83]).

Since protein hydrogen bond networks are not spherically symmetric and force is a vector quantity, we must expect that the mechanical resistance of a given protein varies by varying the orientation of the pulling force vector with respect to the protein. This has been experimentally verified by Brockwell et al. and Carrion-Vazquez et al., pulling the same protein from directions alternative from the canonical N—C direction (9, 13). The unfolding force of the same protein has been shown to vary from less than 20 to 170 pN, depending on the direction of the force. Coarse-grained simulations have also permitted to observe the same pattern on a wide range of proteins, allowing to map the mechanical resistance of a fold in function of the residues on the sequence where force is applied for all the sequence. This analysis predicted that nearly every protein module shows wide variance in its mechanical stability depending on geometry (102).

The capability of pulling proteins in different directions has also been exploited in by H. Dietz and M. Rief to measure intramolecular distances between residues in a folded protein by means of an approach they call "mechanical triangulation" (24). In practice, three residues are chosen on the protein, and the protein is pulled between each combination of the three residues. The measured ΔL_c will depend on both the distances between two residues in the folded and in the unfolded chain. The distance in the folded protein between two residues positioned i-th and j-th in the primary sequence can be retrieved by the formula

$$\mathrm{d}(\mathrm{i},\mathrm{j}) = (j-i) * \mathrm{d}_{\mathrm{aa}} - \Delta L_c(i,j)$$

where $\Delta L_c\ (i, j)$ is the measured ΔL_c when the protein is stretched between the i-th and j-th residue and d_{aa} is the length of a single amino acid.

If the distances between a set of at least three amino acids are determined, the spatial position of those amino acids is known. Dietz and Rief proved this concept by triangulating the positions of three residues on the green fluorescent protein (GFP). The calculated distances correspond to those in the GFP crystal structure within 0.1 nm precision. The approach proved to be capable of giving direct geometrical information on protein structures at the single-molecule level, similarly with FRET, and could find applications in characterizing transient, diverse conformational ensembles.

14.3 THE SMFS METHODOLOGY IN THE STUDIES OF THE MECHANICAL PROPERTIES AND INTERACTIONS OF IDPS

14.3.1 Segments of Mechanical Proteins

The first applications of SMFS to the study of intrinsically disordered polypeptides came from the study of muscle proteins with mechanical function, and specifically titin. This choice was natural given that titin and other mechanical, multimodular proteins were the first natural subjects of SMFS studies (66, 76). Unfolded segments of titin provide a substantial contribution to the passive elasticity of vertebrate muscles. The evolutionary advantage is that the random coil segments can stretch and relax repeatedly with little heat dissipation, behaving as true entropic springs.

The first intrinsically disordered polypeptide studied by SMFS was the human cardiac titin PEVK segment (so called because of the prevalence of those four amino acids in its sequence) in the laboratory of Julio Fernandez in 2001 (48). The PEVK segment was already known to be unfolded by previous structural and microscopic studies (42).

SMFS experiments were performed on an alternate I27-PEVK polyprotein (see section 14.2.3). The experiments confirmed the unfolded state of PEVK and its mostly entropic elasticity: within the force resolution of the system (about 5 pN), no feature due to the detachment of intramolecular interactions was detected. However, conformational diversity was unveiled by persistence length analysis. The ensemble of PEVK segments showed a broad range of persistence lengths, from 0.4 to 2.5 nm, indicating the existence of individual molecules with different elasticity.

To investigate this diversity, single PEVK molecules were stretched and relaxed multiple times, thus allowing to measure the persistence length of the same molecule for each stretching-relaxation cycle. For each single molecule, the distribution of persistence lengths was much narrower than for the whole ensemble of different molecules—that is, each molecule retained its distinctive elasticity. This is visually confirmed by the fact that extension traces of the same PEVK molecule are superimposable, while traces from the extension of different PEVK segments are not (see Fig. 14.7). The existence of PEVK segments with stable different persistence lengths meant that the elasticity of each PEVK segment depends on some kinetically slow process that is substantially unaffected by forces in the tens to hundreds of piconewton range. The proposed explanation was that the elasticity of each PEVK segment depends on the *cis-trans* isomerization state of prolines.

Since PEVK segments were probed at least a few days after expression, one could reasonably expect that, unless the isomerization kinetics is extremely and unusually slow, the population of each proline isomer should equilibrate with each molecule and thus should be statistically similar for each molecule. This would lead to a narrow distribution of PEVK persistence lengths. However, this is not the case: the distribution is approximately flat (see Fig 14.7). Li et al. attribute this to the fact that each isomer influences, in turn, the propensity of neighboring proline to isomerize as well: in particular, if a poly-Pro II helix begins to form, it will cooperatively favor the *trans*-isomerization of neighboring prolines. Therefore, a much broader variety of distinct, varied isoforms of PEVK will form.

An alternative hypothesis to explain the persistence length diversity was, however, that of the "elasticity cassette." Since the PEVK segment is encoded by different exons, it could be that each exon encodes for a segment with different mechanical properties. Alternative splicing could therefore be a mechanism to tune the elasticity of titin in muscles. The hypothesis was not confirmed by subsequent SMFS measurements on constructs containing only one of each PEVK exon. Each exon displayed the same broad diversity of persistence lengths of the whole PEVK segment (81).

In a paper evidencing how essential is a correct data analysis in SMFS, Watanabe et al. (101) also tackled the problem of the elasticity of the PEVK segment, challenging the interpretation of the Fernandez group. They noticed that whenever two molecules were pulled together, the apparent persistence length was half than that measured on curves coming from the stretching of

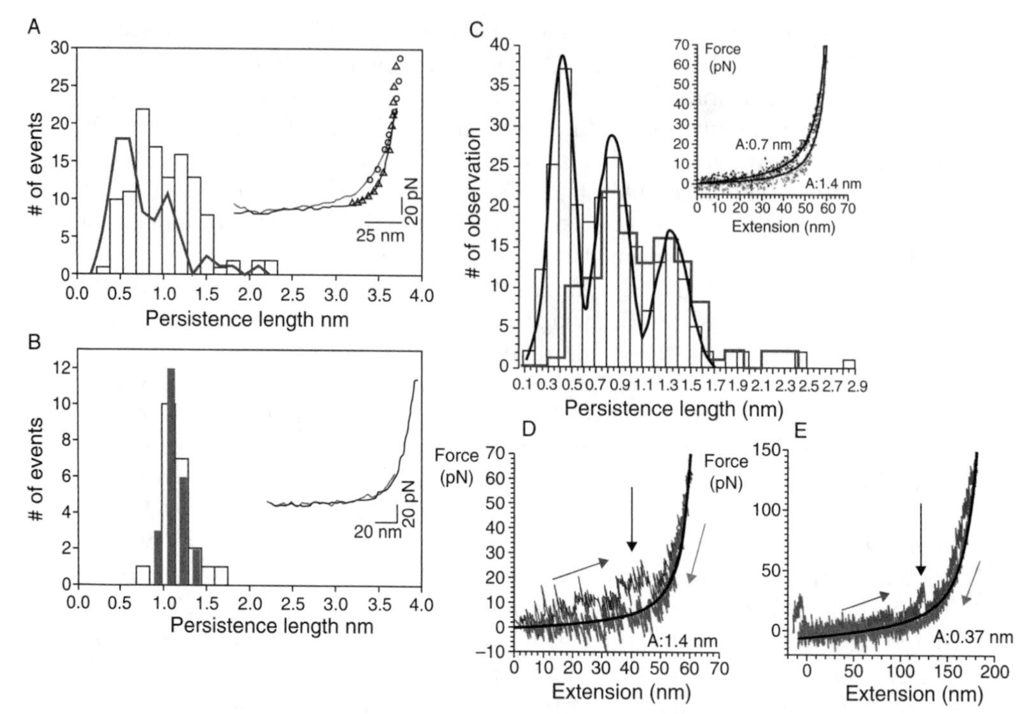

Figure 14.7 (A) Persistence length of the PEVK segment as measured by Li et al. (48) (gray bars) of a set of different single molecules. The WLC function fits the data within experimental noise, thus confirming that the PEVK segment behaves as an entropic spring. In the inset, two representative force curves corresponding to the stretching of a PEVK segment with different persistence lengths are superimposed (solid lines are experimental recordings, open symbols are Levenberg–Marquardt nonlinear fit of WLC model to the individual recordings). The red curve has a $p = 0.40\,\text{nm}$, the black recording has $p = 1.08\,\text{nm}$. (B) Persistence length of the PEVK segments of a single $(\text{I27-PEVK})_3$ molecule during repeated stretch and relaxation cycles. The persistence length is narrowly distributed around 1.1 nm and is the same for the stretch (gray bars) and the relaxation (red bars) traces. In the inset, two consecutive stretch (black) and relaxation (red) recordings. No hysteresis between stretching and relaxation was observed, indicating that this process is fully reversible. From Reference 48, (C) persistence length of PEVK segments as measured by Watanabe et al. (101), with the histogram by Li et al. superimposed (red line). (D) and (E) show peaks possibly coming from the unfolding of transient rarely populated structures (arrows) in the PEVK segment (D) and the N2B segment (E). Taken from Reference 101.

single molecules. In fact, the conformational space available to two molecules under stretching is twice as wide than that of a single molecule, thus dropping the apparent persistence length. By measuring persistence lengths from the pulling of both single and multiple molecules, they found that the persistence length distribution is regularly multimodal (see Fig. 14.7C), and the peak at longest persistence length is that of the single molecule. The persistence length of PEVK, in this case, is thus measured as being about 1.4 nm, with a sharp Gaussian distribution. Remarkably, the whole multimodal histogram overlaps the data of Li et al., thus putting into question the cooperative *trans*-isomerization hypothesis.

In the same study, Watanabe et al. also investigated the elasticity of the unfolded N2B unique sequence (N2B-Us) of titin. The persistence length of this segment was measured as 0.6 nm, less than half that of PEVK. The difference in elasticity between the two segments is attributed to the prevalence of proline residues in PEVK: the polyproline structures increase rigidity.

Intriguingly, Watanabe et al. reported the rare (~5%) but significant occurrence of small peaks in the stretching of both PEVK and N2B segments (see Fig. 14.7D,E). The authors attributed this occurrence to artifacts on the basis that structured conformations giving rise to strong force peaks were not predicted for the two segments and that the events were not always reproducible. By comparison with what we will see below (in section 14.4), one could wonder if these authors were among the first to observe conformational equilibria in unfolded proteins at the single-molecule level, even if they did not recognize it.

SMFS studies have also been reported on the unstructured EH segment of the muscle protein myomesin, which has an analogous function to that of PEVK in titin (6, 82). They found that the segment behaves mostly as an entropic spring. In contrast with PEVK, the persistence lengths of EH is sharply peaked around 0.3 nm, comparable with the length of one amino acid. Analogous to the study of Watanabe et al. reported above, in this case, the rare (~1%) occurrence of small force peaks that may be attributed to the unraveling of intramolecular structures is reported in the unfolding of EH segments.

14.3.2 Intrinsically Disordered Nucleoporins

The group of Ueli Aebi has studied the properties of flexible phenilalanine-glycine nucleoporins (FG domains) by means of several techniques, among those SMFS. FG domains are intrinsically unfolded (22) and form the physical constituents of the selective barrier-gate of nuclear pore complexes in eukaryotic cells. The main structural hypothesis on the function of FG domains is that they act as extended polymer brushes working as an entropic barrier to the nuclear pore, keeping molecules away by means of steric repulsion.

SMFS found that FG domains indeed behave as purely entropic springs upon stretching (52). To investigate their function, FG proteins were functionalized on gold nanodots and the repulsive force needed to indent the protein layer was investigated by means of SMFS. A considerable repulsion force between the tip and the surface was measured, consistent with the presence of a swollen polymer brush on the nanodot surface. The addition of 5% hexanediol to the buffer led to the cancellation of this repulsive force, probably due to the collapse of the individual FG domains to a more compact conformation. The transition was reversible: switching back the buffer to standard PBS restored the repulsive behavior.

In a landmark paper, Lim et al. proved by means of SMFS that the flexibility of FG domains is the fundamental biophysical parameter that nuclear transport proteins modulate to control the selective gating by the nuclear pore complex (51). The karyopherin Kapβ1 is a protein whose role is to facilitate the ingress of protein along the nuclear pore. SMFS demonstrated that addition of Kapβ1 in causes collapse of the FG repeats, thus relieving entropic repulsion. On the other hand, the protein Ran-GTP is known to terminate nuclear import, dissociating Kapβ1 from the FG domains. Subsequent addition of Ran-GTP to the FG-coated surface previously treated with Kapβ1 restored the entropic repulsion of the FG monolayer. If we translate these data in the nuclear pore complex, we can imagine the FG domains as building a gate with the expanded, "brush"-like conformation corresponding to the closed gate, and the collapsed, compact conformation corresponding to the open gate. Thus, SMFS experiments gave a mechanistic explanation of the nuclear pore complex gating, a fundamental problem of cell biology related to biophysical conformational equilibria of IDPs.

A third study investigated the interactions of FG repeats with soluble karyopherins like importin β. Using an importin β-coated tip, FG domains were stretched and a distinct plateau at relatively high forces (~200 pN), arising from the detaching of the interaction between FG and the importin, was detected on these force curves, thus directly monitoring the FG–importin tight binding. The additional work ΔW required to stretch FG domains in the presence of importin β was calculated to be around 1000 k_bT. The fact that a plateau several nanometers long appears, instead of a single peak superimposed on the entropic stretching of FG repeats, suggests that the importin–FG interface is supported by multiple interactions along the protein chain.

14.3.3 Engineered Elastic Polypeptides

Elastin-like polypeptides (ELPs) have been studied by the research group of Stefan Zauscher and coworkers (94, 95). ELPs are unfolded, elastic polypeptides made of consecutive VPGXG pentapeptides, where *x* is any amino acid except proline. ELPs present a lower critical solution temperature (LCST) phase transition. Above the LCST, the ELP is soluble; below it aggregates and precipitates. While other polymers show this kind of transition, the LCST of ELPs can be tuned precisely anywhere between 0 and 100°C. ELPs are of bioengineering interest as force generators, or tunable molecular switches to modulate the hydrophobicity of surfaces, and they can be cheaply and controllably produced by means of synthetic genes.

SMFS has revealed a few fundamental facts on the properties of ELPs. Valiaev et al. (94) showed that at forces between 200 and 260 pN, ELPs undergo a structural transition, monitored by a non-entropic jump in the force curve. Valiaev et al. fitted the curves with the freely jointed chain model (a model similar to the WLC) and found that the jump corresponds to a 0.2 nm increase of the Kuhn length (i.e, equivalent to a 0.1 nm increase in persistence length in the WLC model). Control experiments found that poly-Pro polypeptides featured the same transition under force, while poly-Ile polypeptides did not. Theoretical models of the cis-to-trans isomerization confirmed that the transition is consistent with mechanical force-induced *cis*-to-*trans* proline isomerization.

Recently, ELPs solvent-polymer interactions have been elucidated by means of SMFS (95). Single ELP molecules were stretched at 11, 25, and 42°C. The lower critical solution temperature (see above) of the studied ELP was 41°C. The measured distribution of Kuhn lengths were peaked on 0.33, 0.36, and 0.41 nm, respectively. The shift in the elastic properties of the polymer was attributed to the entropic contribution to hydration-free energy during unfolding: by rising the temperature, the solvent entropy increases and the entropic penalty caused by the solvation of nonpolar groups decreases correspondingly.

On testing different ELPs in widely varied conditions (increasing ionic strength, solvents of different polarity), a strong correlation between the entropic elasticity of the ELPs and the hydrophobic surface exposed to the solvent was unveiled, with Kuhn length decreasing with increasing hydrophobicity. This is due to the effect of the water shell around hydrophobic residues: the more hydrophobic residues are exposed and the larger the ordered water shell around the residues, the higher energy is required to stretch the polymer to a given elongation.

14.3.4 Intermolecular Interactions between IDPs

Force spectroscopy can also be employed to study *inter*-molecular recognitions and interactions between a couple of proteins. One molecule is covalently attached to the surface and the other is covalently linked to the tip, usually (but not always) with appropriate spacers. The two molecules are brought together and then pulled apart. Force peaks (see sections 14.2.1 and 14.2.4) or a plateau (in the case of unzipping geometries, see section 14.2.6) marking the detachment of the noncovalent interactions between the two molecules are recorded. This type of experiments have been conducted to study and explore the energy landscape of the molecular interactions under investigation (12, 46) and for the so-called "recognition imaging," that allows individual molecules to be spotted on the surface of cells (35). The analysis of the interaction force contains information about the mechanical dissociation rate, the mechanical dissociation energy landscape, and the behavior of the individual bond under mechanical force; parameters that are extremely difficult to obtain with bulk techniques. A detailed discussion of these experiments can be found in Reference 96.

Recently, this type of bimolecular force spectroscopy experiment has been applied to study the interaction among disordered, amyloidogenic proteins. Chad Ray and Boris B. Akhremitchev (73) pioneered this approach in 2005, analyzing the force of interaction among α-synuclein fragments. The force distribution at different loading rates cannot be fitted by models implying a single kind of interaction taking place between the peptides. This suggests the existence of different, multiple possible interactions among the peptides. Strikingly, the measured mechanical dissociation barrier heights were comparable to the mechanical unfolding energy barriers of fully formed β-sheet folds like that of titin.

McAllister et al. published a force spectroscopy protein–protein interaction study, comparing the self-dissociation forces for Alzheimer's β-peptide (Aβ), α-synuclein, and lysozyme in different pH conditions (55). For all three proteins, self-interaction force increased characteristically with a pH decrease, reaching an approximate maximum around pH 2–3. The authors observed that, at the same pH and conditions corresponding to the maximum interaction force, the protein acquires a substantial β-structure, as monitored by UV-CD. On the other hand, in control experiments where the protein was tethered only on the surface aspecific tip–protein interactions that contribute to the overall force, signals were recorded whose dependence to the pH, albeit not identical, was similar. These aspecific interactions were not minimized like in the work by Ray and Akhremitchev. In fact, McAllister et al. did not employ a polymeric spacer but instead directly tethered the protein to the AFM tip by glutaraldehyde. As the authors themselves recognize, glutaraldehyde is expected to "glue" covalently the protein aspecifically in multiple points of the sequence. This may strongly interfere with the pH-induced conformational changes under investigations. On the other hand, despite a perhaps nonoptimal tailoring and setup of the experiment, the study is among the first trying to tackle the problem of the conformational heterogeneity of IDPs at the single-molecule level.

More recently, Yu and Lyubchenko analyzed the self-dissociation energy landscape of full α-synuclein molecules at pH 2.7 (107). Dimers detached at forces ranging between 40 and 130 pN, in a loading rate interval between 200 and 100.000 pN/s. By using the classical Evans–Ritchie model (26, 96), they found evidence of two kinetic barriers to mechanical dissociation of α-synuclein dimers, with *k0* of about 4.0 and 0.08 s, respectively. The study suggests that α-synuclein dimers could have a lifetime of the order of 10^{0}–10^{-2} s, and it is a principle demonstration of the possibility to study IDP aggregation processes with single-molecule resolution.

14.4 PROBING THE CONFORMATIONAL EQUILIBRIA OF THE IDPS INVOLVED IN NEURODEGENERATIVE DISEASES: THE CASE OF α-SYNUCLEIN

14.4.1 α-Synuclein and Parkinson's Disease

The IDP α-synuclein is the main constituent of Lewy bodies, intracellular proteinaceous inclusion bodies whose formation is a cellular hallmark of Parkinson's disease (PD) and other neurodegenerative pathologies (89, 91, 105). Fibrils of α-synuclein found in the dopaminergic neurons of PD patients, when reproduced *in vitro* and investigated with spectroscopic techniques, reveal a high content of β-sheet secondary structure (72, 100). Thus, the aggregation process primarily requires the acquiring of β-structure (90).

Point-mutated α-synuclein responsible of familial PD, the A30P, A53T, and E46K mutants, have been reported to show marked differences in their aggregation behaviors when compared with WT α-synuclein, with regard to both fibrillization and oligomerization. Mutations promote the formation of annular protofibrils (A30P), full amyloid fibrils (E46K), or both (A53T) (18–20, 29, 37, 39, 40, 43, 44, 50). Wild-type α-synuclein retains the capability to form annular protofibrils, but only after extended incubation (43).

Quite surprisingly, in ensemble-averaged experiments, such marked contrasts in the aggregation behavior of the different mutants have not been found to correspond to analogously clear-cut differences in their monomeric conformational behavior (10, 20). Only subtle differences were detected at the monomeric level among WT, A30P, A53T, and E46K α-synuclein by means of various spectroscopic techniques in terms of the amount of NAC region shielding, (5, 69) N-terminal α-helix propensity, (11) and β-structure propensity at high concentration (50).

14.4.2 SMFS Experiments on α-Synuclein

We first applied the SMFS methodology to the study of WT α-synuclein. The stretching experiments were carried out on multimodular constructs containing the α-synuclein moiety because handles are needed to connect one end of the protein to the tip and the other to the substrate (see section 14.2.2). To this aim, we followed the design proposed by J. Fernandez for the study of the random-coiled titin N2B segment (96). A chimeric polyprotein composed of a single α-Synuclein module flanked on either side by three tandem I27 domains (see Fig. 14.8A, 3S3) was expressed (23, 24, 75). Two terminal cysteines were added at the C-terminal to promote binding to gold surfaces. The central α-synuclein module had the sequence of either WT α-synuclein or one of its pathogenic mutations A30P, A53T, and E46K.

The first main result of the experiments was that in physiological conditions (Tris 10 mM, pH 7.5), the signals coming from the stretching of α-synuclein are not homogeneous nor broadly diverse, but can be divided in three major,

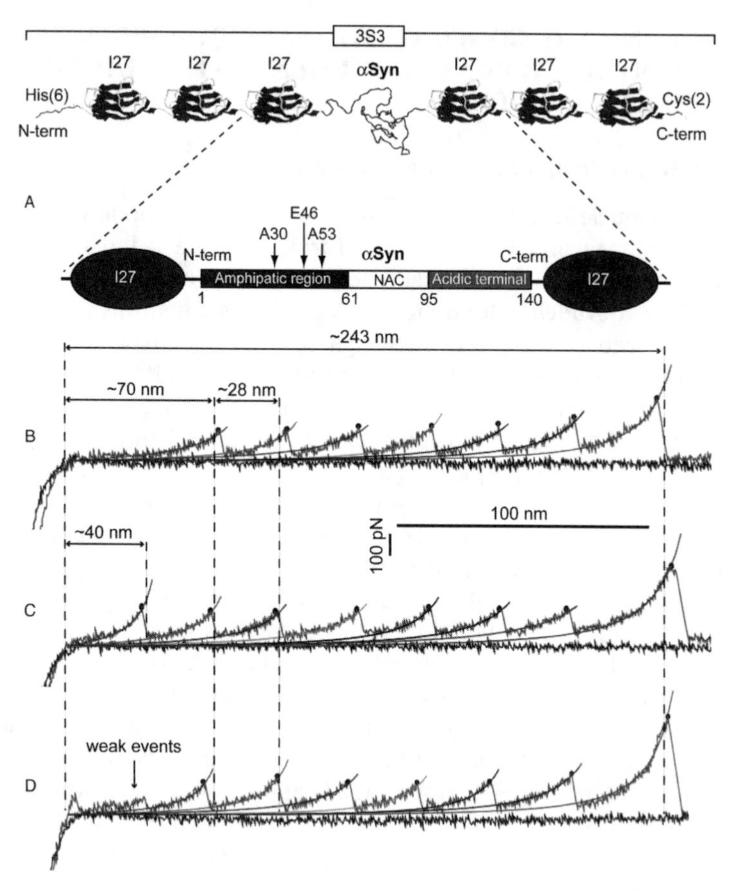

Figure 14.8 (A) Schematic representation of the polyprotein 3S3 constructs containing the α-synuclein sequence flanked on either side by three titin I27 modules, the N-terminal His-tag needed for purification purposes, and the C-terminal Cys-Cys tail needed for covalent attachment to the gold surface. In the α-synuclein moiety (enlarged), three regions are shown: (1) the amphipathic region, prone to fold in α-helical structures when in contact with phospholipid membranes; (2) the fibrillogenic NAC region, characteristic of the fibril core of α-synuclein amyloid; and (3) the acidic C-terminal tail, strongly charged and not prone to fold. The positions of mutation sites A30, E46, and A53 are indicated. (B) Example of curve characterized by a featureless region assigned to the stretching of α-synuclein moiety having, in this case, the mechanical properties of a random coil. This region is followed (from left to right) by six unfolding peaks ~200 pN high, with ~28-nm gaps between each, assigned to the unfolding of I27 domains. The characteristic contour lengths of the fully folded construct, of the I27 modules and of the wholly unfolded construct are shown (dashed lines). In all cases, contour lengths are obtained by fitting a worm-like chain equation with two free parameters (contour and persistence length). (C) Example of the curves featuring the β-like signature of α-synuclein, showing seven practically indistinguishable unfolding events of similar magnitude and spacing. (D) Example of the curves featuring the signature of mechanically weak interactions, showing small peaks preceding the six sawtooth-like peaks. Taken from Reference 10.

discrete classes of mechanically different monomeric α-synuclein conformations (see Fig. 14.8B–D).

The lifetime of these conformers is compatible with the SMFS measurement time scale, that is, longer than 10^{-3} s. SMFS thus showed that monomeric α-synuclein is thus capable to assume multiple, but discrete, structured conformations prior to any aggregation step (78).

Using this methodology, we could quantify the population of these classes of α-synuclein conformations, and how their relative abundance shifts in response to different environments. This allowed to understand the relationship between the observed conformational diversity and the pathogenic potential of α-synuclein In physico-chemical conditions known to induce α-synuclein aggregation, like low concentrations of Cu^{2+}, (8, 93) increased buffer ionic strength (78) experiments detected a substantial perturbation of the α-synuclein conformational equilibrium (see Fig. 14.9).

The mechanical behavior of one class of conformers is compatible with an amount of β-structuring comparable with that found in α-synuclein aggregates and amyloids. The proportion of these possibly β-containing conformations increased markedly in conditions known to be related to PD pathogenesis. This result suggested that the specific ensemble of conformations visited by α-synuclein at the monomeric level, and their relative abundance, can influence the whole aggregation process right at the onset.

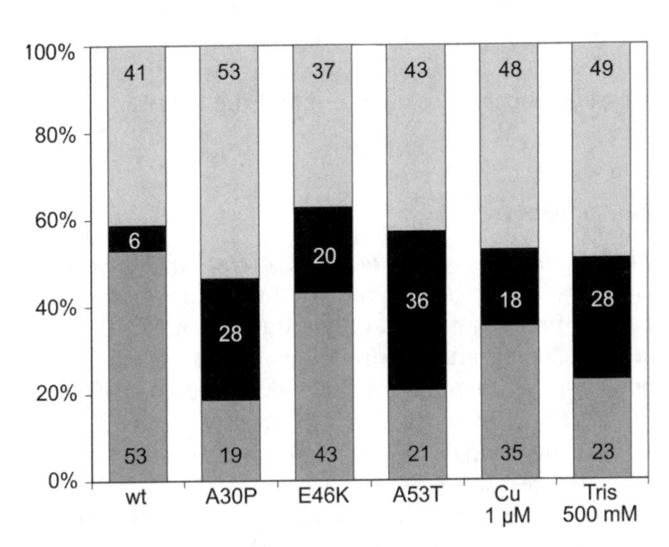

Figure 14.9 Population of α-synuclein conformers in six α-synuclein sequence variants or environmental conditions tested. Percentages observed for each curve type (see Fig. 14.8) of the wild-type protein ($n = 51$), the A30P mutant ($n = 43$), the E46K mutant ($n = 51$), the A53T mutant ($n = 63$), the wild type protein in presence of Cu 1 μM ($n = 34$) and the wild-type protein in Tris 500 mM buffer ($n = 61$). Data from References 78 and 10.

Afterward, we employed the same SMFS methodology to characterize the conformational equilibria of monomeric α-synuclein pathological mutants A30P, A53T, and E46K, and to evaluate the differences with respect to WT and among each other (10). This single-molecule approach monitored marked differences in the conformational behaviors of the mutants with respect to the WT sequence in contrast with the bulk ensemble-averaged spectroscopies. As shown in Figure 14.9, all three mutants show a markedly higher proportion of β-structure than WT α-synuclein, which was observed to have a β-conformation in only 6 ± 2% of the curves. A53T shows the largest β-structure propensity (37 ± 4%) while A30P and E46K have lower values (27 ± 3% and 20 ± 3%, respectively). In general, A30P and A53T show similar conformational equilibria, distinct from both the wild-type and the E46K.

The SMFS single-molecule approach thus proved to provide investigating capabilities that are inaccessible to any of the spectroscopies previously applied to study the structure of α-synuclein and other IDPs. First, working at the strictly single-molecule level ensured that the conformer distribution is detected and quantified without interference from oligomeric forms of the protein. Second, SMFS directly probes conformations with a lifetime longer than 10^{-3} s in the conformational space of the protein under investigation. These long-lived conformers might be the most biologically relevant. Third, the sensitivity of SMFS makes it possible to detect even rare (<10% population) conformers without the need of selectively enhancing their population by adding specific agents, as commonly done with bulk ensemble-averaged experiments. Due to an unexpected coincidence between the shape and height of the signal of one class with those of the I27-linker, our data analysis was carefully tuned up and the details of the such analysis are outlined.

14.4.3 Data Analysis

14.4.3.1 Force Curve Analysis and Interpretation Tens of thousands of force versus extension curves were recorded with an atomic force microscopy (AFM) apparatus for each mutant, only a fraction of which contained single-molecule mechanical unfolding information. After a first automatic data filtering to eliminate curves devoid of any significant number of unfolding peaks, only those traces which could be unambiguously assigned to the complete mechanical unfolding of single 3S3 molecules were tagged for successive measurement, while the rest were discarded.

The remaining curves were then examined and characterized in terms of position, height, and number of their mechanical unfolding peaks. It must be emphasized that the contour length increase caused by each rupture event allows to univocally quantify the span of the amino acidic chain that was involved in that interaction. This analysis evidenced only three homogeneous classes of curves.

One homogeneous group of curves showed six unfolding peaks, evenly spaced by a distance corresponding to the unfolding of an I27 module (see Fig. 14.10). In these curves, the average contour length measured at the first I27 unfolding peak was 68 ± 15 nm, corresponding to that of a fully stretched α-synuclein moiety (49 nm) plus six folded I27 modules (approximately 4–4.5 nm each) (65). This means that in these curves, α-synuclein was captured in a conformation whose extension does not imply the overcoming of measurable unfolding energy barriers, that is, that offered a purely entropic mechanical resistance to pulling. This is a typical behavior of random coils or mostly unfolded conformations, similar to the unfolded segments of mechanical proteins previously studied by SMFS (see sections 14.2.5 and 14.2.6). Thus, we assign this class to mainly unfolded α-synuclein monomer conformations.

In addition to the same entropic behavior previously identified in PEVK, N2B, and ELPs (see sections 14.3.1 and 14.3.3), force curves were recorded by us in the case of α-synuclein that reveal the presence also of variously structured conformers of this protein. These force curves were divided in two classes. A highly homogeneous group of curves showed seven equally spaced peaks (see Fig. 14.11). The average contour length measured at the second peak was 70 ± 10 nm, which is in good accord with the value found at the first

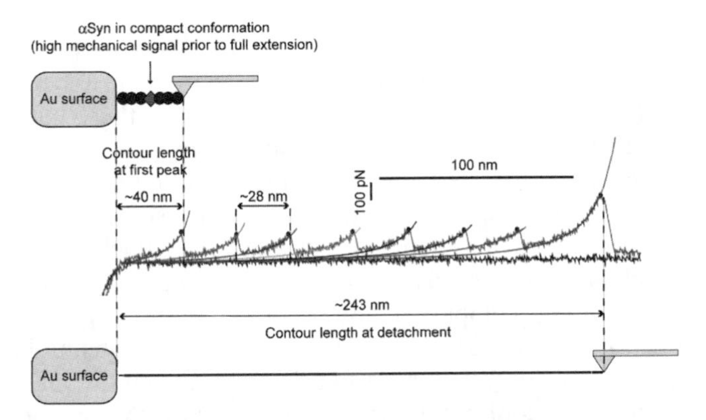

Figure 14.10 Single 3S3 molecule mechanical unfolding trace showing six mechanical events. Blue trace = approach. Green trace = retraction. The force curve shows six clean rupture peaks, separated by ~28 nm. The average contour length measured by WLC and fitted on the first rupture event is ~70 nm. This distance is compatible with the sum of six folded I27 modules (~4.5 nm each[1] = 4.5 × 6 = 27 nm) and one fully unfolded α-synuclein moiety (see cartoon at the top of the figure). The WLC contour length fitted on the last peak (detachment from surface) is ~243 nm, compatible with the expected length of a fully unfolded 3S3 construct (680 AA). Taken from from Reference 10. See color insert.

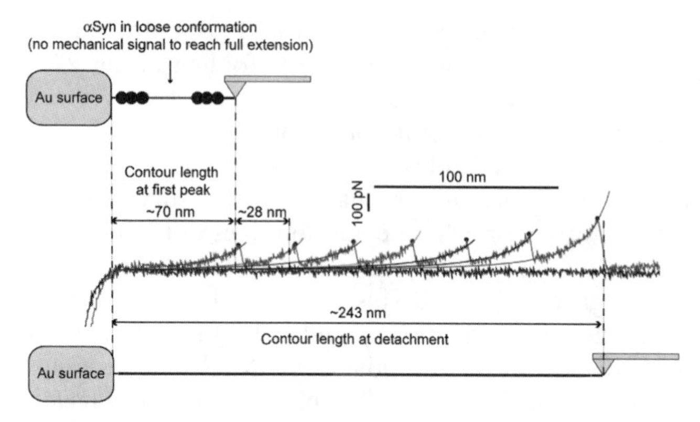

Figure 14.11 Example of a single 3S3 molecule mechanical unfolding traces showing seven mechanical events. Blue trace = approach. Green trace = retraction. The average WLC contour length fitted on the first rupture event is ~40 nm. This distance is compatible with the sum of six folded I27 modules (~4.5 nm each = 4.5 × 6 = 27 nm) and one partially folded α-Synuclein moiety (see cartoon at the top of the figure). The folded portion of α-synuclein has the same length of one I27 module (90AA, ~28 nm), which is compatible with the portion of α-synuclein found to assume a β-conformation in fully formed amyloid fibrils. The WLC contour length fitted on the last peak (detachment from surface) is ~243 nm, compatible with the expected length of a fully unfolded 3S3 construct (680 AA). Taken from Reference 10. See color insert.

I27 peak in the class of curves with six peaks described above. The seventh peak appears in the region that contains no signal when α-synuclein is in random coil. This does not imply that the first peak in these curves can be ascribed to the unfolding of α-synuclein, but rather, that one of the seven peaks must necessarily be due to the mechanical unfolding of an α-synuclein compact conformation since significant mechanical resistance is recorded at elongations which are smaller than a fully stretched α-synuclein. Thus, in these curves, α-synuclein was captured in a conformation that offers a high mechanical resistance to pulling, comparable to that of I27 modules. The only secondary structural motif known to be capable of showing such high unfolding forces is the antiparallel β-sheet (see section 14.2.5), so we assign this class of curves to β-rich α-synuclein conformations. In these curves, the average contour length of the first peak is only 40 ± 7 nm, which corresponds (after subtracting the length of six folded I27 domains) to a synuclein free chain of ~50 amino acids and, consequently, an α-synuclein folded domain of about 90 amino acids. Interestingly, this length matches the portion of α-synuclein that is found to be included in the fibrils after aggregation (72, 100).

Curves not assignable to the previous two classes were grouped in a third class. In all these curves, weak mechanical signals appeared between the first I27 peak and the contact point (see for instance the curve in Fig. 14.8D) signifying that α-synuclein was captured in conformations stabilized by interactions that offered a limited mechanical resistance to unfolding (heretofore mentioned as "mechanically weak interactions," MWI). Similar to what was found in the previously described classes of curves, the distance between the first I27 unfolding peak and contact point is 73 ± 13 nm, corresponding to the length of the totally unfolded α-synuclein plus six folded I27 modules.

The distance between the MWI signal and the contact point corresponds to the 3S3 chain length that is already accessible to extension prior to the mechanical rupture of the MWI. Accordingly, the distance between the MWI signal and the first I27 unfolding peak corresponds to the 3S3 chain length, which was rendered accessible to extension via the rupture of the same MWI. On the basis of the location of the mechanical signals, it is therefore possible to locate the corresponding MWIs along the chain (see Fig. 14.12).

14.4.3.2 Ruling Out Possible Artifacts It is important to note that the occurrence of a seventh peak in the class of curves described above cannot be attributed to the unfolding of additional I27 domains in dimerized 3S3 constructs.

First and foremost, a control experiment performed on WT 3S3 in disulfide-reducing conditions (78) showed the same occurrence of seven-peaked curves as in the experiments conducted without the addition of a reducing agent. In nonreducing conditions, 3S3 can form dimers, and we indeed observed an extremely small proportion of curves (i.e., about 0.002% of the recorded curves) attributable to the mechanical stretching of such molecules. However, these force traces could be easily distinguished from those generated by the unfolding of a monomeric 3S3 construct, and thus discarded.

Most importantly, no geometry of pulling of a 3S3-3S3 dimer can give seven-peaked force traces like those described above. If one such dimer was indeed picked up in a position that resulted in exactly seven I27 modules being trapped between the AFM tip and the surface, then at least one α-synuclein moiety would necessarily be also included (see Fig. 14.13 for an example of such a curve). Due to this, the resulting force trace would inevitably show a distance between the first I27 peak and the contact point equal or longer than the length of a fully extended α-synuclein molecule. Since the first peak in the class of seven-peaked curves consistently show a contour length of 46 ± 13 nm, this can only be ascribed to the unfolding of seven adjacent modules, one of which must necessarily be α-synuclein.

The possibility exists for a 3S3 construct in which α-synuclein is in a compact, "β-rich" conformation to be picked up by the AFM tip at a position which would result in a mechanical unfolding trace showing only six peaks. However, in this case, the contour length of the first peak would not be equal to a fully stretched α-synuclein plus six folded I27 modules as in the first class of curves described above. Rather, it would be significantly shorter (see Fig. 14.14).

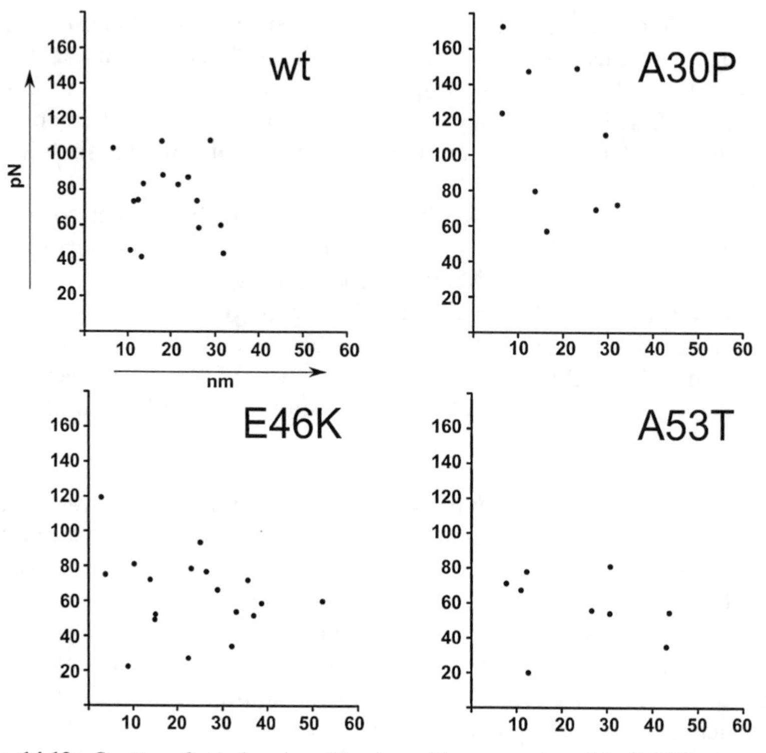

Figure 14.12 Scatter plots showing the size of loops enclosed by MWIs versus the corresponding measured interaction force for each of the four α-synuclein sequence variants. Taken from Reference 10.

The total construct length at detachment would also be noticeably shorter. This kind of signal was indeed observed. Other types of signal artifacts, such as the simultaneous pulling via the AFM tip of more than one 3S3 construct, can also be cogently ruled out as discussed elsewhere (78).

14.4.3.3 Possible Effects of I27 Linkers on the Conformational Equilibria The presence of I27 linkers could have several plausible repercussions on the behavior of the central α-synuclein module. The mere presence of such large objects linked to both termini of α-synuclein could give rise to entropic pulling effects, in turn leading to averagely more extended α-synuclein conformations. "Aspecific" interactions of the I27 linkers with α-synuclein are also plausible. We recently substantiated by CD spectroscopy (78) that the presence of I27 domains in 3S3 construct slightly increases the α-helical content of the central α-synuclein. I27 linkers could also give rise to "crowding effects" (3, 56, 58, 59). Since six I27 modules are present for each α-synuclein, the volume left accessible to α-synuclein conformational changes

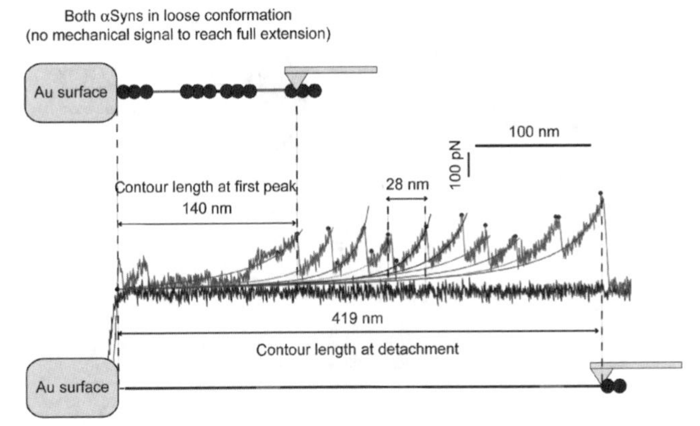

Figure 14.13 Example of one force curve resulting from the pulling of a 3S3-3S3 dimer. Dimers can form via oxidation of the C-terminal cysteines. While other thiol-functionalized proteins we studied in our lab are prone to dimerization via the formation of disulfide bonds, surprisingly low amounts of 3S3 dimers were observed in the present study (about 0.002% of the recorded curves were unambiguously attributed to dimerized 3S3–3S3 constructs). These traces could show up to 12 I27 mechanical rupture events, although we never observed curves with more than 10 peaks. The curve reported here has 10 I27 rupture events, and the contour lengths at first peak and at detachment are in accord with the situation depicted in the cartoon at the top of the figure. Curves with eight or more peaks were immediately recognized as dimers and discarded. Dimerized 3S3 construct could also give mechanical traces showing seven peaks. These cannot in any case be mistaken for the seven-peaked curves described in Figure 14.11 because no pulling geometry of a 3S3–3S3 dimer can give the contour length at first peak observed in monomeric 3S3 seven-peaked curves (please refer to Fig. 14.11). This is due to the fact that, if seven I27 modules of a dimer are stretched, then the mechanical extension trace of at least one α-synuclein moiety must be also included. Due to this, the minimum possible contour length at first peak for a seven-peaked curve generated by the stretching of a dimer is ($4.5 \times 7 = 31.5$ nm for the I27 modules + 49 nm for a fully stretched α-synuclein = ~80 nm). Of course, also the detachment contour length is accordingly higher in this type of curves. Taken from Reference 10. See color insert.

is severely limited. The effect of macromolecular crowding on α-synuclein aggregation properties is well characterized (62, 85, 92). Molecular crowding gives rise to excluded volume effects, which promote reactions with negative activation volumes such as molecular compactions and aggregations. Consequently, α-synuclein compact conformations should be more stable and have significantly longer lifetimes in our 3S3 constructs when compared with α-synuclein dissolved in ideal uncrowded solutions, thus leading us to detect a substantially larger amount of α-synuclein structured conformers. Intriguingly, the populations we estimated are in good agreement with data on monomeric α-synuclein obtained in conditions with high crowding or high α-synuclein

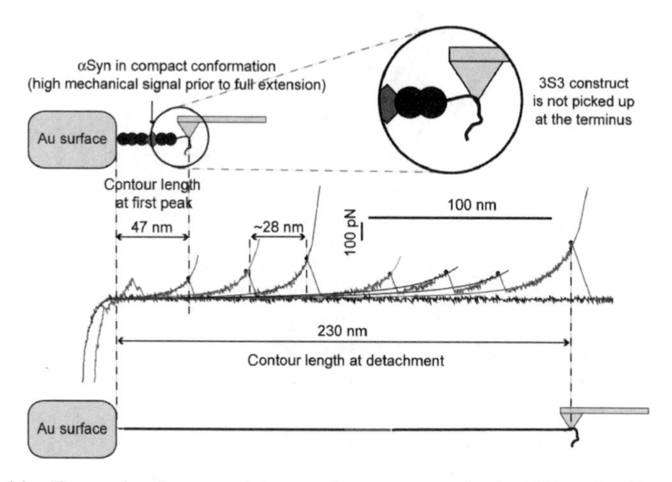

Figure 14.14 Example of one ambiguous force curve: single 3S3 molecule mechanical unfolding trace showing six mechanical unfolding events and a shorter-than-expected contour length for the first peak. As depicted in the cartoon at the top of the figure, an ambiguity is possible regarding force curves showing six peaks. The six mechanical unfolding peaks could arise from six I27 unfolding events (as in the class of curves described in Fig. 14.10), or from five I27 and one α-synuclein unfolding event. In the latter case, final contour length is lower than a full 3S3 construct (see the situation depicted in the cartoon at the top of the figure). Discerning between the two cases must rely on the contour length measurement of the first peak: the expected value for the first case is ~70 nm (see Fig. 14.10), while the expected value in the second case is comprised 35–70 nm. The shorter of the two values is one I27 module less than a fully folded 3S3 (~40–4.5 = ~35 nm, see Fig. 14.11), while the longer is one fully unfolded I27 module plus the unfolded portion of α-synuclein (~50AA). Also due to intrinsic errors of WLC fitting and contact point determination, discerning between the two situations is impossible; this means that a portion of the curves classified in the same class described in Figure 14.10 could be due to 3S3 conformations in which α-synuclein is in a compact conformation. This portion can be expected to be higher in situations in which α-synuclein is more prone to assume compact conformations, thus reducing the average contour length at first peak of curves showing six peaks. This expectation is indeed confirmed by the measured contour lengths (10, 78), thus providing a further confirmation of the internal consistency of our data. Taken from Reference 10. See color insert.

concentration (1, 47, 50). Of course, the fact that the I27 modules are covalently linked to both termini of α-synuclein, rather than floating in solution, introduces further effects, in addition to simple crowding, as it is often the case in natural or chimeric multimodular proteins (3, 4, 33), whose contribution to the final observed conformer population is at the moment not easy to discern.

On the other hand, regardless of how exactly these effects can modify the conformational equilibria of α-synuclein included in the 3S3 construct, being the same protein and its mutants inserted in the same stereochemical context (all 3S3 constructs are identical, except for the relevant α-synuclein point mutations), the differences we observed in the relative abundance of their conformational classes can be considered to be representative of the effect that these mutations have on free monomeric α-synuclein.

14.5 FUTURE PERSPECTIVES

14.5.1 Improving SMFS Sensitivity

The use of SMFS to probe the conformational equilibria of IDPs is just beginning to take place. One of the main limitations of the SMFS is sensitivity: we have seen (see section 14.2.5) that SMFS is sensitive to the secondary structure of the protein being probed, but it still cannot give an amount of structural information comparable to that of in-bulk techniques. The limitation in force resolution is especially problematic for IDPs, where conformers with labile or partially folded structures are likely to form. Two routes are currently in development to overcome these limitations. The first is acquiring a better force resolution, as to be able to solve the unraveling of single hydrogen bonds and, ultimately, the secondary structure of the protein into study. Currently, the only limit to force resolution is cantilever thermal fluctuation due to thermal noise. A reduction of the size of the force sensor, keeping the same stiffness, should improve the signal-to-noise ratio (30). Small cantilevers have been tested and allow a resolution of about 7–10 pN, two to three times better than that achieved with standard cantilevers (83, 99). Usage of these cantilevers is however still extremely limited in the SMFS community due to them still being scarcely available commercially, highly expensive, and technically difficult to manipulate.

An alternative approach to improve the force resolution is lock-in force spectroscopy, recently proposed by Schlierf et al. (81). The basic idea behind this technique is oscillating the tip with a small (~5 nm) amplitude. Recording the amplitude transmission of the oscillation through the polypeptide to the cantilever gives information on the elasticity of the individual molecule and this signal can be directly converted, multiplying it with a reference signal, to a low-noise force curve with resolution of 0.4 pN. This technique allowed to directly observe refolding of folded modules with constant speed force spectroscopy (81).

Another long-sought objective is the coupling optical and force measurements on a single molecule. By using appropriate FRET probes and optical equipment, it could be possible to see what residues are being pulled apart mechanically during an SMFS experiment. The numerous technological challenges still prevent this to be a viable approach, but the first attempts have been made (80).

14.5.2 Improving Data Throughput and Analysis

Current SMFS instrumentation has not been designed for high data throughput. This is a limiting parameter in the study of IDPs by means of SMFS because a full characterization of conformational equilibria in a single experimental condition can still take weeks of experiments. Such timescales are of course unreasonable if SMFS wants to leap toward becoming a routine screening technique for pharmacological agents or mutational analysis.

Three main challenges have to be resolved. The first is for SMFS experiments to be fully automated, gathering data without operators having to supervise the (hours-long) experiment. The second is automatization of data analysis and statistical collection: as described above, SMFS experiments return thousands of raw force curves that have to be sorted, classified, and measured to find the few precious ones corresponding to single-molecule unfolding events. The last challenge is increasing the actual rate of data harvesting that currently cannot exceed much the limit of one force curve per second per instrument.

Automatic classification of force curves is a long-standing problem, on which a few interesting progresses have been made (38, 41, 67) but that currently has no generally accepted solution. Most recently, a promising automated SMFS instrument, with integrated curve selection and analysis, has been presented (86). The instrument automatically aligns and calibrates the cantilever, controls temperature and buffer conditions in the fluid cell, and classifies and measures the force curves in output. The future commercial availability of such greatly simplified setups will be fundamental for the standardization and diffusion of SMFS as a routine biophysical technique in the analysis of IDPs and other proteins.

To increase the data throughput of each single instrument, parallel designs have been proposed, like in the "tip arrays" (57) and the "Millipede" design (98) devices. These approaches could allow "multiple single-molecules" approaches in which different single molecules are picked up by different tips in the same tip approach–retraction cycle, each tip being effectively independent. These methodologies have also the advantage of allowing for "reference cantilever" to be used, effectively removing problems due to mechanical noise and thermal drifts.

14.5.3 Improving Polyprotein Design

It is clear that many more factors than the currently recognized have to be considered for polyprotein design. An ideal marker in fact has to (1) be easily distinguished from the analyte, and (2) interfere as little as possible with the analyte.

Due to the stochastical nature of the bond rupture, the first requirement cannot be fulfilled by a difference in force. The marker requirement about force usually is simply: as strong as possible, to increase the odds of having the analyte signal in a known place in the force curve (i.e., before the first marker unfolding peak). A difference in unfolding length, instead, can be measured with relatively high precision, and can be easily used to distinguish a known marker from the analyte. Unfolding length depends only on the nature of the fold and on the module sequence length, and therefore, does not usually change randomly or depend on buffer conditions. Unfolding length can therefore often be predicted, at least roughly, for the analyte and the marker as well. Unfolding length can also be shortened for a given protein module by engineering disulfide bonds in the module structure, mostly retaining mechanical properties, as successfully demonstrated by the group of Julio Fernandez (104).

The second requirement is more subtle. A polyprotein construct puts the analyte and the marker in extremely close proximity, and it can be expected this alters the analyte surrounding environment and therefore the analyte properties. There is evidence that even modules belonging to naturally multimodular proteins, and that therefore have evolved to be independently folding units, are slightly but measurably influenced by the nature of neighboring modules (32, 49). In any case, the presence of two large moieties flanking the analyte unit has entropic consequences that must be taken into account (84, 96).

It can be possible, however, to design the marker to minimize and control these influencing effects. The most important factor is the electrostatic nature of the marker and analyte surfaces. Markers having an electrostatic charge complementing that of the analyte should be avoided since it is virtually certain that the analyte and the marker will strongly interact, potentially altering the analyte structure. On the same basis, markers and analytes with large exposed hydrophobic patches should be avoided too. It is conceivable that the best combination is that of markers and analyte with roughly the same electrostatic charge, leading to repulsion and minimization of marker–analyte interactions. Tailoring the electrostatic surface of an already known protein module, although requiring substantial work, can be made, as shown by Reference 45.

Advanced hypotheses for marker design include the possibility of building "modular" marker libraries, in which markers are joined to the analyte by means of "click chemistry" (60), so that different suitable markers are exploited for different purposes and proteins, without having to redesign a whole polyprotein each time. Nonprotein markers can also be considered, as for example polysaccharide markers; however, stretching polysaccharides do not give rise to clear-cut peak signals and can easily adhere to proteins, thus making their usefulness as markers still uncertain.

REFERENCES

1. Apetri, M. M., N. C. Maiti, M. G. Zagorski, P. R. Carey, and V. E. Anderson. 2006. Secondary structure of alpha-synuclein oligomers: characterization by Raman and atomic force microscopy. J Mol Biol **355**:63–71.
2. Banavar, J. R., and A. Maritan. 2007. Physics of proteins. Annu Rev Biophys Biomol Struct **36**:261–80.
3. Batey, S., and J. Clarke. 2008. The folding pathway of a single domain in a multidomain protein is not affected by its neighbouring domain. J Mol Biol **378**:297–301.
4. Batey, S., L. G. Randles, A. Steward, and J. Clarke. 2005. Cooperative folding in a multi-domain protein. J Mol Biol **349**:1045–59.
5. Bertoncini, C. W., C. O. Fernandez, C. Griesinger, T. M. Jovin, and M. Zweckstetter. 2005. Familial mutants of alpha-synuclein with increased neurotoxicity have a destabilized conformation. J Biol Chem **280**:30649–52.
6. Bertoncini, P., R. Schoenauer, I. Agarkova, M. Hegner, J. C. Perriard, and H. J. Guntherodt. 2005. Study of the mechanical properties of myomesin proteins using dynamic force spectroscopy. J Mol Biol **348**:1127–37.
7. Best, R. B., B. Li, A. Steward, V. Daggett, and J. Clarke. 2001. Can non-mechanical proteins withstand force? Stretching barnase by atomic force microscopy and molecular dynamics simulation. Biophys J **81**:2344–56.
8. Binolfi, A., R. M. Rasia, C. W. Bertoncini, M. Ceolin, M. Zweckstetter, C. Griesinger, T. M. Jovin, and C. O. Fernandez. 2006. Interaction of alpha-synuclein with divalent metal ions reveals key differences: a link between structure, binding specificity and fibrillation enhancement. J Am Chem Soc **128**:9893–901.
9. Brockwell, D. J., E. Paci, R. C. Zinober, G. S. Beddard, P. D. Olmsted, D. A. Smith, R. N. Perham, and S. E. Radford. 2003. Pulling geometry defines the mechanical resistance of a beta-sheet protein. Nat Struct Biol **10**:731–7.
10. Brucale, M., M. Sandal, S. Di Maio, A. Rampioni, I. Tessari, M. Bisaglia, L. Bubacco, and B. Samorì. 2009. Pathogenic mutations shift the equilibria of Alpha Synuclein single molecules towards structured conformers. ChemBioChem **10**(1):176–83.
11. Bussell, R., and D. Eliezer. 2001. Residual structure and dynamics in Parkinson's disease-associated mutants of alpha-synuclein. J Biol Chem **276**:45996–6003.
12. Bustanji, Y., C. R. Arciola, M. Conti, E. Mandello, L. Montanaro, and B. Samori. 2003. Dynamics of the interaction between a fibronectin molecule and a living bacterium under mechanical force. Proc Natl Acad Sci U S A **100**:13292–7.
13. Carrion-Vazquez, M., H. Li, H. Lu, P. E. Marszalek, A. F. Oberhauser, and J. M. Fernandez. 2003. The mechanical stability of ubiquitin is linkage dependent. Nat Struct Biol **10**:738–43.
14. Carrion-Vazquez, M., A. F. Oberhauser, H. Diez, R. Hervas, J. Oroz, J. Fernandez, and M.-M. D. 2006. Protein Nanomechanics—as Studied by AFM Single-Molecule Force Spectroscopy, pp. 163–245. In J. L. R. A. and A. Alonso (eds.), Advanced Techniques in Biophysics. Springer-Verlag, Berlin.
15. Carrion-Vazquez, M., A. F. Oberhauser, T. E. Fisher, P. E. Marszalek, H. Li, and J. M. Fernandez. 2000. Mechanical design of proteins studied by single-molecule force spectroscopy and protein engineering. Prog Biophys Mol Biol **74**:63–91.

16. Carrion-Vazquez, M., A. F. Oberhauser, S. B. Fowler, P. E. Marszalek, S. E. Broedel, J. Clarke, and J. M. Fernandez. 1999. Mechanical and chemical unfolding of a single protein: a comparison. Proc Natl Acad Sci U S A **96**:3694–9.

17. Cecconi, C., E. A. Shank, F. W. Dahlquist, S. Marqusee, and C. Bustamante. 2008. Protein-DNA chimeras for single molecule mechanical folding studies with the optical tweezers. Eur Biophys J **37**:729–38.

18. Choi, W., S. Zibaee, R. Jakes, L. C. Serpell, B. Davletov, R. A. Crowther, and M. Goedert. 2004. Mutation E46K increases phospholipid binding and assembly into filaments of human alpha-synuclein. FEBS Lett **576**:363–8.

19. Conway, K. A., S. J. Lee, J. C. Rochet, T. T. Ding, J. D. Harper, R. E. Williamson, and P. T. Lansbury, Jr. 2000. Accelerated oligomerization by Parkinson's disease linked alpha-synuclein mutants. Ann N Y Acad Sci **920**:42–5.

20. Conway, K. A., S. J. Lee, J. C. Rochet, T. T. Ding, R. E. Williamson, and P. T. Lansbury, Jr. 2000. Acceleration of oligomerization, not fibrillization, is a shared property of both alpha-synuclein mutations linked to early-onset Parkinson's disease: implications for pathogenesis and therapy. Proc Natl Acad Sci U S A **97**:571–6.

21. Daggett, V., and A. R. Fersht. 2003. Is there a unifying mechanism for protein folding? Trends Biochem Sci **28**:18–25.

22. Denning, D. P., S. S. Patel, V. Uversky, A. L. Fink, and M. Rexach. 2003. Disorder in the nuclear pore complex: the FG repeat regions of nucleoporins are natively unfolded. Proc Natl Acad Sci U S A **100**:2450–5.

23. Dietz, H., M. Bertz, M. Schlierf, F. Berkemeier, T. Bornschlogl, J. P. Junker, and M. Rief. 2006. Cysteine engineering of polyproteins for single-molecule force spectroscopy. Nat Protoc **1**:80–4.

24. Dietz, H., and M. Rief. 2006. Protein structure by mechanical triangulation. Proc Natl Acad Sci U S A **103**:1244–7.

25. Evans, E. 2001. Probing the relation between force-lifetime-and chemistry in single molecular bonds. Annu Rev Biophys Biomol Struct **30**:105–28.

26. Evans, E., and K. Ritchie. 1997. Dynamic strength of molecular adhesion bonds. Biophys J **72**:1541–55.

27. Fernandez-Busquets, X., N. S. de Groot, D. Fernandez, and S. Ventura. 2008. Recent structural and computational insights into conformational diseases. Curr Med Chem **15**:1336–49.

28. Fowler, S. B., R. B. Best, J. L. Toca Herrera, T. J. Rutherford, A. Steward, E. Paci, M. Karplus, and J. Clarke. 2002. Mechanical unfolding of a titin Ig domain: structure of unfolding intermediate revealed by combining AFM, molecular dynamics simulations, nmR and protein engineering. J Mol Biol **322**:841–9.

29. Fredenburg, R. A., C. Rospigliosi, R. K. Meray, J. C. Kessler, H. A. Lashuel, D. Eliezer, and P. T. Lansbury. 2007. The impact of the E46K mutation on the properties of alpha-synuclein in its monomeric and oligomeric states. Biochemistry **46**:7107–18.

30. Gittes, F., and C. F. Schmidt. 1998. Thermal noise limitations on micromechanical experiments. Eur Biophys J **27**:75–81.

31. Grandi, F., M. Sandal, G. Guarguaglini, E. Capriotti, R. Casadio, and B. Samori. 2006. Hierarchical mechanochemical switches in angiostatin. Chembiochem **7**:1774–82.

32. Hamill, S. J., A. E. Meekhof, and J. Clarke. 1998. The effect of boundary selection on the stability and folding of the third fibronectin type III domain from human tenascin. Biochemistry **37**:8071–9.

33. Han, J. H., S. Batey, A. A. Nickson, S. A. Teichmann, and J. Clarke. 2007. The folding and evolution of multidomain proteins. Nat Rev Mol Cell Biol **8**:319–30.

34. Harris, N. C., Y. Song, and C. H. Kiang. 2007. Experimental free-energy surface reconstruction from single-molecule force spectroscopy using Jarzynski's equality. Phys Rev Lett **99**:68101–104.

35. Hinterdorfer, P., and Y. F. Dufrene. 2006. Detection and localization of single molecular recognition events using atomic force microscopy. Nat Methods **3**:347–55.

36. Imparato, A., F. Sbrana, and M. Vassalli. 2008. Reconstructing the free energy landscape of a polyprotein by single-molecule experiments. Europhys Lett **82**:58006.

37. Kamiyoshihara, T., M. Kojima, K. Ueda, M. Tashiro, and S. Shimotakahara. 2007. Observation of multiple intermediates in alpha-synuclein fibril formation by singular value decomposition analysis. Biochem Biophys Res Commun **355**:398–403.

38. Kasas, S., B. M. Riederer, S. Catsicas, B. Cappella, and G. Dietler. 2000. Fuzzy logic algorithm to extract specific interaction forces from atomic force microscopy data. Rev Sci Instrum **71**:2082–6.

39. Koo, H. J., H. J. Lee, and H. Im. 2008. Sequence determinants regulating fibrillation of human alpha-synuclein. Biochem Biophys Res Commun **368**:772–8.

40. Krishnan, S., E. Y. Chi, S. J. Wood, B. S. Kendrick, C. Li, W. Garzon-Rodriguez, J. Wypych, T. W. Randolph, L. O. Narhi, A. L. Biere, M. Citron, and J. F. Carpenter. 2003. Oxidative dimer formation is the critical rate-limiting step for Parkinson's disease alpha-synuclein fibrillogenesis. Biochemistry **42**:829–37.

41. Kuhn, M., H. Janovjak, M. Hubain, and D. J. Muller. 2005. Automated alignment and pattern recognition of single-molecule force spectroscopy data. J Microsc **218**:125–32.

42. Labeit, S., and B. Kolmerer. 1995. Titins: giant proteins in charge of muscle ultrastructure and elasticity. Science **270**:293–6.

43. Lashuel, H. A., D. Hartley, B. M. Petre, T. Walz, and P. T. Lansbury, Jr. 2002. Neurodegenerative disease: amyloid pores from pathogenic mutations. Nature **418**:291.

44. Lashuel, H. A., B. M. Petre, J. Wall, M. Simon, R. J. Nowak, T. Walz, and P. T. Lansbury, Jr. 2002. Alpha-synuclein, especially the Parkinson's disease-associated mutants, forms pore-like annular and tubular protofibrils. J Mol Biol **322**:1089–102.

45. Lawrence, M. S., K. J. Phillips, and D. R. Liu. 2007. Supercharging proteins can impart unusual resilience. J Am Chem Soc **129**:10110–10112.

46. Leckband, D. 2004. Nanomechanics of adhesion proteins. Curr Opin Struct Biol **14**:524–30.

47. Lee, J. C., R. Langen, P. A. Hummel, H. B. Gray, and J. R. Winkler. 2004. Alpha-synuclein structures from fluorescence energy-transfer kinetics: implications for the role of the protein in Parkinson's disease. Proc Natl Acad Sci U S A **101**:16466–71.

48. Li, H., A. F. Oberhauser, S. D. Redick, M. Carrion-Vazquez, H. P. Erickson, and J. M. Fernandez. 2001. Multiple conformations of PEVK proteins detected by single-molecule techniques. Proc Natl Acad Sci U S A **98**:10682–6.

49. Li, H. B., A. F. Oberhauser, S. B. Fowler, J. Clarke, and J. M. Fernandez. 2000. Atomic force microscopy reveals the mechanical design of a modular protein. Proc Natl Acad Sci U S A **97**:6527–31.

50. Li, J., V. N. Uversky, and A. L. Fink. 2002. Conformational behavior of human alpha-synuclein is modulated by familial Parkinson's disease point mutations A30P and A53T. Neurotoxicology **23**:553–67.

51. Lim, R. Y., B. Fahrenkrog, J. Koser, K. Schwarz-Herion, J. Deng, and U. Aebi. 2007. Nanomechanical basis of selective gating by the nuclear pore complex. Science **318**:640–3.

52. Lim, R. Y., N. P. Huang, J. Koser, J. Deng, K. H. Lau, K. Schwarz-Herion, B. Fahrenkrog, and U. Aebi. 2006. Flexible phenylalanine-glycine nucleoporins as entropic barriers to nucleocytoplasmic transport. Proc Natl Acad Sci U S A **103**:9512–7.

53. Lin, J. C., and H. L. Liu. 2006. Protein conformational diseases: from mechanisms to drug designs. Curr Drug Discov Technol **3**:145–53.

54. Marszalek, P. E., H. Lu, H. Li, M. Carrion-Vazquez, A. F. Oberhauser, K. Schulten, and J. M. Fernandez. 1999. Mechanical unfolding intermediates in titin modules. Nature **402**:100–3.

55. McAllister, C., M. A. Karymov, Y. Kawano, A. Y. Lushnikov, A. Mikheikin, V. N. Uversky, and Y. L. Lyubchenko. 2005. Protein interactions and misfolding analyzed by AFM force spectroscopy. J Mol Biol **354**:1028–42.

56. McNulty, B. C., G. B. Young, and G. J. Pielak. 2006. Macromolecular crowding in the Escherichia coli periplasm maintains alpha-synuclein disorder. J Mol Biol **355**:893–97.

57. Minne, S. C., G. Yaralioglu, S. R. Manalis, J. D. Adams, J. Zesch, A. Atalar, and C. F. Quate. 1998. Automated parallel high-speed atomic force microscopy. Appl Phys Lett **72**:2340–2.

58. Minton, A. P. 2000. Implications of macromolecular crowding for protein assembly. Curr Opin Struct Biol **10**:34–39.

59. Morar, A. S., A. Olteanu, G. B. Young, and G. J. Pielak. 2001. Solvent-induced collapse of alpha-synuclein and acid-denatured cytochrome c. Protein Sci **10**:2195–9.

60. Moses, J. E., and A. D. Moorhouse. 2007. The growing applications of click chemistry. Chem Soc Rev **36**:1249–62.

61. Mukhopadhyay, S., R. Krishnan, E. A. Lemke, S. Lindquist, and A. A. Deniz. 2007. A natively unfolded yeast prion monomer adopts an ensemble of collapsed and rapidly fluctuating structures. Proc Natl Acad Sci U S A **104**:2649–54.

62. Munishkina, L. A., A. Ahmad, A. L. Fink, and V. N. Uversky. 2008. Guiding protein aggregation with macromolecular crowding. Biochemistry **47**:8993–9006.

63. Oberhauser, A. F., C. Badilla-Fernandez, M. Carrion-Vazquez, and J. M. Fernandez. 2002. The mechanical hierarchies of fibronectin observed with single-molecule AFM. J Mol Biol **319**:433–47.

64. Oberhauser, A. F., and M. Carrion-Vazquez. 2008. Mechanical biochemistry of proteins one molecule at a time. J Biol Chem **283**:6617–21.

65. Oberhauser, A. F., P. E. Marszalek, M. Carrion-Vazquez, and J. M. Fernandez. 1999. Single protein misfolding events captured by atomic force microscopy. Nat Struct Biol **6**:1025–8.

66. Oberhauser, A. F., P. E. Marszalek, H. P. Erickson, and J. M. Fernandez. 1998. The molecular elasticity of the extracellular matrix protein tenascin. Nature **393**:181–5.

67. Odorico, M., J. M. Teulon, O. Berthoumieu, S. W. Chen, P. Parot, and J. L. Pellequer. 2007. An integrated methodology for data processing in dynamic force spectroscopy of ligand-receptor binding. Ultramicroscopy **107**:887–94.

68. Onuchic, J. N., and P. G. Wolynes. 2004. Theory of protein folding. Curr Opin Struct Biol **14**:70–5.

69. Palecek, E., V. Ostatna, M. Masarik, C. W. Bertoncini, and T. M. Jovin. 2008. Changes in interfacial properties of alpha-synuclein preceding its aggregation. Analyst **133**:76–84.

70. Papoian, G. A. 2008. Proteins with weakly funneled energy landscapes challenge the classical structure-function paradigm. Proc Natl Acad Sci U S A **105**: 14237–8.

71. Pearl, F. M., C. F. Bennett, J. E. Bray, A. P. Harrison, N. Martin, A. Shepherd, I. Sillitoe, J. Thornton, and C. A. Orengo. 2003. The CATH database: an extended protein family resource for structural and functional genomics. Nucleic Acids Res **31**:452–5.

72. Qin, Z., D. Hu, S. Han, D. P. Hong, and A. L. Fink. 2007. Role of different regions of alpha-synuclein in the assembly of fibrils. Biochemistry **46**:13322–30.

73. Ray, C., and B. B. Akhremitchev. 2005. Conformational heterogeneity of surface-grafted amyloidogenic fragments of alpha-synuclein dimers detected by atomic force microscopy. J Am Chem Soc **127**:14739–44.

74. Rief, M., J. M. Fernandez, and H. E. Gaub. 1997. Elastically Coupled Two-Level Systems as a Model for Biopolymer Extensibility. Phys Rev Lett **81**:4764–7.

75. Rief, M., M. Gautel, and H. E. Gaub. 2000. Unfolding forces of titin and fibronectin domains directly measured by AFM. Adv Exp Med Biol **481**:129–41.

76. Rief, M., M. Gautel, F. Oesterhelt, J. M. Fernandez, and H. E. Gaub. 1997. Reversible unfolding of individual titin immunoglobulin domains by AFM. Science **276**:1109–12.

77. Samori, B., G. Zuccheri, and P. Baschieri. 2005. Protein unfolding and refolding under force: methodologies for nanomechanics. Chemphyschem **6**:29–34.

78. Sandal, M., F. Valle, I. Tessari, S. Mammi, E. Bergantino, F. Musiani, M. Brucale, L. Bubacco, and B. Samori. 2008. Conformational equilibria in monomeric alpha-Synuclein at the single-molecule level. PLoS Biol **6**(1):0099–0108.

79. Sarkar, A., S. Caamano, and J. M. Fernandez. 2005. The elasticity of individual titin PEVK exons measured by single molecule atomic force microscopy. J Biol Chem **280**:6261–64.

80. Sarkar, A., R. B. Robertson, and J. M. Fernandez. 2004. Simultaneous atomic force microscope and fluorescence measurements of protein unfolding using a calibrated evanescent wave. Proc Natl Acad Sci U S A **101**:12882–6.

81. Schlierf, M., F. Berkemeier, and M. Rief. 2007. Direct observation of active protein folding using lock-in force spectroscopy. Biophysical Journal **93**:3989–98.

82. Schoenauer, R., P. Bertoncini, G. Machaidze, U. Aebi, J. C. Perriard, M. Hegner, and I. Agarkova. 2005. Myomesin is a molecular spring with adaptable elasticity. J Mol Biol **349**:367–79.

83. Schwaiger, I., C. Sattler, D. R. Hostetter, and M. Rief. 2002. The myosin coiled-coil is a truly elastic protein structure. Nat Mater **1**:232–5.

84. Segall, D., P. Nelson, and R. Phillips. 2006. Volume-exclusion effects in tethered-particle experiments: bead size matters. Phys Rev Lett **96**:088306.

85. Shtilerman, M. D., T. T. Ding, and P. T. Lansbury, Jr. 2002. Molecular crowding accelerates fibrillization of alpha-synuclein: could an increase in the cytoplasmic protein concentration induce Parkinson's disease? Biochemistry **41**:3855–60.

86. Struckmeier, J., R. Wahl, M. Leuschner, J. Nunes, H. Janovjak, U. Geisler, G. Hofmann, T. Jahnke, and D. J. Muller. 2008. Fully automated single-molecule force spectroscopy for screening applications. Nanotechnology **19**:384020.

87. Sulkowska, J. I., and M. Cieplak. 2008. Stretching to understand proteins—a survey of the protein data bank. Biophys J **94**:6–13.

88. Tompa, P., C. Szasz, and L. Buday. 2005. Structural disorder throws new light on moonlighting. Trends Biochem Sci **30**:484–9.

89. Tröster, A. I. 2008. Neuropsychological characteristics of dementia with Lewy bodies and Parkinson's disease with dementia: differentiation, early detection, and implications for "mild cognitive impairment" and biomarkers. Neuropsychol Rev **18**:103–19.

90. Uversky, V. N. 2008. Amyloidogenesis of natively unfolded proteins. Curr Alzheimer Res **5**:260–87.

91. Uversky, V. N. 2007. Neuropathology, biochemistry, and biophysics of alpha-synuclein aggregation. J Neurochem **103**:17–37.

92. Uversky, V. N., E. M. Cooper, K. S. Bower, J. Li, and A. L. Fink. 2002. Accelerated alpha-synuclein fibrillation in crowded milieu. FEBS Lett **515**:99–103.

93. Uversky, V. N., J. Li, and A. L. Fink. 2001. Metal-triggered structural transformations, aggregation, and fibrillation of human alpha-synuclein. A possible molecular NK between Parkinson's disease and heavy metal exposure. J Biol Chem **276**:44284–96.

94. Valiaev, A., D. W. Lim, T. G. Oas, A. Chilkoti, and S. Zauscher. 2007. Force-induced prolyl cis-trans isomerization in elastin-like polypeptides. J Am Chem Soc **129**:6491–7.

95. Valiaev, A., D. W. Lim, S. Schmidler, R. L. Clark, A. Chilkoti, and S. Zauscher. 2008. Hydration and conformational mechanics of single, end-tethered elastin-like polypeptides. J Am Chem Soc **130**:10939–46.
96. Valle, F., M. Sandal, and B. Samori. 2007. The interplay between chemistry and mechanics in the transduction of a mechanical signal into a biochemical function. Phys Life Rev **4**:157–88.
97. van Gunsteren, W. F., D. Bakowies, R. Baron, I. Chandrasekhar, M. Christen, X. Daura, P. Gee, D. P. Geerke, A. Glattli, P. H. Hunenberger, M. A. Kastenholz, C. Oostenbrink, M. Schenk, D. Trzesniak, N. F. van der Vegt, and H. B. Yu. 2006. Biomolecular modeling: goals, problems, perspectives. Angew Chem Int Ed Engl **45**:4064–92.
98. Vettiger, P., G. Cross, M. Despont, U. Drechsler, U. Durig, B. Gotsmann, W. Haberle, M. A. Lantz, H. E. Rothuizen, R. Stutz, and G. K. Binnig. 2002. The "millipede"—nanotechnology entering data storage. IEEE Trans Nanotechnol **1**:39–55.
99. Viani, M. B., T. E. Schaffer, A. Chand, M. Rief, H. E. Gaub, and P. K. Hansma. 1999. Small cantilevers for force spectroscopy of single molecules. J Appl Phys **86**:2258–62.
100. Vilar, M., H. T. Chou, T. Luhrs, S. K. Maji, D. Riek-Loher, R. Verel, G. Manning, H. Stahlberg, and R. Riek. 2008. The fold of alpha-synuclein fibrils. Proc Natl Acad Sci U S A **105**:8637–42.
101. Watanabe, K., P. Nair, D. Labeit, M. S. Kellermayer, M. Greaser, S. Labeit, and H. Granzier. 2002. Molecular mechanics of cardiac titin's PEVK and N2B spring elements. J Biol Chem **277**:11549–58.
102. West, D. K., D. J. Brockwell, P. D. Olmsted, S. E. Radford, and E. Paci. 2006. Mechanical resistance of proteins explained using simple molecular models. Biophys J **90**:287–97.
103. West, D. K., P. D. Olmsted, and E. Paci. 2006. Free energy for protein folding from nonequilibrium simulations using the Jarzynski equality. J Chem Phys **125**:204910.
104. Wiita, A. P., S. R. Ainavarapu, H. H. Huang, and J. M. Fernandez. 2006. Force-dependent chemical kinetics of disulfide bond reduction observed with single-molecule techniques. Proc Natl Acad Sci U S A **103**:7222–7.
105. Windisch, M., H. Wolf, B. Hutter-Paier, and R. Wronski. 2008. The role of alpha-synuclein in neurodegenerative diseases: a potential target for new treatment strategies? Neurodegener Dis **5**:218–21.
106. Yang, G., C. Cecconi, W. A. Baase, I. R. Vetter, W. A. Breyer, J. A. Haack, B. W. Matthews, F. W. Dahlquist, and C. Bustamante. 2000. Solid-state synthesis and mechanical unfolding of polymers of T4 lysozyme. Proc Natl Acad Sci U S A **97**:139–44.
107. Yu, J., and Y. L. Lyubchenko. 2008. Early stages for Parkinson's development: alpha-synuclein misfolding and aggregation. J Neuroimmune Pharmacol **4**:10–16.

PART IV

METHODS TO ASSESS PROTEIN SIZE AND SHAPE

15

ANALYTICAL ULTRACENTRIFUGATION, A USEFUL TOOL TO PROBE INTRINSICALLY DISORDERED PROTEINS

Florence Manon and Christine Ebel

Commissariat à l'Energie Atomique, Institut de Biologie Structurale, Grenoble, France; Centre National de la Recherche Scientifique, Grenoble, France; Université Joseph, Fourier, Grenoble, France

ABSTRACT

Approximately 10% of proteins are predicted to be fully disordered and roughly 40% of eukaryotic proteins possess one or more long internal disordered region. This disorder is now believed to allow intrinsically disordered proteins (IDPs) to better support their functions. Considerable efforts are dedicated nowadays for a better understanding of their mode of action. In this chapter, we indicate how analytical ultracentrifugation (AUC) can be used for probing and characterizing the size and oligomeric state of IDPs. After some basic information about instrumentation, we show how the sedimentation phenomenon is related to the hydrodynamic radius (R_H) and molar mass (M) of the proteins in solution. A section is dedicated to the frictional ratio, which relates R_H and M and differs for IPDs and globular compact proteins. The two usual types of AUC experiments, sedimentation equilibrium and sedimentation velocity, and their analysis are then described in the framework of the IDPs' characterization.

Instrumental Analysis of Intrinsically Disordered Proteins: Assessing Structure and Conformation, Edited by Vladimir Uversky and Sonia Longhi

15.1 INTRODUCTION

Intrinsically disordered proteins (IDPs) are involved in a number of key cellular processes such as regulation, signaling, and control. They mainly function *via* molecular recognition. These proteins lack a well-defined three-dimensional structure under physiological conditions. Nevertheless, they are native and fully functional. Their structural disorder confers functional advantages; IDPs can adopt many different conformations and can recognize very specifically, and with low affinity, different binding partners. Structural disorder thus provides a great versatility in partner binding that enables proteins to have distinct functions (25). Several good reviews have recently been written on the subject (4, 16, 24, 25, 27).

The structural flexibility of IDPs makes three-dimensional high-resolution structure by X-ray crystallography difficult to obtain. Bioinformatical and biophysical methods can be used in combination to assign the intrinsic disorder property of IDPs, and characterize them in solution. The size of IDPs is large compared with their molecular mass and the resulting mass-to-size ratio is unusual. Methods based on size determination (e.g., size exclusion chromatography, native PAGE, DLS) cannot be used to determine the association state of IDPs in solution. Analytical ultracentrifugation (AUC) allows, in a rigorous thermodynamic way, to characterize both the size (hydrodynamic radius), related to the macromolecular conformation, and the molar mass, that is, the association state. AUC is also useful to characterize the sample homogeneity. Furthermore, AUC studies as a function of protein concentrations allow the determination of association constants.

Analytical centrifugation was developed by Svedberg and colleagues in the 1920s (21) and was then broadly used during several decades mainly to assess the sample purity or determine molecular weights. AUC principles and applications were detailed in several historical biophysics textbooks (3, 17, 22, 28).

Analytical ultracentrifuges then almost disappeared from laboratories. However, in 1992, the Beckman Coulter XLA analytical ultracentrifuge with absorbance optics and then the Beckman Coulter XLI with both absorbance and interference optical systems were commercialized. New free and easy-to-use data analysis software became also available. This allowed the resurgence of the method and AUC hasp become once again one of the most widely and powerful biomolecular research tools used to characterize complex biological macromolecules interactions and assemblies. Free software can be found in the Reversible Associations in Structural and Molecular Biology (RASMB) website (http://www.bbri.org/RASMB/rasmb.html). For recent reviews, book chapters, or books, the reader can refer to References 5–7, 11, 13, 19, and 20.

15.2 INSTRUMENTATION AND DATA ANALYSIS SOFTWARE

An analytical ultracentrifuge can be very simply described as a classical preparative ultracentrifuge associated with a special rotor and an optical detec-

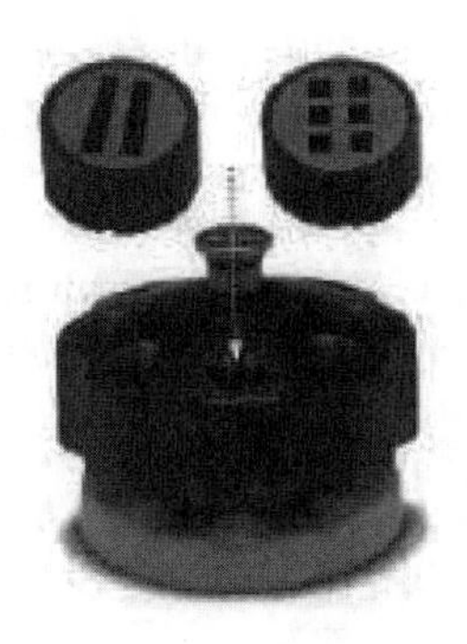

Figure 15.1 The analytical ultracentrifuge. From left to right: the analytical ultracentrifuge; the eight-hole rotor with two- and six-channel 12-mm optical path-length centerpieces; the optical arm used for absorbance and/or interference optics. Origin of the figures: Courtesy of Beckman Coulter, Inc.

tion system (Fig. 15.1). It allows the sample concentration at different radial positions and times to be directly measured inside the centrifuge cell during sedimentation as the centrifuge is running. Centrifugation parameters, such as temperature and rotor speed, and data acquisition are under computer control. For the Beckman Coulter XLA and XLI analytical ultracentrifuges, two different rotors are commercially available: one with four holes and one with eight holes (Fig. 15.1). These rotors contain the centrifuge cells that are composed of a centerpiece (Fig. 15.1) inserted between two quartz or sapphire windows in a cell housing.

There are different types of centerpieces. For example, two-channel 12- or 3-mm optical path-length centerpieces (Fig. 15.1) are used for sedimentation velocity (cf. section 15.6). One sector of the 12-mm optical path-length centerpiece contains typically 420 μL of the reference solvent and the second one contains 420 μL of the sample solution (105 μL in each compartment when using 3-mm optical path-length centerpieces). Six-channel 12-mm optical path-length centerpieces (Fig. 15.1) are used for sedimentation equilibrium (cf. section 15.5). They are loaded with three sample-reference pairs of 120 μL.

The absorbance optics of the Beckman Coulter XLA and XLI analytical ultracentrifuges is composed of a xenon flashlamp and a scanning monochromator. Wavelengths from 230 to ~650 nm can be used to measure the sample concentration. The xenon flashlamp is synchronized with the rotor precession and directs light through an optical arm (Fig. 15.1). The sample and reference compartments are illuminated and intensities are measured as function of the radial position. Absorbances are obtained and can be converted into concentrations using the Beer–Lambert law.

$$A = E_{0.1\%} lc \tag{15.1}$$

where A is the absorbance, $E_{0.1\%}$ the extinction coefficient, l the optical path length and c the concentration.

The Beckman Coulter XLI analytical ultracentrifuge detects both absorbance and interference. The Rayleigh interference optical system uses a coherent laser light at 675 nm. Both the sample and reference solvent compartments are simultaneously illuminated and an interference pattern is detected. This interference pattern consists in a succession of white and black fringes at each radial position. The sample concentration variation is measured on the basis of changes in the refractive index since a variation in macromolecule concentration induces a shift in the fringe positions expressed in fringe units ΔJ. ΔJ is related to the concentration change *via* the wavelength λ and the refractive index increment $(\partial n/\partial c)$ by the following relation:

$$\Delta J = ((\partial n/\partial c)/\lambda) lc \qquad (15.2)$$

A typical value of $\partial n/\partial c$ for proteins is 0.186 mL/g.

The absorbance and the Rayleigh interference optical systems have distinct advantages. The absorbance system is particularly sensitive for the detection of macromolecules containing chromophores. An absorbance between 0.1 and 1.2 can be measured. The interference system requires the analysis of systematic noise. It is a universal optical system particularly well adapted for the analysis of macromolecules lacking strong-enough chromophores, when the buffer components absorb too strongly, or for highly concentrated samples.

15.3 THEORETICAL BACKGROUND—TRANSPORT IN AUC

Sedimentation is a transport method akin to diffusion and sedimentation. Sedimentation can be described as the evolution of the protein weight concentration c, with time t, and radial position, r. The sedimentation is given by the Lamm equation, also referred to as the transport equation. For a homogeneous diluted solution of a noninteracting macromolecule in a dilute buffer (i.e., an ideal solution):

$$(\partial c/\partial t) = -(1/r)\partial/\partial r\left[r\left(cs\omega^2 r - D\,\partial c/\partial r\right)\right] \qquad (15.3)$$

where ω is the angular velocity of the rotor in the centrifuge, s and D the sedimentation and diffusion coefficients of the macromolecule. s is defined as the ratio of the macromolecule velocity (cm/s) to the centrifugal field ($\omega^2 r$ in cm/s^2) ratio. s is expressed in Svedberg unit S (1 S = 10^{-13} s).

The transport of each protein depends on s and D (additional terms are introduced however, linking the concentrations of the different species, for interacting systems). s and D are functions of the molar mass M, the hydrodynamic radius R_H (also referred to as the Stokes radius R_s), and the partial specific volume $\bar{v}$ of the macromolecule. s and D also depend on the solvent density ρ and viscosity η.

The Svedberg equation relates s to R_H (or D), M, and $\bar{v}$:

$$s = M(1-\rho\bar{v})/(N_A 6\pi\eta R_H) = M(1-\rho\bar{v})D/RT \quad (15.4)$$

N_A is Avogadro's number and T the absolute temperature. The sedimentation coefficient is generally expressed as $s_{20,w}$ after correction for solvent density and viscosity in relation to the density and viscosity of water at 20°C ($\rho_{20,w}$ = 0.99832 g/mL; $\eta_{20,w}$ = 1.022 mPa/s):

$$s_{20,w} = s((1-\rho_{20,w}\bar{v})/(1-\rho\bar{v}))(\eta/\eta_{20,w}) \quad (15.5)$$

The Stokes-Einstein equation relates D to R_H:

$$D = RT/(N_A 6\pi\eta R_H) \quad (15.6)$$

D is often expressed as $D_{20,w}$ after correction for temperature and solvent viscosity compared with the conditions of water at 20°C (T_{20} = 293.45 K, $\eta_{20,w}$ = 1.002 mPa/s):

$$D_{20,w} = D(T_{20}/T)(\eta/\eta_{20,w}) \quad (15.7)$$

The molar mass and the partial specific volume $\bar{v}$ of proteins can be estimated from amino acid composition using the program SEDNTERP (free download from http://www.jphilo.mailway.com/). The hydrodynamic radius R_H can be estimated or determined from calibrated size exclusion chromatography or quasi-elastic light scattering experiments. R_H can also be estimated, for a protein of a given molar mass, from the value of the frictional ratio; this is detailed in section 15.4.

The solvent density ρ and viscosity η can be either simply measured experimentally or estimated from the solvent composition using tabulated data with the program SEDNTERP.

Two types of AUC experiments, referred to as sedimentation equilibrium (SE) and sedimentation velocity (SV), will be described below (sections 15.5 and 15.6). SE provides s/D values (thus M values):

$$s/D = M(1-\rho\bar{v})/RT \quad (15.8)$$

SV experiments provide s values (the M/R_H values) and, in the case of a homogeneous solution, D values (thus M and R_H values).

15.4 FRICTIONAL RATIO OF IDPS

We will see in section 15.6 that sedimentation velocity allows the rapid and robust determination of sedimentation coefficients. The values of the sedimentation coefficients will be analyzed using the Svedberg equation in terms of M/R_H (Eq. 15.4). In favorable cases, that is, for homogeneous solutions, M and R_H can be obtained independently from the analysis of SV profiles using the Lamm equation (Eq. 15.3) (assuming the knowledge of ρ and $\bar{v}$). But even in that case, it should be checked that the order of magnitude of R_H value is compatible with the protein molar mass M. This can be done by calculating R_H through the value of the frictional ratio.

The hydrodynamic radius R_H can be compared with the minimum theoretical hydrodynamic radius R_{min} compatible with the nonhydrated volume, V, of the particle through the frictional ratio f/f_{min}.

We have

$$V = (4/3)\pi R_{min}^3 = M\bar{v}/N_A \quad (15.9)$$

$$R_H = (f/f_{min})R_{min} \quad (15.10)$$

The frictional ratio f/f_{min} depends on the hydration, surface roughness, shape, and flexibility of the particle. Uversky (26) has classified IPDs in two classes: coil-like and premolten globule (PMG)-like IPDs, according to their M-R_H characteristics. We have used his data to extract f/f_{min} values. We have considered $\bar{v}$ = 0.73 mL/g, a typical value for proteins. Figure 15.2 presents f/f_{min} values for coil-like IPDs, PMG-like IPDs, and globular compact proteins.

The continuous lines result from the transformation of the linear fits of log M versus log R_H given by Uversky. Figure 15.2 shows that IDPs have extended shapes with large value of f/f_{min}. f/f_{min} increases with the size of IDPs. For example, f/f_{min} is typically 2.1 for a 20-kDa and 3.0 for a 200-kDa coil-like IPD; it is 1.75 for a 20-kDa and 2.05 for a 200-kDa PMG-like IPD. Whatever is the range of the molar mass, these values are significantly larger than those for globular compact proteins, reported also in Figure 15.2. From the data set reported by Tcherkasskaya (23) and fitted by Uversky (26) for globular compact macromolecules, there is a very slight increase of f/f_{min} with M, with, for example, extrapolated values of 1.19 and 1.25 for M of 20 and 200 kDa. In our laboratory, we consider typically for globular compact particles f/f_{min} values of 1.25 ± 0.05 (following the data of Cantor et al. [3] reported in Fig. 15.2).

15.5 SEDIMENTATION EQUILIBRIUM EXPERIMENTS

Because SE leads to M values independently of the protein shape, it is one of the best ways for determining the molecular masses of proteins including IDPs. It is also the method of choice for studying monomer-multimer equilibrium of association (12).

Sedimentation equilibrium experiments generally use absorbance optics and small volumes corresponding to short column (typically 3 mm long). For

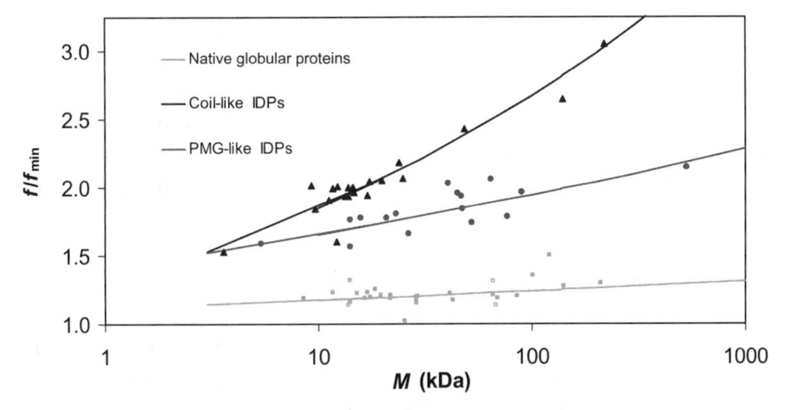

Figure 15.2 Frictional ratios of native globular proteins, PMG-like IDPs, and coil-like IDPs as a function of their molar masses. Semilog representation. Squares, circles, and triangles are for native globular proteins, PMG-like IDPs, and coil-like IDPs, respectively. Lines are derived from the linear fits of log (R_H) versus log (M) given by Uversky (2002) (26). Symbols represent data. Data for filled squares are from Reference 23 and data for open squares from Reference 3. Data for circles and triangles are from Reference 3.

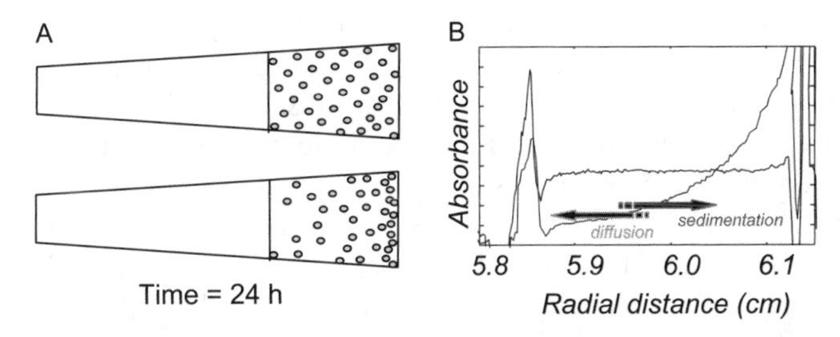

Figure 15.3 Sedimentation equilibrium. (A) Schematic representation of protein (circle) repartition in the sample compartment at the beginning of the experiment (top) and after the equilibrium is reached (bottom). (B) Superimposition of the absorbance profile at the beginning of sedimentation (essentially horizontal) and that obtained after equilibrium (monotonically increasing). The flow of particles due to sedimentation increases with the radial distance. At equilibrium, this flow is reversed by the flow from diffusion which increases with the concentration gradient.

example, 110 µL of sample is required in a two-channel 12-mm optical path-length cell, and 35 µL of sample in a 3-mm path-length cell. The angular velocity used in sedimentation equilibrium experiments is quite low. The solute sediments toward the cell bottom, but the reverse process of diffusion—in response to the concentration variation within the cell—is not negligible (Fig. 15.3). A dynamic state of equilibrium is slowly reached after typically 24 h: the flux of particles due to sedimentation and that caused by diffusion compensate exactly at each radial position.

Usually nine equilibrium profiles are obtained using three different concentrations and three different rotor speeds. It should be noted that systems have to be stable (i.e., neither nonhydrolysis nor aggregation) to achieve the equilibrium conditions. Unstable or absorbing additives should be avoided; the analysis being based on the total signal.

For an ideal homogeneous macromolecule in solution, the Lamm equation (Eq. 15.3) simplifies to

$$c(r) = c_0 exp\left[(\omega^2 M(1-\overline{v}\rho)/2RT)(r^2 - r_0^2)\right] + \delta \quad (15.11)$$

where $c(r)$ is the concentration at radial position r and c_0 is the concentration at radial position r_0, δ represents noise, for example, a baseline related to instrumentation or the presence in the solution of absorbing nonsedimenting impurities. If $\delta = 0$, Equation 15.11 can be transformed into

$$M = [2RT/((1-\overline{v}\rho)\omega^2)]\partial(\ln c)/\partial r^2 \quad (15.12)$$

This representation was the usual way to analyze the data before the 1990s. Nowadays, the different profiles are fitted by appropriate software such as Sedfit (http://www.analyticalultracentrifugation.com), Ultraspin (http://ultraspin.mrc-cpe.cam.ac.uk), WinNONLIN (http://www.biotech.uconn.edu/auf), or Utrascan (http://www.ultrascan.uthscsa.edu). For heterogeneous systems, linear combinations of Equation 15.12 are considered with, in the equilibrium case, additional links between the concentrations (related by equilibrium association constants). However, despite the simplicity of the mathematical analysis of the data, SE experiments are easy to analyze only in the case of well-defined solutions. SV experiments, which are detailed in the following section, are advantageously used before SE experiments, in order to define the number of macromolecular species in solution and the possibility of association equilibrium.

The following simulated data show how SE can unambiguously discriminate between monomer and dimer states. We take the example of a 50-kDa polypeptide chain. The monomer molar mass is thus 50 kDa and that of the dimer 100 kDa (see Fig. 15.4).

The shape of IDPs being anomalous, SE is a method of choice to assess, in a rigorous way, their molar masses—and thus the oligomeric state. Illustrations of the use of SE for the determination of the molar mass of IPDs can be found in recent literature. To illustrate, three examples are given below. The multifunctional nuclear protein, MeCP2, a methylation-dependent transcriptional repressor also involved in the maintenance of condensed chromosomal superstructures and mRNA splicing regulation, was shown to be homogeneous by SV, and a molecular mass within 3% of the mass calculated from the sequence was obtained from SE, demonstrating an extended monomer in solution (1). Glutamic acid-rich proteins (GARPs), which are signaling components required for visual transduction, exhibit a large degree of intrinsic disorder.

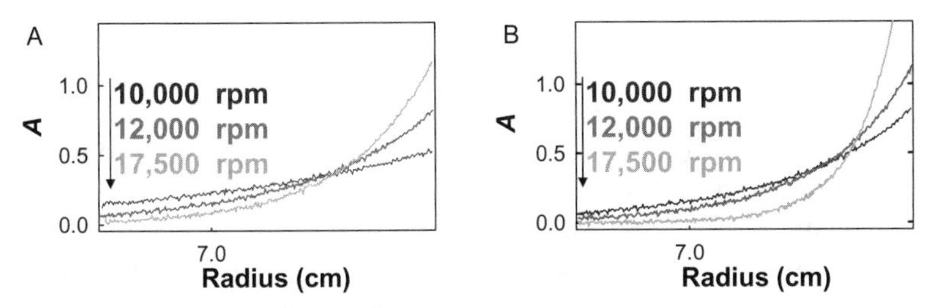

Figure 15.4 Simulated equilibrium sedimentation data for a 50-kDa and a 100-kDa protein. Data were simulated using Sedfit. $\omega_1 = 10{,}000\,\text{rpm}$, $\omega_2 = 12{,}000\,\text{rpm}$, $\omega_3 = 17{,}500\,\text{rpm}$. ω_1, ω_2, and ω_3 have been chosen in such a way that $(\omega_2/\omega_1)^2 = 1.4$ and $(\omega_3/\omega_1)^2 = 3$. These represent the ratios of ω that are recommended to be used for a proper analysis. $\bar{v} = 0.73\,\text{mL/g}$. $\rho = 1\,\text{g/mL}$. The initial absorbance is 0.3, which is a usual starting condition. The curves shown are for 24 h of sedimentation. (A) SE for the 50-kDA protein. (B) SE for the 100-kDa protein.

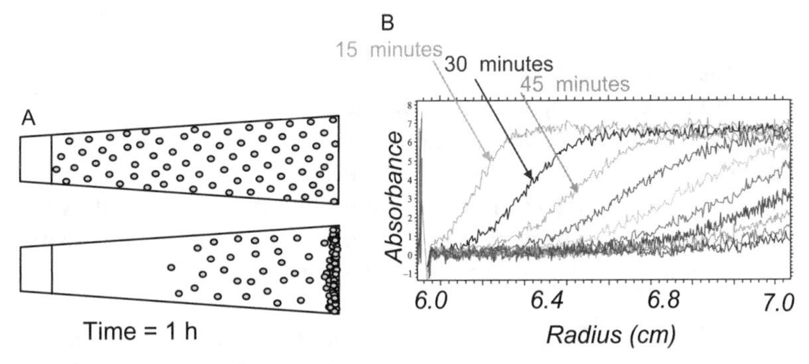

Figure 15.5 Sedimentation velocity. (A) Schematic representation of the protein in the sample compartment at the beginning of the sedimentation velocity experiment (top) and when the particles are sedimenting to the cell bottom upon centrifugation (below). (B) Superimposition of absorbance profiles obtained at different times.

AUC (SV and SE) and chemical cross-linking showed that GARPs exist in a monomer/multimer equilibrium (2). The tomato stress protein ASR1 was found by SE to be monomeric in the absence of zinc and dimeric in the presence of the metal ion—with a transition from a disordered to an ordered state upon the binding of zinc ions (9).

15.6 SEDIMENTATION VELOCITY EXPERIMENTS

SV experiments are performed at a high angular velocity. The angular velocity being much higher than that used in equilibrium experiments, molecules sediment toward the cell bottom in a couple of hours forming a boundary (Fig. 15.5). However, overnight acquisition is recommended for small or noncompact species. This is the procedure we typically follow since we consider sample composition as unknown (possible proteolysis, impurities).

The effect of diffusion is restricted and appears as boundary spreading (Eq. 15.3). The sedimentation profiles are measured by absorbance and/or interference and analyzed as a function of time. The sedimentation boundary motion analysis easily provides precise values, even in the case of complex solutions, of the protein sedimentation coefficient (s, which depends on the ratio M/R_H), and, only in favorable cases (see below) of the diffusion coefficient (D, which depends only on R_H). Because the different species are sedimenting at different velocities (they have different s values), SV is a key technique to ascertain solution homogeneity. As mentioned above, for this reason, SV has to be performed prior to SE.

Nowadays, modern software uses numerical solutions of the Lamm equation (Eq. 15.3) to analyze the SV profiles. The analysis in terms of noninteracting species determines independently s and D from the SV profiles. This is done routinely for homogeneous systems (one particle type in solution) and, in principle, for systems with two or three types of particles. However, the model for the analysis has to be adequately chosen. Particularly, the analysis of the boundary spreading leads to the determination of D, and boundary spreading is affected by sample heterogeneity. Except for homogeneous samples, D obtained by this way is not precise. In the case of unconsidered heterogeneity or interaction in the solution, the apparent D values given by the programs are even wrong. For an equilibrium association, the position and shape of the boundary depend on concentration and the association/dissociation kinetics.

To determine or check the appropriate models of analysis (e.g., the assumption of one or two noninteracting species, required for the analysis in terms of s and D in SV analysis or M in SE experiments), we always use the very powerful $c(s)$ analysis described below. Other authors prefer the Van Holde–Weischet procedure (29) implemented by B. Demeler in the Ultrascan program (http://www.ultrascan.uthscsa.edu/). We typically check the invariance of the species number and of the values of the sedimentation coefficient with concentration, by measuring SV for a set of samples at three concentrations (typically: absorbances of 1.2, 0.6 and 0.3).

The $c(s)$ analysis in the program SEDFIT (freely available at www.analyticalultracentrifugation.com) was described in 2000 by P. Schuck (18). This method deconvolutes the effects of diffusion broadening, which results in a high-resolution sedimentation coefficient distribution. This is done in the program Sedfit by assuming a mathematical relationship between s and D. The M, s, and D values are linked through the values of $\bar{v}$, f/f_{min}, ρ, and η (Eqs. 15.4, 15.6, 15.9, and 15.10). SEDFIT considers the same but reasonable values of $\bar{v}$ and f/f_{min} for all the particles in solution, regardless of their sizes. The program simulates the sedimentation of, for example, 200 particles in solution (characterized by their different s values) and is asked to find their concentrations: a distribution of sedimentation coefficients ($c(s)$) is obtained (Fig. 15.6). If $\bar{v}$ and f/f_{min} are inadequately chosen, the fit will not be very good. An example is shown in Figure 15.6B, where the sedimentation of an IDP was analyzed and where the f/f_{min} values corresponding to a globular compact protein were considered. The mean values of s resulting from the $c(s)$ analysis are always correct but the shape and details of the distribution should not be considered (width of the $c(s)$ peal or minor peaks). The $c(s)$ method allows the assessment of the sample homogeneity in a very powerful way. It determines the value of the sedimentation coefficients in a rather precise way, even in the case of heterogeneous solutions. The concentration-dependent variation of the distribution indicates the possible interactions in solution (in that case, the mean s value increases with the concentration).

The resulting value of the sedimentation coefficient can be then analyzed with the Svedberg equation and the knowledge of the system.

IDPs have more highly elongated shapes than folded globular proteins. They will hence be subjected to more hydrodynamic friction. Their sedimentation coefficients will be smaller than those of folded globular proteins of the same molecular weight. SV appears here as complementary to size exclusion chromatography, a technique that provides apparent molar masses for IDPs that are, due to their large sizes, obviously overestimated. To illustrate these features and the potency of the $c(s)$ analysis, we have calculated R_H for a 70-kDa protein that is either globular compact, PMG-like, or coil-like (section 15.2), and then derived D and s for a PBS buffer at 20°C (Eqs. 15.5 and 15.6). The values are given in Table 15.1, and compared with those of a globular compact dimer. The IPDs appear larger but sediment slower than globular compact proteins. We have simulated the sedimentation velocity profiles in the same conditions of the globular compact and PMG-like proteins (Fig. 15.6).

The behavior of the two species is clearly different, showing that the differences in the s values are easy to measure. We have analyzed the data in terms of a distribution of globular compact species. Boundary spreading is poorly fitted in the case of the PMG-like protein because the frictional ratio of 1.25 used in the analysis is not the real one. Even in the unfavorable conditions of the analysis, the $c(s)$ analysis describes the main features of the sample, that is, its homogeneity and the s value. Additional steps of the analysis would be the fit of f/f_{min} and then, because the sample is homogeneous, the fit of s and D, thus R_H and M, in the model of one noninteracting species. SV is more and more used for IDPs characterization (8, 10, 14, 15).

The three examples below illustrate the recent use of SV for such studies.

1. The protein VPg from potyvirus is required for virus infectivity and is a multifunctional protein. Grezla et al. (10) used structure prediction programs, functional assays, and biochemical and biophysical analyses to show that VPg exists in a natively unfolded conformation. Covalent dimers—through disul-

fide bridges—were evidenced in addition to the monomer for aged samples. Monomer and dimer bind their partner (eIF4E) with similar efficiency. SV was performed to evaluate the shape and oligomeric state of VPg under non-denaturating conditions. Three samples were evaluated: "M": monomer (or mainly monomer), "D": dimer (or mainly dimer), and "M + D": monomer + dimer. The $c(s)$ analysis gives mean $s_{20,w}$ values of 2.0, 3.0, and 2.5 S, respectively. The three data sets were then modeled as a mixture of monomer and dimer, providing: for "M": 90% monomer, for "M+D": equal

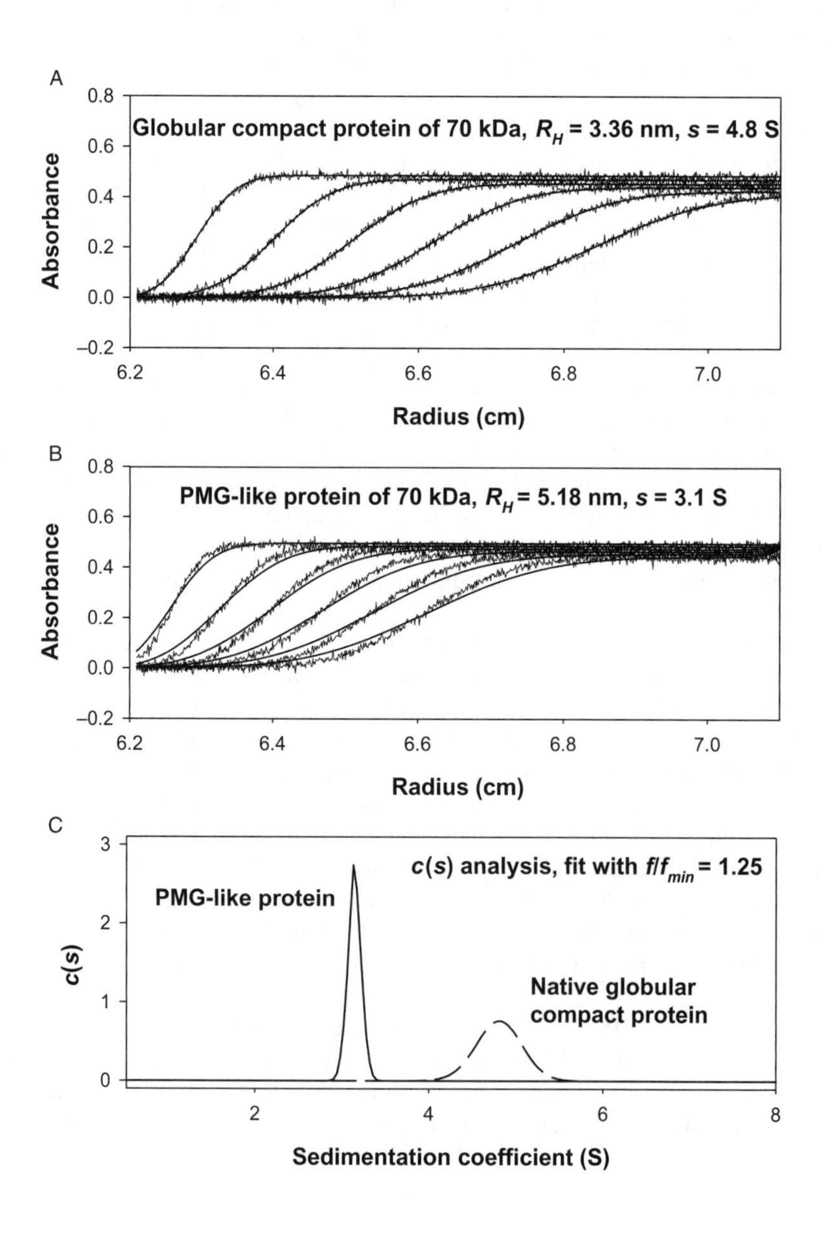

Table 15.1 Hydrodynamic Characteristics, f/f_{min}, R_H, s, $s_{20,w}$, and D, of a 70-kDa Protein in Different Conformations and Association State

	Globular Compact Monomer	PMG-Like IDP Monomer	Coil-Like IDP Monomer	Globular Compact Dimer
f/f_{min}	1.23	1.9	2.52	1.25
R_{H} (nm)	3.36	5.18	6.87	4.29
s (S)	4.8	3.11	2.34	7.5
$s_{20,w}$ (S)	4.98	3.22	2.43	7.78
D ($10^7\,\mathrm{cm^2/s}$)	6.28	4.06	3.06	4.9

We used for the calculations: $\rho = 1.005\,\mathrm{mL/g}$; η (= $1.02\,\mathrm{mPa/s}$; $\bar{v} = 0.73\,\mathrm{mL/g}$.

amounts of the monomer and dimer, and for "D": 90% dimer. The quality of the data did not allow to distinguish between rapid equilibrium (as suggested by the intermediate position of the VPg "M+D" peak in the $c(s)$ analysis) and slow equilibrium (corresponding to an analysis with two species). Nevertheless, the $s_{20,\mathrm{w}}$ values of 1.7 and 3.0 S for VPg monomer and dimer, respectively, are rather small compared with values calculated for globular compact species, namely 2.1 and 3.4S, suggesting an elongated structure for both monomeric and the dimeric VPg (10).

2. In higher plant cells, microtubules play crucial roles in cell morphogenesis, such as cell expansion, cell wall organization, and intracellular transport. Gaillard et al. (8) studied the role of two members of the microtubule-associated protein family MAP65 (AtMAP65-1 and AtMAP65-5) on microtubule

Figure 15.6 Sedimentation velocity and $c(s)$ analysis for a 70-kDa globular compact protein and a 70-kDa PMG-like IDP. (A) Simulation of six sedimentation profiles for a 70-kDa globular compact protein for a total of three and a half hours at 42,000 rpm using the s and D values of Table 15.1. The fitted profiles resulting from the $c(s)$ analysis are given as continuous lines. (B) Simulation of six sedimentation profiles for a 70-kDa PMG-like IDP for a total of three and a half hours at 42,000 rpm using the s and D values of Table 15.1. The $c(s)$ analysis was made using the inappropriate value $f/f_{\mathrm{min}} = 1.25$ corresponding to a globular compact particle. The fitted profiles resulting from the $c(s)$ analysis are given as continuous lines. Clearly, the fit is not good. This figure is for pedagogical purpose: fitting the f/f_{min} value would provide a nice superimposition of the modeled and experimental SV profiles. (C) Results of the $c(s)$ analysis. The analysis was made considering 200 particles between $s = 0.5\,S$ and $s = 8\,S$, with $f/f_{\mathrm{min}} = 1.25$, and a regularization procedure (F ratio = 0.68) leading to smooth distribution. The dashed line corresponds to the analysis of the experiment of Panel (A), the continuous line to that of Panel (B). The f/f_{min} value is not appropriate in the case of IDPs: in Panel B, the fit is not very good. However, the s values from the area under the peak in the $c(s)$ analysis are accurate.

organization. These two MAP65s stimulate the cross-linking of microtubules. Overall, this work favored a microtubule bundling mechanism involving the binding of monomeric *At*MAP65-1 and *At*MAP65-5 to microtubules followed by homodimerization when adjacent microtubules encounter each other. SV was used to investigate the state of the proteins in solution. The sedimentation coefficients $s_{20w} = 3.7 \pm 0.25$ S were significantly smaller than those calculated with values of $f/f_{min} = 1.25$ for hydrated globular compact monomers (4.7 S for *At*MAP65-1 [66 kDa] and 5 S for *At*MAP65-5 [73 kDa]). The molecular mass roughly estimated from SV boundary spreading was 60 kDa for the two proteins, in agreement with the predicted values for monomers. From the s_{20w} values, *At*MAP65-1 and *At*MAP65-5 are monomeric in solution while adopting extended conformations, with derived $R_H \sim 45$ Å, $f/f_{min} \sim 1.6$. The hypothesis of a dimer would lead to $f/f_{min} \sim 2.7$ (Fig. 15.2), slightly higher than for coil-like ($f/f_{min} \sim 2.5$), and rejected since complete unfolding is hardy compatible with a dimer state. Several disordered domains were identified in *At*MAP65-1 and *At*MAP65-5 sequences. The presence of such domains supports the viewpoint that MAP65s undergo conformational changes upon interaction with microtubules (8).

3. The receptor Tir is a key virulence factor of enteropathogenic *Escherichia coli* and related bacteria. The phosphorylation of two serine residues located in the carboxy-terminal tail, Ser-434 and Ser-463, has previously been shown to play an important role in structural/electrostatic changes in the protein and in intermolecular interaction. Race et al. (14) generated wild-type and Ala (unphosphorylatable) and Asp (phosphate-mimics) mutants of the full-length protein and its Cter domain. The full-length protein and the C-ter domain appear to be natively unfolded in solution. For example, SV shows a main species with a sedimentation coefficient $s_{20,w} = 1.2\,S$ for the wild-type C-ter domain of Tir. An $s_{20,w}$ of $2.46\,S$ is expected considering a globular protein of same molar mass with a typical hydration, demonstrating an elongated conformation in solution. Modification of the targeted residues has little effect on the structures at physiological conditions (150 mM NaCl). When the salt concentration of the solvent is lowered to 50 mM for the C-ter domain, AUC evidences changes to compactness (related f/f_{min} value of 1.2) and oligomeric states (dimer and possibly tetramer in slow equilibrium with the monomer). The authors conclude that subtle changes in the composition of the cell medium may affect the structural state of Tir, which may have functional implications.

To summarize, AUC gives information on the shape and molar masses of IDPs. The experimental hydrodynamic radii of IDPs differ significantly from those of globular compact proteins of same molar mass. As a consequence, the frictional ratios of IDPs are large (up to 3!) with values increasing with molar mass, while that of globular compact proteins are between 1.20 and 1.25 in a large range of molar mass. Sedimentation equilibrium experiments allow the determination of the molar mass in solution of IDPs but the analysis is easy only if the system is pure, stable, and well characterized. Sedimentation velocity experiments are advantageously made prior to sedimentation equilibrium experiments. They can ascertain the homogeneity of the preparation and evidence auto-association behavior. The analysis provides in favorable cases the molar mass and hydrodynamic radius or frictional ratio of IDPs in solution. The combination of calibrated size exclusion chromatography or dynamic light scattering (related to size: R_H) and sedimentation velocity (related to M/R_H) is particularly useful and permits to easily distinguish auto-association and disorder.

ACKNOWLEDGMENTS

We are grateful to Dr. David Stroebel, Dr. Frederic Vellieux, and Dr. Kathleen Wood for critical reading of the manuscript.

REFERENCES

1. Adams, V. H., S. J. McBryant, P. A. Wade, C. L. Woodcock, and J. C. Hansen. 2007. Intrinsic disorder and autonomous domain function in the multifunctional nuclear protein, MeCP2. J Biol Chem **282**:15057–64.
2. Batra-Safferling, R., K. Abarca-Heidemann, H. G. Korschen, C. Tziatzios, M. Stoldt, I. Budyak, D. Willbold, H. Schwalbe, J. Klein-Seetharaman, and U. B. Kaupp. 2006. Glutamic acid-rich proteins of rod photoreceptors are natively unfolded. J Biol Chem **281**:1449–60.
3. Cantor, C. R., and P. R. Schimmel. 1980. Techniques for the Study of Biological Structure and Function, pp. 344–846, Biophysical Chemistry. W. H. Freeman, San Franscisco, CA.
4. Dunker, A. K., M. S. Cortese, P. Romero, L. M. Iakoucheva, and V. N. Uversky. 2005. Flexible nets. The roles of intrinsic disorder in protein interaction networks. FEBS J **272**:5129–48.
5. Ebel, C. 2004. Analytical ultracentrifugation for the study of biological macromolecules. Progr Colloid Polym Sci **127**:73–82.
6. Ebel, C. 2007. Analytical ultracentrifugation. State of the art and perspectives, pp. 00–00. In V. Uversky and E. A. Permyakov (eds.), Protein Structures: Methods in Protein Structure and Stability Analysis. Nova Science Publishers, New York.
7. Ebel, C., J. V. Moller, and M. le Maire. 2007. Analytical ultracentrifugation: membrane protein assemblies in the presence of detergent, pp. 91–120. In E. Pebay-Peyroula (ed.), Biophysical Analysis of Membrane Proteins. Investigating Structure and Function. Wiley, pp. 229–60.
8. Gaillard, J., E. Neumann, D. Van Damme, V. Stoppin-Mellet, C. Ebel, E. Barbier, D. Geelen, and M. Vantard. 2008. Two microtubule-associated proteins of arabi-

dopsis MAP65s promote anti-parallel microtubule bundling. Mol Biol Cell **19**:4534–44.

9. Goldgur, Y., S. Rom, R. Ghirlando, D. Shkolnik, N. Shadrin, Z. Konrad, and D. Bar-Zvi. 2007. Desiccation and zinc binding induce transition of tomato abscisic acid stress ripening 1, a water stress- and salt stress-regulated plant-specific protein, from unfolded to folded state. Plant Physiol **143**:617–28.
10. Grzela, R., E. Szolajska, C. Ebel, D. Madern, A. Favier, I. Wojtal, W. Zagorski, and J. Chroboczek. 2008. Virulence factor of potato virus Y, genome-attached terminal protein VPg, is a highly disordered protein. J Biol Chem **283**:213–21.
11. Harding, S. E. 2005. Challenges for the modern analytical ultracentrifuge analysis of polysaccharides. Carbohydr Res **340**:811–26.
12. Laue, T. M. 1995. Sedimentation equilibrium as thermodynamic tool. Methods Enzymol **259**:427–52.
13. Lebowitz, J., M. S. Lewis, and P. Schuck. 2002. Modern analytical ultracentrifugation in protein science: a tutorial review. Protein Sci **11**:2067–79.
14. Race, P. R., A. S. Solovyova, and M. J. Banfield. 2007. Conformation of the EPEC Tir protein in solution: investigating the impact of serine phosphorylation at positions 434/463. Biophys J **93**:586–96.
15. Chenal, A., J. I. Guijarro, B. Raynal, M. Delepierre, and D. Ladant. 2009. RTX calcium binding motifs are intrinsically disordered in the absence of calcium: implication for protein secretion. J Biol Chem **284**:1781–9.
16. Receveur-Brechot, V., J. M. Bourhis, V. N. Uversky, B. Canard, and S. Longhi. 2006. Assessing protein disorder and induced folding. Proteins **62**:24–45.
17. Schachman, H. K. 1959. Ultracentrifugation in Biochemistry. Academic Press, New York.
18. Schuck, P. 2000. Size-distribution analysis of macromolecules by sedimentation velocity ultracentrifugation and lamm equation modeling. Biophys J **78**:1606–19.
19. Scott, D. J., S. E. Harding, and A. J. Rowe (eds.). 2006. Modern Analytical Ultracentrifugation: Techniques and Methods. The Royal Society of Chemistry, Cambridge, UK.
20. Serdyuk, I. N., N. R. Zaccai, and G. Zaccai. 2007. Analytical ultracentrifugation, pp. 229–387. Methods in Molecular Biophysics: Structure, Dynamics, Function, first edition. Cambridge University Press, Cambridge, UK.
21. Svedberg, T., and K. O. Pedersen. 1940. The Ultracentrifuge. Oxford Clarendon Press, London.
22. Tanford, C. 1961. Physical Chemistry of Macromolecules.
23. Tcherkasskaya, O., and V. N. Uversky. 2001. Denatured collapsed states in protein folding: example of apomyoglobin. Proteins **44**:244–54.
24. Tompa, P. 2002. Intrinsically unstructured proteins. Trends Biochem Sci **27**:527–33.
25. Tompa, P., C. Szasz, and L. Buday. 2005. Structural disorder throws new light on moonlighting. Trends Biochem Sci **30**:484–9.
26. Uversky, V. N. 2002. What does it mean to be natively unfolded? Eur J Biochem **269**:2–12.

27. Uversky, V. N., C. J. Oldfield, and A. K. Dunker. 2005. Showing your ID: intrinsic disorder as an ID for recognition, regulation and cell signaling. J Mol Recognit **18**:343–84.
28. Van Holde, K. E. 1985. Physical Biochemistry. Prentice-Hall, Englewoods Cliffs, NJ.
29. Van Holde, K. E., and W. O. Weischet. 1975. Boundary analysis of sedimentation velocity experiments with monodisperse and paucidisperse solutes. Biopolymers **17**:1387–1403.

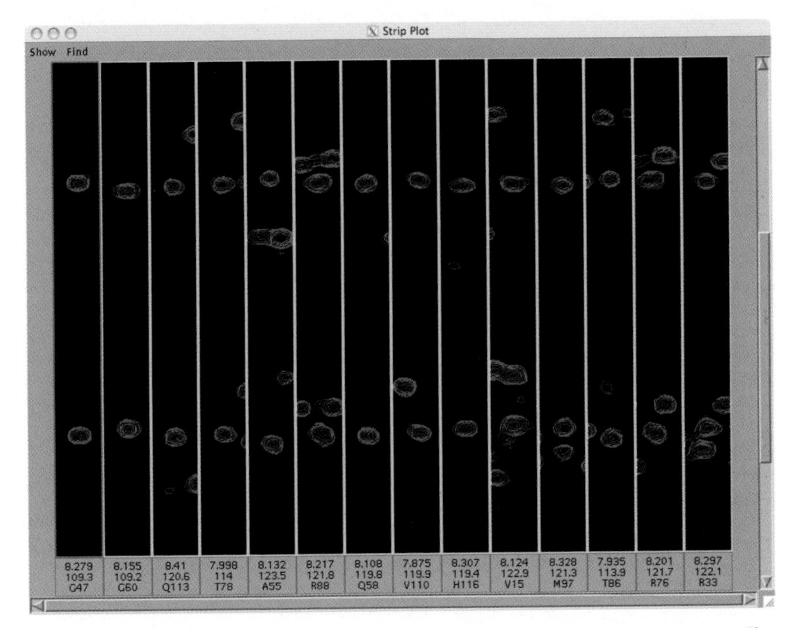

Figure 3.9 Strip plots for 14 of 16 leucine residues of Nlg3cyt, correlating $^{13}C\alpha/^{13}C\beta$ pairs with the backbone ^{1}H of the next residue, indicated at the bottom. Leucine residues that are followed by proline do not yield any cross-peaks. In case where (^{15}N,^{1}H) correlations (nearly) overlap, cross-peaks of other spin systems are observable in addition.

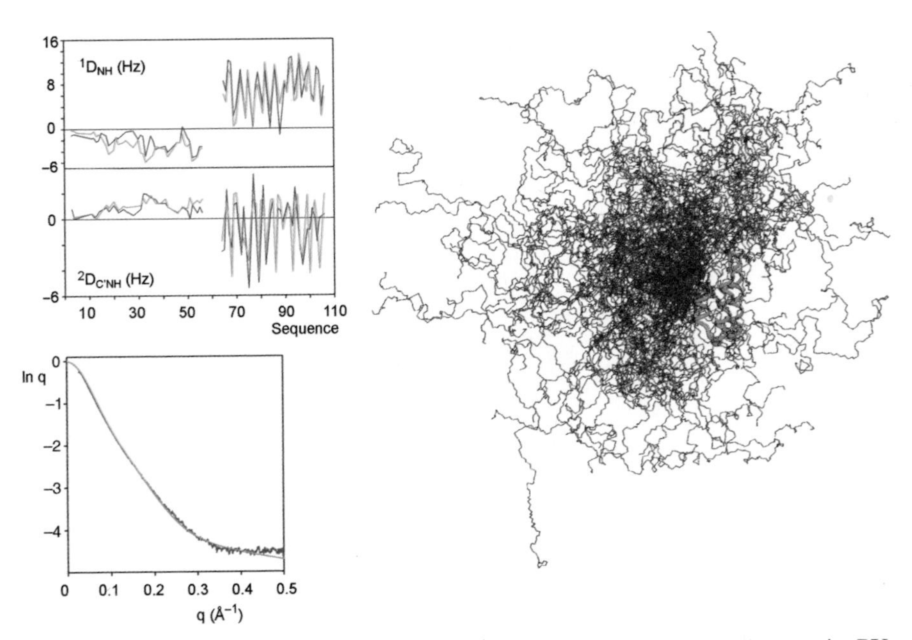

Figure 4.1 Top left: experimental $^{1}D_{NH}$ and $^{2}D_{C'NH}$ from the two-domain protein, PX, from the Sendai virus. $^{1}D_{NH}$ and $^{2}D_{C'NH}$ RDCs are evidently accurately reproduced from throughout the protein using the explicit ensemble flexible-Meccano approach. Experimental values are shown in blue. Small-angle X-ray scattering data are reproduced from the same ensemble (from Reference 5).

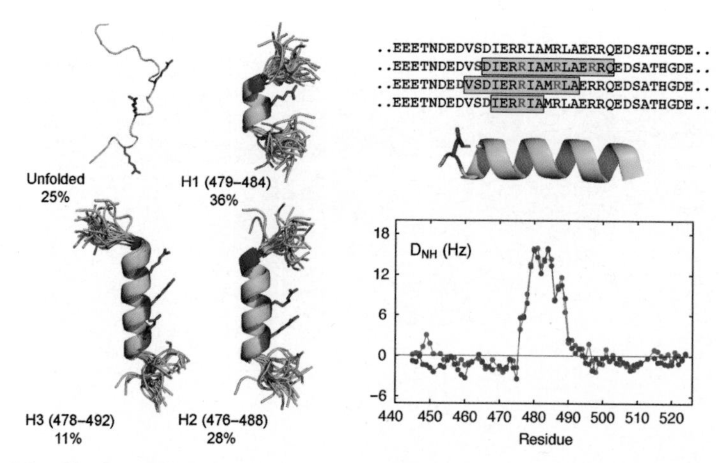

Figure 4.3 Conformational behavior of the molecular recognition element of N-tail protein from the Sendai virus. (See text for full caption)

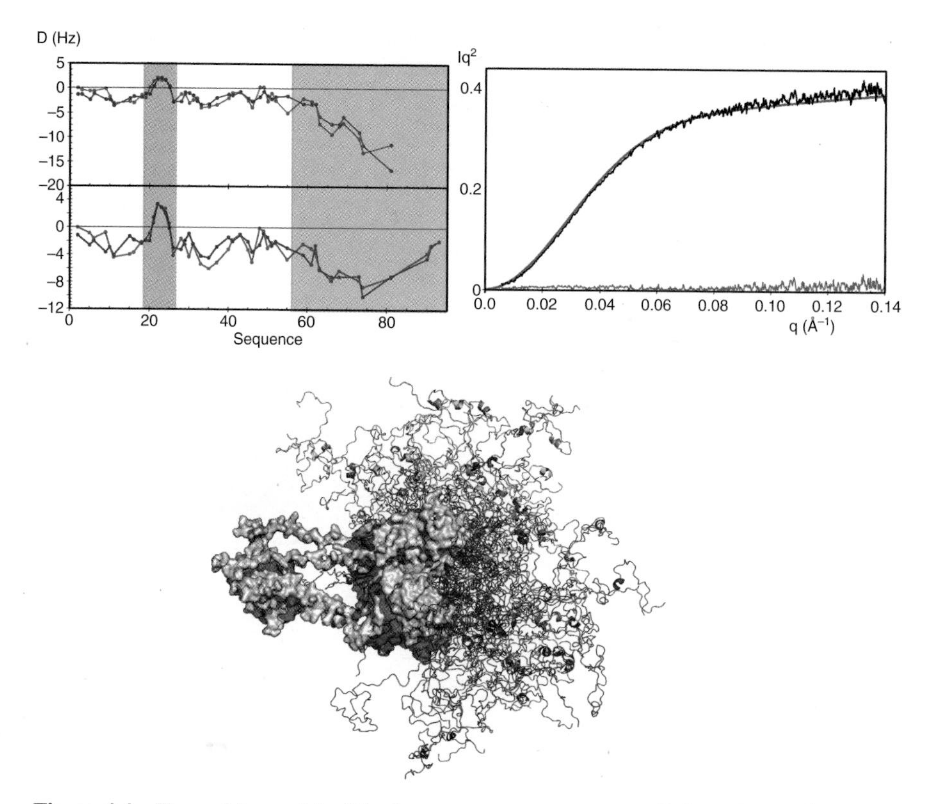

Figure 4.4 Ensemble model of the human tumor repressor p53 from RDCs (top left), SAXS (top right), MD simulation, and *FM* (bottom). RDCs were measured in the isolated TAD and in the full-length DNA-bound and free forms of the protein. Nascent helical propensity was clearly visible in the binding site of the TAD prior to interaction with Mdm2 (red shading) and enhanced rigidity in the proline-rich region that projects the TAD from the surface of the folded core domain of the protein (blue shading) (from Reference 63).

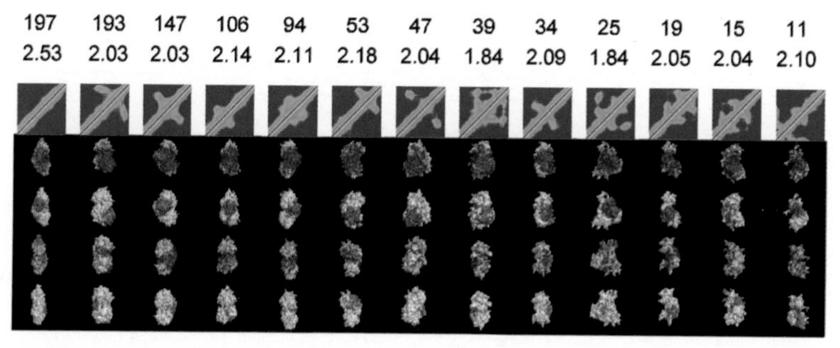

Figure 5.7 Average contact maps and surface representations for clusters of structures identified using PCA. (See text for full caption)

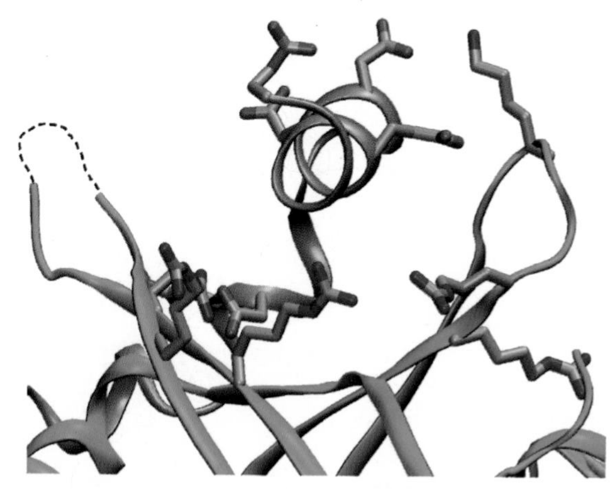

Figure 5.8 Ribbon diagram showing asymmetric distribution of charge for p53TAD when bound to RPA70. This figure was made using coordinates in the PDB file 2B3G and the visual molecular dynamics molecular graphics package (12, 31). Side chains show GLU and ASP residues in p53TAD, and ARG and LYS residues in RPA70.

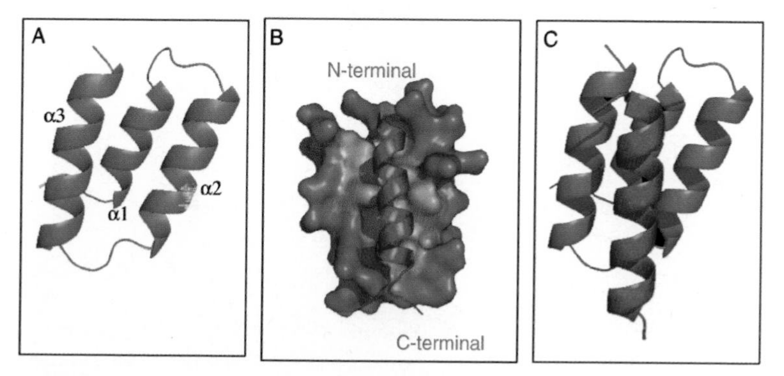

Figure 6.10 (A) Ribbon representation of the XD structure. (B) Model of the complex between XD and the α-MoRE of N-tail according to Reference 41. Hydrophobic residues are shown in orange. (See text for full caption)

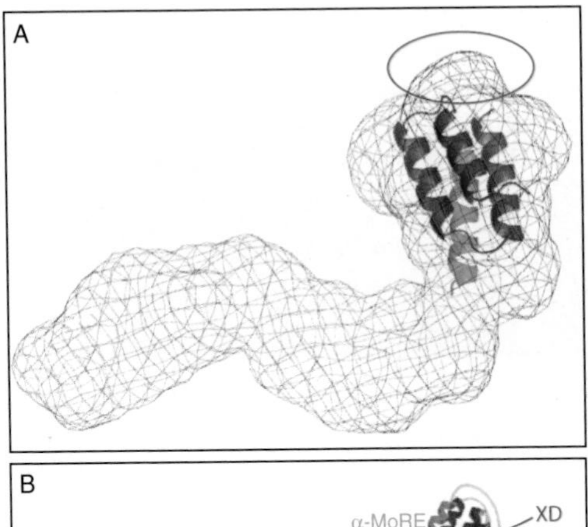

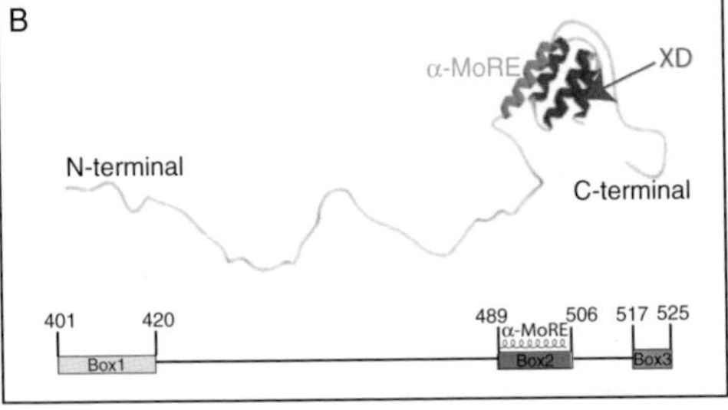

Figure 6.11 (A) Global shape of the N-tail–XD complex as derived by small-angle X-ray scattering studies (12). (See text for full caption)

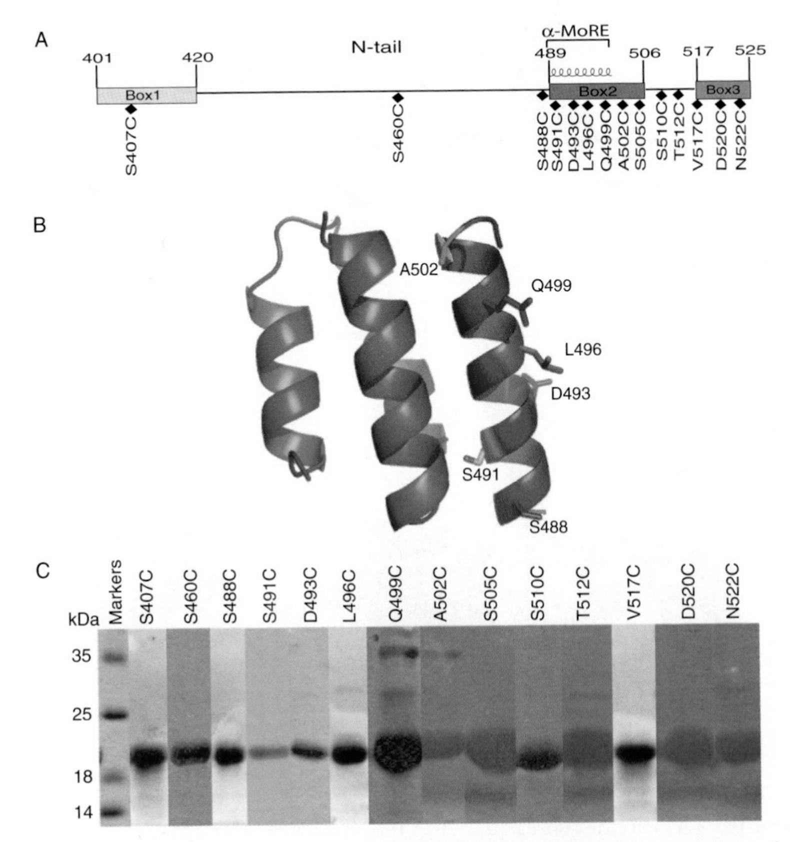

Figure 6.12 (A) Schematic representation of positions targeted for cysteine substitution and spin labeling (black diamonds). (See text for full caption)

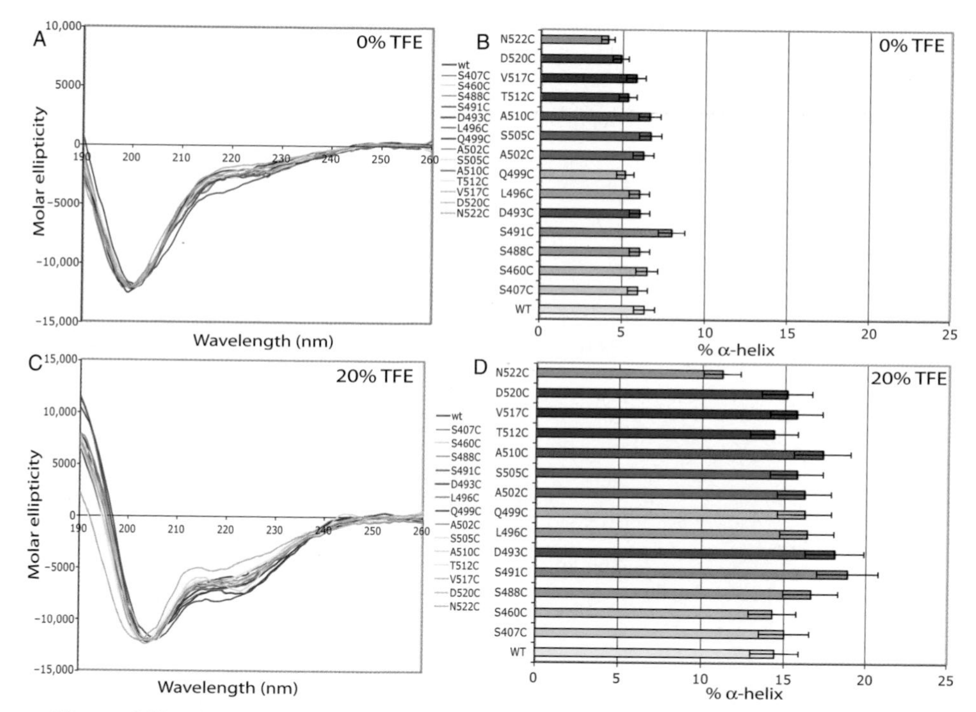

Figure 6.13 CD spectra of spin labeled and wt N-tail proteins at 0% (A) and 20% TFE (C). α-Helical content of spin labeled and wt N-tail proteins at 0% (B) and 20% (D) TFE. Each spectrum is the mean of three independent acquisitions. The error bar (10% of the value) corresponds to the experimentally determined standard deviation from three independent experiments. The α-helical content was derived from the ellipticity at 222 nm as described in Reference 62. Data were taken from Reference 5.

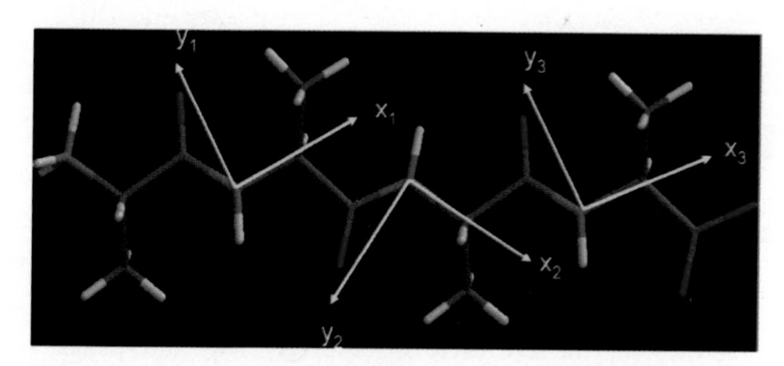

Figure 7.21 Planar structure of tetraalanine (ϕ = 180°, ψ = 180°). The coordinate systems S1(x1, y1, z1), S2(x2, y2, z2), and S3(x3, y3, z3) were used to express the Raman tensors of the individual, uncoupled amide I modes and their transition dipole moments (the z-component for S2 has been omitted for the sake of clarity). The structure was obtained by using the program TITAN from Schrödinger, Inc.

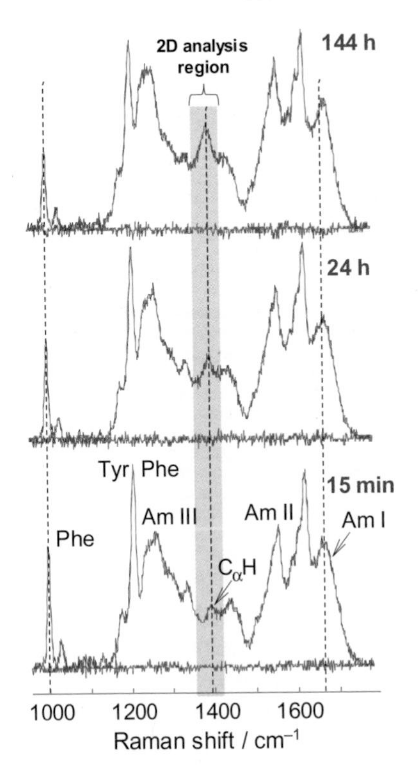

Figure 9.1 Experimental (blue), modeled (red), and deference (green) DUVRR spectra of lysozyme incubated for various times. The experimental Raman spectra were fitted with three pure component spectra, that is, the spectra of "nucleus" β-sheet and partially unfolded intermediate calculated by the independent component analysis, and the experimental spectrum of native lysozyme. A mixed soft-hard modeling approach provided the refined DUVRR spectra of β-sheet and partially unfolded intermediate, kinetic profiles for all three species and the characteristic times for each step of lysozyme transformation (157). The highlighted C_{α}–H bending region of the spectra has been used for the initial 2D correlation analysis (157). Adopted from Shashilov et al. (157) with permission from the American Chemical Society.

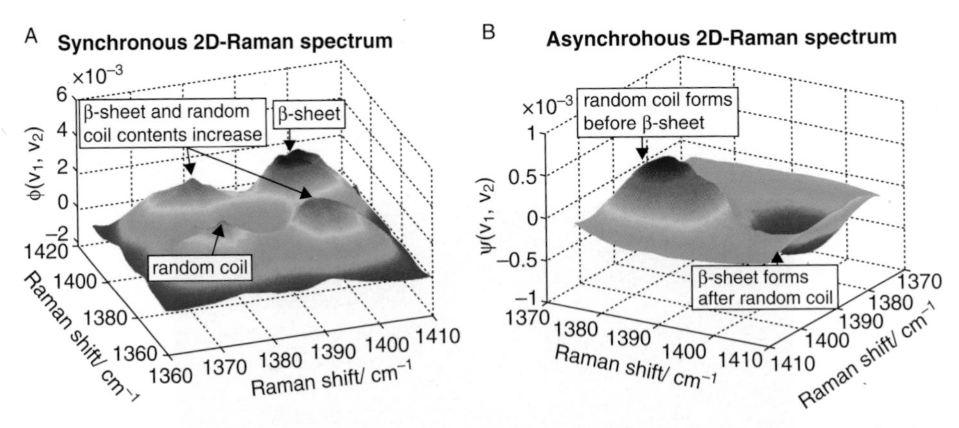

Figure 9.2 Synchronous (A) and asynchronous (B) 2D correlation spectra of the CαH bending region of lysozyme DUVRR spectra. (A) Two auto-peaks at 1387/1387 and 1396/1396 cm^{-1} correspond to random coil and β-sheet, respectively, and two positive cross-peaks at 1387/1396 cm^{-1} and 1396/1387 cm^{-1} show positive correlation between β-sheet and random coil changes. (B) Two opposite-sign areas at 1387/1396 cm^{-1} and 1396/1387 cm^{-1} indicate asynchronous formation of random coil and β-sheet; valley at 1396/1387 cm^{-1} shows that β-sheet appeared after random coil.

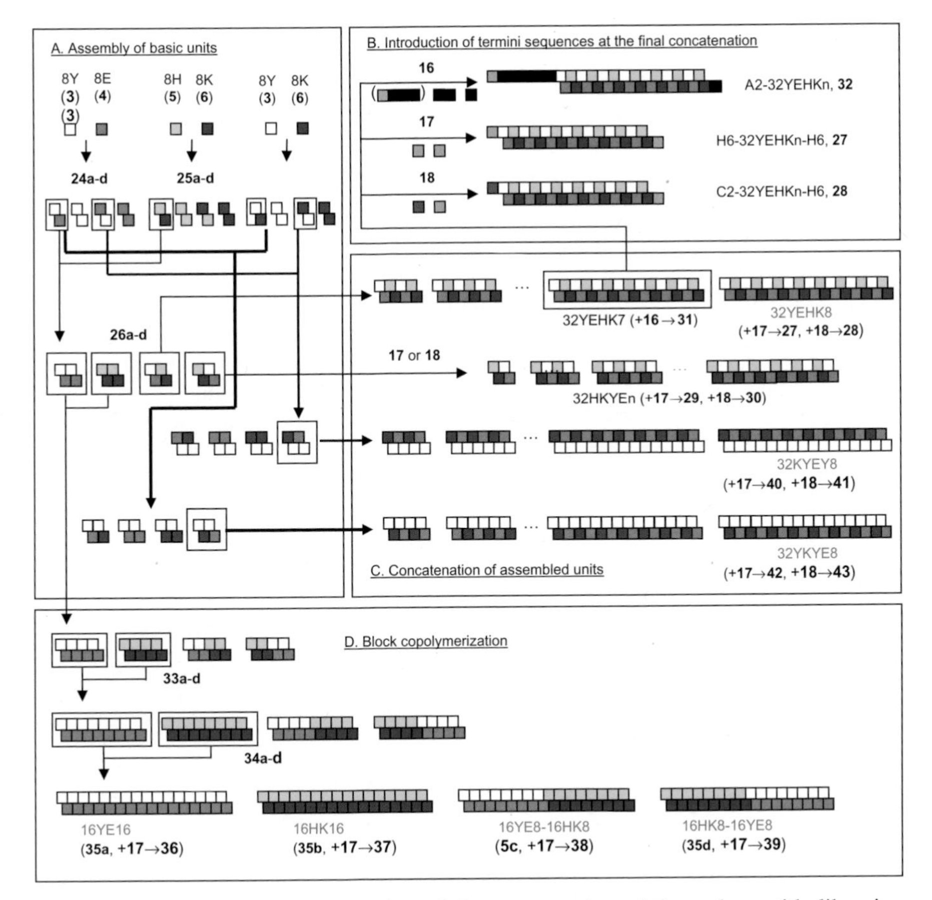

Figure 9.8 Schematic representation of the construction of the polypeptide libraries. Adopted from Biomacromolecules 2007, **8**, 1487–97, page 1492.

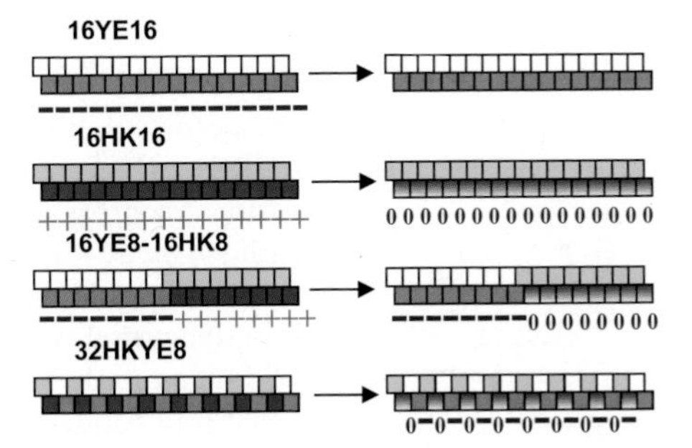

Figure 9.14 The schematic representation of the polypeptide charge distribution before and after posttranslational chemical modification.

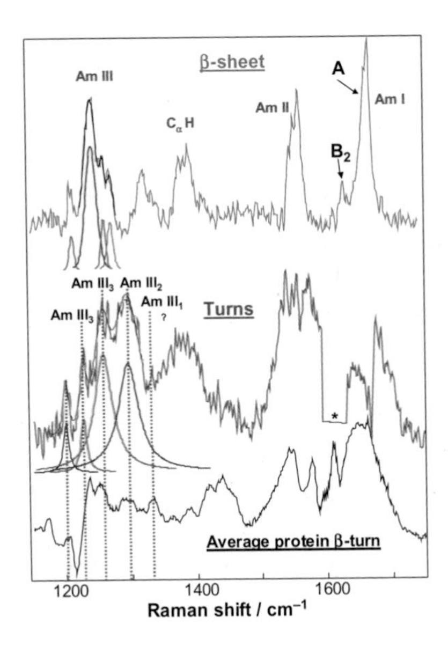

Figure 9.27 DUVRR spectra of YEHK fibril β-sheet and turns, and the average spectrum of native protein β-turns adopted from Reference 77. * = Spectral range of strong contribution of aromatic amino acids. Adopted from Sikirzhytsli et al. (166) with permission from the American Chemical Society.

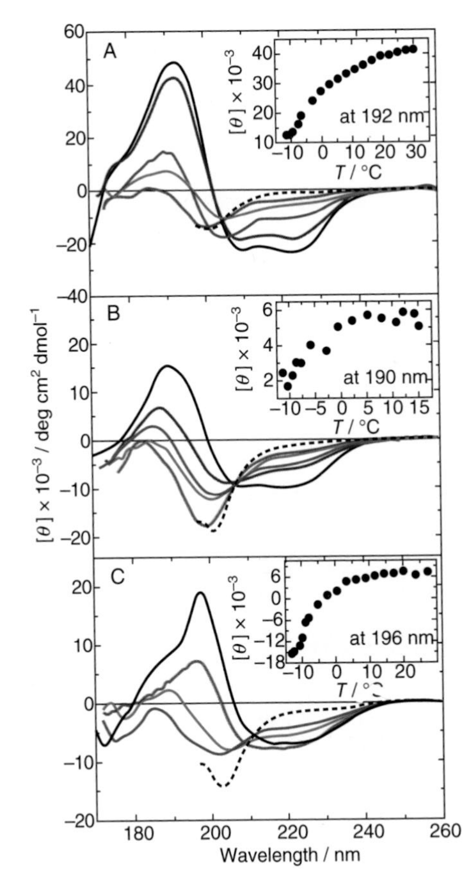

Figure 10.5 Vacuum-UV CD spectra of metmyoglobin (A), staphylococcal nuclease (B), and thioredoxin (C) in various conformational states: native (solid black line), heat-denatured (red line), cold-denatured (blue line), acid-denatured (green line), partially acid-denatured (brown line), and urea-denatured (dotted black line) (35). The inset shows the temperature dependence of the molar ellipticity at the indicated wavelength for each protein. (From Reference 35, with permission.)

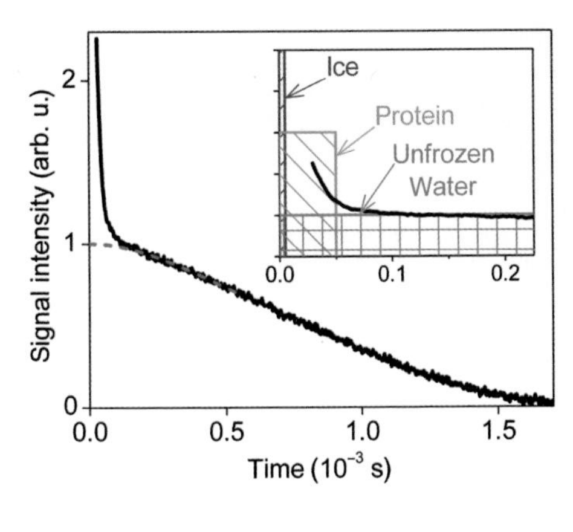

Figure 12.1 Illustration of the method applied to measure the fraction of the unfrozen water component x_{unfrozen}. (See text for full caption)

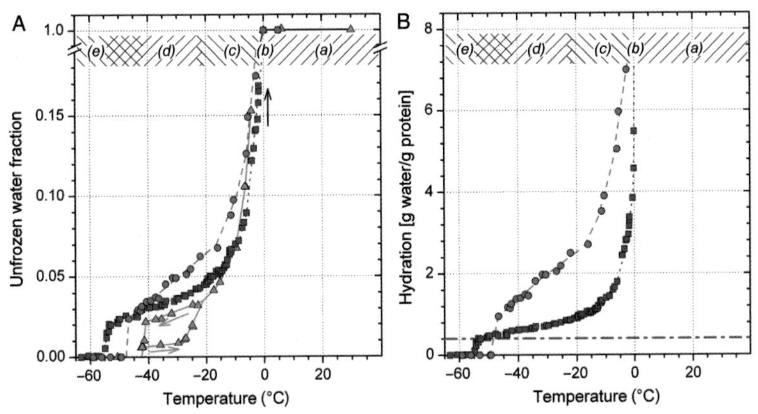

Figure 12.3 (A) Unfrozen water fraction as a function of temperature for buffer (green circles), BSA-buffer (blue squares), and ERD10-buffer (red triangles) solutions as measured by extrapolated FID and CPMG techniques. (See text for full caption)

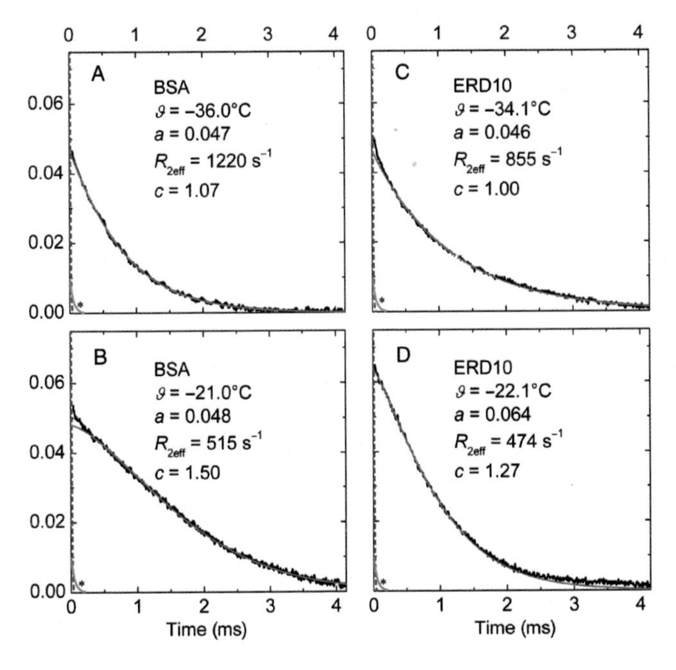

Figure 12.4 FID signals (thin line) of protons in BSA-buffer (A and B) and ERD10-buffer solutions (C and D) at two characteristic temperatures. (See text for full caption)

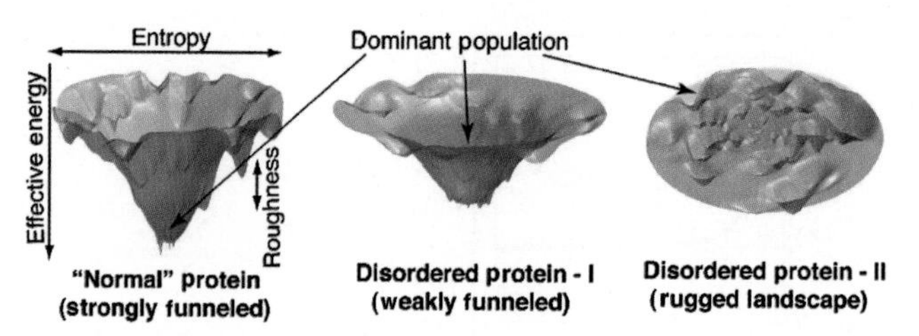

Figure 14.1 Schematic representation of the interplay between protein's energy landscape and its folding. (See text for full caption)

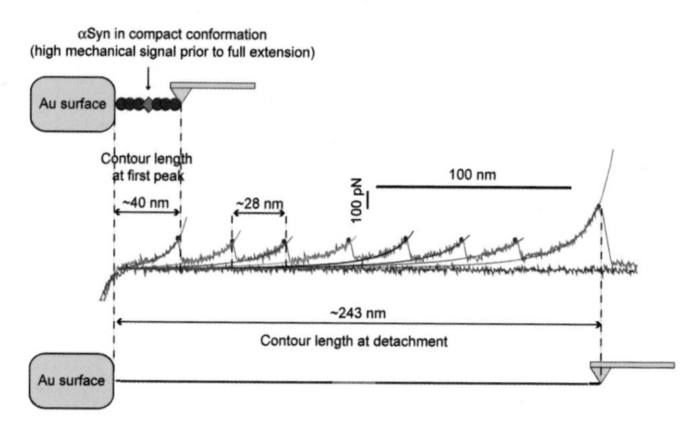

Figure 14.10 Single 3S3 molecule mechanical unfolding trace showing six mechanical events. (See text for full caption)

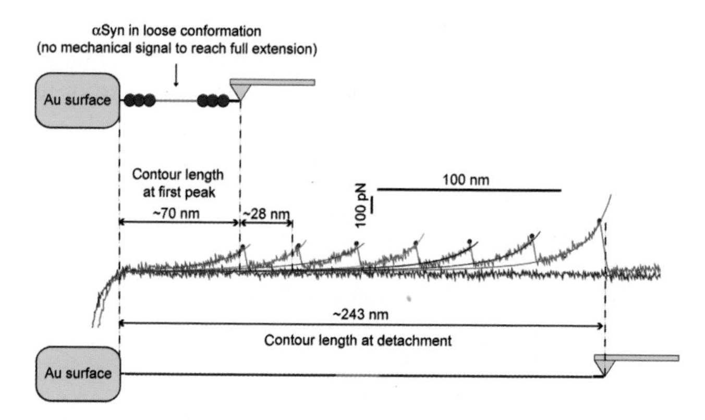

Figure 14.11 Example of a single 3S3 molecule mechanical unfolding traces showing seven mechanical events. Blue trace = approach. Green trace = retraction. The average WLC contour length fitted on the first rupture event is ~40 nm. This distance is compatible with the sum of six folded I27 modules (~4.5 nm each = 4.5 × 6 = 27 nm) and one partially folded α-Synuclein moiety (see cartoon at the top of the figure). The folded portion of α-synuclein has the same length of one I27 module (90AA, ~28 nm), which is compatible with the portion of α-synuclein found to assume a β-conformation in fully formed amyloid fibrils. The WLC contour length fitted on the last peak (detachment from surface) is ~243 nm, compatible with the expected length of a fully unfolded 3S3 construct (680 AA). Taken from Reference 10.

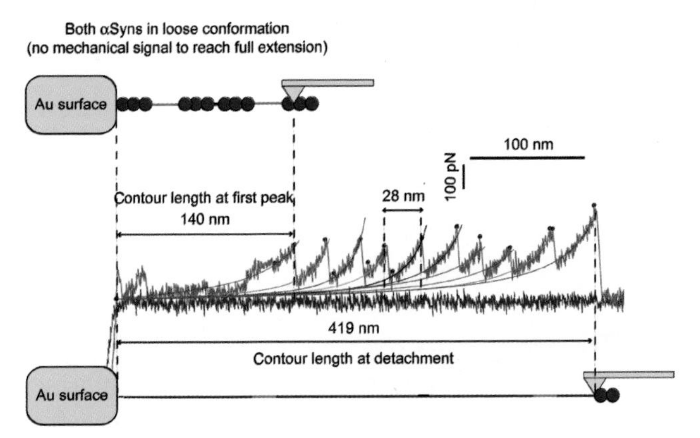

Figure 14.13 Example of one force curve resulting from the pulling of a 3S3-3S3 dimer. Dimers can form via oxidation of the C-terminal cysteines. (See text for full caption)

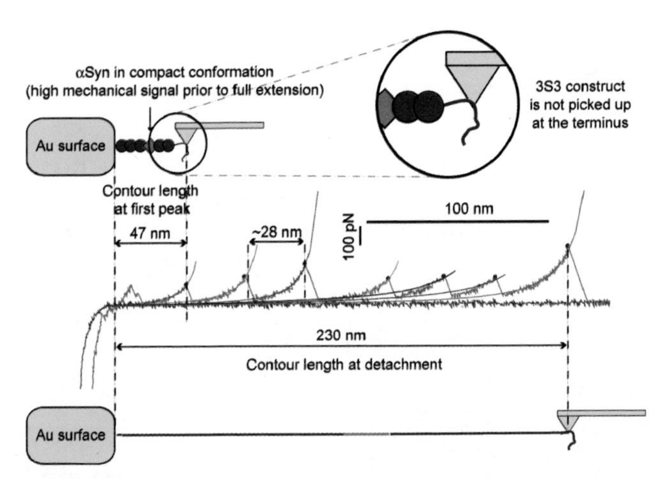

Figure 14.14 Example of one ambiguous force curve: single 3S3 molecule mechanical unfolding trace showing six mechanical unfolding events and a shorter-than-expected contour length for the first peak. (See text for full caption)

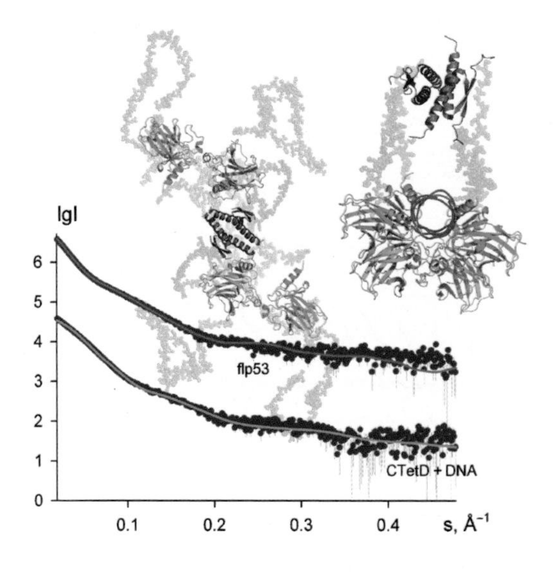

Figure 16.5 The models of tumor suppressor p53 generated from the SAXS curves (dots: experimental data; solid lines: computed patterns). (See text for full caption)

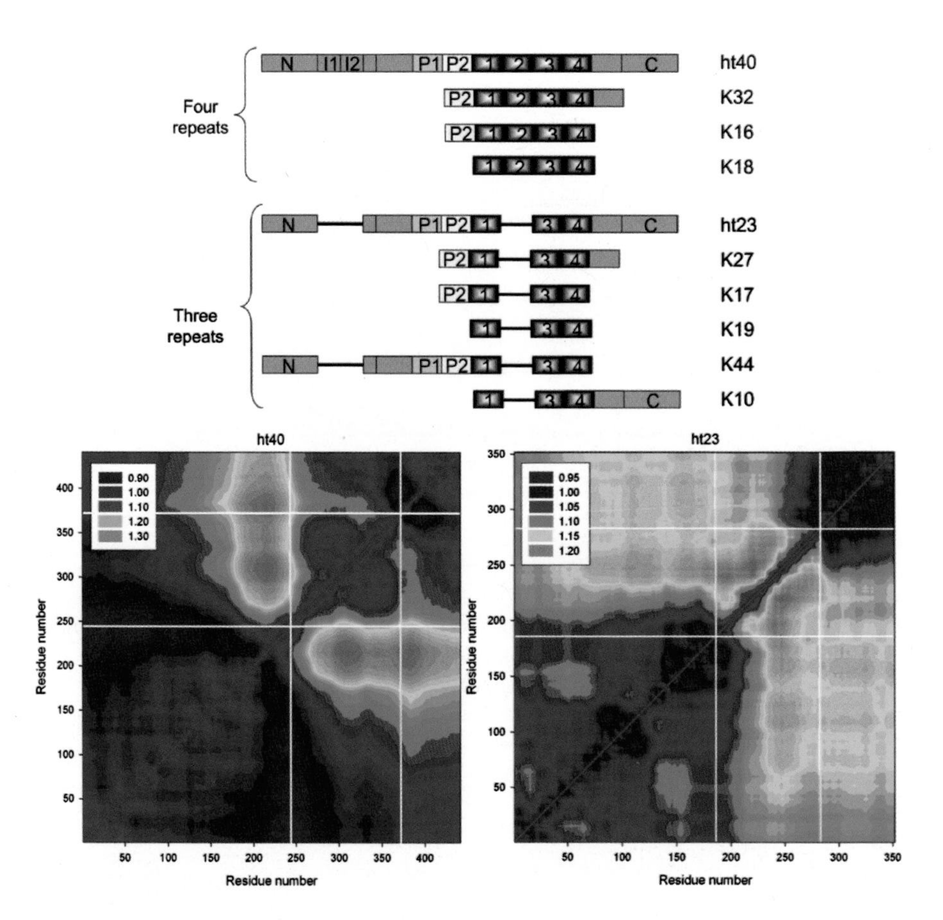

Figure 16.8 Bar diagrams of isoforms of tau protein used for the simultaneous SAXS curve analysis with EOM (top panel). (See text for full caption)

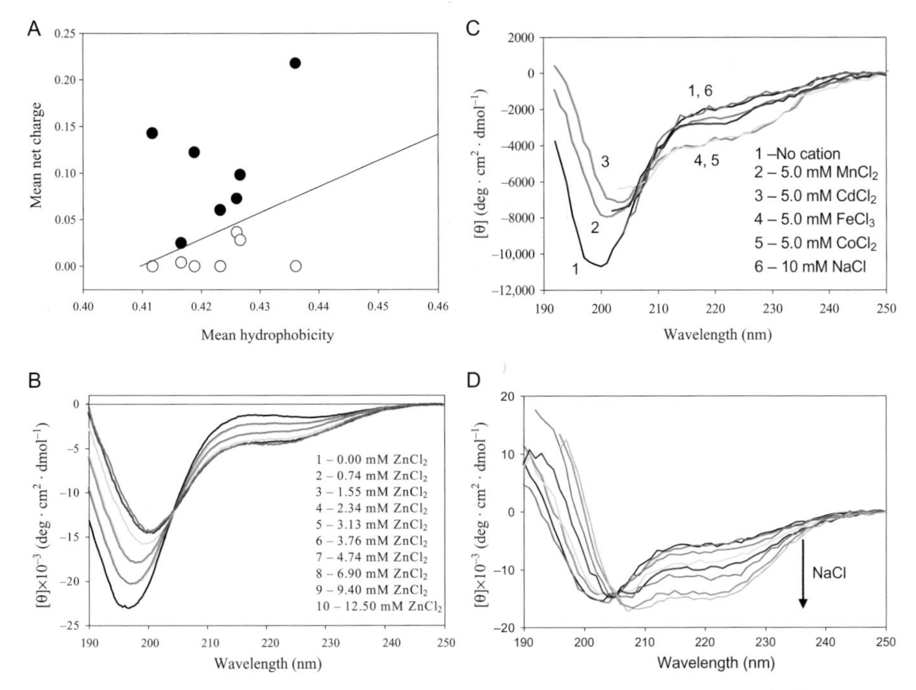

Figure 19.5 The effect of counterions on structural properties of IDPs. (A) Charge–hydropathy plot for seven extended IDPs, osteocalcin, osteonectin, α-casein, HPV16E7 protein, calsequestrin, manganese-stabilizing protein, and HIV-1 integrase in their apo- and metal-saturated forms. The far-UV CD spectra of prothymosin α (plot B), α-synuclein (plot C), and core histones (plot D) measured in the absence or presence of various metal ions.

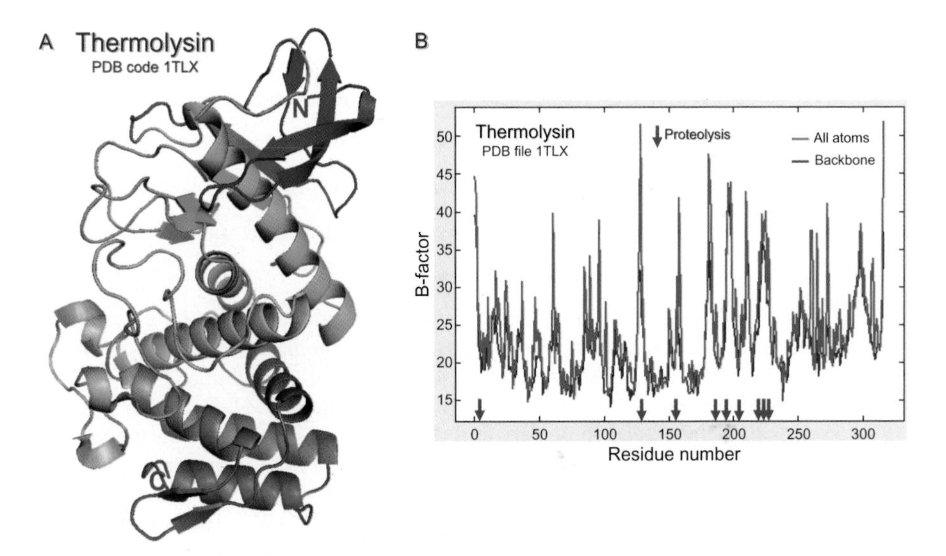

Figure 20.2 Correlation between sites of enhanced chain flexibility (segmental mobility) and sites of limited proteolysis in thermolysin. (See text for full caption)

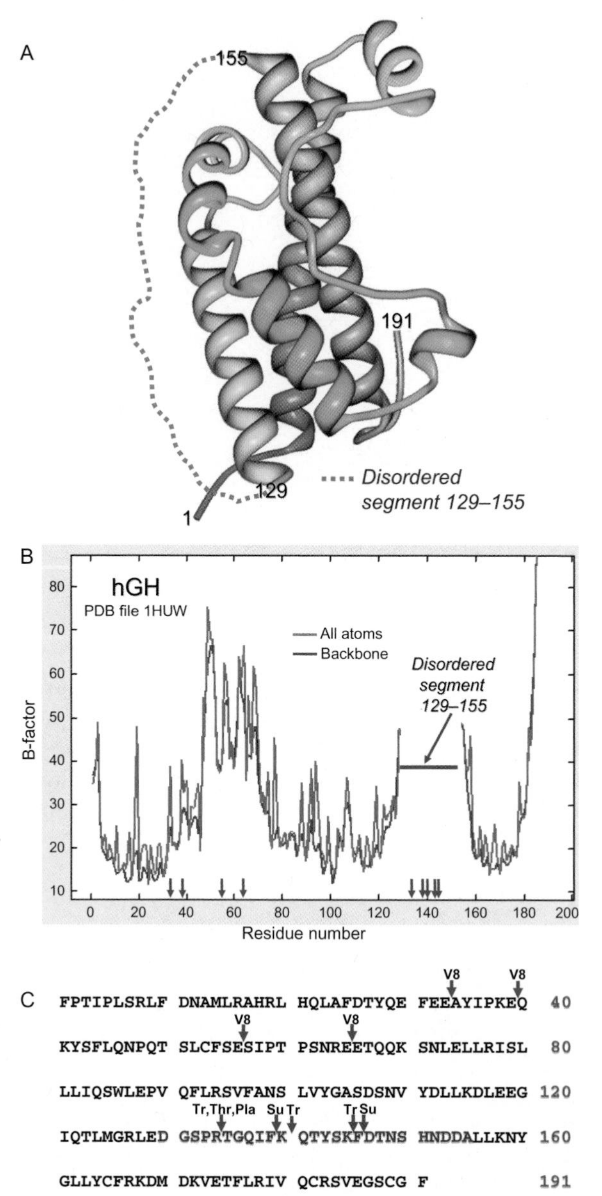

Figure 20.8 Limited proteolysis of human growth hormone (hGH) by a variety of proteases occurs at disordered/mobile regions of the protein. (A) A 3D model of the 191-residue chain of hGH (PDB code 1HUW). The structure is that of an hGH mutant, but this is essentially identical to that of the wild-type protein (241). The chain segment 129–155 does not show electron density and thus is considered to be disordered. This segment is arbitrarily drawn with a red line. The 3D model of the hormone was generated with the MBT software (47) available at PDB. (B) The B-factor profile along the 191-residue chain of hGH taken from the PDB file (code 1HUW). The chain segment 129–155 is disordered and does not display electron density. Arrows indicate sites of proteolysis by several proteolytic enzymes. (C) Amino acid sequence of the 191-residue chain of hGH. The amino acid residues located in the disordered segment 129–155 are colored in red. Arrows indicate the sites of proteolysis by staphylococcal V8-protease (V8), trypsin (Tr), thrombin (Thr), subtilisin (Su), and plasmin (Pla).

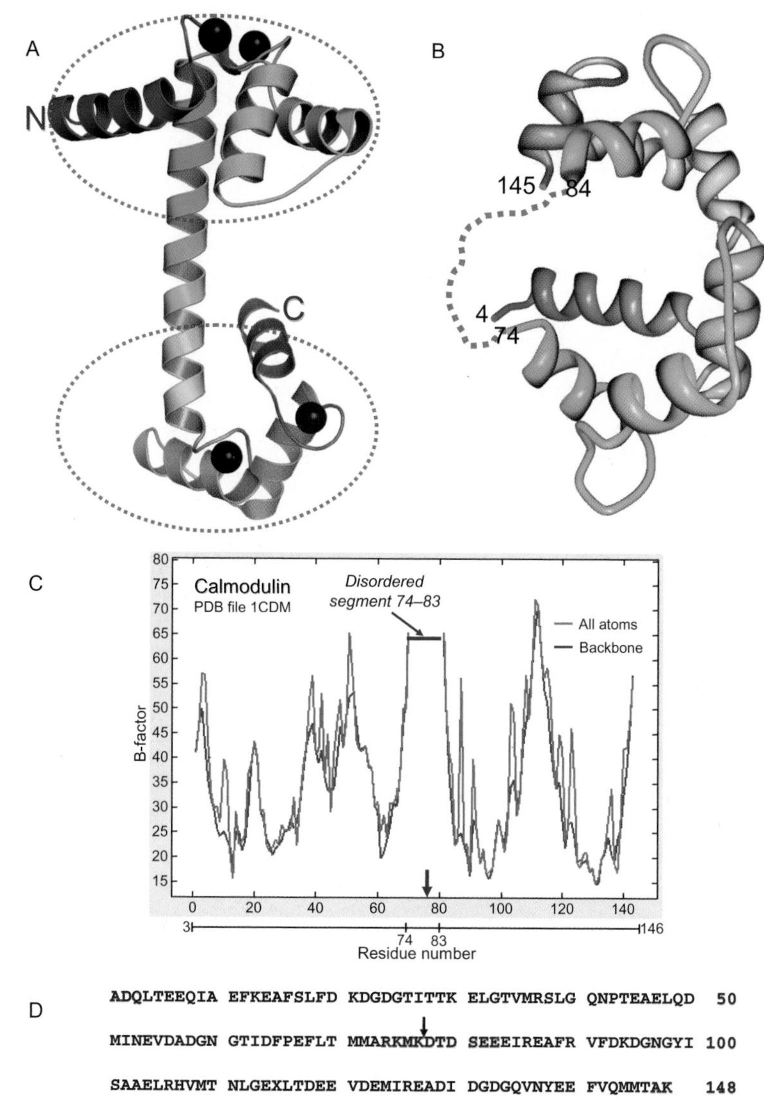

Figure 20.9 Dissection of calmodulin (CaM) into two domains by tryptic hydrolysis. (A) The most commonly represented dumbbell structure of CaM highlighting its two-domain topology. The two domains or lobes bind each two calcium ions. (B) Structure of CaM complexed with a peptide ligand. The segment 74–83, arbitrarily drawn by a red line, does not show electron density and thus is disordered. Also, short segments at the N- and C-terminal ends of the protein appear to be disordered. The 3D model was constructed with the MBT software (47) available in PDB. The chain is with rainbow colors from the N-terminus (blue) to the C-terminus (red). (C) The B-factor profile along the 148-residue chain of CaM (PDB code 1CDM). There is a discontinuity in the profile at segment 74–83 due to missing electron density and thus this chain segment is disordered. Note that the actual residue numbers of the CaM polypeptide chain are those indicated at the bottom of the diagram. (D) Amino acid sequence of CaM. Arrow indicates the site of proteolytic cleavage by trypsin cutting at the Lys77–Asp78 peptide bond (53). The disordered residues in the crystal structure of the protein are colored in red in the amino acid sequence.

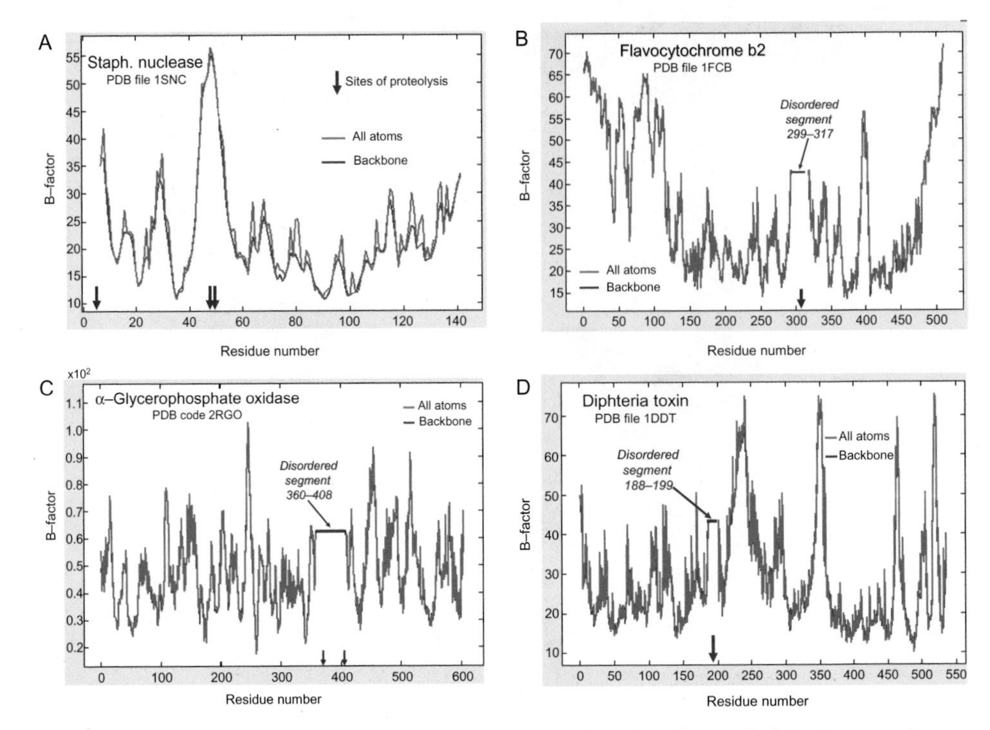

Figure 20.10 Limited proteolysis occurs at disordered regions of globular proteins. (See text for full caption)

16

STRUCTURAL INSIGHTS INTO INTRINSICALLY DISORDERED PROTEINS BY SMALL-ANGLE X-RAY SCATTERING

PAU BERNADÓ[1] AND DMITRI I. SVERGUN[2]

[1]*Institute for Research in Biomedicine, Parc Científic de Barcelona, Barcelona, Spain*
[2]*European Molecular Biology Laboratory, Hamburg Outstation, Hamburg, Germany*

ABSTRACT

Small-angle X-ray scattering (SAXS) is widely employed to study the overall solution structure and structural transitions of folded biologic macromolecules. At the same time, SAXS is one of the very few techniques yielding structural information about flexible macromolecules, unfolded and intrinsically unfolded proteins. In the past, SAXS was more often employed to qualitatively monitor the folding/unfolding processes by a few overall parameters; modern modeling methods allow one to extract much more detailed quantitative information for a better understanding of the biologic role of these proteins. In this chapter, after a brief introduction to the technical and experimental details of SAXS, its applications to the IDPs are described. Classical approaches based on the analysis of overall parameters are presented, and a recent development, the ensemble optimization method, is also explained in detail. The latter approach takes the conformational plasticity of IDPs into account by allowing for coexistence of multiple protein conformations in solution, and the analysis of the ensembles provides a new source of structural information.

Instrumental Analysis of Intrinsically Disordered Proteins: Assessing Structure and Conformation, Edited by Vladimir Uversky and Sonia Longhi

This approach and the combination of SAXS with other biophysical techniques, such as NMR, Föster Resonance Energy Transfer (FRET), and molecular simulations, promise structural insights into the detection of structural disorder and into monitoring of structural perturbations of IDPs upon environmental changes or binding to biologic partners.

16.1 INTRODUCTION

Diffraction methods are among the most powerful techniques to study macromolecular structure up to the atomic resolution. Large-scale structure determination initiatives using X-ray crystallography (41) yielded unprecedented numbers of high-resolution models for isolated proteins and/or their domains, and these numbers are expected to grow rapidly in the coming years. Much more difficulties have been encountered in the study of the proteins like intrinsically disordered proteins (IDPs). The inherent structural flexibility makes IDPs difficult (or impossible) to crystallize and also limits the amount of structural information, which can be obtained from solution nuclear magnetic resonance (NMR). The interest for this family of proteins has been enhanced by the discovery that they are fully active and play important biologic roles in several metabolic and signal transduction pathways even though they lack a permanent tertiary structure (14, 18, 19, 76, 87). In addition, some IDPs and nonnative states of proteins have been linked to the formation of toxic aggregates that cause important degenerative diseases (16).

The present chapter describes one of the most efficient methods for structural characterization of macromolecules in solution, a diffraction technique called small-angle X-ray scattering (SAXS). This method allows one to study the low-resolution structure of native particles in solutions and to analyze structural changes in response to variation of external conditions. Computational approaches are well established to retrieve low-resolution, three-dimensional structural models of proteins and complexes, and they are widely used in structural biology (59, 61, 72). Importantly, unlike most other structural methods, SAXS is applicable to equilibrium and nonequilibrium mixtures and monitor kinetic processes such as (dis)assembly (1) and (un)folding (17). In particular, the technique is effectively used to quantitatively characterize the overall structure of IDPs, and recent development of novel data analysis methods made it possible to also describe the flexibility of IDP ensembles in solution based on SAXS data. In this chapter, after a short reminder of the main theoretical and experimental aspects of X-ray scattering from macromolecular solutions, applications of SAXS to the analysis of IDPs will be presented.

16.2 BASICS OF X-RAY SCATTERING

This section will briefly describe the basic theoretical and experimental aspects of SAXS as applied to solutions of biologic macromolecules. The reader is

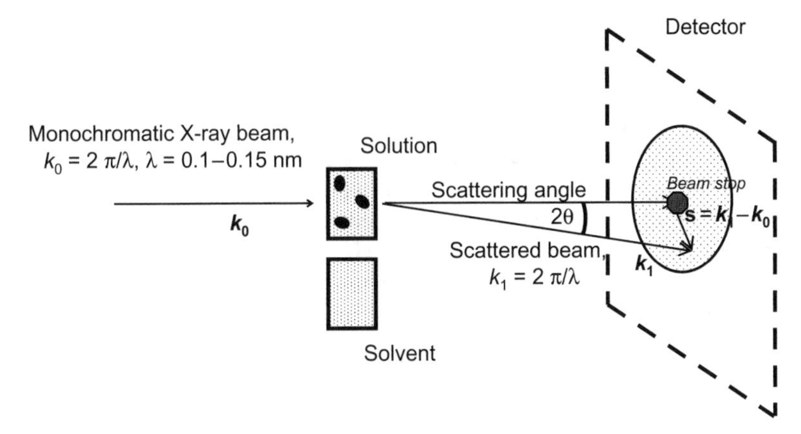

Figure 16.1 General scheme of a SAXS experiment.

referred to textbooks (20) or to recent reviews (36, 44, 73) for more detailed information.

In a SAXS experiment, a solution of macromolecules placed in a sample container (e.g., capillary or cuvette) is illuminated by a collimated monochromatic X-ray beam, and the isotropic intensity of the scattered beam is measured as a function of the scattering angle (Fig. 16.1). The solvent scattering is also measured and subtracted from that of the macromolecular solution. The difference is due to the dissolved particles and contains information about their structure. The SAXS experiments are usually performed using synchrotron radiation, and all major synchrotrons, like ESRF (Grenoble), ANL (Argonne), DESY (Hamburg), and Spring-8 (Himeji), have biologic SAXS beamlines, but home X-ray sources, although yielding much lower flux, can in some cases also be used. For structure analysis (shape, quaternary structure, flexibility studies), solutions containing single molecular species without aggregates are required. Sample monodispersity should usually be better than 90% and must be verified by other methods like native gel filtration, dynamic light scattering (DLS), or analytical ultracentrifugation (AUC) before going to a large-scale facility. Typical protein concentrations required for SAXS are in the range of about 0.5–10 mg/mL, and a concentration series is usually measured to get rid of interparticle interference. Knowledge about the absolute value of the solute concentration, with accuracy better than 10%, is required to estimate the molecular mass. The sample volume per measurement is about 20–100 μL depending on the setup so that about 2–5 mg of purified material is typically required for a complete structural study. Radiation damage is an issue for synchrotron studies, and the measured samples cannot be recovered and used for other experiments.

In biologic SAXS, relatively hard X-ray photons with wavelength λ, about 0.10–0.15 nm are typically employed. When the sample is illuminated by a monochromatic X-ray plane wave with wave vector $k_0 = |\mathbf{k_0}| = 2\pi/\lambda$, the electrons within the object interacting with the incident radiation become sources of spherical waves. For structural studies, elastic scattering effects are relevant, where the modulus of the scattered wave $k_1 = |\mathbf{k_1}|$ is equal to k_0. The amplitude of the wave scattered by each atom is described by its scattering length f, which is proportional to the number of electrons.

The scattering process involves a transformation from the "real" space (coordinates $\mathbf{r}$ of the object) to the "reciprocal" space, that is, coordinates of scattering vectors $\mathbf{s} = (s,\Omega) = \mathbf{k_1} - \mathbf{k_0}$. Here, the momentum transfer is $s = 4\pi\ \lambda^{-1} \sin(\theta)$, where 2θ is the scattering angle and Ω is the direction of the scattering vector. The transformation is mathematically described by the Fourier operator

$$A(\mathbf{s}) = \Im[\rho(\mathbf{r})] = \int \rho(\mathbf{r}) \exp(i\mathbf{sr})\, d\mathbf{r}, \tag{16.1}$$

where the function $\rho(\mathbf{r})$ is the electron density distribution in the sample and $A(\mathbf{s})$ is the scattering amplitude. Experimentally, one measures the scattering intensity, $I(\mathbf{s}) = [A(\mathbf{s})]^2$. The real space resolution of the scattering data can be estimated as $d = 2\pi/s$, which is a counterpart of the well-known Bragg equation in crystallography. The scattering at small angles, that is, at small s, therefore provides information about large distances (much larger than the wavelength), that is, about overall particle structure.

In the present chapter, we consider solution scattering studies where the intensity is isotropic and depends only on the momentum transfer s. To obtain the net signal from the macromolecules, the solvent scattering is subtracted from that of the solution, and the measured intensity corresponds to the excess density distribution $\Delta\rho(\mathbf{r}) = \langle\rho(\mathbf{r})\rangle - \rho_s$, where ρ_s is the electron density of the solvent. Two major cases can be distinguished for biologic structural studies in solution:

1. The most important idealized case represents a dilute solution of randomly distributed noninteracting particles, all having the same structure. The intensity from the entire ensemble is proportional to the scattering from a single particle averaged over all orientations: $I(s) = \langle I(\mathbf{s})\rangle_\Omega = [F[\Delta\rho(\mathbf{r})]^2]_\Omega$. Typical objects of this kind are monodisperse dilute solutions of folded purified proteins, nucleic acids, or macromolecular complexes. Novel methods of SAXS data analysis allow for structure determination of such systems at low resolution (1–2 nm) (58, 70, 74). Note that for nonideal solutions of identical particles, which can interact with each other, the measured scattering intensity can be written as $I_S(s) = I(s) \times S(s)$. Here, $S(s)$ is the "structure factor" related to the interparticle interactions, which can, in most cases, be eliminated by measurements at different concentrations and extrapolation to infinite dilution.
2. Mixtures of particles of different types, where the scattering intensity will be a linear combination of their contributions

$$I(s) = \sum_{k=1}^{K} \nu_k I_k(s), \tag{16.2}$$

where K is the number of distinct particle types (components), and ν_k and $I_k(s)$ are the volume fraction and the scattering intensity from the kth component, respectively. Typical objects of this kind are equilibrium systems like oligomeric mixtures of proteins or nonequilibrium systems in the process of (dis) assembly or (un)folding. SAXS is employed to determine the volume fractions of the components in the system, if their scattering patterns are available, or to find the number of components based on the measurements at different conditions. IDPs, as we shall see below, represent a specific type of mixture with many components ($K \gg 1$) due to different configurations of individual molecules in solution.

It should also be noted that the basic equations and concepts valid for SAXS are also applicable for small-angle neutron scattering (SANS). The latter technique is also used in the structural studies of biomolecular solutions, but mostly for the analysis of complexes, where neutrons are able to provide additional information thanks to their sensitivity to hydrogen/deuterium composition of the solvent and of the solute. For IDPs, SAXS is more often used, which is faster, requires less material, and provides more precise data at higher scattering angles.

16.3 SCATTERING PROPERTIES OF IDPS

16.3.1 SAXS Profiles and Kratky Plots

The scattering profile measured for an IDP is the average of all these arising from the astronomical number of conformations that the protein adopts in solution. This particularity provides to the experimental SAXS data with very special trends. Figure 16.2A displays the synthetic SAXS curves for 10 randomly selected 100-residue-long polyalanine conformations from a large pool of 10,000 of them. The individual conformations present a large number of features along the simulated momentum transfer range. The initial part of the curve, describing the overall size of the conformation, also presents distinct slopes indicating a large variety of possible sizes that an unstructured chain can adopt. The average SAXS profile, obtained from the individual 10,000 conformations of the pool, presents a smoother behavior, with essentially no features.

Traditionally, Kratky plots ($I(s)\cdot s^2$ as a function of s) have been used to identify disordered states and distinguish them from globular particles. The Kratky representation enhances the particular features observed in scattering profiles allowing an easier identification of different kinds of particles (17). The scattering intensity of a globular protein behaves approximately as $1/s^4$ conferring a bell-shaped Kratky plot with a well-defined maximum. Conversely,

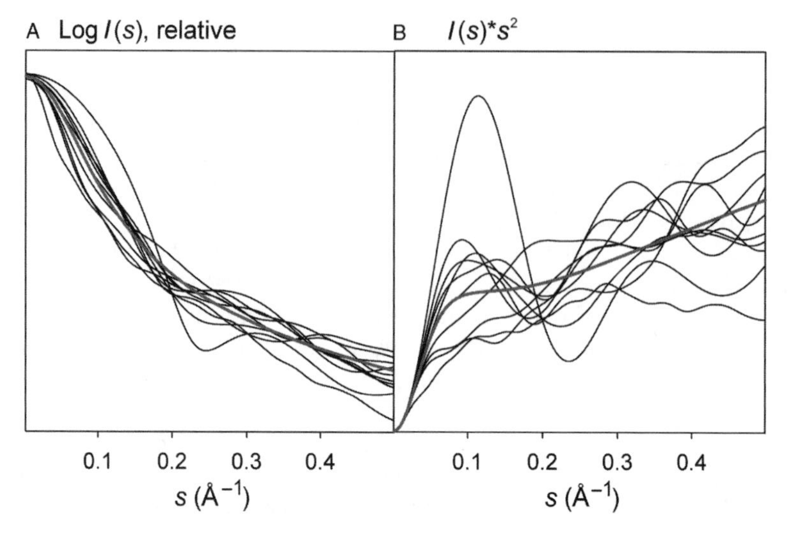

Figure 16.2 Effect of the conformational polydispersity of IDPs on SAXS profiles and their Kartky representations. (A) SAXS profiles (black) of 10 randomly selected 100-residue polyalanine chain conformations from a 10,000 pool of them. The average of the SAXS intensities from the 10,000 conformations is displayed with a thick gray line. (B) Kratky representation of the same data.

an ideal Gaussian chain has a $1/s^2$ dependence of $I(s)$ and, therefore presents a plateau at large s values. In the case of a chain with no thickness, the Kratky plot also presents a plateau over a specific range of s, which is followed by a monotonic increase. This last behavior is normally observed experimentally in unfolded proteins. To exemplify these results, we present the Kratky plots of the above-mentioned 10 individual SAXS profiles compared with the averaged one (Fig. 16.2B). A huge diversity of the Kratky plots for the different conformations is observed. Whereas some of them present a clear maximum, indicating a compact conformation, others present less evident shapes. The average SAXS profile presents a monotonic increase as expected for an unstructured system.

16.3.2 Radius of Gyration: A Single Parameter to Identify IDPs

Unstructured proteins, due to the possibility of adopting highly extended conformations, are characterized by large average sizes compared with globular proteins. One of the best known size characteristic of a protein, its radius of gyration R_g can be directly obtained from SAXS data using a classical Guinier approximation (28). The quantitative comparison of the experimentally measured R_g with these predicted by theoretical models, mainly derived from polymer physics, is used as a diagnostic of the unstructured nature of a protein. In this section, we briefly describe the model values of R_g used for IDPs.

In the Guinier approximation, assuming that at very small angles ($s < 1.3/R_g$) the intensity is represented as $I(s) = I(0)\exp(-(sR_g)^2/3)$, and the R_g is obtained by a simple linear fit in the logarithmic scale (28). For unstructured proteins, alternative Debye approximation (13) and the transformation mentioned in Equation 16.2 (69) are also used. These two methods may provide more accurate R_g estimates for unstructured systems (57). The experimental R_g is a single value representation of the size of the molecule, which for the scattering from an ensemble of IDPs yields an average (so-called z-average, [20]) over all accessible conformations in solution. The most common quantitative R_g interpretation for unfolded proteins is based on Flory's equation (22), which relates it to the length of the protein chain through a power law

$$R_g = R_0 \cdot N^{\nu}, \tag{16.3}$$

where N is the number of residues in the polymer chain, a constant R_0 depends on several factors, in particular, on the persistence length, and ν is an exponential scaling factor. For an excluded-volume polymer, Flory estimated ν to be approximately 3/5, and more accurate estimates established a value of 0.588 (40). A recent compilation of R_g values measured for 26 chemically denatured proteins sampling broad range of chain lengths found a ν value of 0.598 ± 0.028 (37), in good agreement with hydrodynamic data (86), and an R_0 value of 1.927 ± 0.27. These results suggest the random coil nature of the chemically denatured proteins, at least in terms of the R_g parameter.

The comparison of R_gs measured for proteins with threshold values derived from different Flory's equation parameterizations is a very common strategy to identify IDPs. However, the question whether the conformational sampling in the denatured state is equivalent to that found in intrinsically disordered states is still unclear. Potential bias from the effect of denaturants such as urea or guanidinium chloride on the Ramachandran description of the backbone conformations has been postulated (67 and references therein). As a consequence of the potential differences at residue level in both environments, a modification of the overall properties could be envisioned, leading to differences in the SAXS properties. A recent NMR study based on the measurement of several residual dipolar couplings (RDCs) along the ubiquitin backbone indicated that the conformational sampling in chemically denatured proteins could be different from that found in IDPs (47). These results suggest that even in the absence of transiently structured regions, present Flory's law parameterizations may not be appropriate to estimate R_g for IDPs from the length of the protein, although more work on this field is needed to derive clearer conclusions.

16.4 APPLICATIONS OF SAXS TO IDPS

16.4.1 Biophysical Characterization of IDPs Using SAXS

SAXS has since decades been widely used to characterize unstructured protein states and IDPs. Normally, SAXS curves are analyzed in combination with other experimental techniques and bioinformatics tools to identify unstructured regions in proteins. Circular dichroism (CD), NMR, fluorescence spectroscopy, and hydrodynamic techniques such as size exclusion chromatography, AUC, or DLS have been used in combination with SAXS to identify proteins as IDPs. Since the pioneering study of prothymosin α (25), several other IDPs have been biophysically classified: C-terminal measles virus nucleoprotein (45), PIR domain (49), neuroligin 3 (56), synthetic resilin (54), Ki-1/57 (12), FEZ1 (39), and HrpO (26). In these cases, the monotonic increase of Kratky plots, as well as the large R_g values, and expanded $p(r)$ functions with too large D_{max} values for folded particles were the key features useful to identify them as unstructured proteins.

An excellent example of the synergy between SAXS and other experimental and computational tools to structurally characterize IDPs is the C-terminal measles virus nucleoprotein case, N-tail (45), a 139-residue-long polypeptide. The random coil CD spectra and the low dispersion of the ^{1}H NMR spectra clearly suggested the unstructrured nature of the N-tail. These observations were in agreement with the notably large R_g obtained from Guinier analysis of the small-angle region, 27.5 ± 0.7 Å. However, N-tail Kratky plot presented a clear bump at s = 0.08 Å^{-1} followed by a flat region (Fig. 16.3). This dual behavior indicates a certain degree of compactness in a concrete region of the chain. A bioinformatics analysis predicts a 16-residue-long fragment to have a strong tendency to form an α-helix. In addition, an X-ray structure of this

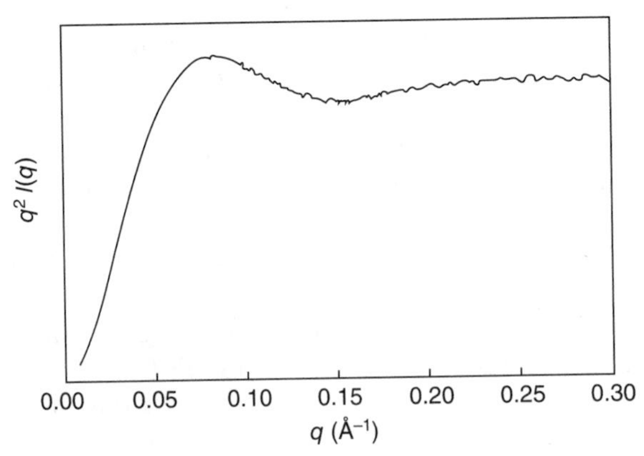

Figure 16.3 Kratky plot of the measles virus N-tail at 9 mg/mL. The bump at $q \approx 0.08$ Å^{-1} suggests the presence of a partially folded region within the disordered protein. Figure extracted with permission from Longhi et al. [45].

fragment with its natural partner, the C-terminal region of the phosphoprotein, was obtained in a helical conformation, confirming this stretch as a molecular recognition element (35). Interestingly, the same region of the protein belonging to the family-related Sendai virus was identified to have a large population, up to 75%, of helical conformations (31, 32), substantiating the SAXS results. Importantly, populations of nascent secondary structural elements such as the one found in the N-tail are often related with biomolecular recognition events involved in regulation and signaling (2, 23, 68).

There have been attempts to reconstruct low-resolution structures from SAXS data measured for disordered chains using standard programs. The resulting ab initio reconstructions display highly elongated shapes, dictated by the large D_{max} of IDPs. The validity of this single conformation approach to describe a highly plastic protein is uncertain, although it clearly helps in visualizing a largely diffused molecule.

Intrinsically disordered fragments are often attached to or tether folded domains. This architecture provides advantages in recognition events (18). There are several SAXS studies of partially folded proteins such as the transcriptional repressor CtBP (55), the C-terminus of Sendai virus phosphoprotein (5), α4 (66), N-terminus of Msh6 (64), p53 (75), Prion protein (43), or NEIL1 (78). The presence of highly dynamic fragments attached to globular particles induces a dual behavior in the Kratky representation that presents a maximum, corresponding to the folded part of the protein and a contribution with a continuous rise from the unfolded region. The relative number of amino acids from both distinct fragments dictates the actual features of the Kratky plot. When comparing the experimental scattering profile to the high-resolution structure of the globular domain, if available, normally a bad fit is obtained (55). When performing ab initio reconstructions from these mixed proteins, an additional density not corresponding to the folded part is obtained, which can be assigned to the flexible region (66, 78). Whether this density reflects the space sampling of the unstructured region is not clear.

16.4.2 Monitoring Environmental Changes in IDPs with SAXS

IDPs are often involved in signaling processes and must change their global properties upon environmental modifications within the cell in order to bind to or release from their natural partners. SAXS is a well-suited tool to rapidly monitor structural changes in proteins upon such environmental modifications. The changes, associated with varying pH (38), ionic strength (52), presence of specific ions (29), phosphorylation (29), or additives (30), must induce global size variation in IDPs in order to be monitored by SAXS. These global alterations are reflected again in the apparent R_g, D_{max}, and the appearance of Kratky plots.

Prothymosin α is an example of a protein studied by SAXS in different conditions (Fig. 16.4). Out of its 109 residues, roughly half are aspartic and glutamic acids, such that the protein is predicted to have a global charge of

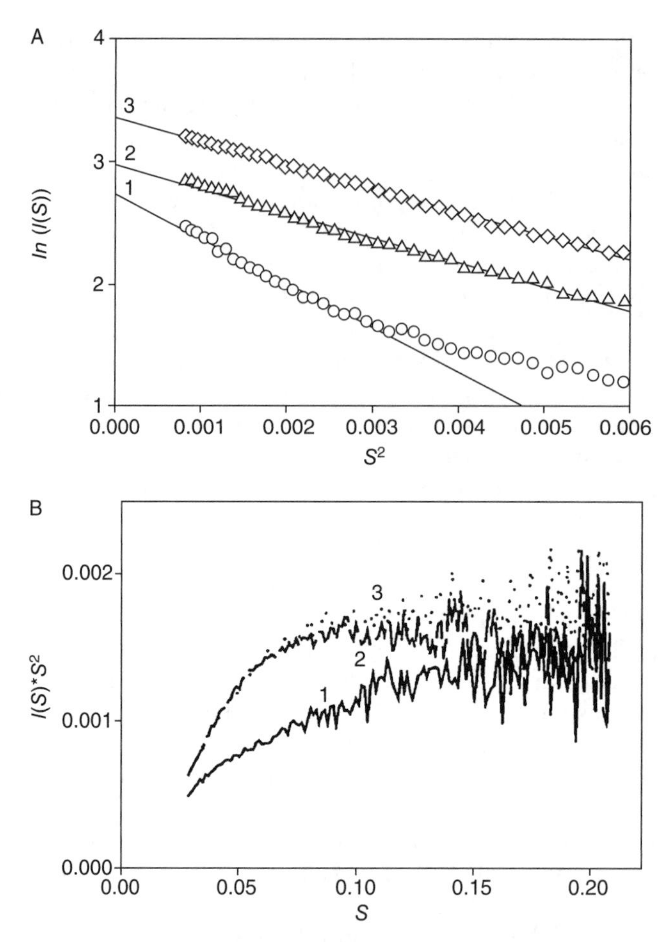

Figure 16.4 Guinier (A) and Kratky plot (B) representations of the SAXS data measured for prothymosin α under different experimental conditions: (1) pH 7.5, (2) pH 2.5, (3) pH 7.5, and 15 mM of $ZnCl_2$. Guinier plots have been scaled arbitrarily for a better visualization (note that the momentum transfer has been expressed units of $S = s/2\pi$). Reprinted from Uversky et al. [79], Copyright (2009), with permission from Elsevier.

–54 at neutral pH. SAXS curves of prothymosin α have been measured at nearly neutral, 7.5, and at acidic, 2.5, pHs (80). Despite Kratky plots indicating that the system was unstructured in both conditions, a dramatic R_g reduction was observed upon pH decrease, from 37.8 ± 0.9 to 27.6 ± 0.9 Å. Interestingly, a similar level of compaction, down to R_g of 28.1 ± 0.8 Å, was observed when, at neutral pH, 15 mM of Zn^{2+} was added to the sample (79). These dramatic changes in protein size can be explained by the decrease of the electrostatic repulsion within the chain at low pH, and with the presence of cations bound to the prothymosin α.

α-Synuclein (αS), a protein found in the aggregates involved in Parkinson's disease and other neurodegenerative pathologies, is a 140-amino acid IDP that has been widely studied and biophysically characterized in different conditions. Some biologic data indicate that the presence of certain metal ions accelerates fibrillation kinetics of αS. The direct interaction of this IDP with Cu^{2+} was confirmed by NMR (62), but the effect of this process on the structure of αS was unclear. SAXS experiments on αS in the presence or absence of Cu^{2+} revealed no global changes upon metal binding, indicating that the recognition process only induced local modifications (10). Through SAXS measurements, it was also demonstrated that the enhanced propensity to form aggregates of the two naturally occurring point mutants A30P and A53T was not provoked by distinct global properties in solution with respect to the wild type (42). Conversely, in acidic conditions, the R_g decreases from 40 ± 1 to 30 ± 1 Å (42), suggesting the presence of either partial folding or the formation of long-range contacts between remote regions of the chain as suggested in different NMR studies (3, 9).

16.4.3 A SAXS Perspective of Biomolecular Interactions of IDPs

The large conformational sampling observed by IDPs provides them with very specific binding properties and consequently pivotal roles in several signaling pathways (19). Interaction with a partner normally requires freezing the highly plastic IDP in a single conformation leading to a so-called folding upon binding (88). The entropic cost of this process makes IDPs very well suited for transient interactions required for signal transduction. Additionally, the large surface exposed by these proteins and the adaptability to different environments make IDPs highly selective or promiscuous partners, depending on the case. Studying the biomolecular recognition processes for IDPs is thus crucial for understanding of the biologic role of this family of proteins, and SAXS has been an important tool for this goal. Several studies were devoted to the interactions of different IDPs with other proteins (11, 24, 64) and with DNA (43, 75). We will give a more detailed explanation of two of them, Msh6 and p53, to exemplify the role that SAXS can have in this field.

The complex of Msh2 and Msh6 recognizes mismatched bases in DNA during mismatch repair. The N-terminal region of Msh6, a 304-residue-long IDP, recognizes PCNA, a homotrimeric protein that controls processivity of DNA polymerases. Shell and coworkers demonstrated this direct interaction with SAXS (64). A comparison of the R_g, Kratky plots, and $p(r)$ functions of the isolated partners and the complex showed that PCNA does not induce substantial structure to the N-terminal region of Msh6, which remains mainly disordered and proteolytically accessible upon binding. The interaction of the Msh2–Msh6 complex with PCNA was also addressed by SAXS. The interaction was shown to produce a heterocomplex that could be considered as a flexible dumbbell, where both globular regions are tethered by the N-terminal Msh6 fragment that acts as a molecular leash. These observations where further confirmed in an additional experiment with a biologically active deletion mutant of Msh6 with a notably shorter N-terminal tail. In these conditions, the important size changes upon binding were enhanced and easily monitored by the $p(r)$ and D_{max} derived from the SAXS profiles.

The tumor suppressor p53 is a multifunctional protein that plays a crucial role in the processes like apoptosis control and DNA repair p53 is a homotetramer with two folded domains that are tethered and flanked by unstructured regions. In fact, 37% of the whole sequence can be considered an IDP, and, as a consequence, p53 is a highly flexible protein as shown by the SAXS data of the full-length and different constructs (75, 85). In the modeling of p53 against the SAXS data, high-resolution structures of the folded domains from crystallography and NMR were used as rigid bodies, and the unfolded domains were represented by chains of dummy residues following a well-established method (58). The free p53 is a rather open cross-like tetrameric assembly, which collapses in the presence of DNA to tightly embrace the latter (Fig. 16.5). This is how the flexibility of IDPs helps the protein to fulfil its function of constantly watching over the DNA in the cell. Interestingly, the SAXS model of the complex without the disordered N-terminal transactivation domains is in excellent agreement with an independent electron microscopy map of the same molecule (75). In a subsequent study (85), a yet more detailed

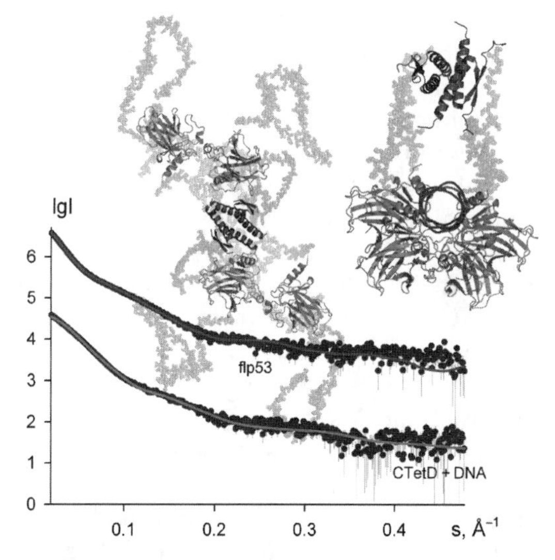

Figure 16.5 The models of tumor suppressor p53 generated from the SAXS curves (dots: experimental data; solid lines: computed patterns). In the free form (left), the four domains of p53 recognizing DNA (green and cyan) are arranged in two separated dimers, which wrap around DNA (pink) upon binding (right). The available high-resolution structures of the domains employed in the modeling are displayed as ribbons, the flexible portions of p53 as semitransparent beads. The N-terminal transactivation domains in the free full-length protein flp53 (tomato beads) are truncated in the complex with DNA (CTeD + DNA). See color insert.

information about the flexibility of the isolated transactivation domain and its structure within the full-length p53 complexed coactivator was obtained by a joint use of SAXS and NMR.

16.4.4 SAXS as a Tool to Validate Computational Models of Unstructured Proteins

The development of realistic ensemble models of unstructured states of proteins has been an important subject of research for many years (91). Initially, these models were focused on protein states under denaturing conditions for a better understanding of the folding process (65). In addition, molecular properties straightforwardly derived from these ensembles such as solvent accessibility serve as a basis for a quantitative interpretation of protein thermodynamics and stability (4, 63). More recently, structural models of IDPs have been developed in order to structurally characterize this family of proteins (5).

The quality of in silico structural models has been evaluated by their ability to properly reproduce experimental data measured for disordered proteins. SAXS has arguably been the most popular technique, although hydrodynamic measurements and NMR have also been used (48). Several types of models with different levels of structural resolution have been validated with previously mentioned compilations of R_g for denatured proteins (15, 21, 27, 33, 77, 83, 84, 90). In these studies, a remarkable agreement with the experimental data is obtained. However, for most of the models, an additional parameterization accounting for the exclusion or the solvation terms is required.

In more advanced studies, complete scattering profiles were considered. Extending the momentum transfer range simulated the development of more adequate ensemble models, which has to be tested at higher resolution and in a more stringent context. Zagrovic et al. addressed the conformational study of a synthetic peptide by combining state-of-the-art molecular dynamic simulations with the SAXS profile experimentally measured (89). In another example, scattering patterns of the reduced ribonuclease A in different denaturing conditions were well reproduced with a proper selection of the solvation term (83). A very simplistic structural model based on polymer theory was used to describe the SANS curve of denatured phosphoglycerate kinase (60).

Fewer studies of this kind were performed for IDPs. To explain the SAXS curve measured for the complex of $p27^{kip1}$/Cdk2/Cyclin A, mobility of the system was explicitly accounted for (24). Hundreds of snapshots of the molecular dynamics simulation of the complex were collected, and their theoretical SAXS profile computed and averaged yielded a curve with a better description of the experimental data than other single-conformation models.

The program flexible Meccano (FM) has been the most tested structural model for IDPs. FM assembles peptidic units, considered as rigid entities, in a consecutive way (5). The force field used for this algorithm includes a coil description of the residue-specific Ramachandran space sampled by the amino acids, and a coarse-grained description of the side chains that avoids collapse within the chain. This program has been tested for a large number of IDPs, and it has successfully described SAXS and several NMR observables measured for these proteins. A detailed description of the algorithm and its applications to IDPs can be found in chapter 21. FM was used to describe the SAXS profile of Protein X, a three-helix bundle that has a 60-amino-acid-long tail attached at the N-terminus (5). The biophysical characterization of the transactivation domain of p53 was also addressed in the context of the full-length protein and when isolated (85). Importantly, in these cases, the same structural model was able to simultaneously describe NMR properties, which mainly report on the conformational (local) properties and SAXS curves that report on the size and shape (global) characteristics of the proteins. These results suggest the general applicability of the FM structural model to fully describe IDPs.

16.5 THE ENSEMBLE OPTIMIZATION METHOD (EOM)

The EOM has been developed as a general approach to address the structural characterization of IDPs by SAXS (6). Using this strategy, the experimental SAXS data are assumed to arise from a (undetermined a priori) number of coexisting conformational states generalizing the traditional "single-state" methods. The conformations belonging to the final ensemble are selected by a Monte-Carlo-type search based on the scattering patterns computed from a large representative random pool, becoming a data-driven optimized ensemble strategy. The price to pay in order to optimize the large number of degrees of freedom by EOM is that assumptions are required to generate the pool of unfolded chains. The randomization model is also used to establish a threshold for the later quantitative interpretation of the results in terms of individual selected structures, and in this sense the EOM is not entirely model independent. It is worth mentioning that EOM provides a new source of structural information for disordered and also for flexible multidomain proteins (7, 8, 81, 82). The major features of the EOM will be discussed in the following sections.

16.5.1 The EOM Algorithm

In the algorithm, schematically depicted in Figure 16.6, the solute is represented by an ensemble containing N different conformations of the same molecule (in practice, N is about 20–50). The appropriate ensemble is selected from a pool containing $M \gg N$ conformers, which should cover the conformational space of the molecule. A genetic algorithm (GA) (34) is then employed to select subsets of configurations that, after averaging their individual scattering profiles, fit the experimental data.

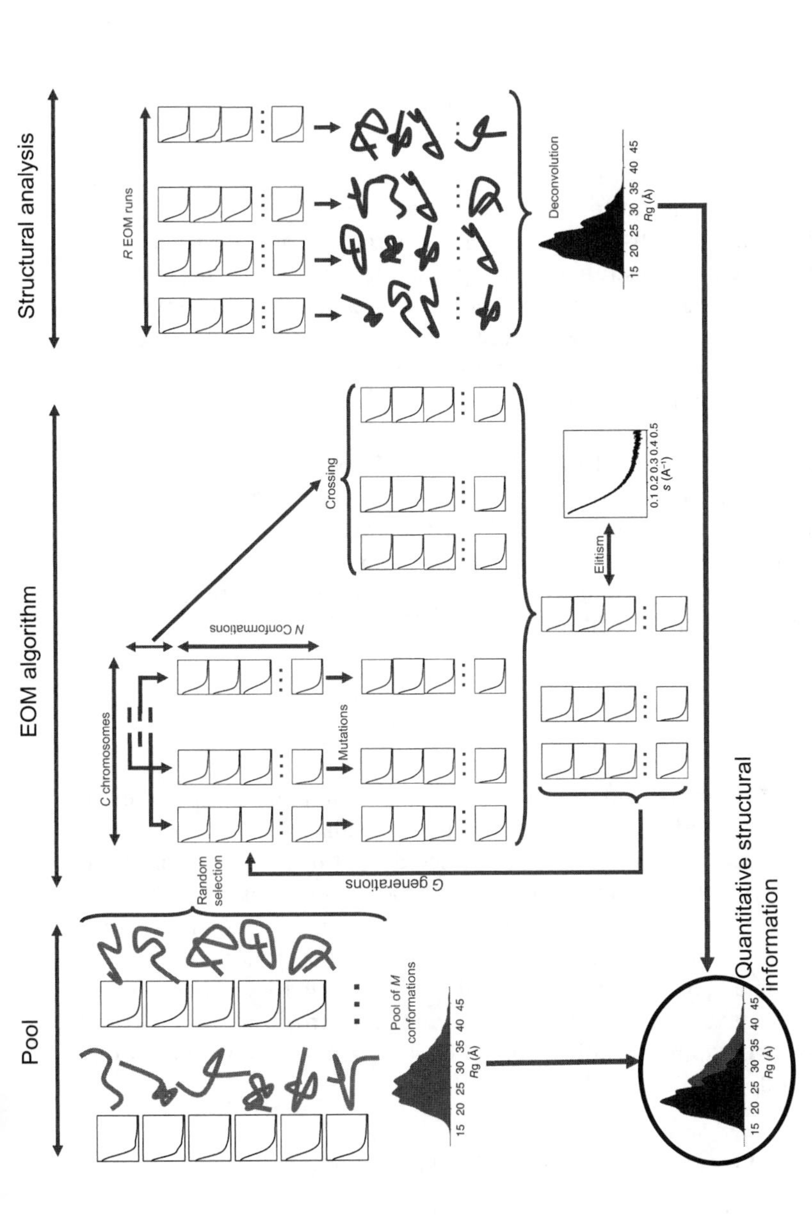

Figure 16.6 Schematic representation of the EOM scheme for the analysis of SAXS data in terms of R_g distributions. The *M* conformations/curves of the pool (random distribution), left part of the figure, are used to generate the initial *C* chromosomes and to feed the genetic operators (*mutations*, *crossing*, and *elitism*) along the GA process that runs for *G* generations. The complete process is repeated *R* independent times, and each run provides *N* selected structures/curves that fit the experimental profile. The structural analysis of the resulting conformations is displayed on the right part of the scheme; the R_g distribution of the selected $(N \times R)$ conformations is compared with that derived from the pool that is considered as a complete conformational freedom scenario. From this, comparison is possible to derive a quantitative structural estimation of the IDP conformations coexisting in solution.

The scattering data from such an ensemble is computed by summing up the individual scattering patterns as in Equation 16.2. We can assume that the subsets are uniformly populated, so that the intensity of a subset $I(s)$ containing N conformers is

$$I(s) = \frac{1}{N} \sum_{n=1}^{N} I_n(s), \tag{16.4}$$

where $I_n(s)$ is the scattering from the nth conformer. The scattering curves from all the structures in the pool are first precomputed using the program CRYSOL (71), and the subsequent selection operators are preformed using these patterns, which significantly speeds up the calculations. The subset selection is performed using a GA, where each subset (called chromosome) contains N scattering profiles (genes) from different conformers. In the first generation, $C = 50$ chromosomes are generated by selecting the conformations randomly from the pool. In each generation G, these C chromosomes are submitted to *mutation* and *crossing*. In *mutation*, the genes of each chromosome are randomly exchanged for others either from the pool or from the chromosomes of the same generation. In the *crossing*, genes of two randomly selected chromosomes are exchanged. After the two genetic operations, the population is composed of $3C$ chromosomes. For each chromosome, the average (Eq. 16.4) of its individual SAXS profiles is compared with the experimental scattering yielding the fitness function (Eq. 16.5)

$$\chi^2 = \frac{1}{K-1} \sum_{j=1}^{K} \left[\frac{\mu I(s_j) - I_{\exp}(s_j)}{\sigma(s_j)} \right]^2, \tag{16.5}$$

where $I_{\exp}(s)$ is the experimental scattering, K is the number of experimental points, $\sigma(s)$ is the standard deviation, and μ is a scaling factor. The best C chromosomes yielding the lowest χ^2 are selected for further evolution in an *elitism* fashion, typically for up to G = 1000–5000 generations. The chromosome with the best agreement to the experimental data is collected for further structural analysis. The complete process is repeated R times in order to collect $R \times N$ final conformations. The distributions of the low-resolution parameters, R_g, maximum distance (D_{max}), and anisotropy in the selected conformations provide information about the structure of the solute, as discussed in the following sections.

16.5.2 Structural Information Coded in EOM Ensembles

It may seem that a subpopulation with $N \approx 50$ is far too small to explain the conformational behavior of a flexible protein sampling an astronomical amount of conformations in solution. However, these subpopulations are sufficient to depict the global properties of the protein in terms of size and shape (i.e., R_g, D_{max}, and anisotropy) rather than conformational details at residue level. The structures selected by the EOM are neither the most populated conformations, not even are they claimed to really exist in solution. Instead, the EOM ensemble is a tool to describe the size and shape distributions sampled by the unfolded molecule. Selected ensembles derived from repeated EOM runs starting from different random seeds normally contain different conformations, but they all provide similar R_g, D_{max}, and anisotropy distributions. Therefore, the algorithm is able to find equivalent minima in terms of distributions but of course not in terms of individual molecular configurations: the latter is not identifiable given the low resolution of SAXS. Still, the distributions provided by EOM yield a major improvement over traditional approaches that condense all structural characteristics in the averaged R_g value. One may also argue that due to the restricted information content of the SAXS data, the EOM model containing a relatively small number of conformers adequately describe the entire system. Similar approaches using higher-resolution techniques (e.g., RDCs in NMR) require much larger ensembles to reach reasonable levels of convergence (5, 33).

It is very instructive to compare the EOM-derived distributions with those of the random ensembles (e.g., initial pools) to potentially detect the presence of nascent secondary structural elements or low populations of transient long-range contacts as alterations of the distributions with respect to the pool. Local transient structures display broader distributions of R_g and normally shifted toward larger values, whereas long-range contacts and residual tertiary structure provide more compact distributions than the random coils.

The EOM has been applied to the biophysical characterization of intrinsically unfolded neuligin 3 (56) and HrpO (26). For the neuroligin 3 case, the EOM suggests the presence of two different families of conformations coexisting in solution (Fig. 16.7). The first family includes relatively compact structures, with less conformational diversity than the starting pool. A second family of structures displayed extremely large values of R_g that were attributed to either a population of highly extended molecules or the presence of oligomers. Interestingly, this bimodal distribution disappeared upon the addition of urea, where a unimodal and highly extended ensemble of conformations was depicted by the EOM.

16.5.3 Multiple SAXS Curve Fitting with EOM

The EOM provides overall structural information, whereas the exact regions responsible for conformational restriction remains elusive. It appears, however, that using deletion mutants, more detail could be obtained even with such a low-resolution method as SAXS. Indeed, the scattering profiles from different deletion mutants of the full-length protein can be fitted simultaneously with EOM, and the sum of the individual figures of merit drives the optimization (6). In this way, contributions from different chain fragments to the SAXS data can be distinguished to identify the regions responsible for the increase or decrease in compactness of the protein. The SAXS resolution is therefore improved by the addition of multiple experimental curves (of course, this

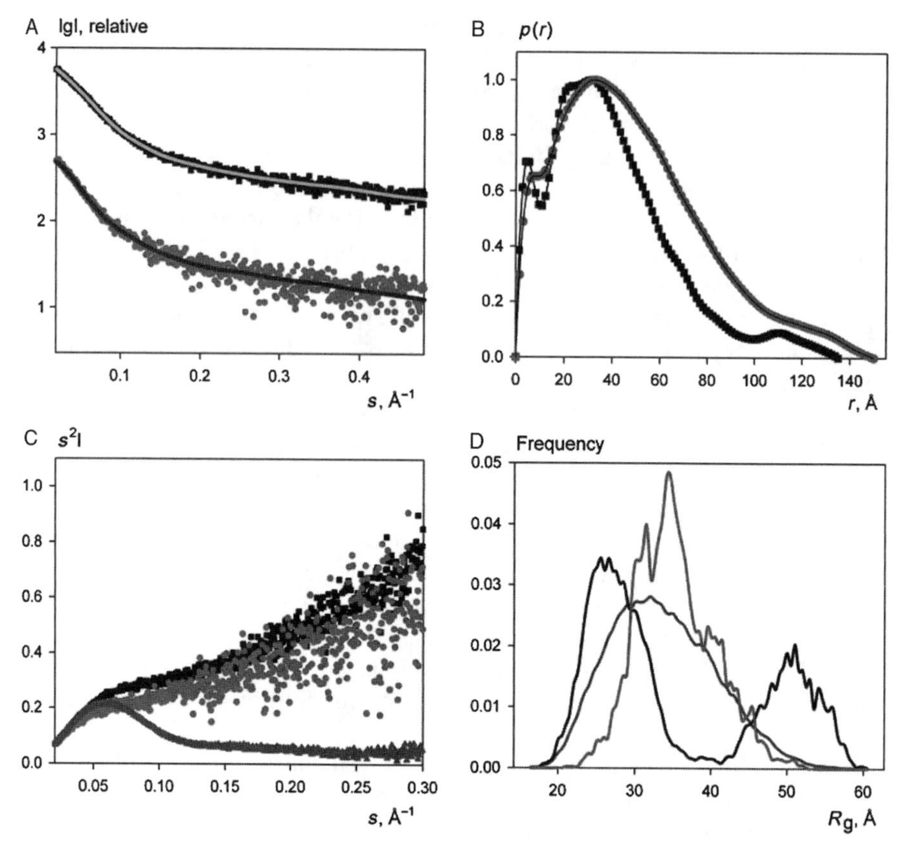

Figure 16.7 SAXS measurements and analysis on neuroligin 3. (A) X-ray scattering intensities in native (black squares) and denaturing (gray circles) conditions. EOM fits to the profiles are displayed in solid lines. (B) $p(r)$ distribution functions for both data sets. (C) Kratky plots of both data sets (same color code) compared with that obtained for the globular protein BSA (triangles). (D) R_g distributions derived from the EOM analysis of both data sets (same color code). The dual and the monotonic behavior of neuroligin 3 in native and denaturing conditions respectively are observed when compared from the distribution of the randomly generated conformations of the pool. Figure reprinted with permission from Paz et al. [56].

strategy is only valid if the structural elements of the full-length protein remain intact in the deletion mutants).

Mylonas et al. (53) applied the multiple SAXS curve fitting with EOM to the SAXS analysis of tau, an IDP involved in neuronal microtubule stabilization, and found abnormal deposits in the brain of Alzheimer's disease patients (46). Two tau isoforms were studied, and SAXS data for the full-length and several different deletion mutants for each isoform were used (Fig. 16.8). The EOM unambiguously identifies the so-called repeat region as the source of residual secondary structure in tau, in perfect agreement with previous NMR

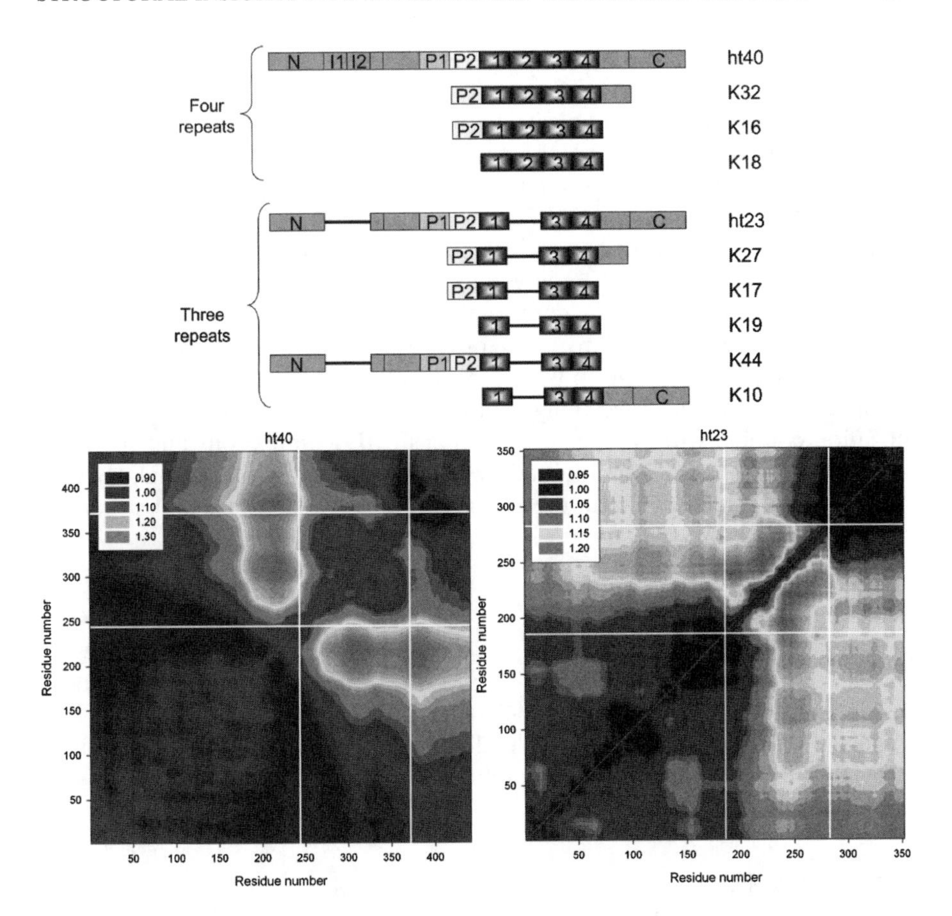

Figure 16.8 Bar diagrams of isoforms of tau protein used for the simultaneous SAXS curve analysis with EOM (top panel). Two families are depicted with four (ht40) and three (ht23) repeat regions. Cα–Cα distance plot of ht40 (middle panel) and ht23 (bottom panel) using multiple curve fitting. Each plot point shows the ratio of the average distance of the selected structures with that obtained from the pool. White lines indicate the residues belonging to the repeat domain. Figure extracted with permission from Mylonas et al. [53]. See color insert.

data indicating the presence of turns and extended fragments in this region (50, 51). The multiple curve fitting, through the averaged Cα–Cα inter-residue distance matrix identifies a distinct conformational behavior depending on the number of repeats present in the isoforms. For the isoform ht23, with three repeats, the maximum separation is found within the repeat domain itself. The full-length ht40 isoform with four domains reveals an enhanced separation between the repeat domain and the preceding region. These results suggest that the different number of turns (one per repeat) may lead to different global arrangements of the chain in that region enhancing or shortening the average interdomain distances expected from a random coil.

16.6 CONCLUSIONS

High-resolution methods are barely applicable to study flexible molecules like IDPs in solution, and among the low-resolution structural techniques, not many can be used. Due to its versatility, speed, and ability to deal with disordered matter and mixtures, SAXS is one of the major structural players in the IDP field. For decades, SAXS was employed to unfolded proteins at a rather simplistic level, by qualitative analysis using Kratky plots and the only parameter, R_g being quantitatively determined. Although even these approaches yielded valuable information about numerous IDPs in solution, a genuine breakthrough was brought up by recent developments in SAXS allowing for comprehensive quantitative analysis of the scattering patterns. Ensemble optimization coupled with Molecular Dynamics (MD) simulations and the joint use with local information obtained by NMR or FRET promises make SAXS an extremely effective tool in the characterization of structure and dynamics of IDPs in solution.

ACKNOWLEDGMENTS

Financial support form the Spanish Ministry of Education-FEDER (BIO2007-63458) is gratefully acknowledged. P. B. holds a Ramón y Cajal contract that is partially financed by the Spanish Ministry of Education and the by funds provided to the Institute for Research in Biomedicine by the Generalitat de Catalunya.

REFERENCES

1. Akiyama, S., A. Nohara, and Y. Maéda. 2008. Assembly and disassembly dynamics of the cyanobacterial periodosome. Mol Cell **29**:703–16.
2. Baker, J. M. R., R. P. Hudson, V. Kanelis, W.-Y. Choy, P. H. Thibodeau, P. J. Thomas, and J. D. Forman-Kay. 2007. CFTR regulatory region interacts with NBD1 predominantly via multiple transient helices. Nature Struct Mol Biol **14**:738–45.
3. Bernadó, P., C. W. Bertoncini, C. Griesinger, M. Zweckstetter, and M. Blackledge. 2005. Defining long-range order and local disorder in native α-synuclein using residual bipolar couplings. J Am Chem Soc **127**:17968–9.
4. Bernadó, P., M. Blackledge, and J. Sancho. 2006. Sequence-specific solvent accessibilities of protein residues in unfolded protein ensambles. Biophys J **91**:4536–43.
5. Bernadó, P., L. Blanchard, P. Timmins, D. Marion, R. W. H. Ruigrok, and M. Blackledge. 2005. A structural model for unfolded proteins from residual dipolar couplings and small-angle X-ray scattering. Proc Natl Acad Sci U S A **102**:17002–7.
6. Bernadó, P., E. Mylonas, M. V. Petoukhov, M. Blackledge, and D. I. Svergun. 2007. Structural characterization of flexible proteins using small-angle X-ray scattering. J Am Chem Soc **129**:5656–64.

7. Bernadó, P., Y. Pérez, D. I. Svergun, and M. Pons. 2008. Structural characterization of the active and inactive states of Src kinase in solution by small-angle X-ray scattering. J Mol Biol **376**:492–505.
8. Bertini, I., V. Calderone, M. Fragai, R. Jaiswal, C. Luchinat, M. Melikian, E. Mylonas, and D. I. Svergun. 2008. Evidence of reciprocal reorientation of the catalytic and hemopexin-like domains of full-length MMP-12. J Am Chem Soc **130**:7011–21.
9. Bertoncini, C. W., Y.-S. Jung, C. O. Fernández, W. Hoyer, C. Griesinger, T. M. Jovin, and M. Zweckstetter. 2005. Release of long-range tertiary interactions potentiates aggregation of natively unstructured α-synuclein. Proc Natl Acad Sci U S A **102**:1430–5.
10. Binolfi, A., R. M. Rasia, C. W. Bertoncini, M. Ceolin, M. Zweckstetter, C. Griesinger, T. M. Jovin, and C. O. Fernández. 2006. Interaction of α-synuclein with divalent metal ions reveals key differences: a link between structure, binding specificity and fibrillation enhancement. J Am Chem Soc **128**:9893–901.
11. Bourhis, J.-M., V. Receveur-Bréchot, M. Oglesbee, X. Zhang, M. Buccellato, H. Darbon, B. Cannard, S. Finet, and S. Longhi. 2005. The intrinsically disordered C-terminal domain of the measles virus nucleoprotein interacts with the C-terminal domain of the phosphoprotein via two distinct sites and remains predominantly unfolded. Protein Sci **14**:1975–92.
12. Bressan, G. C., J. C. Silva, J. C. Borges, D. O. dos Passos, C. H. I. Ramos, I. L. Torriani, and J. Kobarg. 2008. Human regulatory protein Ki-1/57 has characteristics of an intrinsically unstructured protein. J Proteome Res **7**:4465–74.
13. Calmettes, P., D. Durand, M. Desmadril, V. Receveur, and J. C. Smith. 1994. How random is a highly denatured protein? Biophys Chem **53**:105–14.
14. Dafforn, T. R., and C. J. I. Smith. 2004. Natively unfolded domains in endicytosis: hooks, lines and linkers. EMBO Rep **5**:1046–52.
15. Ding, F., R. K. Jha, and N. V. Dokholyan. 2005. Scaling behavior and structure of denatured proteins. Structure **13**:1047–54.
16. Dobson, C. M. 2004. Experimental investigation of protein folding and misfolding. Methods **34**:4–14.
17. Doniach, S. 2001. Changes in biomolecular conformation seen by small angle X-ray scattering. Chem Rev **101**:1763–78.
18. Dunker, A. K., C. J. Brown, J. D. Lawson, L. M. Iakoucheva, and Z. Obradovic. 2002. Intrinsic disorder and protein function. Biochemistry **41**:6573–82.
19. Dyson, H. J., and P. E. Wright. 2005. Intrinsically unstructured proteins and their functions. Nature Rev **6**:197–208.
20. Feigin, L. A., and D. I. Svergun. 1987. Structure Analysis by Small-Angle X-ray and Neutron Scattering. Plenum Press, New York.
21. Fitzkee, N. C., and G. D. Rose. 2004. Reassessing random-coil statistics in unfolded proteins. Proc Natl Acad Sci U S A **101**:12497–502.
22. Flory, P. J. 1953. Principles of Polymer Chemistry. Cornell University Press, Ithaca, NY.
23. Fuxreiter, M., I. Simon, P. Friedrich, and P. Tompa. 2004. Preformed structural elements feature in partner recognition by intrinsically unstructured proteins. J Mol Biol **338**:1015–26.

24. Galea, C. A., A. Nourse, Y. Wang, S. G. Sivakolundu, W. T. Séller, and R. W. Kriwacki. 2008. Role of intrinsic flexibility in signal transduction mediated by the cell cycle regulator, p27Kip1. J Mol Biol **376**:827–38.

25. Gast, K., H. Damaschun, H. K. Eckert, K. Schulze-Forster, H. R. Maurer, M. Müller-Frohne, D. Zirwer, J. Czarnecki, and G. Damaschun. 1995. Prothymosin α: a biologically active protein with random coil conformation. Biochemistry **34**:13211–8.

26. Gazi, A. D., M. Bastaki, S. N. Charola, E. A. Gkougkoulia, E. A. Kapellios, N. J. Panapoulos, and M. Kokkinidis. 2008. Evidence for a coiled-coil interaction mode of disordered proteins from bacterial type III secretion systems. J Biol Chem **283**:34062–8.

27. Goldenberg, D. P. 2003. Computational simulation of the statistical properties of unfolded proteins. J Mol Biol **326**:1615–33.

28. Guinier, A. 1939. La diffraction des rayons X aux trés petits angles: application a l'étude de phenomenes ultramicroscopiques. Ann Phys (Paris) **12**:161–237.

29. He, G., A. Ramachandran, T. Dahl, S. George, D. Schultz, D. Cookson, A. Veis, and A. George. 2005. Phosphorylation of phosphophoryn is crucial for its function as a mediator of biomineralization. J Biol Chem **280**:33109–14.

30. Hong, D.-P., A. L. Fink, and V. N. Uversky. 2008. Structural characteristics of α-synuclein oligomers stabilized by flavonoid baicalein. J Mol Biol **383**:214–23.

31. Houben, K., D. Marion, N. Tarbouriech, R. W. H. Ruigrok, and L. Blanchard. 2007. Interaction of the C-terminal domains of Sendai virus N and P proteins: comparison of polymerase-nucleocapsid interactions within the paramyxovirus family. J Virol **81**:6807–16.

32. Jensen, M. R., K. Houben, E. Lescop, L. Blanchard, R. W. H. Ruigrok, and M. Blackledge. 2008. Quantitative conformational analysis of partially folded proteins from residual dipoar couplings: application to the molecular recognition element of Sendai virus nucleoprotein. J Am Chem Soc **130**:8055–61.

33. Jha, A. K., A. Colubri, K. F. Freed, and T. R. Sosnick. 2005. Statistical coil model of the unfolded state: resolving the reconciliation problem. Proc Natl Acad Sci U S A **102**:13099–104.

34. Jones, G. 1998. Genetic and evolutionary algorithms. In Encyclopedia of Computational Chemistry. Wiley, Chichester, UK.

35. Kingston, R. L., D. J. Hamel, L. S. Gay, F. W. Dahlquist, and B. W. Matthews. 2004. Structural basis for the attachment of a paramyxoviral polymerase to its template. Proc Natl Acad Sci U S A **101**:8301–6.

36. Koch, M. H. J., P. Vachette, and D. I. Svergun. 2003. Small-angle scattering: a view on the properties, structures and structural changes of biological macromolecules in solution. Q Rev Biophys **36**:147–227.

37. Kohn, J. E., I. S. Millet, J. Jacob, B. Zagrovic, T. M. Dillon, N. Cingel, R. S. Dothager, S. Seifert, P. Thiyagarajan, T. R. Sosnick, M. Z. Hasan, V. S. Pande, I. Ruczinski, S. Doniach, and K. W. Plaxco. 2004. Random coil behaviour and the dimensions of chemically unfolded proteins. Proc Natl Acad Sci U S A **101**: 12491–6.

38. Konno, T., N. Tanaka, M. Kataoka, E. Takano, and M. Maki. 1997. A circular dichroism study of preferential hydration and alcohol effects on a denatured protein, pig calpastatin domain I. Biochim Biophys Acta **1342**:73–82.

39. Lanza, D. C. F., J. C. Silva, E. M. Assmann, A. J. C. Quaresma, G. C. Bressan, I. L. Torriani, and J. Kobarg. 2009. Human FEZ1 has characteristics of a natively unfolded protein and dimerizes in solution. Proteins **74**:104–21.

40. LeGuillou, J. C., and J. Zinn-Justin. 1977. Critical exponents for the n-vector model in three dimensions from field theory. Phys Rev Lett **39**:95–8.

41. Levitt, M. 2007. Growth of novel protein structural data. Proc Natl Acad Sci U S A **104**:3183–8.

42. Li, J., V. N. Uversky, and A. L. Fink. 2002. Conformational behavior of human α-synuclein is modulated by familial Parkinson's disease point mutations A30P and A53T. Neurotoxicology **23**:553–67.

43. Lima, L. M. T. R., Y. Cordeiro, L. W. Tinoco, A. F. Marques, C. L. P. Oliveira, S. Sampath, R. Kodali, G. Choi, D. Foguel, I. Torriani, B. Caughey, and J. L. Silva. 2006. Structural insights into the interaction between prion protein and nucleic acid. Biochemistry **45**:9180–7.

44. Lipfert, J., and S. Doniach. 2007. Small-angle X-ray scattering from RNA, proteins, and protein complexes. Annu Rev Biophys Biomol Struct **36**:307–27.

45. Longhi, S., V. Receveur-Brechot, D. Karlin, K. Johansson, H. Darbon, D. Bhella, R. Yeo, S. Finet, and B. Canard. 2003. The C-terminal domain of the measles virus nucleoprotein is intrinsically disordered and folds upon binding to the C-terminal moiety of the phosphoprotein. J Biol Chem **278**:18638–48.

46. Mandelkow, E. M., and E. Mandelkow. 1998. Tau in Alzheimer's disease. Trends Cell Biol **8**:425–7.

47. Meier, S., S. Grzesiek, and M. Blackledge. 2007. Mapping the conformational landscape of urea-denatured ubiquitin using residual dipolar couplings. J Am Chem Soc **129**:9799–807.

48. Mittag, T., and J. D. Forman-Kay. 2007. Atomic-level characterization of disordered protein ensembles. Curr Opin Struct Biol **17**:3–14.

49. Moncoq, K., I. Broutin, C. T. Craescu, P. Vachette, A. Ducruix, and D. Durand. 2004. SAXS study of the PIR domain from the Grb14 molecular adaptor: a natively unfolded protein with transient structure primer? Biophys J **87**:4056–64.

50. Mukrasch, M. D., J. Biernat, M. von Bergen, C. Griesinger, E. Mandelkow, and M. Zweckstetter. 2005. Sites of tau important for aggregation populate β-structure and bind to microtubules and polyanions. J Biol Chem **280**:24978–86.

51. Mukrasch, M. D., P. Markwick, J. Biernat, M. von Bergen, P. Bernadó, C. Griesinger, E. Mandelkow, M. Zweckstetter, and M. Blackledge. 2007. Highly populated turn conformations in natively unfolded tau protein identified from residual dipolar couplings and molecular simulation. J Am Chem Soc **129**:5235–43.

52. Munishkina, L. A., A. L. Fink, and V. N. Uversky. 2004. Conformational prerequisites for formation of amyloid fibrils from histones. J Mol Biol **342**:1305–24.

53. Mylonas, E., A. Hascher, P. Bernadó, M. Blackledge, E. Mandelkow, and D. I. Svergun. 2008. Domain conformation of tau protein studied by solution small-angle X-ray scattering. Biochemistry **47**:10345–53.

54. Nairn, K. M., R. E. Lyons, R. J. Mulder, S. T. Mudie, D. J. Cookson, E. Lesieur, M. Kim, D. Lau, F. H. Scholes, and C. M. Elvin. 2008. A synthetic resilin is largely unstructured. Biophys J **95**:3358–65.

55. Nardini, M., D. Svergun, P. V. Konarev, S. Spano, M. Fasano, C. Bracco, A. Pesce, A. Donadini, C. Cericola, F. Secundo, A. Luini, D. Corda, and M. Bolognesi. 2006. The C-terminal domain of the transcriptional corepressor CtBP is intrinsically unstructured. Protein Sci **15**:1042–50.

56. Paz, A., T. Zeev-Ben-Mordehai, M. Lundqvist, E. Sherman, E. Mylonas, L. Weiner, G. Haran, D. I. Svergun, F. A. A. Mulder, J. L. Sussman, and I. Silman. 2008. Biophysical characterization of the unstructured cytoplasmatic domain of the human neuronal adhesion protein neuroligin 3. Biophys J **95**:1928–44.

57. Pérez, J., P. Vachette, D. Russo, M. Desmandril, and D. Durand. 2001. Heat-induced unfolding of neocarzinostatin, a small all-β protein investigated by small-angle X-ray scattering. J Mol Biol **308**:721–43.

58. Petoukhov, M. V., and D. I. Svergun. 2005. Global rigid body modeling of macromolecular complexes against small-angle scattering data. Biophys J **89**:1237–50.

59. Petoukhov, M. V., and D. I. Svergun. 2007. Analysis of X-ray and neutron scattering from biomolecular solutions. Curr Opin Struct Biol **17**:562–71.

60. Petrescu, A.-J., V. Receveur, P. Calmettes, D. Durand, M. Desmandril, B. Roux, and J. C. Smith. 1997. Small-angle neutron scattering by strongly denatured protein: analysis using random polymer theory. Biophys J **72**:335–42.

61. Putnam, C. D., M. Hammel, G. L. Hura, and J. A. Tainer. 2007. X-ray solution scattering (SAXS) combined with crystallography and computation: defining achúrate macromolecular structure, conformations and assemblies in solution. Q Rev Biophys **40**:191–285.

62. Rasia, R. M., C. W. Bertoncini, D. Marsh, W. Hoyer, D. Cherny, M. Zweckstetter, C. Griesinger, T. M. Jovin, and C. O. Fernández. 2005. Structural characterization of copper(II) binding to α-synuclein: insights into the bioinorganic chemistry of Parkinson's disease. Proc Natl Acad Sci U S A **102**:4294–9.

63. Robertson, A. D., and K. P. Murphy. 1997. Protein structure and the energetics of protein stability. Chem Rev **97**:1251–67.

64. Shell, S. S., C. D. Putnam, and R. D. Kolodner. 2007. The N terminus of *Saccharomyces cerevisiae* Msh6 is an ustructured tether to PCNA. Mol Cell **26**:565–78.

65. Shortle, D. 1996. The denatured state (the other half of the holding equation) and its role in protein stability. FASEB J **10**:27–34.

66. Smetana, J. H. C., C. L. P. Oliveira, W. Jablonka, T. A. Perthinez, F. R. G. Carneiro, M. Montero-Lomeli, I. Torriani, and N. I. T. Zanchin. 2006. Low resolution structure of the human α4 protein (IgBP1) and studies on the stability of α4 and of its yeast ortholog Tap42. Biochim Biophys Acta **1764**:724–34.

67. Stumpe, M. C., and H. Grubmüller. 2007. Interaction of urea with amino acids: implications for urea-induced protein denaturation. J Am Chem Soc **129**:16126–31.

68. Sugase, K., H. J. Dyson, and P. E. Wright. 2007. Mechanism of coupled folding of an intrinsically disordered protein. Nature **447**:1021–5.

69. Svergun, D. I. 1992. Determination of the regularization parameter in indirect-transform methods using perceptual criteria. J Appl Crystallogr **25**:495–503.

70. Svergun, D. I. 1999. Restoring low resolution structure of biological macromolecules from solution scattering using simulated annealing. Biophys J **76**:2879–86.

71. Svergun, D. I., C. Barberato, and M. H. J. Koch. 1995. CRYSOL—a program to evaluate X-ray solution scattering of biological macromolecules from atomic coordinates. J Appl Cryst **28**:768–73.

72. Svergun, D. I., and M. H. J. Koch. 2002. Advances in structure analysis using small-angle scattering in solution. Curr Opin Struct Biol **12**:654–660.

73. Svergun, D. I., and M. H. J. Koch. 2003. Small-angle scattering studies of biological macromolecules in solution. Rep Prog Phys **66**:1735–82.

74. Svergun, D. I., Petoukhov, M. V., and M. H. J. Koch. 2001. Determination of domain structure of proteins from X-ray solution scattering. Biophys J **80**:2946–53.

75. Tidow, H., R. Melero, E. Mylonas, S. M. V. Freund, J. G. Grossmann, J. M. Carazo, D. I. Svergun, M. Valle, and A. R. Fersht. 2007. Quaternary structure of tumor suppressor p53 and a specific p53-DNA complex. Proc Natl Acad Sci U S A **104**:12324–9.

76. Tompa, P. 2002. Intrinsically unstructured proteins. Trends Biochem Sci **27**:527–33.

77. Tran, H. T., X. Wang, and R. V. Pappu. 2005. Reconciling observations of sequence-specific conformational propensities with the generic polymeric behaviour of denatured proteins. Biochemistry **44**:11369–80.

78. Tsutakawa, S. E., G. L. Hura, K. A. Frankel, P. K. Cooper, and J. A. Tainer. 2007. Structural analysis of flexible proteins in solution by small angle X-ray scattering combined with crystallography. J Struct Biol **158**:214–23.

79. Uversky, V. N., J. R. Gillespie, I. S. Millett, A. V. Khodyakova, R. N. Vasilenko, A. M. Vasiliev, I. L. Rodionov, G. D. Kozlovskaya, D. A. Dolgikh, A. L. Fink, S. Doniach, E. A. Permyanov, and V. M. Abramov. 2000. Zn^{2+}-mediated structure formation and compactation of the "natively unfolded" human prothymosin α. Biochem Biophys Res Commun **267**:663–8.

80. Uversky, V. N., J. R. Gillespie, I. S. Millett, A. V. Khodyakova, A. M. Vasiliev, T. V. Chernovskaya, R. N. Vasilenko, G. D. Kozlovskaya, D. A. Dolgikh, A. L. Fink, S. Doniach, and V. M. Abramov. 1999. Natively unfolded human prothymosin α adopts partially folded collapsed conformation at acidic pH. Biochemistry **38**:15009–16.

81. VanOudenhove, J., E. Anderson, S. Krueger, and J. L. Cole. 2009. Analysis of PKR structure by small-angle scattering. J Mol Biol **387**:910–20.

82. Wang, S., M. T. Overgaard, Y. X. Hu, and D. B. McKay. 2008. The *Bacillus subtilis* RNA helicase YxiN is distended in solution. Biophys J **94**:L01–3.

83. Wang, Y., J. Trewhella, J., and D. P. Goldenberg. 2008. Small-angle X-ray scattering of reduced ribonuclease A: effects of solution conditions and comparisons with a computational model of unfolded proteins. J Mol Biol **377**:1576–92.

84. Wang, Z., K. W. Plaxco, and D. E. Makarov. 2007. Influence of loca and residual structures on the scaling behavior and dimensions of unfolded proteins. Biopolymers **68**:321–8.

85. Wells, M., H. Tidow, T. J. Rutherford, P. Markwick, M. R. Jensen, E. Mylonas, D. I. Svergun, M. Blackledge, and A. R. Fersht. 2008. Structure of tumor suppressor p53 and its intrinsically disordered N-terminal transactivation domain. Proc Natl Acad Sci U S A **105**:5762–7.
86. Wilkins, D. K., S. B. Grimshaw, V. Receveur, C. M. Dobson, J. A. Jones, and L. J. Smith. 1999. Hydrodynamic radii of native and denatured proteins measured by pulse field gradient NMR techniques. Biochemistry **38**:16424–31.
87. Wright, P. E., and H. J. Dyson. 1999. Intrinsically unstructured proteins: re-assessing the protein structure-function paradigm. J Mol Biol **293**:321–31.
88. Wright, P. E., and H. J. Dyson. 2009. Linking holding and binding. Curr Opin Struct Biol **19**:31–8.
89. Zagrovic, B., J. Lipfert, E. J. Soria, I. S. Millett, W. F. van Gunsteren, S. Doniach, and V. S. Pande. 2005. Unusual compactness of a polyproline type II structure. Proc Natl Acad Sci U S A **102**:11698–703.
90. Zhou, H.-X. 2002. Dimensions of denatured protein chains from hydrodynamic data. J Phys Chem B **106**:5769–75.
91. Zhou, H.-X. 2004. Polymer models of protein stability, folding, and interactions. Biochemistry **43**:2141–54.

17

DYNAMIC AND STATIC LIGHT SCATTERING

KLAUS GAST

Universität Potsdam, Institut für Biochemie und Biologie, Physikalische Biochemie, Potsdam-Golm, Germany

ABSTRACT

This chapter illustrates how molecular parameters as size, molar mass, and intermolecular interactions, which are important to identify and characterize intrinsically disordered proteins (IDPs), can be obtained from light scattering measurements. The physical basis of light scattering, experimental techniques, sample treatment, and data evaluation schemes are outlined, with special emphasis on studies on proteins. Static light scattering (SLS) and dynamic light scattering (DLS) yield different physical quantities of macromolecules. SLS is capable of measuring molar masses within the range 10^3–10^8 g/mol and is therefore ideal for determining the state of association of proteins in solution. Since proteins are, in general, too small to obtain the geometric radius of gyration R_G from SLS, it is more useful to determine the hydrodynamic Stokes radius, R_S, which can be obtained easily and quickly from DLS experiments. Accordingly, DLS is an appropriate technique to monitor expansion or compaction of protein molecules. This is especially important for IDPs, which can be recognized and characterized by comparing the measured Stokes radii with those calculated for particular reference states, such as the compactly folded and the fully unfolded states. The combined application of DLS and SLS improves measurements of the molar mass and is essential when

Instrumental Analysis of Intrinsically Disordered Proteins: Assessing Structure and Conformation, Edited by Vladimir Uversky and Sonia Longhi

changes in the molecular dimensions and molecular association/dissociation take place simultaneously.

17.1 INTRODUCTION

The experimental discovery of intrinsically disordered proteins (IDPs) was based mainly on two observations: lack of ordered secondary structure and atypically large molecular dimensions of particular proteins under native conditions, which were found to be fully biologically active. For an exploration of the molecular dimensions of IDPs, coupled with thorough control of their monomeric state, the combination of dynamic light scattering (DLS) and static light scattering (SLS) is one of the methods of choice.

It is worth mentioning that the unexpected results of size exclusion chromatography (SEC) led to much confusion before the protein community became aware of the extraordinary properties of IDPs. Since a compact globular structure was expected, the apparently large molecular masses were interpreted assuming an oligomeric structure for the protein under study. Despite some earlier work pointing to the unfolded nature of those proteins (56, 57, 81, 87, 115, 128), the beginning of extensive investigations leading to the realization of the importance of this unusual behavior can be dated roughly to the middle of the 1990s (42, 106, 130). The number of measurements of the molecular dimensions of IDPs has since increased considerably. Some of the experimental approaches are discussed briefly in relation to DLS in section 17.4.

DLS and SLS measurements yield different macromolecular quantities. From DLS experiments, hydrodynamic parameters of macromolecules in solution, for example, the Stokes radius R_S, can be obtained. An advantage over other hydrodynamic techniques is that DLS is fast and entirely noninvasive, and studies can be done under a wide variety of solvent conditions. The curiosity of being a hydrodynamic rather than an optical method results from the fact that DLS analyzes the temporal fluctuations of the light scattering intensity caused by hydrodynamic motions in solution. The connection to the light scattering process itself, which had already been studied for a long time by measuring time-integrated or static light scattering (SLS), is important for several reasons. First, SLS enables the mass of the scattering particles to be determined, which is a useful complement to measurements of the hydrodynamic dimensions. Second, since the scattering effect is something like a weighting factor in a DLS experiment, its knowledge is essential for proper application of the method. Therefore, we shall discuss the basics of light scattering before turning to the application of DLS to studying the dimensions and conformational transitions of proteins.

SLS, which obtains molecular parameters from the angular and concentration dependence of the time-averaged scattering intensity, was one of the main tools for the determination of molecular masses in the middle of the last century. The advent of DLS in the early 1970s led to great expectations concerning the study of dynamic processes in solution. Up to the 1990s, however, the method appeared to be experimentally more difficult than SLS. Laboratory-built or commercially available DLS photometers consisted of gas lasers and expensive, huge correlation electronics, and subsequent data evaluation was tedious because of severe requirements on computation: powerful online computers were not generally available. The first compact DLS devices (particle sizers) were only applicable to strongly scattering systems. This situation has changed during the last 15 years. A sophisticated single-board correlator, which can be easily plugged into a desktop computer, is essentially the only additional element needed to upgrade an SLS to a DLS instrument. It is noteworthy that the popularity of SLS itself has also increased now due mostly to significant technical improvements. Well-calibrated instruments equipped with lasers and sensitive solid-state detectors allow precise measurements over a wide range of particle masses (10^3–10^8 kDa) using sample volumes of the order of 10 μL. As a consequence of these developments, light scattering can now be considered as a standard laboratory method. Several compact instruments are on the market that are valuable tools for protein analysis. Nevertheless, for special applications including particular studies of protein folding and aggregation, a more flexible laboratory-built setup assembled from commercially available components should be preferred. It is worth mentioning that SLS measurements are performed using conventional fluorescence spectrometers, including stopped-flow fluorescence devices in the case of kinetic experiments. Measuring SLS and DLS in one and the same experiment is particularly useful for studying the molecular dimensions of IDPs or macromolecular association and aggregation. Further improvements have been achieved by combining light scattering and particle separation techniques.

However, some limitations and some drawbacks have to be noted. Both DLS and SLS measurements must be done with solutions that are free of undesired large particles and bubbles. Care has to be taken with those proteins that absorb light in the visible region.

After dealing with the basic principles of SLS and DLS (section 17.2) the general benefits, difficulties, and limits of light scattering investigations on proteins are outlined in section 17.3. Section 17.4 provides the results for two typical IDPs: prothymosin α (ProTα) and α-synuclein (αS). Since a considerable proportion of IDPs is involved in protein aggregation and amyloid formation, section 17.5 concentrates on the light scattering studies of aggregation phenomena.

17.2 LIGHT SCATTERING: BASIC PRINCIPLES AND INSTRUMENTATION

17.2.1 Light Scattering from Macromolecular Solutions

SLS has been used for a long time and much of the experimental methodology, skillful practice, and basic theoretical approaches had been developed in the first half of the last century. The advent of DLS can be dated to the middle of the 1960s, and the next decade yielded much of the fundamental groundwork on the principles and practice of DLS from researchers in the fields of physics, macromolecular biophysics, biochemistry, and biology. The following short introduction to both methods is an attempt to supply readers who are not familiar with these techniques, with an intuitive understanding of the basic principles and the applicability to protein folding and aggregation problems. We will therefore only recall some of the more fundamental equations. A clearer and more detailed understanding can be achieved on the basis of numerous excellent textbooks (8, 11, 16, 59, 69, 102).

The theory of light scattering can be approached from either the so-called single-particle analysis or the density fluctuation viewpoint. The single-particle analysis approach, which will be used here, is simpler to visualize and is adequate for studying the structure and dynamics of macromolecules in dilute solutions. The fluctuation approach is appropriate for studying light scattering in liquids. It is useful to distinguish between light scattering from small particles or molecules ($d < \lambda/20$) and from large particles whose maximum dimensions d are comparable to or larger than the incident wavelength λ (the wavelength of blue-green light is about 500 nm). The former case is easier to handle, but much information about the structure is not accessible. Proteins used for folding studies entirely fulfill the conditions of the first case, with the exception of large protein complexes and aggregates. The concepts of light scattering from small particles in vacuum can be applied to the scattering of macromolecules in solution by considering the excess of various quantities (e.g., light scattering, polarizability, refractive index) of macromolecules over that of the solvent.

The strength of the scattering effect depends primarily on the polarizability α of a small scattering element, which might be the molecule itself if it is sufficiently small. Larger molecules are considered as consisting of several scattering elements. The oscillating electric field vector $\vec{E}_0(t,\vec{r}) = \vec{E}_0 \cdot e^{-i(\omega t - \vec{k}_0 \cdot \vec{r})}$ of the incident light beam induces a small oscillating dipole with the dipole moment $\vec{p}(t) = \alpha \cdot \vec{E}(t)$. The variable ω is the angular frequency, and $\vec{k}_0$ is the wave vector with the magnitude $|\vec{k}_0| = 2\pi \cdot n_0/\lambda$. This oscillating dipole reemits electromagnetic radiation, which has the same wavelength in the case of elastic scattering. The intensity of scattered light at distance r is

$$I_S = \frac{16\pi^4}{\lambda^4} \cdot \frac{\alpha^2}{r^2} \cdot I_0, \qquad (17.1)$$

where λ is the wavelength in vacuum, n_0 is the refractive index of the surrounding medium, and $I_0 = |\vec{E}_0|^2$ is the intensity of the incident vertically polarized laser beam. Equation 17.1 differs only by a constant factor for the case of unpolarized light. In general, the scattered light is detected at an angle θ with respect to the incident beam in the plane perpendicular to the polarization of the beam (Fig. 17.1). The wave vector pointing in this direction is $\vec{k}_S$, where $\vec{k}_0$ and $\vec{k}_S$ have the same magnitude.

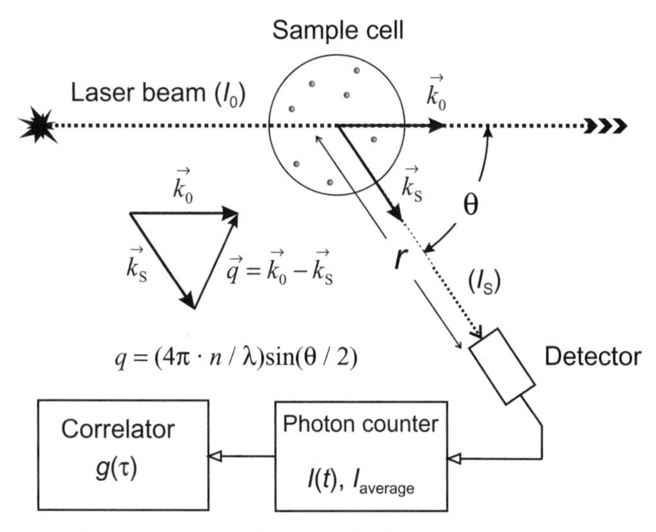

Figure 17.1 Schematic diagram of a light scattering instrument.

An important quantity is the vector difference $\vec{q} = \vec{k}_0 - \vec{k}_S$, the so-called scattering vector, which determines the spatial distribution of the phases $\phi_i = \vec{q} \cdot \vec{r}_i$ of the scattered light wave emitted by individual scattering elements i (Eq. 17.1 and Fig. 17.1). The magnitude of $\vec{q}$ is $q = (4\pi \cdot n/\lambda)\sin(\theta/2)$. The phases play an essential role for the total instantaneous intensity, which results from the superposition of light waves emitted by all scattering elements within the scattering volume v defined by the primary beam and the aperture of the detector. The instantaneous intensity fluctuates in time for nonfixed particles, like macromolecules in solution, due to phase fluctuations $\phi_i(t) = \vec{q} \cdot \vec{r}_i(t)$ caused by changes in their location $\vec{r}_i(t)$.

SLS measures the time average of the intensity; thus, the term "time-averaged light scattering" would be more appropriate, although "static light scattering" appears the accepted convention now. SLS becomes q-dependent for large particles, when light waves emitted from scattering elements within an individual particle have distinct phase differences.

DLS analyzes the above-mentioned temporal fluctuations of the instantaneous intensity of scattered light. Accordingly, DLS can measure several dynamic processes in solution. It is evident that only those changes in the location of scattering elements lead to intensity fluctuations that produce a sufficiently large phase shift. This is always the case for the translational diffusion of macromolecules. The motion of segments of chain molecules and rotational motion can be studied for large structures. Rotational motion of monomeric proteins and chain dynamics of unfolded proteins are, in practice, not accessible by DLS.

17.2.2 SLS

The light scattering intensity from a macromolecular solution can be calculated using Equation 17.1 and summing over the contributions of all macromolecules in the scattering volume v. It is useful to substitute the polarizability α by appropriate physical parameters. This can be done by applying the Clausius–Mosotti equation to macromolecular solutions, which leads to $n^2 - n_0^2 = 4\pi \cdot N' \cdot \alpha$, where n and n_0 are the refractive indices of the solution and the solvent, respectively. N' is the number of particles (molecules) per unit volume and can be expressed by $N' = N_A \cdot c/M$. N_A is Avogadro's number, and c and M are the weight concentration and the molecular mass (molar mass) of the macromolecules, respectively. Using the approximation $n + n_0 \sim 2n_0$, we obtain $n_0^2 - n^2 \sim 2n_0(\partial n/\partial c) \cdot c$. $\partial n/\partial c$ is the specific refractive index increment of the macromolecules in the particular solvent. The excess scattering of the solution over that of the solvent $I_{ex} = I_{solution} - I_{solvent}$ of noninteracting small molecules is then

$$I_{ex} = \frac{4\pi^2 n_0^2 (\partial n/\partial c)^2}{\lambda^4 N_A} \cdot \frac{v}{r^2} \cdot c \cdot M \cdot I_0 = H \cdot \frac{v}{r^2} \cdot c \cdot M \cdot I_0 \tag{17.2}$$

The optical constant $H = 4\pi^2 n_0^2 (\partial n/\partial c)^2 / \lambda^4 N_A$ depends only on experimental parameters and the scattering properties of the molecules in the particular solvent, which is reflected by $\partial n/\partial c$. The exact knowledge of $\partial n/\partial c$, the dependence of n on protein concentration in the present case, is very important for absolute measurements of the molecular mass. For proteins in aqueous solvents of low ionic strength, it is about $0.19\,\mathrm{cm^3/g}$ and, in practice, does not depend significantly on the amino acid sequence. However, it is markedly different in solvents containing high concentrations of denaturants.

One can eliminate the instrument parameters in Equation 17.2 by using the Rayleigh (excess) ratio $R_q = (I_{ex}/I_0) \cdot (r^2/v)$. In practice, R_q of an unknown sample is calculated from the scattering intensities of the sample and that of a reference sample of known Rayleigh ratio R_{ref} by

$$R_q = R_{ref} \cdot f_{corr} \cdot (I_{ex}/I_{ref}). \tag{17.3}$$

The variable f_{corr} is an experimental correction factor accounting for differences in the refractive indices of sample and reference sample (16). In the more general case, including intermolecular interactions and large molecules that have a refractive index n_p and satisfy the condition $4\pi(n_p - n_0)d/\lambda << 1$ of the Rayleigh–Debye approximation, SLS data are conveniently presented by the relation

$$\frac{H \cdot c}{R_q} = \frac{1}{M \cdot P(q)} + 2A_2 \cdot c. \tag{17.4}$$

$P(q)$ is the particle scattering function, which is mainly expressed in terms of the product $q \cdot R_G$, where $q \cdot R_G = \langle R^2 \rangle^{1/2}$ is the root mean square radius of gyration of the particles. Analytic expressions for $P(q)$ are known for different particle shapes, whereas other characteristic size-dependent parameters are used instead of R_G (e.g., length L for rods or cylinders). In the limit $q \cdot \bar{R}_G << 1$, the approximation $P(q)^{-1} = 1 + (q \cdot \bar{R}_G)^2/3$ can be used to estimate $\bar{R}_G$ from the angular dependence of R_q. A perceptible angular dependence of the scattering intensity can only be expected for particles with $R_G > 10\,nm$. Thus, light scattering is not an appropriate method for monitoring changes of R_G during unfolding and refolding of small monomeric proteins. However, a substantial angular dependence is observed for large protein aggregates.

The concentration dependence of the right-hand side of Equation 17.4 yields the second virial coefficient A_2, which reflects the strength and type of intermolecular interactions. The usefulness of measuring A_2 will be discussed below. A_2 is positive for predominantly repulsive (covolume and electrolyte effects) and negative for predominantly attractive intermolecular interactions.

In general, both extrapolation to zero concentration and to zero scattering angle ($q = 0$) are done in a single diagram (Zimm plot) for the calculations of M from Equation 17.4. The primary mass "moment" or "average" obtained is the weight-average molar mass

$$M = \sum_i c_i M_i / \sum_i c_i$$

in the case of polydisperse systems. This has to be taken into consideration for proteins when both monomers and oligomers are present. Molar masses of proteins used in folding studies (i.e., $M < 50{,}000\,g/mol$) can be determined at 90° scattering angle because the angular dependence of R_q is negligible. Measurements at different concentration are mandatory, however, because remarkable electrostatic and hydrophobic interactions may exist under particular environmental conditions. In the following, the values of parameters measured at finite concentration are termed apparent values, for example, M_{app}.

17.2.3 DLS

Information about dynamic processes in solution is contained primarily in the temporal fluctuations of the scattered electric field $\vec{E}_s(t)$. The time characteristics of these fluctuations can be described by the first-order time autocorrelation function

$$g^{(1)}(\tau) = \left\langle \vec{E}_s(t) \cdot \vec{E}_s(t+\tau) \right\rangle.$$

The brackets denote an average over many products of $\vec{E}_s(t)$, with its value after a delay time τ. $g^{(1)}(\tau)$ is only accessible in the heterodyne detection mode, where the scattered light is mixed with a small fraction of the incident beam on the optical detector. This experimentally complicated detection method must be used if particle motions relative to the laboratory frame, for example, the electrophoretic mobility in an external electric field, is to be measured. The less complicated homodyne mode, where only the scattered light intensity is measured, is normally the preferred optical scheme. This has the consequence that only the second-order intensity correlation function

$$g^{(2)}(\tau) = \langle I(t) \cdot I(t+\tau) \rangle$$

is directly available. Under particular conditions, which are met in the case of light scattering from dilute solutions of macromolecules, the Siegert relation

$$g^{(2)}(\tau) = 1 + \left| g^{(1)}(\tau) \right|^2$$

can be used to obtain the normalized first-order correlation function $g^{(1)}(\tau)$ from the measured $g^{(2)}(\tau)$. Analytic forms of $g^{(1)}(\tau)$ have been derived for different dynamic processes in solution. As we have already indicated above, essentially only translational diffusional motion contributes to the fluctuations of the scattered light in the case of monomeric proteins, and we can reasonably neglect rotational effects. $g^{(1)}(\tau)$ for identical particles with a translational diffusion coefficient D has the form of an exponential

$$g^{(1)}(\tau) = e^{-q^2 \cdot D \cdot \tau}. \tag{17.5}$$

D is related to the hydrodynamic Stokes radius R_S by the Stokes–Einstein equation

$$R_S = \frac{k \cdot T}{6\pi \cdot \eta \cdot D}, \tag{17.6}$$

where k is Boltzmann's constant, T is the temperature in K, and η is the solvent viscosity. $g^{(1)}(\tau)$ for a polydisperse solution containing L different macromolecular species (or aggregates) with masses M_i, diffusion coefficients D_i, and weight concentrations c_i is

$$g^{(1)}(\tau) = \frac{1}{S} \sum_{i=1}^{L} a_i \cdot e^{-q^2 \cdot D_i \cdot \tau}, \tag{17.7a}$$

where $S = \sum_{i=1}^{L} a_i$ is the normalization factor. The weights $a_i = c_i \cdot M_i = n_i \cdot M_i^2$ reflect the $c \cdot M$ dependence of the scattered intensity (see Eq. 17.2). The variable n_i is the number concentration (molar concentration) of the macromolecular species. Accordingly, even small amounts of large particles are considerably represented in the measured $g^{(1)}(\tau)$. The general case of an arbitrary size distribution, which results in a distribution of D, can be treated by an integral

$$g(\tau) = \frac{1}{S} \int a(D) \cdot e^{-q^2 \cdot D \cdot \tau} dD. \tag{17.7b}$$

Equation 17.7b has the mathematical form of a Laplace transformation of the distribution function *a*(*D*). Thus, an inverse Laplace transformation is needed to reconstruct *a*(*D*) or the related distribution functions *c*(*D*) and *n*(*D*). This is an ill-conditioned problem from the mathematical point of view because of the experimental noise in the measured correlation function. However, numerical procedures exist, termed "regularization," which allow stabilized, "smoothed" solutions to be obtained. A widely used program package for this purpose is "CONTIN" (96). Nevertheless, the distributions obtained can depend sensitively on the experimental noise and parameters used in the data evaluation procedure in special cases. Thus, it might be more appropriate to use simpler but more stable data evaluation schemes like the method of cumulants (68), which yields the z-averaged diffusion coefficient and higher moments reflecting the width and asymmetry of the distribution. can be obtained simply from the limiting slope of the logarithm of $g^{(1)}(\tau)$, viz

$$\bar{D} = -q^{-2} \cdot \frac{d}{d\tau}\left(\ln\left|g^{(1)}(\tau)\right|\right)_{\tau\to 0}.$$

This approach is very useful for rather narrow distributions.

D, like *M*, is concentration dependent, usually written in the form

$$D(c) = D_0(1 + k_D \cdot c), \tag{17.8}$$

where k_D is the diffusive concentration dependence coefficient, which can be used to characterize intermolecular interactions. The variable k_D can vary considerably in magnitude and sign, and in dependence on solvent, conformational state, and net charge of the protein. Some data are shown in Table 17.1. However, k_D differs from A_2. The concentration dependence of *D* and other macromolecular parameters has been discussed in more detail by Harding and Johnson (54). Extrapolation to zero protein concentration, yielding D_0, is essential in order to calculate the hydrodynamic dimensions in terms of R_S for individual protein molecules.

17.2.4 Running Light Scattering Experiments

17.2.4.1 Experimental Setup and Sample Requirements The experimental setup for light scattering measurements in macromolecular solutions is relatively simple (Fig. 17.1). A beam from a continuous wave laser is focused into a temperature-stabilized cuvette. Temperature stability better than ±0.1°C is required for DLS experiments because of the temperature dependence of the solvent viscosity. The scattered light intensity is detected at one or various scattering angles using either photomultiplier tubes or avalanche photodiode detectors (APDs). Though photomultiplier tubes have still some advantages, an important feature of APDs is the much higher quantum yield (>50%) within the wavelength range of laser light between 400 and 900 nm.

TABLE 17.1 Stokes Radii R_S, Relative Compactness $R_S/R_{S,native}$, and Diffusive Concentration Dependence Coefficients k_D Measured by DLS for Selected Proteins in Differently Folded States

Protein/State	R_S (nm)	$R_S/R_{S,nat}$	k_D (mL/g)	Reference
Barstar				
Native	1.72		5	(44)
Unfolded 6 M urea	2.83	1.65	50	(44)
Cold denatured	2.50	1.45	−3	(44)
RNase T1				
Native	1.74		49	(46)
Unfolded 5.3 M GdmCl	2.40	1.38	–	(46)
Heat denatured, 60°C	2.16	1.24	87	(46)
RNase A				
Native	1.90		≈0	(89)
Unfolded 6 M GdmCl	2.60	1.37	–	(89)
Unfolded 6 M GdmCl, red.	3.14	1.65	–	(89)
TFE-state	2.28	1.20	15	(47)
Apomyoglobin				
Native state	2.09		26	(43)
Acid denatured	4.29	2.05	520	(43)
Molten globule	2.53	1.21	104	(43)
Bovine α-lactalbumin				
Native (holo)	1.88		−7	(48)
Unfolded 5 M GdmCl	2.46	1.31	−9	(48)
A-state (pH 2)	2.08	1.11	15	(48)
Molten globule (pH 7)	2.04	1.09	62	(48)
Molten globule (TFE)	2.11	1.12	−100	(47)
TFE-state (40% v/v)	2.25	1.20	−10	(47)
Kinetic molten globule	1.99	1.06	−60	(48)
PGK				
Native state	2.97		0.8	(21)
Unfolded 2 M GdmCl	5.66	1.91	–	(21)
Cold-denatured state	5.10	1.72	–	(21)
Acid-denatured state	7.42	2.50	552	(22)
TFE-state (50% v/v, pH 2)	7.76	2.61	1030	(22)

Experimental errors omitted for brevity. Errors in R_S are typically less than ±2%. Errors in k_D are of the order of ±10% except for k_D values close to zero.
For the specific solvent conditions, we refer to the original literature.

The expense of a light scattering system depends essentially on the type of laser and the complexity of the detection system. For example, simple systems detect the light only at 90° or in the nearly backward scattering direction, while sophisticated devices are able to monitor the scattered light simultaneously at many selectable scattering angles. An important feature is the construction of the sample cell holder, which should allow the use of different cell types. For folding studies and molecular weight determinations of proteins, detection at one scattering angle is sufficient. Multiangle detection is recommended for studies of protein assembly, when the size of the formed particles exceeds about 50 nm. A criterion for choosing the appropriate cell type is the decision between batch and flow experiments and the available amount of protein. Batch experiments require the smallest amount of substance, since standard microfluorescence cells with volumes of the order of 10 μL can be used. These cells have the further advantage that the protein concentration can be measured by UV absorption in the same cell. Flow-through cells, which are essential for stopped-flow experiments and on-line coupling to particle separation devices, like HPLC, and field-flow fractionation (FFF), are also useful for batch experiments in the following respect. Direct connection of these cells to a filter unit is the best way to obtain perfectly cleaned samples. The cells must be cleaned first by flushing a large amount of water and solvent through the filter and the cell. The protein sample is then applied without removing the filter unit. Several cell types can be used including standard fluorescence flow-through cells, sample cells used in stopped-flow systems, or special flow-through cells provided by manufacturers of light scattering equipment. Most of these flow cells have the disadvantage of a restricted angular range. An exception is the cell used in the Wyatt multiangle laser light scattering device. Cylindrical cells placed in an index matching bath have to be used for precise measurements at different, particularly small scattering angles. Special experimental equipment has been developed for measurements at very low angles.

In the case of SLS experiments, measurements of the photocurrent, with subsequent analog-to-digital conversion as well as photon-counting techniques, are used (Fig. 17.1). The latter technique is preferred in DLS experiments. Appropriate correlation electronics is no longer a problem for modern DLS devices since practically ideal digital correlators can be built up.

The minimum protein concentration needed in a DLS experiment depends on the molecular mass and the optical quality of the instrument, and particularly on the available laser power. But, even at sufficiently high laser power of about 0.5 W, experiments become impractical below a certain concentration. This happens when the excess scattering of the protein solution is weaker than the scattering of pure water, for example, for a protein with $M = 10\,\text{kDa}$ at concentrations below 0.25 mg/mL. Thus, DLS experiments can be done without special efforts at concentrations satisfying the condition $c \cdot M \geq 5\,\text{kDa} \cdot \text{mg/mL}$. For SLS experiments, the value of $c \cdot M$ can be about one order of magnitude lower.

Removal of undesirable contaminants, such as dust, bubbles, or other large particles, is a crucial step in preparing samples for light scattering experiments. Large particles can readily be removed by membrane filters having pore sizes of about 0.1 μm, or by centrifugation at 10,000 g or higher, while SEC or similar separation techniques have to be applied when a protein solution contains oligomers or small aggregates only slightly larger than the monomeric protein. Fluids used to clean the scattering cell must also be free of all particulates.

In an SLS experiment, the time-averaged scattering intensities of the solvent, solution, and reference sample (scattering standard) have to be measured. Data have to be evaluated according to Equations 17.3 and 17.4. Commercial instruments are, in general, equipped with software packages that generate Zimm plots according to Equation 17.4. The software also contains the Rayleigh ratios for scattering standards at the wavelength of the instrument. A widely used scattering standard is toluene. The Rayleigh ratios of toluene and other scattering standards at different wavelengths are given in Reference 16. Instrument calibration can also be done using an appropriate protein solution. However, the protein must be absolutely monomeric under the chosen solvent conditions.

DLS measurements demand slightly more experience concerning experiment duration and data evaluation. Calibration is not needed. The basis for obtaining stable and reliable results is an adequately measured correlation function. Visual inspection of these primary data is recommended before any data evaluation is done. The signal-to-noise ratio (S/N) that is required depends on the complexity of the system under study. For example, short measurement times are acceptable only in the case of unimodal, narrow size distributions, where a noisy correlation function can be well fitted by a single exponential. In practice, two data evaluation schemes are usually used, the method of cumulants (68), which yields an average Stokes radius, and the second and third moments of the size distribution and/or the approximate reconstruction of the size distribution by an inverse Laplace transformation, usually using the program CONTIN (96). The corresponding software packages are available from the author (96) or are provided by the manufacturers of the instruments. CONTIN enables the reconstruction of a distribution function in terms of scattering power, weight concentration, or number (molar) concentration. Though the latter two types are more instructive, they should only be calculated when a reasonable particle shape can be anticipated within the considered size range.

17.2.4.2 Supplementary Measurements and Accessory Devices Additional physical quantities are required for estimating the molecular mass from SLS, and the diffusion coefficient and the Stokes radius from DLS. To calculate the optical constant in Equation 17.2, the refractive index of the solvent and the refractive index increment of the protein in the particular solvent must be known. The former can be measured easily with an Abbe refractometer, whereas a differential refractometer is needed for the precise measurements of $\partial n/\partial c$. Measurements of $\partial n/\partial c$ are often more expensive than the light scattering experiment itself. A comprehensive collection of $\partial n/\partial c$ values can be found in Reference 114. $\partial n/\partial c = 0.19\,\mathrm{mL/g}$ is a good approximation for proteins in aqueous solution at moderate (<200 mM) salt concentrations.

For calculating R_S from DLS data by Equation 17.6, the dynamic viscosity η of the solvent must be known. It can be obtained from the kinematic viscosity ν (e.g., measured by an Ubbelohde-type viscometer) and the density ρ (measured by a digital densitometer) by $\eta = \nu \cdot \rho$.

A very useful option for light scattering instruments is on-line coupling with FPLC, HPLC, or FFF and a concentration detector, which can be a UV absorption monitor or a refractive index detector. This allows direct measurements of M during the flow. In the case of known $\partial n/\partial c$ for the particular solvent conditions, the molecular mass can be obtained from the output signals of the SLS and refractive index detectors by $M = k_e \cdot (\partial n/\partial c)^{-1} \cdot (output)_{SLS}/(output)_{RI}$, where k_e is the instrument calibration constant. A parallel estimation of R_S is possible when the scattering is strong enough, allowing a sufficiently precise DLS experiment within a few minutes.

A further option is the coupling to a mixer allowing kinetic light scattering experiments. Kinetic light scattering is discussed in more detail in section 17.3.10. Two schemes can be used. Either a mixing device is coupled to a light scattering apparatus equipped with a flow cell (9) or the light scattering apparatus is built up around a stopped-flow device originally constructed for fluorescence detection (89).

17.2.4.3 Light Scattering Instruments Several companies produce light scattering instruments and optional units. Many of them have taken into consideration the special demands of studies on protein solutions. Accordingly, most of the systems can handle very small sample volumes, have the sensitivity to study low concentrations of monomeric proteins, and can work in batch and flow mode. For those interested in applying SLS and DLS to folding and aggregation studies, the following companies and the corresponding websites are listed in alphabetic order:

1. ALV-Laservertriebsgesellschaft Germany: http://www.alvgmbh.de,
2. Brookhaven Instruments Corporation, USA: http://www.bic.com,
3. Malvern Instruments Ltd., UK: http://www.malvern.co.uk, and
4. Wyatt Technology Corporation, USA: http://www.wyatt.com or http://www.wyatt.de.

17.2.4.4 Advantages of Simultaneous DLS and SLS Experiments Many of the modern light scattering instruments allow both DLS and SLS to be measured in one and the same experiment. The combination of the two experimental procedures involves some problems, since the optimum optical schemes for DLS and SLS are different. Briefly, DLS needs focused laser beams and only a much smaller detection aperture can be employed because of the required spatial coherence of the scattered light. This reduces the SLS signal and demands higher beam stability.

Combined DLS/SLS offers two main advantages for studies of the conformational states of proteins.

The first concerns the reliability of measurements of the hydrodynamic dimensions and plays an important role during folding or unfolding investigations and, particularly, for studies of IDPs. The observed changes of R_S or unexpected large values of R_S measured by DLS could partly or entirely

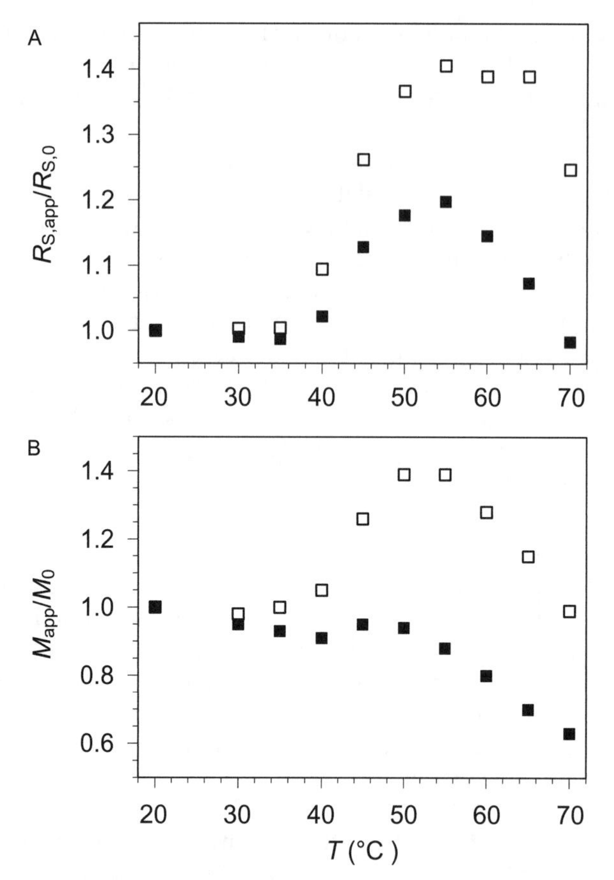

Figure 17.2 Heat-induced unfolding and association/dissociation of Cro repressor wild type at 1.8 mg/mL (■) and 5.8 mg/mL (□) in 10 mM sodium cacodylate, pH 5.5. The reference values for the relative changes in $R_{S,app}$ (A) and M_{app} (B) were taken at 20°C.

result from an accompanying aggregation reaction. Such effects can clearly be recognized when the molecular mass is determined in the same experiment by SLS.

Such a situation will be demonstrated by the complex thermal unfolding behavior of wild type λ Cro repressor (Cro-WT). The native Cro-WT is active in the dimeric form. Spectroscopic and calorimetric studies (29 and references cited therein) revealed that thermal unfolding proceeds via an intermediate state. The resultant changes of the relative apparent Stokes radius, $R_{S,rel}$, and of the relative average mass, M_{rel}, are quite different at low- and high-protein concentrations as shown in Figure 17.2A,B, respectively. At low concentration, Cro-WT first unfolds partially, according to the increase in $R_{S,rel}$, between 40 and 55°C, but remains essentially in the dimeric form as substantiated by the nearly constant M_{rel}. Above 55°C, both $R_{S,rel}$ and M_{rel} decrease due to the dissociation of the dimer. At high concentration, the increase of $R_{S,rel}$ during the first unfolding step is much stronger and is accompanied by an increase in M_{rel}. This is due to the dimerization of partly unfolded dimers. At temperatures above 55°C, dissociation into monomers is indicated by the decrease of both $R_{S,rel}$ and M_{rel}. The observed transient population of a tetramer is a peculiarity of the thermal unfolding of Cro-WT.

The second advantage of combined DLS/SLS is its ability to measure correct molar masses of proteins in imperfectly clarified solutions that contain protein monomers and an unavoidable small amount of aggregates. This situation is often met, for example, during a folding reaction or when the amount of protein is too small for appropriate purification procedures. In such a case, SLS alone would measure a meaningless weight average mass. The Stokes radii of monomers and aggregates are frequently well separated, allowing the weighting, a_i, for monomers and aggregates to be estimated by fitting the measured $g^{(1)}(\tau)$ by Equation 17.7a. Knowledge of the weighting allows the contributions of monomers and aggregates to the total scattering intensity to be distinguished and thereby, the correct molecular mass of the protein to be obtained. This situation is illustrated in Figure 17.3, which shows, as an example, the size distribution of Stokes radii for recombinant Syrian hamster prion protein at pH 4.2. In this case, only 82% of the total scattering intensity arises from monomeric protein, and 18% is caused by a very small amount of aggregate. Subtracting the contribution of aggregates from the scattering intensity gives a molar mass of 18,000 g/mol, which is close to the 16,400 g/mol calculated from the amino acid composition. Without correction, a molar mass of about 22,000 g/mol would be obtained.

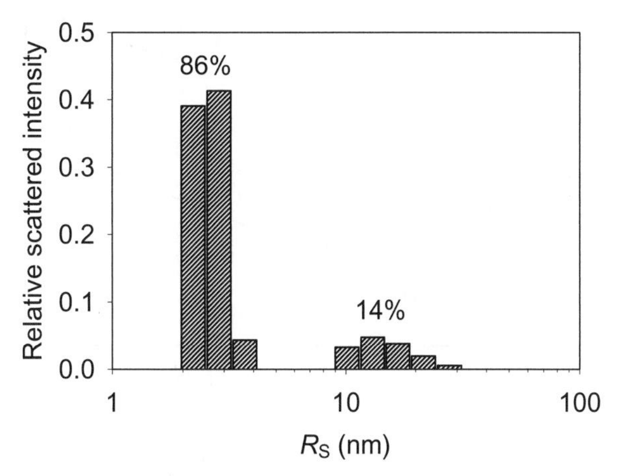

Figure 17.3 Size distribution in terms of relative scattered intensity versus R_S for ShaPrP$^{90-232}$, $c = 1.1$ mg/mL in the native folded state in 20 mM sodium acetate, pH 4.2, 50 mM NaCl.

17.3 ACCESSING CONFORMATION AND CONFORMATIONAL TRANSITIONS OF PROTEINS BY LIGHT SCATTERING

17.3.1 The Compactly Folded State—General Remarks

The molecular parameter of interest in studying conformations of proteins is primarily the hydrodynamic Stokes radius R_S. Additional direct estimations of the molecular mass from SLS data are recommended to check whether the protein is and remains in the monomeric state during conformational transitions. Formation of a dimer of two globular subunits could easily be misinterpreted as a swollen monomer since its Stokes radius is larger by a factor of only 1.39 than that of the monomer (39). Hydrodynamic radii in the native state have been measured for many proteins. For globular proteins, a good correlation with the mass or the number of amino acids was found (17) (see also section 17.3.8). This has encouraged some researchers and manufacturers of DLS devices to estimate M from R_S in general. This procedure has to be applied with great care. A fixed relation between M and R_S is at variance with changes of the compactness during folding/unfolding transitions (see section 17.3.8).

For correct estimation of molecular dimensions in terms of R_S before and after a folding transition, extrapolation of D to zero protein concentration should be done before applying Equation 17.6. The concentration dependence of D indicated by k_D possibly changes during unfolding/refolding transitions as is shown in Figure 17.4 for thermal unfolding of RNase T1. Large changes in k_D are often observed in the case of pH-induced denaturation. Apparent Stokes radii, $R_{S,app}$, are measured if extrapolation to zero protein concentration is not possible due to certain experimental limitations.

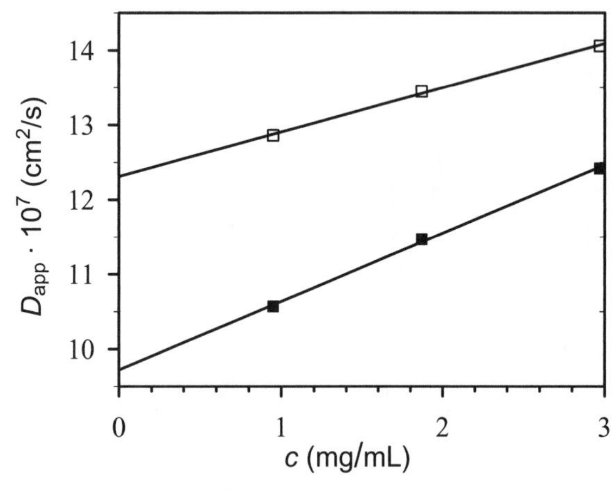

Figure 17.4 Concentration dependence of the diffusion coefficient D for ribonuclease T1 in 10 mM sodium cacodylate, pH 7, 1 mM EDTA at 20°C (□) and 60°C (■). Linear fits to the data yield the diffusion coefficient at zero protein concentration D_0 and the diffusive concentration dependence coefficient k_D.

17.3.2 Heat-Induced Unfolding

The first DLS studies of the thermal denaturation of proteins were reported by Nicoli and Benedek (86). These authors investigated heat-induced unfolding of lysozyme in the pH range between 1.2 and 2.3 and at different ionic strength. The size transition coincided with the unfolding curve recorded by spectroscopic probes. The average radius increased by 18% from 1.85 to 2.18 nm. The same size increase was also found for ribonuclease A (RNase A) at high ionic strength. The proteins had intact disulfide bonds.

A good candidate for thermal unfolding/refolding studies is RNase T1, since the transition is completely reversible and aggregation does not occur (46). The unfolding transition curves measured at three different protein concentrations at pH 7 are shown in Figure 17.5. The corresponding concentration dependences of D in the native and unfolded states are shown in Figure 17.4. The increase of the diffusive concentration dependence coefficient k_D from 49 to 87 mL/g on the transition to the unfolded state indicates a strengthening of the repulsive intermolecular interactions. The Stokes radii obtained after extrapolation to zero protein concentration are 1.74 and 2.16 nm for the native and unfolded states, respectively. The 24% increase in R_S on unfolding is somewhat larger than those for lysozyme and RNase A. This could be due to the fact that RNase T1 differs from lysozyme and RNase A in the number and position of the disulfide bonds.

Unfortunately, many proteins aggregate upon heat denaturation, thus preventing reliable measurements of the dimensions.

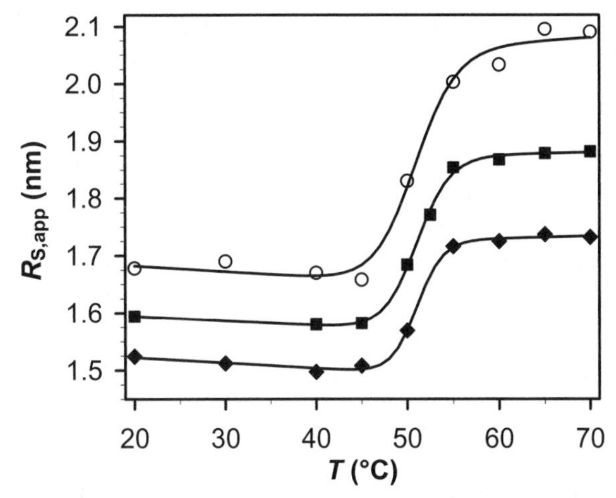

Figure 17.5 Heat-induced unfolding of ribonuclease T1 in 10 mM sodium cacodylate, pH 7, 1 mM EDTA, monitored by the temperature-dependent changes of the apparent Stokes radius at concentrations of 0.8 mg/mL (○), 1.9 mg/mL (■), and 3.9 mg/mL (◆). The continuous lines were obtained by nonlinear least squares fits according to a two-state transition yielding $T_m = 51.0 \pm 0.5$°C and $\Delta H_m = 497 \pm 120$ kJ/mol (average over all three fits).

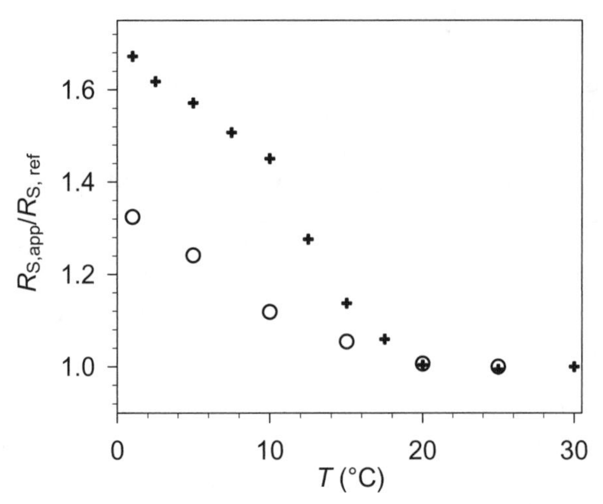

Figure 17.6 Relative expansion, $R_{S,app}/R_{S,ref}$, upon cold denaturation for barstar (pseudowild type), 2.4 mg/mL, in 50 mM Tris/HCl, pH 8, 0.1 M KCl, 2.2 M urea (○); and phosphoglycerate kinase from yeast, 0.97 mg/mL, in 20 mM sodium phosphate, pH 6.5, 10 mM EDTA, 1 mM dithiothreitol (DTT), 0.7 M GdmCl (+). $R_{S,ref}$ is the Stokes radius in the reference state at the temperature of maximum stability under slightly destabilizing conditions.

17.3.3 Cold Denaturation

Since the stability curves for proteins have a maximum at a characteristic temperature, unfolding in the cold is a general phenomenon (95). However, easily attainable unfolding conditions in the cold exist only for a few proteins. Cold denaturation under destabilizing conditions was reported for phosphoglycerate kinase from yeast (19, 51) and barstar (1, 132). The increases in R_S upon cold denaturation for PGK (19) and barstar (unpublished results) are shown in Figure 17.6. The size increase follows a three-state transition in the case of PGK, because of the independent unfolding of the two domains in the cold. Both proteins aggregate on heating, preventing comparison of the cold-denaturation results with those for heat-induced unfolding.

17.3.4 Denaturant-Induced Unfolding

Before the advent of DLS, most data relating to the hydrodynamic dimensions of proteins under strongly denaturing conditions had been obtained by measuring intrinsic viscosities (112). In a pioneering DLS experiment on protein denaturation, Dubin et al. (26) studied the unfolding transition of lysozyme in guanidinium chloride (GdmCl). These authors meticulously analyzed the transition at 31 GdmCl concentrations between 0 and 6.4 M in 100 mM acetate buffer, pH 4.2, at a protein concentration of 10 mg/mL. It was found that $R_{S,app}$ increases during unfolding by 45% and 86% in the case

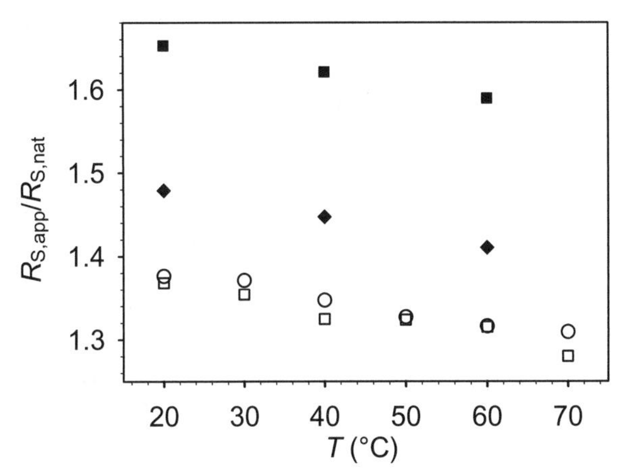

Figure 17.7 Temperature dependence of the relative expansion $R_{S,app}/R_{S,nat}$ of RNase A and RNase T1 in the unfolded state at high concentrations of GdmCl. RNase T1, 2.2 mg/mL, in 10 mM sodium cacodylate, pH 7, 1 mM EDTA, 5.3 M GdmCl (○); RNase A, 2.9 mg/mL, in 50 mM MES, pH 5.7, 1 mM EDTA, 6 M GdmCl (□); and RNase A (broken disulfide bonds), 2.9 mg/mL, in 50 mM MES, pH 6.5, 1 mM EDTA, 6 M GdmCl (■). RNase A, 2.5 mg/mL, with broken disulfide bonds is highly unfolded even in the absence of GdmCl (◆).

of intact and reduced disulfide bonds, respectively. Since then, the hydrodynamic dimensions at high concentrations of GdmCl have been measured for many proteins lacking disulfide bonds by using DLS. The results will be discussed in section 17.3.8 in connection with scaling (i.e., power) laws, which can be derived from these data. Here, we consider briefly the influence of disulfide bonds and temperature on the dimensions of proteins in the highly unfolded state. Such experiments were done with RNase A (88). Figure 17.7 shows the temperature dependence of the relative dimensions, $R_{S,rel} = R_{S,app}/R_{S,native}$, for RNase A with intact and with reduced disulfide bonds. The slight compaction with increasing temperature is typical of proteins in highly unfolded states: for example, similar results are obtained with RNase T1 (Fig. 17.7). Surprisingly, RNase A without disulfide bonds is already unfolded in the absence of GdmCl and has larger dimensions than RNase A with intact disulfide bonds in 6 M GdmCl.

17.3.5 Acid-Induced Unfolding

Many proteins can be denatured by extremes of pH, and most of such studies have focused on using acid as denaturant. Proteins respond very differently to acidic pH. For example, lysozyme remains native-like, some other proteins adopt the molten globule conformation with hydrodynamic dimensions about 10% larger than in the native state, and many proteins unfold into an expanded conformation. Fink et al. (34) introduced a classification scheme for the unfolding behavior under acidic conditions. The molecular mechanisms of acid denaturation have been studied in the case of apomyoglobin by Baldwin and coworkers (6). The dimensions of selected acid-denatured proteins are listed below in Table 17.1. Some proteins are more expanded in the acid-denatured than in the chemical-denatured state, for example, PGK: $R_{S,acid\ denat}/R_{S,native} = 2.50$ and $R_{S,unf}/R_{S,native} = 1.91$. Furthermore, acid-denatured proteins exhibit strong repulsive intermolecular interactions, as reflected by large values of k_D (e.g., 550 mL/g for PGK at pH 2). This means, that D_{app} is twice as large as D_0 at a concentration of about 2 mg/mL. This underlines the importance of extrapolation of the measured quantity to zero protein concentration.

17.3.6 Partially Folded States—Molten Globule and Fluoroalcohol-Induced States

From the wealth of partly folded states existing for many proteins under different destabilizing conditions, DLS data obtained for some particular states will be considered here. Molten globule states (3, 73, 97) have been most extensively studied. DLS data on the hydrodynamic dimensions were published for the molten globule states of two widely studied proteins: α-lactalbumin (48) and apomyoglobin (43). The results including data for the kinetic molten globule of α-lactalbumin are shown in Table 17.1. Data also exist on the geometric dimensions in terms of the radius of gyration R_G for both α-lactalbumin (64) and apomyoglobin (43, 65) in different conformational states.

Only a few DLS studies on the hydrodynamic dimension of proteins in aqueous solvents containing specific structure promoting substances like trifluoroethanol (TFE) and hexafluoroisopropanol have been reported so far. The Stokes radii at high fluoroalcohol content, at which the structure stabilizing effect saturates, are very different for various proteins. An increase in R_S of only about 20% was found for α-lactalbumin and RNase A (47) with intact disulfide bonds. In contrast, PGK at pH 2 and 50% (v/v) TFE has a Stokes radius $R_S = 7.76$ nm, exceeding that of the native state by a factor of 2.6 (22). According to the high helical content estimated from CD data and the high charge according to the large value of k_D, it is conceivable that the entire PGK molecule consists of a long flexible helix. Fluoroalcohols may have a different effect at low volume fractions. In the case of α-lactalbumin, a molten globule-like state was found at 15% (v/v) TFE (47). This state possessed native-like secondary structure and a Stokes radius 10% larger than that of native protein. Furthermore, strong attractive intermolecular interactions were indicated by a large negative diffusive concentration dependence coefficient, $k_D = -100$ mL/g, which result presumably from the exposure of hydrophobic patches in this state. k_D was close to zero in both the absence and high percentage of TFE.

17.3.7 Comparison of the Dimensions of Proteins in Different Conformational States

The hydrodynamic dimensions in different equilibrium states of five proteins frequently used in folding studies are summarized in Table 17.1. Some general rules become evident from these data; for example, heat-denatured proteins are relatively compact as compared with those denatured by high concentration of denaturants or cold-denatured proteins. These considerations are particularly interesting in connection with the dimensions of IDPs.

17.3.8 Scaling Laws for the Native and Highly Unfolded States

A systematic dependence of the hydrodynamic dimensions on the number of amino acids N or the relative molecular mass M has been established for the dimensions of globular proteins in the native state and also for proteins lacking disulfide bonds in the unfolded state induced by high concentrations of GdmCl or urea. The scaling laws can be written in two different forms:

$$R_S = R_{0,N} \cdot N^{\nu} \quad \text{or} \quad R_S = R_{0,M} \cdot M^{\nu}.$$

where ν is the scaling exponent.

Scaling (i.e., "power") laws for native globular proteins have been published by several authors. Damaschun et al. (20) obtained the relation $R_S[\text{Å}] = 3.62 \cdot N^{1/3}$ on the basis of the Stokes radii calculated for more than 50 globular proteins deposited in the protein data bank. Uversky (118) found the dependence $R_S[\text{Å}] = 0.557 \cdot M^{0.369}$ using experimental data from the literature and Stokes radii measured with a carefully calibrated size exclusion column. The relation between R_S and N determined by Wilkins et al. (131) using pulsed field gradient NMR (PFG NMR) is $R_S[\text{Å}] = 4.75 \cdot N^{0.29}$. An advantage of DLS in measuring the hydrodynamic dimensions is that no calibration of the method is needed. The scaling laws for the native state are useful for the classification of proteins for which the high-resolution structure is not known or cannot be obtained. A summary of published data for native folded proteins is given in Table 17.2. Deviation of the experimentally measured Stokes radius from the scaling law indicates that the protein has no globular shape or does not adopt a compactly folded structure as in the case of IDPs.

In contrast to the native state, proteins in unfolded states populate a large ensemble of configurations, which are restricted only by the allowed ϕ-ψ-angles and nonlocal interactions of the polypeptide chains. The conformational properties of the polypeptide chains can be described by the rotational isomeric state theory (38). The first systematic analysis of the chain length dependence of hydrodynamic dimensions was done by Tanford et al. (113) by measuring the intrinsic viscosities of unfolded proteins. The Stokes radii measured by DLS (21) for 12 proteins denatured by high concentrations of GdmCl are shown in Figure 17.8. From the data, the scaling law $R_S[\text{Å}] = 0.28 \cdot N^{0.5}$

TABLE 17.2 Scaling Laws for Proteins in Native Folded and Chemically Unfolded States

State, Conditions	Exponent ν	$R_{0,N}$ (Å)	$R_{0,M}$ (Å)	Reference
Native folded				
	0.29	4.75		(131)
	0.369		0.557	(118)
	0.357		0.625	(120)
	0.33	3.62		(20)
Unfolded				
GdmCl, urea	0.57	2.21		(131)
GdmCl	0.502		0.286	(118)
Urea	0.524		0.22	(118)
GdmCl	0.543		0.189	(120)
Urea	0.521		0.224	(120)
GdmCl (Fig. 17.7)	0.50	2.80		(21)
	0.522	2.518		(135)

The scaling laws are given by either $R_S = R_{0,N} \cdot N^{\nu}$ or $R_S = R_{0,M} \cdot M^{\nu}$, where N is the number of amino acid residues, M is the relative molecular mass in dalton, and R_S is obtained in nanometer. Experimental errors are omitted.

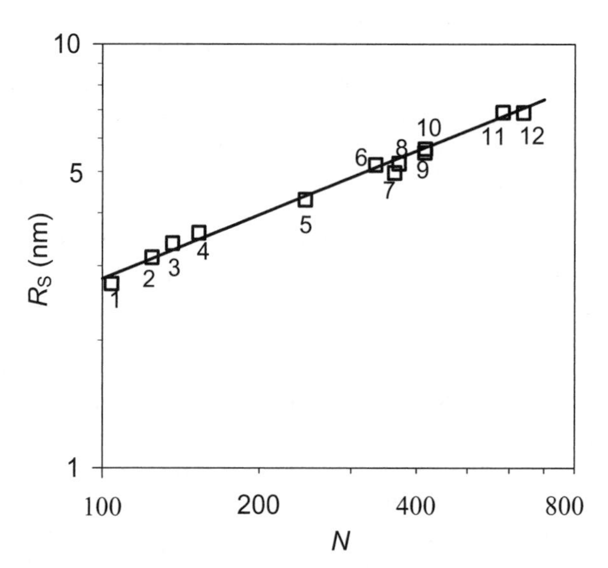

Figure 17.8 Stokes radii R_S of 12 proteins unfolded by 6 M GdmCl and their dependence on the number of amino acid residues N. The linear fit yields the scaling law $R_S[\text{Å}] = 2.8 \cdot N^{0.5}$. The unit Å is used here for an easier comparison with data from the literature. 1, apocytchrome *c*; 2, staphylokinase; 3, RNase A; 4, apomyoglobin; 5, chymotrypsinogen A; 6, glyceraldehyde-3-phosphate dehydrogenase; 7, aldolase; 8, pepsinogen; 9, streptokinase; 10, phosphoglycerate kinase; 11, bovine serum albumin; 12, DnaK. Disulfide bonds, where they exist in the native proteins, were reduced.

was obtained. Our data and the scaling laws for R_S in highly unfolded states published by different authors using SEC (118) or PFG NMR (131) are summarized in Table 17.2. Power laws can also be established by measuring radii of gyration by small-angle X-ray scattering (SAXS) (67).

Whether or not unfolded proteins are indeed true random chains is a matter of debate (5). Though all experimental results concerning the dimensions are consistent with random chain behavior, this is not a sufficient criterion to exclude the existence of remaining ordered structure within the polypeptide chain. Fitzkee and Rose (36) have introduced the "rigid segment model" in which rigid segments of native protein structure are interconnected by flexible hinge residues. They demonstrated by Monte Carlo simulations that this model can reproduce random chain behavior even in the extreme case of a high content of ordered structure. Nevertheless, the chain-length dependence of the hydrodynamic dimension in the form of scaling laws is an important experimental basis for the characterization of unfolded polypeptide chains. The differences in the experimentally determined exponents in Table 17.2 may reflect different protein solvent interactions existing in the protein systems under study.

Empirical scaling laws for intermediate conformations between those of the native and the fully unfolded states were introduced by Uversky (118, 119) (see section 17.4.1).

17.3.9 Hydrodynamic Modeling and Hydration

Two aspects, which are closely related to measurements of the hydrodynamic dimensions of proteins are hydrodynamic modeling (14, 41) and the hydration problem (52, 53). For globular proteins in particular, both are important for the link between high-resolution data from X-ray crystallography or NMR and hydrodynamic data. Protein hydration appears as an adjustable parameter in recent hydrodynamic modeling procedures (40). Values averaged over different proteins are 0.3 g water/g protein or, in terms of the hydration shell, 0.12 nm. Model calculations are also in progress for unfolded proteins (135).

17.3.10 Protein Folding Kinetics

17.3.10.1 General Considerations and Attainable Time Range An important characteristic of the folding process is the relation between compaction and structure formation. Therefore, it is intriguing to monitor compactness during folding by measuring the Stokes radius using DLS, which is the fastest of the hydrodynamic methods. Furthermore, R_S and the radius of gyration R_G are direct measures of the overall molecular compactness. The first studies of this kind were measurements of the compaction during lysozyme refolding by continuous-flow DLS (31). In general, the basic experimental techniques of continuous- and stopped-flow experiments can also be applied for laser light scattering.

For SLS, there are no restrictions from the general physical principles concerning the accessible time range. The data acquisition time T_A that is needed to obtain a sufficiently high (S/N), depends solely on the photon flux and can be minimized by increasing the light level.

In the case of DLS, however, there are two fundamental processes that limit the accessible timescales. These problems were discussed in detail (45) and will be outlined only briefly in this section.

First, unlike other methods including SLS, enhancement of the signal is not sufficient to increase the (S/N), since the measured signal is itself a statistical quantity. At light levels above the shot-noise limit, the (S/N) of the measured time-correlation function depends on the ratio of T_A to the correlation time τ_C of the diffusion process by $(S/N) = (T_A/\tau_C)^{1/2}$. The variable τ_C is of the order of tens of microseconds for proteins leading to a total acquisition time $T_A \sim 5$–10s in order to achieve an acceptable (S/N). This time can be shortened by splitting it into N_S "shots" in a stopped-flow experiment. $N_S = 100$ is an acceptable value leading to acquisition times of 50–100ms during one "shot." This corresponds to the time resolution in the case of the stopped-flow technique.

The second limiting factor becomes evident when the continuous-flow method is used. Excellent time resolution has been achieved with this method in the case of fluorescence measurements (108) and SAXS (93, 107). Continuous-flow experiments are especially useful for slow methods, since the time resolution depends only on the aging time between mixer and detector and not on the speed of data acquisition. Increasing the flow speed in order to get short aging times also reduces the resting time of the protein molecules within the scattering volume. However, to avoid distortion of the correlation functions, this time must be longer than the correlation time corresponding to the diffusional motion of the molecules. This limits the flow speed to about 15cm/s under typical experimental conditions (45) resulting in aging times of about 100ms. The benefit of continuous-flow measurements over stopped-flow measurements does not exist for DLS. Thus, DLS is only suitable for studying rather slow processes, for example, late stages of protein folding in the time range >100ms. However, this time range is important for folding and oligomerization or misfolding and aggregation (see section 17.5). Two examples for the compaction of monomeric protein molecules during folding will be considered now.

17.3.10.2 Hydrodynamic Dimensions of the Kinetic Molten Globule of Bovine α-Lactalbumin The hydrodynamic dimensions of equilibrium molten globule states are well characterized (Table 17.1). It is interesting to compare the dimensions in these states with that of the kinetic counterpart. Stopped-flow DLS investigations (48) were done with the Ca^{++}-free apoprotein, which folds more slowly than the holoprotein. The time course of the changes of the apparent Stokes radius at a protein concentration of 1.35mg/mL is shown in Figure 17.9. Kinetic measurements at different protein concentrations revealed the "sticky" nature of the kinetic molten globule as

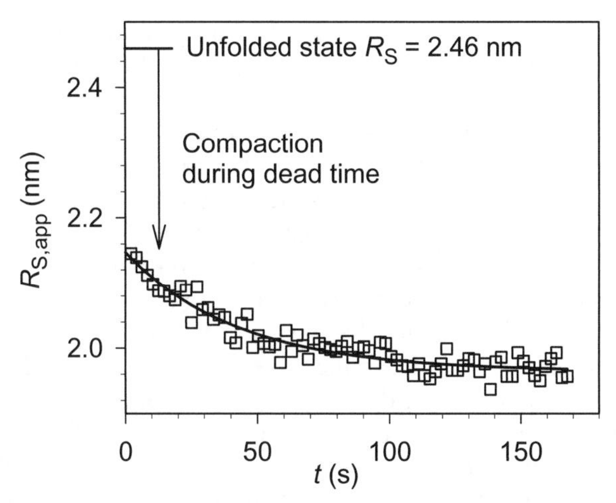

Figure 17.9 Kinetics of compaction of bovine α-lactalbumin after a GdmCl concentration jump from 5 to 0.5 M, final protein concentration 1.35 mg/mL, 50 mM sodium cacodylate, pH 7, 50 mM NaCl, 2 mM ethylene glycol tetraacetic acid (EGTA), $T = 20°C$. The data can be fitted by a single exponential with a decay time $\tau = 43 \pm 6$ s.

compared with the completely unfolded and refolded states. This becomes evident from the extremely negative value of k_D in the kinetic molten globule state (see Table 17.1). The kinetic molten globule appears to be slightly more compact than the equilibrium molten globules. The dimensions of the natively folded apo- and the holoprotein are practically the same. The changes of the dimensions of holo α-lactalbumin were measured later by time-resolved X-ray scattering (2).

17.3.10.3 RNase A Is Only Weakly Collapsed during the Burst Phase of Folding In the previous section, it was demonstrated that α-lactalbumin collapses very rapidly into the molten globule with dimensions close to that of the native state. A different folding behavior was observed for RNase A (89). The Stokes radius of RNase is 2.56 nm in the unfolded state at 6 M GdmCl and decreases only to 2.34 nm during the burst phase. Most of the compaction toward the native state occurs at later stages in parallel with the final arrangement of secondary structure and the formation of tertiary structure.

17.4 LIGHT SCATTERING STUDIES OF IDPs

17.4.1 Experimental Identification and Characterization of IDPs by Simultaneous DLS and SLS Experiments

Based on the demonstration of the capabilities of light scattering in studying protein conformations in general, we now exemplify the application to IDPs.

The achievable quantities of interest are the molecular dimensions or the (equivalent) overall compactness in terms of the hydrodynamic Stokes radius R_S obtained from DLS (the notation compactness is particularly useful for comparing the dimensions of different conformational states). The simultaneous SLS experiment is essential to ensure that the results indeed refer to the expected state of association of the protein molecules (see section 17.2.4.4). Estimations of the compactness by DLS have the following advantages: the experimental procedure does not put any stress on the sample, and the environmental conditions and protein concentrations can easily be adapted to special requirements. This could be particularly useful for IDPs in order to check which conditions lead to an increase in compaction of the polypeptide chain. The concentration dependence is important for estimating the influence of thermodynamic nonideality or crowding effects at high solute concentrations. Two kinds of experiment are conceivable in general: (1) to decide whether a protein under study can be considered to be intrinsically disordered or (2) to study the molecular compactness of a known IDP in more detail:

1. The first step in analyzing the conformational type is to compare the measured Stokes radius with the hydrodynamic dimensions of the compact globular state on the one hand and with the random coil dimensions on the other, calculated for a polypeptide chain of the same amino acid composition or equivalent mass. This can be done on the basis of empirical relationships (scaling laws) introduced in section 17.3.8, yielding an idea of how the dimensions are related to these limiting cases of the conformational state. In addition to these extremes, Uversky (118) introduced scaling laws for intermediate compactness, for example, the molten globule state or the premolten globule state (118, 119). Furthermore, it is instructive to estimate the content of ordered secondary structure by CD (see chapter 8) or other spectroscopic methods in order to distinguish disordered structures from highly elongated structures with a large amount of ordered secondary structure (21) (see section 17.3.6).
2. More detailed DLS studies of IDPs are essentially based on the experience of and data evaluation procedures for unfolded proteins. To obtain deeper insights, a combined approach with other biophysical techniques is extremely helpful. Some general aspects of DLS studies on IDPs are considered in the following.

A very useful approach is to compare the hydrodynamic Stokes radius R_S with the geometric radius of gyration R_G. R_G can be obtained from the angular dependence of the light scattering intensity if the particle size exceeds about 10 nm. Since this is not fulfilled for monomeric proteins with relative molecular masses $M < 100$ kDa, R_G has to be measured by SAXS. SAXS experiments are more expensive than DLS measurements, but small-angle scattering curves yield far more information on the protein conformation in different states than R_G alone (see chapter 16). A useful parameter is the ratio between R_G and R_S, the ρ-factor, where $\rho = R_G/R_S$ (13). This factor is sensitive to conformational changes of proteins (21).

The Stokes radius can also be determined by other experimental methods including gel filtration (chapter 18), ultracentrifugation (chapter 15), PFG NMR (131), and fluorescence correlation spectroscopy (100). All these techniques have particular advantages and limitations that should be taken into consideration. In order to evaluate the reliability in estimating R_S of IDPs, it is useful to compare the results of some of these methods in particular cases. This will be done for ProTα in section 17.4.2. Wilkins et al. (131) introduced a compaction factor C based on the experimental R_S and the calculated reference values of R_S in the compactly folded state, $R_{S,C}$, and the fully unfolded state, $R_{S,U}$,

$$C = \frac{R_{S,U} - R_S}{R_{S,U} - R_{S,C}}$$

Measurements of the hydrodynamic dimensions are implicitly connected with the solvation problem (section 17.3). This is particularly important for IDPs because their polypeptide chains are highly exposed to solvent. Therefore, the results of direct estimations of hydration by other methods such as NMR (chapter 10) are important for improving the interpretation of hydrodynamic data.

Since hydrodynamic parameters reflect only global compactness, it is valuable to relate the experimental values of R_S to results of methods that measure residual structure and local compactness, for example, NMR and approaches for determining intramolecular distances, distance distributions, and chain dynamics (33, 103, 104) (see chapter 12).

The variations of compactness with solvent conditions and temperature are important properties of IDPs. IDPs are often highly charged at neutral pH. Accordingly, changes in compactness can be expected and have been measured in special cases upon variation of pH. This can be simply monitored by DLS (section 17.4.2).

The extent of intermolecular interactions can be explored by measuring the concentration dependences of both $1/M_{app}$ (SLS) and D_{app} (DLS).

In the following section, some of the above-mentioned aspects are exemplified in the case of two representative members of the IDP family, namely ProTα and αS.

17.4.2 ProTα Reveals Typical Properties of IDPs

ProTα is one of the first polypeptides to be shown to be intrinsically disordered. Human ProTα is a small acidic protein (calculated $pI = 3.5$) of 110 amino acid residues that shows high evolutionary conservation and a wide tissue distribution in mammals. In early attempts to measure the molecular mass in solution by SEC (18, 55), the results were interpreted as indicating a

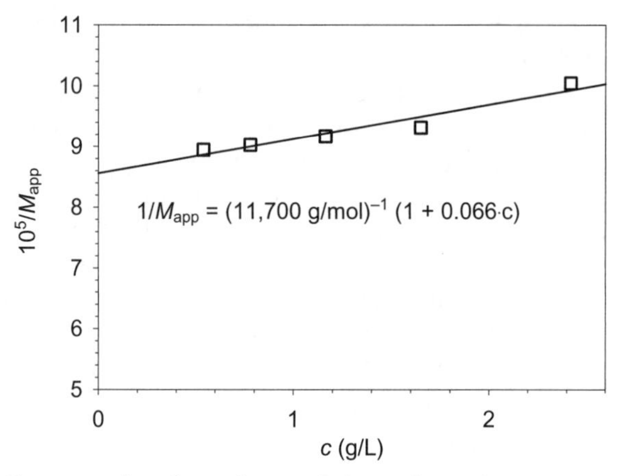

Figure 17.10 Concentration dependence of the reciprocal apparent molecular mass of prothymosin α in PBS, pH 7.4 at 20°C.

molecular mass fivefold larger than that calculated from the amino acid sequence, $M = 11{,}983$ g/mol. Equilibrium sedimentation measurements yielded $M = 12{,}800$ g/mol (55). The first indications that ProTα and related polypeptides do not adopt folded conformations under physiological conditions came from spectroscopic investigations by Watts et al. (128). Later, it was shown (42) that ProTα satisfies all criteria of being an IDP, albeit that term was not coined at this time. These observations were substantiated by additional biophysical investigations (123). It was shown that ProTα undergoes partial compaction at low pH (42, 123). Interestingly, ProTα adopts a more compact state with some ordered secondary structure based on CD and NMR measurements in the presence of Zn^{2+} (122, 133), while titration of Ca^{2+} or Mg^{2+} hardly affect the disordered conformation and the molecular dimensions.

The results of simultaneous SLS and DLS experiments on ProTα are shown in Figures 17.10 and 17.11, respectively. The molecular mass of 12,700 g/mol obtained by extrapolation of $1/M_{app}$ to zero protein concentration reveals that the protein is a monomer under physiological conditions (PBS, pH 7.4). Though ProTα is highly charged at pH 7.4, the concentration dependences of $1/M_{app}$ (Fig. 17.10) and D_{app} (Fig. 17.11) are only moderate due to the presence of salt. The large Stokes radius $R_S = 3.07$ nm calculated from $D(c = 0) = 8.15 \cdot 10^{-7}\,\text{cm}^2/\text{s}$ using the Stokes–Einstein equation (Eq. 17.6) is a strong indication that ProTα is disordered. This becomes evident by comparing the Stokes radius with those of compactly folded and highly unfolded polypeptide chains of the same mass as shown in Table 17.3. The hydrodynamic radii obtained under comparable conditions by SEC (123) and PFG NMR (133) are also included in this table. The disordered conformation of ProTα at pH 7.4 is confirmed by the corresponding CD spectrum (42). Further support comes from the ρ-factor 1.55, which is close to the ρ-factor 1.51 for a

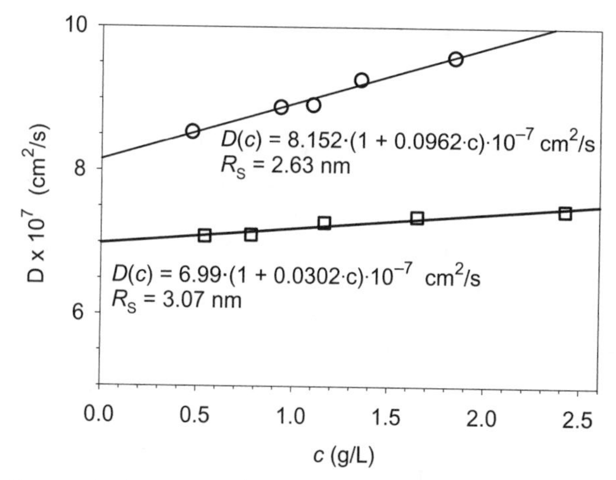

Figure 17.11 Concentration dependence of the translational diffusion coefficient D of prothymosin α in PBS, pH 7.4 (□) and in 10 mM glycine-HCl, pH 2.45 (○) at 20°C.

TABLE 17.3 Hydrodynamic Dimensions of Prothymosin α under Different Environmental Conditions Compared with Those Calculated for the Compactly Folded and Highly Unfolded States

State, Condition	R_S (nm)	$R_S/R_{S,C}$	Reference
PBS, pH 7.4	3.07	1.72	
10 mM glycine/HCl, pH 2.45	2.63	1.48	
Calculated data			
Compactly folded	1.87	1	
Highly unfolded (GdmCl)	3.19	1.79	
Obtained by other methods			
SEC (pH 7.5)	3.14	1.76	(123)
SEC (pH 2.2)	2.49	1.40	(123)
PFG NMR pH 7	3.37	1.80	(133)
PFG NMR pH 7, Zn^{2+}	2.41	1.35	(133)

$R_S = R_{0,M} \cdot M^{\nu}$ (compactly folded: $R_{0,M} = 0.0557$, $\nu = 0.369$, unfolded in GdmCl: $R_{0,M} = 0.0286$, $\nu = 0.502$; see Table 17.2). Experimental errors are omitted.

random coil under theta-conditions (13). The ρ-factor was calculated from $R_S = 3.07$ nm and $R_G = 4.76$ nm obtained from SAXS experiments (42).

The effect on D_{app} of lowering the pH from 7.4 to 2.45 is shown in Figure 17.11. Extrapolation of the data to zero protein concentration showed a compaction from $R_S = 3.07$ nm to $R_S = 2.63$ nm. The value at pH 2.45 is in fair agreement with $R_S = 2.49$ nm measured by SEC (123) at pH 2.2.

Taken together, these results demonstrate that ProTα is a prototype of an IDP. It is worth mentioning that ProTα forms fibrillar structures by agitation at pH 2.5, at high concentrations and $T = 37°C$ (92).

17.4.3 SLS and DLS Studies of Human αS

αS, a protein consisting of 140 amino acid residues that is linked to a group of neurodegenerative disorders termed synucleinopathies (127) is a further prototype of an IDP. Weinreb et al. (130) were the first to discover its unusual properties, for example, the large hydrodynamic radius of 3.4 nm. They created the term "natively unfolded protein" for an emerging but at that time still small family of proteins with similar properties. The involvement of αS aggregates in neurodegenerative disorders has tremendously stimulated investigations of the initial monomeric state and transiently formed oligomeric states (124, 125) as well as the conditions that favor the formation of oligomers, fibrils, and intracellular inclusions (25, 121). Additionally, it was shown that specific mutations in the αS gene corresponding to A53T and A30P substitutions accelerate the formation of oligomers and fibrils (10, 74, 76). Interestingly, the conformation of the initial monomeric species has recently been studied at the single-molecule level (101) (see also chapter 13).

The following light scattering results were obtained under conditions where αS is essentially in a low molecular weight, nonfibrillar state. We have investigated wild type (wt) and A53T αS by SLS and DLS at neutral pH, and temperatures of 20 and 37°C (unpublished results).

Wt αS was found to be a monomer according to the measured molecular mass of 12,200 g/mol, which is close to the calculated M = 14,460 g/mol. The Stokes radius $R_S = 2.91$ nm derived from D extrapolated to zero protein concentration (Figure 17.12) is much larger than the calculated $R_{S,C} = 1.91$ nm but somewhat smaller than $R_{S,U} = 3.43$ nm for the highly unfolded state (126). This is in accordance with the findings (126) that αS is intrinsically disordered but slightly more compact than a fully unfolded chain.

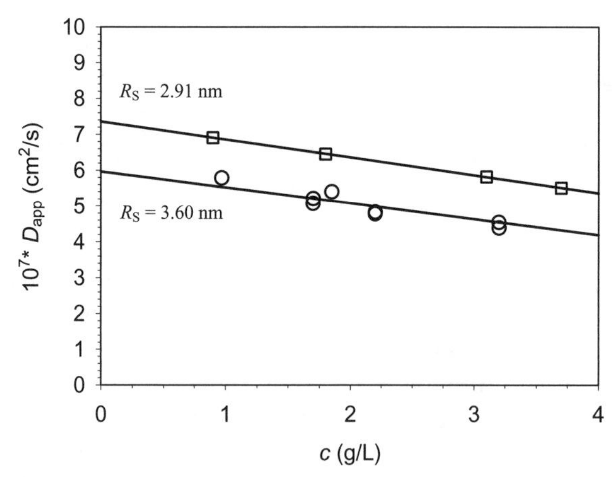

Figure 17.12 Concentration dependence of the translational diffusion coefficient D of α-synuclein wt (□) and A53T (○) in 9.5 mM Tris buffer, pH 7.4, 47.5 mM NaCl, 0.1 mM EDTA, 5% dimethyl sulfoxide (DMSO) at 20°C.

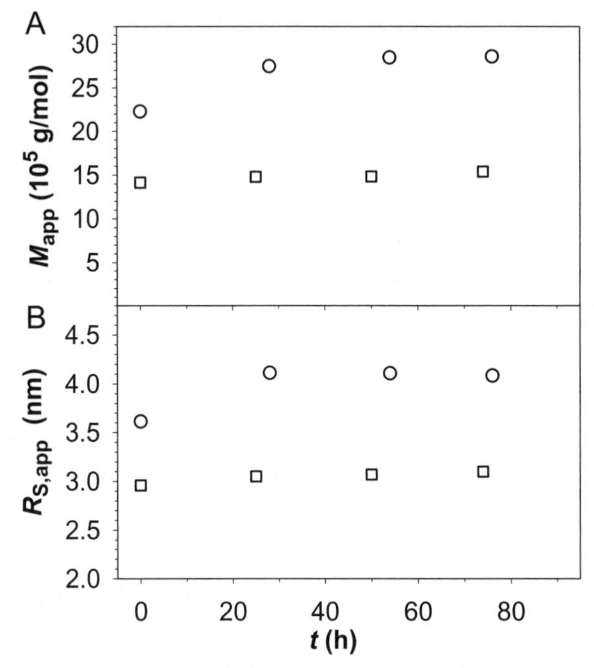

Figure 17.13 Time dependence of the apparent molecular mass M_{app} (A) and the apparent Stokes radius $R_{S,app}$ (B) of α-synuclein wt (□) and A53T (○) in 9.5 mM Tris buffer, pH 7.4, 47.5 mM NaCl, 0.1 mM EDTA, 5% DMSO at 37°C: protein concentration is 0.97 mg/mL.

The situation is clearly different for A53T. For a freshly prepared sample, a molecular mass of 21,300 g/mol and a Stokes radius R_S = 3.6 nm were found at 20°C. This means that the majority of molecules must be in a dimeric state. The effect of subsequent incubation of both wt and A53T αS at 37°C without agitation on the apparent molecular mass and R_S is shown in Figure 17.13. While the wt remained in the monomeric state on incubation for about 80 h and showed a practically constant $R_{S,app}$ of about 3 nm, the transition of A53T to dimers was completed by approaching M_{app} = 28,000 g/mol. The measured Stokes radius R_S = 4.13 nm is somewhat larger than that reported for the dimer by Uversky et al. (124).

17.4.4 Remarks on Further Applications

IDPs like ProTα and αS (sections 17.4.2. and 17.4.3.) have large expansion factors under physiological conditions and may show a considerable compaction under specific conditions, for example at low pH or in the presence of Zn^{2+} in the case of ProTα.

However, DLS measurements are also useful in more complicated situations. First, many protein sequences have been found that contain both intrinsically disordered and folded regions. Second, induced folding of IDPs may be restricted to particular regions of the polypeptide chain. In both cases, changes in compactness, which have to be detected during conformational transitions, may be less pronounced. However, DLS experiments are still advantageous if they are done in combination with methods that probe local structural changes. By this approach, the effect of local or domain folding on the global dimensions can be estimated. Such studies are particularly interesting since a correlation between the degree of compaction and the amount of ordered secondary structure has been observed (see Reference 23 and references cited herein).

SLS is also useful in the case of proteins containing regions of intrinsic disorder for monitoring the state of association during induced folding events. Furthermore, complex formation during interaction with partner molecules (peptides, nucleic acids) can be followed.

The above-mentioned studies underline the importance of the multiple technique approach for studying IDPs (98). References concerning light scattering and IDPs can be found by using the database DisProt on the website http://www.disprot.org/index.php.

17.5 ASSOCIATION, AGGREGATION, AND AMYLOID FORMATION STUDIED BY SLS AND DLS

Many of the IDPs including both ProTα and αS considered in section 17.4 are prone to oligomerization, aggregation, and fibril formation. Therefore, the potential for and typical examples of studying protein association/aggregation by SLS and DLS will be discussed.

17.5.1 Association and Aggregation of Proteins

Association and aggregation phenomena comprise inevitably elementary bimolecular reaction steps. The rate of diffusion-limited reactions, the fastest bimolecular reactions in solution, is given by the Smoluchowski rate (109)

$$k_s = \frac{8k_B T}{3\eta}, \tag{17.9}$$

which is equal to $k_s = 6.6 \times 10^9\,M/s$ using the viscosity of water at $T = 20°C$. The time resolution of SLS devices coupled with a stopped-flow apparatus is about one-tenth of a second (9) and thus within the range of DLS as discussed in the previous section. This resolution lies far above the one required to study diffusion-limited reactions for proteins at concentrations needed to get a reasonable SLS signal ($c \times M \geq 0.5\,kDa/mg/mL$). Protein association phenomena,

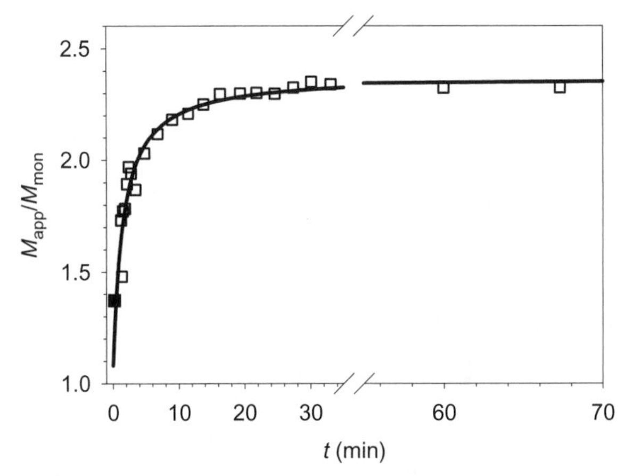

Figure 17.14 Association kinetics of pyruvate decarboxylase (PDC). The protein concentration was 0.033 mg/mL in 50 mM MES, pH 6.4, 5 mM $MgSO_4$ at 20°C. The filled square marks the value of the apparent mass related to the mass of monomeric PDC before the reaction was initiated by addition of ThDP to a final concentration of 1 mM. The continuous line indicates the fit (correlation coefficient $R = 0.94$) to a bimolecular reaction according to Equation 17.10, with $\tau = (1.6 \pm 0.7)$ min translating to a rate constant of 2×10^4 M/s. The other parameters are $a = 1.3 \pm 0.3$ and $b = 1.1 \pm 0.3$.

such as antibody–antigen complexation, occur with a bimolecular rate constant of the order of 10^6 M/s, which is three orders of magnitude slower than the Smoluchowski rate. These rates are at the limit of the time resolution (millisecond time range) achieved by using a stopped-flow fluorescence device for measuring the SLS signal (35). With this time resolution, special cases, like the association of ribosomal subunits, can be studied (49). But apart from these exceptional studies, light scattering investigations are restricted to slower processes in the minutes to hours range.

As a typical example, the association kinetics of pyruvate decarboxylase (PDC) is shown in Figure 17.14. PDC is an enzyme on the glycolytic pathway in fermenting cells. The native enzyme is a dimer of dimers in a hetereotetrameric state (66). The monomer has a molecular mass of 60 kDa. The initial state shown in Figure 17.9 is a mixture of predominantly monomers and dimers, whereas the final state is a mixture of predominantly dimers and tetramers. The kinetics after the addition of the cofactor thiamine diphosphate (ThDP) could be fitted according to

$$\frac{M_{\text{app}}(t)}{M_{\text{mon}}} = a \cdot \left(1 - \frac{1}{1 + \frac{t}{\tau}} \right) + b, \tag{17.10}$$

which represents a simple two-state model for a bimolecular reaction (84). The true mechanism is more complex in detail. But nevertheless, this example demonstrates that such changes in the average molecular mass can be monitored well by SLS. SLS is more appropriate for studying molecular association than DLS, since dimerization leads to a doubling in the average mass, whereas the diffusion coefficient only increases by a factor of 1.39 in the case of globular proteins (39).

17.5.2 Protein Aggregation Studies by Light Scattering

Investigations of aggregation phenomena in solution by SLS have their historic roots in the fields of polymer (37) and colloid chemistry (70) in the first half of the twentieth century. The development of the DLS technique, the discovery of the fractal geometry of colloidal aggregates, and the advance of computer simulations of such systems led to intensive studies in this field (77).

Some proteins show aggregation behavior analogous to colloidal systems, as summarized in Table 17.4. A typical example is barstar, the bacterial inhibitor of barnase (44). Its aggregation can be induced through raised levels of salt at low pH, as shown in Figure 17.15. The initial state of barstar is a symmetrical 16-mer with a molten globule-like conformation (63). The aggregates are amorphous or granular in their appearance (44). The time course of the average mass M and hydrodynamic radius R follow the dependence expected for colloidal aggregation (90) given by

$$M(t) = M(0)(1 + t/\tau)^z \qquad (17.11a)$$

TABLE 17.4 Proteins Showing Colloidal-Like Aggregation Behavior as Reported in the Literature by Light Scattering Methods

Protein	Aggregation Conditions	d	Reference
IgG	Heat induced	2.56 ± 0.3	(30)
Casein	Ca^{2+} induced	2.2–2.3	(58)
Lysozyme	Salt induced	1.8 ± 0.1 2.0 ± 0.1 2.24 ± 0.15*	(117)
Transferrin receptor	Heat induced/low pH and salt	2.18 ± 0.03	(105)
β-Lactoglobulin	Heat induced	1 (pH 2)/2 (pH 7)	(27)
BSA	Heat induced	$d > 2.1$	(71)
Ovalbumin	Heat induced	1.7 no salt 2 with salt	(129)
Patatin	Heat induced	n.d.	(94)
Barstar	Low pH, salt induced	2.1 ± 0.1	(44)

*Depending on the exact solution conditions.
n.d., not determined.

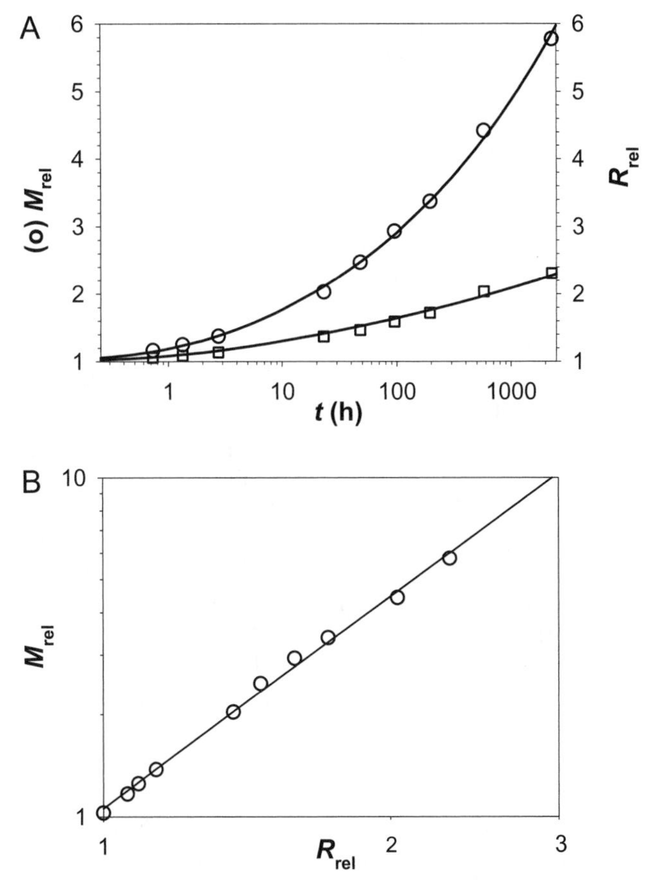

Figure 17.15 Amorphous growth of the A-state of barstar. Protein concentration was $c = 1.0$ mg/mL in 10 mM HCl, pH 2 at 20°C. Aggregation was induced by raising the salt concentration above 75 mM NaCl, where the solution is stable, to a final NaCl concentration of 200 mM. (A) Relative increases in mass $M_{rel} = M(t)/M(0)$ (○) and Stokes radius $R_{rel} = R_S(t)/R_S(0)$ (□). The continuous lines are fits according to Equations 17.11a and Equations 17.11b (correlation coefficients of $R = 0.999$ and 0.997). The derived parameters are $\tau = (0.85 \pm 0.08)$ h, $z = 0.223 \pm 0.003$, and $z/d = 0.104 \pm 0.003$, yielding $d = 2.14 \pm 0.09$. (B) Dimensionality plot, M_{rel} versus R_{rel}, according to the mass scaling law for the data shown in (A). The slope of the regression curve equals $d = 2.08 \pm 0.04$ ($R = 0.998$).

and

$$R(t) = R(0)(1 + t/\tau)^{z/d}. \tag{17.11b}$$

The exponent z is related to the aggregation rates and determines the aggregation regime between the limiting cases of reaction-limited and diffusion-limited aggregation. The exponent d describes the morphology of the aggregates according to the mass scaling law $M \propto R^d$. The variable d corresponds to $1/\nu$ in the notation of the scaling laws used in section 17.3.8. Linear aggregates are characterized by $d = 1$ and spherical ones by $d = 3$. Fractional values in between are interpreted as fractal dimensions. But care has to be taken to interpret such aggregates as true fractal objects. The power-law behavior of the mass scaling law has to be verified over several orders of magnitude, which is certainly not fulfilled in the example shown in Figure 17.15B.

Many light scattering investigations have been made in connection with heat-induced aggregation of proteins relevant for the food industry (see Table 17.4). β-Lactoglobulin is the most intensively investigated food protein in this respect (4, 7, 75) (for a review see Reference 50). These studies are examples of very careful light scattering investigations of protein aggregation. In many of these studies, the consumption of the initial species (mostly monomers) has been followed additionally by SEC. Consumption of the initial species is a very important characteristic of the aggregation process (see below), but it cannot be derived easily from light scattering data, since the measured weight averaged mass is dominated by the growing large particles. Jossang et al. (62) studied the aggregation kinetics of human immunoglobulins by measuring the increase in the relative mass and the relative Stokes radius. From the changes of both quantities, the authors obtained detailed information about the aggregation process, which could be consistently described by Smoluchowski's coagulation theory (110).

Aggregation of proteins has been regarded for a long time as undesirable side reaction in molecular biology, and the importance of these processes became evident only recently in connection with recombinant protein technology and the observation that the development of particular diseases correlates with protein misfolding and misassembly (60). The basic principles of protein aggregation were considered for example by De Young et al. (24).

Typical experiments studying protein aggregation as a side reaction during refolding were reported by Zettlmeissl et al. (134), who studied the aggregation kinetics of lactate dehydrogenase using a stopped-flow laser light scattering apparatus. These authors thoroughly investigated the effect of protein concentration on the kinetics of aggregation. From the concentration dependence of the initial slopes of the light scattering signal, they obtained an apparent reaction order of 2.5.

If the protein aggregates, or if the polymerization products become large enough, further structural information about the growing species can be obtained from the angular dependence of light scattering intensity. A careful analysis of fibrin formation using stopped-flow multiangle light scattering was reported by Bernocco et al. (9).

An interesting question is how can protein aggregation be influenced? Suppression of citrate synthase aggregation and facilitation of correct folding were demonstrated by Buchner et al. (12) using a commercial fluorometer as a light scattering device.

17.5.3 Investigations of Amyloid Formation by SLS and DLS

17.5.3.1 Initial States, Critical Oligomers, Protofibrils, and Fibrils Conformational conversion of proteins into misfolded structures with accompanying assembly into large particles, mostly of fibrillar structure called amyloid, is presently an expanding field of research. In this section, we will try to sketch the potentials and limits of light scattering for elucidating these processes. It is already clear that SLS and DLS are best applied for this purpose in conjunction with other methods, particularly with those sensitive to changes in secondary structure (CD, FTIR, and NMR spectroscopy) and methods providing evidence of the morphology of the growing species, such as electron microscopy and/or atomic force microscopy. Nevertheless, we will concentrate upon the contribution of the light scattering studies.

Amyloid formation comprises different stages of particle growth. The information that can be obtained from light scattering experiments depends strongly on the quality of the initial state (32). A starting solution containing essentially the conversion competent monomeric species is a prerequisite for studies of early stages involving the formation of defined oligomers, later called critical oligomers, and protofibrillar structures. These structures became interesting recently because of their possible toxic role (15). The size distributions and the kinetics of formation of these rather small particles can be characterized very well by light scattering. A careful experimental design, for example, multiangle light scattering or model calculations regarding the angular dependence of the scattered light intensity, is needed for the characterization of late products such as large protofibrils or long "mature" fibrils. The problems involved in light scattering from long fibrillar structures were discussed in a review article by Lomakin et al. (78). Quantitative light scattering studies of fibril formation have been reported for a few protein systems. Some of them will be discussed now.

17.5.3.2 Aggregation Kinetics of Aβ Peptides Aβ peptide is the major protein component found in amyloid deposits of Alzheimer's disease patients and comprises 39–43 amino acids. The first light scattering investigations of Aβ aggregation kinetics were reported by Tomski and Murphy (116) for synthetic Aβ(1–40). According to their SLS and DLS data, these authors modeled the assembly kinetics as the formation of rods with a diameter of 5 nm built up from spontaneously formed small cylinders consisting of eight Aβ monomers. The kinetic aggregation model was based on Smoluchowski's coagulation theory. Later, the same group (85) studied the concentration dependence of the aggregation process. They found that a high molecular weight species is formed rapidly when Aβ(1–40) is diluted from 8 M urea to physiological conditions. The size of this species was largest and constant at the lowest concentration (70 μM), while it was smaller and grew with time at higher concentrations. Furthermore, dissociation of Aβ monomers from preformed fibrils was observed.

Detailed light scattering investigations of Aβ(1–40) fibrillogenesis in 0.1 M HCl were reported by Lomakin et al. (72, 79, 80). Aβ(1–40) fibrillization was found to be highly reproducible at low pH. In a review article (78), these authors thoroughly analyzed the requirements for obtaining good quantitative data on the fibril length from the measured hydrodynamic radius. Adequate measurements can be made if the fibril length does not exceed about 150 nm. Rate constants for fibril nucleation and elongation were determined from the measured time evolution of the fibril length distribution. The initial fibril elongation rate varied linearly with the initial peptide concentration c_0 below a critical concentration $c^* = 0.1$ mM, and was constant above c^*. From these observations, particular mechanisms for Aβ fibril nucleation were derived. Homogeneous nucleation within small ($d \sim 14$ nm) Aβ micelles was proposed for $c_0 > c^*$ and heterogeneous nucleation on seeds for $c < c_0$. From the temperature dependence of the elongation rate, an activation energy of 23 kcal/mol was estimated for the proposed monomer addition to fibrils.

Finally, it should be noted that reproducibility of Aβ fibrillogenesis *in vitro* is, in general, a serious problem due to possible variations in the starting conformation and the assembly state of the Aβ preparations (32).

17.5.3.3 Kinetics of Oligomer and Fibril Formation of PGK Presently, more than 60 proteins and peptides are known that form fibrillar structures under appropriate environmental conditions. Many of these proteins are not related to any disease. Particularly interesting are those that have served as model proteins in folding studies and could also be good candidates for studying basic principles of structure conversion and amyloid formation. The potential of light scattering for this purpose will be demonstrated here for PGK from yeast. A similar example for the recombinant prion protein can be found in the literature (111). A well-defined, conversion competent state, free of any "seeds" can be achieved for PGK under specific environmental conditions, thus allowing studies of early stages of misfolding and aggregation.

Conversion of PGK starts from a partially folded state at pH 2, at room temperature and in the presence of defined amounts of salt (22). The time dependence of the relative increases in average mass and average Stokes radius after adding NaCl and sodium trichloroacetate (Na-TCA), respectively, are shown in Figure 17.16A,B. Two growth steps are clearly visible in Figure 17.16B, whereas the growth curves in Figure 17.16A increase strictly monotonously. Kinks in the dimensionality plot (Figure 17.16C,D) are also observed when various growth stages, because of comparable growth rates, are not clearly separated in plots of M and R_S versus time (Figure 17.16A,C) (82). This is a clear advantage of this data representation. However, one must be careful in relating the (apparent) dimensionality derived from these plots to simple geometric models.

The mass and size transition curves scale linearly with concentration (83). Thus, the growth process is a second-order reaction. Both growth steps can be well described within the framework of Smoluchowski's coagulation theory.

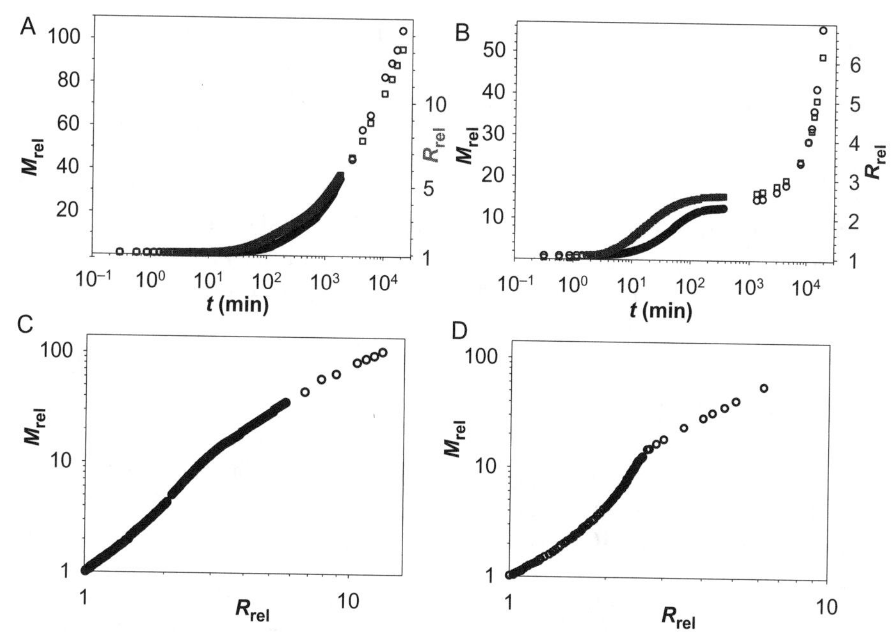

Figure 17.16 Amyloid assembly of PGK. Relative increases in mass M_{rel} (○) and Stokes radius R_{rel} (□) during assembly of PGK, $c = 2.7\,mg/mL$, into fibrillar structures in 10 mM HCl, pH 2, $T = 20°C$. Structural conversion was initiated by adding either 190 mM NaCl (A) or 9.1 mM Na-TCA (B), respectively. (C) and (D) Corresponding dimensionality plots, M_{rel} versus R_{rel} for the data shown in (A) and (B).

During the first stage, clustering occurs toward oligomers consisting of about 10 in the case of NaCl and 11–13 PGK molecules in the case of TCA (82). These oligomers, termed "critical oligomers," assemble into protofibrillar structures during the second stage. Such behavior could be inferred from the relation between mass and size (Figure 17.16C,D), which yields an idea of the dimensionality of the growth process. EM at selected growth stages has confirmed this idea (83). The mutual processes of growth and secondary structure conversion were analyzed by relating the increase in mass to the changes in the far UV CD (83) and amide I region (82).

17.5.3.4 Misfolding and Misassembly: General Remarks Viewed from the perspective of chemical kinetics amyloid formation is a complex polymerization reaction, since assembly into amyloid fibrils is accompanied by secondary structure rearrangement into a predominantly β-sheet structure. β-Sheet strands are needed for association at the edges of monomeric subunits to build up the fibrillar end products (99). This is in marked contrast to polymerization reactions of low molecular weight compounds, where the monomers already possess the functional groups for polymerization at the initiation of the reaction (37), and to polymerization of proteins like actin, microtubulin (91), and sickle-cell hemoglobin (28). In the latter cases, the proteins possess the assembly-competent structure immediately after the start of the reaction. These processes are usually interpreted in the framework of a nucleation polymerization mechanism (28, 91). This interpretation has been taken over to rationalize the kinetics of amyloid formation (61). However, the processes of amyloid formation are far more complex. In particular, the initial steps, the appearance of productive intermediates for fibril formation, and the problem of how to redirect the process into nontoxic dead-end aggregation products are topics of intense research.

ACKNOWLEDGMENTS

The author is grateful to Ralph Golbik (University Halle/Wittenberg) for supporting the experiments with PDC and to Dagmar Ehrenhöfer and Erich Wanker (Max-Delbrück-Centrum Berlin-Buch) for providing samples of wt and A53T αS.

The author would like to thank Roger Pain (Jožef Stefan Institute, Ljubljana) and Stefanie Barbirz (University of Potsdam) for critically reading the manuscript and many valuable suggestions.

REFERENCES

1. Agashe, V. R., and J. B. Udgaonkar. 1995. Thermodynamics of denaturation of barstar: evidence for cold denaturation and evaluation of the interaction with guanidine hydrochloride. Biochemistry **34**:3286–99.
2. Arai, M., K. Ito, T. Inobe, M. Nakao, K. Maki, K. Kamagata, H. Kihara, Y. Amemiya, and K. Kuwajima. 2002. Fast compaction of alpha-lactalbumin during folding studied by stopped-flow X-ray scattering. J Mol Biol **321**:121–32.
3. Arai, M., and K. Kuwajima. 2000. Role of the molten globule state in protein folding. Adv Protein Chem **53**:209–82.
4. Aymard, P., T. Nicolai, D. Durand, and A. Clark. 1999. Static and dynamic scattering of beta-lactoglobulin aggregates formed after heat-induced denaturation at pH 2. Macromolecules **32**:2542–52.
5. Baldwin, R. L., and B. H. Zimm. 2000. Are denatured proteins ever random coils? Proc Natl Acad Sci U S A **97**:12391–2.
6. Barrick, D., F. M. Hughson, and R. L. Baldwin. 1994. Molecular mechanism of acid denaturation—the role of histidine residues in the partial unfolding of apomyoglobin. J Mol Biol **237**:588–601.
7. Bauer, R., R. Carrotta, C. Rischel, and L. Ogendal. 2000. Characterization and isolation of intermediates in beta-lactoglobulin heat aggregation at high pH. Biophys J **79**:1030–8.
8. Berne, B. J., and R. Pecora. 2000. Dynamic Light Scattering with Applications to Chemistry, Biology, and Physics. Dover Publications, Mineola, New York.

9. Bernocco, S., F. Ferri, A. Profumo, C. Cuniberti, and M. Rocco. 2000. Polymerization of rod-like macromolecular monomers studied by stopped-flow, multiangle light scattering: set-up, data processing, and application to fibrin formation. Biophys J **79**:561–83.
10. Bertoncini, C. W., C. O. Fernandez, C. Griesinger, T. M. Jovin, and M. Zweckstetter. 2005. Familial mutants of alpha-synuclein with increased neurotoxicity have a destabilized conformation. J Biol Chem **280**:30649–52.
11. Brown, W. 1993. Dynamic Light Scattering. Claredon Press, Oxford.
12. Buchner, J., M. Schmidt, M. Fuchs, R. Jaenicke, R. Rudolph, F. X. Schmid, and T. Kiefhaber. 1991. GroE facilitates refolding of citrate synthase by suppressing aggregation. Biochemistry **30**:1586–91.
13. Burchard, W., M. Schmidt, and W. H. Stockmayer. 1980. Information on polydispersity and branching from combined quasi-elastic and integrated scattering. Macromolecules **13**:1265–72.
14. Byron, O. 2000. Hydrodynamic bead modeling of biological macromolecules, pp. 278–304. In M. L. Johnson and L. Brands (eds.), Methods in Enzymology, vol. 321. Academic Press, New York.
15. Caughey, B., and P. T. Lansbury. 2003. Protofibrils, pores, fibrils, and neurodegeneration: separating the responsible protein aggregates from the innocent bystanders. Annu Rev Neurosci **26**:267–98.
16. Chu, B. 1991. Laser Light Scattering. Academic Press, New York.
17. Claes, P., M. Dunford, A. Kenney, and V. Penny. 1992. An on-line dynamic light scattering instrument for macromolecular characterization, pp. 66–76. In S. E. Harding, D. B. Sattelle, and V. A. Bloomfield (eds.), Laser Light Scattering in Biochemistry. Royal Society of Chemistry, Cambridge, UK.
18. Cordero, O. J., C. S. Sarandeses, J. L. Lopez, and M. Nogueira. 1992. On the anomalous behaviour on gel-filtration and SDS-electrophoresis of prothymosin-alpha. Biochem Int **28**:1117–24.
19. Damaschun, G., H. Damaschun, K. Gast, R. Misselwitz, J. J. Muller, W. Pfeil, and D. Zirwer. 1993. Cold denaturation-induced conformational changes in phosphoglycerate kinase from yeast. Biochemistry **32**:7739–46.
20. Damaschun, G., H. Damaschun, K. Gast, R. Misselwitz, D. Zirwer, K. H. Guhrs, M. Hartmann, B. Schlott, H. Triebel, and D. Behnke. 1993. Physical and conformational properties of staphylokinase in solution. Biochim Biophys Acta **1161**: 244–8.
21. Damaschun, G., H. Damaschun, K. Gast, and D. Zirwer. 1998. Denatured states of yeast phosphoglycerate kinase. Biochemistry (Mosc) **63**:259–75.
22. Damaschun, G., H. Damaschun, K. Gast, and D. Zirwer. 1999. Proteins can adopt totally different folded conformations. J Mol Biol **291**:715–25.
23. Daughdrill, G. W., G. J. Pielak, V. N. Uversky, M. S. Cortese, and A. K. Dunker. 2005. Natively disordered proteins, pp. 271–353. In J. Buchner and T. Kiefhaber (eds.), Protein Folding Handbook. Wiley-VCH, Weinheim, Germany.
24. De Young, L. R., K. A. Dill, and A. L. Fink. 1993. Aggregation of globular proteins. Acc Chem Res **26**:614–20.
25. Ding, T. T., S. J. Lee, J. C. Rochet, and P. T. Lansbury. 2002. Annular alpha-synuclein protofibrils are produced when spherical protofibrils are incubated in solution or bound to brain-derived membranes. Biochemistry **41**:10209–17.

26. Dubin, S. B., G. Feher, and G. B. Benedek. 1973. Study of the chemical denaturation of lysozyme by optical mixing spectroscopy. Biochemistry **12**:714–9.
27. Durand, D., J. C. Gimel, and T. Nicolai. 2002. Aggregation, gelation and phase separation of heat denatured globular proteins. Physica A **304**:253–65.
28. Eaton, W. A., and J. Hofrichter. 1990. Sickle cell hemoglobin polymerization. Adv Protein Chem **40**:63–279.
29. Fabian, H., K. Falber, K. Gast, D. Reinstadler, V. V. Rogov, D. Naumann, D. F. Zamyatkin, and V. V. Filimonov. 1999. Secondary structure and oligomerization behavior of equilibrium unfolding intermediates of the lambda Cro repressor. Biochemistry **38**:5633–42.
30. Feder, J., T. Jossang, and E. Rosenqvist. 1984. Scaling behavior and cluster fractal dimension determined by light scattering from aggregating proteins. Phys Rev Lett **53**:1403–6.
31. Feng, H. P., and J. Widom. 1994. Kinetics of compaction during lysozyme refolding studied by continuous-flow quasielastic light scattering. Biochemistry **33**:13382–90.
32. Fezoui, Y., D. M. Hartley, J. D. Harper, R. Khurana, D. M. Walsh, M. M. Condron, D. J. Selkoe, P. T. Lansbury, A. L. Fink, and D. B. Teplow. 2000. An improved method of preparing the amyloid beta-protein for fibrillogenesis and neurotoxicity experiments. Amyloid **7**:166–78.
33. Fierz, B., K. Joder, F. Krieger, and T. Kiefhaber. 2007. Using triplet-triplet energy transfer to measure conformational dynamics in polypeptide chains. Methods Mol Biol **350**:169–87.
34. Fink, A. L., L. J. Calciano, Y. Goto, T. Kurotsu, and D. R. Palleros. 1994. Classification of acid denaturation of proteins: intermediates and unfolded states. Biochemistry **33**:12504–11.
35. Finke, J. M., M. Roy, B. H. Zimm, and P. A. Jennings. 2000. Aggregation events occur prior to stable intermediate formation during refolding of interleukin 1 beta. Biochemistry **39**:575–83.
36. Fitzkee, N. C., and G. D. Rose. 2004. Reassessing random-coil statistics in unfolded proteins. Proc Natl Acad Sci U S A **101**:12497–502.
37. Flory, P. J. 1953. Principles of Polymer Chemistry. Cornell University Press, Ithaca, NY.
38. Flory, P. J. 1969. Statistical Mechanics of Chain Molecules. John Wiley & Sons, New York.
39. Garcia Bernal, J. M., and J. G. de la Torre. 1981. Transport properties of oligomeric subunit structures. Biopolymers **20**:129–39.
40. Garcia de la Torre, J. 2001. Hydration from hydrodynamics. General considerations and applications of bead modelling to globular proteins. Biophys Chem **93**:159–70.
41. Garcia de la Torre, J., M. L. Huertas, and B. Carrasco. 2000. Calculation of hydrodynamic properties of globular proteins from their atomic-level structure. Biophys J **78**:719–30.
42. Gast, K., H. Damaschun, K. Eckert, F. K. Schulze, H. R. Maurer, F. M. Muller, D. Zirwer, J. Czarnecki, and G. Damaschun. 1995. Prothymosin alpha: a biologically active protein with random coil conformation. Biochemistry **34**:13211–8.

43. Gast, K., H. Damaschun, R. Misselwitz, F. M. Muller, D. Zirwer, and G. Damaschun. 1994. Compactness of protein molten globules: temperature-induced structural changes of the apomyoglobin folding intermediate. Eur Biophys J **23**:297–305.

44. Gast, K., A. J. Modler, H. Damaschun, R. Krober, G. Lutsch, D. Zirwer, R. Golbik, and G. Damaschun. 2003. Effect of environmental conditions on aggregation and fibril formation of barstar. Eur Biophys J **32**:710–23.

45. Gast, K., D. Zirwer, and G. Damaschun. 2000. Time-resolved dynamic light scattering as a method to monitor compaction during protein folding, pp. 205–20. In M. Helmstedt and K. Gast (eds.), Data Evaluation in Light Scattering of Polymers. Wiley-VCH, Weinheim, Germany.

46. Gast, K., D. Zirwer, H. Damaschun, U. Hahn, F. M. Muller, M. Wirth, and G. Damaschun. 1997. Ribonuclease T1 has different dimensions in the thermally and chemically denatured states: a dynamic light scattering study. FEBS Lett **403**:245–8.

47. Gast, K., D. Zirwer, M. Muller Frohne, and G. Damaschun. 1999. Trifluoroethanol-induced conformational transitions of proteins: insights gained from the differences between alpha-lactalbumin and ribonuclease A. Protein Sci **8**:625–34.

48. Gast, K., D. Zirwer, M. Mullerfrohne, and G. Damaschun. 1998. Compactness of the kinetic molten globule of bovine alpha-lactalbumin: a dynamic light scattering study. Protein Sci **7**:2004–11.

49. Görisch, H., D. J. Goss, and L. J. Parkhurst. 1976. Kinetics of ribosome dissociation and subunit association studied in a light-scattering stopped-flow apparatus. Biochemistry **15**:5743–53.

50. Gosal, W. S., and S. B. Ross-Murphy. 2000. Globular protein gelation. Curr Opin Colloid Interface Sci **5**:188–94.

51. Griko, Y. V., S. Y. Venyaminov, and P. L. Privalov. 1989. Heat and cold denaturation of phosphoglycerate kinase (interaction of domains). FEBS Lett **244**:276–8.

52. Halle, B., and M. Davidovic. 2003. Biomolecular hydration: from water dynamics to hydrodynamics. Proc Natl Acad Sci U S A **100**:12135–40.

53. Harding, S. E. 2001. The hydration problem in solution biophysics: an introduction. Biophys Chem **93**:87–91.

54. Harding, S. E., and P. Johnson. 1985. The concentration dependence of macromolecular parameters. Biochem J **231**:543–7.

55. Haritos, A. A., P. P. Yialouris, E. P. Heimer, A. M. Felix, E. Hannappel, and M. A. Rosemeyer. 1989. Evidence for the monomeric nature of thymosins. FEBS Lett **244**:287–90.

56. Hemmings, H. C., Jr., A. C. Nairn, D. W. Aswad, and P. Greengard. 1984. DARPP-32, a dopamine- and adenosine 3′:5′-monophosphate-regulated phosphoprotein enriched in dopamine-innervated brain regions. II. Purification and characterization of the phosphoprotein from bovine caudate nucleus. J Neurosci **4**:99–110.

57. Hernandez, M. A., J. Avila, and J. M. Andreu. 1986. Physicochemical characterization of the heat-stable microtubule-associated protein MAP2. Eur J Biochem **154**:41–8.

58. Horne, D. S. 1987. Determination of the fractal dimension using turbidimetric techniques. Faraday Discuss **83**:259–70.
59. Huglin, M. 1972. Light Scattering from Polymer Solutions. Academic Press, New York.
60. Jaenicke, R., and R. Seckler. 1997. Protein misassembly in vitro. Adv Protein Chem **50**:1–59.
61. Jarrett, J. T., and P. T. Lansbury, Jr. 1993. Seeding "one-dimensional crystallization" of amyloid: a pathogenic mechanism in Alzheimer's disease and scrapie? Cell **73**:1055–8.
62. Jossang, T., J. Feder, and E. Rosenqvist. 1985. Heat aggregation kinetics of human IgG. J Chem Phys **82**:574–89.
63. Juneja, J., N. S. Bhavesh, J. B. Udgaonkar, and R. V. Hosur. 2002. NMR identification and characterization of the flexible regions in the 160kDa molten globule-like aggregate of barstar at low pH. Biochemistry **41**:9885–99.
64. Kataoka, M., K. Kuwajima, F. Tokunaga, and Y. Goto. 1997. Structural characterization of the molten globule of alpha lactalbumin by solution X-ray scattering. Protein Sci **6**:422–30.
65. Kataoka, M., I. Nishii, T. Fujisawa, T. Ueki, F. Tokunaga, and Y. Goto. 1995. Structural characterization of the molten globule and native states of apomyoglobin by solution X-ray scattering. J Mol Biol **249**:215–28.
66. Killenberg, J. M., G. Kern, G. Hubner, and R. Golbik. 2002. Folding and stability of different oligomeric states of thiamin diphosphate dependent homomeric pyruvate decarboxylase. Biophys Chem **96**:259–71.
67. Kohn, J. E., I. S. Millett, J. Jacob, B. Zagrovic, T. M. Dillon, N. Cingel, R. S. Dothager, S. Seifert, P. Thiyagarajan, T. R. Sosnick, M. Z. Hasan, V. S. Pande, I. Ruczinski, S. Doniach, and K. W. Plaxco. 2004. Random-coil behavior and the dimensions of chemically unfolded proteins. Proc Natl Acad Sci U S A **101**:12491–6.
68. Koppel, D. E. 1972. Analysis of macromolecular polydispersity in intensity correlation spectroscopy: the method of cumulants. J Chem Phys **57**:4814–20.
69. Kratochvil, P. 1987. Classical Light Scattering from Polymer Solutions. Elsevier, Amsterdam.
70. Kruyt, H. (ed.). 1952. Colloid Science. Elsevier Publishing Company, Amsterdam.
71. Kumagai, H., T. Matsunaga, and T. Hagiwara. 1999. Effect of salt addition on the fractal structure of aggregates formed by heating dilute BSA solutions. Biosci Biotechnol Biochem **63**:223–5.
72. Kusumoto, Y., A. Lomakin, D. B. Teplow, and G. B. Benedek. 1998. Temperature dependence of amyloid beta-protein fibrillization. Proc Natl Acad Sci U S A **95**:12277–82.
73. Kuwajima, K., and M. Arai. 2000. The molten globule state: the physical picture and biological significance, pp. 138–74. In R. H. Pain (ed.), Mechanisms of Protein Folding. Oxford University Press, Oxford.
74. Lashuel, H. A., B. M. Petre, J. Wall, M. Simon, R. J. Nowak, T. Walz, and P. T. Lansbury. 2002. Alpha-synuclein, especially the Parkinson's disease-associated mutants, forms pore-like annular and tubular protofibrils. J Mol Biol **322**:1089–102.

75. Le Bon, C., T. Nicolai, and D. Durand. 1999. Kinetics of aggregation and gelation of globular proteins after heat-induced denaturation. Macromolecules **32**: 6120–7.

76. Li, J., V. N. Uversky, and A. L. Fink. 2001. Effect of familial Parkinson's disease point mutations A30P and A53T on the structural properties, aggregation, and fibrillation of human alpha-synuclein. Biochemistry **40**:11604–13.

77. Lin, M. Y., H. M. Lindsay, D. A. Weitz, R. C. Ball, R. Klein, and P. Meakin. 1989. Universality in colloid aggregation. Nature **339**:360–2.

78. Lomakin, A., G. B. Benedek, and D. B. Teplow. 1999. Monitoring protein assembly using quasielastic light scattering spectroscopy, pp. 429–59. Methods in Enzymology, vol. **309**. Academic Press, New York.

79. Lomakin, A., D. S. Chung, G. B. Benedek, D. A. Kirschner, and D. B. Teplow. 1996. On the nucleation and growth of amyloid beta-protein fibrils: detection of nuclei and quantitation of rate constants. Proc Natl Acad Sci U S A **93**:1125–9.

80. Lomakin, A., D. B. Teplow, D. A. Kirschner, and G. B. Benedek. 1997. Kinetic theory of fibrillogenesis of amyloid beta-protein. Proc Natl Acad Sci U S A **94**:7942–7.

81. Lynch, W. P., V. M. Riseman, and A. Bretscher. 1987. Smooth muscle caldesmon is an extended flexible monomeric protein in solution that can readily undergo reversible intra- and intermolecular sulfhydryl cross-linking. A mechanism for caldesmon's F-actin bundling activity. J Biol Chem **262**:7429–37.

82. Modler, A. J., H. Fabian, F. Sokolowski, G. Lutsch, K. Gast, and G. Damaschun. 2004. Polymerization of proteins into amyloid protofibrils shares common critical oligomeric states but differs in the mechanisms of their formation. Amyloid **11**:215–31.

83. Modler, A. J., K. Gast, G. Lutsch, and G. Damaschun. 2003. Assembly of amyloid protofibrils via critical oligomers—a novel pathway of amyloid formation. J Mol Biol **325**:135–48.

84. Moore, J. W., and R. G. Pearson. 1981. Kinetics and Mechanism. Wiley, New York.

85. Murphy, R. M., and M. R. Pallitto. 2000. Probing the kinetics of beta-amyloid self-association. J Struct Biol **130**:109–22.

86. Nicoli, D. F., and G. B. Benedek. 1976. Study of the thermal denaturation of lysozyme and other globular proteins by light-scattering spectroscopy. Biopolymers **15**:2421–37.

87. Nimmo, G. A., and P. Cohen. 1978. The regulation of glycogen metabolism. Purification and characterisation of protein phosphatase inhibitor-1 from rabbit skeletal muscle. Eur J Biochem **87**:341–51.

88. Nöppert, A., K. Gast, M. Mullerfrohne, D. Zirwer, and G. Damaschun. 1996. Reduced-denatured ribonuclease A is not in a compact state. FEBS Lett **380**:179–82.

89. Nöppert, A., K. Gast, D. Zirwer, and G. Damaschun. 1998. Initial hydrophobic collapse is not necessary for folding RNase A. Fold Des **3**:213–21.

90. Olivier, B. J., and C. M. Sorensen. 1990. Variable aggregation rates in colloidal gold—kernel homogeneity dependence on aggregant concentration. Phys Rev A **41**:2093–100.

91. Oosawa, F., and S. Asakura. 1975. Thermodynamics of the Polymerization of Protein. Academic Press, London.
92. Pavlov, N. A., D. I. Cherny, G. Heim, T. M. Jovin, and V. Subramaniam. 2002. Amyloid fibrils from the mammalian protein prothymosin alpha. FEBS Lett **517**:37–40.
93. Pollack, L., M. W. Tate, N. C. Darnton, J. B. Knight, S. M. Gruner, W. A. Eaton, and R. H. Austin. 1999. Compactness of the denatured state of a fast-folding protein measured by submillisecond small-angle X-ray scattering. Proc Natl Acad Sci U S A **96**:10115–7.
94. Pots, A. M., E. T. Grotenhuis, H. Gruppen, A. G. J. Voragen, and K. G. de Kruif. 1999. Thermal aggregation of patatin studied in situ. J Agric Food Chem **47**:4600–5.
95. Privalov, P. L. 1990. Cold denaturation of proteins. Crit Rev Biochem Mol Biol **25**:281–305.
96. Provencher, S. W. 1982. CONTIN—a general-purpose constrained regularization program for inverting noisy linear algebraic and integral equations. Comput Phys Commun **27**:229–42.
97. Ptitsyn, O. B. 1995. Molten globule and protein folding. Adv Protein Chem **47**:83–229.
98. Receveur-Brechot, V., J. M. Bourhis, V. N. Uversky, B. Canard, and S. Longhi. 2006. Assessing protein disorder and induced folding. Proteins **62**:24–45.
99. Richardson, J. S., and D. C. Richardson. 2002. Natural beta-sheet proteins use negative design to avoid edge-to-edge aggregation. Proc Natl Acad Sci U S A **99**:2754–9.
100. Rigler, R., and E. S. Elson. 2001. Fluorescence Correlation Spectroscopy: Theory and Applications. Springer, New York.
101. Sandal, M., F. Valle, I. Tessari, S. Mammi, E. Bergantino, F. Musiani, M. Brucale, L. Bubacco, and B. Samori. 2008. Conformational equilibria in monomeric alpha-synuclein at the single-molecule level. PLoS Biol **6**:e6.
102. Schmitz, K. S. 1990. An Introduction to Dynamic Light Scattering by Macromolecules. Academic Press, New York.
103. Schuler, B. 2007. Application of single molecule Forster resonance energy transfer to protein folding. Methods Mol Biol **350**:115–38.
104. Schuler, B., and W. A. Eaton. 2008. Protein folding studied by single-molecule FRET. Curr Opin Struct Biol **18**:16–26.
105. Schuler, J., J. Frank, W. Saenger, and Y. Georgalis. 1999. Thermally induced aggregation of human transferrin receptor studied by light-scattering techniques. Biophys J **77**:1117–25.
106. Schweers, O., E. Schonbrunn-Hanebeck, A. Marx, and E. Mandelkow. 1994. Structural studies of tau protein and Alzheimer paired helical filaments show no evidence for beta-structure. J Biol Chem **269**:24290–7.
107. Segel, D. J., A. Bachmann, J. Hofrichter, K. O. Hodgson, S. Doniach, and T. Kiefhaber. 1999. Characterization of transient intermediates in lysozyme folding with time-resolved small-angle X-ray scattering. J Mol Biol **288**:489–99.
108. Shastry, M. C. R., S. D. Luck, and H. Roder. 1998. A continuous-flow capillary mixing method to monitor reactions on the microsecond time scale. Biophys J **74**:2714–21.

109. Smoluchowski, M. V. 1916. Drei Vorträge über Diffusion, Brownsche Molekularbewegung und Koagulation von Kolloidteilchen. Physik Zeitschr **XVII**: 557–99.

110. Smoluchowski, M. V. 1917. Versuch einer mathematischen Theorie der Koagulationskinetik kolloider Lösungen. Z Phys Chem **92**:129–68.

111. Sokolowski, F., A. J. Modler, R. Masuch, D. Zirwer, M. Baier, G. Lutsch, D. A. Moss, K. Gast, and D. Naumann. 2003. Formation of critical oligomers is a key event during conformational transition of recombinant Syrian hamster prion protein. J Biol Chem **278**:40481–92.

112. Tanford, C. 1968. Protein denaturation. Adv Protein Chem **23**:121–282.

113. Tanford, C., K. Kawahara, and S. Lapanje. 1966. Proteins in 6M guanidine hydrochloride. Demonstration of random coil behavior. J Biol Chem **241**:1921–3.

114. Theisen, A., C. Johann, M. P. Deacon, and S. E. Harding. 2000. Refractive Increment Data-Book. Nottingham University Press, Nottingham.

115. Thomas, J., S. M. Van Patten, P. Howard, K. H. Day, R. D. Mitchell, T. Sosnick, J. Trewhella, D. A. Walsh, and R. A. Maurer. 1991. Expression in *Escherichia coli* and characterization of the heat-stable inhibitor of the cAMP-dependent protein kinase. J Biol Chem **266**:10906–11.

116. Tomski, S. J., and R. M. Murphy. 1992. Kinetics of aggregation of synthetic beta-amyloid peptide. Arch Biochem Biophys **294**:630–638.

117. Umbach, P., Y. Georgalis, and W. Saenger. 1998. Time-resolved small-angle static light scattering on lysozyme during nucleation and growth. J Am Chem Soc **120**:2382–90.

118. Uversky, V. N. 1993. Use of fast protein size-exclusion liquid-chromatography to study the unfolding of proteins which denature through the molten globule. Biochemistry **32**:13288–98.

119. Uversky, V. N. 2002. Natively unfolded proteins: a point where biology waits for physics. Protein Sci **11**:739–56.

120. Uversky, V. N. 2002. What does it mean to be natively unfolded? Eur J Biochem **269**:2–12.

121. Uversky, V. N. 2003. A protein-chameleon: conformational plasticity of alpha-synuclein, a disordered protein involved in neurodegenerative disorders. J Biomol Struct Dyn **21**:211–34.

122. Uversky, V. N., J. R. Gillespie, I. S. Millett, A. V. Khodyakova, R. N. Vasilenko, A. M. Vasiliev, I. L. Rodionov, G. D. Kozlovskaya, D. A. Dolgikh, A. L. Fink, S. Doniach, E. A. Permyakov, and V. M. Abramov. 2000. Zn2+-mediated structure formation and compaction of the "natively unfolded" human prothymosin alpha. Biochem Biophys Res Commun **267**:663–8.

123. Uversky, V. N., J. R. Gillespie, I. S. Millett, A. V. Khodyakova, A. M. Vasiliev, T. V. Chernovskaya, R. N. Vasilenko, G. D. Kozovskaya, D. A. Dolgikh, A. L. Fink, S. Doniach, and V. M. Abramov. 1999. Natively unfolded human prothymosin alpha adopts partially folded collapsed conformation at acidic pH. Biochemistry **38**:15009–16.

124. Uversky, V. N., H. J. Lee, J. Li, A. L. Fink, and S. J. Lee. 2001. Stabilization of partially folded conformation during alpha-synuclein oligomerization in both purified and cytosolic preparations. J Biol Chem **276**:43495–8.

125. Uversky, V. N., J. Li, and A. L. Fink. 2001. Evidence for a partially folded intermediate in alpha-synuclein fibril formation. J Biol Chem **276**:10737–44.
126. Uversky, V. N., J. Li, P. Souillac, I. S. Millett, S. Doniach, R. Jakes, M. Goedert, and A. L. Fink. 2002. Biophysical properties of the synucleins and their propensities to fibrillate—inhibition of alpha-synuclein assembly by beta- and gamma-synucleins. J Biol Chem **277**:11970–8.
127. Uversky, V. N., C. J. Oldfield, and A. K. Dunker. 2008. Intrinsically disordered proteins in human diseases: introducing the D2 concept. Annu Rev Biophys **37**:215–46.
128. Watts, J. D., P. D. Cary, P. Sautiere, and C. Crane-Robinson. 1990. Thymosins: both nuclear and cytoplasmic proteins. Eur J Biochem **192**:643–51.
129. Weijers, M., R. W. Visschers, and T. Nicolai. 2002. Light scattering study of heat-induced aggregation and gelation of ovalbumin. Macromolecules **35**:4753–62.
130. Weinreb, P. H., W. G. Zhen, A. W. Poon, K. A. Conway, and P. T. Lansbury. 1996. NACP, a protein implicated in Alzheimer's disease and learning, is natively unfolded. Biochemistry **35**:13709–15.
131. Wilkins, D. K., S. B. Grimshaw, V. Receveur, C. M. Dobson, J. A. Jones, and L. J. Smith. 1999. Hydrodynamic radii of native and denatured proteins measured by pulse field gradient NMR techniques. Biochemistry **38**:16424–31.
132. Wong, K. B., S. M. V. Freund, and A. R. Fersht. 1996. Cold denaturation of barstar: H-1, N-15 and C-13 NMR assignment and characterisation of residual structure. J Mol Biol **259**:805–18.
133. Yi, S., B. L. Boys, A. Brickenden, L. Konermann, and W. Y. Choy. 2007. Effects of zinc binding on the structure and dynamics of the intrinsically disordered protein prothymosin alpha: evidence for metalation as an entropic switch. Biochemistry **46**:13120–30.
134. Zettlmeissl, G., R. Rudolph, and R. Jaenicke. 1979. Reconstitution of lactic dehydrogenase. Noncovalent aggregation vs. reactivation. 1. Physical properties and kinetics of aggregation. Biochemistry **18**:5567–71.
135. Zhou, H. X. 2002. Dimensions of denatured protein chains from hydrodynamic data. J Phys Chem **106**:5769–75.

18

ANALYZING INTRINSICALLY DISORDERED PROTEINS BY SIZE EXCLUSION CHROMATOGRAPHY

VLADIMIR N. UVERSKY

Institute for Intrinsically Disordered Protein Research, Center for Computational Biology and Bioinformatics, and Department of Biochemistry and Molecular Biology, Indiana University School of Medicine, Indianapolis, IN
Institute for Biological Instrumentation, Russian Academy of Sciences, Pushchino, Moscow Region, Russia

ABSTRACT

This chapter is dedicated to the gel-filtration chromatography, also known as size exclusion chromatography (SEC) or gel-permeation chromatography, and its applications for the various analyses of intrinsically disordered proteins (IDPs). SEC can be used for the estimation of the hydrodynamic dimensions of a given IDP, for evaluation of the association state, for the analysis of IDP interactions with binding partners, and for the induced folding studies. It can also be used to physically separate IDP conformers based on their hydrodynamic dimensions, thus providing a unique possibility for the independent analysis of their physicochemical properties.

18.1 INTRODUCTION

Chromatography, a broad range of physical methods for separation and analysis of complex mixtures, occupies a unique position among various analytic

Instrumental Analysis of Intrinsically Disordered Proteins: Assessing Structure and Conformation, Edited by Vladimir Uversky and Sonia Longhi

techniques of modern biochemistry, biophysics, and molecular biology. In chromatography, the components to be separated are generally partitioned between two phases: a stationary phase and a mobile phase. Stationary phase is typically packed into a column, whereas mobile phase usually contains dissolved sample and percolates through the stationary phase in a definite direction. The phases are chosen such that components of the sample differ in their capabilities to interact with the stationary and mobile phases. The components should interact with the stationary phase to be retained and separated by it, as a component with higher affinity to the stationary phase will travel longer through the column than a component with lower affinity. These differences in the mobility of sample components determine their separation as they travel through the stationary phase. Importantly, various retention mechanisms based on reversible physical interactions can be utilized (e.g., adsorption at a surface, absorption in an immobilized solvent layer, and electrostatic interactions). Furthermore, more than one type of interaction may contribute simultaneously to the separation mechanism and various means may be employed to achieve the reversibility of the component interaction with the stationary phase. All this defines the uniqueness of chromatography as an exceptionally useful and pliable analytic tool with almost endless applications.

Let us have a closer look at one of the chromatography types, size exclusion chromatography (SEC), also known as molecular exclusion chromatography, gel-filtration or gel-permeation chromatography (GPC), which is commonly used for the separation of biomolecules, including proteins, based on their hydrodynamic dimensions. Separation in SEC is achieved via the use of the porous beads with a well-defined range of pore sizes as the stationary phase. Therefore, the separation mechanism of gel-filtration is nonadsorptive, independent of the eluent system used, and very gentle.

In SEC, molecules in the mobile phase pass by a number of these porous beads while flowing through the column (Fig. 18.1). Molecules whose hydrodynamic dimensions are smaller than a particular limit can fit inside all the pores in the beads. They will be drawn in pores by the force of diffusion, where they will stay for a short time and then will move out. These molecules are totally included as they have total access to all the mobile phase inside and between the beads. They have the largest retention and therefore will elute last during the gel-filtration separation. On the other hand, large molecules that are too massive to fit inside any pore will have access only to the mobile phase between the beads. These molecules are excluded as they just follow the solvent flow and reach the end of the column before molecules with smaller size. Finally, molecules of intermediate size are partially included as they can fit inside some but not all of the pores in the beads and therefore possess an intermediate retention and elute between the large ("excluded") and small ("totally included") molecules. Within the fractionation range chosen, molecules are eluted in order of decreasing size. It is important to remember that all the molecules larger than all the pores in a matrix will elute

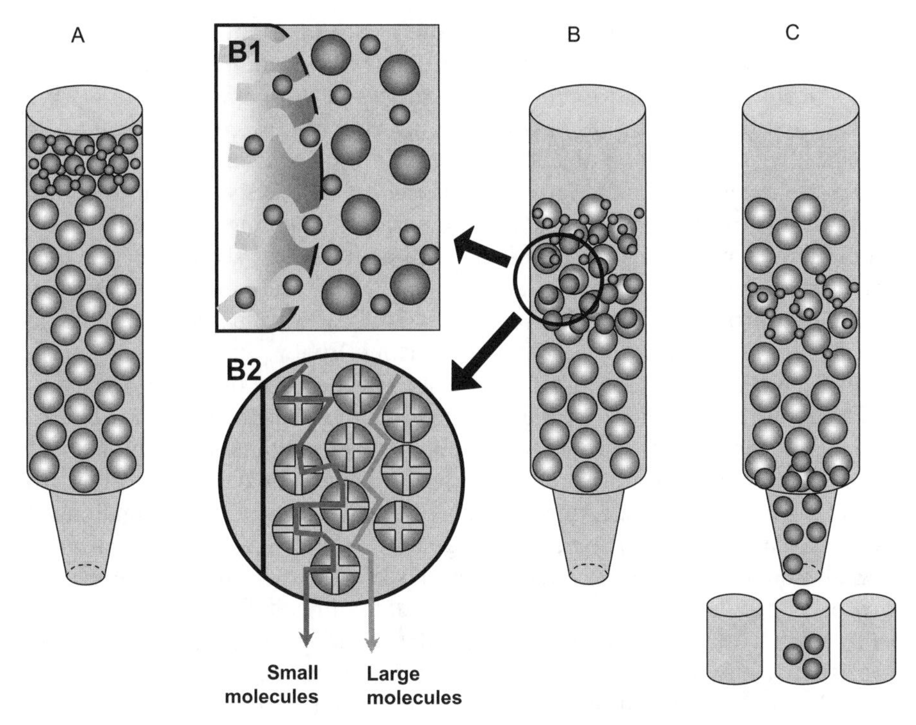

Figure 18.1 Schematic representation of the physical principles of molecule separation by size exclusion chromatography. Porous column matrix is shown as light gray spheres. Small and large molecules are shown as dark gray spheres. (A) Beginning of the chromatographic process, where a mixture of large and small molecules is loaded into a column containing porous matrix. (B) Beginning of the chromatographic separation, where large molecules migrate faster through the column matrix. (C) End of the chromatographic separation, where large molecules reach the end of the column and already start to elute, whereas small molecules continue to wonder through the column matrix. (B1) Part of the porous bead from the column matrix. Small molecules can diffuse in and out of the pores (shown as curved tunnels), whereas large molecules are mostly excluded from the pores. (B2) Migration trajectories of large (light gray line) and small (dark gray line) molecules inside the porous column matrix. As small molecule can diffuse inside the pores, their trajectories are longer, and therefore, they spent more time migrating through the column. This difference in the trajectory length represents a basis for the chromatographic separation.

together regardless of their size. Likewise, any molecules that can fit into all the pores in the beads will elute at the same time.

In the protein field, the most frequent uses of SEC are separation of proteins based on their size and estimation of their molecular masses. Formally, SEC is a separation technique based on hydrodynamic radius (see below); however, for similarly shaped molecules, hydrodynamic radius is proportional to molecular mass. Therefore, we can talk about SEC as a mass-based separation, even though this is not strictly true.

The hydrodynamic volume is one of the most important and fundamental structural parameters of a protein molecule. Hydrodynamic volume changes dramatically during the protein denaturation and unfolding (36, 41, 44, 45), and evaluation of the protein hydrodynamic dimensions (compact, extended, or partially swollen) is an absolute prerequisite for an accurate classification of a protein conformation. Many experimental techniques were elaborated to estimate the protein hydrodynamic dimensions, including viscometry, sedimentation, dynamic light scattering, small-angle X-ray scattering (SAXS), small-angle neutron scattering, and so on. Peculiarities of the intrinsically disordered protein (IDP) analysis by analytic ultracentrifugation are outlined in chapter 15 by Florence Manon and Christine Ebel.

Although many of the listed above approaches are based on the well-developed theories, all of them have some difficulties and pitfalls. Some hydrodynamic techniques require large protein quantities; others use complex and expensive equipment and sophisticated approaches for data processing. All techniques based on scattering (dynamic light scattering, SAXS, small-angle neutron scattering) require very homogeneous samples, as presence of even a small fraction of the aggregated material is known to dramatically affect the scattering profile, making interpretation of data difficult. Furthermore, practically all these methods consume a lot of time for sample preparation and precise measurements and have some limitations in the variation of experimental conditions. Application of SEC allows researchers to overcome many of these experimental difficulties. For example, the protein concentration can be decreased up to 0.001 mg/mL by using the 226-nm filter in the optical registration system or even to the nanogram level if the radiolabeled proteins are studied.

In comparison with the classical hydrodynamic methods such as viscometry and sedimentation, the use of SEC as a technique for the macromolecular dimension evaluation is a novel approach. In 1959, it was recognized that the SEC-based fractionation of macromolecules is determined by their molecular sizes (29). This brought the molecular sieve hypothesis of the gel-forming polymer action to the existence. SEC now is considered as a general separation technique where size and shape of molecules are the prime separation parameters (28). Therefore, the elution behavior of proteins on the SEC column is determined by their Stokes radii rather than by molecular masses (2, 3, 5, 8, 11, 21, 22, 30, 41, 45).

Originally, a set of SEC applications was rather limited because of the poor chemical and physical properties of early column matrixes that possessed high resistance to flow and were mechanically and chemically unstable. This unfavorable situation dramatically changed by the middle of the 1970s due to the elaboration of the new column packing materials (silanized silica) and the introduction of the high-pressure liquid chromatography (HPLC) concept. These inventions really revolutionized the field, and by the 1980s, HPLC became one of the most widely utilized techniques for compound separation.

Currently, SEC-HPLC fast protein liquid chromatography (FPLC) is commonly used as a convenient tool for the estimation of molecular dimensions and analysis of their changes under a variety of conditions. For example, the processes of protein denaturation and unfolding are often analyzed by SEC either in terms of changes in the retention time, which correlate with the changes of Stokes radii of the protein conformers (6–9, 20, 41, 45), or by following the appearance of a new elution peak (7–9, 14, 41, 45, 54–57). Hydrodynamic dimensions of IDPs are also studied by SEC (34, 49, 51–53). This chapter describes the peculiarities of the SEC application for evaluation of protein hydrodynamic dimensions, for conformational classification of IDPs, for separation of different conformational states of a protein by their dimensions, and for the independent structural characterization of these separated conformers.

18.2 ESTIMATION OF THE HYDRODYNAMIC DIMENSIONS BY SEC

18.2.1 Calibration of the SEC Column

Each of the numerous different chromatographic materials suitable for gel filtration is characterized by an "exclusion limit." This parameter defines an approximate upper limit for the size of molecules that can be separated using a given column matrix. Gel-filtration columns are characterized by two parameters: the void volume (V_O) and the total volume (V_T). V_O is essentially the volume of the mobile phase between the beads of the chromatographic medium. Molecules larger than the exclusion limit, that is, excluded molecules, elute in the V_O. V_T is the volume of all of the liquid within the column (i.e., both within the porous beads, as well as between them). The smallest, or included, molecules appear in the V_T.

Calibration of a gel-filtration column represents a crucial primary step in obtaining the quantitative information on the protein molecular dimensions by SEC. Column calibration implies the determination of a correlation between the parameters characterizing the column permeation properties (or the retention capability) and the protein hydrodynamic dimensions. In SEC, the retention of solute molecules by the column depends on their continuous exchange between the mobile phase and the stagnant mobile phase within the pores of the column matrix. This exchange is an equilibrium entropy-controlled process, as enthalpic processes such as adsorption are undesirable in SEC. Thus, the SEC retention volume (V_r) is expressed by the following equation (25):

$$V_r = V_O + V_P K_{SEC} + V_S K_{LC}, \tag{18.1}$$

where V_O is the void volume and V_P is the pore volume, whereas V_S is the stationary phase volume, K_{SEC} corresponds to the SEC solute distribution coefficient, and K_{LC} is the coefficient characterizing the liquid chromatography solute distribution. As it has been already mentioned, the ideal SEC retention has to be governed only by entropic contributions (i.e., it has to exclude both specific and nonspecific interactions of solute molecules with the column matrix). That is why the column packing material/eluent combination has to be chosen such that K_{LC} is minimized or ideally is equal to zero (25).

The value of K_{SEC} for peaks eluting in the region resolvable by the SEC column is $0 < K_{SEC} < 1$. Note, as the largest species is entirely excluded from the pores in the column matrix, its retention volume is equal to the void volume, in which case K_{SEC} is zero. On the other hand, the smallest molecule permeates all of the pores within the SEC column, and its retention volume equals the sum of the void volume and the pore volume, that is, the total volume or total permeation limit (V_T). The value of K_{SEC} for species eluting at the total volume is 1. Finally, the intermediate-size molecules can permeate the pores to some extent and thus, can be separated according to their respective hydrodynamic volumes. The retention of a given molecule by a SEC column can be described by the column partition coefficient, K_d, which is determined from the elution profiles by the following equation:

$$K_d = \frac{V_x - V_O}{V_T - V_O}, \tag{18.2}$$

where V_O and V_T are void and total solvent-accessible column volumes, respectively, whereas V_x is the elution (or retention) volume of a given molecule under given conditions.

The dependence of the retention volume of a solute on its hydrodynamic dimension represents the SEC calibration curve. A starting point of column calibration is an injection of a series of well-characterized SEC standards, proteins with known hydrodynamic dimensions, followed by the determination of the corresponding retention volumes. This information is then used for the conversion of the retention volume axis in SEC to a hydrodynamic dimension axis (i.e., calibration), which can be accomplished in a number of ways. Ackers developed one of the first calibration approaches assuming that the distribution of column matrix pore size followed the Gaussian law (2, 3). This assumption has led to the prediction that there is a linear correlation between the molecule Stokes radius (R_S) and the inverse error function complement of the column partition coefficient K_d (2, 3). The linearity of $\mathrm{erf}^{-1}(1 - K_d)$ versus R_S dependence was observed for several different column matrixes (16, 26). However, in many cases, this dependence was shown to be nonlinear (8, 21–23), suggesting that the assumption on the Gaussian-like distribution of the pore sizes is not generally correct. Based on these observations, it has been concluded that the use of a complex $\mathrm{erf}^{-1}(1 - K_d)$ versus R_S dependence did not provide obvious advantages over a direct R_S versus K_d plot (22, 40). Finally, it has been shown that when the column permeation properties are independent of the experimental conditions (i.e., when $V_T - V_O = const$ for all conditions and buffers used), the simplified calibration procedure, plotting the migration rate ($1000/V_{el}$) versus R_S can be used (8).

The accuracy of the R_S estimations by SEC depends significantly on the number of proteins used for the column calibration. Furthermore, if the determination of hydrodynamic dimensions for denatured and unfolded proteins is planned, then a set of denatured and unfolded proteins with known R_S values should be used for column calibration (41). This is also applicable for the evaluation of dimensions of IDPs. However, in the case of IDPs, a set of globular proteins with known hydrodynamic dimensions in various denatured and unfolded conformations can be used for calibration.

18.2.2 Measuring Hydrodynamic Dimensions of Proteins by SEC

Thorough SEC-based analysis of several proteins, whose hydrodynamic dimensions in different conformational states were estimated by other hydrodynamic techniques, revealed that the SEC-determined R_S values were in good agreement with those obtained by traditional hydrodynamic methods such as viscometry, sedimentation, and dynamic light scattering (8, 41, 45). The accuracy of SEC measurements was high enough to obtain the reliable information on the hydrodynamic dimensions of a protein in different conformational states. In fact, even molten globules, whose hydrodynamic dimensions are very close to the respective values of the globular ordered proteins, were reliably discriminated form the corresponding folded proteins (41, 45).

SEC was shown to represent a very useful tool to follow changes in the hydrodynamic dimensions accompanying denaturation and unfolding of globular proteins (7, 8, 41, 45). Importantly, it has been established that the unfolding curve retrieved for a given protein by SEC coincides with the unfolding curves measured for this protein by other techniques. This clearly indicated that the reliable R_S measurements can be done not only under the conditions preceding and following the conformational transition, but also within the transition region (7, 8, 41, 45). In general, SEC is an "inert" technique, which does not shift the equilibrium between the conformers and, therefore, can be used for a quantitative study of processes involved protein conformational rearrangements (7, 8, 41, 45).

18.2.3 Evaluating Hydrodynamic Dimensions of Proteins in Different Conformational States by SEC

Molecular density, and hence hydrodynamic dimensions, is one of the most unambiguous characteristics of a polymer molecule. Additional knowledge can be gained via the analysis of the molecular mass dependence of the molecular density for a polymer in different conformational states. In fact, the density of a globule is expected to be independent of the chain length, whereas the density of a partially collapsed or swelled macromolecules depends on both the chain length, and therefore on its molecular weight M, and on the nonspecific interactions of the monomer units with the solvent (13). Keeping this in mind, data retrieved by SEC for several proteins in different conformational states can be utilized for finding a potential correlation between the hydrodynamic dimensions of a protein molecule in a variety of conformational states and the length of the polypeptide chain (1, 37–39, 43, 44, 46). The analyzed proteins were grouped in the following classes: native globular proteins with nearly spherical shapes; equilibrium molten globules and equilibrium premolten globule states in the presence of strong denaturants; denaturant-unfolded proteins without cross-links; and natively unfolded proteins. Figure 18.2 represents the results of this analysis and clearly shows that in all cases studied, an excellent correlation between the apparent molecular density (determined as $\rho = M/(4\pi R_S^3/3)$, where M is a molecular mass and R_S is a hydrodynamic radius of a given protein) and molecular mass was observed.

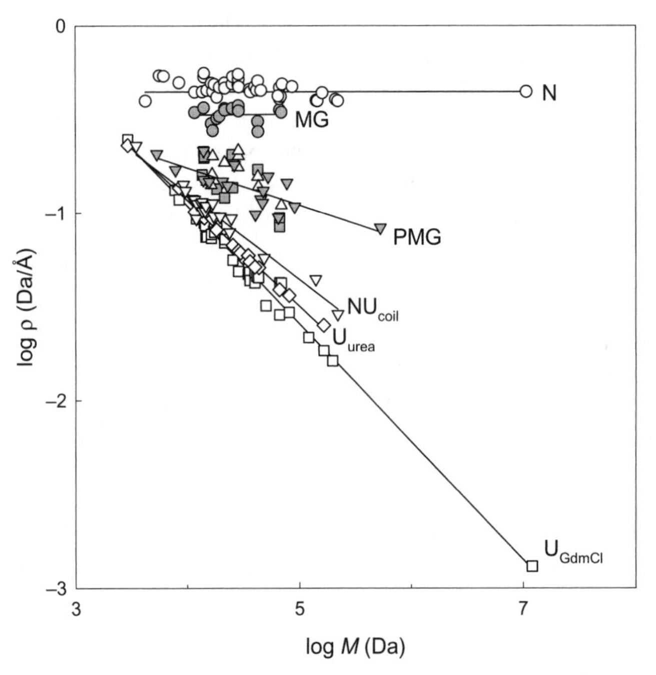

Figure 18.2 Variation of the density of protein molecules, ρ, with protein molecular weight, M, for a number of thermodynamically stable conformational states: N, ordered globular protein; MG, molten globule; PMG, partially folded and partially collapsed conformations (native premolten globules are shown as reversed triangles; proteins with intact disulfate bridges in 8M urea or 6M GdmHCl are shown as squares; intermediates accumulated during the unfolding by urea or GdmCl are shown as circles); NU_{coil}, native coil-like proteins under the physiological conditions; U_{urea}, unfolded in 8M urea (proteins without cross-links or with reduced cross-links); and U_{GdmCl}, unfolded in 6M GdmHCl (proteins without cross-links or with reduced cross-links). Lines represent the best fits. Modified from Reference 39.

Thus, regardless of the differences in the amino acid sequences and biologic functions, protein molecules behave as polymer homologues in a number of conformational states (1, 37–39, 43, 44, 46).

This analysis gave rise to a set of the standard equations for a polypeptide chain in a number of conformational states (46):

$$\log\left(R_S^N\right) = -(0.204 \pm 0.023) + (0.357 \pm 0.005) \cdot \log(M), \quad (18.3)$$

$$\log\left(R_S^{MG}\right) = -(0.053 \pm 0.094) + (0.334 \pm 0.021) \cdot \log(M), \quad (18.4)$$

$$\log\left(R_S^{PMG}\right) = -(0.21 \pm 0.18) + (0.392 \pm 0.041) \cdot \log(M), \quad (18.5)$$

$$\log\left(R_S^{U(urea)}\right) = -(0.649 \pm 0.016) + (0.521 \pm 0.004) \cdot \log(M), \quad (18.6)$$

$$\log\left(R_S^{U(GdmCl)}\right) = -(0.723 \pm 0.033) + (0.543 \pm 0.007) \cdot \log(M), \quad (18.7)$$

$$\log\left(R_S^{NU(coil)}\right) = -(0.551 \pm 0.032) + (0.493 \pm 0.008) \cdot \log(M), \quad (18.8)$$

$$\log\left(R_S^{NU(PMG)}\right) = -(0.239 \pm 0.055) + (0.403 \pm 0.012) \cdot \log(M), \quad (18.9)$$

where N, MG, PMG, U(urea), and U(GdmCl) correspond to the native, molten globule, premolten globule, urea-, and GdmCl-unfolded globular proteins, respectively, whereas NU(coil) and NU(PMG) correspond to native coil-like and native premolten globule-like proteins, respectively.

Importantly, statistical analysis has revealed that the relative errors of the recovered approximations exhibit random distribution over the wide range of chain lengths and do not generally exceed 10% (37). This means that the effective protein dimensions in a variety of conformational states can be predicted based on the chain length with an accuracy of 10%. In other words, this set of equations can be used to estimate the R_S value for any protein with known molecular mass M in any conformational state. Another important point is that having the R_S measured by SEC and knowing the molecular mass of the protein, one can understand what conformational state the studied protein is in under the given conditions.

18.3 CONFORMATIONAL CLASSIFICATION OF IDPS BY SEC

18.3.1 Evidence for Dual Personality of Natively Unfolded Proteins

The fact that the hydrodynamic data for natively unfolded proteins (i.e., IDPs, which, in their native states, are highly extended and do not have noticeable amounts of ordered secondary structure) could not be fit by one dependence but require two rather different curves (see Fig. 18.2) clearly shows that these proteins do not possess uniform structural properties, as expected for the members of a single thermodynamic entity (43). In fact, based on the data of SEC analysis, natively unfolded proteins were divided into two structurally different groups: intrinsic coils and premolten globules (43). Proteins from the first group had hydrodynamic dimensions typical of random coils in poor solvent and did not possess any (or almost any) ordered secondary structure. Proteins from the second group were essentially more compact, exhibiting some amount of residual secondary structure, although they were still less densely pact than ordered globular proteins or molten globule proteins (43).

The validity of this assumption was supported by comparing the results of the hydrodynamic studies with the data on secondary structure evaluation. Earlier, the analysis of data on the conformational characteristics of 41 globular proteins in native and partially folded conformational states revealed that for a polypeptide chain, a good correlation exists between relative decrease in hydrodynamic volume and increase in secondary structure content (47). The results of this analysis are presented in Figure 18.3A as R_S^U/R_S (relative compactness) versus $[\theta]^{222}/[\theta]_{222}^{U}$ (relative content of ordered secondary structure). In this plot, open and closed circles correspond to the data for native globular proteins and their partially folded intermediates, respectively. Data for both classes of conformations (native states and various intermediate states) were described by the expression (correlation coefficient $r^2 = 0.97$) (47):

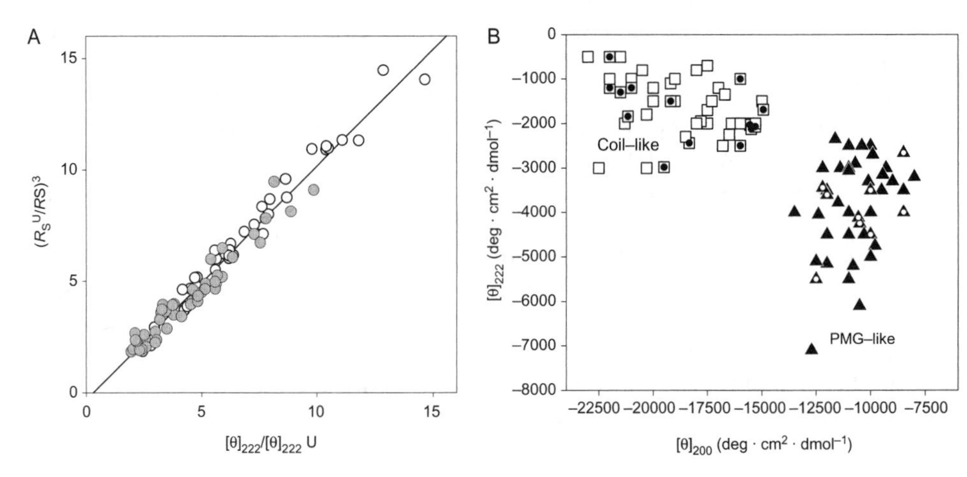

Figure 18.3 (A) Correlation between the degree of compactness (calculated for different conformational states as the decrease in hydrodynamic volume relative to the volume of the unfolded conformation) and amount of ordered secondary structure (calculated for different conformational states from their far-UV CD spectra as the increase in negative ellipticity at 222 nm, $[\theta]_{222}$, relative to that of the unfolded conformation). Open and gray symbols correspond to the data for native globular proteins and their partially folded intermediates, respectively. (B) Analysis of far-UV CD spectra in terms of double wavelength plot, $[\theta]_{222}$ versus $[\theta]_{200}$, allows the natively unfolded proteins division on native coils (open squares) and native premolten globules (black triangles). Native premolten globules and native coils with known hydrodynamic parameters are marked by white-dotted and black-dotted symbols, respectively.

$$\left(\frac{R_S^U}{R_S}\right)^3 = (1.047 \pm 0.010) \cdot \frac{[\theta]_{222}}{[\theta]_{222}^U} - (0.31 \pm 0.12) \tag{18.10}$$

This meant that the degree of compactness and the amount of ordered secondary structure were highly correlated. In other words, no compact intermediates lacking secondary structure, or highly ordered noncompact species, were found.

Unfolded polypeptide chains are characterized by very specific shapes of their far-UV circular dichroism (CD) spectrum, with an intensive minimum in the vicinity of 200 nm and an ellipticity close to zero in the vicinity of 222 nm (4, 10, 18, 19, 31, 34, 42, 58). This is a very useful spectroscopic criterion for the selection of natively unfolded proteins (48). More information on the peculiarities of the IDP analysis by CD can be found in chapter 10 by Robert Woody.

A coil-like shape of far-UV CD spectrum has been reported for ~100 proteins, which is almost threefold larger than the number of proteins shown to be unfolded by hydrodynamic techniques (43). Figure 18.3B represents a "double wavelength" plot, $[\theta]_{222}$ versus $[\theta]_{200}$, that assorted natively unfolded proteins into two nonoverlapping groups (43). Fifty-one proteins were characterized by far-UV CD spectra characteristic of almost completely unfolded polypeptide chains: with $[\theta]_{200} = -(18,900 \pm 2800)$ deg·cm^2·dmol^{-1} and $[\theta]_{222} = -(1700 \pm 700)$ deg·cm^2·dmol^{-1}. On the other hand, 44 other protein spectra were consistent with the existence of some residual secondary structure, possessing a shape typical of the premolten globule state of globular proteins (with $[\theta]_{200} = -(10,700 \pm 1300)$ deg·cm^2·dmol^{-1} and $[\theta]_{222} = -(3900 \pm 1100)$ deg·cm^2·dmol^{-1}).

The difference in the shape of far-UV CD spectra along does not allow the unambiguous discrimination between the two conformations. However, among more than 100 reported cases, 23 proteins were simultaneously characterized by CD and hydrodynamic methods, making classification more certain. Intrinsic premolten globules and intrinsic coils studied by both techniques are indicated in Figure 18.3B as white-dotted and black-dotted symbols, respectively. These data were consistent with the important conclusions that more compact polypeptides (with PMG-like hydrodynamic characteristics) possess larger amounts of ordered secondary structure than less compact coil-like natively unfolded proteins. Thus, the simultaneous application of CD and hydrodynamic techniques left no doubts that natively unfolded proteins should be subdivided into two structurally distinct groups: intrinsic coils and intrinsic premolten globules.

18.3.2 Knowing IDP Conformation from SEC Measurements

As it follows from the discussion above, SEC is very useful in ascertaining the degree of compactness of a protein and can distinguish between partially and

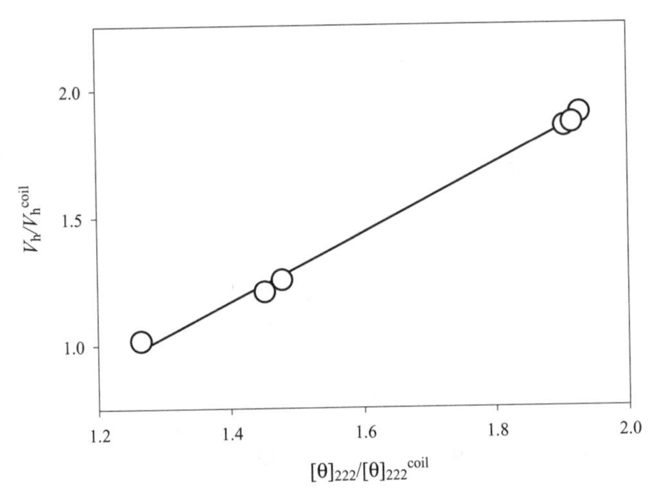

Figure 18.4 Comparison of the degree of compactness, measured as the ratio V_h^{coil}/V_h, where V_h^{coil} and V_h correspond to hydrodynamic volume calculated for random coil of given molecular mass and the measured hydrodynamic volume, respectively, determined for α-, β-, and γ-synucleins under different experimental conditions with the amount of ordered secondary structure determined as $[\theta]_{222}/[\theta]_{222}^{coil}$.

fully unfolded states, since an increase in the hydrodynamic volume is associated with unfolding. Transformation of a typical globular protein into a molten globule state results in a ~15–20% increase in its hydrodynamic radius (32, 33, 39, 41, 44, 45). The relative increase in hydrodynamic volume of less folded intermediates is even larger (33, 37, 39, 44, 54, 55). Figure 18.2 shows that the folded and unfolded conformations of globular proteins possess very different molecular mass dependencies of their hydrodynamic radii, R_S (37–39, 43, 44). Therefore, equilibrium conformations of globular proteins (native, molten globule, unfolded states, etc.) can easily be discriminated by the degree of compactness of the polypeptide chain. Figure 18.3 illustrates that SEC can also be used to differentiate between differently collapsed forms of IDPs.

The hydrodynamic properties of the members of the synuclein family, α-, β-, and γ-synucleins, were studied under the variety of experimental conditions by SEC and SAXS (52). Figure 18.4 illustrates that the results of the chromatographic analysis agreed well with the far-UV CD data and shows that under the conditions of neutral pH, β-synuclein was slightly more extended than α- and γ-synucleins, whereas in acidic solutions all three proteins possessed the same degree of compaction (52).

In fact, at neutral pH, the hydrodynamic dimensions of β-synuclein were typical of a completely unfolded polypeptide chain, whereas α- and γ-synucleins were more compact than expected for a random coil. This conclusion followed from the comparison of measured Stokes radius values with those calculated for a completely unfolded polypeptide chain of the appropriate molecular mass. In the case of β-synuclein, the experimentally determined value (33.9 ± 0.4 Å) perfectly matched the calculated one (34.1 Å), but the Stokes radii measured for α- (31.8 ± 0.4 Å) and γ-synucleins (30.4 ± 0.4 Å) were notably lower than the corresponding calculated values (34.3 and 32.8 Å, respectively) (51). The conclusion on the partially collapsed form in α-synuclein was further confirmed by the measurement of its R_S in the presence of 8 M urea, where the protein behaved as a random coil (R_S = 34.5 ± 0.4 Å) (51).

Importantly, SEC analysis revealed that the decrease in pH was accompanied by the formation of partially folded conformation in all three synucleins as evidenced by substantial decrease in their hydrodynamic dimensions (R_S = 27.9 ± 0.4, 27.5 ± 0.4, and 26.5 ± 0.4 Å for α-, β-, and γ-synuclein, respectively). As these data were in perfect agreement with the values calculated from Equation 18.9, it has been concluded that at acidic pH, all three proteins formed premolten globule-like conformation (52).

Figure 18.4 compares hydrodynamic volumes measured for synucleins under different experimental conditions with the results of far-UV CD analysis. Comparison of the degree of compactness (measured as the ratio V_h^{coil}/V_h, where V_h^{coil} and V_h correspond to hydrodynamic volume calculated for random coil of given molecular mass and measured hydrodynamic volume, respectively) with the amount of secondary structure (determined as $[\theta]_{222}/[\theta]_{222}^{coil}$) shows an excellent correlation. These data illustrated that under physiological conditions, β-synuclein was more unfolded than the α- and γ-synucleins, but at acidic pH, all three proteins were in the premolten globular form (52).

SEC analysis of another IDP, inhibitory γ subunit of the cGMP phosphodiesterase (PDEγ) revealed that at neutral pH in the absence and in the presence of 8 M urea the hydrodynamic dimensions of PDEγ at neutral pH were close to those measured in the presence of 8 M urea (R_S = 24.8 ± 0.8 and 29.1 ± 0.8 Å, respectively) (53). Both values were very close to R_S calculated for a completely unfolded protein in 8 M urea with a molecular mass of 9669 Da (R_S = 26.7 Å) or to a native coil with this molecular mass (R_S = 25.9 Å). Based on these observations, it has been concluded that under conditions studied PDEγ *was* essentially unfolded even in the absence of a denaturant (53). In fact, this close proximity of SEC and calculated values showed that PDEγ is a native coil.

Results of the chromatographic studies on the C-terminal domain of chicken gizzard caldesmon (CaD136, residues 636–771) agreed well with the data of far-UV CD, SAXS, and intrinsic fluorescence, and showed that this domain was essentially unfolded under the conditions of neutral pH (27). In fact, the hydrodynamic dimensions of CaD136 were relatively close to those measured in the presence of 6 M GdmCl (R_S = 28.1 ± 0.8 and 35.3 ± 0.8 Å, respectively), confirming the fact that CaD136 is essentially unfolded even in the absence of a denaturant. Comparison of these measured values with R_S calculated using Equations 18.4, 18.7, 18.8, and 18.9 for a protein with a molecular mass of 14,514 Da (19.1, 21.7, 34.4, 31.7, and 27.4 Å for N, MG, U, NU(coil), and

NU(PMG), respectively) suggested that CaD136 belonged to the native premolten globule class (27).

18.4 ASSESSING INDUCED FOLDING AND ASSOCIATION OF IDPS BY SEC

18.4.1 Analysis of Induced Folding

The presented data above for the members of the synuclein family illustrate that SEC is a useful technique for the evaluation of the IDP partial folding induced by changes in the environment. Similar results on the pH-induced gaining of partially folded conformation were obtained for another typical IDP, prothymosin α (49). The hydrodynamic dimensions of this protein at neutral pH were close to those measured in the presence of 8 M urea (31.4 ± 0.3 and 32.8 ± 0.3 Å, respectively). The small difference between the two values was explained by swelling of the unfolded polypeptide chain in a good solvent. Both values were virtually indistinguishable from R_S, calculated for the completely unfolded protein with a molecular mass of 12.21 kDa (31.3 Å). A decrease in pH led to a pronounced decrease of the prothymosin α hydrodynamic dimension (R_S = 24.9 ± 0.3 Å) (49). It should be emphasized that this R_S value determined at pH 2.5 was still far from that expected for a globular protein of 12 kDa, but correlated well with the dimensions of the premolten globule (25.5 Å). Although we considered here only the data on the pH-induced partial folding of IDPs, SEC can be utilized for the analysis of folding induced in IDPs by any other environmental factors.

18.4.2 Studies of IDP Association and Association-Induced Folding

The situation when IDP partially folds while it oligomerizes is more complex, as here we are dealing with two opposite effects—decrease in the hydrodynamic dimensions induced by folding and increase in the molecular mass (and consequently hydrodynamic volume) caused by oligomerization. However, even in this case, very useful information can be extracted both on the polypeptide conformation and on its oligomerization state. For a protein with a molecular mass of M, this information can be retrieved using Equations 18.3, 18.4, and 18.7–18.9 utilizing a set of molecular masses $n \times M$, with $n = 1, 2, 3, \ldots, N$ being the oligomerization state.

Analysis of the heat-induced dimerization of α-synuclein represents an illustrative example of this approach. Incubation of α-synuclein at high temperature induced partially folded conformation (51). This structural transformation was completely reversible, when the heat treatment was transient. It has been hypothesized that if the partially folded conformation serves as an intermediate of the fibril assembly, then populating this structure for a longer period of time should induce the self-assembly and might trap the structure in oligomeric forms (50). To test this hypothesis and see if the sustained heat treatment, thereby sustained partially folded structure, can stabilize the structure and initiate the oligomerization, purified wild-type human recombinant α-synuclein was incubated at different temperatures for up to 3 days. This incubation resulted in a temperature-dependent, progressive aggregation. The incubation at 65°C showed small oligomers (mostly dimers) at day 1, and at day 3, larger aggregates were detected with increased amount of small oligomers (dimers). Whereas the oligomerization at 50°C was slower, but apparent at day 3, no oligomers of any size were detectable at 37°C or room temperature for up to 3 days (50).

At the next stage, hydrodynamic properties of different associated forms of α-synuclein have been analyzed by GPC (50). This analysis revealed that the initial conformation of α-synuclein was essentially unfolded polypeptide chain with the Stokes radius R_S = 31.3 Å, whereas trapped dimeric form, being characterized by R_S = 36.3 Å, had to be comprised of more compact protein molecules. These conclusions followed from the comparison of the measured Stokes radius, R_S, values with those calculated for native coil or native premolten globule with a molecular mass of 14,460 kDa. In the case of initial α-synuclein conformation, the experimentally determined value perfectly matched the value calculated for the native coil of 14,460 Da (R_S = 31.6 Å). R_S measured for the trapped conformation coincided with expected dimensions of the premolten globule protein with molecular mass of 28,920 Da (R_S = 36.2 Å). In other words, incubation of α-synuclein at elevated temperatures for prolonged periods of time may induce formation and trapping the stable dimers comprised of partially folded premolten globule-like intermediates (50). In other words, partially folded premolten globule-like conformation of α-synuclein seems to be stabilized as the protein undergoes a highly selective self-assembly process during prolonged incubation at elevated temperatures.

18.5 SEC-BASED PHYSICAL SEPARATION OF IDP CONFORMERS AND THEIR INDEPENDENT ANALYSIS

An exceptional advantage of SEC, in comparison with the vast majority of traditional structural methods, is its capability of physical separation of protein conformers, which are different in their hydrodynamic dimensions. Although such separation takes place only under particular conditions (e.g., under the conditions that are favoring slow conformational exchange between these species or upon the formation of stable oligomeric forms), this property of SEC allows one to perform an independent investigation of different physical properties of compact and less compact or monomeric and oligomeric conformations. Various traditional spectroscopic techniques, being combined with the chromatographic facilities, can be used for such structural characterization.

This property of SEC was successfully applied for studying the formation of baicalein-stabilized oligomers of α-synuclein. Baicalein is the main component of a traditional Chinese herbal medicine *Scutellaria baicalensis* and has multiple biologic activities including antiallergic, anticarcinogenic, and anti-HIV properties (12, 17, 24, 35). Furthermore, baicalein was shown to markedly inhibit α-synuclein fibrillation *in vitro* (15, 59). This inhibition occurred via inducing the specific oligomerization. This ability of baicalein to effectively induce oligomerization of α-synuclein was shown using SEC-HPLC. After incubation for 2 days with 100 μM baicalein, the HPLC profile of α-synuclein showed a new peak with a retention time of 11.5 min, indicating formation of the stable oligomeric species (15). The peak corresponding to the monomeric protein was also observed in the elution profile. Purified samples eluting from the HPLC column were monitored by UV spectroscopy to confirm the baicalein binding (15). The baicalein has three characteristic maxima in the absorption spectrum, at 216, 277, and 324 nm. Zhu et al. (59) showed that when the baicalein was oxidized, the absorbance at 324 nm disappeared, whereas when it was bound to α-synuclein, a new peak at ~360 nm was observed. The UV absorption spectrum of α-synuclein oligomer showed an absorbance at around 360 nm, suggesting the effective baicalein binding. Interestingly, the peak in the HPLC profile corresponding to the monomeric α-synuclein coincubated with baicalein also showed an absorbance at 360 nm, indicating baicalein binding (15).

These two samples separated by SEC were used for the detailed biophysical analysis, including atomic force and electron microscopy, SAXS, FTIR, and far-UV CD (15). Furthermore, thermodynamic stability of the baicalein-stabilized oligomers was evaluated via the analysis of their GdmCl-induced unfolding (15). The purified baicalein-stabilized oligomers were incubated at 37°C for 1 month. No fibrils were formed, and no dissociation was observed after this prolonged incubation, suggesting the high stability of the oligomers. Inhibitory effects of these oligomers on α-synuclein fibrillation were also evaluated. Finally, the effect of the baicalein-stabilized oligomers on the integrity of lipid membranes was evaluated (15). All these very important studies became possible due to the ability of SEC to physically separate monomeric and oligomeric α-synucelin species.

18.6 CONCLUSIONS

The major advantages of SEC application for the analysis of IDPs and induced folding are summarized below:

1. The use of SEC with its capability to use very low protein concentrations might exclude the experimental artifact associated with protein association and aggregation.

2. Highly reliable and accurate data on the hydrodynamic dimensions of a given protein under the variety of conditions as well as information on their changes can be retrieved within a short time period.
3. These data can be used to understand the conformational state of a given IDP.
4. SEC can be utilized to evaluate the association state of an IDP and its conformation within the oligomer.
5. SEC allows real physical separation of IDP conformers with different hydrodynamic dimensions (or different oligomeric species). This opens the unique possibility for the independent structural analysis of these conformers/oligomers preseparated by gel-filtration column.

ACKNOWLEDGMENTS

This work was supported in part by the Program of the Russian Academy of Sciences for the "Molecular and cellular biology," and by grants R01 LM007688-01A1 and GM071714-01A2 from the National Institutes of Health. We gratefully acknowledge the support of the IUPUI Signature Centers Initiative.

REFERENCES

1. Abramov, V. M., A. M. Vasiliev, V. S. Khlebnikov, R. N. Vasilenko, N. L. Kulikova, I. V. Kosarev, A. T. Ishchenko, J. R. Gillespie, I. S. Millett, A. L. Fink, and V. N. Uversky. 2002. Structural and functional properties of *Yersinia pestis* Caf1 capsular antigen and their possible role in fulminant development of primary pneumonic plague. J Proteome Res **1**:307–15.
2. Ackers, G. K. 1970. Analytical gel chromatography of proteins. Adv Protein Chem **24**:343–446.
3. Ackers, G. K. 1967. Molecular sieve studies of interacting protein systems. I. Equations for transport of associating systems. J Biol Chem **242**:3026–34.
4. Adler, A. J., N. J. Greenfield, and G. D. Fasman. 1973. Circular dichroism and optical rotatory dispersion of proteins and polypeptides. Methods Enzymol **27**:675–735.
5. Andrews, P. 1965. The gel-filtration behaviour of proteins related to their molecular weights over a wide range. Biochem J **96**:595–606.
6. Brems, D. N., S. M. Plaisted, H. A. Havel, E. W. Kauffman, J. D. Stodola, L. C. Eaton, and R. D. White. 1985. Equilibrium denaturation of pituitary- and recombinant-derived bovine growth hormone. Biochemistry **24**:7662–8.
7. Corbett, R. J., and R. S. Roche. 1983. The unfolding mechanism of thermolysin. Biopolymers **22**:101–5.
8. Corbett, R. J., and R. S. Roche. 1984. Use of high-speed size-exclusion chromatography for the study of protein folding and stability. Biochemistry **23**:1888–94.

9. Endo, S., Y. Saito, and A. Wada. 1983. Denaturant-gradient chromatography for the study of protein denaturation: principle and procedure. Anal Biochem **131**:108–20.
10. Fasman, G. D. 1996. Circular Dichroism and Conformational Analysis of Biomolecules. Plenum Press, New York.
11. Fish, W. W., J. A. Reynolds, and C. Tanford. 1970. Gel chromatography of proteins in denaturing solvents. Comparison between sodium dodecyl sulfate and guanidine hydrochloride as denaturants. J Biol Chem **245**:5166–8.
12. Gao, Z., K. Huang, and H. Xu. 2001. Protective effects of flavonoids in the roots of *Scutellaria baicalensis* Georgi against hydrogen peroxide-induced oxidative stress in HS-SY5Y cells. Pharmacol Res **43**:173–8.
13. Grossberg, A. Y., and A. R. Khohlov. 1989. Statistical Physics of Macromolecules. Nauka, Moscow.
14. Gupta, B. B. 1983. Determination of native and denatured milk proteins by high-performance size exclusion chromatography. J Chromatogr **282**:463–75.
15. Hong, D. P., A. L. Fink, and V. N. Uversky. 2008. Structural characteristics of alpha-synuclein oligomers stabilized by the flavonoid baicalein. J Mol Biol **383**:214–23.
16. Horiike, K., H. Tojo, T. Yamano, and M. Nozaki. 1983. Interpretation of the Stokes radius of macromolecules determined by gel filtration chromatography. J Biochem **93**:99–106.
17. Ikezoe, T., S. S. Chen, D. Heber, H. Taguchi, and H. P. Koeffler. 2001. Baicalin is a major component of PC-SPES which inhibits the proliferation of human cancer cells via apoptosis and cell cycle arrest. Prostate **49**:285–92.
18. Johnson, W. C., Jr. 1988. Secondary structure of proteins through circular dichroism spectroscopy. Annu Rev Biophys Chem **17**:145–66.
19. Kelly, S. M., and N. C. Price. 1997. The application of circular dichroism to studies of protein folding and unfolding. Biochim Biophys Acta **1338**:161–85.
20. Lau, S. Y., A. K. Taneja, and R. S. Hodges. 1984. Synthesis of a model protein of defined secondary and quaternary structure. Effect of chain length on the stabilization and formation of two-stranded alpha-helical coiled-coils. J Biol Chem **259**:13253–61.
21. Le Maire, M., L. P. Aggerbeck, C. Monteilhet, J. P. Andersen, and J. V. Moller. 1986. The use of high-performance liquid chromatography for the determination of size and molecular weight of proteins: a caution and a list of membrane proteins suitable as standards. Anal Biochem **154**:525–35.
22. le Maire, M., A. Ghazi, J. V. Moller, and L. P. Aggerbeck. 1987. The use of gel chromatography for the determination of sizes and relative molecular masses of proteins. Interpretation of calibration curves in terms of gel-pore-size distribution. Biochem J **243**:399–404.
23. le Maire, M., E. Rivas, and J. V. Moller. 1980. Use of gel chromatography for determination of size and molecular weight of proteins: further caution. Anal Biochem **106**:12–21.
24. Li, B. Q., T. Fu, W. H. Gong, N. Dunlop, H. Kung, Y. Yan, J. Kang, and J. M. Wang. 2000. The flavonoid baicalin exhibits anti-inflammatory activity by binding to chemokines. Immunopharmacology **49**:295–306.

25. Meunier, D. M. 1997. Molecular weight determinations, pp. 853–66. In F. Settle (ed.), Handbook of Instrumental Techniques for Analytical Chemistry. Prentice Hall, Upper Saddle River, NJ.
26. Nozaki, Y., N. M. Schechter, J. A. Reynolds, and C. Tanford. 1976. Use of gel chromatography for the determination of the Stokes radii of proteins in the presence and absence of detergents. A reexamination. Biochemistry **15**:3884–90.
27. Permyakov, S. E., I. S. Millett, S. Doniach, E. A. Permyakov, and V. N. Uversky. 2003. Natively unfolded C-terminal domain of caldesmon remains substantially unstructured after the effective binding to calmodulin. Proteins **53**:855–62.
28. Porath, J. 1968. Molecular sieving and adsorption. Nature **218**:834–8.
29. Porath, J., and P. Flodin. 1959. Gel filtration: a method for desalting and group separation. Nature **183**:1657–9.
30. Potschka, M. 1987. Universal calibration of gel permeation chromatography and determination of molecular shape in solution. Anal Biochem **162**:47–64.
31. Provencher, S. W., and J. Glockner. 1981. Estimation of globular protein secondary structure from circular dichroism. Biochemistry **20**:33–7.
32. Ptitsyn, O. B. 1995. Molten globule and protein folding. Adv Protein Chem **47**:83–229.
33. Ptitsyn, O. B., V. E. Bychkova, and V. N. Uversky. 1995. Kinetic and equilibrium folding intermediates. Philos Trans R Soc Lond B Biol Sci **348**:35–41.
34. Receveur-Brechot, V., J. M. Bourhis, V. N. Uversky, B. Canard, and S. Longhi. 2006. Assessing protein disorder and induced folding. Proteins **62**:24–45.
35. Shieh, D. E., L. T. Liu, and C. C. Lin. 2000. Antioxidant and free radical scavenging effects of baicalein, baicalin and wogonin. Anticancer Res **20**:2861–5.
36. Tanford, C. 1968. Protein denaturation. Adv Protein Chem **23**:121–282.
37. Tcherkasskaya, O., E. A. Davidson, and V. N. Uversky. 2003. Biophysical constraints for protein structure prediction. J Proteome Res **2**:37–42.
38. Tcherkasskaya, O., and V. N. Uversky. 2001. Denatured collapsed states in protein folding: example of apomyoglobin. Proteins **44**:244–54.
39. Tcherkasskaya, O., and V. N. Uversky. 2003. Polymeric aspects of protein folding: a brief overview. Protein Pept Lett **10**:239–45.
40. Ui, N. 1979. Rapid estimation of the molecular weights of protein polypeptide chains using high-pressure liquid chromatography in 6 M guanidine hydrochloride. Anal Biochem **97**:65–71.
41. Uversky, V. N. 1994. Gel-permeation chromatography as a unique instrument for quantitative and qualitative analysis of protein denaturation and unfolding. Int J Bio-Chromatogr **1**:103–14.
42. Uversky, V. N. 1999. A multiparametric approach to studies of self-organization of globular proteins. Biochemistry (Mosc) **64**:250–66.
43. Uversky, V. N. 2002. Natively unfolded proteins: a point where biology waits for physics. Protein Sci **11**:739–56.
44. Uversky, V. N. 2003. Protein folding revisited. A polypeptide chain at the folding-misfolding-nonfolding cross-roads: which way to go? Cell Mol Life Sci **60**:1852–71.

45. Uversky, V. N. 1993. Use of fast protein size-exclusion liquid chromatography to study the unfolding of proteins which denature through the molten globule. Biochemistry **32**:13288–98.
46. Uversky, V. N. 2002. What does it mean to be natively unfolded? Eur J Biochem **269**:2–12.
47. Uversky, V. N., and A. L. Fink. 2002. The chicken-egg scenario of protein folding revisited. FEBS Lett **515**:79–83.
48. Uversky, V. N., J. R. Gillespie, and A. L. Fink. 2000. Why are "natively unfolded" proteins unstructured under physiologic conditions? Proteins **41**:415–27.
49. Uversky, V. N., J. R. Gillespie, I. S. Millett, A. V. Khodyakova, A. M. Vasiliev, T. V. Chernovskaya, R. N. Vasilenko, G. D. Kozlovskaya, D. A. Dolgikh, A. L. Fink, S. Doniach, and V. M. Abramov. 1999. Natively unfolded human prothymosin alpha adopts partially folded collapsed conformation at acidic pH. Biochemistry **38**:15009–16.
50. Uversky, V. N., H. J. Lee, J. Li, A. L. Fink, and S. J. Lee. 2001. Stabilization of partially folded conformation during alpha-synuclein oligomerization in both purified and cytosolic preparations. J Biol Chem **276**:43495–8.
51. Uversky, V. N., J. Li, and A. L. Fink. 2001. Evidence for a partially folded intermediate in alpha-synuclein fibril formation. J Biol Chem **276**:10737–44.
52. Uversky, V. N., J. Li, P. Souillac, I. S. Millett, S. Doniach, R. Jakes, M. Goedert, and A. L. Fink. 2002. Biophysical properties of the synucleins and their propensities to fibrillate: inhibition of alpha-synuclein assembly by beta- and gamma-synucleins. J Biol Chem **277**:11970–8.
53. Uversky, V. N., S. E. Permyakov, V. E. Zagranichny, I. L. Rodionov, A. L. Fink, A. M. Cherskaya, L. A. Wasserman, and E. A. Permyakov. 2002. Effect of zinc and temperature on the conformation of the gamma subunit of retinal phosphodiesterase: a natively unfolded protein. J Proteome Res **1**:149–59.
54. Uversky, V. N., and O. B. Ptitsyn. 1996. Further evidence on the equilibrium "pre-molten globule state": four-state guanidinium chloride-induced unfolding of carbonic anhydrase B at low temperature. J Mol Biol **255**:215–28.
55. Uversky, V. N., and O. B. Ptitsyn. 1994. "Partly folded" state, a new equilibrium state of protein molecules: four-state guanidinium chloride-induced unfolding of beta-lactamase at low temperature. Biochemistry **33**:2782–91.
56. Uversky, V. N., G. V. Semisotnov, R. H. Pain, and O. B. Ptitsyn. 1992. "All-or-none" mechanism of the molten globule unfolding. FEBS Lett **314**:89–92.
57. Withka, J., P. Moncuse, A. Baziotis, and R. Maskiewicz. 1987. Use of high-performance size-exclusion, ion-exchange, and hydrophobic interaction chromatography for the measurement of protein conformational change and stability. J Chromatogr **398**:175–202.
58. Woody, R. W. 1995. Circular dichroism. Methods Enzymol **246**:34–71.
59. Zhu, M., S. Rajamani, J. Kaylor, S. Han, F. Zhou, and A. L. Fink. 2004. The flavonoid baicalein inhibits fibrillation of alpha-synuclein and disaggregates existing fibrils. J Biol Chem **279**:26846–57.

PART V

CONFORMATIONAL STABILITY

19

CONFORMATIONAL BEHAVIOR OF INTRINSICALLY DISORDERED PROTEINS: EFFECTS OF STRONG DENATURANTS, TEMPERATURE, PH, COUNTERIONS, AND MACROMOLECULAR CROWDING

VLADIMIR N. UVERSKY

Institute for Intrinsically Disordered Protein Research, Center for Computational Biology and Bioinformatics, and Department of Biochemistry and Molecular Biology, Indiana University School of Medicine, Indianapolis, IN

Institute for Biological Instrumentation, Russian Academy of Sciences, Pushchino, Moscow Region, Russia

ABSTRACT

Intrinsically disordered proteins (IDPs) differ from "normal" ordered proteins at several levels: structural, functional, and conformational. Amino acid biases characteristic for IDPs (e.g., low content of hydrophobic amino acid residues combined with high relative content of charged and polar residues) determine their structural variability and lack of rigid well-folded structure. This structural plasticity is necessary for the unique functional repertoire of IDPs, which is complementary to the catalytic activities of ordered proteins. Amino acid biases also drive atypical responses of IDPs to changes in their environment. The conformational behavior of IDPs is characterized by the

Instrumental Analysis of Intrinsically Disordered Proteins: Assessing Structure and Conformation, Edited by Vladimir Uversky and Sonia Longhi

low cooperativity (or the complete lack thereof) of the denaturant-induced unfolding, lack of the measurable excess heat absorption peaks characteristic for the melting of ordered proteins, "turned out" response to heat and changes in pH, the ability to gain structure in the presence of various counterions, and the unique response to macromolecular crowding. This chapter describes some of the most characteristic features of the IDP conformational behavior.

19.1 INTRODUCTION: AMINO ACID CODE FOR INTRINSIC DISORDER

A class of biologically active proteins lacking ordered structure under physiological conditions is rapidly growing. These molecules, known as intrinsically disordered proteins (IDPs) among other names, exist as highly dynamic ensembles in which the atom positions and backbone Ramachandran angles vary significantly over time with no specific equilibrium values. They carry numerous vital biologic functions, thus challenging the century-old paradigm that a unique biologic function is a specific property of a unique polypeptide chain specifically folded into a unique 3D structure, that is, challenging a lock-and-key model proposed by Emil Fisher to describe the astonishing specificity of well-ordered enzymes. Both ordered and disordered proteins are polypeptides. Why then are IDPs disordered and flexible, contrasting the realm of well-folded globular proteins, each possessing unique and highly specific structure? To answer this question, we should consider ordered globular proteins first.

The unique 3D structure of a globular protein is stabilized by noncovalent interactions (conformational forces) of different natures. This includes hydrogen bonds, hydrophobic interactions, electrostatic interactions, van der Waals interactions, and so on. Complete (or almost complete) disruption of all these interactions can be achieved in concentrated solutions of strong denaturants (such as urea or guanidinium chloride [GdmCl]). Here, an initially folded protein molecule unfolds, being transformed into a highly disordered random coil-like conformation (2, 58, 61, 79). However, some environmental changes can decrease (or even completely eliminate) only a part of the conformational interactions, whereas the remaining interactions could stay unchanged (or could even be intensified). Very often, a protein will lose its biologic activity under these conditions, thus becoming denatured (79). It is important to remember that denaturation is not necessarily accompanied by the unfolding of a protein, but rather might result in the appearance of various partially folded conformations with properties intermediate between those of the folded (ordered) and the completely unfolded states.

Globular proteins are known to exist in at least four different equilibrium conformations: folded (ordered), molten globule, premolten globule, and unfolded (68, 70, 80, 84, 88, 92, 98, 99), and the ability of a globular protein to adopt different stable conformations is believed to be an intrinsic property of a polypeptide chain. The structural properties of a typical globular protein in the molten globule and the premolten globule conformations are outlined below.

The molten globular protein is denatured and therefore has no (or has only a trace of) rigid cooperatively melted tertiary structure. Small-angle X-ray scattering analysis reveals that it has a globular structure typical of folded globular proteins (27, 45, 46, 75, 92). The 2D NMR coupled with hydrogen–deuterium exchange shows that the molten globule is characterized not only by the native-like secondary structure content, but also by the native-like folding pattern (3, 7, 8, 10, 13, 28, 43, 104). A considerable increase in the accessibility of a protein molecule to proteases is noted as a specific property of the molten globule (32, 53). The transformation into this intermediate state is accompanied by a considerable increase in the affinity of a protein molecule to the hydrophobic fluorescence probes (such as 8-anilinonaphthalene-1-sulfonate [ANS]), and this behavior is a characteristic property of the molten globules (76, 100). Finally, on the average, the hydrodynamic radius of the molten globule is increased by no more than 15% compared with that of the folded state, which corresponds to the volume increase of ~50% (85).

The globular protein in the premolten globule state is also denatured. It has a considerable amount of residual secondary structure, which is much less pronounced than that of the native or the molten globule protein. The premolten globule form is considerably less compact than the molten globule or folded states, being still noticeably more compact than the random coil. The premolten globule can effectively interact with ANS, though essentially weaker than the molten globule. The premolten globule has no globular structure (84, 88, 92), suggesting that the premolten globule probably represents a "squeezed" and partially collapsed and partially ordered form of a coil. Finally, the premolten globule was shown to be separated from the molten globule by an all-or-none transition (68, 84, 98, 99), suggesting that the molten globule and the premolten globule are different thermodynamic (phase) states of a globular protein.

By analogy with the above-mentioned partially folded conformations of ordered globular proteins, IDPs are grouped into three structurally different subclasses: native molten globules (so-called collapsed IDPs), native premolten globules, and native coils (both are known as extended IDPs). It is clear now that the above-mentioned structure-function paradigm, which emphasizes that ordered 3D structures represent the indispensable prerequisite to the effective protein functioning, should be redefined to include IDPs (19–24, 72, 83, 84, 86, 95, 103). According to this redefined paradigm, native proteins (or their functional regions) can exist in any of the known conformational states, ordered, molten globule, premolten globule, and coil. Function can arise from any of these conformations and transitions between them. It has also been proposed that in addition to the "protein folding" problem, where the correct folding of a globular protein into the rigid biologically active conformation is determined by its amino acid sequence (1), the "protein nonfolding problem" does exists too, where the lack of a rigid globular structure in a given IDP may be encoded in some specific features of its amino acid sequence.

In an attempt to understand the relationship between the amino acid sequence and protein intrinsic disorder, a set of experimentally characterized IDPs and intrinsically disordered regions (IDRs) was systematically compared with a set of ordered globular proteins (22, 72, 74, 102). This analysis revealed that IDPs and IDRs differ from structured globular proteins and domains with regard to many attributes, including amino acid composition, sequence complexity, hydrophobicity, charge, and flexibility. IDPs were shown to be significantly depleted in a number of so-called order-promoting residues, including bulky hydrophobic (Ile, Leu, and Val) and aromatic amino acid residues (Trp, Tyr, and Phe), which would normally form the hydrophobic core of a folded globular protein, and IDPs/IDRs also possess low content of Cys and Asn residues. On the other hand, IDPs were substantially enriched in the so-called disorder-promoting amino acids: Ala, Arg, Gly, Gln, Ser, Pro, Glu, and Lys (22, 72, 74, 102). This is illustrated in Figure 19.1A, which represents the average amino acid frequencies in two IDP data sets in comparison with the average frequencies found in ordered globular proteins (16, 101). Data for this analysis were extracted from the DisProt database (78) that currently assembles 520 experimentally verified disordered proteins.

Almost simultaneously and completely independently from these studies, it has been established that the combination of low mean hydrophobicity and relatively high net charge constitutes an important prerequisite for the lack of compact structure in extended (coil-like and premolten globule-like) IDPs (89). Figure 19.1B shows that such extended IDPs are specifically localized within a unique region of the charge-hydrophobicity phase space, being separated from well-folded ordered proteins by a boundary, satisfying the following relationship (89):

$$\langle H \rangle_{\text{boundary}} = \frac{\langle R \rangle + 1.151}{2.785}. \tag{19.1}$$

This equation estimates the "boundary" mean hydrophobicity value, $<H>_{\text{boundary}}$, below which a polypeptide chain with a given mean net charge $<R>$ will most probably be disordered. Overall, Figure 19.1 clearly shows that the amino acid sequences of IDPs and ordered proteins are very different, supporting a hypothesis that the propensity of a polypeptide chain to fold or stay disordered is encoded in its amino acid sequence. For example, extended IDPs are disordered under physiological conditions because of the strong electrostatic repulsion (due to their high net charges) and weak hydrophobic attraction (due to their low contents of hydrophobic residues). These data also suggest that IDPs (especially extended IDPs) with their highly biased amino acid sequences might possess obscure conformational responses to changes in their environment.

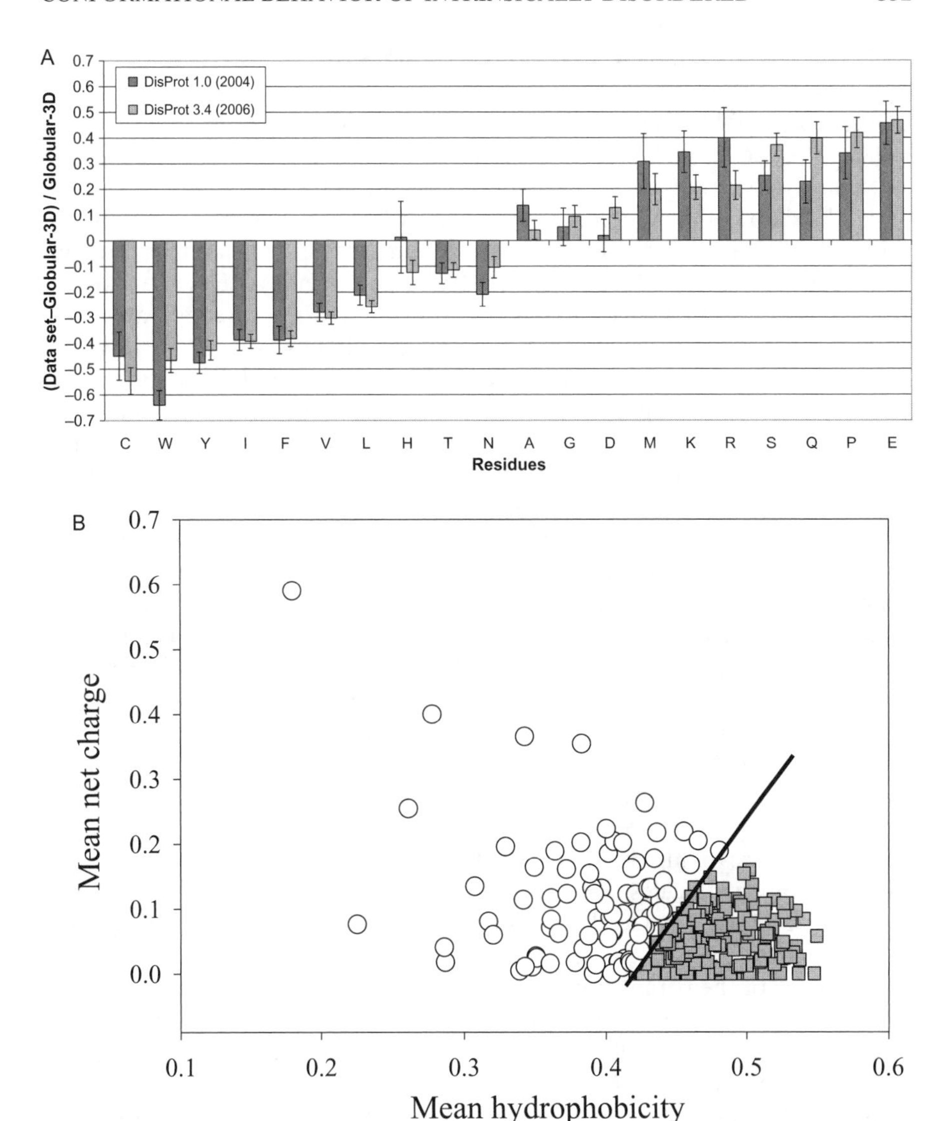

Figure 19.1 Peculiarities of the amino acid sequences of intrinsically disordered proteins. (A) Amino acid composition, relative to the set of globular proteins Globular-3D, of intrinsically disordered regions 10 residues or longer from the DisProt database. Dark gray indicates DisProt 1.0 (152 proteins), whereas light gray indicates DisProt 3.4 (460 proteins). Amino acid compositions were calculated per disordered regions and then averaged. The arrangement of the amino acids is by peak height for the DisProt 3.4 release. Confidence intervals were estimated using per-protein bootstrapping with 10,000 iterations. (B) Mean net charge versus mean hydropathy plot (charge–hydropathy plot) for the set of 275 folded (squares) and 91 natively unfolded proteins (circles).

19.2 EFFECTS OF STRONG DENATURANTS

IDPs, being highly dynamic, are characterized by low conformational stability, which is reflected in low steepness of the transition curves describing their unfolding induced by strong denaturants or even in the complete lack of the sigmodal shape of the unfolding curves. This is in strict contrast to the solvent-induced unfolding of ordered globular proteins, which is known to be a highly cooperative process. In fact, we can find here an extreme case of the cooperative transition, which is an all-or-none transition where a cooperative unit includes the whole molecule, that is, no intermediate states can be observed in the transition region. Based on the analysis of the unfolding transitions in ordered globular proteins, it has been concluded that the steepness of urea- or GdmCl-induced unfolding curves depends strongly on whether a given protein has a rigid tertiary structure (i.e., it is ordered) or is already denatured and exists as a molten globule (71, 97). In fact, urea-induced or GdmCl-induced protein unfolding of globular proteins often involves at least two steps: the ordered (native) state to molten globule (N↔MG) and the molten globule to unfolded state (MG↔U) transitions (12, 49, 68–70, 73). Both transitions are rather cooperative (as they follow an S-shaped curve). However, for a long time, the dimensions of the cooperative units for these transitions were unknown, and, as a consequence, it was unclear whether or not these transitions have a real all-or-none nature.

The usual method for estimation of the cooperativity of transition is the measurement of the slope of the transition curve at its middle point. This slope is proportional to the change of the thermodynamic quantity conjugated with the variable provoking the transition, that is, to the difference in the numbers of denaturant molecules "bound" to the initial and final states in the urea-induced or GdmCl-induced transitions, $\Delta\nu_{eff}$. To understand the thermodynamic nature of solvent-induced transitions in globular proteins, the dependence of their transition slopes on the protein molecular mass (M) was analyzed (71, 97). This analysis was based on the hypothesis that the slope of a phase transition in small systems depends on the dimensions of this system (39, 42). In the case of first-order phase transition, the slope increases proportionally to the number of units in a system (42), whereas the slope for second-order phase transition is proportional to the square root of this number (39). Therefore, it is possible to distinguish between phase and nonphase intramolecular transitions by measuring whether their slopes depend on molecular weight.

Figure 19.2 represents the results of such an analysis of the available at that moment experimental data on urea-induced and GdmCl-induced N↔U, N↔MG, and MG↔U transitions in small globular proteins (71, 97). Here, the cooperativity was measured based on the corresponding $\Delta\nu_{eff}$ values, which were obtained from the equilibrium constants K_{eff} as (71, 97):

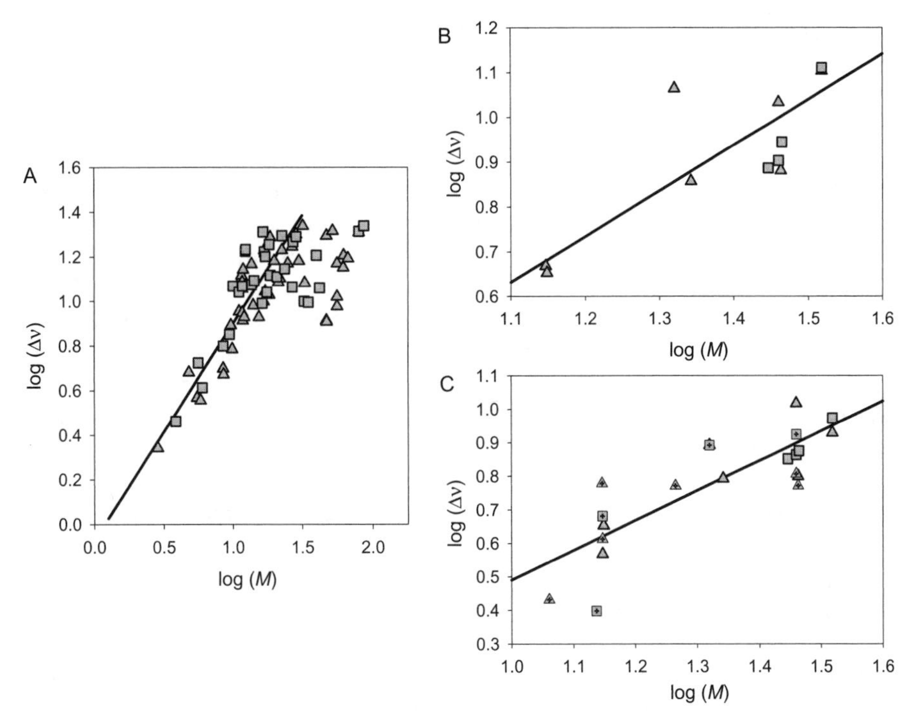

Figure 19.2 Molecular mass dependence of cooperativity parameters for urea- and GdmCl-induced transitions in globular proteins. Plot (A) represents $\log(\Delta\nu_{\text{eff}}^{\text{N}\leftrightarrow\text{U}})$ versus $\log(M)$ dependence. Plot (B) illustrates $\log(\Delta\nu_{\text{eff}}^{\text{N}\leftrightarrow\text{MG}})$ versus $\log(M)$ dependence, whereas $\log(\Delta\nu_{\text{eff}}^{\text{MG}\leftrightarrow\text{U}})$ versus $\log(M)$ dependence is shown in plot (C). Squares and triangles correspond to urea- and GdmCl-induced transitions, respectively. Crossed symbols in plot (C) correspond to the unfolding of acid-induced molten globules.

$$\Delta\nu_{\text{eff}} = \left(\frac{\partial \ln K_{\text{eff}}}{\partial \ln a}\right)_{a=a_t} = 4a_t\left(\frac{\partial \Theta}{\partial a}\right)_{a=a_t}, \tag{19.2}$$

where Θ is the fraction of molecules in one of the states separated by the all-or-none transition, a is the activity of a denaturing agent, and a_t is its activity at the middle transition point. The activities of urea and GdmCl as functions of their molar concentrations (m) can be calculated by the empirical equations (63)

$$a_{\text{urea}} = 0.9815\,(m) - 0.02978\,(m)^2 + 0.00308\,(m)^3 \tag{19.3}$$

and

$$a_{\text{GdmCl}} = 0.6761\,(m) - 0.1468\,(m)^2 + 0.02475\,(m)^3 + 0.00132\,(m)^4. \tag{19.4}$$

Figure 19.2A shows that the $\Delta\nu_{eff}$ (M) curve comprises two parts: for small globular proteins (with $M < 25$–$30\,kDa$), cooperativity of unfolding transition increases with M, whereas for large proteins (with $M > 25$–$30\,kDa$), cooperativity does not depend on M. The existence of pronounced molecular mass dependence for the degree of cooperativity shows that denaturant-induced unfolding of small globular proteins exhibits the characteristics of phase transition. On the other hand, the independence of the cooperativity of unfolding from M for large proteins can be related to their multidomain organization. Based on the fact that $\Delta\nu_{eff}$ (M) dependencies for N↔U transitions induced by urea and GdmCl coincide within experimental errors, it was proposed that the number of binding sites or the area of exposed residue surface (but not the energy of the binding!) is the same for urea and GdmCl (71, 97). Alternatively, both denaturants may change the structure of a solvent in a similar manner.

The analysis of the $\log(\Delta\nu_{eff}^{N\leftrightarrow U})$ versus $\log(M)$ dependence for solvent-induced N↔U transitions in small globular proteins revealed that it can be described as

$$\log(\Delta\nu_{eff}^{N\leftrightarrow U}) = 0.97\log(M) - 0.07, \tag{19.5}$$

with the root mean square deviation (*rmsd*) of 0.112 and the correlation coefficient (r) of 0.87 (71, 97). As the proportionality coefficient in the above equation is equal to 0.97 ± 0.15, we can conclude that $\Delta\nu_{eff}^{N\leftrightarrow U}$ is proportional to M, clearly showing that the urea- and GdmCl-induced unfolding of small globular proteins is an all-or-none transition, that is, an intramolecular analog of first-order phase transition in macroscopic systems.

The results of the analogous analyses for the denaturant-induced N↔MG and MG↔U transitions are shown in Figure 19.2B,C, respectively. Despite a much smaller number of experimental points, the slopes of the N↔MG and MG↔U transition curves clearly increase with the molecular mass of a protein. These increases were approximated by the following equations (71, 97):

$$\log(\Delta\nu_{eff}^{N\leftrightarrow MG}) = 1.02\log(M) - 0.49, \tag{19.6}$$

with $rmsd = 0.090$ and $r = 0.82$; and

$$\log(\Delta\nu_{eff}^{MG\leftrightarrow U}) = 0.89\log(M) - 0.40, \tag{19.7}$$

with $rmsd = 0.092$ and $r = 0.84$.

These data clearly show that all denaturant-induced transitions in small globular proteins can be described in terms of all-or-none transitions (71, 97).

In application to IDPs, it has been proposed that this type of analysis can be used to differentiate whether a given protein has ordered (rigid) structure or exists as a native molten globule (83). In fact, the comparison of Equations 19.5 and 19.7 suggests that the slope of the N↔U transition for a protein with a molecular mass of M is more than twice as steep as the slope of the MG↔U transition. For example, for a protein with the molecular mass of 30 kDa, $\Delta\nu_{eff}^{N\leftrightarrow U} = 23.1$, whereas $\Delta\nu_{eff}^{MG\leftrightarrow U} = 8.2$. Therefore, to extend this type of analysis to IDP, the corresponding $\Delta\nu_{eff}$ value should be determined from the denaturant-induced unfolding experiments. Then this quantity should be compared with the $\Delta\nu_{eff}^{N\leftrightarrow U}$ and $\Delta\nu_{eff}^{MG\leftrightarrow U}$ values corresponding to the N↔U and MG↔U transitions in globular protein of a given molecular mass, evaluated by Equations 19.5 and 19.7, respectively.

Although the denaturant-induced unfolding of a native molten globule can be described by a shallow sigmoidal curve (e.g., see Reference 62), urea- or GdmCl-induced structural changes in native premolten globules or native coils are noncooperative and typically seen as monotonous featureless changes in the studied parameters. This is due the low content of the residual structure in these species.

19.3 EFFECT OF TEMPERATURE

According to the modern classification of phase transitions, the first-order phase transitions are those that involve a latent heat. During such a transition, a system either absorbs or releases a fixed amount of energy, which can be directly observed by means of thermophysical methods of investigation. Therefore, the presence of specific heat absorption peaks in melting curve measured by the differential scanning calorimetry is a simple and convenient criterion indicating the presence of a rigid tertiary structure in a given protein (66, 68, 82). On the other hand, the lack of such cooperative heat sorption peaks is the reflection of a highly dynamic structure. Thus, differential scanning microcalorimetry can be used to find whether a given protein has an ordered (rigid) structure. For example, the scanning microcalorimetry data obtained for metal-depleted forms of pike parvalbumin (64) did not reveal any measurable excess heat absorption peaks characteristic for the first-order thermal transitions. Furthermore, in these studies, the specific heat capacity of apoprotein, $C_{p,apo}$, significantly exceeded the values observed for the ordered Ca^{2+}-saturated form of this protein. The $C_{p,apo}$ values were closer to the heat capacity of fully unfolded parvalbumin, as estimated according to Hackel et al. (40). Therefore, the protein specific heat capacity absolute values, which can be regarded as a measure of hydration of its amino acid residues (67), also suggested that apo-parvalbumin was essentially solvated and, consequently, lacked ordered and rigid tertiary structure (64).

The analysis of the temperature effects on structural properties of several extended IDPs revealed that native coils and native premolten globules possess so-called turned out response to heat. This is illustrated by Figure 19.3, where the temperature-induced changes in the far-UV circular dichroism

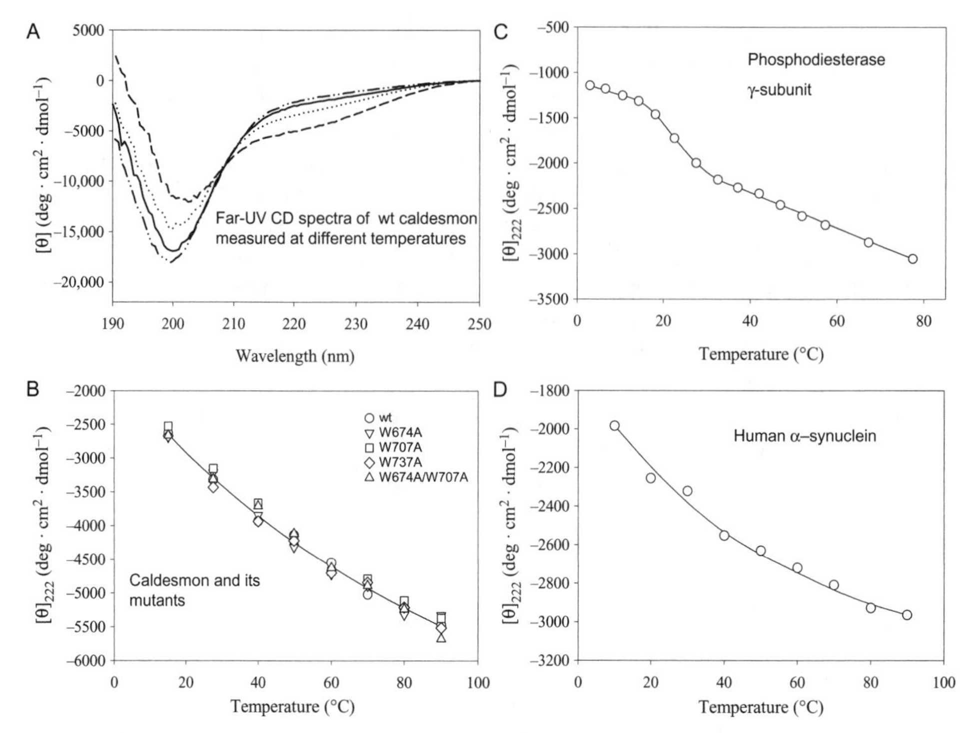

Figure 19.3 The effect of temperature on structural properties of IDPs. The temperature-induced changes in the far-UV CD spectrum of caldesmon (plot A) and the temperature-dependence of $[\theta]_{222}$ for several extended IDPs: caldesmon and its mutants (plot B), phosphodiesterase γ-subunit (plot C), and human α-synuclein (plot D).

(CD) spectrum of α-synuclein (cf. Reference 93) (Fig. 19.3A) and the temperature dependence of $[\theta]_{222}$ for several extended IDPs (Fig. 19.3B) are shown. As it follows from Figure 19.3A, at low temperatures, α-synuclein possessed a far-UV CD spectrum typical of an unfolded polypeptide chain. As the temperature was increased, the shape of spectrum changed, reflecting the temperature-induced formation of secondary structure. This temperature-induced folding is further illustrated in Figure 19.4B, which represents the $[\theta]_{222}$ temperature dependence for α-synuclein (93), caldesmon 636–771 fragment (65), and phosphodiesterase γ-subunit (96). Here, the major spectral changes occurred over the range of 3 to 30–50°C, and further heating was accompanied by less pronounced effects. Analogous temperature dependencies indicative of heat-induced structure formation have been reported for the receptor extracellular domain of nerve growth factor (81) and α_s-casein (47). The structural heating-induced changes in all these IDPs were completely reversible. Thus, an increase in temperature induces the partial folding of intrinsically unstructured proteins, rather than the unfolding typical of ordered

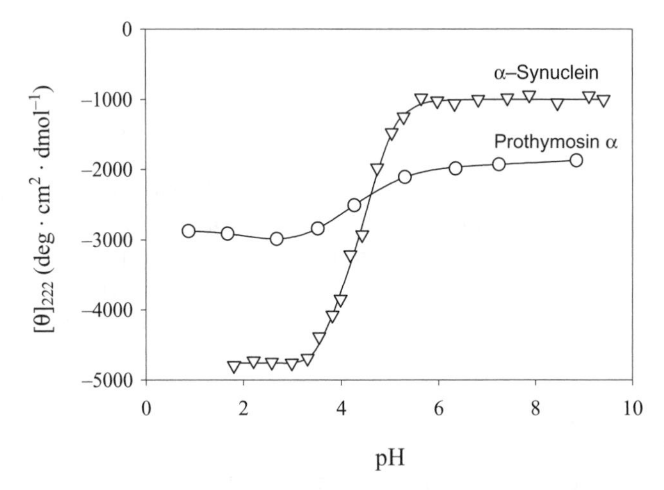

Figure 19.4 The effect of pH on structural properties of IDPs. The pH dependence of $[\theta]_{222}$ for α-synuclein and prothymosin α.

globular proteins. The effects of elevated temperatures may be attributed to the increased strength of the hydrophobic interaction at higher temperatures, leading to a stronger hydrophobic attraction, which is the major driving force for folding.

19.4 EFFECT OF PH

Extended IDPs are also characterized by the "turned out" response to changes in pH. Figure 19.4 represents the pH dependence of $[\theta]_{222}$ for α-synuclein (93) and prothymosin α (91). It is clearly seen that there was little or no change in the far-UV CD spectra between pH ~9.0 and ~5.5. However, a decrease in pH from 5.5 to 3.0 resulted in a substantial increase in negative intensity in the vicinity of 220 nm. It has also been established that the pH-induced changes in the far-UV CD spectrum of these two proteins were completely reversible and consistent with the formation of partially folded premolten globule-like conformation (91, 93).

Similar pH-induced structural transformations have been described for such extended IDPs as pig calpastatin domain I (48), histidine-rich protein II (51), and naturally occurring human peptide LL-37 (44). These observations show that a decrease (or increase) in pH induces partial folding of extended IDPs due to the minimization of their large net charge present at neutral pH, thereby decreasing charge/charge intramolecular repulsion and permitting hydrophobic-driven collapse to the partially folded conformation.

19.5 EFFECT OF COUNTERIONS

Under physiological pH, extended IDPs are essentially unfolded because of the strong electrostatic repulsion between the noncompensated charges of the same sign and because of the low hydrophobic attraction. To some extent, this strong uncompensated electrostatic repulsion leading to the substantial unfolding resembles the situation occurring for many ordered proteins at extremely low or high pH. Many of the pH-unfolded globular proteins were transformed into more ordered conformations when the electrostatic repulsion was reduced by binding of oppositely charged ions (30, 35–37). Similar mechanisms may be expected for extended IDPs, and the metal ion-stimulated conformational changes have been indeed described for several of them. Furthermore, any interaction of extended IDPs with natural ligand that will affect its mean net charge, mean hydrophobicity, or both, may change net charge and hydrophobicity of a system in such a way that these parameters will approach those typical of ordered proteins.

The validity of this hypothesis is illustrated in Figure 19.5A, which represents the data for seven extended IDPs, osteocalcin, osteonectin, α-casein, HPV16E7 protein, calsequestrin, manganese-stabilizing protein, and HIV-1 integrase (89). In this figure, black circles correspond to the ligand-free proteins, whereas open circles describe proteins complexed with particular metal cations. The solid line represents the border between extended IDPs and ordered proteins. Thus, the interaction of at least these seven proteins with their natural ligands (metal ions) results in a shift of their charge-hydropathy parameters to those typical for ordered proteins (89).

As a further illustration, Figure 19.5B–D represents the far-UV CD spectra of prothymosin α (Fig. 19.5B), α-synuclein (Fig. 19.5C), and core histones (Fig. 19.5D) measured in the absence or presence of various metal ions. An increase in the Zn^{2+} content was accompanied by essential changes in the shape of the far-UV CD spectra of prothymosin α (Fig. 19.5B), reflecting partial folding of the protein (90). Similarly, many cations (monovalent, bivalent, and trivalent) were shown to induce conformational changes in α-synuclein and transformed this extended IDP into a partially folded conformation (Fig. 19.5C) (94). The folding strength of cations increases with the increase in the ionic charge density (94), reflecting that the effective screening of the Coulombic charge/charge repulsion represents a major driving force for such folding. For polyvalent cations, an additional important factor could be the potential capability for cross-linking or bridging between two or more carboxylates (94). Figure 19.5 shows that the degree of cation-induced folding of extended IDPs varies in a wide range (from subtle ordering of prothymosin α [Fig. 19.5B] and α-synuclein [Fig. 19.5C] to a complete folding of core histones [Fig. 19.5D]). Finally, human antibacterial protein LL-37, a natively unfolded protein with extremely basic net charge, was shown to be essentially folded in the presence of several anions (44).

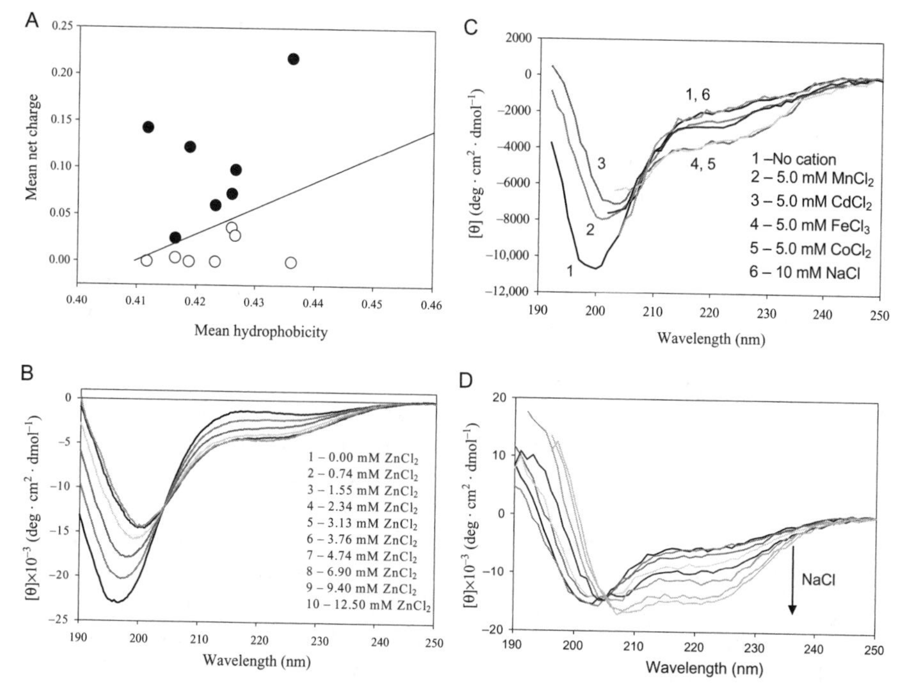

Figure 19.5 The effect of counterions on structural properties of IDPs. (A) Charge–hydropathy plot for seven extended IDPs, osteocalcin, osteonectin, α-casein, HPV16E7 protein, calsequestrin, manganese-stabilizing protein, and HIV-1 integrase in their apo- and metal-saturated forms. The far-UV CD spectra of prothymosin α (plot B), α-synuclein (plot C), and core histones (plot D) measured in the absence or presence of various metal ions. See color insert.

19.6 EFFECT OF MACROMOLECULAR CROWDING

Up to date, the vast majority of *in vitro* experiments on various IDPs were performed under the relatively ideal thermodynamic conditions of low protein and moderate salt concentrations. However, proteins have evolved to function within cells, where the concentration of macromolecules, including proteins, nucleic acids, and carbohydrates, within a cell can be as high as 400 g/L (106), creating a crowded medium, with considerably restricted amounts of free water (29, 33, 55, 57, 105, 106). Such media are referred as "crowded" rather than "concentrated," as, in general, no individual macromolecular species may be present at high concentration, and a special term "molecular crowding" has been introduced to describe the effect of high solute concentrations on chemical reactions (54, 55). Obviously, the volume occupied by solutes is unavailable to other molecules because two molecules cannot be in the same place at the same time. The thermodynamic consequences of the unavailable volume are called excluded volume effects (56, 105). Volume exclusion in biologic fluids may have large effects on both stability of biologic macromolecules (6, 25, 26, 54), and macromolecular equilibrium, including alteration of protein–protein interactions (54, 59) and modulation of the rate and extent of amyloid formation (41, 77, 87). In particular, the addition of high concentrations of polymers (proteins, polysaccharides, polyethyleneglycols) dramatically accelerated α-synuclein fibrillation *in vitro* (87).

The view that macromolecular crowding is an important yet a neglected variable in biochemical studies has gained attention (29, 105). The effect of excluded volume on macromolecules may be examined experimentally by using concentrated solutions of a model "crowding agent" such as polyethylene glycol, dextran, Ficoll, or inert proteins (41, 56). The effect of high concentrations of crowders on structural properties of several IDPs was analyzed. For example, Flaugh and Lumb (31) established that molecular crowding modeled by the high concentrations (of up to 250 g/L) of the dextrans of average molecular weights 9.5, 37.5, and 77 kDa and Ficoll 70 did not induce significant folding in two IDPs: FosAD and p27ID. FosAD corresponds to the C-terminal activation domain of human c-Fos (residues 216–310) and is functional for interacting with transcription factors in whole-cell extract (11). p27ID corresponds to the cyclin-dependent kinase inhibition domain of the cell-cycle inhibitor human $p27^{Kip1}$ (residues 22–97) and is active as a cyclin A-Cdk2 inhibitor (4). Both protein domains were shown to be intrinsically disordered as judged by CD spectra that were characteristic of the unfolded polypeptide chains, lack of ^{1}H chemical-shift dispersion and negative ^{1}H-^{15}N nuclear Overhauser effects (4, 11). In the presence of macromolecular crowding agents, none of these IDPs underwent any significant conformational change reflected in noticeable changes in either CD or fluorescence spectra. Based on these observations, it has been concluded that molecular crowding effects are not necessarily sufficient to induce ordered structure in IDPs (31). Similarly, α-synuclein was shown to preserve its mostly unfolded conformation in the presence of several crowding agents (60) and even in the periplasm of the bacterial cells (52). The analysis of FlgM, which is a 97-residue IDP from *Salmonella typhimurium* that regulates flagellar synthesis by binding the transcription factor σ^{28}, revealed that approximately half of this IDP gained structure in the crowded environment (18). Importantly, although free FlgM was mostly unstructured in the dilute solutions, its C-terminal half formed a transient helix in the unbound form (15), became structured on binding to σ^{28} (14), and was shown to be folded in the crowded environment (18). Therefore, IDPs could be grouped into two classes, foldable and nonfoldable, based on their response to the crowded environment. Foldable IDPs can gain structure in crowded environment (and, thus, inside the living cells) likely due to the crowding-induced formation of a hydrophobic core. Nonfoldable IDPs remain mostly unstructured at the crowded conditions. Some of these nonfoldable by crowding IDPs may require another protein (or DNA, or RNA, or some other natural binding partners) to provide a framework for structure formation. FlgM clearly represents a unique case of the two-faced Janus, where the first face exemplified by the C-terminal half of FlgM is structured in the crowded environment, whereas the second face exemplified by the N-terminal half of FlgM does not become structured at physiologically relevant solute concentrations (18).

One of the experimental approaches to model the confined intracellular space is to encapsulate studied proteins in silica glass using the sol-gel techniques to create a crowded microenvironment. It is believed though that the fraction of the total volume excluded by the silica matrix is less than the fractional volume occupied by macromolecules in a living cell (25, 26). However, the size of protein-occupied pores in these gels is believed to have the same order of magnitude as the diameter of protein, and it has been suggested that the protein itself may dictate the pore size during the gelatio process (50). Proteins are not bound covalently to the silica matrix, but the matrix substantially impedes the rotational freedom of the protein (38). The mild conditions of sol-gel glass encapsulation have proven to be compatible with the folded structure and function of several ordered proteins (9, 17, 34). The optically transparent glass products may be analyzed by the majority of spectroscopic techniques used to monitor the structure of proteins in dilute solutions. This includes fluorescence (5, 9) and CD (25, 26). Furthermore, the highly porous character of the glass allows easy exchange of the solvent that permeates the silica matrix, but the encapsulated macromolecules are unable to escape the glass under most solvent conditions (25). Using this approach, solvent effects on the secondary structure of several ordered proteins have been analyzed by CD following encapsulation in the hydrated pores of a silica glass matrix by the sol-gel method (25, 26). Unfortunately, this attractive technique was not applied for the IDP analysis as of yet.

19.7 CONCLUSIONS

Some of the most typical features of IDP conformational behavior are summarized below:

1. The low cooperativity (or the complete lack thereof) of the denaturant-induced unfolding. The denaturant-induced unfolding of native molten globules can be described by shallow sigmoidal curves, whereas urea- or GdmCl-induced unfolding of native premolten globules or native coils is a noncooperative process and is typically seen as monotonous featureless changes in the studied parameters.
2. The lack of the measurable excess heat absorption peaks characteristic for the melting of ordered proteins in the calorimetric experiments. This is typically combined with high values of the specific heat capacity, which significantly exceed the values observed for the ordered proteins, being close to the heat capacity of fully unfolded polypeptide chain.

3. "Turned out" response to heat and changes in pH, where high temperatures or extreme pH values induce partial folding of the extended IDPs.
4. The intriguing ability of extended IDPs (native premolten globules and native coils) to gain structure in the presence of various counterions. The degree of this counterion-induced folding of different IDPs varies in a wide range, starting from subtle conformational rearrangements in some IDPs and ending with the complete folding of other IDPs.
5. Based on their response to the crowded environment, IDPs form two classes: foldable and nonfoldable. Foldable IDPs can gain structure, whereas nonfoldable IDPs remain mostly unstructured in a crowded environment.

The above-mentioned observations of the obscure response of IDPs to changes in their environment definitely should be taken into account while discussing conformational behavior of these proteins.

ACKNOWLEDGMENTS

This work was supported in part by the Program of the Russian Academy of Sciences for the "Molecular and cellular biology" and by grants R01 LM007688-01A1 and GM071714-01A2 from the National Institutes of Health. We gratefully acknowledge the support of the Indiana University Purdue University in Indianapolis (IUPUI) Signature Centers Initiative.

REFERENCES

1. Anfinsen, C. B., E. Haber, M. Sela, and F. H. White, Jr. 1961. The kinetics of formation of native ribonuclease during oxidation of the reduced polypeptide chain. Proc Natl Acad Sci U S A **47**:1309–14.
2. Anson, M. L., and A. E. Mirsky. 1932. The effect of denaturation on the viscosity of protein systems. J Gen Physiol **15**:341–50.
3. Baum, J., C. M. Dobson, P. A. Evans, and C. Hanley. 1989. Characterization of a partly folded protein by NMR methods: studies on the molten globule state of guinea pig alpha-lactalbumin. Biochemistry **28**:7–13.
4. Bienkiewicz, E. A., J. N. Adkins, and K. J. Lumb. 2002. Functional consequences of preorganized helical structure in the intrinsically disordered cell-cycle inhibitor p27(Kip1). Biochemistry **41**:752–9.
5. Bismuto, E., and G. Irace. 2001. The effect of molecular confinement on the conformational dynamics of the native and partly folded state of apomyoglobin. FEBS Lett **509**:476–80.
6. Bismuto, E., P. L. Martelli, A. De Maio, D. G. Mita, G. Irace, and R. Casadio. 2002. Effect of molecular confinement on internal enzyme dynamics: frequency domain fluorometry and molecular dynamics simulation studies. Biopolymers **67**:85–95.

7. Bose, H. S., R. M. Whittal, M. A. Baldwin, and W. L. Miller. 1999. The active form of the steroidogenic acute regulatory protein, StAR, appears to be a molten globule. Proc Natl Acad Sci U S A **96**:7250–5.
8. Bracken, C. 2001. NMR spin relaxation methods for characterization of disorder and folding in proteins. J Mol Graph Model **19**:3–12.
9. Brennan, J. D. 1999. Using fluorescence to investigate proteins entrapped in sol-gel derived materials. Appl Spectrosc **53**:106A–21A.
10. Bushnell, G. W., G. V. Louie, and G. D. Brayer. 1990. High-resolution three-dimensional structure of horse heart cytochrome c. J Mol Biol **214**: 585–95.
11. Campbell, K. M., A. R. Terrell, P. J. Laybourn, and K. J. Lumb. 2000. Intrinsic structural disorder of the C-terminal activation domain from the bZIP transcription factor Fos. Biochemistry **39**:2708–13.
12. Christensen, H., and R. H. Pain. 1991. Molten globule intermediates and protein folding. Eur Biophys J **19**:221–9.
13. Chyan, C. L., C. Wormald, C. M. Dobson, P. A. Evans, and J. Baum. 1993. Structure and stability of the molten globule state of guinea-pig alpha-lactalbumin: a hydrogen exchange study. Biochemistry **32**:5681–91.
14. Daughdrill, G. W., M. S. Chadsey, J. E. Karlinsey, K. T. Hughes, and F. W. Dahlquist. 1997. The C-terminal half of the anti-sigma factor, FlgM, becomes structured when bound to its target, sigma 28. Nat Struct Biol **4**:285–91.
15. Daughdrill, G. W., L. J. Hanely, and F. W. Dahlquist. 1998. The C-terminal half of the anti-sigma factor FlgM contains a dynamic equilibrium solution structure favoring helical conformations. Biochemistry **37**:1076–82.
16. Daughdrill, G. W., G. J. Pielak, V. N. Uversky, M. S. Cortese, and A. K. Dunker. 2005. Natively disordered proteins, pp. 271–353. In J. Buchner and T. Kiefhaber (eds.), Handbook of Protein Folding. Wiley-VCH, Verlag GmbH & Co., Weinheim, Germany.
17. Dave, B. C., B. Dunn, J. S. Valentine, and J. I. Zink. 1994. Sol-gel encapsulation methods for biosensors Anal Chem **66**:1120–7.
18. Dedmon, M. M., C. N. Patel, G. B. Young, and G. J. Pielak. 2002. FlgM gains structure in living cells. Proc Natl Acad Sci U S A **99**:12681–4.
19. Dunker, A. K., C. J. Brown, J. D. Lawson, L. M. Iakoucheva, and Z. Obradovic. 2002. Intrinsic disorder and protein function. Biochemistry **41**:6573–82.
20. Dunker, A. K., C. J. Brown, and Z. Obradovic. 2002. Identification and functions of usefully disordered proteins. Adv Protein Chem **62**:25–49.
21. Dunker, A. K., M. S. Cortese, P. Romero, L. M. Iakoucheva, and V. N. Uversky. 2005. Flexible nets. The roles of intrinsic disorder in protein interaction networks. FEBS J **272**:5129–48.
22. Dunker, A. K., J. D. Lawson, C. J. Brown, R. M. Williams, P. Romero, J. S. Oh, C. J. Oldfield, A. M. Campen, C. M. Ratliff, K. W. Hipps, J. Ausio, M. S. Nissen, R. Reeves, C. Kang, C. R. Kissinger, R. W. Bailey, M. D. Griswold, W. Chiu, E. C. Garner, and Z. Obradovic. 2001. Intrinsically disordered protein. J Mol Graph Model **19**:26–59.
23. Dunker, A. K., and Z. Obradovic. 2001. The protein trinity—linking function and disorder. Nat Biotechnol **19**:805–6.

24. Dunker, A. K., C. J. Oldfield, J. Meng, P. Romero, J. Y. Yang, J. W. Chen, V. Vacic, Z. Obradovic, and V. N. Uversky. 2008. The unfoldomics decade: an update on intrinsically disordered proteins. BMC Genomics **9** (Suppl 2):S1.

25. Eggers, D. K., and J. S. Valentine. 2001. Crowding and hydration effects on protein conformation: a study with sol-gel encapsulated proteins. J Mol Biol **314**:911–22.

26. Eggers, D. K., and J. S. Valentine. 2001. Molecular confinement influences protein structure and enhances thermal protein stability. Protein Sci **10**:250–61.

27. Eliezer, D., K. Chiba, H. Tsuruta, S. Doniach, K. O. Hodgson, and H. Kihara. 1993. Evidence of an associative intermediate on the myoglobin refolding pathway. Biophys J **65**:912–7.

28. Eliezer, D., J. Yao, H. J. Dyson, and P. E. Wright. 1998. Structural and dynamic characterization of partially folded states of apomyoglobin and implications for protein folding. Nat Struct Biol **5**:148–55.

29. Ellis, R. J. 2001. Macromolecular crowding: obvious but underappreciated. Trends Biochem Sci **26**:597–604.

30. Fink, A. L., L. J. Calciano, Y. Goto, T. Kurotsu, and D. R. Palleros. 1994. Classification of acid denaturation of proteins: intermediates and unfolded states. Biochemistry **33**:12504–11.

31. Flaugh, S. L., and K. J. Lumb. 2001. Effects of macromolecular crowding on the intrinsically disordered proteins c-Fos and p27(Kip1). Biomacromolecules **2**:538–40.

32. Fontana, A., P. Polverino de Laureto, and V. De Philipps. 1993. Molecular aspects of proteolysis of globular proteins, pp. 101–10. In W. van den Tweel, A. Harder, and M. Buitelear (eds.), Protein Stability and Stabilization. Elsevier, Amsterdam.

33. Fulton, A. B. 1982. How crowded is the cytoplasm? Cell **30**:345–7.

34. Gill, I., and A. Ballestros. 2000. Bioencapsulation within synthetic polymers (part 1): sol-gel encapsulated biologicals. Trends Biotechnol **18**:282–96.

35. Goto, Y., L. J. Calciano, and A. L. Fink. 1990. Acid-induced folding of proteins. Proc Natl Acad Sci U S A **87**:573–7.

36. Goto, Y., and A. L. Fink. 1990. Phase diagram for acidic conformational states of apomyoglobin. J Mol Biol **214**:803–5.

37. Goto, Y., N. Takahashi, and A. L. Fink. 1990. Mechanism of acid-induced folding of proteins. Biochemistry **29**:3480–8.

38. Gottfried, D. S., A. Kagan, B. M. Hoffman, and J. M. Friedman. 1999. Impeded rotation of a protein in sol-gel matrix. J Phys Chem B **103**:2803–7.

39. Grosberg, A. Y., and A. R. Khokhlov. 1989. Statistical Physics of Macromolecules. Nauka, Moscow.

40. Hackel, M., H. J. Hinz, and G. R. Hedwig. 1999. A new set of peptide-based group heat capacities for use in protein stability calculations. J Mol Biol **291**:197–213.

41. Hatters, D. M., A. P. Minton, and G. J. Howlett. 2002. Macromolecular crowding accelerates amyloid formation by human apolipoprotein C-II. J Biol Chem **277**:7824–30.

42. Hill, T. L. 1963–1964. Thermodynamics of the Small Systems. Wiley, New York.

43. Jeng, M. F., S. W. Englander, G. A. Elove, A. J. Wand, and H. Roder. 1990. Structural description of acid-denatured cytochrome c by hydrogen exchange and 2D NMR. Biochemistry **29**:10433–7.
44. Johansson, J., G. H. Gudmundsson, M. E. Rottenberg, K. D. Berndt, and B. Agerberth. 1998. Conformation-dependent antibacterial activity of the naturally occurring human peptide LL-37. J Biol Chem **273**:3718–24.
45. Kataoka, M., Y. Hagihara, K. Mihara, and Y. Goto. 1993. Molten globule of cytochrome c studied by small angle X-ray scattering. J Mol Biol **229**: 591–6.
46. Kataoka, M., K. Kuwajima, F. Tokunaga, and Y. Goto. 1997. Structural characterization of the molten globule of alpha-lactalbumin by solution X-ray scattering. Protein Sci **6**:422–30.
47. Kim, T. D., H. J. Ryu, H. I. Cho, C. H. Yang, and J. Kim. 2000. Thermal behavior of proteins: heat-resistant proteins and their heat-induced secondary structural changes. Biochemistry **39**:14839–46.
48. Konno, T., N. Tanaka, M. Kataoka, E. Takano, and M. Maki. 1997. A circular dichroism study of preferential hydration and alcohol effects on a denatured protein, pig calpastatin domain I. Biochim Biophys Acta **1342**:73–82.
49. Kuwajima, K. 1989. The molten globule state as a clue for understanding the folding and cooperativity of globular-protein structure. Proteins **6**:87–103.
50. Lan, E. H., B. C. Dave, J. M. Fukuto, B. Dunn, J. I. Zink, and J. S. Valentine. 1999. Synthesis of sol-gel encapsulated heme proteins with chemical sensing properties. J Mater Chem **9**:45–53.
51. Lynn, A., S. Chandra, P. Malhotra, and V. S. Chauhan. 1999. Heme binding and polymerization by *Plasmodium falciparum* histidine rich protein II: influence of pH on activity and conformation. FEBS Lett **459**:267–71.
52. McNulty, B. C., G. B. Young, and G. J. Pielak. 2006. Macromolecular crowding in the *Escherichia coli* periplasm maintains alpha-synuclein disorder. J Mol Biol **355**:893–7.
53. Merrill, A. R., F. S. Cohen, and W. A. Cramer. 1990. On the nature of the structural change of the colicin E1 channel peptide necessary for its translocation-competent state. Biochemistry **29**:5829–36.
54. Minton, A. P. 2000. Implications of macromolecular crowding for protein assembly. Curr Opin Struct Biol **10**:34–9.
55. Minton, A. P. 1997. Influence of excluded volume upon macromolecular structure and associations in 'crowded' media. Curr Opin Biotechnol **8**:65–9.
56. Minton, A. P. 2001. The influence of macromolecular crowding and macromolecular confinement on biochemical reactions in physiological media. J Biol Chem **276**:10577–80.
57. Minton, A. P. 2000. Protein folding: thickening the broth. Curr Biol **10**:R97–9.
58. Mirsky, A. E., and L. Pauling. 1936. On the structure of native, denatured and coagulated proteins. Proc Natl Acad Sci U S A **22**:439–47.
59. Morar, A. S., A. Olteanu, G. B. Young, and G. J. Pielak. 2001. Solvent-induced collapse of alpha-synuclein and acid-denatured cytochrome c. Protein Sci **10**:2195–9.

60. Munishkina, L. A., E. M. Cooper, V. N. Uversky, and A. L. Fink. 2004. The effect of macromolecular crowding on protein aggregation and amyloid fibril formation. J Mol Recognit **17**:456–64.
61. Neurath, H., J. P. Greenstein, F. W. Putnam, and J. O. Erickson. 1944. The chemistry of protein denaturation. Chem Rev **34**:157–265.
62. Neyroz, P., B. Zambelli, and S. Ciurli. 2006. Intrinsically disordered structure of *Bacillus pasteurii* UreG as revealed by steady-state and time-resolved fluorescence spectroscopy. Biochemistry **45**:8918–30.
63. Pace, C. N. 1986. Determination and analysis of urea and guanidine hydrochloride denaturation curves. Methods Enzymol **131**:266–80.
64. Permyakov, S. E., A. G. Bakunts, A. I. Denesyuk, E. L. Knyazeva, V. N. Uversky, and E. A. Permyakov. 2008. Apo-parvalbumin as an intrinsically disordered protein. Proteins **72**:822–36.
65. Permyakov, S. E., I. S. Millett, S. Doniach, E. A. Permyakov, and V. N. Uversky. 2003. Natively unfolded C-terminal domain of caldesmon remains substantially unstructured after the effective binding to calmodulin. Proteins **53**:855–62.
66. Privalov, P. L. 1979. Stability of proteins: small globular proteins. Adv Protein Chem **33**:167–241.
67. Privalov, P. L., and A. I. Dragan. 2007. Microcalorimetry of biological macromolecules. Biophys Chem **126**:16–24.
68. Ptitsyn, O. B. 1995. Molten globule and protein folding. Adv Protein Chem **47**:83–229.
69. Ptitsyn, O. B. 1987. Protein folding: hypotheses and experiments. J Prot Chem **6**:273–93.
70. Ptitsyn, O. B. 1995. Structures of folding intermediates. Curr Opin Struct Biol **5**:74–8.
71. Ptitsyn, O. B., and V. N. Uversky. 1994. The molten globule is a third thermodynamical state of protein molecules. FEBS Lett **341**:15–8.
72. Radivojac, P., L. M. Iakoucheva, C. J. Oldfield, Z. Obradovic, V. N. Uversky, and A. K. Dunker. 2007. Intrinsic disorder and functional proteomics. Biophys J **92**:1439–56.
73. Rodionova, N. A., G. V. Semisotnov, V. P. Kutyshenko, V. N. Uverskii, and I. A. Bolotina. 1989. Staged equilibrium of carbonic anhydrase unfolding in strong denaturants. Mol Biol (Mosk) **23**:683–92.
74. Romero, P., Z. Obradovic, X. Li, E. C. Garner, C. J. Brown, and A. K. Dunker. 2001. Sequence complexity of disordered protein. Proteins **42**:38–48.
75. Semisotnov, G. V., H. Kihara, N. V. Kotova, K. Kimura, Y. Amemiya, K. Wakabayashi, I. N. Serdyuk, A. A. Timchenko, K. Chiba, K. Nikaido, T. Ikura, and K. Kuwajima. 1996. Protein globularization during folding. A study by synchrotron small-angle X-ray scattering. J Mol Biol **262**:559–74.
76. Semisotnov, G. V., N. A. Rodionova, O. I. Razgulyaev, V. N. Uversky, A. F. Gripas, and R. I. Gilmanshin. 1991. Study of the "molten globule" intermediate state in protein folding by a hydrophobic fluorescent probe. Biopolymers **31**:119–28.
77. Shtilerman, M. D., T. T. Ding, and P. T. Lansbury, Jr. 2002. Molecular crowding accelerates fibrillization of alpha-synuclein: could an increase in the

cytoplasmic protein concentration induce Parkinson's disease? Biochemistry **41**:3855–60.

78. Sickmeier, M., J. A. Hamilton, T. LeGall, V. Vacic, M. S. Cortese, A. Tantos, B. Szabo, P. Tompa, J. Chen, V. N. Uversky, Z. Obradovic, and A. K. Dunker. 2007. DisProt: the database of disordered proteins. Nucleic Acids Res **35**:D786–93.
79. Tanford, C. 1968. Protein denaturation. Adv Protein Chem **23**:121–282.
80. Tcherkasskaya, O., and V. N. Uversky. 2001. Denatured collapsed states in protein folding: example of apomyoglobin. Proteins **44**:244–54.
81. Timm, D. E., P. Vissavajjhala, A. H. Ross, and K. E. Neet. 1992. Spectroscopic and chemical studies of the interaction between nerve growth factor (NGF) and the extracellular domain of the low affinity NGF receptor. Protein Sci **1**:1023–31.
82. Uversky, V. N. 1999. A multiparametric approach to studies of self-organization of globular proteins. Biochemistry (Mosc) **64**:250–66.
83. Uversky, V. N. 2002. Natively unfolded proteins: a point where biology waits for physics. Protein Sci **11**:739–56.
84. Uversky, V. N. 2003. Protein folding revisited. A polypeptide chain at the folding-misfolding-nonfolding cross-roads: which way to go? Cell Mol Life Sci **60**:1852–71.
85. Uversky, V. N. 1993. Use of fast protein size-exclusion liquid chromatography to study the unfolding of proteins which denature through the molten globule. Biochemistry **32**:13288–98.
86. Uversky, V. N. 2002. What does it mean to be natively unfolded? Eur J Biochem **269**:2–12.
87. Uversky, V. N., E. M. Cooper, K. S. Bower, J. Li, and A. L. Fink. 2002. Accelerated alpha-synuclein fibrillation in crowded milieu. FEBS Lett **515**:99–103.
88. Uversky, V. N., and A. L. Fink. 1999. Do protein molecules have a native-like topology in the pre-molten globule state? Biochemistry (Mosc) **64**:552–5.
89. Uversky, V. N., J. R. Gillespie, and A. L. Fink. 2000. Why are "natively unfolded" proteins unstructured under physiologic conditions? Proteins **41**:415–27.
90. Uversky, V. N., J. R. Gillespie, I. S. Millett, A. V. Khodyakova, R. N. Vasilenko, A. M. Vasiliev, I. L. Rodionov, G. D. Kozlovskaya, D. A. Dolgikh, A. L. Fink, S. Doniach, E. A. Permyakov, and V. M. Abramov. 2000. Zn(2+)-mediated structure formation and compaction of the "natively unfolded" human prothymosin alpha. Biochem Biophys Res Commun **267**:663–8.
91. Uversky, V. N., J. R. Gillespie, I. S. Millett, A. V. Khodyakova, A. M. Vasiliev, T. V. Chernovskaya, R. N. Vasilenko, G. D. Kozlovskaya, D. A. Dolgikh, A. L. Fink, S. Doniach, and V. M. Abramov. 1999. Natively unfolded human prothymosin alpha adopts partially folded collapsed conformation at acidic pH. Biochemistry **38**:15009–16.
92. Uversky, V. N., A. S. Karnoup, D. J. Segel, S. Seshadri, S. Doniach, and A. L. Fink. 1998. Anion-induced folding of staphylococcal nuclease: characterization of multiple equilibrium partially folded intermediates. J Mol Biol **278**:879–94.
93. Uversky, V. N., J. Li, and A. L. Fink. 2001. Evidence for a partially folded intermediate in alpha-synuclein fibril formation. J Biol Chem **276**:10737–44.

94. Uversky, V. N., J. Li, and A. L. Fink. 2001. Metal-triggered structural transformations, aggregation, and fibrillation of human alpha-synuclein. A possible molecular NK between Parkinson's disease and heavy metal exposure. J Biol Chem **276**:44284–96.
95. Uversky, V. N., C. J. Oldfield, and A. K. Dunker. 2005. Showing your ID: intrinsic disorder as an ID for recognition, regulation and cell signaling. J Mol Recognit **18**:343–84.
96. Uversky, V. N., S. E. Permyakov, V. E. Zagranichny, I. L. Rodionov, A. L. Fink, A. M. Cherskaya, L. A. Wasserman, and E. A. Permyakov. 2002. Effect of zinc and temperature on the conformation of the gamma subunit of retinal phosphodiesterase: a natively unfolded protein. J Proteome Res **1**:149–59.
97. Uversky, V. N., and O. B. Ptitsyn. 1996. All-or-none solvent-induced transitions between native, molten globule and unfolded states in globular proteins. Fold Des **1**:117–22.
98. Uversky, V. N., and O. B. Ptitsyn. 1996. Further evidence on the equilibrium "pre-molten globule state": four-state guanidinium chloride-induced unfolding of carbonic anhydrase B at low temperature. J Mol Biol **255**:215–28.
99. Uversky, V. N., and O. B. Ptitsyn. 1994. "Partly folded" state, a new equilibrium state of protein molecules: four-state guanidinium chloride-induced unfolding of beta-lactamase at low temperature. Biochemistry **33**:2782–91.
100. Uversky, V. N., S. Winter, and G. Lober. 1996. Use of fluorescence decay times of 8-ANS-protein complexes to study the conformational transitions in proteins which unfold through the molten globule state. Biophys Chem **60**:79–88.
101. Vacic, V., V. N. Uversky, A. K. Dunker, and S. Lonardi. 2007. Composition profiler: a tool for discovery and visualization of amino acid composition differences. BMC Bioinformatics **8**:211.
102. Williams, R. M., Z. Obradovi, V. Mathura, W. Braun, E. C. Garner, J. Young, S. Takayama, C. J. Brown, and A. K. Dunker. 2001. The protein non-folding problem: amino acid determinants of intrinsic order and disorder. Pac Symp Biocomput **6**:89–100.
103. Wright, P. E., and H. J. Dyson. 1999. Intrinsically unstructured proteins: re-assessing the protein structure-function paradigm. J Mol Biol **293**:321–31.
104. Wu, L. C., P. B. Laub, G. A. Elove, J. Carey, and H. Roder. 1993. A noncovalent peptide complex as a model for an early folding intermediate of cytochrome c. Biochemistry **32**:10271–6.
105. Zimmerman, S. B., and A. P. Minton. 1993. Macromolecular crowding: biochemical, biophysical, and physiological consequences. Annu Rev Biophys Biomol Struct **22**:27–65.
106. Zimmerman, S. B., and S. O. Trach. 1991. Estimation of macromolecule concentrations and excluded volume effects for the cytoplasm of *Escherichia coli*. J Mol Biol **222**:599–620.

20

DETECTING DISORDERED REGIONS IN PROTEINS BY LIMITED PROTEOLYSIS

Angelo Fontana, Patrizia Polverino de Laureto, Barbara Spolaore, Erica Frare, and Marcello Zambonin
CRIBI Biotechnology Centre, University of Padua, Padua, Italy

ABSTRACT

The limited proteolysis technique can be used to analyze protein structure and dynamics and, in particular, to identify disordered sites or regions within otherwise folded globular proteins. The approach relies on the fact that the proteolysis of a polypeptide substrate requires its binding and adaptation at the protease's active site and thus enhanced backbone flexibility or local unfolding of the site of proteolytic attack. A striking correlation was found between sites of limited proteolysis and sites of enhanced chain flexibility of the polypeptide chain, this last evaluated by the crystallographically determined B-factor. It is herewith shown that limited proteolysis often occurs at chain regions characterized by missing electron density, thus indicating that protein disorder occurs at these regions. It is concluded that limited proteolysis is a very useful and reliable experimental technique that can be used to probe protein structure and dynamics and, in particular, to detect sites of disorder in proteins, thus complementing the results that can be obtained by the use of other physicochemical and computational approaches. The peculiar advantages of this simple biochemical technique include the requirement of

Instrumental Analysis of Intrinsically Disordered Proteins: Assessing Structure and Conformation, Edited by Vladimir Uversky and Sonia Longhi

very minute amounts of protein sample, thus when other experimental techniques cannot be used.

20.1 INTRODUCTION

In the past few years, it became clear that a large number of proteins are devoid of structure or contain disordered regions along their polypeptide chain (57, 58, 62, 79, 199, 236, 242, 243, 249, 275). Intrinsically disordered (ID) proteins are those lacking stable secondary and tertiary structure under physiological conditions and in the absence of a binding partner/ligand. Biophysical properties of several ID proteins appear to be consistent with a fully unfolded molecule, but evidence has been provided that they may contain some local regions of residual structure. It has been shown that many ID proteins interact with target proteins, nucleic acids, and other ligands and that their binding to their cognate partners result in a conformational transition from a disordered to an ordered state (264, 277, 278). Evidence is accumulating that ID proteins comprise a large fraction of eukaryotic proteins, since possibly as many as 30% are either completely or at least partially disordered (60, 192, 237, 267). The abundance of ID proteins raises the important question of their biologic function. Indeed, the classical notion that proteins require a well-defined globular structure to be functional has been challenged (55, 56, 59, 61, 65, 250, 263). Thus, it has been proposed that it is time to move beyond the limited view that only a defined 3D structure is a requirement for protein function and to acknowledge the growing body of evidence that ID proteins exist and have functions (55, 59, 275). The role of ID proteins in mediating protein interactions has been considered, and nowadays, a favored view is that disordered proteins are more malleable, leading to advantages with respect to regulation and binding of diverse ligands (55, 56, 59, 61, 65, 236, 250, 263, 275). Of interest, it has been observed that a significant number of amyloidogenic proteins involved in severe and devastating diseases, such as Parkinson's, Alzheimer's, and prion diseases, are ID proteins (116, 247, 251). Therefore, the structural characterization of fully or partly disordered proteins appears to be relevant for analyzing the fundamental problem of protein folding and stability and the correlations between structure and biologic function, as well as for gaining insights into the molecular features of severe diseases involving protein aggregation (37, 49, 218, 247).

Proteins can display different levels of order and disorder. Proteins can be fully folded into a fixed, native-like 3D structure, and these proteins are the most well-characterized and abundant in the Protein Data Bank (PDB) database (18). However, even otherwise folded globular proteins can have chain regions of moderate length (10–30 residues) that are disordered, as given by the lack of backbone coordinates in their X-ray structures (142) or diminished chemical shift dispersion in nuclear magnetic resonance (NMR) measurements (17, 63, 64, 121). Moreover, folded proteins normally contain rather flexible or disordered chain segments at their N- and C-terminal ends. In the case of multidomain proteins, flexible linker segments are joining the domains, thus allowing the domain movements that determine protein dynamics and function (16). Finally, a protein can adopt a partly folded state, nowadays named molten globule (MG), defined as retaining significant secondary structure, but lacking tertiary interactions (140, 189, 190). Therefore, proteins appear to exhibit a continuum of structures, ranging from a tightly and compactly folded state, an otherwise folded state but containing local flexible segments or disordered regions, an MG state, and highly extended states as that displayed by ID proteins. An understanding of the molecular features of these various protein states would pave the way for the elucidation of structure–function relationships in proteins and for analyzing the mechanism of protein folding, considering that the acquisition of the final stable and ordered protein structure starting from its denatured state requires folding intermediates that need to be characterized (9, 72, 78, 245).

The identification and analysis of the molecular properties of fully or partly denatured proteins is extremely interesting and timely. Indeed, techniques have only recently been developed to analyze the structural features of unfolded and partly folded proteins in solution (63, 64, 72, 206). While X-ray crystallography and NMR spectroscopy are the main tools in 3D-structure determination of globular proteins at atomic resolution, these techniques suffer from limitations if applied to unravel molecular aspects of protein disorder. The use of X-ray crystallography for analyzing fully disordered proteins is precluded, since these proteins cannot be crystallized due to their inherent flexibility. However, X-ray methods can also provide information regarding the dynamics of a protein (94, 198, 219). Indeed, the crystallographically determined temperature factor, usually called B-factor, can be used as a reliable source of information of protein dynamics (19, 139, 212). The profile of B-factor values along the protein polypeptide chain can be used to identify regions that display enhanced values of B-factors, thus identifying the most flexible or disordered chain regions. Moreover, missing electron density in protein structures very likely corresponds to disordered regions, since the absence of interpretable electron density for some sections of the structure is associated with the increased mobility of atoms in these regions, which leads to the noncoherent X-ray scattering and makes atoms invisible. Therefore, the lack of backbone coordinates of a specific region of the polypeptide indicates that this region is highly flexible or disordered.

Spectroscopic methods such as NMR can be successfully used for the characterization of unfolded states of proteins, including MG states and ID proteins in particular. Several reviews have focused on the use of NMR to characterize the structure and dynamics of nonnative states of globular proteins, including ID proteins (17, 63, 64, 121). However, NMR measurements require substantial amounts of a nonaggregating protein sample at millimolar concentration. Other methods for analyzing protein disorder include circular dichroism (CD), fluorescence, Raman spectroscopy, and Fourier-transform infrared spectroscopy (FTIR), but these methods can be used for analyzing the average conformation of a protein and do not provide sequence-specific information. The hydrogen–deuterium (H/D) exchange technique is a useful technique for analyzing protein dynamics (71, 135, 195). Regions of intrinsic disorder can be established by analyzing the efficiency of H/D exchange using NMR, FTIR, and mass spectrometry (MS) experiments. The technique involves H/D exchange of protein backbone amide hydrogens and subsequent detection of the deuterated regions of the protein sample after its proteolytic digestion and subsequent analysis of deuterated protein fragments by MS techniques (266). The amount of deuterium incorporated by a fragment reflects the mobility of the corresponding region in the intact protein, considering that chain segments hydrogen bonded in a regular secondary structure (helices, strands) allow much less H/D exchange than unfolded/disordered regions (195, 266). Size exclusion chromatography, small-angle X-ray scattering, and hydrodynamic measurements can provide insights into the dimensions and globularity of a protein, and indeed these methods have been applied for characterizing ID proteins. Even sodium dodecyl sulfate (SDS)-gel electrophoresis can be used to detect ID proteins, since it has been found that these proteins display an aberrant mobility (117). However, all experimental methods can have advantages and drawbacks, and the results obtained by each technique can complement those obtained by other techniques. Some of these methods are highly demanding in terms of amounts of purified protein, special instrumentation, and expertise, and they can also be time-consuming. For a list and brief description of the various techniques that can be used for analyzing protein disorder, see the website of DisProt (www.disprot.org) (210).

Finally, we must mention herewith that recently a number of algorithms have been developed for predicting, based on the protein amino acid sequences, whether a protein is ID or contains disordered chain segments/regions (23, 77, 170). These algorithms aim at predicting the ordered/disordered status of a given sequence based on the physicochemical properties of the protein sequence alone, thus not taking into account the folded structure of the mature protein. A curated database of disordered proteins can be found in DisProt, including the associated functions for many of them. It was shown that these algorithms can produce satisfactory predictions, but still they can generate false positives and negatives (77). Nevertheless, there is a need of methods for a direct experimental identification of ID proteins and disordered chain regions.

Here, we focus on the use of limited proteolysis yielding a reliable identification of disordered or at least highly flexible regions in otherwise folded proteins (83–87, 89, 257). We highlight that this technique can provide very useful structural information that complement the results obtained using other physicochemical and computational methods. In particular, we show how clean the correlation is between the results of proteolysis experiments and those derived from X-ray crystallography. No attempt has been made to cover all aspects of the use of limited proteolysis approach for unraveling features of structure and dynamics of proteins in their native, partly, or fully denatured states. The reader, therefore, may find here some personal selection of issues covered and examples of experimental studies herewith presented and commented. Our laboratory was involved since two decades in studying molecular aspects of the limited proteolysis phenomenom and, when possible, we will refer to the results of experiments conducted in our own laboratory.

20.2 THE LIMITED PROTEOLYSIS APPROACH

20.2.1 General

The term "limited proteolysis" was coined by Linderstrøm-Lang to indicate the specific fission of only one peptide bond (or a few) among the many present in a protein molecule substrate (145). Limited proteolysis phenomena occur in a variety of physiologically relevant processes, such as blood coagulation, fertilization, protein transport across membranes, posttranslational protein processing, peptide hormones production from their larger protein precursors, zymogen activation, and others (149, 166–168). Often, limited proteolysis of a globular protein substrate results in cleaving only one peptide bond, thus leading to a nicked protein given by two fragments remaining associated in a stable and often functional complex. The most classical example is that of bovine ribonuclease A, which is selectively cleaved by subtilisin at peptide bond Ala20–Ser21, thus leading to the nicked ribonuclease S constituted by peptide S (fragment 1–20) and protein S (fragment 21–124) (196). Of course, the fact that a rather unspecific protease as subtilisin cuts the 124-residue chain of ribonuclease A at a single peptide bond is intriguing. Clearly, it is immediately evident that the higher-order structure of the protein and not its amino acid sequence is dictating this extraordinary selectivity of proteolytic attack. Therefore, an understanding of the molecular features underlying the biorecognition phenomenon between a protease and a polypeptide substrate leading to specific peptide fissions is of paramount importance for shedding light into the molecular aspects and mechanisms of a variety of physiologically relevant processes (149, 203).

A widely used nomenclature to describe the interaction of a polypeptide substrate at the protease's active site is that introduced by Schechter and Berger (205). As shown in Figure 20.1, it is considered that the amino acid residues of the polypeptide substrate bind at subsites of the active site. By convention, these subsites on the protease are called S (for subsites) and the substrate amino acid residues are called P (for peptide). The amino acid residues of the N-terminal side of the scissile bond are numbered P3, P2, and P1, and those residues of the C-terminal side are numbered P1′, P2′, P3′. ... The P1 or P1′ residues are those residues located near the scissile bond. The substrate residues around the cleavage site can then be numbered up to P8. The

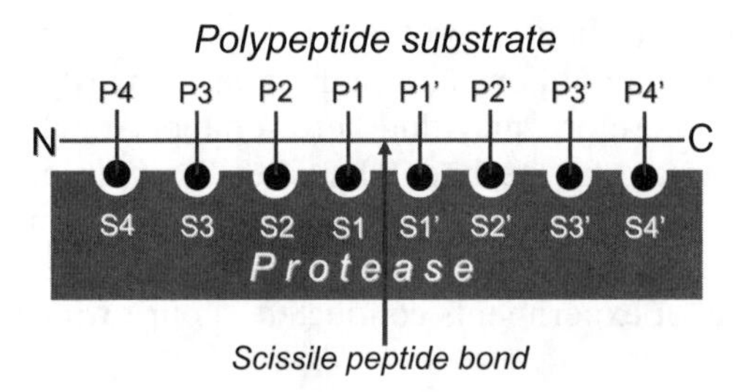

Figure 20.1 Schematic representation of the binding of a polypeptide substrate at the active site of a protease. An eight-residue segment of a peptide chain interacts with its side-chain residues (P) at a series of subsites (S) of the protease. The interaction of the substrate at the active site of the protease requires a specific stereochemical adaptation of the substrate, and thus likely a significant degree of chain mobility, in order to form the idealized transition state of the hydrolytic reaction. Up to 12 residues of the substrate can be involved in the interaction. The P1 side-chain residue interacting with the S1 binding site of the protease is usually the major determinant for the protease's specificity, but is not unique. Usually, proteases cleave at the carboxyl-terminal side of the scissile bond, while thermolysin instead cleaves at the amino side. The figure and nomenclature are adapted from Schechter and Berger (205).

subsites on the protease that complement the substrate binding residues are numbered S3, S2, S1, S1′, S2′, S3′. ... In several studies, it has been established that the protease–peptide/protein substrate interaction involves a stretch of up to 12 amino acid residues (112, 114, 115).

Enhanced chain flexibility (segmental mobility) is the key feature of the sites of limited proteolysis of a globular protein substrate, as clearly emphasized for the first time by the results of systematic proteolysis and autolysis experiments conducted on the thermophilic protease thermolysin (83). As shown in Figure 20.2, it was demonstrated that there is a clear-cut correlation between sites of limited proteolytic attack and sites of enhanced chain flexibility, the latter reflected by the crystallographic temperature factors (B-values). Indeed, the crystal structure of a protein includes information about the thermal and other fluctuations of the atoms in a protein crystal. To each atom, a Debye–Waller temperature factor or B-factor can be assigned, with the latter being proportional to the mean square amplitude of the fluctuations (19, 94, 139, 198, 212, 219). Although important, B-factors can be an imperfect measure of the true chain flexibility or segmental mobility of a protein, since they result from both thermal motions and static disorder (198). Moreover, reduced B-factors can derive from intermolecular crystal packing interactions, which reduce the flexibility of those residues involved in packing and, therefore, damp down the segmental mobility. In agreement with this, different crystal forms of proteins have been shown to display different B-factor profiles along their chain (19, 139, 212). Nevertheless, our systematic analysis of numerous limited proteolysis data in terms of structure and dynamics of pro-

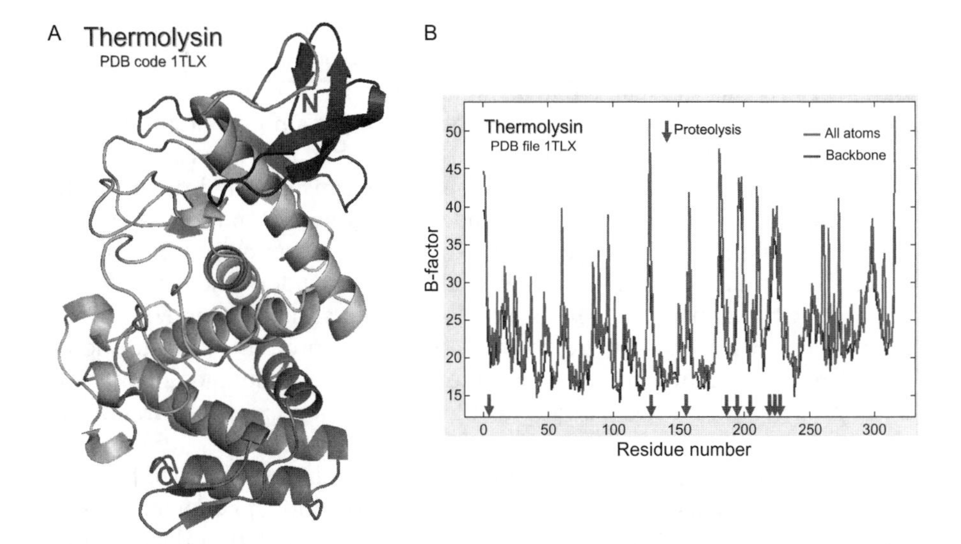

Figure 20.2 Correlation between sites of enhanced chain flexibility (segmental mobility) and sites of limited proteolysis in thermolysin. (A) 3D structure of thermolysin taken from the RCSB Protein Data Bank (PDB code 1TLX). The protein model was prepared using the PyMol molecular graphics system (www.pymol.org) (47). (B) The profile of the B-factor along the 316-residue chain of thermolysin. The sites of limited proteolysis or autolysis along the protein chain are indicated by arrows. Proteolysis events resulted in nicked protein species constituted by two or even three fragments associated in rather stable protein complexes (83). See color insert.

teins of known 3D structure determined crystallographically revealed that enhanced chain flexibility or protein disorder detected by high B-factor values or missing electron density, respectively, are reliable parameters of protein flexibility/disorder (83–87, 89, 257).

Chain flexibility of the site of limited proteolytic attack is in keeping with the idea that the protein substrate must undergo considerable structural changes for its binding and adaptation at the precise 3D structure of the protease's active site in order to form the optimal transition state needed for hydrolysis (83, 87). Indeed, modeling studies of the conformational changes required for proteolytic cleavage indicate that the sites of limited proteolysis must suffer a large conformational change (local unfolding) of a stretch of up to 12 residues (112, 114, 115). Limited proteolysis of a globular protein, therefore, occurs at flexible loops, and so chain segments in a regular secondary structure (such as helices) are not sites of limited proteolysis. As we will show below, the technique is eminently suitable to detect sites of chain flexibility or protein disorder.

The key role of flexibility in the digestion of a polypeptide substrate is substantiated by the recent systematic analysis of the recognition mechanism of proteases for polypeptide substrates and inhibitors (239, 240). This analysis, made possible by the recent availability of many protease-inhibitor crystallographic structures, revealed that an almost universal recognition mechanism by all proteolytic enzymes implies that the binding of a polypeptide substrate at the active site of the protease occurs in an extended β-strand conformation (240). This implies that a polypeptide substrate can bind and adapt at the active site of a protease only in its extended conformation and that, therefore, a protease can select only the extended strand from a conformational ensemble of polypeptide structures. Consequently, we may understand why a folded and quite rigid globular protein is usually rather resistant to proteolysis, and we may anticipate that for proteolysis to occur the protein substrate should suffer significant conformational transition in order to bind and adapt at the enzyme's active site a stretch of up to 12 residues in an extended strand conformation (115).

The implication of the fact that substrate binding in the extended β-strand conformation appears to be a universal biorecognition phenomenon with proteases (240) is that the rate of proteolysis is much faster after denaturation of the protein substrate, as demostrated by a large variety of experimental studies. For example, the largely unfolded apocytochrome *c* obtained by the removal of the covalently bound heme moiety is degraded more easily and faster by proteases than the holo protein or when embedded in a folded, noncovalent complex with heme (214). In particular, the very different rate of proteolysis of a globular protein substrate in respect to that of an unfolded/disordered substrate has been demonstrated by determining the kinetics of the selective trypsin hydrolysis of the Arg77–Val78 peptide bond in the folded versus unfolded ribonuclease T1 (173). This study was made possible by the fact that (1) ribonuclease T1 contains only two peptide bonds (Lys41–Thr42 and Arg77–Val78) susceptible to tryptic hydrolysis and cleavage at Lys41 can be blocked by selective Nε-acetylation and (2) this protein unfolds completely when the two disulfide bonds are broken by reduction. The results of this elegant study indicated that the Arg77–Val78 peptide bond is cleaved at least 1700 times more rapidly in the unfolded than in the folded species of ribonuclease T1.

Accepting the view that proteolysis overhelmingly requires a polypeptide substrate in an extended β-strand conformation (240), we may predict that proteolysis of an otherwise globular protein can occur only at the flexible/disordered loops connecting elements of secondary structure (helices, strands) (83–87, 89, 257). Indeed, as we will show below, highly flexible loops or disordered chain segments in globular proteins are the most suitable targets for limited or site-specific proteolysis, since these are able to properly bind and adapt at the protease's active site in an extended conformation. Often in past and current literature, limited proteolysis phenomena are wrongly explained in terms of surface "exposure" of the site of proteolytic attack, but this terminology of exposure, accessibility, or protrusion is fully insufficient to describe the selective proteolytic event, since it is evident that even in a small globular protein, there are many exposed sites which could be targets of proteolysis (87).

20.2.2 Methodology

The limited proteolysis approach for probing protein structure and dynamics implies that the proteolytic event should be determined by the stereochemistry and flexibility of the protein substrate and not by the specificity of the attacking protease (86, 87). Clearly, by using the substrate-specific proteases as trypsin or V8-protease, cleavages are not expected even with an unfolded polypeptide substrate, if this last does not contain Lys or Arg residues (trypsin) or Glu residues (V8-protease) (54). Therefore, these specific proteases will not be useful to detect flexible/disordered sites of a protein. The most suitable proteases are those displaying broad substrate specificity, such as subtilisin (154), thermolysin (107), proteinase K (66), and pepsin (95). These endopeptidases display only a moderate preference for hydrolysis at hydrophobic or neutral amino acid residues, but often cleavages occur at other residues as well (21). The reader is referred to the website of the Peptide Cutter for a comparative analysis of substrate specificity of several proteases (www.expasy.ch/tools/peptidecutter).

In order to facilitate the analysis of the resulting proteolytic fragments, it is advisable to conduct initial experiments using proteases of narrow specificity, that is, with proteolytic activity against one or two amino acid types. The most used proteolytic enzymes are endoproteases as trypsin or V8-protease. Then, the information obtained with specific proteases will facilitate the identification of sites of cleavage obtained by using less specific proteases such as proteinase K or subtilisin. Usually, exopeptidases such as amino- or carboxypeptidases are not used for limited proteolysis, even if few specific applications of these proteases have been described in the literature to remove short-chain segments at the amino- or carboxy-terminal ends of proteins (255).

It is not easy to predict in advance the most useful experimental conditions for conducting a limited proteolysis experiment, since these actually depend on the structure, dynamics, and stability/rigidity properties of the protein substrate, as well as the actual aim of the experiment, for example, isolation of the rigid core of the protein or location of the sites of protein flexibility/disorder (86). The most commonly used enzyme:substrate (E:S) ratio to be employed is 1:50 or 1:100, but E:S ratios as low as 1:1000 or 1:5000 are required if the protein is very labile and thus proteolysis occurs very quickly. Similarly, the reaction time can vary between one minute and several hours or even days (86).

In order to correlate limited proteolysis data with the structure and dynamics of the intact native protein, as deduced for example by X-ray crystallography, it is very important to rank proteolysis events into the initial and subsequent cleavages (86). The most informative sites of proteolysis are those that can be classified as initial sites, since subsequent cleavages occur at a perturbed protein substrate not necessarily retaining the overall structure and dynamics of the intact native protein. Therefore, limited proteolysis experiments should be devised in order to monitor the kinetics of proteolysis. It is clear that if one waits long enough, many peptide bond fissions occur, and the resulting proteolysis is not at all limited or selective. Under these conditions, proteolysis experiments are clearly less informative.

The usual way to analyze the time course of proteolysis is by SDS-polyacrylamide gel electrophoresis and reverse-phase HPLC. These analyses can provide not only the first indication of the time course but also the extent of the proteolysis reaction. Since limited proteolysis is a kinetic process involving a bimolecular reaction between the protein substrate and the protease, it is possible to predict, on the basis of initial experiments, how to adjust the E:S ratio, reaction time, protein substrate concentration, and incubation temperature in order to either limit or enhance the extent of protein digestion. Usually, this is most interesting and useful for detecting the initial peptide bond fissions of the protein substrate, for example, in order to identify the most flexible or disordered sites of the protein chain (83, 86, 87).

Protein chemistry methods combining electrophoresis or chromatography with N-terminal sequencing by the Edman technique can be used to establish the identity of the polypeptide fragments and thus to identify nicksites along the protein chain. These methods can be quite labor intensive and relatively highly demanding in terms of protein sample requirements. On the other hand, nowadays the identification of protein fragments can be made much more easily by MS techniques. In this case, the analysis can require minute amounts of protein sample (1–10 ng, femtomoles) and can be completed within minutes (1, 3, 28, 52, 153, 270, 281). Indeed, MS methods provide a powerful method due to the mass accuracy that allows a precise fragment identification and, using tandem MS, they can also be used for the partial sequencing of peptides. Of course, the time-consuming steps of electrophoretic and chromatographic separation of protein fragments and their subsequent N-terminal sequencing can be eliminated, since it is possible to analyze directly by MS methods proteolysis mixtures and perform partial sequencing using tandem MS (3, 52, 270).

It is likely that the recent developments of MS techniques will prompt a more general and widespread application of limited proteolysis in protein research (28, 52, 281). Indeed, MS methods can be used effectively for protein fragment identification, even when they are present in a mixture and in very minute amounts. MS data allow an easy definition of the sites of cleavage if the amino acid sequence of the protein is known. A specific advantage of ESI-MS is that the separated protein fragments being eluted from an HPLC column can be directly injected into the ion source of the MS instrument. Indeed, the combined use of limited proteolysis and MS techniques has already been described in numerous recent papers. In a recent study, it was demonstrated that the MS analysis of the time course of limited proteolysis reactions can provide information that is self-sufficient to identify all proteolytic fragments, thus leading to a significant improvement over the traditional protein chemistry methods relying on the N-terminal sequencing method (231).

20.3 ANALYSIS OF PARTLY FOLDED STATES

In recent years, numerous studies have been conducted on the partially folded or MG states that proteins can adopt under specific solvent conditions, such as at low pH, in the presence of moderate concentrations of denaturants or after the removal of protein-bound metal ions or cofactors (140, 189, 190). The interest in MGs resides in the fact that they are considered protein folding intermediates. The key characteristics of an MG include a native-like secondary structure, lack of specific tertiary interactions, and a more expanded and flexible structure than that of the native protein (140, 189, 190). The structural analysis of MGs was expected to be difficult, since these states are not only flexible but can also be an ensemble of conformations, so elucidating their structure by either X-ray crystallography or NMR could be difficult if not impossible (72). Nevertheless, recent developments in H/D exchange, combined with two-dimensional NMR spectroscopy, have provided quite a detailed structural information for MGs (63, 64, 72).

In our laboratory, we have shown that limited proteolysis experiments can be very useful in analyzing the structural features of MGs of several model proteins, such as α-lactalbumin (α-LA), lysozyme, apomyoglobin (apoMb), and cytochrome *c*. Overall, the chain regions identified as mobile or unfolded by proteolysis closely correlated with those detected by using other physicochemical and spectroscopic measurements. In the following, we will discuss in detail the results of limited proteolysis experiments aimed to unravel conformational features of the MGs adopted by the calcium-free α-LA (apo-LA) when exposed to mildly denaturing conditions in acid solution or in the presence of trifluoroethanol or oleic acid.

20.3.1 The MG of α-LA

The conformational state of α-LA exposed to acid pH 2.0 (A-state) has been investigated in great detail using a variety of experimental approaches and techniques by numerous investigators (4, 51, 141, 175). As a result of these efforts, nowadays the A-state of α-LA is regarded as a prototype protein MG (141). These studies were conducted on bovine, human, goat, and guinea pig α-LA, but the conformational features of these homologous proteins are very similar, and thus results obtained with α-LA from different sources and their interpretations can likely be used interchangeably (141). NMR and H/D exchange measurements revealed that α-LA in acid solution adopts a partly folded or MG state characterized by a disordered β-domain, whereas the α-domain maintains substantial, albeit dynamic, helical secondary structure (2,

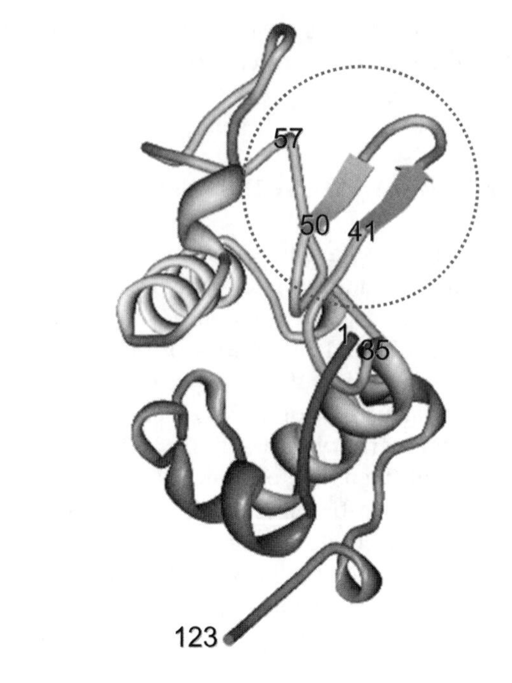

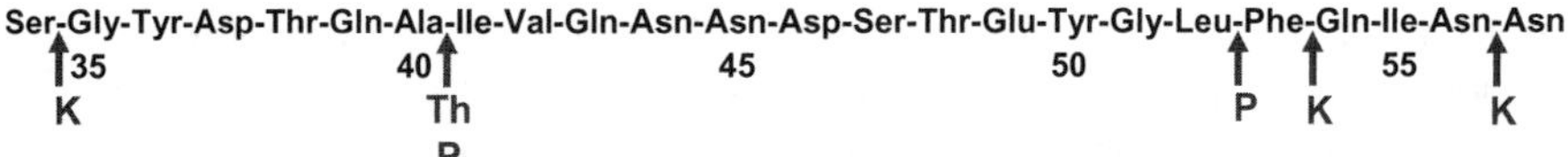

Figure 20.3 Probing local unfolding in the α-lactalbumin's (α-LA) molten globule by proteolysis. Schematic representation of the crystal structure of the 123-residue chain of α-LA. The model was produced with the MBT software (161) available in PDB (code 1HFZ). The chain segment of α-LA in its MG state that suffers limited proteolysis is circled by a dashed line. The amino acid sequence of the chain region (residues 34–57) of α-LA is explicitly shown. Arrows indicate the sites of peptide bond fission by proteinase K (K), pepsin (P), and thermolysin (Th). Limited proteolysis of α-LA exposed to acid pH, in the presence of trifluoroethanol or by addition of oleic acid occurs at the level of the β-domain. This implies that essentially, only this region of the 123-residue chain of α-LA becomes rather flexible under different protein denaturing conditions and thus amenable to proteolytic attack. Moreover, proteolysis data indicate that the α-domain retains sufficient structure and rigidity to resist to proteolytic attacks. The local unfolding of the β-domain in the molten globule state of α-LA has been observed also by means of other spectroscopic measurements (see text).

42, 191). The α-domain is a discontinuous domain given by the N-terminal segment 1–37 and the C-terminal segment 85–123 and comprises all helical segments of the protein, whereas the β-domain is given by the remainder of the protein chain encompassing the β-strands and a coil region (179) (see Fig. 20.3).

The calcium-depleted form of α-LA, as obtained by dissolving the protein at neutral pH in the presence of EDTA, also adopts an MG state at neutral pH, but a moderate heating is required (141, 175). There are discrepancies in the reported experimental results and conflicting proposals regarding the conformational state of apo-LA, ranging from a classical MG devoid of a cooperative thermal transition to a partly folded state with some native-like properties and displaying instead cooperativity (141). Often, it was assumed that apo-LA, as obtained for example by dissolving the protein at room temperature in Tris buffer, pH 8.0, containing a calcium-chelating agent, adopts an MG state, but this may not be true without specifying ionic strength and temperature of the protein solution. It became clear that the conformational state adopted by apo-LA is strongly influenced by the specific solvent conditions (141, 175).

With the view to deduce conformational features of α-LA in its MG or partly folded state, in our laboratory, we have conducted a series of limited proteolysis experiments on α-LA exposed to various mild perturbing conditions, such as exposing α-LA in acid solution (181) or at neutral pH in the presence of EDTA, trifluoroethanol, or oleic acid (183). We investigated the effect of the fatty acid on the structural features of α-LA, since it has been reported that an ill-defined oleic acid/α-LA complex displays the unusual, but interesting, property of killing tumor but not healthy cells by an apoptosis-like mechanism (102, 225). It was found that various MGs of α-LA obtained under different solvent conditions all suffered limited proteolysis at a rather short portion of the 123-residue chain of the protein (see Fig. 20.3). The conclusion reached from our studies (85, 87, 181, 183) was that the chain region approximately from residue 34 to 56, encompassing most of the β-domain, was disordered, while the rest of the protein chain remains folded and sufficiently rigid to resist proteolysis (184, 188). Therefore, proteolysis data are in agreement with the results of other physicochemical measurements indicating that the MG of α-LA is characterized by an unfolded β-domain and a native-like α-domain (2, 4, 42, 191, 207) (see Fig. 20.3). Since the protein in its MG state retains a rather large portion of its polypeptide chain stable and rigid enough to resist proteolysis, it was possible to remove selectively the β-domain from the 123-residue chain of α-LA and to isolate a stable, folded "gapped" protein species given by fragment 1–34 covalently linked to fragment 54/57–123 by disulfide bridges (188).

The strong reduction of the signal in the near-UV CD when α-LA adopts the MG state, as for example when exposed in acid solution (A-state), was taken as evidence that the tertiary structure of the protein is lost (51). However, this assumption was shown later not to be true, on the basis of results of a variety of studies, mostly conducted by 2D-NMR and H/D exchange measurements (207). Indeed, the MG of α-LA produced in acid solution or at neutral pH in the presence of a chelating agent and at moderately high temperature, reveals that some tertiary structure of the native protein is still present (2, 141, 207). Also, the results of proteolysis experiments clearly indicate that the gross overall structure of the protein is retained in the protein MG, with the exception of the β-domain region (181, 183, 184, 188). Therefore, the MG of α-LA does not adhere to the original definition of the MG being a collapsed conformational state retaining a native-like content of secondary structure, but fully lacking tertiary interactions (189).

Nowadays, it is clear that proteins can display a great variety of partly folded states, ranging from those highly denatured to those much similar to the native state. As a matter of fact, protein MGs are considered generic protein folding intermediates. It is our belief that it is more straight and simple to relate the partly folded states of apo-LA to the native than to the MG state. Indeed, as we will see below, proteolysis experiments conducted on α-LA are not much different from those obtained with other protein systems. In the following, we will discuss and interpret proteolysis experiments that lead to a specific nick or to a removal of a segment of a protein chain without the need to invoke an explanation of these results in terms of an MG state of proteins.

20.4 REMOVING LOOSE, FLEXIBLE PARTS FROM A PROTEIN

The N- and C-terminal ends of proteins are usually more flexible than the rest of the chain, as shown by the results of X-ray analysis of proteins in the solid state, as well as in solution, by utilizing spectroscopic methods. A loose and flexible N- or C-terminal stretch sequence will be more sensitive to adventitious proteolysis during expression and purification. This explains the fact that several native proteins possess ragged N- and C-terminal ends, as a result of a relatively easy proteolysis at these chain sites mostly by the action of exopeptidases. Even well-structured proteins may contain disordered tails that can inhibit protein crystallization (133). Limited proteolysis appears to be the technique of choice for removing these loose disordered segments in order to produce protein samples that can form useful crystals. This approach has been used often for obtaining crystals that better diffract and lead consequently to higher-resolution data (97). The possibility of removing the loose parts from a protein moiety by proteolysis has been tested in a number of cases. Below we will comment a few specific examples examined in our laboratory.

20.4.1 Fragment 1–44 of Human Growth Hormone (hGH)

Fragment 1–44 of hGH, prepared *in vitro* by limited proteolysis of the hormone with pepsin at low pH (215, 216), encompasses in full the N-terminal helix 9–34 of this four-helix bundle protein (46, 241). This fragment, which is rather disordered in isolation, can bind heme and acquires a helical secondary structure upon heme binding (215). Heme appears to be bound to the fragment in a stereospecific way, since an induced dichroic signal is observed in the Soret region of the CD spectrum. It was proposed that the heme-binding properties

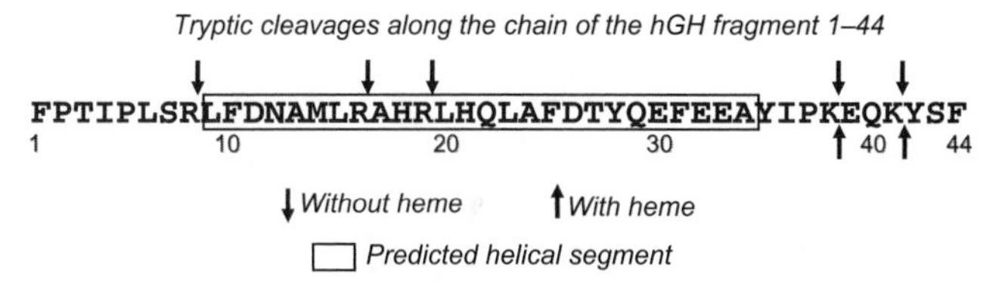

Figure 20.4 Proteolysis with trypsin of the human growth hormone (hGH) fragment 1–44 in the absence or presence of heme. The amino acid sequence of the fragment is shown, and the location of the N-terminal helix observed in intact hGH is indicated by a box (residues 9–34). Arrows indicate the sites of cleavage by trypsin in the absence or presence of heme (215).

of the fragment reside in the amphipathic character of the helix adopted by the fragment upon binding of the heme moiety, as well as in the presence in its polypeptide chain of His18, His21, and Met14 (see Fig. 20.4). These residues can act as specific ligands for the heme iron, as observed with cytochromes (215).

The hGH fragment 1–44 alone is much more susceptible to tryptic digestion than the heme/fragment complex, implying a more folded and rigid structure of this last species (87). In the absence of heme, the tryptic peptides produced from fragment 1–44 derive from cleavages at the Lys and Arg residues of the fragment. While the free fragment is fully digested by trypsin after a few minutes of reaction, in the presence of heme, the fragment is almost fully resistant to proteolysis and trypsin cuts are observed only at its C-terminal end, which is expected to be disordered, if the fragment complexed to heme attains a native-like secondary structure from Leu9 to Ala34 (Fig. 20.4). Therefore, the proteolytic probe substantiates that the heme fragment is in a quite rigid, hydrogen-bonded, helical secondary structure, likely encompassing the chain segment 9–34 as in the intact hormone (46, 241). Of note, proteolysis at Arg8 does not occur in the heme complex, since this Arg residue is likely rigidified by the first turn of the helix (see Fig. 20.4).

20.4.2 Hirudin

Hirudin is a thrombin inhibitor isolated from leeches (220). The most studied proteins are those isolated from *Hirudo medicinalis* (variant HV1) (50) and *Hirudo manillensis* (variant HM2) (204). The inhibitor is constituted by 64 or 65 residues and held compact by three disulfide bridges (Fig. 20.5). Hirudin has been intensively investigated as a potential protein drug with anticoagulant activity, since this mini-protein is the most potent and specific inhibitor of thrombin, a key enzyme in the coagulation cascade. Hirudin has a tadpole-like shape, being constituted of an N-terminal disulfide cross-linked core

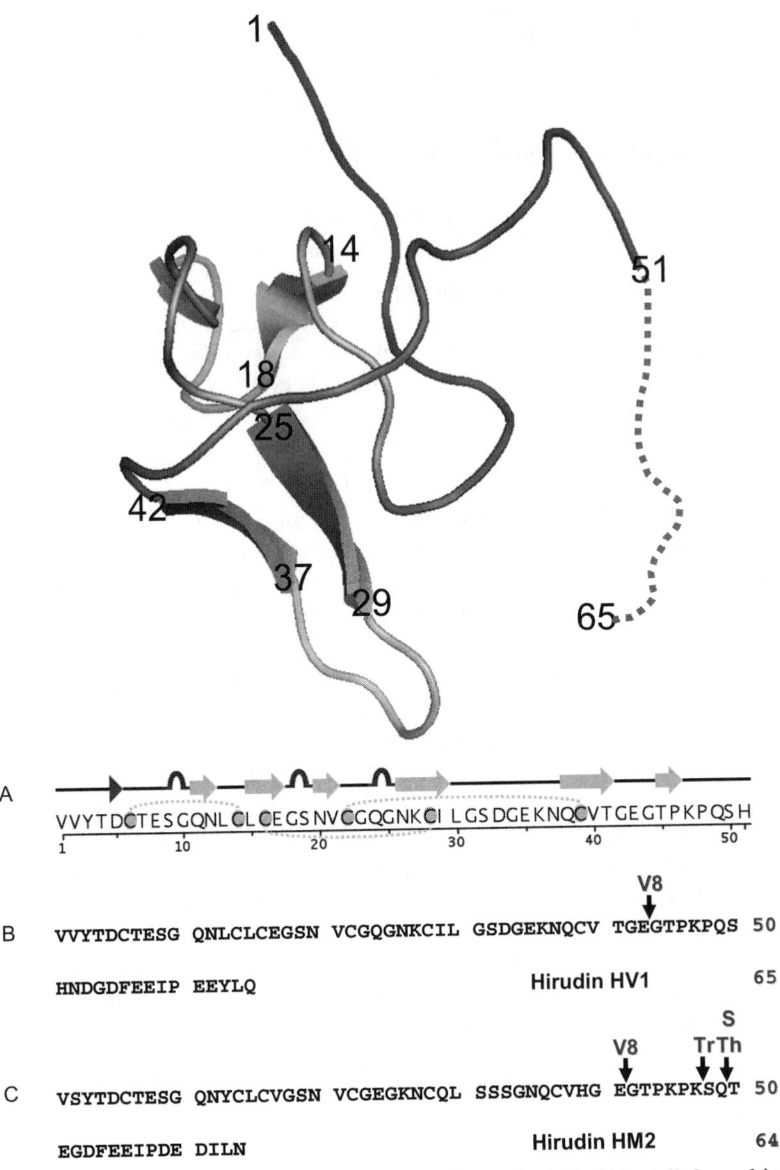

Figure 20.5 Limited proteolysis removes the C-terminal flexible tail from hirudin. (Top) Schematic three-dimensional (3D) solution structure of chain segment 1–51 of hirudin from *Hirudinaria medicinalis* HV1 as obtained by NMR measurements (228) (PDB code 1HIC). The model was constructed with the PyMol software (161) (www.pymol.org). The C-terminal tail 51–65 is not determined by NMR and thus is disordered. (Bottom) The amino acid sequence of the N-terminal 51-residue chain of hirudin HV1 is shown (A). Arrows indicate the location and dimensions of β-sheet secondary structure. The amino acid sequences of the full polypeptide chains of hirudin HV1 from *Hirudinaria manillensis* (B) and hirudin HM2 from *Hirudo medicinalis* (C) are shown. Hirudin HV1 and hirudin HM2 are a 65- and 64-residue chain, respectively. The amino acid sequences of the two hirudin variants are highly similar (75%). The disulfide bonding of hirudin HV1 (A) is indicated by dashed lines connecting pairs of Cys residues. The sites of limited proteolysis of hirudin HV1 (34) by the Glu-specific protease from *Staphylococcus aureus* V8 and hirudin HM2 (256) by subtilisin (S), thermolysin (Th), trypsin (Tr), and V8-protease (V8) are indicated by arrows.

domain and a C-terminal tail. Both X-ray (124, 201) as well as NMR (105, 227, 228) data are available for the crystal and solution structure of hirudin, respectively. A dual binding of hirudin to thrombin was elucidated by X-ray analysis of the complex hirudin–thrombin (124, 201, 227). While the N-terminal core domain of hirudin binds at the active site of thrombin, its C-terminal tail binds at the fibrinogen binding exosite of thrombin and contributes to the formation of a very tight protein complex. NMR measurements conducted on hirudin in solution revealed that the C-terminal tail is disordered, while a well-defined structure was obtained for the N-terminal fragment 1–51 (105, 228). Of interest, the solution structure of the core fragment 1–51 is very similar to that in the corresponding chain region in the hirudin–thrombin complex determined crystallographically (227).

Detailed studies of limited proteolysis of hirudin HV1 and hirudin HM2 revealed that proteolysis occurs only at the level of the C-terminal tail, thus allowing the isolation of N-terminal core fragments (see Fig. 20.5). Several interesting fragments were isolated from both HV1 and HM2 hirudin, and they were shown to maintain inhibitory action toward thrombin by binding at its active site (33, 34, 256). Of interest, a Trp3 analogue of the N-terminal fragment 1–47 was synthesized by solid-phase chemical methods and shown to be even more potent inhibitor than the natural Tyr3 species (45). Therefore, the loose flexible tail of hirudin can be removed neatly from the hirudin chain by proteolysis, in agreement with the view that proteolysis of globular proteins occur only at disordered regions.

20.4.3 Fragments of Thermolysin

Several C-terminal fragments of thermolysin are capable of folding into a native-like structure independently from the rest of the polypeptide chain, thus possessing protein domain properties (44, 75, 258–261). These fragments of thermolysin were prepared by cyanogen bromide cleavage of the protein at the level of methionine residues in positions 120 and 205 of the 316-residue chain of thermolysin (261). Of course, the dimensions of the fragments (1–120, 121–205, and 206–316) were dictated by the location of the Met residues in the protein chain. With the aim to define the minimum size of a C-terminal fragment capable to acquire a stable, native-like conformation, the 111-residue fragment 206–316 was digested by several proteases (44) (Fig. 20.6A). From the kinetics of proteolysis digestion and analysis of the isolated subfragments, it was found that proteolysis occurred only at the N-terminal region of fragment 206–316. Proteolysis occurred in a stepwise mode from the N-terminal region and the rather short fragment 255–316 was quite resistant to further proteolysis with subtilisin, implying a tightly folded conformation (86, 87). Indeed, NMR measurements provided clear-cut evidence that this small, single-chain thermolysin fragment (62 residues, lacking disulfide bridges) acquires in solution a stable, highly helical, and native-like structure (197) (Fig. 20.6B).

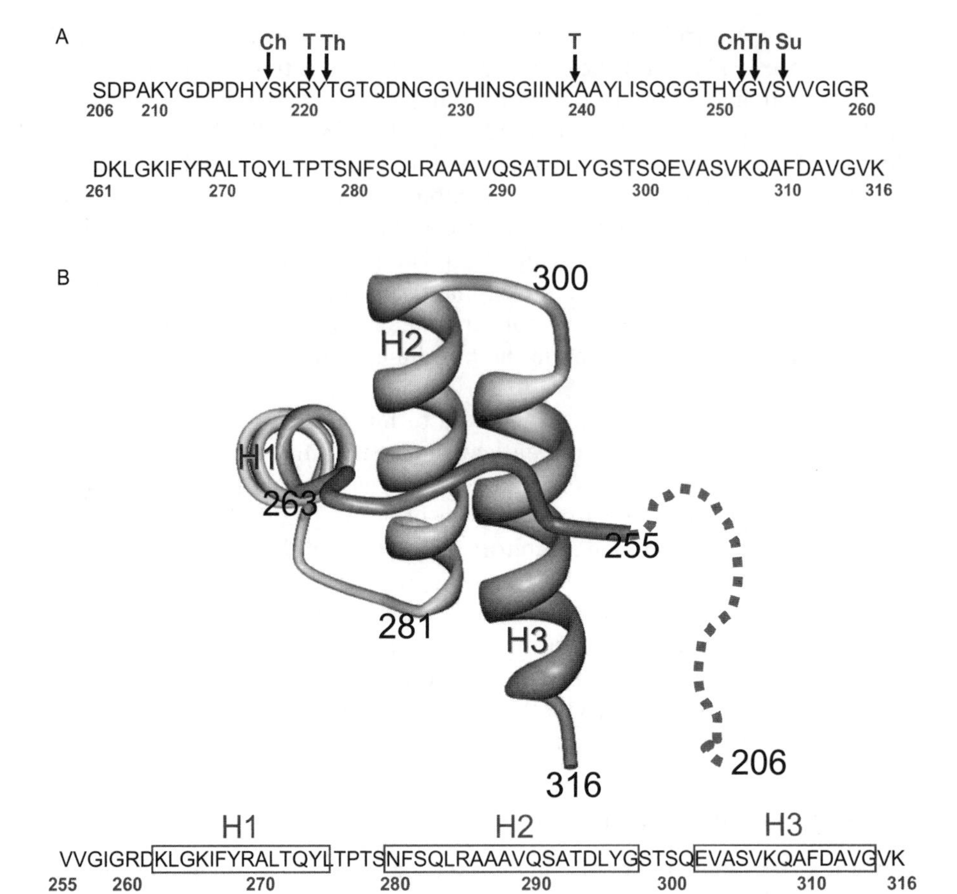

Figure 20.6 Removal of the N-terminal disordered tail from the thermolysin C-terminal fragment 206–316. (A) Amino acid sequence of BrCN-fragment 206–316 of thermolysin. This fragment is a 111-residue polypeptide chain without stabilizing effectors as disulfide bridges, metal ions, or bound cofactors (260). Limited proteolysis of the fragment allows the selective and stepwise removal of its disordered N-terminal region (44). Arrows indicate sites of proteolysis along the fragment chain by chymotrypsin (Ch), trypsin (T), thermolysin (Th), and subtilisin (Su). The fragment most resistant to further proteolysis by subtilisin was fragment 255–316. This fragment was isolated to homogeneity and studied for its conformational and stability properties. (B) 3D structure of thermolysin fragment 255–316 determined by NMR (197) (PDB code 1TRL). Numbers near the polypeptide chains refer to residue numbers of the 62-residue chain of fragment 255–316. The helical segments (H1, H2, and H3) indicated by boxed sequences below the 3D structure of the fragment almost coincide with the helical segments observed in the same chain region in intact thermolysin. This implies that fragment 255–316 displays protein domain characteristics, that is, the capability of an autonomous folding. Moreover, it was found that fragment 255–316 shows cooperative unfolding transitions mediated by heat or urea in much analogy to those displayed by small globular proteins.

20.5 IDENTIFICATION OF DISORDERED REGIONS IN PROTEINS

Limited proteolysis of a globular protein occurs at those loops or regions that display inherent conformational flexibility, whereas the protein core remains quite rigid and thus resistant to proteolysis (83, 86, 87). Usually, it is a region, rather than a specific site, which is the target of limited proteolysis. Indeed, if several proteases are used, it can be observed that cleavage takes place over a stretch of nearby peptide bonds (84, 86, 87). Our view is that a mechanism of local unfolding can be used to explain limited proteolysis phenomena in proteins, since only an unfolded polypeptide substrate can bind and adapt to the protease's active site in an extended conformation (83, 87) (see above). Here, we will be using the B-factor profile along the polypeptide chain of a protein as giving a graphic image of the degree of chain flexibility of a protein and thus as a means of predicting or explaining the sites prone to limited proteolysis (83). We will show below that limited proteolysis can be very selective especially when the site of proteolysis is embedded in a disordered chain region, this last identified by missing electron density.

Other efforts aimed at discovering structural parameters that dictate the limited proteolysis phenomenon placed emphasis on features of the protein substrate such as solvent accessibility, surface exposure, or protrusion index (113, 180). An algorithm, based on these parameters, named NICKPRED, was developed in order to predict the nicksites in proteins (113). However, when predictions were compared with experimental results of actual limited proteolysis experiments, the correlation between predictions and experimental nicksites was moderate only. Moreover, usually NICKPRED and related computational methods (180) predict numerous sites of limited proteolysis, besides the one (or very few) observed experimentally. This can be explained by considering that, if one observes a 3D model of a globular protein, clearly he or she can identify, simply by visual inspection of a globular protein model, many exposed sites located at the whole surface of the protein. Therefore, we may conclude that the utility of the protrusion index as a predictor of sites of limited proteolysis appears to be dubious.

In our analysis of the structural and dynamic features of protein substrates dictating limited proteolysis events, occasionally, it was difficult to reconcile the predicted flexibility/disorder of protein chain regions detected by proteolysis experiments with similar features of the corresponding chain segment in the crystal structure of the protein substrate (84, 87). However, in a number of cases, we observed that this correlation was instead evident by considering the structure of the apo form of the protein substrate, that is, in its free state without substrates, inhibitors, cofactors, or metal ions bound. Moreover, in our analysis, it was very critical to consider only the "initial sites" of proteolysis, since subsequent cleavages occur in a nicked or gapped protein substrate not necessarily retaining the structure and dynamics of the intact native protein. Indeed, initial sites of proteolysis are most informative for locating the sites of enhanced flexibility of the protein chain. Clearly, if the two complementary fragments comprising the entire chain are identified in the proteolysis mixture, they clearly correspond to the initial, very selective peptide bond fission (86).

Herewith, we will examine few selected cases of limited proteolysis experiments conducted on proteins of known 3D structure and dynamics, this last mostly deduced from crystallographic data. These examples will demonstrate that there is a strict correlation between sites of protein disorder and sites of limited proteolysis.

20.5.1 Apomyoglobin (apoMB)

apoMb, myoglobin without the heme, is a small monomeric protein of 153 amino acid residues that since decades has been used as a model protein for studies of protein folding and stability (80, 100, 208, 233, 252). The structure of apoMb has not been determined by X-ray methods and, in the past, for simplicity, the structure of apoMb has often been assumed to be similar to that of heme-containing holo myoglobin (holoMb). Indeed, the results of a variety of spectroscopic studies have indicated that, in solution and at neutral pH, apoMb retains much of the secondary structure observed in the crystal structure of holoMb. Nonetheless, CD measurements (104, 109) provided evidence that the helical content of apoMb is reduced in respect to that of holoMb, which in the crystal state is highly helical and constituted by eight helices (named A through H) (73, 280) (see Fig. 20.7).

With the view to probe the structure and dynamics of apoMb utilizing the limited proteolysis approach, apoMb was subjected to a series of proteolysis experiments utilizing a variety of proteolytic enzymes (90, 178). At neutral pH and at 25°C, these proteases initially cleave horse (90) or sperm-whale (178) apoMb at the chain region encompassing helix F (chain segment 82–97) of native holoMb (73, 280) (see Fig. 20.7). When reacted under identical experimental conditions, holoMb was instead fully resistant to proteolysis. From the proteolytic mixture of apoMb obtained using thermolysin, the two complementing fragments 1–88 and 89–153 were isolated and shown to be able to associate in a stable protein complex, named nicked apoMb (164). The results of limited proteolysis experiments, therefore, indicate that apoMb in solution at neutral pH adopts a well-folded conformation, with the exception of the highly flexible, disordered chain segment corresponding to helix F in holoMb (90, 178).

The conclusions reached from proteolysis experiments of apoMb are nicely and precisely supported by a number of additional experimental and theoretical studies on the structure and dynamics of apoMb. First of all, the unfolding of the segment encompassing helix F has been documented by analyzing the structure and dynamics of apoMb by NMR spectroscopy (68, 69). Hydrogen exchange experiments by NMR fail to detect H/D exchange protection in the chain segment encompassing helix F, whereas the rest of the helical core of apoMb shows structural and dynamic properties similar to those of native

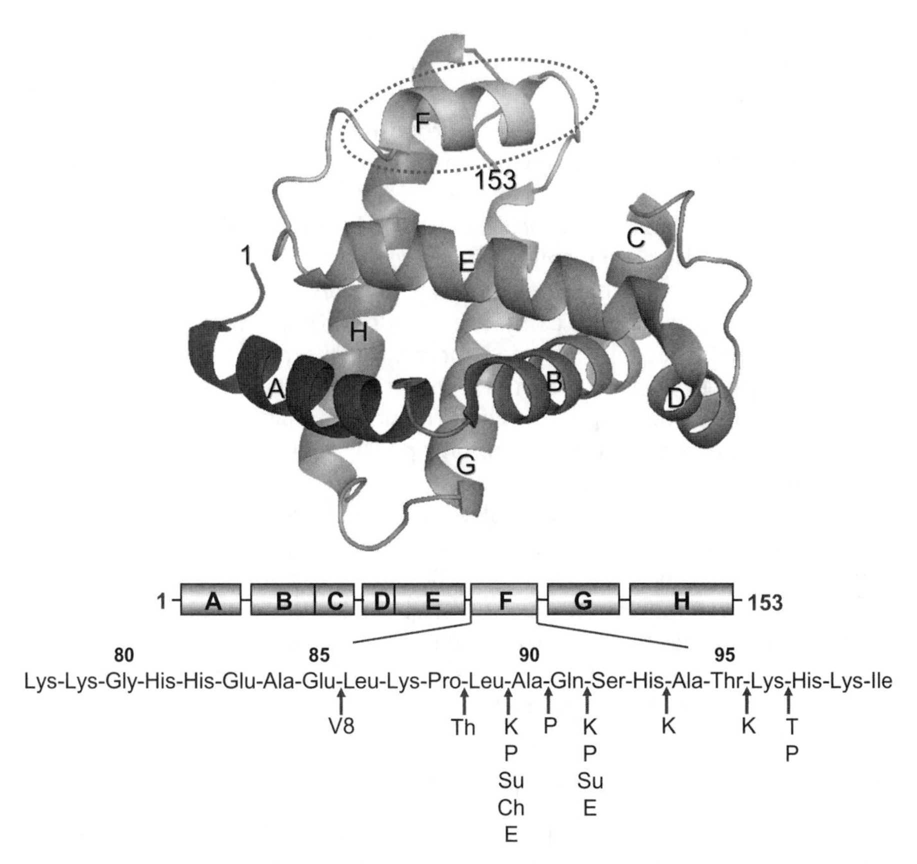

Figure 20.7 Proteolysis experiments reveal that helix F is unfolded in apomyoglobin (apoMb). (Top) Three-dimensional model of myoglobin (PDB code 1MBO) produced with the PyMol software (161) (www.pymol.org). The eight (A through H) helices of the protein are labeled. The heme noncovalently bound to the protein is not shown. The segment of helix F, circled by a dashed line, is disrupted/unfolded in apoMb, that is, by removing the heme moiety from the holo protein (holoMb). (Bottom) The eight helical segments of the protein are indicated by boxes, and below them the amino acid sequence of the chain segment comprising helix F (residues 82–97) is given. The sites of proteolytic cleavage of apoMb by thermolysin (Th), subtilisin (Su), trypsin (T), papain (P), proteinase K (K), staphylococcal V8-protease (V8), and chymotrypsin (Ch) are indicated by arrows (90, 178).

holoMb. Moreover, the results of molecular dynamics simulations conducted on apoMb provided evidence that helix F is substantially more mobile than the core of the protein (29, 171, 235). It is therefore quite reassuring that the results of proteolysis experiments are in full agreement with the results obtained from other physicochemical studies, emphasizing therefore the reliability of proteolytic probes of protein structure and dynamics.

Upon examination of the amino acid sequence of the chain segment encompassing helix F (see Fig. 20.7), it is obvious to note that this chain segment, otherwise helical in the holo protein, contains the strong helix breaker Pro88 residue. Therefore, it is reasonable to propose that there are several interactions between the heme and side chains of amino acid residues belonging to helix F that can stabilize this helix in the holo protein and thus counterbalance the helix-breaking effect of Pro88. Indeed, helix F in holoMb is strongly stabilized by the hydrophobic environment created by the nearby heme moiety and, in particular, by the covalent bond between the pyrrole nitrogen of His93 and the heme iron, as well as by some hydrophobic interactions (73, 280). Therefore, it is expected that, upon removal of the heme from holoMb, the heme-free apoMb acquires a more flexible state and the chain segment 82–97 becomes disordered, adopting a conformation compatible with its intrinsic conformational propensity. Indeed, an apoMb mutant with a Pro88→Ala substitution is fairly well protected against proteolytic attack (178). This is explained by considering that the helix-breaking Pro residue has been replaced by the helix-forming Ala residue, thus leading to a quite rigid and hydrogen-bonded helical structure in the Pro88Ala mutant of apoMb, thus making this mutant protein quite resistant to proteolysis. Therefore, the unfolding of helix F in wild-type apoMb is clearly documented by the results of proteolysis experiments, in agreement with the results obtained by using other spectroscopic and computational approaches.

20.5.2 Human Growth Hormone (hGH)

hGH was subjected to the action of a variety of proteolytic enzymes with the aim to prepare hGH fragments maintaining some of the biologic properties of the intact hormone (144). These studies identified a sequence region (residues 130–150) in the hGH molecule highly susceptible to limited proteolysis by trypsin, plasmin, thrombin, and subtilisin (see Fig. 20.8). Recently, the action of kallikreins on hGH was also studied in detail, and it was found that these proteases preferentially cleave the protein substrate at chain region 134–150, in particular cleaving peptide bonds Arg134–Thr135 and Tyr143–Ser144 (132). The determination of the crystal structure of hGH permits a structural evaluation of the proteolysis data. The structure of hGH is a four-helix bundle (residues 9–34, 72–92, 106–128, and 155–184) with three short helices (46, 241). Crystallographic data indicated a quite disordered segment connecting helices 3 (residues 106–128) and 4 (residues 155–184). In particular, the chain segment 129–155 was not included in the final model of hGH, since this region is not visible in the electron density map and thus is disordered (Fig. 20.8B). It is striking to observe, therefore, that all sites of limited proteolysis of hGH by a variety of proteases all occur at the chain segment which appears to be disordered in the crystal structure of the protein. Therefore, the same disordered chain region in hGH is identified by both limited proteolysis experiments and X-ray analysis. On the other hand, the

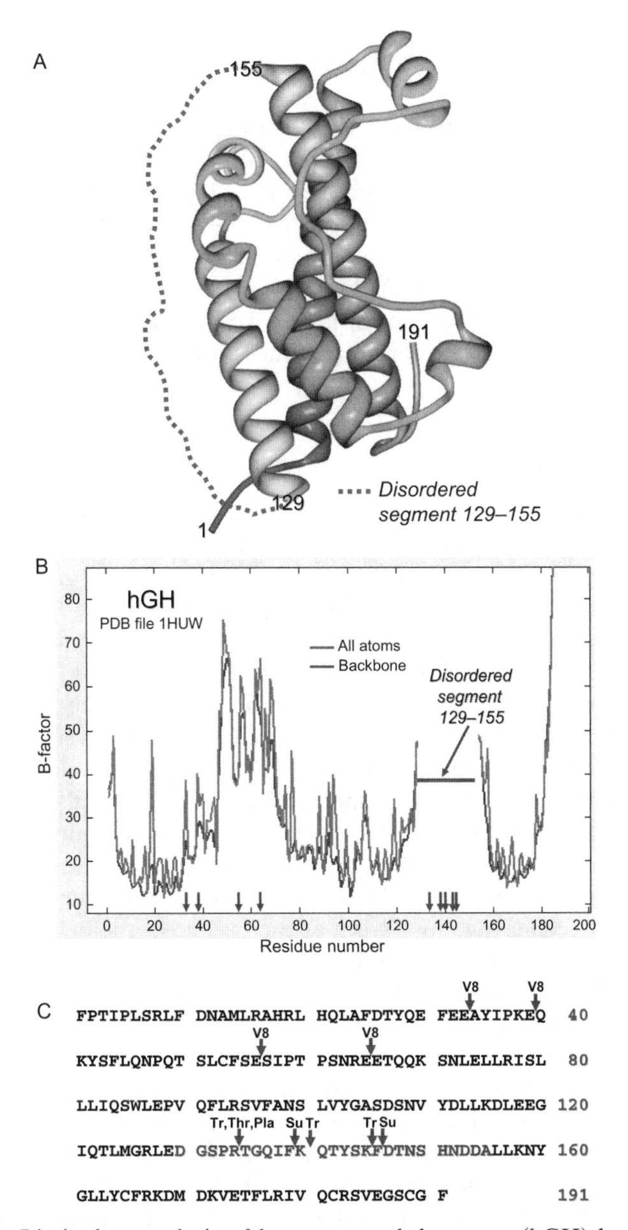

Figure 20.8 Limited proteolysis of human growth hormone (hGH) by a variety of proteases occurs at disordered/mobile regions of the protein. (A) A 3D model of the 191-residue chain of hGH (PDB code 1HUW). The structure is that of an hGH mutant, but this is essentially identical to that of the wild-type protein (241). The chain segment 129–155 does not show electron density and thus is considered to be disordered. This segment is arbitrarily drawn with a red line. The 3D model of the hormone was generated with the MBT software (47) available at PDB. (B) The B-factor profile along the 191-residue chain of hGH taken from the PDB file (code 1HUW). The chain segment 129–155 is disordered and does not display electron density. Arrows indicate sites of proteolysis by several proteolytic enzymes. (C) Amino acid sequence of the 191-residue chain of hGH. The amino acid residues located in the disordered segment 129–155 are colored in red. Arrows indicate the sites of proteolysis by staphylococcal V8-protease (V8), trypsin (Tr), thrombin (Thr), subtilisin (Su), and plasmin (Pla). See color insert.

Glu-specific V8-protease does not cleave at the level of the disordered region 129–155, since at this region there are no Glu residues (see Fig. 20.8C) to be attacked by the Glu-specific V8-protease (54). However, this protease preferentially attacks at Glu residues, which are embedded in chain regions that are anyway rather flexible, as given by the B-factor profile (186) (Fig. 20.8B). A kinetic analysis of the proteolytic digestion process by V8-protease has shown that, among the 14 possible sites of proteolysis of the hGH chain, Glu33 and Glu39 were the sites of preferential attack, followed by Glu56 and Glu66 (186).

20.5.3 Calmodulin (CaM)

CaM is a 148-amino acid protein belonging to a family of homologous proteins that bind calcium through similar structural domains. CaM acts as the primary receptor for intracellular calcium in all eukaryotic cells and mediates a variety of physiological processes in a calcium-dependent manner (96, 193, 226, 274, 283). More than 20 enzymes are activated by CaM, and in several instances, the mode of enzyme–CaM interaction has been elucidated. Detailed studies of the structural features of CaM were conducted using both X-ray and NMR (5, 6, 12, 36, 120, 155). The crystal structure of CaM has been solved and shown to contain two globular lobes (domains) each binding two calcium ions (5, 6, 36, 155). The two domains are connected by a single solvent-exposed α-helix of about eight turns (see Fig. 20.9). In the presence of calcium, trypsin cleaves selectively CaM at the level of this interconnecting helix into two fragments of about equal size, residues 1–77 and 78–148 (53). It is immediately evident that the selective cleavage by trypsin at the middle of the domain-connecting helix in CaM is an odd result, considering that only an extended β-strand is the required conformationals feature for a polypeptide substrate to bind at the protease's active site for a proteolysis to occur (240) (see above).

In the context of the present discussion, the tryptic cleavage of the central helix of CaM appears to be unusual and clearly contradicts the possible general conclusion that flexible loops are the favored sites of limited proteolysis and, in particular, that helices are not cleaved (83, 87). However, this apparent discrepancy can be clarified by a careful reading of the original publication on the X-ray structure of CaM at 3 Å resolution. In fact, it was stated that "regions 20–25, 75–80, and 120–125 show a weak electron density or are invisible in the density map" (6). Thus, even if in the three-dimensional model of CaM the central helix is usually depicted as a clearly defined element of secondary structure (see Fig. 20.9A), the helix turn comprised by residues 75–80 should be disordered or at least somewhat flexible. Subsequently, the structure of CaM was refined at 2.2 Å resolution, and a more accurate model was proposed (5). The overall features of the CaM structure previously determined were confirmed, including the unusual feature of the long central helix (residues 65–92) between the two calcium-binding domains. However, it was reported that "the phi and psi dihedral angles of residues 79–81 show significant devia-

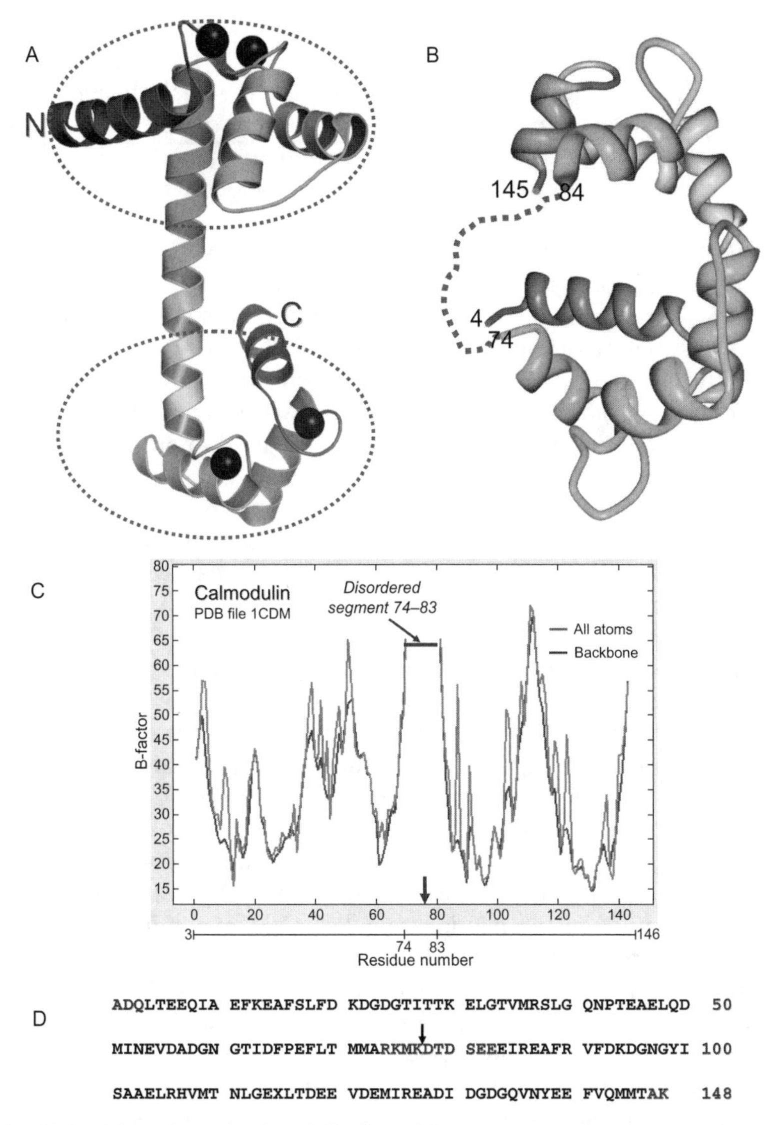

Figure 20.9 Dissection of calmodulin (CaM) into two domains by tryptic hydrolysis. (A) The most commonly represented dumbbell structure of CaM highlighting its two-domain topology. The two domains or lobes bind each two calcium ions. (B) Structure of CaM complexed with a peptide ligand. The segment 74–83, arbitrarily drawn by a red line, does not show electron density and thus is disordered. Also, short segments at the N- and C-terminal ends of the protein appear to be disordered. The 3D model was constructed with the MBT software (47) available in PDB. The chain is with rainbow colors from the N-terminus (blue) to the C-terminus (red). (C) The B-factor profile along the 148-residue chain of CaM (PDB code 1CDM). There is a discontinuity in the profile at segment 74–83 due to missing electron density and thus this chain segment is disordered. Note that the actual residue numbers of the CaM polypeptide chain are those indicated at the bottom of the diagram. (D) Amino acid sequence of CaM. Arrow indicates the site of proteolytic cleavage by trypsin cutting at the Lys77–Asp78 peptide bond (53). The disordered residues in the crystal structure of the protein are colored in red in the amino acid sequence. See color insert.

tions from ideal α-helical geometry," the central helix actually being "unwound and somewhat strained in the middle" (5). Moreover, residues 76–82 have the highest temperature factors of the all polypeptide chain of CaM, suggesting that this chain region is highly flexible (see crystallographic data, PDB code 3CLN).

Studies on the structure of CaM in solution provided evidence of the dynamic features of the central helix of CaM (81, 176). For example, X-ray solution scattering data gave indications of differences between the solution and crystal structure of CaM and, in particular, that the interconnecting helix is "bent" in solution, thereby bringing the two calcium-binding domains in closer contact (106). The conformation of CaM was examined by CD measurements in the presence of the glycols, which were used to promote crystallization of CaM. It was found that CaM in aqueous buffer at pH 5–7 shows reduced percentage helicity in respect to that calculated from the crystal structure of CaM and that helicity is significantly increased in the presence of polyols and other organic solvents. The conclusion reached in these studies was that the central helix in CaM is induced to form under the experimental conditions employed for growing crystals of the protein (15).

Additional studies on the solution structure of CaM in complex with CaM-binding peptides using NMR also revealed significant distortion or disruption of the central helix (12, 120). Clear-cut evidence of the dynamic nature of the chain segment connecting the two calcium-binding domains of CaM came from NMR and X-ray studies of the complex formed by CaM and peptides. The 3D structure of calcium–CaM complexed, with a 26-residue synthetic peptide comprising the CaM-binding region (residues 577–602) of skeletal muscle myosin light chain kinase, has been solved by heteronuclear NMR spectroscopy (120). In the peptide complex, CaM maintains in solution its two-domain topology of the crystalline state, but "the long central helix (residues 65–93), which connects the two domains, is disrupted into two helices connected by a long flexible loop (residues 74–82)," thereby enabling the two domains to clamp the bound peptide, which adopts a helical conformation (120). A similar structure of CaM bound to the peptide corresponding to the CaM-binding domain of brain CaM-dependent protein kinase II-α has been found by X-ray methods (155). The crystal structure of the peptide–CaM complex solved at 2.0 Å resolution revealed that the central helix is missing electron density and that the two calcium-binding lobes interact and form a compact globular protein with a shape of an ellipsoid (PDB code 1CDM).

The structural and dynamic features of CaM have been analyzed by several authors utilizing molecular dynamics simulations (253, 274, 279). The general conclusion from all these studies was that the central helix functions as a flexible tether connecting the two calcium-binding domains of CaM. The simulations predict an actual structure of CaM more compact than the dumbbell structure seen in the crystal state of the protein and a kinking or bending in the central part of the helix tether, bringing the two domains into a closer proximity (see Fig. 20.9B). The flexibility of the central helix of CaM has been investigated also by a variety of computational techniques (molecular dynamics simulations, normal mode analysis, and essential dynamics analysis) (279).

The CaM case convincingly documents the great potential and reliability of proteolytic enzymes as probes of structure and dynamics of globular proteins and adds strength to the view that flexible loops only, and not helices, are the sites of proteolytic attack (83, 85, 87). There has been considerable controversy on the presence, role, and structure of the "long central helix" of CaM (81, 176). The results of a simple and clear-cut experiment of selective tryptic hydrolysis of CaM have anticipated the key conclusions on the structure of CaM achieved by NMR, X-ray, small-angle X-ray scattering, time-resolved fluorescence, and chemical cross-linking experiments, as well as mutagenesis studies of the CaM molecule (262).

20.5.4 Staphylococcal Nuclease

Nuclease from *Staphylococcus aureus* has been used extensively as a model protein in fundamental studies of protein structure, function, and stability. Staphylococcal nuclease is cleaved by trypsin at peptide bonds 5–6, 48–49, and 49–50, giving the two-fragment complex nuclease-T (230). The structure of native nuclease has been refined at 1.65 Å resolution and the crystallographic temperature factors (B-values) of the main chain were determined (148). The first amino acid for which electron density was observed is Leu7, indicating that the N-terminal segment of six residues is highly disordered. As seen in Figure 20.10A, the specific tryptic cleavages of nuclease occur at the level of the disordered amino-terminus and at the chain segment showing a maximum of mobility, as given by the figures of B-factors. Moreover, the two sites of tryptic cleavage are located outside chain segments of regular secondary structure (helices and strands). Hence, the structural and dynamic characteristics of the sites of proteolytic cleavage in staphylococcal nuclease document that the limited proteolysis events occur at highly flexible or disordered sites of the protein.

20.5.5 Flavocytochrome b2

Selective proteolysis of baker's yeast flavocytochrome b2 by staphylococcal protease, chymotrypsin, and yeast protease occurs in a restricted segment (region 301–310) of the 57.5-kDa polypeptide chain of the enzyme (99) (Fig. 20.10B). Cleavage does not result in disruption of the tetrameric structure of the enzyme containing one heme and one flavin moiety per subunit, but does lead to an alteration of its kinetic properties. The 3D structure of yeast flavocytochrome b2 allows an explanation of the high susceptibility to proteolysis of region 301–310 along the 511-residue chain of the enzyme. The flavin-binding domain of the enzyme forms an eight-stranded β/α-barrell, interrupted by an excursion of the chain by about 40 residues (between β4 and α4) (276). This excursion contains two helices (E and F) and a fully disordered

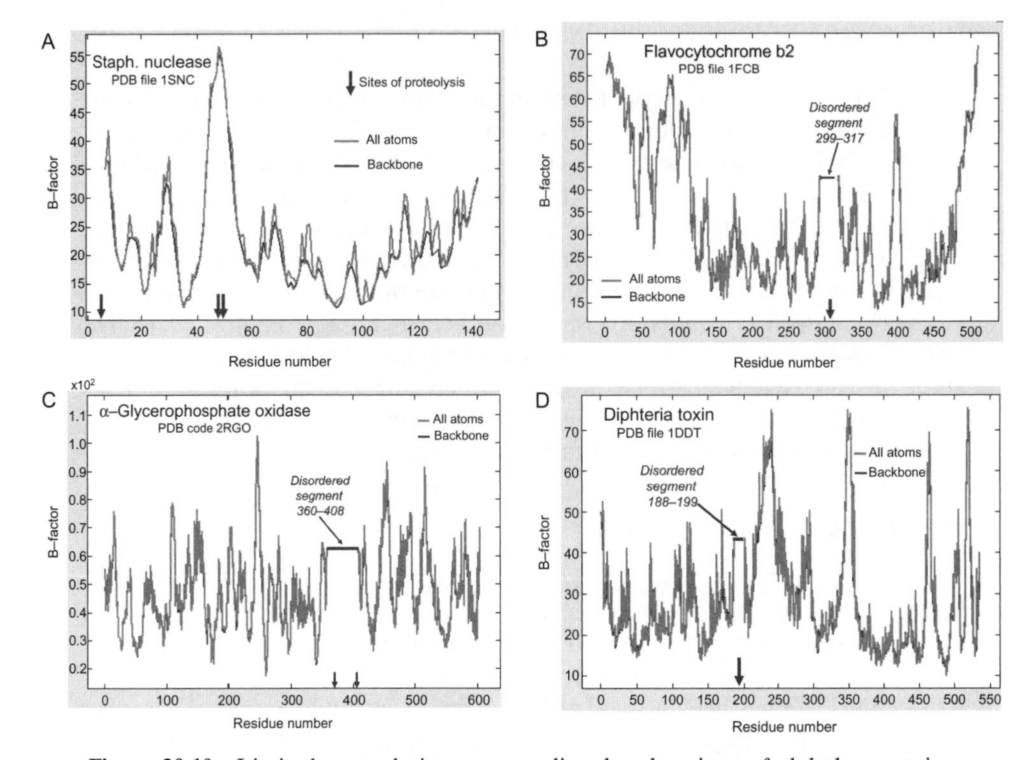

Figure 20.10 Limited proteolysis occurs at disordered regions of globular proteins. (A) The B-factor profile along the 149-residue chain of staphylococcal nuclease taken from PDB (code 1SNC). The graph contains figures for the B-factor from Leu7 to Ser141. Thus, the N-terminal segment 1–6 and the C-terminal segment 142–149 are not visible in the electron density map of the protein and thus are disordered. Arrows indicate sites of tryptic hydrolysis at Lys6, located at the disordered N-terminal segment, and at Lys48 and Lys49, which are embedded in the most flexible region of the protein. Note that the disordered C-terminal tail 142–149 does not contain basic residues (Lys and Arg) and, therefore, trypsin does not cleave the protein at the C-terminal region. (B) The B-factor profile of the 511-residue chain of flavocytochrome b2 taken from PDB (code 1FCB). The graph shows a discontinuity at chain segment 299–317 due to missing electron density, and thus this segment is disordered. The sites of limited proteolysis of flavocytochrome b2 by yeast protease, chymotrypsin, and staphylococcal V8-protease are encompassed by chain segment 305–315. (C) The B-factor profile of the 607-residue chain of α-glycerophosphate oxidase (GlpO) from *Streptococcus* sp. taken from PDB (code 2RGO). The chain segment 360–408 is missing electron density, thus indicating protein disorder. This chain region corresponds to the 50/52-residue insert that occurs in GlpO from a number of enterococcal and streptococcal sources. On limited proteolysis with trypsin, the GlpO from *Streptococcus* sp. is readily converted to two major fragments Met1–Lys368 and Ala405–Lys607. Hence, limited proteolysis excises from intact GlpO most of the chain segment that is disordered, as given by X-ray data. (D) The B-factor profile along the 544-residue chain of diphtheria toxin taken from the PDB (code 1DDT). For the chain segment 188–199, electron density is missing and thus this segment is disordered. Limited proteolysis of the toxin by trypsin occurs at Arg residues in chain positions 190, 192, and 193. Cleavages occur at the chain segment containing the Cys186–Cys201 bridge and thus the fragments of the nicked toxin remain covalently bound by this disulfide bond. See color insert.

299–317 segment, since electron density is missing for this segment (Fig. 20.10B). Therefore, limited proteolysis by several proteases occurs at the level of the single disordered chain region of the protein.

20.5.6 α-Glycerophosphate Oxidase (GlpO)

The flavoprotein GlpO from a number of enterococcal and streptococcal sources contains a conserved 50-residue insert that is completely absent in the homologous α-glycerophosphate dehydrogenases from other sources. On limited proteolysis with trypsin, the GlpO from *Streptococcus* sp. (67.6 kDa) is readily converted to two major fragments corresponding to masses of approximately 40 and 23 kDa (35). The combined application of sequence and mass spectrometric analyses demonstrates that the 40-kDa fragment represents the N-terminus of intact GlpO (Met1–Lys368; 40.5 kDa), while the 23-kDa band represents a C-terminal fragment (Ala405–Lys607; 22.9 kDa). Hence, limited proteolysis excises most of the GlpO insert (Ser355–Lys404), indicating that this represents a flexible/unfolded region of the protein amenable to proteolytic attack. Attempts to identify the expected 4.2-kDa protein fragment corresponding to Ser369–Lys404 were not successful, suggesting that this fragment dissociates from the core protein and suffers further proteolysis. This is largely plausible, considering that there are five internal Lys + Arg in this chain segment (35).

The active site and other spectroscopic properties of the nicked enzyme (1–368/405–607) remain essentially unaltered in respect to those of the intact enzyme, including both flavin and tryptophan fluorescence spectra, titration behavior with both dithionite and sulfite, and preferential binding of the oxidized flavin. The structure of the 607-residue chain of intact *Streptococcus* sp. GlpO has been determined using multiwavelength anomalous dispersion data and refined at 2.4 Å resolution (43). The B-factor profile along the protein chain of GlpO shows a discontinuity in the chain region 360–410, due to missing electron density (see Fig. 20.10C). Hence, this region is disordered on the basis of crystallographic criteria and thus easily removed by tryptic proteolysis from the protein core. The insert of 50-residue segment in streptococcal GlpO can be removed without impairing the structure and stability of the enzyme, as clearly demonstrated by the X-ray structure of a recombinant GlpO mutant and largely lacking this insert (43).

20.5.7 Diphteria Toxin (DT)

Limited proteolysis of the 535-residue chain of DT and subsequent reduction generates two fragments, the N-terminal fragment A (21 kDa) and the C-terminal fragment B (37 kDa), as a result of cleavage at Arg residues in positions 190, 192, or 193 (163) (see Fig. 20.10D). There is evidence that *in vivo* nicking plays a role in the cytotoxic action of DT and that the nicked protein separates into the two individual fragments when the disulfide bond

Cys186–Cys201 is exposed to the reducing environment of the cytoplasm during membrane translocation of the toxin. The highly selective proteolysis of DT by trypsin can be interpreted on the basis of the crystal structure of the toxin (40). The site of tryptic cleavage is located at a 14-residue loop (residues 187–200) connecting fragment A (the catalytic domain) to fragment B (consisting of two domains). This loop appears to be disordered, based on the fact that for the chain segment 188–199, electron density is missing (see Fig. 20.10C).

20.5.8 Other Proteins

We have analyzed many more results of limited proteolysis experiments conducted on proteins of known 3D structure determined by X-ray methods. In general, in analogy to the cases herewith examined, we have observed a clear-cut correlation of sites of limited proteolysis with sites or regions which appear to be disordered by crystallography, that is, chain regions of missing electron density. We may mention here that this correlation was observed with β-galactosidase, actin, α-thrombin, myosin, protein kinase-2, phaseolin, luciferase, thymidylate synthase, aspartate oxidase, superoxide dismutase, ferredoxin, xanthine oxidase, ovalbumin, factor XIII, tryptophan synthase, interleukin-6, and others. Therefore, this comparative analysis of proteolysis versus crystallographic data firmly establishes that limited proteolysis data can be the experimental technique of choice for identifying disordered regions in proteins, especially when very minute amounts of protein sample are available.

20.6 CUTTING AT FLEXIBLE HINGES BETWEEN PROTEIN DOMAINS

Relatively large globular proteins are assemblies of compactly folded substructures, usually called domains or modules. Protein domains are almost invariably seen in proteins of more than about 100 amino acid residues and often their existence can be recognized simply by visual inspection of the 3D model of the protein molecule (13, 103, 200, 268, 272, 282). Computer algorithms utilizing the X-ray coordinates of protein structure have been developed for the identification and analysis of protein domains (32, 122, 147, 172, 211, 232). These algorithms, based on principles such as interface area minimization, plane cutting, clustering, distance mapping, specific volume minimization, and compactness, allow a description of globular proteins in terms of a hierarchic architecture given by elements of secondary structure (helices, strands), subdomains (supersecondary structures or folding units), domains, and whole protein molecule. The concept of domain, therefore, appears to be a convenient way to simplify the description and classification of protein structures. There is no strict, universally accepted definition of a protein domain, but a consensus view of a domain involves a compact, local, and independent structural unit, relatively well separated from the rest of the protein molecule (110, 254, 268).

Identification of domains or modules in relatively large proteins can be reached also from sequence analysis of the databases of amino acid sequences (147, 172). In particular, these sequence-based methods can allow the definition of domain boundaries by estimation of sequence homology, since the connecting segments between domains usually show higher sequence variability. Indeed, domain borders may be estimated by computational approaches based on multiple sequence alignments and/or homology-modeling studies (110, 147, 172, 254). These analyses were applied to families of multimodule proteins involved in cell adhesion, clotting, fibrinolysis, and signaling, such as fibronectins, kringles, SK2 and SK3 domains, SH3 domains, epidermal growth factor (EGF)-containing proteins, and immunoglobulins. However, when a limited number of homologous sequences are available and in the absence of structural data, the precision of these theoretical approaches is limited and experimental data are required.

Protein domains were proposed to be protein substructures that can act as intermediates in the folding process of a globular protein from the unfolded to the native state (268, 272). It is conceivable to suggest that, in a multidomain protein, specific segments of the unfolded polypeptide chain first refold to individual domains and that they subsequently associate and interact with each other to give the final tertiary structure of the protein, much the same as do subunits in oligomeric proteins. The major implication of this model of protein folding by a mechanism of modular assembly is that isolated protein fragments corresponding to domains in the intact protein are expected to fold into a native-like structure, thus resembling in their properties a small globular protein (258).

The size of an individual domain has limits and a large proportion of them are given by approximately 100 residues (147, 172). Very short domains constituted by less than 50 residues usually require the stabilizing effect of bound metal ions or disulfide bonds. For example, the 64/65-residue chain of hirudin is folded into the N-terminal core domain of about 50 residues cross-linked by three disulfide bridges, while the C-terminal tail is flexible and disordered (see Fig. 20.5). Usually, protein domains are "continuous" and given by one segment of the polypeptide chain folding into a compact module or lobe. For example, simple visual inspection of the thermolysin molecule reveals that the 316-residue chain of the protein folds into two structural domains of about equal size (see Fig. 20.2). However, "discontinuous" domains as in lysozyme or α-LA also exist (see the model of α-LA in Fig. 20.3), meaning that more than one chain segment is required to form the domain.

The current rapid growth in the number of rather large protein structures solved by X-ray methods, together with the fact that usually domains are the constituent building blocks of these proteins, has prompted an intensive experimental research on protein domains (110, 254). Efforts are being directed toward the experimental dissection of rather large proteins into their constituent domains (97, 133). Limited proteolysis was shown to be the best experimental technique for splicing out a protein fragment that can fold autonomously (20, 41, 97, 133, 221). The success of the technique resides in the fact that limited (specific) proteolysis of a multidomain globular protein occurs at the connecting segments between domains. These "hinge" regions are usually more flexible than the rest of the polypeptide chain forming the globular domains and thus are easily and selectively attacked by a protease. The results of domain definition by the limited proteolysis approach parallel those that can be reached by the use of H/D exchange measurements for identifying the flexible hinges between domains (174, 217).

The chain segments linking protein domains display varying degrees of flexibility connected with their amino acid compositions and lengths. Statistical studies of linkers in a variety of multidomain proteins revealed that there is a preference of Pro, Arg, Phe, Thr, Glu, and Gln in order of decreasing preference (98, 158). Proline is the most preferred amino acid type of the linker region, since it has no amide hydrogen to donate in hydrogen bonding, and therefore, it can serve to structurally isolate the linker from the domains. Moreover, Pro cannot fit into a regular structure of either the α-helix or β-sheet and is a common "breaker" of secondary structure (98). Linker regions usually adopt a coil structure, but even helical segments are found at the interface between domains. However, sometimes these helical segments contain a central Pro residue in order to determine a kink in the helix and allow some motion of the linker. A database of interdomain linkers has been constructed and is available at http://www.ibi.vu.nl/programs/linkerdbwww.

Limited proteolysis has been widely used to delineate the structural organization of multidomain proteins (11, 97, 123, 133, 221, 265). It was shown that the technique can offer an experimental route for the determination of exact boundaries of the domains combining limited proteolysis with electrospray ionization or MALDI-TOF MS (221). The strategy aims at identifying the proteolysis-resistant core, considering that an undigested protein species very likely corresponds to a compact folded domain (41, 44). The technique allowed the isolation of fragments capable of independent folding into a native-like structure, displaying some of the activities of the parent protein and thus helping to clarify features of structure and biologic activity of complex proteins. Numerous interesting applications of the limited proteolysis/MS approach for producing protein domains from multifunctional proteins have been described in the literature (11, 123, 265).

The limited proteolysis/MS technique for identification and production of protein domains is having a systematic use in the laboratories belonging to the New York Structural GenomiX Research Consortium (http://www.nysgxrc.org) aimed to conduct a large-scale structural analysis of proteins. The process leading to X-ray structure determination is time-consuming and labor intensive, since it requires choosing appropriate strategies for protein production by recombinant methods and protein purification, crystallization, and structure determination. Clearly, the overall strategy would significantly benefit from a prior knowledge of the domain architecture of target proteins. Automated protocols for proteolytic digestion of protein samples have been developed with the view to systematically define domain boundaries in proteins. The results from this work have already demonstrated that proteins yielding useful crystals are typically resistant to proteolysis and that crystallizability can be significantly improved by removing the flexible/disordered tails of proteins (20).

20.7 LIMITED PROTEOLYSIS IN THE CONTEXT OF AMYLOIDOGENIC PROTEINS

Protein aggregation causes the formation of fibrillar aggregates or well-ordered amyloid precipitates that characterize the group of human diseases known as amyloidoses (37, 48, 49, 76, 213, 218). Indeed, more than 20 human disorders involve the conversion of a specific protein or protein fragment from a soluble state into insoluble amyloid fibrils that are deposited in a variety of organs and tissues (37). Among the amyloidogenic proteins are the Aβ peptide, involved in Alzheimer's disease, the islet amyloid polypeptide in type II diabetes, α-synuclein (α-syn) in Parkinson's disease, and the prion proteins in the transmissible spongiform encephalopathies. Although the proteins associated with amyloid diseases differ in their primary and tertiary structures, as well as their size and function, all of them form elongated amyloid fibrils characterized by a common cross-β structural motif, having β-strands oriented perpendicular to the fibril axis (152, 194, 209, 224, 238, 269). More recently, it has been found that several proteins not associated with human disease can be induced to form amyloid fibrils *in vitro* (38, 39, 74). On this basis, the ability to form amyloid fibrils appears to be a general property of the polypeptide chain and, therefore, an increasingly adopted view is that amyloid fibrils can be formed by a large variety of proteins and peptides under suitable experimental conditions (38, 39, 74).

An aspect that does not appear to be sufficiently emphasized in current literature is the fact that a large proportion of physiologically relevant amyloid deposits in tissues are made up of protein fragments derived from relatively larger protein precursors (37). For example, the fibrils associated with Alzheimer's disease are formed by peptide fragments deriving from limited proteolysis of the amyloid precursor protein by secretases (273). Other examples of amyloid deposits given by protein fragments include serum amyloid A protein, gelsolin, apolipoprotein A1, prolactin, calcitonin, transthyretin, medin, and fibrinogen among others (37). Protein fragments derived by limited proteolysis of protein precursors are particularly prone to aggregation, since they can usually only adopt, at most, partly folded states and cannot establish the long-range interactions present in the intact native protein. In particular, a property of protein fragments is that they may contain clusters of hydrophobic residues that can trigger protein aggregation by intermolecular hydrophobic interactions. Therefore, protein fragments appear to be key players in the mechanism of protein aggregation, and, consequently, protein fragmentation by limited proteolysis appears to be an important phenomenon underlying a substantial proportion of amyloid diseases and perhaps even a causative mechanism of them.

Proteolysis experiments were used to prepare protein fragments in order to study the propensity for amyloid formation of different regions of a protein chain. Limited proteolysis of lysozyme was used to produce amyloidogenic fragments, and among them, the most amyloidogenic was fragment 57–107 (93). Of interest, this fragment encompasses the β-domain of the protein and corresponds to the chain region that more easily unfolds in mutants of human lysozyme that cause a form of a systematic non-neuropathic amyloidosis (22). Proteolysis of α-LA was used to prepare nicked and gapped protein species that were shown to aggregate at low pH much faster than the intact protein (182). This result was interpreted as indicating that the MG state of α-LA in acid solution needs to adopt a more relaxed and flexible structure in order to form the amyloid precipitate. Limited proteolysis experiments were also used to analyze molecular features of intermediates and protofibrils of the SH3 domain (185) and HypF-N protein (31). As an additional example of protein fragmentation, β2-microglobulin was dissected into 11 fragments using the Arg-specific *Achromobacter* protease, and fragment 20–41 was found to be the most amyloidogenic (134).

Recently, in our laboratory, we have subjected the 153-residue chain of apoMb to limited proteolysis with pepsin at low pH, and from the digest, fragment 1–29 was isolated to homogeneity. This fragment displays amyloidogenic properties and aggregates into fibrils much more easily than intact apoMb (177). Of interest, fibrils formed by fragment 1–29 in acid solution redissolve at neutral pH forming mostly monomeric fragment, which can be induced to aggregate again by acidification of the solution. The easy dissolution of the fibrils of fragment 1–29 contrasts with the much more drastic solvent conditions required for dissolving fibrils or amyloid precipitates of other proteins requiring the use of 6 M guanidine hydrochloride or neet dimethylsulfoxide (111). Analogues of the apoMb fragment 1–29 prepared by solid-phase synthesis were used to investigate the role of hydrophobic clusters in protein aggregation phenomena (159).

20.7.1 Amyloidogenicity of ID Proteins

A significant portion of known amyloidogenic proteins belong to the class of ID proteins, among them are α-syn, tau protein, and huntingtin (37, 247). Clearly, it is reasonable to propose that such proteins are well suited for amyloidogenesis, as they lack significant structure and conformational constraints and thus are able to polymerize more readily than tightly packed globular proteins (247). While native globular proteins need to relax at least partly their rigid structure in order to aggregate, substantial evidence suggests that the fibrillation of ID proteins requires their partial refolding (247, 248). As a specific and well-studied example, we may mention here the case of the ID protein α-syn, an abundant protein found in brain tissues and occurring in the pathological intracellular protein deposits in Parkinson's diseases (244, 246). Evidence was accumulated to indicate that the formation of a folded intermediate represents the critical step of α-syn fibrillogenesis. This intermediate can be stabilized by numerous factors, including high temperatures, low pH, and the presence of several common pesticides and herbicides, metal ions, or moderate concentrations of trimethylamine-N-oxide or organic solvents (244, 246). Under all these conditions, α-syn was shown to undergo significantly enhanced fibrillation. In contrast, fibril formation by α-syn was considerably slowed or inhibited under conditions favoring formation of more folded conformations. For an excellent discussion of the fibrillogenesis phenomena of α-syn (246) and other ID proteins (247), the reader is referred to recent reviews.

As expected for an ID protein, α-syn is readily digested by proteolytic enzymes, and indeed it has been found that proteolytic cleavages of this protein affect its aggregation behavior process. Truncated forms of α-syn have a higher tendency to assemble into aggregates than the wild-type protein, and, moreover, they can seed protein aggregation of the intact protein (246). These truncated protein species prepared *in vitro* likely have some role in the protein aggregation phenomena *in vivo*, considering that approximately 15% of α-syn in Lewy bodies is C-terminally truncated. Furthermore, mutations in the α-syn gene (A30P and A53T) have been reported to increase the amount of C-terminally truncated α-syn. The proteolysis of α-syn *in vitro* with several proteases was investigated in the presence of micellar SDS, used as a membrane-mimetic environment (187). While α-syn at neutral pH is easily degraded to a variety of small fragments, as expected for a largely unfolded protein substrate, in the presence of micellar SDS, the proteolysis of α-syn is rather selective and the complementary fragments 1–111 and 112–140, 1–113 and 114–140, and 1–123 and 124–140 are obtained when thermolysin, proteinase K, and V8-protease, respectively, are used.

Recently, the role of limited proteolysis of α-syn with several matrix metalloproteases, including MMP-1 and MMP-3, has been analyzed, and it was also found that these proteases lead to C-truncated versions of α-syn displaying enhanced tendency to aggregate (143). Using MMP-1 and MMP-3, α-syn is specifically cleaved into fragments 1–90 and 91–140. From these studies, it was concluded that specific proteolytic cleavages can remove the highly negatively charged and aggregation-inhibiting C-terminal domain, approximately from residue 96 to 140, enhancing the aggregation process of the pro-aggregating NAC domain (chain segment 61–95) of α-syn. Therefore, limited proteolysis of α-syn may influence the pathogenesis of Parkinson's disease *in vivo* by generation of specific aggregation-enhancing α-syn fragments resulting from limited proteolysis of the protein (143).

20.7.2 Analysis of Amyloid Precipitates by Proteolysis

The molecular features of protein aggregation processes have been intensively investigated in recent years, but a detailed molecular description of intermediate protein aggregates and mature fibrils or amyloids is still lacking. Indeed, the intrinsic heterogeneous, transient, and insoluble nature of the amyloid aggregates makes the application of high resolution structural techniques difficult, such as NMR spectroscopy and X-ray diffraction techniques (152, 194, 209, 224, 238, 269). Useful information regarding the molecular features of amyloid deposits have been obtained by cryo-electron microscopy, FTIR, solid-state NMR, and H/D exchange analyzed by NMR or MS (127, 128). Recently, the limited proteolysis technique has been successfully exploited to analyze molecular aspects of protein fibrils, including those of Aβ peptide (129), Ure2p (14, 26, 27), α-syn (157), β2-microglobulin (160, 165), the HET-s prion protein of the fungus *Podospora anserina* (10), apolipoprotein C-II (271), lysozyme (91), and others. The rationale of this approach resides in the fact that hydrolysis sites by proteases occur at flexible chain regions devoid of hydrogen-bonded regular secondary structure, and, therefore, the prospects are to remove the flexible parts or tails from the amyloid core.

Limited proteolysis experiments have shown that the entire 130-residue chain of human lysozyme is not involved in the β-sheet fibril core, since pepsin cleaves off the N- and C-terminal segments of the protein embedded in the amyloid fibrils (91). The conclusion reached in this study was that the lysozyme amyloid is given by a core structure comprising mostly of the chain region 32–108, flanked by flexible tails. Also, the initial aggregates formed by human lysozyme have been analyzed recently by a combination of biophysical techniques and limited proteolysis experiments (92). It was found that protein aggregates and protofibrils are more easily digested than the native protein or its final amyloid precipitate. Thus, the relative flexibility of the initial aggregates may explain why these species are more toxic than the final well-formed amyloid (30), since they can easily enter into protein–protein interactions at the cellular level.

The fibrils formed by the 289-residue HET-s prion protein when digested with proteinase K lead to the proteolysis-resistant C-terminal fragment 218–289, implying that this region of the protein chain constitutes the core of the fibrils (10). Similarly, the amyloid precipitate formed by apolipoprotein C-II contains core regions, while the flexible parts of the fibrils can be removed by proteolysis (271). The fibrils are extensively digested by trypsin, but fragment 56–76 was quite resistant to further hydrolysis, suggesting that it likely forms the core of the fibrils. Indeed, synthetic peptides covering the chain segment 60–70 of apolipoprotein C-II were found to easily form amyloid aggregates (271). The polymorphism of amyloid precipitates formed by β2-microglobulin in acid solution at pH 2.5 (long fibrils) or 3.6 (short fibrils, less than 500 nm long) was analyzed by proteolysis experiments using pepsin as probe (165).

The results of these analyses indicated different conformational features of the protein when embedded in long or short fibrils.

20.8 DISORDERED REGIONS OF PROTEINS AS SITES OF OTHER ENZYMATIC ATTACKS

Since an enzyme that must act on polypeptide substrates has not been designed by nature to interact only with a single specific substrate, it may be well that the enzyme's active site should have the intrinsic capability to interact with a variety of potential substrates. Therefore, a flexible or disordered polypeptide on one side can have the structural plasticity to adopt the extended conformation that enables a stereospecific binding at the enzyme's active site and on the other can allow the enzyme to interact with a number of potential substrates (55, 56, 59, 65, 85, 263). It is tempting to propose that the molecular features herewith emphasized for a productive interaction between a protease and a polypeptide substrate can be extended also to other enzymatic reactions occurring with polypeptide substrates, that is, to other posttranslational protein modifications, such as phosphorylation, glycosylation, acetylation, methylation, hydroxylation, amidation, sulfation, and others.

The importance of protein disorder in a variety of biomolecular recognition processes of proteins has been emphasized (see Introduction). For example, it has been shown that the phosphorylation sites in proteins mostly occur in protein regions characterized by an intrinsic disorder (118). Indeed, also the reaction of a kinase with a polypeptide substrate requires a flexible/disordered substrate. In a number of X-ray structures of kinases bound to peptide substrates or inhibitors, it has been found that the ligands in their protein-bound state have extended, irregular conformations (24, 151). The bound peptide substrates or inhibitors do not show intramolecular backbone hydrogen bonding, while they have instead extensive hydrogen bonding between the peptide backbones and the backbones and side chains of the kinase partners (24, 151). There are other examples of peptide substrates that are bound to the kinases in an extended, irregular conformation (see structures in PDB, codes 1ATP, 1IR3, 1O6I, 1QMZ, 1PHK, 1O6K, 1GY3, 1CDK, and 1JBP). Clearly, the final structure of a kinase/peptide complex requires a disordered structure of the peptide ligand prior to its association with the enzyme, so that all potential hydrogen bonds of the peptide substrate are available for hydrogen bonding. Therefore, the peptide should be disordered and not intramolecularly hydrogen bonded in a regular secondary structure (helices, β-strands). The structural details of the complexes between kinases and peptide ligands parallel those observed in the analogous complexes of proteases (239, 240) (see above). We may note that the implications of these crystallographic data in terms of disordered chain segments as a requirement for the enzymatic reactions involving polypeptide substrates have been clearly overlooked (see Reference 118 for additional comments).

Recently, we have shown that the action of transglutaminase (TGase) on protein substrates is dictated by the same molecular mechanism herewith described for proteases (88, 156). TGase catalyzes acyl transfer reactions between the γ-carboxamide groups of protein-bound glutamine (Gln) residues, which serve as acyl donors, and primary amines, resulting in the formation of new γ-amides of glutamic acid and the release of ammonia (82, 101, 125, 150). In a number of cases, it has been demonstrated that the TGase-mediated reactions at glutamine (Gln) residues in proteins can be very selective (88). For example, TGase mediates the conjugation of an amino derivative of poly(ethylene glycol) (PEG-NH_2) at Gln91 in apoMb and at Gln40 and Gln141 in hGH, despite these proteins having many more Gln residues (156). Of note, as shown above, these proteins suffered highly selective limited proteolysis phenomena at the same flexible/disordered chain regions being attacked by TGase (see Fig. 20.7 for apoMb and Fig. 20.8 for hGH). Of interest, granulocyte colony-stimulating factor is selectively modified by TGase at Gln134 (156), whereas the natural form of this protein is glycosilated at the nearby Thr133. Thus, both sites of specific enzymatic modifications of this protein factor are located within the chain region 126–137 shown by crystallography to be disordered (108) (see the crystal structure in PDB, code 1RHG). Therefore, in analogy to proteases, TGase also requires a flexible polypeptide substrate in order to perform its catalytic event. This correlates with the fact that TGase can be considered a "reverse protease," in the sense that it can catalyze the synthesis instead of hydrolysis of an amide bond. Indeed, in analogy to cysteine proteases, the active site of TGase contains similar active site residues (Cys64, His274, and Asp255 in microbial TGase) (125).

20.8.1 Actin

As an additional example, we may mention here that the 375-residue chain of actin can be selectively cleaved at peptide bonds Gly42–Val43 and Met47–Gly48 by digestion with subtilisin, a protease from *Escherichia coli* and other proteases (126, 130, 162, 222). Moreover, actin can be selectively labeled at Gln41 by a TGase-mediated reaction using dansyl-ethylenediamine or dansylcadaverine used as an amino donor (229). Also the 43-residue chain of thymosin β4 can be covalently bound to Gln41 of actin via an isopeptide bond involving mostly Lys38 of thymosin β4 by means of TGase (202). That chain disorder controls the selectivity of the reactions on actin by both proteases and TGase is substantiated by the X-ray structure of this protein. Indeed, the B-factor profile along the polypeptide chain of actin shows a discontinuity at chain region 39–51 due to missing electron density, indicative of a disordered chain segment (131) (PDB code 1QZ6).

Since TGase catalyses the formation of isopeptide bonds between the side chains of Gln and Lys residues in proteins, thus leading to inter- or intramolecular protein cross-linking, perhaps both protein-bound acyl donor (Gln) and amino donor (Lys) should be characterized by sufficient flexibility in order to facilitate the TGase-mediated reaction. Indeed, TGase can form an intramolecular cross-link between Gln41 and Lys50, both residues being located in the disordered chain segment 39–51 (67). When a mixture of actin and myosin subfragment 1 (S1) was reacted with bacterial TGase, a selective covalent, intermolecular isopeptide bond was formed between Gln41 of actin and the Lys-rich loop (residues 636–642, Gly635-KKGGKKK-Gly646) of myosin S1 (70). Both residues forming the intermolecular isopeptide bond are embedded in disordered regions, namely region 39–51 in actin (PDB code 1QZ6) and region 627–646 in myosin S1 (PDB code 2MYS).

The disordered region 40–50 of actin is involved in the binding of DNase-I (126, 130), as well as in other enzymatic reactions. The actin of the *Dictyostelium*, highly homologous to the rabbit actin, was shown to become selectively phosphorylated at Tyr53 during the development cycle of amoebae (7, 146). Recently, the crystal structures of both native and phosphorylated actin were solved (7). The 40–50 loop is partially ordered and stabilized by hydrogen-bonding contacts with the phosphate group of Tyr53 in the phosphorylated actin, whereas it is fully disordered in nonphosphorylated actin. Four residues of the loop 40–50 were resolved in the electron density map (Gly42, Met47, Gly48, Gln49) and the average B-factor for these residues is 79.4 $Å^2$, compared with ~30 $Å^2$ for the rest of the protein chain. This is an indication that in the phosphorylated actin, the loop is more constrained but still quite dynamic. Of interest, the rate of proteolytic digestion of the peptide bond Met47–Gly48 by subtilisin is 50% reduced in respect to that observed with the native actin (7), in keeping with the view that flexibility/disorder is required for proteolysis. Finally, we mention that Lys50 of actin is involved in the selective isopeptide bond formation catalyzed by the RTX toxin of the bacterium *Vibrio cholerae* (138). Overall, these various observations indicate that the disordered loop 40–50 of actin is involved in several other enzymatic reactions, besides limited proteolysis events.

20.9 OUTLOOK

The examples we have presented here establish that the sites of limited proteolysis (nicksites) in globular proteins of known 3D structure are characterized by enhanced chain flexibility or segmental mobility, as reflected by the B-factor profile along the protein chain. Proteolysis can be extremely selective especially when nicksites occur at regions for which no recognizable signals appear in the electron density maps and thus are disordered. Therefore, a mechanism of local unfolding is explaining the often observed exquisite selectivity of the limited proteolysis phenomenon. Indeed, this notion is in agreement with the results of modeling experiments aimed to unravel details of the molecular recognition phenomenon between a protease and a protein substrate. It was found that the prime determinant of the site-specific proteolytic event is the ability of a chain segment of up to 12 residues to unfold locally without perturbing the overall protein conformation (85, 112, 115). Moreover, flexibility or unfolding of a polypeptide substrate enables the adoption of the extended conformation invariably required for its binding to the protease's active site (239, 240). We emphasize here again that the parameters of surface accessibility, solvent exposure, or protrusion are not at all sufficient to explain the selectivity of the limited proteolysis phenomenon. Hopefully, the use of the misleading terminology of "surface exposure," "accessibility," "protein topology," or "topography" for explaining molecular aspects of limited proteolysis phenomena will be abandoned.

Proteolytic enzymes can be used to identify and isolate folded proteins in libraries of proteins produced by the phage-display technology (8, 136) and to pinpoint, in globular proteins, sites (loops or turns) characterized by local unfolding or regions of protein disorder (85, 137). In the past, the technique was rather difficult to apply, since the analytic methods required to isolate and characterize protein fragments were labor intensive and not sufficiently sensitive. No doubt that limited proteolysis has attracted the interest on numerous investigators since the beginning of protein science several decades ago. These continuing efforts have produced a vast amount of experimental studies, as documented by the fact that a literature search on PubMed for the key word "limited proteolysis" produces 4601 publications. The recent dramatic advances in MS techniques in analyzing and even sequencing peptides and proteins (1, 28, 52, 281) will likely prompt a more systematic use of the limited proteolysis/MS approach as a simple first step in the elucidation of structure–dynamics–function relationships of a novel and rare protein, especially if available in minute amounts. A specific advantage of using proteolytic probes is that they provide data on the solution structure of a protein, even if these data do not reach the high-resolution level provided by X-ray or NMR techniques. Challenging applications of the limited proteolysis/MS approaches are in the analysis of protein assemblies in virus (25, 169, 234) and ribosomes (223). A curated website, named PMAP, has been very recently established with the view to aid the large scientific community interested in proteolytic events and pathways (119).

The experimental data herewith presented and discussed document that even the difficult problem of analyzing the molecular properties of ID proteins can benefit from the use of the limited proteolysis technique, shown here to provide a very reliable identification of disordered regions in proteins. Considering that the identification and analysis of these disordered regions in otherwise folded proteins can be achieved only with moderate success using physicochemical, spectroscopic and computational techniques, alternative experimental methods should be investigated. In conclusion, we suggest that protein disorder should be investigated by taking advantage of a variety of experimental techniques, including the simple biochemical approach relying on the use of proteolytic probes of protein structure and dynamics.

ACKNOWLEDGMENTS

This work was supported in part by the Italian Ministry of University and Research (PRIN-2003, PRIN-2006, and FIRB-2003 No. RBNEOPX83 Projects on Protein Folding, Misfolding and Aggregation). The authors wish to thank Ms. Barbara Sicoli for the expert typing of the manuscript.

REFERENCES

1. Aebersold, R., and M. Mann. 2003. Mass spectrometry-based proteomics. Nature **422**:198–207.
2. Alexandrescu, A. T., P. A. Evans, M. Pitkeathly, J. Baum, and C. M. Dobson. 1993. Structure and dynamics of the acid-denatured molten globule state of α-lactalbumin: a two-dimensional NMR study. Biochemistry **32**:1707–18.
3. Andersen, J. S., B. Svensson, and P. P. Roepstorff. 1996. Electrospray ionization and matrix assisted laser desorption/ionization mass spectrometry: powerful analytical tools in recombinant protein chemistry. Nat Biotechnol **14**:449–57.
4. Arai, M., and K. Kuwajima. 2000. Role of the molten globule state in protein folding. Adv Protein Chem **53**:209–82.
5. Babu, Y. S., C. E. Bugg, and W. J. Cook. 1988. Structure of calmodulin refined at 2.2 Å resolution. J Mol Biol **204**:191–204.
6. Babu, Y. S., J. S. Sack, T. J. Greenhough, C. E. Bugg, A. R. Means, and W. J. Cook. 1985. Three-dimensional structure of calmodulin. Nature **315**:37–40.
7. Baek, K., X. Liu, F. Ferron, S. Shu, E. D. Korn, and R. Dominguez. 2008. Modulation of actin structure and function by phosphorylation of Tyr-53 and profilin binding. Proc Natl Acad Sci U S A **105**:11748–53.
8. Bai, Y., and H. Feng. 2004. Selection of stably folded proteins by phage-display with proteolysis. Eur J Biochem **271**:1609–14.
9. Baldwin, R. L. 2008. The search for folding intermediates and the mechanism of protein folding. Annu Rev Biophys **37**:1–21.
10. Balguerie, A., S. Dos Reis, B. Coulary-Salin, S. Chaignepain, M. Sabourin, J. M. Schmitter, and S. J. Saupel. 2004. The sequences appended to the amyloid core region of the HET-s prion protein determine higher-order aggregate organization *in vivo*. J Cell Sci **117**:2599–610.
11. Bantscheff, M., V. Weiss, and M. O. Glocker. 1999. Identification of linker regions and domain borders of the transcription activator protein NtrC from *Escherichia coli* by limited proteolysis, in-gel digestion and mass spectrometry. Biochemistry **38**:11012–20.
12. Barbato, G., M. Ikura, L. E. Kay, R. W. Pastor, and A. Bax. 1992. Backbone dynamics of calmodulin studied by 15N-relaxation using inverse detected two-dimensional NMR spectroscopy: the central helix is flexible. Biochemistry **31**:5269–78.
13. Baron, M., D. G. Norman, and I. D. Campbell. 1991. Protein modules. Trends Biochem Sci **16**:13–7.

14. Baxa, U., K. L. Taylor, J. S. Wall, M. N. Simon, N. Cheng, R. B. Wickner, and A. C. Steven. 2003. Architecture of Ure2p prion filaments: the N-terminal domains form a central core fiber. J Biol Chem **278**:43717–27.

15. Bayley, P. M., and S. R. Martin. 1992. The alpha-helical content of calmodulin is increased by solution conditions favouring protein crystallisation. Biochim Biophys Acta **1160**:16–21.

16. Bennett, W. S., and R. Huber. 1984. Structural and functional aspects of domain motions in proteins. CRC Crit Rev Biochem **15**:291–384.

17. Berjanskii, M. V., and D. S. Wishart. 2005. A simple method to predict protein flexibility using secondary chemical shifts. J Am Chem Soc **127**:14970–1.

18. Berman, H. M., J. Westbrook, Z. Feng, G. Gilliland, T. N. Bhat, H. Weissig, I. N. Shindyalov, and P. E. Bourne. 2000. The Protein Data Bank. Nucleic Acids Res **28**:235–42.

19. Bhalla, J., G. B. Storchan, C. M. McCarthy, V. N. Uversky, and O. Tcherkasskaya. 2006. Local flexibility in molecular function paradigm. Mol Cell Proteomics **5**:1212–23.

20. Bonanno, J. B., S. C. Almo, A. Bresnick, M. R. Chance, A. Fiser, S. Swaminathan, J. Jiang, F. W. Studier, L. Shapiro, C. D. Lima, T. M. Gaasterland, A. Sali, K. Bain, L. Feil, X. Gao, D. Lorimer, A. Ramos, J. M. Sauder, S. R. Wasserman, S. Emtage, K. L. D'Amico, and S. K. Burley. 2005. New York-Structural GenomiX Research Consortium (NYSGXRC): a large scale center for the protein structure initiative. J Struct Funct Genomics **6**:225–32.

21. Bond, J. S. 1990. Commercially available proteases, pp. 232–40. In R. J. Beynon and J. S. Bond (eds.), Proteolytic Enzymes: A Practical Approach. IRL Press, Oxford.

22. Booth, D. R., M. Sunde, V. Bellotti, C. V. Robinson, W. L. Hutchinson, P. E. Fraser, P. N. Hawkins, C. M. Dobson, S. E. Radford, C. C. Blake, and M. B. Pepys. 1997. Instability, unfolding and aggregation of human lysozyme variants underlying amyloid fibrillogenesis. Nature **27**:787–93.

23. Bordoli, L., F. Kiefer, and T. Schwede. 2007. Assessment of disorder predictions in CASP7. Proteins **69**:129–36.

24. Bossemeyer, D., R. A. Engh, V. Kinzel, H. Ponstingl, R. Huber. 1993. Phosphotransferase and substrate binding mechanism of the cAMP-dependent protein kinase catalytic subunit from porcine heart as deduced from the 2.0 Å structure of the complex with Mn2+ adenylyl imidodiphosphate and inhibitor peptide PKI(5-24). EMBO J. **12**:849–59.

25. Bothner, B., X. F. Dong, L. Bibbs, J. E. Johnson, and G. Siuzdak. 1998. Evidence of viral capsid dynamics using limited proteolysis and mass spectrometry. J Biol Chem **273**:673–6.

26. Bousset, L., F. Briki, J. Doucet, and R. Melki. 2003. The native-like conformation of Ure2p in fibrils assembled under physiologically relevant conditions switches to an amyloid-like conformation upon heat-treatment of the fibrils. J Struct Biol **141**:132–42.

27. Bousset, L., V. Redeker, P. Decottignies, S. Dubois, P. Le Maréchal, and R. Melki. 2004. Structural characterization of the fibrillar form of the yeast *Saccharomyces cerevisiae* prion Ure2p. Biochemistry **43**:5022–32.

28. Bradshaw, R. A., and A. L. Burlinghame. 2005. From proteins to proteomics. IUBMB Life **57**:267–72.

29. Brooks, C. L. 1992. Characterization of “native” apomyoglobin by molecular dynamics simulation. J Mol Biol **227**:375–80.

30. Bucciantini, M., E. Giannoni, F. Chiti, F. Baroni, L. Formigli, J. Zurdo, N. Taddei, G. Ramponi, C. M. Dobson, and M. Stefani. 2002. Inherent toxicity of aggregates implies a common mechanism for protein misfolding diseases. Nature **416**:507–11.

31. Campioni, S., M. F. Mossuto, S. Torrassa, G. Calloni, P. Polverino de Laureto, A. Relini, A. Fontana, and F. Chiti. 2008. Conformational properties of the aggregation precursor state of HypF-N. J Mol Biol **379**:554–67.

32. Carugo, O. 2007. Identification of domains in protein crystal structures. J Appl Crystallogr **40**:778–81.

33. Chang, J.-Y. 1990. Production, properties and thrombin inhibitory mechanism of hirudin amino-terminal core fragments. J Biol Chem **265**:22159–66.

34. Chang, J.-Y., J.-M. Schlaeppi, and S. R. Stone. 1990. Antithrombin activity of the hirudin N-terminal core domain residues 1–43. FEBS Lett **260**:209–12.

35. Charrier, V., J. Luba, D. Parsonage, and A. Claiborne. 2000. Limited proteolysis as a structural probe of the soluble α-glycerophosphate oxidase from *Streptococcus* sp. Biochemistry **39**:5035–44.

36. Chattopadhyaya, R., W. E. Meador, A. R. Means, and F. A. Quiocho. 1992. Calmodulin structure refined at 1.7 Å resolution. J Mol Biol **228**:1177–92.

37. Chiti, F., and C. M. Dobson. 2006. Protein misfolding, functional amyloid and human disease. Annu Rev Biochem **75**:333–66.

38. Chiti, F., N. Taddei, F. Baroni, C. Capanni, M. Stefani, G. Ramponi, and C. M. Dobson. 2002. Kinetic partitioning of protein folding and aggregation. Nat Struct Biol **9**:137–43.

39. Chiti, F., P. Webster, N. Taddei, A. Clark, M. Stefani, G. Ramponi, and C. M. Dobson. 1999. Designing conditions for *in vitro* formation of amyloid protofilaments and fibrils. Proc Natl Acad Sci U S A **96**:3590–4.

40. Choe, S., M. J. Bennett, G. Fujii, P. M. Curmi, K. A. Kantardjieff, R. J. Collier, and D. Eisenberg. 1992. The crystal structure of diphtheria toxin. Nature **357**:216–22.

41. Christ, D., and G. Winter. 2006. Identification of protein domains by shotgun proteolysis. J Mol Biol **358**:364–71.

42. Chyan, C. L., C. Wormald, C. M. Dobson, P. A. Evans, and J. Baum. 1993. Structure and stability of the molten globule state of guinea-pig α-lactalbumin: a hydrogen exchange study. Biochemistry **32**:5681–91.

43. Colussi, T., D. Parsonage, W. Boles, T. Matsuoka, T. C. Mallett, P. A. Karplus, and A. Claiborne. 2008. Structure of α-glycerophosphate oxidase from *Streptococcus* sp.: A template for the mitochondrial α-glycerophosphate dehydrogenase. Biochemistry **47**:965–77.

44. Dalzoppo, D., C. Vita, and A. Fontana. 1985. Folding of thermolysin fragments: identification of the minimum size of a carboxyl-terminal fragment that can fold into a stable native-like structure. J Mol Biol **182**:331–40.

45. De Filippis, V., A. Vindigni, L. Altichieri, and A. Fontana. 1995. Core domain of hirudin from the leech *Hirudinaria manillensis*: Chemical synthesis, purification, and characterization of a Trp3 analog of fragment 1–47. Biochemistry **34**:9552–64.

46. de Vos, A. M., M. H. Ultsch, and A. A. Kossiakoff. 1992. Human growth hormone and extracellular domain of its receptor: crystal structure of the complex. Science **255**:306–12.

47. DeLano, W. L. 2002. The PyMol Molecular Graphics System. DeLano Scientific, San Carlos, CA. www.pymol.org.

48. Dobson, C. M. 2003. Protein folding and diseases: a view from the First Horizon Symposium. Nat Rev Drug Discov **2**:154–60.

49. Dobson, C. M. 2006. Protein aggregation and its consequences for human diseases. Protein Pept Lett **13**:219–27.

50. Dodt, J., U. Seemuller, R. Maschler, and H. Fritz. 1985. The complete covalent structure of hirudin: localization of the disulfide bonds. Biol Chem Hoppe-Seyler **366**:379–85.

51. Dolgikh, D. A., R. L. Gilmanshin, E. V. Brazhnikov, V. E. Bychkova, G. V. Semisotnov, S. Y. Venyaminov, and O. B. Ptitsyn. 1981. α-Lactalbumin: compact state with fluctuating tertiary structure? FEBS Lett **136**:311–5.

52. Domon, B., and R. Aebersold. 2006. Mass spectrometry and protein analysis. Science **312**:212–7.

53. Draibikowski, W., H. Brzeska, and S. Y. Venyaminov. 1982. Tryptic fragments of calmodulin. J Biol Chem **257**:11584–90.

54. Drapeau, G. R. 1977. Cleavage at glutamic acid with staphylococcal protease. Methods Enzymol **47**:189–91.

55. Dunker, A. K., C. J. Brown, J. D. Lawson, L. M. Iakoucheva, and Z. Obradovic. 2002. Intrinsic disorder and protein function. Biochemistry **41**:6573–82.

56. Dunker, A. K., C. J. Brown, and Z. Obradovic. 2002. Identification and functions of usefully disordered proteins. Adv Protein Chem **62**:25–49.

57. Dunker, A. K., E. Garner, S. Guilliot, P. Romero, K. Albrecht, J. Hart, Z. Obradovic, C. Kissinger, and J. E. Villafranca. 1998. Protein disorder and the evolution of molecular recognition: Theory, predictions and observations. Pac Symp Biocomput **3**:473–84.

58. Dunker, A. K., J. D. Lawson, C. J. Brown, R. M. Williams, P. Romero, J. S. Oh, C. J. Oldfield, A. M. Campen, C. M. Ratliff, K. W. Hipps, J. Ausio, M. S. Nissen, R. Reeves, C. Kang, C. R. Kissinger, R. W. Bailey, M. D. Griswold, W. Chiu, E. C. Garner, and Z. Obradovic. 2001. Intrinsically disordered proteins. J Mol Graph Model **19**:26–59.

59. Dunker, A. K., and Z. Obradovic. 2001. The protein trinity: linking function and disorder. Nat Biotechnol **19**:805–6.

60. Dunker, A. K., Z. Obradovic, P. Romero, E. C. Garner, and C. J. Brown. 2000. Intrinsic protein disorder in complete genomes. Genome Inform Ser Workshop Genome Inform **11**:161–71.

61. Dunker, A. K., I. Silman, V. N. Uversky, and J. L. Sussman. 2008. Function and structure of inherently disordered proteins. Curr Opin Struct Biol **18**:756–64.

62. Dyson, H. J., and P. E. Wright. 2002. Coupling of folding and binding for unstructured proteins. Curr Opin Struct Biol **12**:54–60.
63. Dyson, H. J., and P. E. Wright. 2002. Insights into the structure and dynamics of unfolded proteins from nuclear magnetic resonance. Adv Protein Chem **62**:311–40.
64. Dyson, H. J., and P. E. Wright. 2004. Unfolded proteins and protein folding studied by NMR. Chem Rev **104**:3607–22.
65. Dyson, H. J., and P. E. Wright. 2005. Intrinsically unstructured proteins and their functions. Nature Rev Mol Cell Biol **6**:197–208.
66. Ebeling, W., N. Hennrich, M. Klockow, H. Meta, D. Orth, and H. Lang. 1974. Proteinase K from *Tritirachium album Limber*. Eur J Biochem **47**:91–7.
67. Eli-Berchoer, L., G. Hegyi, A. Patthy, E. Reisler, and A. Muhlrad. 2000. Effect of intramolecular cross-linking between glutamine-41 and lysine-50 on actin structure and function. J Muscle Res Cell Motil **21**:405–14.
68. Eliezer, D., and P. E. Wright. 1996. Is apomyoglobin a molten globule? Structural characterization by NMR. J Mol Biol **263**:531–8.
69. Eliezer, D., J. Yao, H. J. Dyson, and P. E. Wright. 1998. Structural and dynamic characterization of partially folded states of apomyoglobin and implications for protein folding. Nature Struct Biol **5**:148–55.
70. Eligula, L., L. Chuang, M. L. Phillips, M. Motoki, K. Seguro, and A. Muhlrad. 1998. Transglutaminase-induced cross-linking between subdomain 2 of G-actin and the 636–642 lysine-rich loop of myosin subfragment 1. Biophys J **74**:953–63.
71. Englander, S. W. 2000. Protein folding intermediates and pathways studied by hydrogen exchange. Annu Rev Biophys Biomol Struct **29**:213–38.
72. Evans, P. A., and S. E. Radford. 1994. Probing the structure of folding intermediates. Curr Opin Struct Biol **4**:100–6.
73. Evans, S. V., and G. D. Brayer. 1990. High-resolution study of the three-dimensional structure of horse heart metmyoglobin. J Mol Biol **213**:885–97.
74. Fändrich, M., M. A. Fletcher, and C. M. Dobson. 2001. Amyloid fibrils from muscle myoglobin. Nature **410**:165–6.
75. Fassina, G., C. Vita, D. Dalzoppo, M. Zamai, M. Zambonin, and A. Fontana. 1986. Autolysis of thermolysin: isolation and characterization of a folded three-fragment complex. Eur J Biochem **156**:221–8.
76. Fernàndez-Busquets, X., N. S. de Groot, D. Fernandez, and S. Ventura. 2008. Recent structural and computational insights into conformational diseases. Curr Med Chem **15**:1336–49.
77. Ferron, F., S. Longhi, B. Canard, and D. Karlin. 2006. A practical overview of protein disorder prediction methods. Proteins **65**:1–14.
78. Fink, A. L. 1995. Compact intermediate states in protein folding. Annu Rev Biomol Struct **24**:495–522.
79. Fink, A. L. 2005. Natively unfolded proteins. Curr Opin Struct Biol **15**:35–41.
80. Fink, A. L., K. A. Oberg, and S. Seshadri. 1998. Discrete intermediates versus molten globule models for protein folding: characterization of partially folded intermediates of apomyoglobin. Fold Des **3**:19–25.

81. Finn, B. E., and S. Forsén. 1995. The evolving model of calmodulin structure, function and activation. Structure **3**:7–11.

82. Folk, J. E., and J. S. Finlayson. 1977. The ε-(γ-glutamyl)lysine crosslink and the catalytic role of transglutaminases. Adv Protein Chem **31**:1–133.

83. Fontana, A., G. Fassina, C. Vita, D. Dalzoppo, M. Zamai, and M. Zambonin. 1986. Correlation between sites of limited proteolysis and segmental mobility in thermolysin. Biochemistry **25**:1847–51.

84. Fontana, A., P. Polverino de Laureto, and V. De Filippis. 1993. Molecular aspects of proteolysis of globular proteins, pp. 101–10. In W. van den Tweel, A. Harder, and M. Buitelear (eds.), Protein Stability and Stabilization. Elsevier Science Publishing, Amsterdam.

85. Fontana, A., P. Polverino de Laureto, V. De Filippis, E. Scaramella, and M. Zambonin. 1997. Probing the partly folded states of proteins by limited proteolysis. Fold Des **2**:R17–26.

86. Fontana, A., P. Polverino de Laureto, V. De Filippis, E. Scaramella, and M. Zambonin. 1999. Limited proteolysis in the study of protein conformation, p. 257–284. *In* E. E. Sterchi, and W. Stöcker (ed.), Proteolytic Enzymes: Tools and Targets, Springer Verlag, Heidelberg.

87. Fontana, A., P. Polverino de Laureto, B. Spolaore, E. Frare, P. Picotti, and M. Zambonin. 2004. Probing protein structure by limited proteolysis. Acta Biochim Pol **51**:299–321.

88. Fontana, A., B. Spolaore, A. Mero, and F. M. Veronese. 2008. Site-specific modification and PEGylation of pharmaceutical proteins mediated by transglutaminase. Adv Drug Deliv Rev **60**:13–28.

89. Fontana, A., C. Vita, D. Dalzoppo, and M. Zambonin. 1989. Limited proteolysis as a tool to detect structure and dynamic features of globular proteins: studies on thermolysin, pp. 315–24. In B. Wittman-Liebold (ed.), Methods in Protein Sequence Analysis. Springer-Verlag, Berlin.

90. Fontana, A., M. Zambonin, P. Polverino de Laureto, V. De Filippis, A. Clementi, and E. Scaramella. 1997. Probing the conformational state of apomyoglobin by limited proteolysis. J Mol Biol **266**:223–30.

91. Frare, E., M. F. Mossuto, P. Polverino de Laureto, M. Dumoulin, C. M. Dobson, and A. Fontana. 2006. Identification of the core structure of lysozyme amyloid fibrils by proteolysis. J Mol Biol **361**:551–61.

92. Frare, E., M. F. Mossuto, P. Polverino de Laureto, S. Tolin, L. Menzer, M. Dumoulin, C. M. Dobson, and A. Fontana. 2009. Characterization of oligomeric species on the aggregation pathway of human lysozyme. J Mol Biol **387**:17–27.

93. Frare, E., P. Polverino de Laureto, J. Zurdo, C. M. Dobson, and A. Fontana. 2004. A highly amyloidogenic region of hen lysozyme. J Mol Biol **23**:1153–65.

94. Frauenfelder, H., G. A. Petsko, and D. Tsernoglou. 1979. Temperature-dependent X-ray diffraction as a probe of protein structural dynamics. Nature **280**:558–63.

95. Fruton, J. S. 1970. The specificity and mechanism of pepsin action. Adv Enzymol **33**:401–43.

96. Gallo, D., Y. Jacquot, G. Laurent, and G. Leclercq. 2008. Calmodulin, a regulatory partner of the estrogen receptor alpha in breast cancer cells. Mol Cell Endocrinol **291**:20–6.

97. Gao, X., K. Bain, J. B. Bonanno, M. Buchanan, D. Henderson, D. Lorimer, C. Marsh, J. A. Reynes, J. M. Sauder, K. Schwinn, C. Thai, and S. K. Burley. 2005. High-throughput limited proteolysis/mass spectrometry for protein domain elucidation. J Struct Funct Genomics **6**:129–34.
98. George, R. A., and J. Heringa. 2002 An analysis of protein domain linkers: their classification and role in protein folding. Protein Eng **15**:871–9.
99. Ghrir, R., and F. Lederer. 1981. Study of a zone highly sensitive to proteases in flavocytochrome b2 from *Saccharomyces cerevisiae*. Eur J Biochem **120**:279–87.
100. Gilmanshin, R., R. B. Dyer, and R. H. Callender. 1997. Structural heterogeneity of the various forms of apomyoglobin: implications for protein folding. Protein Sci **6**:2134–42.
101. Griffin, R., R. Casadio, and C. M. Bergamini. 2002. Transglutaminases: nature's biological glues. Biochem J **368**:377–96.
102. Håkansson, A., B. Zhivotovsky, S. Orrenius, H. Sabharwal, and C. Svanborg. 1995. Apoptosis induced by a human milk protein. Proc Natl Acad Sci U S A **92**:8064–8.
103. Han, J. H., S. Batey, A. A. Nickson, S. A. Teichmann, and J. Clarke. 2007. The folding and evolution of multidomain proteins. Nat Rev Mol Cell Biol **8**:319–30.
104. Harrison, S. C., and E. R. Blout. 1965. Reversible conformational changes of myoglobin and apomyoglobin. J Biol Chem **240**:299–303.
105. Haruyama, H., and K. Wütrich. 1989. Conformation of recombinant desulfatohirudin in aqueous solution determined by nuclear magnetic resonance. Biochemistry **28**:4301–12.
106. Heidorn, D. B., and J. Trewhella. 1988. Comparison of the crystal and solution structures of calmodulin and troponin *c*. Biochemistry **27**:909–15.
107. Heinrikson, R. L. 1977. Applications of thermolysin in protein structural analysis. Methods Enzymol **47**:175–89.
108. Hill, C. P., T. D. Osslund, and D. Eisenberg. 1993. The structure of granulocyte-colony-stimulating factor and its relationship to other growth factors. Proc Natl Acad Sci U S A **90**:5167–71.
109. Hirst, J. D., and C. L. Brooks. 1994. Helicity, circular dichroism and molecular dynamics of proteins. J Mol Biol **243**:173–8.
110. Holland, T. A., S. Veretnik, I. N. Shindyalov, and P. E. Bourne. 2006. Partitioning protein structures into domains: why is it so difficult? J Mol Biol **361**:562–90.
111. Hoshino, M., H. Katou, Y. Hagihara, K. Hasegawa, H. Naiki, and Y. Goto. 2002. Mapping the core of the β2-microglobulin amyloid fibril by H/D exchange. Nat Struct Biol **9**:332–6.
112. Hubbard, S. J. 1998. The structural aspects of limited proteolysis of native proteins. Biochim Biophys Acta **1382**:191–206.
113. Hubbard, S. J., R. J. Beynon, and J. M. Thornton. 1998. Assessment of conformational parameters as predictors of limited proteolytic sites in native protein structures. Protein Eng **11**:349–59.
114. Hubbard, S. J., S. F. Campbell, and J. M. Thornton. 1991. Molecular recognition: conformational analysis of limited proteolytic sites and serine proteinase protein inhibitors. J Mol Biol **220**:507–30.

115. Hubbard, S. J., F. Eisenmenger, and J. M. Thornton. 1994. Modeling studies of the change in conformation required for cleavage of limited proteolytic sites. Protein Sci **3**:757–68.

116. Iakoucheva, L. M., C. J. Brown, J. D. Lawson, Z. Obradovic, and A. K. Dunker. 2002. Intrinsic disorder in cell-signaling and cancer-associated proteins. J Mol Biol **323**:573–84.

117. Iakoucheva, L. M., A. L. Kimzey, C. D. Masselon, R. D. Smith, A. K. Dunker, and E. J. Ackerman. 2001. Aberrant mobility phenomena of the DNA repair protein XPA. Protein Sci **10**:1353–62.

118. Iakoucheva, L. M., P. Radivojac, C. J. Brown, T. R. O'Connor, J. G. Sikes, Z. Obradovic, and A. K. Dunker. 2004. The importance of intrinsic disorder for protein phosphorylation. Nucleic Acids Res **32**:1037–49.

119. Igarashi, Y., E. Heureux, K. S. Doctor, P. Talwar, S. Gramatikova, K. Gramatikoff, Y. Zhang, M. Blinov, S. S. Ibragimova, S. Boyd, B. Ratnikov, P. Ciepla, A. Godzik, J. W. Smith, A. L. Osterman, and A. M. Eroshkin. 2009. PMAP: databases for analyzing proteolytic events and pathways. Nucleic Acids Res **37**:D611–8. www.proteolysis.org.

120. Ikura, M., G. M. Clore, A. M. Gronenborn, G. Zhu, C. B. Klee, and A. Bax. 1992. Solution structure of a calmodulin-target peptide complex by multidimensional NMR. Science **256**:632–8.

121. Ishima, R., and D. A. Torchia. 2000. Protein dynamics from NMR. Nat Struct Biol **7**:740–3.

122. Islam, S. A., J. Luo, and M. J. Sternberg. 1995. Identification and analysis of domains in proteins. Protein Eng **8**:513–25.

123. Jawhari, A., S. Boussert, V. Lamour, R. A. Atkinson, B. Kieffer, O. Poch, N. Potier, A. van Dorsselaer, D. Moras, and A. Poterszman. 2004. Domain architecture of the p62 subunit from the human transcription/repair factor TFIIH deduced by limited proteolysis and mass spectrometry analysis. Biochemistry **43**:14420–30.

124. Johnson, P. H., and B. F. P. Edwards. 1992. The structure of a complex of bovine α-thrombin and recombinant hirudin at 2.8 Å resolution. J Biol Chem **267**:17670–8.

125. Kashiwagi, T., K. Yokoyama, K. Ishikawa, K. Ono, D. Ejima, H. Matui, and E. Suzuki. 2002. Crystal structure of microbial transglutaminase from *Streptoverticillium mobaraense*. J Biol Chem **277**:44252–60.

126. Khaitlina, S. Y., J. Moraczewska, and H. Strzelecka-Gołaszewska. 1993. The actin/actin interactions involving the N-terminus of the DNase-I-binding loop are crucial for stabilization of the actin filament. Eur J Biochem **218**:911–20.

127. Kheterpal, I., K. D. Cook, and R. Wetzel. 2006. Hydrogen/deuterium exchange mass spectrometry analysis of protein aggregates. Methods Enzymol **413**:140–66.

128. Kheterpal, I., and R. Wetzel. 2006. Hydrogen/deuterium exchange mass spectrometry: a window into amyloid structure. Acc Chem Res **39**:584–93.

129. Kheterpal, I., A. Williams, C. Murphy, B. Bledsoe, and R. Wetzel. 2001. Structural features of the Abeta amyloid fibril elucidated by limited proteolysis. Biochemistry **40**:11757–67.

130. Kiessling, P., W. Jahn, G. Maier, B. Polzar, and H. G. Mannherz. 1995. Purification and characterization of subtilisin cleaved actin lacking the segment of residues 43–47 in the DNase-I binding loop. Biochemistry **34**:14834–42.

131. Klenchin, V. A., J. S. Allingham, R. King, J. Tanaka, G. Marriott, and I. Rayment. 2003. Trisoxazole macrolide toxins mimic the binding of actin-capping proteins to actin. Nat Struct Biol **10**:1058–63.

132. Komatsu, N., K. Saijoh, N. Otsuki, T. Kishi, I. P. Micheal, C. V. Obiezu, C. A. Borgono, K. Takehara, A. Jayakumar, H. K. Wu, G. L. Clayman, and E. P. Diamandis. 2007. Proteolytic processing of human growth hormone by multiple tissue kallikreins and regulation by the serine protease inhibitor Kazal-Type5 (SPINK5) protein. Clin Chim Acta **377**:228–36.

133. Koth, C. M., S. M. Orlicky, S. M. Larson, and A. M. Edwards. 2003. Use of limited proteolysis to identify protein domains suitable for structural analysis. Methods Enzymol **368**:77–84.

134. Kozhukh, G. V., Y. Hagihara, T. Kawakami, K. Hasegawa, H. Naiki, and Y. Goto. 2002. Investigation of a peptide responsible for amyloid fibril formation of β_2-microglobulin by *Achromobacter* protease. J Biol Chem **277**:1310–15.

135. Krishna, M. M., L. Hoang, Y. Lin, and S. W. Englander. 2004. Hydrogen exchange methods to study protein folding. Methods **34**:51–64.

136. Kristensen, P., and G. Winter. 1998. Proteolytic selection for protein folding using filamentous bacteriophages. Fold Des **3**:321–8.

137. Kriwacki, R. W., J. Wu, L. Tennant, P. E. Wright, and G. Siuzdak. 1997. Probing protein structure using biochemical and biophysical methods: proteolysis, matrix-assisted laser desorption/ionization mass spectrometry, high-performance liquid chromatography and size-exclusion chromatography of p21Waf1/Cip1/Sdi1. J Chromatogr A **777**:23–30.

138. Kudryashov, D. S., Z. A. Durer, A. J. Ytterberg, M. R. Sawaya, I. Pashkov, K. Prochazkova, T. O. Yeates, R. R. Loo, J. A. Loo, K. J. Satchell, and E. Reisler. 2008. Connecting actin monomers by iso-peptide bond is a toxicity mechanism of the *Vibrio cholerae* MARTX toxin. Proc Natl Acad Sci U S A **105**:18537–42.

139. Kundu, S., J. S. Melton, D. C. Sorensen, and G. N. Phillips, Jr. 2002. Dynamics of proteins in crystals: comparison of experiment with simple models. Biophys J **83**:723–32.

140. Kuwajima, K. 1989. The molten globule state as a clue for understanding the folding and cooperativity of globular-protein structure. Proteins **6**:87–103.

141. Kuwajima, K. 1996. The molten globule state of α-lactalbumin. FASEB J. **10**:102–9.

142. Le Gall, T., P. R. Romero, M. S. Cortese, V. N. Uversky, and A. K. Dunker. 2007. Intrinsic disorder in the Protein Data Bank. J Biomol Struct Dyn **24**:325–42.

143. Levin, J., A. Giese, K. Boetzel, I. Israel, T. Högen, G. Nübling, H. Kretzschmar, and S. Lorenzl. 2009. Increased α-synuclein aggregation following limited cleavage by certain matrix metalloproteinases. Exp Neurol **215**:201–8.

144. Li, C. H. 1982. Human growth hormone: 1974–1981. Mol Cell Biochem **46**:31–41.

145. Linderstrøm-Lang, K. 1950. Structure and enzymatic breakdown of proteins. Cold Spring Harbor Symp Quant Biol **14**:117–26.

146. Liu, X., S. Shu, M. S. Hong, R. L. Levine, and E. D. Korn. 2006. Phosphorylation of actin Tyr-53 inhibits filament nucleation and elongation and destabilizes filaments. Proc Natl Acad Sci U S A **103**:13694–9.

147. Lo Conte, L., S. E. Brenner, T. J. Hubbard, C. Chothia, and A. G. Murzin. 2002. SCOP database in 2002: refinements accommodate structural genomics. Nucleic Acids Res **30**:264–7.

148. Loll, P. J., and E. E. Lattman. 1989. The crystal structure of the ternary complex of staphylococcal nuclease, Ca2+, and the inhibitor pdTp, refined at 1.65 Å. Proteins **5**:183–201.

149. López-Otín, C., and J. S. Bond. 2008. Proteases: multifunctional enzymes in life and disease. J Biol Chem **283**:30433–7.

150. Lorand, L., and S. M. Conrad. 1984. Transglutaminases. Mol Cell Biochem **58**:9–35.

151. Lowe, E. D., M. E. Noble, V. T. Skamnaki, N. G. Oikonomakos, D. J. Owen, and L. N. Johnson. 1997. The crystal structure of a phosphorylase kinase peptide substrate complex: kinase substrate recognition. EMBO J **16**:6646–58.

152. Makin, O. S., and L. C. Serpell. 2005. Structures for amyloid fibrils. FEBS J **272**:5950–61.

153. Mann, M., and M. Wilm. 1995. Electrospray mass spectrometry for protein characterization. Trends Biochem Sci **20**:219–24.

154. Markland, F. S., and E. L. Smith. 1971. Subtilisins: primary structure, chemical and physical properties, pp. 561–608. In P. D. Boyer (ed.), The Enzymes, third edition, vol. **3**. Academic Press, New York.

155. Meador, W. E., A. R. Means, and F. A. Quiocho. 1993. Modulation of calmodulin plasticity in molecular recognition on the basis of X-ray structures. Science **262**:1718–21.

156. Mero, A., B. Spolaore, F. M. Veronese, and A. Fontana. 2009. Transglutaminase-mediated PEGylation of proteins: direct identification of the sites of protein modification by mass spectrometry using a novel monodisperse PEG. Bioconjug Chem **20**:384–9.

157. Miake, H., H. Mizusawa, T. Iwatsubo, and M. Hasegawa. 2002. Biochemical characterization of the core structure of α-synuclein filaments. J Biol Chem **277**:19213–9.

158. Miyazaki, S., Y. Kuroda, and S. Yokoyama. 2006. Identification of putative domain linkers by a neural network: application to a large sequence database. BMC Bioinformatics **7**:323–31.

159. Monsellier, E., M. Ramazzotti, P. Polverino de Laureto, G. G. Tartaglia, N. Taddei, A. Fontana, M. Vendruscolo, and F. Chiti. 2007. The distribution of residues in a polypeptide sequence is a determinant of aggregation optimized by evolution. Biophys J **93**:4382–91.

160. Monti, M., A. Amoresano, S. Giorgetti, V. Bellotti, and P. Pucci. 2005. Limited proteolysis in the investigation of β_2-microglobulin amyloidogenic and fibrillar states. Biochim Biophys Acta **1753**:44–50.

161. Moreland, J. L., A. Gramada, O. V. Buzko, Q. Zhang, and P. E. Bourne. 2005. The Molecular Biology Toolkit (MBT): a modular platform for developing

molecular visualization applications. BMC Bioinformatics **6**:21–7. http://mbt.sdsc.edu.

162. Mornet, D., and K. Ue. 1984. Proteolysis and structure of skeletal muscle actin. Proc Natl Acad Sci U S A **81**:3680–4.
163. Moskaug, J. O., K. Sletten, K. Sandvig, and S. Olsnes. 1989. Translocation of diphtheria toxin A-fragment to the cytosol: role of the site of inter-fragment cleavage. J Biol Chem **264**:15709–13.
164. Musi, V., B. Spolaore, P. Picotti, M. Zambonin, V. De Filippis, and A. Fontana. 2004. Nicked apomyoglobin: a noncovalent complex of two polypeptide fragments comprising the entire protein chain. Biochemistry **43**:6230–40.
165. Myers, S. L., N. H. Thomson, S. E. Radford, and A. E. Ashcroft. 2006. Investigating the structural properties of amyloid-like fibrils formed *in vitro* from β_2-microglobulin using limited proteolysis and electrospray ionisation mass spectrometry. Rapid Commun Mass Spectrom **20**:1628–36.
166. Neurath, H. 1980. Limited proteolysis, protein folding and physiological regulation, pp. 501–4. In R. Jaenicke (ed.), Protein Folding. Elsevier/North Holland Biomedical Press, Amsterdam and New York.
167. Neurath, H. 1989. Proteolytic processing and physiological regulation. Trends Biochem Sci **14**:268–71.
168. Neurath, H., and K. A. Walsh. 1976. Role of proteolytic enzymes in biological regulation. Proc Natl Acad Sci U S A **73**:3825–32.
169. O'Brien, J. A., J. A. Taylor, and A. R. Bellamy. 2000. Probing the structure of rotavirus NSP4: a short sequence at the extreme C terminus mediates binding to the inner capsid particle. J Virol **74**:5388–94.
170. Oldfield, C. J., Y. Cheng, M. S. Cortese, C. J. Brown, V. N. Uversky, and A. K. Dunker. 2005. Comparing and combining predictors of mostly disordered proteins. Biochemistry **44**:1989–2000.
171. Onufriev, A., D. A. Case, and D. Bashford. 2003. Structural details, pathways, and energetics of unfolding apomyoglobin. J Mol Biol **325**:555–67.
172. Orengo, C. A., A. D. Michie, S. Jones, D. T. Jones, M. B. Swindells, and J. M. Thornton. 1997. CATH: a hierarchic classification of protein domain structures. Structure **5**:1093–108.
173. Pace, N., and A. J. Barrett. 1984. Kinetics of tryptic hydrolysis of the arginine-valine bond in folded and unfolded ribonuclease T1. Biochem J **219**:411–7.
174. Pantazatos, D., J. S. Kim, H. E. Klock, R. C. Stevens, I. A. Wilson, S. A. Lesley, and V. L. Woods, Jr. 2004. Rapid refinement of crystallographic protein construct definition employing enhanced hydrogen/deuterium exchange MS. Proc Natl Acad Sci U S A **101**:751–6.
175. Permyakov, E. A., and L. J. Berliner. 2000. α-Lactalbumin: structure and function. FEBS Lett **473**:269–74.
176. Persechini, A., and R. H. Kretsinger. 1988. The central helix of calmodulin functions as a flexible tether. J Biol Chem **263**:12175–8.
177. Picotti, P., G. De Franceschi, E. Frare, B. Spolaore, M. Zambonin, F. Chiti, P. Polverino de Laureto, and A. Fontana. 2007. Amyloid fibril formation and disag-

gregation of fragment 1–29 of apomyoglobin: insights into the effect of pH on protein fibrillogenesis. J Mol Biol **367**:1237–45.

178. Picotti, P., A. Marabotti, A. Negro, V. Musi, B. Spolaore, M. Zambonin, and A. Fontana. 2004. Modulation of the structural integrity of helix F in apomyoglobin by single amino acid replacements. Protein Sci **13**:1572–85.
179. Pike, A. C. W., K. Brew, and K. R. Acharya. 1996. Crystal structures of guinea-pig, goat and bovine α-lactalbumin highlight the enhanced conformational flexibility of regions that are significant for its action in lactose synthase. Structure **4**:691–703.
180. Pintar, A., O. Carugo, and S. Pongor. 2002. CX, an algorithm that identifies protruding atoms in proteins. Bioinformatics **18**:980–984.
181. Polverino de Laureto, P., V. De Filippis, M. Di Bello, M. Zambonin, and A. Fontana. 1995. Probing the molten globule state of α-lactalbumin by limited proteolysis. Biochemistry **34**:12596–12604.
182. Polverino de Laureto, P., E. Frare, F. Battaglia, M. F. Mossuto, V. N. Uversky, and A. Fontana. 2005. Protein dissection enhances the amyloidogenic properties of α-lactalbumin. FEBS J **272**:2176–88.
183. Polverino de Laureto, P., Frare, E., R. Gottardo, and A. Fontana. 2002. Molten globule of bovine α-lactalbumin at neutral pH induced by heat, trifluoroethanol and oleic acid: a comparative analysis by circular dichroism spectroscopy and limited proteolysis. Proteins **49**:385–97.
184. Polverino de Laureto, P., E. Scaramella, M. Frigo, F. G. Wondrich, V. De Filippis, M. Zambonin, and A. Fontana. 1999. Limited proteolysis of bovine α-lactalbumin: isolation and characterization of protein domains. Protein Sci **8**:2290–303.
185. Polverino de Laureto, P., N. Taddei, E. Frare, C. Capanni, S. Costantini, J. Zurdo, F. Chiti, C. M. Dobson, and A. Fontana. 2003. Protein aggregation and amyloid fibril formation by an SH3 domain probed by limited proteolysis. J Mol Biol **334**:129–41.
186. Polverino de Laureto, P., S. Toma, G. Tonon, and A. Fontana. 1995. Probing the structure of human growth hormone by limited proteolysis. Int J Pept Protein Res **45**:200–8.
187. Polverino de Laureto, P., L. Tosatto, E. Frare, O. Marin, V. N. Uversky, and A. Fontana. 2006. Conformational properties of the SDS-bound state of α-synuclein probed by limited proteolysis: unexpected rigidity of the acidic C-terminal tail. Biochemistry **45**:11523–11531.
188. Polverino de Laureto, P., D. Vinante, E. Scaramella, E. Frare, and A. Fontana. 2001. Stepwise proteolytic removal of the β-subdomain in α-lactalbumin: the protein remains folded and can form the molten globule in acid solution. Eur J Biochem **268**:4324–33.
189. Ptitsyn, O. B. 1995. Molten globule and protein folding. Adv Protein Chem **47**:183–229.
190. Ptitsyn, O. B. 1995. Structures of folding intermediates. Curr Opin Struct Biol **5**:74–8.
191. Quezada, C. M., B. A. Schulman, J. J. Froggatt, C. M. Dobson, and C. Redfield. 2004. Local and global cooperativity in the human alpha-lactalbumin molten globule. J Mol Biol **338**:149–58.

192. Radivojac, P., L. M. Iakoucheva, C. J. Oldfield, Z. Obradovic, V. N. Uversky, and A. K. Dunker. 2007. Intrinsic disorder and functional proteomics. Biophys J **92**:1439–56.

193. Radivojac, P., S. Vucetic, T. R. O'Connor, V. N. Uversky, Z. Obradovic, and A. K. Dunker. 2006. Calmodulin signaling: analysis and prediction of a disorder-dependent molecular recognition. Proteins **63**:398–410.

194. Rambaran, R. N., and L. C. Serpell. 2008. Amyloid fibrils: abnormal protein assembly. Prion **2**:112–7.

195. Raschke, T. M., and S. Marqusee. 1998. Hydrogen exchange studies of protein structure. Curr Opin Biotechnol **9**:80–6.

196. Richards, F. M., and P. J. Vithayathil. 1959. The preparation of subtilisin-modified ribonuclease and the separation of the peptide and protein components. J Biol Chem **234**:1459–64.

197. Rico, M., M. A. Jiménez, C. González, V. De Filippis, and A. Fontana. 1994. NMR solution structure of the C-terminal fragment 253–316 of thermolysin: a dimer formed by subunits having the native structure. Biochemistry **33**:14834–47.

198. Ringe, D., and G. A. Petsko. 1985. Mapping protein dynamics by X-ray diffraction. Prog Biophys Mol Biol **45**:197–235.

199. Romero, P., Z. Obradovic, C. R. Kissinger, J. E. Villafranca, E. Garner, S. Guilliot, and A. K. Dunker. 1998. Thousands of proteins likely to have long disordered regions. Pac Symp Biocomput **3**:437–48.

200. Rossman, M. G., and A. Liljas. 1974. Recognition of structural domains in globular proteins. J Mol Biol **85**:177–81.

201. Rydel, T. J., A. Tulinski, W. Bode, and R. Huber. 1991. Refined structure of the hirudin-thrombin complex. J Mol Biol **221**:583–601.

202. Safer, D., T. R. Sosnick, and M. Elzinga. 1997. Thymosin β4 binds actin in an extended conformation and contacts both the barbed and pointed ends. Biochemistry **36**:5806–16.

203. Salvesen, G. S. 2006. New perspectives on proteases. Nature Rev Drug Discov **5**:1–6.

204. Scacheri, E., G. Nitti, B. Valsasina, G. Orsini, C. Visco, M. Ferreia, R. T. Sawyer, and P. Sarmientos. 1993. Novel hirudin variants from the leech *Hirudinaria manillensis*: amino acid sequence, cDNA cloning and genomic organization. Eur J Biochem **214**:295–304.

205. Schechter, I., and A. Berger. 1967. On the size of the active site in proteases. I. Papain. Biochem Biophys Res Comm **27**:157–62.

206. Schulman, B. A., P. S. Kim, C. M. Dobson, and C. Redfield. 1997. A residue-specific NMR view of the non-cooperative unfolding of a molten globule. Nat Struct Biol **4**:630–4.

207. Schulman, B. A., C. Redfield, Z. Y. Peng, C. M. Dobson, and P. S. Kim. 1995. Different subdomains are most protected from hydrogen exchange in the molten globule and native states of human alpha-lactalbumin. J Mol Biol **253**:651–7.

208. Schwarzinger, S., R. Mohana-Borges, G. J. Kroon, H. J. Dyson, and P. E. Wright. 2008. Structural characterization of partially folded intermediates of apomyoglobin H64F. Protein Sci **17**:313–21.

209. Serpell, L. C. 2000. Alzheimer's amyloid fibrils: structure and assembly. Biochim Biophys Acta **1502**:16–30.

210. Sickmeier, M., J. A. Hamilton, T. LeGall, V. Vacic, M. S. Cortese, A. Tantos, B. Szabo, P. Tompa, J. Chen, V. N. Uversky, Z. Obradovic, and A. K. Dunker. 2007. DisProt: the database of disordered proteins. Nucleic Acids Res **35**:D786–93.

211. Siddiqui, A. S., and G. J. Barton. 1995. Continuous and discontinuous domains: an algorithm for the automatic generation of reliable protein domain definitions. Protein Sci **4**:872–84.

212. Smith, D. K., P. Radivojac, Z. Obradovic, A. K. Dunker, and G. Zhu. 2003. Improved amino acid flexibility parameters. Protein Sci **12**:1060–72.

213. Soto, C., and L. D. Estrada. 2008. Protein misfolding and neurodegeneration. Arch Neurol **65**:184–9.

214. Spolaore, B., R. Bermejo, M. Zambonin, and A. Fontana. 2001. Protein interactions leading to conformational changes monitored by limited proteolysis: apo form and fragments of horse cytochrome *c*. Biochemistry **40**:9460–8.

215. Spolaore, B., V. De Filippis, and A. Fontana. 2005. Heme binding by the N-terminal fragment 1–44 of human growth hormone. Biochemistry **44**:16079–89.

216. Spolaore, B., P. Polverino de Laureto, M. Zambonin, and A. Fontana. 2004. Limited proteolysis of human growth hormone at low pH: isolation, characterization, and complementation of the two biologically relevant fragments 1–44 and 45–191. Biochemistry **43**:6576–86.

217. Spraggon, G., D. Pantazatos, H. E. Klock, I. A. Wilson, V. L. Woods, Jr., and S. A. Lesley. 2004. On the use of DXMS to produce more crystallizable proteins: structures of the *T. maritima* proteins TM0160 and TM1171. Protein Sci. **13**:3187–99.

218. Stefani, M., and C. M. Dobson. 2003. Protein aggregation and aggregate toxicity: new insights into protein folding, misfolding diseases and biological evolution. J Mol Med **81**:678–99.

219. Sternberg, M. J. E., D. E. P. Grace, and D. C. Phillips. 1979. Dynamic information from protein crystallography: an analysis of temperature factors from refinement of the hen egg-white lysozyme. J Mol Biol **130**:231–53.

220. Stringer, K. A., and J. Lindenfeld. 1992. Hirudins: antithrombin anticoagulants. Ann Pharmacother **26**:1535–1540.

221. Stroh, J. G., P. Loulakis, A. J. Lanzetti, and J. Xie. 2005. LC-mass spectrometry analysis of N- and C-terminal boundary sequences of polypeptide fragments by limited proteolysis. J Am Soc Mass Spectrom **16**:38–45.

222. Strzelecka-Gołaszewska, H., J. Moraczewska, S. Y. Khaitlina, and M. Mossakowska. 1993. Localization of the tightly bound divalent-cation-dependent and nucleotide-dependent conformation changes in G-actin using limited proteolytic digestion. Eur J Biochem **211**:731–42.

223. Suh, M. J., S. Pourshahian, and P. A. Limbach. 2007. Developing limited proteolysis and mass spectrometry for the characterization of ribosome topography. J Am Soc Mass Spectrom **18**:1304–17.

224. Sunde, M., and C. Blake. 1997. The structure of amyloid fibrils by electron microscopy and X-ray diffraction. Adv Protein Chem **50**:123–59.

225. Svensson, M., H. Sabharwal, A. Håkansson, A. K. Mossberg, P. Lipniunas, H. Leffler, C. Svanborg, and S. Linse. 1999. Molecular characterization of α-lactalbumin folding variants that induce apoptosis in tumor cells. J Biol Chem **274**:6388–96.

226. Swulius, M. T., and M. N. Waxham. 2008. Ca(2+)/calmodulin-dependent protein kinases. Cell Mol Life Sci **65**:2637–57.

227. Szyperski, S., P. Güntert, S. R. Stone, A. Tulinski, W. Bode, R. Huber, and K. Wüthrich. 1992. Impact of protein-protein contacts on the conformation of thrombin-bound hirudin studied by comparison with the nuclear magnetic resonance solution structure of hirudin (1–51). J Mol Biol **228**:1206–11.

228. Szyperski, S., P. Güntert, S. R. Stone, and K. Wüthrich. 1992. Nuclear magnetic resonance solution structure of hirudin (1–51) and comparison with corresponding three-dimensional structures determined using the complete 65-residue hirudin polypeptide chain. J Mol Biol **228**:1193–205.

229. Takashi, R. 1988. A novel actin label: a fluorescent probe at glutamine-41 and its consequences. Biochemistry **27**:938–43.

230. Taniuchi, H., and C. B. Anfinsen. 1968. Steps in the formation of active derivatives of staphylococcal nuclease during trypsin digestion. J Biol Chem **243**:4778–86.

231. Tao, L., S. E. Kiefer, D. Xie, J. W. Bryson, S. A. Hefta, and M. L. Doyle. 2008. Time-resolved limited proteolysis of mitogen-activated protein kinase-activated protein kinase-2 determined by LC/MS only. J Am Soc Mass Spectrom **19**:841–54.

232. Taylor, W. 1999. Protein structure domain identification. Protein Eng **12**:203–16.

233. Tcherkasskaya, O., and V. N. Uversky. 2001. Denatured collapsed states in protein folding: example of apomyoglobin. Proteins **44**:244–54.

234. Thomas, J. J., R. Bakhtiar, and G. Siuzdak. 2000. Mass spectrometry in viral proteomics. Acc Chem Res **33**:179–87.

235. Tirado-Rives, J., and W. L. Jorgensen. 1993. Molecular dynamics simulations of the unfolding of apomyoglobin in water. Biochemistry **32**:4175–84.

236. Tompa, P. 2002. Intrinsically unstructured proteins. Trends Biochem Sci **27**:527–33.

237. Tompa, P., Z. Dosztanyi, and I. Simon. 2006. Prevalent structural disorder in *E. coli* and *S. cerevisiae* proteomes. J Proteome Res **5**:1996–2000.

238. Tycko, R. 2004. Progress towards a molecular-level structural understanding of amyloid fibrils. Curr Opin Struct Biol **14**:96–103.

239. Tyndall, J. D. A., and D. P. Fairlie. 1999. Conformational homogeneity in molecular recognition by proteolytic enzymes. J Mol Recognit **12**:363–70.

240. Tyndall, J. D. A., T. Nall, and D. P. Fairlie. 2005. Proteases universally recognize beta strands in their active site. Chem Rev **105**:973–99.

241. Ultsch, M. H., W. Somers, A. A. Kossiakoff, and A. M. de Vos. 1994. The crystal structure of affinity-matured human growth hormone at 2Å resolution. J Mol Biol **236**:286–99.

242. Uversky, V. N. 2002. Natively unfolded proteins: a point where biology waits for physics. Protein Sci **11**:739–56.

243. Uversky, V. N. 2002. What does it mean to be natively unfolded? Eur J Biochem **269**:2–12.
244. Uversky, V. N. 2003. A protein-chameleon: conformational plasticity of α-synuclein, a disordered protein involved in neurodegenerative disorders. J Biomol Struct Dyn **21**:211–34.
245. Uversky, V. N. 2003. Protein folding revisited, a polypeptide chain at the folding-misfolding-nonfolding cross-roads: which way to go? Cell Mol Life Sci **60**:1852–71.
246. Uversky, V. N. 2008. Alpha-synuclein misfolding and neurodegenerative diseases. Curr Protein Pept Sci **9**:507–40.
247. Uversky, V. N. 2008. Amyloidogenesis of natively unfolded proteins. Curr Alzheimer Res **5**:260–87.
248. Uversky, V. N., and A. L. Fink. 2004. Conformational constraints for amyloid fibrillation: the importance of being unfolded. Biochim Biophys Acta **1698**:131–53.
249. Uversky, V. N., J. R. Gillespie, and A. L. Fink. 2000. Why are natively unfolded proteins unstructured under physiological conditions? Proteins **41**:415–27.
250. Uversky, V. N., C. J. Oldfield, and A. K. Dunker. 2005. Showing your ID: intrinsic disorder as an ID for recognition, regulation and cell signaling. J Mol Recognit **18**:343–84.
251. Uversky, V. N., C. J. Oldfield, and A. K. Dunker. 2008. Intrinsically disordered proteins in human diseases: introducing the D2 concept. Annu Rev Biophys **37**:215–46.
252. Uzawa, T., C. Nishimura, S. Akiyama, K. Ishimori, S. Takahashi, H. J. Dyson, and P. E. Wright. 2008. Hierarchical folding mechanism of apomyoglobin revealed by ultra-fast H/D exchange coupled with 2D NMR. Proc Natl Acad Sci U S A **105**:13859–64.
253. van der Spoel, D., B. L. de Groot, S. Hayward, H. J. Berendsen, and H. J. Vogel. 1996. Bending of the calmodulin central helix: a theoretical study. Protein Sci **5**:2044–53.
254. Veretnik, S., P. E. Bourne, N. N. Alexandrov, and I. N. Shindyalov. 2004. Toward consistent assignment of structural domains in proteins. J Mol Biol **339**:647–78.
255. Villanueva, J., V. Villegas, E. Querol, F. X. Avilés, and L. Serrano. 2002. Protein secondary structure and stability determined by combining exoproteolysis and matrix-assisted laser desorption/ionization time-of-flight mass spectrometry. J Mass Spectrom **37**:974–84.
256. Vindigni, A., V. De Filippis, G. Zanotti, C. Visco, G. Orsini, and A. Fontana. 1994. Probing the structure of hirudin from *Hirudinaria manillensis* by limited proteolysis: isolation, characterization and thrombin-inhibitory properties of N-terminal fragments. Eur J Biochem **226**:323–33.
257. Vita, C., D. Dalzoppo, and A. Fontana. 1988. Limited proteolysis of globular proteins: molecular aspects deduced from studies on thermolysin, pp. 57–68. In I. Chaiken, E. Chiancone, A. Fontana, and P. Neri (eds.), Macromolecular Biorecognition. Humana Press, Clifton, NJ.
258. Vita, C., and A. Fontana. 1982. Domain characteristics of the carboxyl-terminal fragment 206–316 of thermolysin: unfolding thermodynamics. Biochemistry **21**:5196–202.

259. Vita, C., A. Fontana, and I. M. Chaiken. 1985. Folding of thermolysin fragments: correlation between conformational stability and antigenicity of carboxyl-terminal fragments. Eur J Biochem **151**:191–6.

260. Vita, C., A. Fontana, and R. Jaenicke. 1989. Folding of thermolysin fragments: hydrodynamic properties of isolated domains and subdomains. Eur J Biochem **183**:513–8.

261. Vita, C., A. Fontana, J. R. Seeman, and I. M. Chaiken. 1979. Conformational and immunochemical analysis of the cyanogen bromide fragments of thermolysin. Biochemistry **18**:3023–31.

262. Vogel, H. J., and M. Zhang. 1995. Protein engineering and NMR studies of calmodulin. Mol Cell Biochem **150**:3–15.

263. Vucetic, S., C. J. Brown, A. K. Dunker, and Z. Obradovic. 2003. Flavors of protein disorder. Proteins **52**:573–84.

264. Vucetic, S., H. Xie, L. M. Iakoucheva, C. J. Oldfield, A. K. Dunker, Z. Obradovic, and V. N. Uversky. 2007. Functional anthology of intrinsic disorder. 2. Cellular components, domains, technical terms, developmental processes and coding sequence diversities correlated with long disordered regions. J Proteome Res **6**:1899–916.

265. Wagenführ, K., S. Pieper, P. Mackeldanz, M. Linscheid, D. H. Krüger, and M. Reuter. 2007. Structural domains in the type III restriction endonuclease EcoP15I: characterization by limited proteolysis, mass spectrometry and insertional mutagenesis. J Mol Biol **366**:93–102.

266. Wang, L., H. Pan, and D. L. Smith. 2002. Hydrogen exchange-mass spectrometry: optimization of digestion conditions. Mol Cell Proteomics **1**:132–8.

267. Ward, J. J., J. S. Sodhi, L. J. McGuffin, B. F. Buxton, and D. T. Jones. 2004. Prediction and functional analysis of native disorder in proteins from the three kingdoms of life. J Mol Biol **337**:635–45.

268. Wetlaufer, D. 1973. Nucleation, rapid folding, and globular intra-chain regions in proteins. Proc Natl Acad Sci U S A **70**:697–701.

269. Wetzel, R., S. Shivaprasad, and A. D. Williams. 2007. Plasticity of amyloid fibrils. Biochemistry **46**:1–10.

270. Wilm, M., A. Shevchenko, T. Houthaeve, S. Breit, L. Schweigerer, T. Fotsis, and M. Mann. 1996. Femtomole sequencing of proteins from polyacrylamide gels by nano-electrospray mass spectrometry. Nature **379**:466–9.

271. Wilson, L. M., Y. F. Mok, K. J. Binger, M. D. Griffin, H. D. Mertens, F. Lin, J. D. Wade, P. R. Gooley, and G. J. Howlett. 2007. A structural core within apolipoprotein C-II amyloid fibrils identified using hydrogen exchange and proteolysis. J Mol Biol **366**:1639–51.

272. Wodak, S. J., and J. Janin. 1981. Location of structural domains in protein. Biochemistry **20**:6544–52.

273. Wolfe, M. S. 2008. Gamma-secretase inhibition and modulation for Alzheimer's disease. Curr Alzheimer Res **5**:158–64.

274. Wriggers, W., E. Mehler, F. Pitici, H. Weinstein, and K. Schulten. 1998. Structure and dynamics of calmodulin in solution. Biophys J **74**:1622–39.

275. Wright, P. E., and H. J. Dyson. 1999. Intrinsically unstructured proteins: re-assessing the protein structure–function paradigm. J Mol Biol **293**:321–31.

276. Xia, Z. X., N. Shamala, P. H. Bethge, L. W. Lim, H. D. Bellamy, N. H. Xuong, F. Lederer, and F. S. Mathews. 1987. Three-dimensional structure of flavocytochrome b2 from baker's yeast at 3.0 Å resolution. Proc Natl Acad Sci U S A **84**:2629–33.

277. Xie, H., S. Vucetic, L. M. Iakoucheva, C. J. Oldfield, A. K. Dunker, Z. Obradovic, and V. N. Uversky. 2007. Functional anthology of intrinsic disorder. 3. Ligands, post-translational modifications and diseases associated with intrinsically disordered proteins. J Proteome Res **6**:1917–32.

278. Xie, H., S. Vucetic, L. M. Iakoucheva, C. J. Oldfield, A. K. Dunker, V. N. Uversky, and Z. Obradovic. 2007. Functional anthology of intrinsic disorder. 1. Biological processes and functions of proteins with long disordered regions. J Proteome Res **6**:1882–98.

279. Yang, C., G. S. Jas, and K. Kuczera. 2004. Structure, dynamics and interaction with kinase targets: computer simulations of calmodulin. Biochim Biophys Acta **1697**:289–300.

280. Yang, F., and G. N. Phillips, Jr. 1996. Crystal structures of CO-, deoxy- and met-myoglobins at various pH values. J Mol Biol **256**:762–74.

281. Yates, J. R. 1998. Mass spectrometry and the age of proteomics. J Mass Spectrom **33**:1–19.

282. Zehfus, M. H., and G. D. Rose. 1986. Compact units in proteins. Biochemistry **25**:5759–65.

283. Zhang, M., and T. Yuan. 1998. Molecular mechanisms of calmodulin's functional versatility. Biochem Cell Biol **76**:313–23.

PART VI

MASS SPECTROMETRY

21

MASS SPECTROMETRY TOOLS FOR THE INVESTIGATION OF STRUCTURAL DISORDER AND CONFORMATIONAL TRANSITIONS IN PROTEINS

Mária Šamalíková, Carlo Santambrogio, and Rita Grandori
Department of Biotechnology and Biosciences, University of Milano-Bicocca, Milan, Italy

ABSTRACT

Electrospray ionization mass spectrometry has recently developed into a central tool of structural biology, which allows the investigation of the structure and the dynamics of proteins and protein assemblies. This contribution describes the principles of the main experimental approaches that have been applied to the study of intrinsically disordered proteins and describes some recent examples from the literature. In particular, we distinguish between methods that are based on maintenance of noncovalent interactions under electrospray conditions, such as charge-state-distribution analysis and ion mobility, from methods based on isotope exchange, in which the protein can be denatured or even digested before mass spectrometry analysis. All together, these methods offer useful structural information, which is complementary to that gained by other biophysical techniques. Particularly interesting is the possibility to directly monitor distinct conformers in complex mixtures, free from averaging over the molecular population. Although this is a very young field,

Instrumental Analysis of Intrinsically Disordered Proteins: Assessing Structure and Conformation, Edited by Vladimir Uversky and Sonia Longhi

the examples discussed in this chapter illustrate the contribution that mass spectrometry methods can give to the structural characterization of intrinsically disordered proteins in the absence and in the presence of interactors.

21.1 INTRODUCTION

In the last decades, mass spectrometry (MS) has evolved into one of the central techniques for protein studies. This process was accompanied by technological and methodological developments that not only improved the potential of MS for analytic purposes but also transformed it into a novel tool for structural biology. In the attempt to describe current MS methods for protein conformational studies, we distinguish between two main categories. On one side, there are approaches that exploit the intrinsic dependence of these techniques on protein conformation. These are charge-state distribution (CSD) analysis and ion mobility (IM) measurements, both coupled to electrospray ionization (ESI). On the other side, there are indirect approaches that make use of MS just for accurate mass determination. In this case, MS is employed in order to characterize the products of conformation-dependent reactions that affect protein mass in some way. The most widely used methods to this purpose are hydrogen/deuterium exchange (HX) and limited proteolysis. Since limited proteolysis is treated in detail in another chapter of this book, we limit the discussion of indirect approaches to HX.

21.2 CSDs

Any MS measurement consists of the following three steps: production of gas-phase ions of the analytes, separation of the ions based on their mass-to-charge ratio (m/z), and ion detection. Accordingly, the three main components of a mass spectrometer are the ion source, where gas-phase ions are generated, the mass analyzer, where ions are sorted, and a detector, where ions are counted (68). Several different ways to ionize and desolvate analyte molecules are available, but the most widely employed methods for proteins and peptides are ESI (18) and matrix-assisted laser desorption/ionization (MALDI) (42, 74). As underscored by the Nobel Prize to John B. Fenn and Koichi Tanaka in 2002, these ionization techniques have brought about a revolution in the application of MS to the analysis of biologic systems, thanks to the possibility to detect intact macromolecules, limiting or even suppressing their fragmentation during the ionization/desolvation process.

Further developments in methods and instrumentation were led in the 1990s to another important achievement, namely preservation of noncovalent interactions, besides maintenance of intact covalent structures, during MS detection of macromolecules (10, 43). Thus, not only the chemical identity but also the *behavior* of molecules of biologic interest could become the object of investigation by MS. This possibility has opened the way to the analysis of protein folding transitions and protein–ligand interactions by MS. Although detection of noncovalent complexes by MALDI-MS has been reported in the literature (36, 83), ESI remains the milder technique and, therefore, appears to be more appropriate for the analysis of weak interactions. Furthermore, the peculiar feature of ESI to produce multiply charged protein ions is fundamental to structural analysis, as discussed in detail below. MALDI, instead, produces predominantly singly charged ions and cannot deliver structural information based on protein ionization behavior.

21.2.1 The Electrospray Process

During ESI, a liquid sample is sprayed through a thin capillary under the presence of a strong electric field, generating small, charged droplets containing analyte molecules/ions (45). Solvent evaporation causes the droplets to shrink and leads, consequently, to an increase of the surface charge density. When electrostatic repulsions at the droplet's surface are no longer offset by surface tension, physical instability emerges and the droplet undergoes fission into smaller daughter droplets (Coulomb explosion). The threshold charge (q_R) for Coulomb explosion is a function of the droplet's radius (R) and surface tension (γ), as expressed by the Rayleigh equation

$$q_R = 8\pi\left(\varepsilon_0 \gamma R^3\right)^{1/2},$$

where ε_0 is the permittivity of vacuum. Cycles of solvent evaporation and droplet fission eventually lead to production of gas-phase ions of the analyte. It is therefore understandable that the original size of the droplets has a strong influence on the efficiency of the process. The innovation introduced by the so-called nanospray sources (78) is to reduce the diameter of the capillary emitter tip and, as a consequence, that of the first-generation droplets. Therefore, nano-ESI sources allow for milder desolvation conditions than conventional ESI sources, especially in terms of voltage and temperature, and are particularly well suited to the study of noncovalent interactions.

21.2.2 Conformational Effects on CSDs

As said before, a peculiar feature of ESI is the production of multiply charged protein ions. The experimentally observed values of net charge depend on several parameters like protein size, conformational state, solution conditions, and instrumental setting. Protein ionization is typically mediated by proton-transfer reactions with the solvent, resulting in protonated or deprotonated protein ions, respectively, in positive- or negative-ion mode (46). The groups carrying positive charges are mainly the side chains of basic residues (Arg, Lys, and His) and the N-terminus, while those carrying negative charges are mainly the side chains of acidic residues (Asp and Glu) and the C-terminus.

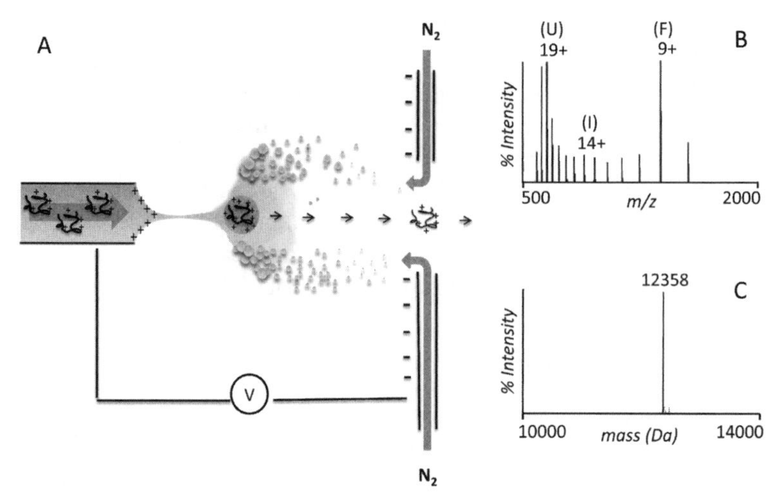

Figure 21.1 (A) Schematic representation of the ESI process. (B) Nano-ESI-MS spectrum of cytochrome *c* sprayed at pH 3 (water/acetate), 25% methanol (22). Labels indicate the main charge state and the conformational state assigned to the different peak envelopes: F, folded; I, intermediate; U, unfolded. (C) Deconvoluted spectrum.

However, results obtained with basic proteins in negative-ion mode or acidic proteins in positive-ion mode provide evidence that ionization at other sites can occur (39). The typical spectrum of a pure protein consists in an envelope of signals where adjacent peaks represent ionization states differing by one charge. These envelopes are bell-shaped, spanning a range of values of net charge around the most populated charge state (most intense peak), and are referred to as CSDs. Deconvolution of such spectra provides the zero-charge mass value for each detectable component (Fig. 21.1).

The reason why ESI-MS spectra of proteins contain structural information is that CSDs are sensitive to the conformation held by proteins at the moment of transfer to the gas phase (10). A given protein in a compact conformation will acquire fewer charges than the same protein in the denatured state. This effect typically results in a major shift of the CSD toward lower *m/z* values upon denaturation. Thus, spectra obtained from a protein in different conformational states are multimodal and typically display well-resolved envelopes of the distinct components (24). Figure 21.1 illustrates the example of cytochrome *c*, a protein that undergoes acid-induced denaturation with a steep unfolding transition between pH 3 and pH 2 (21) and populates a well-characterized, methanol-induced partially folded form. In conclusion, although the mass of different protein conformers is identical, these can be distinguished by ESI-MS, thanks to their different ionization properties. This effect appears to be general and has been observed for proteins of different sizes and isoelectric points. Furthermore, the effect is observable both in positive- and negative-ion modes (49), although it can disappear when measuring basic proteins in negative-ion mode or acidic proteins in positive-ion mode (5, 22, 53, 82).

Although several factors, depending on the employed solvent and instrumental conditions, can affect protein CSDs, conformational effects are largely predominant and can be interpreted confidently when operating under accurately controlled conditions (73). Furthermore, they result in major and abrupt changes in ESI-MS spectra, which reflect the biochemical transition and are distinguishable from the minor, monotonic, nonspecific effects due to the direct influence of the physicochemical conditions on the ionization process. It should be noted that conformational effects on ESI-MS protein spectra depend on preservation of noncovalent interactions under electrospray conditions, but do not depend on maintenance of folded conformations after the desolvation process has been completed. Indeed, the experimentally observed charge states are mainly determined by proton-transfer reactions still involving the solvated protein inside the ESI droplets (46). Gas-phase reactions can, nevertheless, alter charge state distrubutions during dissociation of protein complexes (30).

The structural feature that seems to be captured by ESI is compactness (48, 76). Indeed, tertiary more than secondary structure affects protein ionization during electrospray, as indicated by the fact that the molten-globule state of several proteins can be effectively discriminated from both the native and the fully unfolded forms, for either mainly alpha or mainly beta architectures (7, 10, 23, 32, 39, 49). On the other hand, the coil-to-helix transition per se can be even undetectable by this method, as shown by experimental systems that allow uncoupling of secondary from tertiary structure formation (26, 48, 59). Thus, CSD analysis is nicely complementary to infrared spectroscopy and far-UV circular dichroism (CD) for the investigation of protein conformational transitions and seems to offer a rather sensitive and global probe for structural compactness. Analytic tools have been developed in order to deconvolute spectra with nonresolved peak envelopes and to quantify the relative amounts of the different components (7, 14).

Until now, we have discussed only issues concerning the position of CSDs. In other words, we mainly referred to the average charge state characterizing different conformational states. Another aspect that should be taken into consideration is the width of the distribution. Two main factors contribute in determining this feature: random variations in the proton transfer reactions and structural heterogeneity of the molecular ensemble. As long as structural heterogeneity implies fluctuation in protein structure compactness, it will also affect the extent of ionization and will, therefore, result in a wider range of charge states populated by the protein ions produced by electrospray. It is expected that such an effect be more pronounced in the denatured state. Consistent with these arguments, CSDs of unfolded proteins are always broader than the respective ones under nondenaturing conditions (24, 39). For average size proteins, the latter can consist of only three to five peaks, with a single charge state strongly predominant over the others. On the other hand, tens of peaks can compose the broad and relatively flat CSDs of unfolded proteins. Such an effect on CSD width can be observed also comparing the conformational states of α-synuclein, the primary fibrillar component of Lewy bodies in Parkinson's disease patients. In this case, the protein has a more compact conformation at acidic pH than at pH 7, and thus, it can be concluded that conformational effects on CSDs persist even when the pH dependence of the conformational transition is opposite than usual (5, 22).

21.2.3 Origin of Conformational Effects

Four main hypotheses have been formulated in order to explain conformational effects on protein CSDs. These hypotheses are not mutually exclusive and, rather, highlight mechanisms that could all contribute to the observed phenomenon.

1. *Electrostatic repulsions.* As postulated in Fenn's pioneering work, folded protein structures may result in lower charge states than unfolded ones, due to the smaller distances among ionizable residues in the former relative to the latter (17). In this view, electrostatic repulsions are the dominant force acting on gas-phase protein ions produced by electrospray.
2. *Zwitterionic structure.* According to this hypothesis, the specific interactions within folded protein structures may stabilize charges of opposite sign relative to the net charge of the ion (25, 44). These would be, instead, prone to be neutralized in unfolded structures, with consequent increase in the net charge of the protein ions. In this view, attractive electrostatic forces can survive in protein ions produced by electrospray under nondenaturing conditions.
3. *Solvent accessibility.* Another structural feature that changes during protein unfolding is the degree of exposure of ionizable residues to the solvent (44). This factor has been invoked to explain more extensive ionization of unfolded proteins, due to higher accessibility of groups involved in proton-transfer reactions.
4. *Shape.* In this hypothesis, the protein charge observed by ESI-MS reflects the charge of the precursor ESI droplet, which is in turn assumed to be close to the Rayleigh limit (12). Unfolded proteins would acquire more charges because their precursor ESI droplet would contain more charges. This effect is thought to derive from an elongated shape of the unfolded protein that forces the droplet into a highly nonspherical shape, supporting a larger surface charge than predicted by the Rayleigh limit for a droplet of equal volume.

These hypotheses are not strictly alternative to each other. Different factors likely contribute to the highly complex process of protein ionization by electrospray, as also suggested by recent theoretical (47) and experimental (75) studies.

21.2.4 Size Effect

Besides conformational effects, there is a well-documented effect of protein size on the extent of ionization by electrospray. By comparing the average charge states obtained with folded globular proteins of different sizes, a general trend is observed in which the net charge of the protein increases progressively with protein size (29, 30, 40, 57). This effect is likely related to the conformational effects discussed above and may well similarly arise from the geometric and electrostatic features of protein ions produced by electrospray. Interestingly, the experimental values of mean charge are very close (between 70% and 100%) to the Rayleigh-limit charge calculated for a water droplet of equal radius as the globular protein structure (approximated to a sphere). By assuming that a protein has the same density as water, the Rayleigh equation can be expressed as

$$q_R = 0.078M^{1/2},$$

which, thus, seems to represent the upper limit to the experimentally observed charge state of folded globular proteins as a function of the protein mass M (in daltons). Under the above assumptions on shape and density, a correlation with mass implies a correlation with surface. However, when nonspherical or cave structures are also considered, it turns out that the correlation with surface holds, rather than that with mass, according to the equation

$$q/e = AS^{\alpha},$$

where e is the elementary charge, S the surface area, and A and α are coefficients (40). This empirical power law implies a linear dependence of the ln(average charge) on the ln(surface area) for protein structures observed by nondenaturing ESI-MS. The derived value for α is 0.69. Such a linear dependence appears to hold in general, also considering oligomeric structures and a range of different experimental conditions. These results imply that it is possible to use empirical calibration functions in order to calculate the surface area of unknown structures in solution based on ESI-MS data. This method has not been applied yet to comparative analysis of folded and unfolded structures, but, since accessible surface area is one of the structural features that discriminate the two states, it is expectable that such an analysis will also offer a new tool for detection and characterization of intrinsically disordered proteins (IDPs).

21.2.5 Mechanism of Protein Ionization by Electrospray

The above discussion should be put in the frame of the current models for the process of transfer of protein ions to the gas phase by electrospray. The actual mechanisms responsible for this process are still object of debate. According to the so-called *charged residue model* (15), progressive reduction of droplet size through cycles of solvent evaporation and Coulomb explosions will eventually produce droplets containing a single analyte molecule. Complete solvent evaporation from such a droplet would result in a gas-phase analyte ion. An alternative model, the so-called *ion evaporation model* (34), suggests that electrostatic repulsions at the droplet's surface can lift analyte ions promoting their departure from the droplet. Detailed thermodynamic calculations show that the Gibbs free energy variation, ΔG, for ion evaporation under electrospray conditions can be negative. However, this kind of evidence is available only for small molecules. On the other hand, the above reported correlation between observed protein mean charge and calculated Rayleigh limit charge is an evidence supporting the charged residue model, which, based on this argument, has been considered as the most probable one for protein electrospray (12, 30, 57, 62). Nevertheless, such a correlation does not seem to hold when solvents with different surface tension than water are used (71, 72). Indeed, solvent conditions that drastically lower surface tension result in almost unaltered CSDs, as long as the protein does not unfold. On the opposite, the Rayleigh equation predicts a decrease of the limit charge of ESI droplets as the surface tension decreases. Finally, a recent paper reported on the experimental evidence supporting the ion evaporation model for desolvation of amino acids and peptides by electrospray (58). In this study, it was found that the addition of water vapor to the nitrogen gas bath of the instrument interface enhance the signals of amino acids and peptides contained in the sprayed solution. The opposite would be predicted by the charged residue model, since the presence of water vapor can only decrease the rate of solvent evaporation from the charged droplets. It is, therefore, concluded that the gas-phase ions are most likely produced by the ion-evaporation mechanism even in the case of peptides. The question remains open as far as intact proteins are concerned (75). This discussion might not appear directly related to the application of ESI-MS to protein conformational studies. However, a deep understanding of the electrospray mechanism would be important for confident interpretation of protein ionization behavior and its response to solvent conditions and protein structural properties.

21.2.6 Applications to the Study of IDPs

Although the physicochemical mechanisms underlying the ESI process are still the subject of debate, application of CSD analysis to protein conformational studies has been widely reported in the literature (39). When studying natively folded proteins, it is possible to establish the reference spectra of the folded and fully unfolded protein by making the experimental conditions progressively milder or progressively harsher (67). In order to identify the spectrum of the folded protein, pure water or millimolar concentrations of volatile buffers, typically ammonium acetate, can be used. Then, the source and interface temperature and voltages are lowered progressively in order to identify ranges of values in which instrumental setting does not become a limiting factor for protein conformational stability during electrospray. In such a range, the CSD must remain stable in its most right-shifted position in the spectrum. In order to identify the spectrum of the fully unfolded protein, the shift of the CSD toward smaller m/z values is monitored under progressively more denaturing conditions. Organic solvents, alcohols, and volatile acids or bases can be used as denaturants. Alternatively, source or interface temperature can be increased in order to induce thermal unfolding (33, 54).

IDPs are expected to deviate from this general behavior. Thus, CSD analysis could be used for the assessment of structural disorder, based on the evidence that the protein does not display the typical unfolding transition going from physiological conditions of solvent, pH, and temperature to strong denaturing conditions. Considering that conformational effects on CSDs can disappear when analyzing basic proteins in negative-ion mode or acidic ones in positive-ion mode, knowledge of the amino acid sequence would help in planning the experiment and interpreting the results. Alternatively, data collection both in positive- and negative-ion modes is recommendable.

As mentioned above, studies of this kind have been carried out on the IDP α-synuclein (5). The results of CSD analysis reveal that the protein at pH 7 cannot be unfolded and, rather, acquire a more compact conformation going from pH 7 to pH 2.5, a behavior that had been previously documented by other biophysical methods. Such an effect can be explained by titration of carboxylate groups in this highly acidic protein and consequent reduction of intramolecular electrostatic repulsions. The α-synuclein case study also offers an example of the fact that conformational effects on CSDs can depend on the polarity employed. This acidic protein, indeed, displays stronger conformational effects in negative- (5, 22) than in positive-ion mode (82). Measurements in negative-ion mode have been carried out in order to investigate the interaction of α-synuclein with the polyamine spermine, a naturally occurring compound found in neuronal cells, which is known to increase the aggregation rate of α-synuclein. The ESI-MS spectrum of α-synuclein in the presence of spermine at neutral pH reveals a bimodal CSD, reflecting a dramatic shift in the average charge state of the protein upon binding to spermine (22). These results indicate that the protein engaged in the complex is folded into a more compact conformation than the free protein. This partially folded protein conformation, thus, seems to be an on-pathway intermediate of fibril formation.

The cyclin-dependent kinase inhibitor Sic1 of the yeast *Saccharomyces cerevisiae* is a key regulator of cell cycle progression and its coordination to cell growth (3). The structural properties and conformational transitions of this protein have been studied by several different techniques, including ESI-MS (8). In agreement with sequence analysis, CD, and limited proteolysis, ESI-MS suggests that the pure protein in water is structurally disordered, revealing a broad CSD centered on the 26+ ion (Fig. 21.2), while the size-charge correlation for folded globular proteins would predict an average charge state of 14+

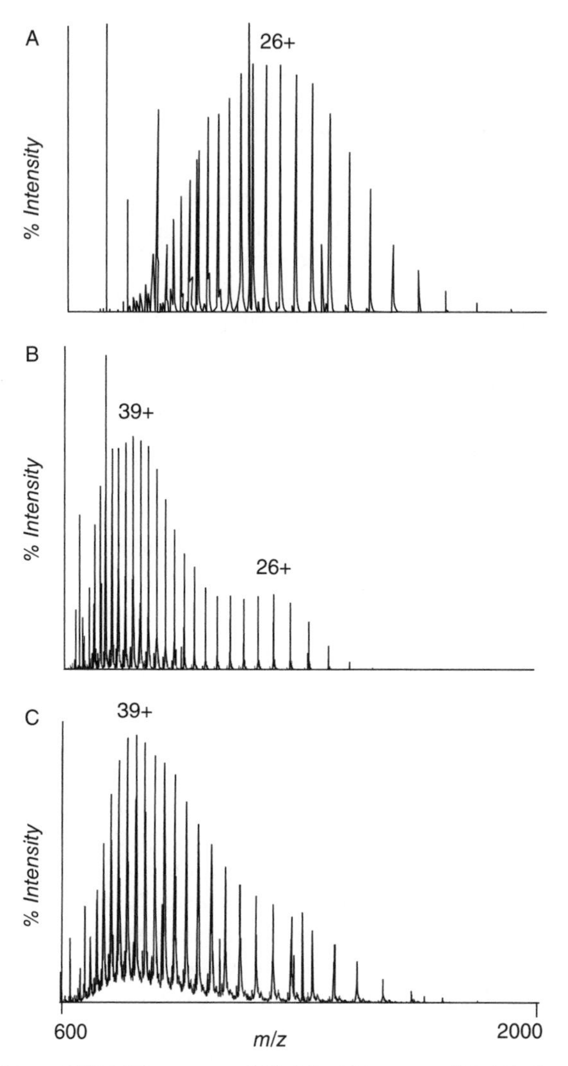

Figure 21.2 Nano-ESI-MS spectra of Sic1 in (A) water, (B) 1% formic acid, and (C) 1% formic acid, 50% acetonitrile. Labels show the main charge state of each peak envelope.

for this protein size. At the same time, the ESI-MS results show that the pure protein in the absence of interactors is in a partially collapsed structure that can be unfolded to a completely denatured state (main charge state 39+) by the addition of acids and organic solvents (Fig. 21.2). Thus, the bimodal distribution observed in 1% formic acid could be ascribed to a partially folded and the completely unfolded form of the protein (Fig. 21.2). The fact that the protein in the absence of denaturing agents has a considerable degree of tertiary structure could be confirmed by gel filtration (8).

Another IDP that has been studied by ESI-MS is prothymosin α, a pleiotropic regulator involved in cell proliferation, apoptosis, gene transcription, and oxidative-stress response (85). This protein is characterized by a bimodal CSD at neutral pH, indicative of coexistence of two distinct conformers. In the presence of metal ions, it is possible to identify the bound species of both forms. These studies reveal a selectivity of the protein toward Zn^{2+}, compared with Ca^{2+} and Mg^{2+}. The ESI data, complemented by nuclear magnetic resonance (NMR) studies, indicate that Zn^{2+} binding results in increased protein compactness inducing folding of the C-terminal half of the protein. Based on these and other results, the authors formulate the hypothesis that the conformational and dynamic changes upon zinc binding may act as an entropic switch that facilitates binding to other proteins.

21.3 IM

The CSDs of individual components of protein mixtures often cover the same narrow m/z range so that charge states of compounds with different molecular weights may generate overlapping peaks. It can also happen that protein ions with the same mass, yet possessing different shape, share the same charge state. These ions, however, can still be distinguished by their mobility in gas phase, that is, by the so-called ion mobility spectrometry (IMS).

IMS was developed by Cohen and Karasek in 1970 (11). It allows to distinguish different conformers among ions of same m/z and can help reduce the spectral complexity of even complicated mixtures (31). The ions, formed by ESI or MALDI, are injected into a cell filled with drift gas (helium) and separated on the basis of their different velocity in such a buffer gas under the influence of an electric field. The velocity (v[cm/s]) of gas-phase ions through the drift cell is directly proportional to the electric field (E [V/cm]), and the proportionality constant is called *mobility* of the ion (K[$cm^2/(s{\cdot}V)$]):

$$v = KE.$$

The larger the ions (and, therefore, their collision cross section), the more hindered they will be moving through the buffer gas. Thus, the time required to traverse the drift tube is directly affected by the ion size and shape. The plot of ion intensity as a function of drift time is called arrival-time distribution (ATD) (Fig. 21.3). A value for the *collision cross section* can be derived from the drift time. Such a value is directly related to the shape and overall topology of the ion, yielding an information about the compactness of the observed structure. However, structural assignments based on mobility results generally require comparison with model structures obtained by molecular-dynamics simulations or molecular modeling (37, 38, 79, 80).

The ion velocity v is proportional to the field strength at low electric fields (e.g., 200 V/cm). Thus, under these conditions, K is independent of E. At high

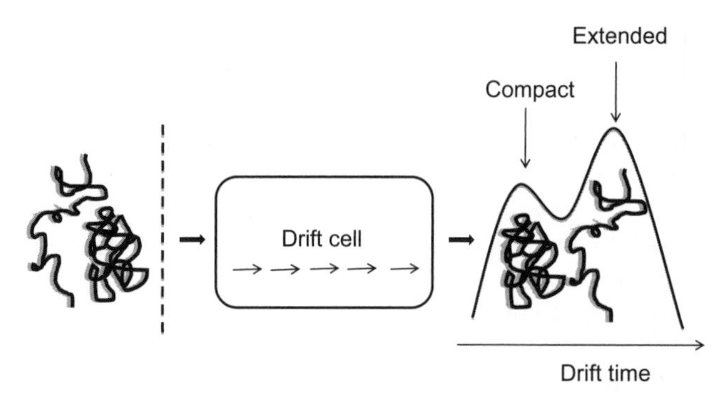

Figure 21.3 Schematic representation of the ion mobility technique.

electric fields (e.g., 10,000V/cm), K starts being dependent on E in a compound-dependent manner, which is the principle of ion separation in differential mobility spectrometry or high-field asymmetric waveform ion mobility spectrometry (63).

21.3.1 Applications to the Study of IDPs

IMS coupled to ESI-MS provides us with a unique tool for the study of molecular conformations, allowing the separation of isomers from very complex mixtures. Nevertheless, it can be argued that gas-phase protein structures are not representative of the conformational properties in solution, since noncovalent interactions are radically affected by the presence of an aqueous solvent. Rapidly growing evidence in the literature shows, instead, that major features of protein structures and protein assemblies survive in the gas phase, at least long enough to allow species separation by IMS (19, 41, 66). In agreement with experimental evidence, molecular-dynamics simulations indicate persistence of compact conformations upon desolvation of folded protein structures. These studies show that desolvated proteins tend to shrink into "inside out" structures exposing hydrophobic residues and engaging polar groups in intramolecular hydrogen bonds and salt bridges (35, 61). Thus, structures do change, but distinct conformational states in solution generally give rise to discernible drift times in IMS measurements.

ESI-IMS-MS has been employed in order to probe the conformation and oligomerization state of amyloidogenic proteins. Back to the α-synuclein case, IM data complement CSD analysis showing that the compact conformation at low pH has an average cross section of 1690 Å^2 in the gas phase, while the same measurement for the protein sprayed at neutral pH gives a value of 2530 Å^2 (5). This result indicates that distinct solution conformers can maintain structural differences in the gas phase, at least on the timescale of IMS measurements, allowing their discrimination based on geometric properties. This conclusion is supported also by measurements on the ensemble of folded and unfolded conformations of the reference protein cytochrome c at pH 3. The application of this technique to the study of α-synuclein–spermine interaction shows that ligand binding induces folding of the protein at neutral pH into an even more compact (1430 $Å^2$) conformation than observed for the protein alone at low pH (22). The unfolding of the amyloidogenic protein β2-microglobulin induced by acidification has also been investigated by this technique (4, 7). The analysis of drift times for each charge state led to the identification of four distinct conformers. Two of them coexist in the folded state, representing a native-like conformation and a slightly more relaxed conformation. The other two correspond to a partially folded state, which starts accumulating at pH < 5, and an extended conformation, which starts accumulating at pH < 3.5.

The amyloidogenic protein Aβ42, responsible for fibril formation in Alzheimer's disease, is mostly disordered in its monomeric state (60–80% random coil), according to NMR and CD. The conformational properties of monomeric Aβ42 have been investigated by a combination of ESI-MS, IMS, and molecular dynamics in the presence and absence of water. This approach allowed to generate molecular models of the most representative structures in solution and in gas phase and to relate these to each other, justifying the observed ATD curves (4). These results led to formulate a testable hypothesis on the mechanism of Aβ42 self-assembly. The C-terminal region of the peptide displays a propensity to form an α-helix in apolar environments. Such a structure could act as nucleation site in the formation of oligomeric intermediates of the fibrillation process. The work on Aβ42 also provides an interesting example of detection of oligomeric intermediates of the fibrillation process. The addition of a second dimension (IMS) to MS analysis enhances tremendously the resolution power in the characterization of complex, heterogeneous mixtures. So dimers, tetramers, hexamers, and dodecamers of Aβ42 could be identified as the first species formed toward protofibril formation (6).

21.4 H/X

From the early 1950s to present days, HX has become one of the most powerful techniques for the study of protein structure and dynamics. The main process involved in this method is the interchange between hydrogens of the polypeptide with deuterium atoms provided in high concentration in the solution. Since the characteristic time for the exchange of amide hydrogens depends on the local conformation, the amount of incorporated deuterium over time offers a probe to analyze protein conformational states. In the last decades, HX has been successfully coupled to MS, shedding light on particular characteristics of protein structures and protein complexes that are hardly investigable by other approaches.

21.4.1 General Strategy

Hydrogens contained in protein structures exchange in a continuous and reversible way with those of the solvent. The important feature of this process is that protein hydrogens are not all equivalent for the exchange reaction (28, 50). Several factors affect the propensity for the swap. It is possible to identify three main categories of hydrogen atoms in proteins according to their exchange behavior. The first group includes hydrogens bound to electronegative atoms (O, N, S) in side chains and at the N- and C-termini. These hydrogens exchange too fast to allow monitoring the progress of the reaction experimentally (28). A second group is formed by carbon-bound aliphatic and aromatic hydrogens, which, on the opposite, exchange too slowly (28). The only group whose exchange takes place on a timescale compatible with realistic experimental approaches is that of backbone amide hydrogens.

The other useful feature of backbone amide hydrogen exchange is that the rate is strongly dependent on local conformational properties, like solvent accessibility and involvement in H bonds. Due to these factors, the exchange rates can vary from milliseconds to years (28). Thus, the degree of protection of amide hydrogens within protein structures, relative to model peptide compounds, offers useful structural information.

Exchange reactions can be made detectable by transferring a protonated protein in deuterated water (D_2O) or vice versa. Incorporation or loss of deuterium can be monitored by changes in proton signals in NMR spectra (16) or by the mass shift in MS spectra (16). MS offers the advantage of low sample consumption and no limitation on protein size, but has the disadvantage to give information at the molecule level and not at residue level. However, the combination with proteolytic digestion and MS/MS protocols can provide site-specific information also when MS is used to characterize the exchange products (69). Furthermore, the peculiar feature of MS data of being averaged over the atoms of each molecule but not among molecules can represent an advantage interpreting exchange kinetics (51).

21.4.2 Exchange and Folding

A precise formalism has been developed to relate the thermodynamic behavior of a protein to the correspondent observed HX rates (50). Hydrogens protected by structure can exchange with the solvent only after a dynamic event that reversibly and transiently unfolds the local structure. Thus, the exchange rates of structured or buried regions are governed by random fluctuations that occasionally expose the exchangeable sites to the solvent. The situation described above can be represented by the following model, referring to each single residue of the protein:

$$F \underset{k_{\mathrm{cl}}}{\overset{k_{\mathrm{op}}}{\rightleftarrows}} U \xrightarrow{k_{\mathrm{in}}} U_e$$

where F stands for folded, U for unfolded, subscripte e for exchanged, and k_{op}, k_{cl}, and k_{in} are the opening, closing, and intrinsic exchange rates, respectively. Intrinsic rate means the rate of exchange for a particular amino acid within a totally random-coil polypeptide. These values vary between 1 and $100\,s^{-1}$ and are tabulated (2, 81). In this scheme, the observed exchange rate, that is, the measured rate of deuterium loss or incorporation, can be expressed as

$$k_{obs} = \frac{k_{op} \cdot k_{in}}{k_{in} + k_{cl} + k_{op}}.$$

Two distinct regimes are commonly identified, depending on the relative rates of isotope exchange and conformational transitions. In the limit case of $k_{cl} \gg k_{in}$ and $k_{cl} \gg k_{op}$ the above equation can be rewritten as

$$k_{obs} \cong \frac{k_{op} \cdot k_{in}}{k_{cl}} \Rightarrow \frac{k_{obs}}{k_{in}} = \frac{k_{op}}{k_{cl}} = K = \frac{1}{P},$$

where K is the unfolding equilibrium constant, given by the ratio between k_{op} and k_{cl}, and P is the *protection factor*, the ratio between intrinsic and observed exchange rates. This regime is called EX2 mechanism and offers a direct link to the free energy variation associated to local unfolding:

$$\Delta G_u^0 = -RT \ln K = -RT \ln \frac{k_{obs}}{k_{in}}.$$

where R is the universal gas constant and T is the absolute temperature.

The opposite limit situation ($k_{in} \gg k_{cl}$ and $k_{in} \gg k_{op}$) is called EX1 mechanism and refers to the case in which exchange is much faster than conformational changes. This behavior is usually observed under high pH conditions, when the intrinsic rate of hydrogen exchange is very fast (84). In such a regime, the observed rate is given by $k_{obs} \cong k_{op}$, directly translating exchange data to the rate-limiting step, that is, local unfolding.

If the exchange is being monitored by MS at the level of the whole protein, the deconvoluted spectra will display different features under the two limit scenarios. The EX2 mechanism will produce a broad, single peak corresponding to an ensemble of partially exchanged molecules that shifts progressively toward the mass of the fully exchanged form. The EX1 mechanism, instead, will generate distinct peaks, corresponding to the distinct conformers, whose relative intensity progressively change as isotope exchange proceeds (50, 55).

21.4.3 Experimental Design

There are two main kinds of experimental scheme for HX: continuous and pulse labeling (77). These are briefly described here, in their association to MS as a detection method, referring to the *exchange-in* mode, in which a protonated protein is labeled in a deuterated solvent. In the first case, the protein is transferred to a deuterated solution that is continuously sprayed into the spectrometer. Since hydrogen concentration in the solvent is very low, once a site is labeled, the probability of back exchange is practically negligible, although it can become appreciable in nano-ESI sources. The mass spectra will show incorporation of deuterium in real time. A pulse-labeling experiment, instead, is based on exposure of a protein to deuterium for a controlled, brief time interval, followed by transfer to quenching conditions (low temperature and low pH), under which the exchange rate of backbone atoms is drastically reduced. If the quenching conditions are such that H_2O is reintroduced in the system, side chains will back exchange to hydrogen. Thus, this kind of protocols yield exchange data for the backbone, free from the background of side chains. However, the problem of undesired back exchange of backbone sites typically alters the results. Control experiments can be performed in order to provide correction factors accounting for back exchange (56). The described protocols can be reversed into the *exchange-off* mode, where the protein is first completely deuterated by prolonged incubation in D_2O, and then the progressive mass reduction in protonated solvents is monitored.

Continuous labeling is suitable for equilibrium conditions, while pulse labeling allows kinetic monitoring of folding and unfolding transitions. In this case, the pulse is applied after increasing time intervals from the beginning of the reaction. Fast-mixing devices have been developed in order to perform isotopic pulse labeling on-line to ESI-MS (20). This kind of experiments has provided rich information on the mechanism of folding and unfolding reactions, detecting intermediates, depicting the time evolution of species distributions, and describing the distinct species from the structural point of view (50).

In order to get site-specific information, exchanged samples can be digested under quenching conditions generating peptides that can then be analyzed by (LC)-MS/MS. The ideal enzyme for this purpose is pepsin (84) because it is active at the low pH of the quenching conditions (around pH 2). The nonspecific cleavage of pepsin can produce even more than 100 peptides from a single protein, making the assignment sometimes extremely laborious. Correct peptide identification and sequence reconstruction is required to cover the entire sequence of the protein with the highest possible number of overlapping peptides. In this way, the exchange rates at nearly single amino acid level can be obtained (13). From the progressive mass increment of each peptide, it is possible to assess the different degrees of protection of distinct regions within a protein structure. Even finer mapping of protection factors can be achieved by gas-phase fragmentation techniques (16, Note Added in Proof).

If temperature and pH are maintained constant during the experiment, each amide hydrogen follows first-order kinetics (1), resulting in the following global exchange kinetics:

$$P = N_{\mathrm{f}} e^{-k_{\mathrm{f}} t} + N_{\mathrm{s}} e^{-k_{\mathrm{s}} t},$$

where P is the number of unexchanged hydrogens in the peptide, N_{f} and N_{s} are the numbers of fast and slow-exchanged hydrogens, with exchange rate constant of k_f and k_s, respectively. The analysis is carried out by a multiexponential fit, in which additional terms can be added for intermediate classes.

21.4.4 Applications to the Study of IDPs

By offering the possibility of distinguishing structured from unstructured regions, HX-MS is clearly a powerful tool for the analysis of IDPs. Identification of structured subdomains within largely unstructured proteins is a very challenging task, but at the same time, it is necessary in order to understand the mechanism of action of these complex molecular switches. In particular, locally structured regions frequently represent interaction nuclei that could drive intermolecular recognition and binding-induced folding (52).

The application of HX-MS to IDPs is complicated by several factors. One of these is the fast exchange that the great majority of peptides will show in a largely unstructured protein. The use of near-quenching conditions throughout all the experiment has proven useful in such cases. In this slow-exchange regime, it is possible to detect moderate protection factors that would be missed under exchange conditions typically used for folded proteins (13). Another problem arises with polyproline helices. Since HX is blind to them, these structures generate artifacts in the analysis of protection factors (52).

The characterization of large unstructured regions in proteins by HX-MS has also been used in order to design deletions aimed at improving protein crystallization (60, 70). Such an approach is expected to give significant contributions to large-scale crystallography projects and to structural characterization of ordered domains containing or associated to large disordered regions.

As an example of structural characterization of an IDP by HX-MS, we mention the case of the tyrosine kinase-interacting protein Tip-C484 of *Herpesvirus saimiri* (52). HX-MS has been applied to conformational analysis of the pure recombinant protein. The rate of deuterium incorporation in the intact protein indicates a mostly disordered structure. However, ~10% of the amide hydrogen atoms are resistant to exchange, revealing the existence of small, ordered regions. Pepsin digestion followed by MS/MS experiments allowed mapping this intrinsic structure within specific sequence stretches (peptides 27–62 and 145–147 and the N-terminal His_6 affinity tag).

HX-MS can also be applied to the identification of regions in IDPs that become structured and/or buried upon binding to other proteins, offering an invaluable tool for the investigation of the molecular mechanisms underlying their biochemical activity (27, 64). The conformational properties of the ligand-dependent transcription factor PPARγ, which is involved in glucose homeostasis, have been investigated by this method in the presence of different agonists and antagonists (27). The binding pocket of PPARγ is significantly more dynamic than the rest of the ligand-binding domain, in order to accomodate different ligands. The HX-MS measurements reveal specific conformational changes elicited by each ligand, locally and globally, that is, in the whole binding domain. The change in protein dynamics induced by ligand binding reveals allosteric effects that likely mediate the biologic activity of PPARγ.

In the same way, regions promoting aggregation in amyloid fibrils have been characterized (9, 65). In this case, the exchange experiment is performed directly on the insoluble fibrils, followed by digestion and solubilization of the labeled material. This kind of studies allows mapping the most protected regions of a self-assembled polypeptide within the fibrillar structure. An example is offered again by the protein α-synuclein (13). Preformed α-synuclein fibrils are resuspended in deuterated buffer to follow the time course of isotope exchange. At different time points, the reaction is quenched, fibrils are dissociated into monomers, and the samples are subjected to pepsin digestion and MS/MS analysis to localize the deuterated regions. The results show that the central part of the α-synuclein molecule (residues 39–101) is the region involved in the cross-β structure of the fibrils, whereas the C- and N-termini remain unprotected. The data also suggest a possible model for the fine structure of the fibrils.

In conclusion, the possibility to probe protein structure and dynamics by MS methods is giving a significant contribution to the characterization of proteins in disordered conformations, which is required in order to understand conformational stability in the folded state. In particular, for IDPs, the elucidation of the folding transitions induced by ligand binding and protein–protein interactions will shed light on the mechanisms of this fascinating case of molecular recognition.

NOTE ADDED IN PROOF

After submission of the present contribution, several papers relevant to this topic have been published. We mention, in particular:

Frimpong, A. K., R. R. Abzalimov, V. N. Uversky, and I. A. Kaltashov. 2009. Characterization of intrinsically disordered proteins with electrospray ionization mass spectrometry: conformational heterogeneity of alpha-synuclein. Proteins Sep **11**. [Epub ahead of print].

Pan, J., J. Han, C. H. Borchers, L. Konermann. 2009. Hydrogen/deuterium exchange mass spectrometry with top-down electron capture dissociation for characterizing structural transitions of a 17 kDa protein. J Am Chem Soc **131**:12801–08.

Rand, K. D., M. Zehl, O. N. Jensen, and T. J. Jørgensen. 2009. Protein hydrogen exchange measured at single-residue resolution by electron transfer dissociation mass spectrometry. Anal Chem **81**:5577–84.

REFERENCES

1. Alomirah, H., I. Alli, and Y. Konishi. 2003. Charge state distribution and hydrogen/deuterium exchange of alpha-lactalbumin and beta-lactoglobulin preparations by electrospray ionization mass spectrometry. J Agric Food Chem **51**:2049–57.

2. Bai, Y., J. S. Milne, L. Mayne, and S. W. Englander. 1993. Primary structure effects on peptide group hydrogen exchange. Proteins **17**:75–86.
3. Barberis, M., E. Klipp, M. Vanoni, and L. Alberghina. 2007. Cell size at S phase initiation: an emergent property of the G1/S network. PLoS Comput Biol **3**:e64.
4. Baumketner, A., S. L. Bernstein, T. Wyttenbach, G. Bitan, D. B. Teplow, M. T. Bowers, and J. E. Shea. 2006. Amyloid beta-protein monomer structure: a computational and experimental study. Protein Sci **3**:420–8.
5. Bernstein, S. L., D. Liu, T. Wyttenbach, M. T. Bowers, J. C. Lee, H. B. Gray, and J. R. Winkler. 2004. Alpha-synuclein: stable compact and extended monomeric structures and pH dependence of dimer formation. J Am Soc Mass Spectrom **15**:1435–43.
6. Bernstein, S. L., T. Wyttenbach, A. Baumketner, J. E. Shea, G. Bitan, D. B. Teplow, and M. T. Bowers. 2005. Amyloid beta-protein: monomer structure and early aggregation states of Abeta42 and its Pro19 alloform. J Am Chem Soc **127**:2075–84.
7. Borysik, A. J., S. E. Radford, and A. E. Ashcroft. 2004. Co-populated conformational ensembles of beta2-microglobulin uncovered quantitatively by electrospray ionization mass spectrometry. J Biol Chem **279**:27069–77.
8. Brocca, S., M. Šamalikova, V. N. Uversky, M. Lotti, M. Vanoni, L. Alberghina, and R. Grandori. 2009. Order propensity of an intrinsically disordered protein, the cyclin-dependent-kinase inhibitor Sic1. Proteins **76**:731–46.
9. Caddy, G. L., and C. V. Robinson. 2006. Insights into amyloid fibril formation from mass spectrometry. Protein Pept Lett **13**:255–60.
10. Chowdhury, S. K., V. Katta, and B. T. Chait. 1990. Probing conformational changes in proteins by mass spectrometry. J Am Chem Soc **112**:9012–3.
11. Cohen, M.J., and F.W. Karasek. 1970. Plasma chromatography—a new dimension for gas chromatography and mass spectrometry. J Chromatogr Sci **8**:330.
12. de la Mora, J. F. 2000. Electrospray ionization of large multiply charged species proceeds via Dole's charged residue mechanism. Anal Chim Acta **406**:93–104.
13. Del Mar, C., E. A. Greenbaum, L. Mayne, S. W. Englander, and V. L. J. Woods. 2005. Structure and properties of alpha-synuclein and other amyloids determined at the amino acid level. Proc Natl Acad Sci U S A **102**:15477–82.
14. Dobo, A., and I. A. Kaltashov. 2001. Detection of multiple protein conformational ensembles in solution via deconvolution of charge-state distributions in ESI MS. Anal Chem **73**:4763–73.
15. Dole, M., L. L. Mack, R. L. Hines, R. C. Mobley, L. D. Ferguson, and M. B. Alice. 1968. Molecular beams of macroions. J Chem Phys **49**:2240–9.
16. Eyles, S. J., and I. A. Kaltashov. 2004. Methods to study protein dynamics and folding by mass spectrometry. Methods **34**:88–99.
17. Fenn, J. B. 1993. Ion formation from charged droplets: roles of geometry, energy, and time. J Am Soc Mass Spectrom **4**:524–35.
18. Fenn, J. B., M. Mann, C. K. Meng, S. F. Wong, and C. M. Whitehouse. 1989. Electrospray ionization for mass spectrometry of large biomolecules. Science **246**:64–71.
19. Fenn, L. S., and J. A. McLean. 2008. Biomolecular structural separations by ion mobility-mass spectrometry. Anal Bioanal Chem **391**:905–9.

20. Ferguson, P. L., J. Pan, D. J. Wilson, B. Dempsey, G. Lajoie, B. Shilton, and L. Konermann. 2007. Hydrogen/deuterium scrambling during quadrupole time-of-flight MS/MS analysis of a zinc-binding protein domain. Anal Chem **79**:153–60.
21. Goto, Y., Y. Hagihara, D. Hamada, M. Hoshino, and I. Nishii. 1993. Acid-induced unfolding and refolding transitions of cytochrome c: a three-state mechanism in H_2O and D_2O. Biochemistry **32**:11878–85.
22. Grabenauer, M., S. L. Bernstein, J. C. Lee, T. Wyttenbach, N. F. Dupuis, H. B. Gray, J. R. Winkler, and M. T. Bowers. 2008. Spermine binding to Parkinson's protein alpha-synuclein and its disease-related A30P and A53T mutants. J Phys Chem B **112**:11147–54.
23. Grandori, R. 2002. Detecting equilibrium cytochrome c folding intermediates by electrospray ionization mass spectrometry: two partially folded forms populate the molten-globule state. Protein Sci **11**:453–8.
24. Grandori, R. 2003. Electrospray-ionization mass spectrometry for protein conformational studies. Curr Org Chem **7**:1589–603.
25. Grandori, R. 2003. Origin of the conformation dependence of protein charge-state distributions in electrospray-ionization mass spectrometry. J Mass Spectrom **38**:11–5.
26. Grandori, R., I. Matecko, and N. Müller. 2001. Uncoupled analysis of secondary and tertiary protein structure by circular dichroism and electrospray ionization mass spectrometry. J Mass Spectrom **37**:191–6.
27. Hamuro, Y., S. J. Coales, J. A. Morrow, K. S. Molnar, S. J. Tuske, M. R. Southern, and P. R. Griffin. 2006. Hydrogen/deuterium-exchange (H/D-Ex) of PPARgamma LBD in the presence of various modulators. Protein Sci **15**: 1883–92.
28. Hamuro, Y., S. J. Coales, M. R. Southern, J. F. Nemeth-Cawley, D. D. Stranz, and P. R. Griffin. 2003. Rapid analysis of protein structure and dynamics by hydrogen/deuterium exchange mass spectrometry. J Biomol Tech **14**:171–82.
29. Hautreux, M., N. Hue, A. Du Fou de Kerdaniel, A. Zahir, V. Malec, and O. Laprévote. 2004. Under non-denaturing solvent conditions, the mean charge state of a multiply charged protein ion formed by electrospray is linearly correlated with the macromolecular surface. Int J Mass Spectrom **231**:131–7.
30. Heck, A. J., and R. H. van den Heuvel. 2004. Investigation of intact protein complexes by mass spectrometry. Mass Spectrom Rev **23**:368–89.
31. Henderson, S. C., S. J. Valentine, A. E. Counterman, and D. E. Clemmer. 1999. ESI/ion trap/ion mobility/time-of-flight mass spectrometry for rapid and sensitive analysis of biomolecular mixtures. Anal Chem **71**:291–301.
32. Invernizzi, G., and R. Grandori. 2007. Detection of the equilibrium folding intermediate of beta-lactoglobulin in the presence of trifluoroethanol by mass spectrometry. Rapid Commun Mass Spectrom **21**:1049–52.
33. Invernizzi, G., M. Šamalikova, S. Brocca, M. Lotti, H. Molinari, and R. Grandori. 2006. Comparison of bovine and porcine beta-lactoglobulin: a mass spectrometric analysis. J Mass Spectrom **41**:717–27.
34. Iribarne, J. V., and B. A. Thomson. 1976. On the evaporation of small ions from charged droplets. J Chem Phys **64**:2287–94.
35. Jarrold, M. F. 2000. Peptides and proteins in the vapor phase. Annu Rev Phys Chem **51**:179–207.

36. Juskowiak, B., M. Chudaka, and S. Takenaka. 2003. Detection of noncovalent interactions of hairpin oligonucleotide with stilbazolium ligands by MALDI TOF mass spectrometry. Int J Mass Spectrom **3**:225–30.

37. Kaddis, C. S., S. H. Lomeli, S. Yin, B. Berhane, M. I. Apostol, V. A. Kickhoefer, L. H. Rome, and J. A. Loo. 2007. Sizing large proteins and protein complexes by electrospray ionization mass spectrometry and ion mobility. J Am Chem Soc **18**:1206–16.

38. Kaddis, C. S., and J. A. Loo. 2007. Native protein MS and ion mobility large flying proteins with ESI. Anal Chem **79**:1778–84.

39. Kaltashov, I. A., and R. R. Abzalimov. 2008. Do ionic charges in ESI MS provide useful information on macromolecular structure? J Am Soc Mass Spectrom **19**:1239–46.

40. Kaltashov, I. A., and A. Mohimen. 2005. Estimates of protein surface areas in solution by electrospray ionization mass spectrometry. Anal Chem **77**: 5370–9.

41. Kanu, B., P. Dwivedi, M. Tam, L. Matz, and H. H. J. Hill. 2008. Ion mobility-mass spectrometry. J Mass Spectrom **43**:1–22.

42. Karas, M., and F. Hillenkamp. 1988. Laser desorption ionization of proteins with molecular masses exceeding 10000 daltons. Anal Chem **60**:2299–301.

43. Katta, V., and B. T. Chait. 1991. Conformational changes in proteins probed by hydrogen-exchange electrospray-ionization mass spectrometry. Rapid Commun Mass Spectrom **5**:214–7.

44. Katta, V., and B. T. Chait. 1991. Observation of the heme-globin complex in native myoglobin by electrospray-ionization mass spectrometry. J Am Chem Soc **113**:8534–5.

45. Kebarle, P., and Y. Ho. 1997. On the mechanism of electrospray mass spectrometry, pp. 3–63. In R. B. Cole (ed.), Electrospray Ionization Mass Spectrometry. John Wiley & Sons, New York.

46. Kebarle, P., and L. Tang. 1993. From ions in solution to ions in the gas phase. Anal Chem **65**:972A–86A.

47. Konermann, L. 2007. A minimalist model for exploring conformational effects on the electrospray charge state distribution of proteins. J Phys Chem B **111**:6534–43.

48. Konermann, L., and D. J. Douglas. 1997. Acid-induced unfolding of cytochrome c at different methanol concentrations: electrospray ionization mass spectrometry specifically monitors changes in the tertiary structure. Biochemistry **36**: 12296–302.

49. Konermann, L., and D. J. Douglas. 1998. Equilibrium unfolding of proteins monitored by electrospray ionization mass spectrometry: distinguishing two-state from multi-state transitions. Rapid Commun Mass Spectrom **12**:435–42.

50. Konermann, L., X. Tong, and Y. Pan. 2008. Protein structure and dynamics studied by mass spectrometry: H/D exchange, hydroxyl radical labeling, and related approaches. J Mass Spectrom **43**:1021–36.

51. Miranker, A., C. V. Robinson, S. E. Radford, R. T. Aplin, and C. M. Dobson. 1993. Detection of transient protein folding populations by mass spectrometry. Science **262**:896–900.

52. Mitchell, J. L., R. P. Trible, L. A. Emert-Sedlak, D. D. Weis, E. C. Lemer, J. J. Applen, B. M. Sefton, T. E. Smithgall, and J. R. Engen. 2007. Functional characterization and conformational analysis of the *Herpesvirus saimiri* Tip-C484 protein. J Mol Biol **366**:1282–93.

53. Murphy, P. W., E. E. Rowland, and D. M. Byers. 2007. Electrospray ionization mass spectra of acyl carrier protein are insensitive to its solution phase conformation. J Am Chem Soc **18**:1525–32.

54. Natalello, A., S. M. Doglia, J. Carey, and R. Grandori. 2007. Role of flavin mononucleotide in the thermostability and oligomerization of *Escherichia coli* stress-defense protein WrbA. Biochemistry **46**:543–53.

55. Nazabal, A., S. Dos Reis, M. Bonneu, S. J. Saupe, and J. M. Schmitter. 2003. Conformational transition occurring upon amyloid aggregation of the HET-s prion protein of *Podospora anserina* analyzed by hydrogen/deuterium exchange and mass spectrometry. Biochemistry **42**:8852–61.

56. Nazabal, A., M. L. Maddelein, M. Bonneu, S. J. Saupe, and J. M. Schmitter. 2005. Probing the structure of the infectious amyloid form of the prion-forming domain of HET-s using high resolution hydrogen/deuterium exchange monitored by mass spectrometry. J Biol Chem **280**:13220–8.

57. Nesatyy, V. J. 2001. Gas-phase binding of non-covalent protein complexes between bovine pancreatic trypsin inhibitor and its target enzymes studied by electrospray ionization tandem mass spectrometry. J Mass Spectrom **36**:950–9.

58. Nguyen, S., and J. B. Fenn. 2007. Gas-phase ions of solute species from charged droplets of solutions. Proc Natl Acad Sci U S A **104**:1111–1117.

59. Pan, X. M., X. R. Sheng, and J. M. Zhou. 1997. Probing subtle acid-induced conformational changes of ribonuclease A by electrospray mass spectrometry. FEBS Lett **402**:25–7.

60. Pantazatos, D., J. S. Kim, H. E. Klock, R. C. Stevens, I. A. Wilson, S. A. Lesley, and V. L. J. Woods. 2004. Rapid refinement of crystallographic protein construct definition employing enhanced hydrogen/deuterium exchange MS. Proc Natl Acad Sci U S A **101**:751–6.

61. Patriksson, A., E. Marklund, and D. Van der Spoel. 2007. Proteins structures under electrospray conditions. Biochemistry **46**:933–45.

62. Peschke, M., U. H. Verkerk, and P. Kebarle. 2004. Prediction of the charge states of folded proteins in electrospray ionization. Eur J Mass Spectrom **10**:993–1002.

63. Purves, R. W., and Guevremont R. 1999. Electrospray ionization—high-field asymmetric waveform ion mobility spectrometry—mass spectrometry. Anal Chem **71**:2346–57.

64. Receveur-Brechot, V., J. M. Bourhis, V. N. Uversky, B. Canard, and S. Longhi. 2006. Assessing protein disorder and induced folding. Proteins **62**:24–45.

65. Redeker, V., F. Halgand, J. P. Le Caer, L. Bousset, O. Laprévote, and R. Melki. 2007. Hydrogen/deuterium exchange mass spectrometric analysis of conformational changes accompanying the assembly of the yeast prion Ure2p into protein fibrils. J Mol Biol **369**:1113–25.

66. Ruotolo, B. T., S. J. Hyung, P. M. Robinson, K. Giles, R. H. Bateman, and C. V. Robinson. 2007. Ion mobility-mass spectrometry reveals long-lived, unfolded inter-

mediates in the dissociation of protein complexes. Angew Chem Int Ed Engl **42**:8001–4.

67. Sanglier, S., C. Atmanene, G. Chevreux, and A. V. Dorsselaer. 2008. Nondenaturing mass spectrometry to study noncovalent protein/protein and protein/ligand complexes: technical aspects and application to the determination of binding stoichiometries. Methods Mol Biol **484**:217–43.
68. Siuzdak, G. 2003. The Expanding Role of Mass Spectrometry in Biotechnology. MCC Press, San Diego, CA.
69. Sours, K. M., S. C. Kwok, T. Rachidi, T. Lee, A. Ring, A. N. Hoofnagle, K. A. Resing, and N. G. Ahn. 2008. Hydrogen-exchange mass spectrometry reveals activation-induced changes in the conformational mobility of p38alpha MAP kinase. J Mol Biol **379**:1075–93.
70. Spraggon, G., D. Pantazatos, H. E. Klock, I. A. Wilson, V. L. J. Woods, and S. A. Lesley. 2004. On the use of DXMS to produce more crystallizable proteins: structures of the *T. maritima* proteins TM0160 and TM1171. Protein Sci **13**:3187–99.
71. Šamalikova, M., and R. Grandori. 2003. Role of opposite charges in protein electrospray-ionization mass spectrometry. J Mass Spectrom **38**:941–7.
72. Šamalikova, M., and R. Grandori. 2005. Testing the role of solvent surface tension in protein ionization by electrospray. J Mass Spectrom **40**:503–10.
73. Šamalikova, M., I. Matecko, N. Müller, and R. Grandori. 2004. Interpreting conformational effects in protein nano-ESI-MS spectra. Anal Bioanal Chem **378**:1112–23.
74. Tanaka, K., H. Waki, Y. Ido, S. Akita, Y. Yoshida, and T. Yohida. 1988. Protein and polymer analyses up to m/z 100,000 by laser ionization time-of-flight mass spectrometry. Rapid Commun Mass Spectrom **2**:151–3.
75. Verkerk, U. H., and P. Kebarle. 2005. Ion-ion and ion-molecule reactions at the surface of proteins produced by nanospray. Information on the number of acidic residues and control of the number of ionized acidic and basic residues. J Am Chem Soc **16**:1325–41.
76. Vis, H., U. Heinemann, C. M. Dobson, and C. V. Robinson. 1998. Detection of a monomeric intermediate associated with dimerization of protein Hu by mass spectrometry. J Am Chem Soc **120**:6427–8.
77. Wales, T. E., and J. R. Engen. 2006. Hydrogen exchange mass spectrometry for the analysis of protein dynamics. Mass Spectrom Rev **25**:158–70.
78. Wilm, M., and M. Mann. 1996. Analytical properties of the nanoelectrospray ion source. Anal Chem **68**:1–8.
79. Wyttenbach, T., and M. T. Bowers. 2003. Gas-phase conformations: the ion mobility/ion chromatography method. Top Curr Chem **225**:207–32.
80. Wyttenbach, T., and M. T. Bowers. 2007. Intermolecular interactions in biomolecular systems examined by mass spectrometry. Annu Rev Phys Chem **58**:511–33.
81. Xiao, H., J. K. Hoerner, S. J. Eyles, A. Dobo, E. Voigtman, A. I. Mel'cuk, and I. A. Kaltashov. 2005. Mapping protein energy landscapes with amide hydrogen exchange and mass spectrometry: I. A generalized model for a two-state protein and comparison with experiment. Protein Sci **14**:543–57.

82. Xie, Y., J. Zhang, S. Yin, and J. A. Loo. 2007. Top-down ESI-ECD-FT-ICR mass spectrometry localizes noncovalent protein-ligand binding sites. J Am Chem Soc **128**:14432–3.

83. Yanes, O., A. Nazabal, R. Wenzel, R. Zenobi, and F. X. Aviles. 2006. Detection of noncovalent complexes in biological samples by intensity fading and high-mass detection MALDI-TOF mass spectrometry. J Proteome Res **5**:2711–9.

84. Yao, Z. P., P. Tito, and C. V. Robinson. 2005. Site-specific hydrogen exchange of proteins: insights into the structures of amyloidogenic intermediates. Methods Enzymol **402**:389–402.

85. Yi, S., B. L. Boys, A. Brickenden, L. Konermann, and W. Y. Choy. 2007. Effects of zinc binding on the structure and dynamics of the intrinsically disordered protein prothymosin alpha: evidence for metalation as an entropic switch. Biochemistry **46**:13120–30.

PART VII

EXPRESSION AND PURIFICATION OF IDPS

22

RECOMBINANT PRODUCTION OF INTRINSICALLY DISORDERED PROTEINS FOR BIOPHYSICAL AND STRUCTURAL CHARACTERIZATION

Dmitri Tolkatchev, Josee Plamondon, Richard Gingras, Zhengding Su, and Feng Ni

Biomolecular NMR and Protein Research Group, Biotechnology Research Institute, National Research Council of Canada, Montreal, Quebec, Canada

ABSTRACT

We have developed a procedure for recombinant expression of intrinsically disordered proteins (IDPs) in *Escherichia coli*. It can be used to produce unfolded proteins with a yield comparable to and even higher than that for well-folded proteins. Our strategy is to express IDPs/IUPs fused to a carrier protein that accumulates in inclusion bodies in the cell, where the precipitated fusion protein can be protected from proteolytic degradation. We have successfully prepared a large number of polypeptides using carrier proteins derived from the N-terminal oligonucleotide-binding domain of staphylococcal nuclease. The lengths of these peptides ranged from a few to ~80 amino acids with consistently high yields. In this chapter, we summarize the practical experience and related experimental details used for the production of a number of IDPs/IUPs.

Instrumental Analysis of Intrinsically Disordered Proteins: Assessing Structure and Conformation, Edited by Vladimir Uversky and Sonia Longhi

22.1 INTRODUCTION

There has been a growing interest in the studies of intrinsically disordered or intrinsically unstructured proteins (IDPs/IUPs), ranging from activity assays to detailed physicochemical and structural characterization. Comprehensive investigations require the production of these proteins or polypeptides in highly purified forms at reasonable quantities. The latter is often not a trivial task, considering the intrinsic instability of unfolded polypeptide chains resulting from either aggregation tendencies or proteolytic degradation. As well, only short polypeptides (<30 aa) can be prepared cost-effectively via chemical (solid-phase) synthesis. The yield of synthetic peptides longer than 30–40 residues may be so low that it may become prohibitive for the intended studies or applications. Biophysical and structural (NMR) studies often require uniform enrichment of the IDPs/IUPs with ^{15}N and/or ^{13}C isotopes, which elevates further the costs of chemical synthesis and imposes additional limitations on the length of accessible polypeptide materials. Recombinant overexpression of IDPs/IUPs is also challenging even when protease-deficient *Escherichia coli* strains are utilized to avoid degradation. As a result, elaborate fusion proteins are often tried for a given IDP/IUP to overcome problems related to endogenous degradation and proteolysis.

Since IDPs/IUPs do not have a uniquely folded three-dimensional structure, a natural strategy is to choose a carrier protein to direct them to inclusion bodies in the cell, where the precipitated fusion protein can be protected from proteolytic degradation. Expressed polypeptides can subsequently be extracted from insoluble inclusion bodies under denaturing conditions. Several such protein carriers have been described so far, especially for the production of small peptides (8–11). In our laboratory, we have successfully produced a large number of polypeptides derived from IDPs/IUPs using carrier proteins developed from the N-terminal oligonucleotide-binding domain of staphylococcal nuclease (SNase) (17, 26). In this chapter, we summarize the practical experience and some related experimental details used for the preparation of these representative IDPs/IUPs.

22.2 GENERAL PROPERTIES OF CARRIER PROTEINS

Figure 22.1 shows the DNA and amino acid sequences of carrier proteins used for the production of unfolded polypeptides. The SFC120 protein is derived from the first 117 residues in the N-terminal fragment of SNase. A methionine-free version of SFC120, that is, MFH, was generated by replacing all the methionine residues by leucine (underlined) and with the addition of a poly-His purification tag to the C-terminus. A leucine residue would occupy a van der Waals volume similar to that of methionine, and the two residues would also have a similar hydrophobicity. In general, Met-->Leu mutations are not required for high-yield production of IDPs/IUPs, as will be shown in an example below. These mutations do not visibly affect the ability of the SNase fragment to accumulate in inclusion bodies. However, the methionine-free version of the carrier protein is often advantageous if the IDP/IUP is to be released from the carrier protein by means of cyanogen bromide cleavage. The "wild-type" carrier sequence (i.e., SFC120, Fig. 22.1) contains four methionine residues each of which upon treatment with CNBr would be hydrolyzed on the C-terminal side, producing small peptide fragments. Separation of these peptides from the product of interest may represent a practical challenge.

The carrier protein has an unfolded conformation as judged by the heteronuclear single-quantum coherence (HSQC) spectrum of uniformly ^{15}N-enriched MFH (Fig. 22.2), (which, in this preparation, was fused with an unstructured peptide derived from the C-terminal fragment of hirudin), with a narrow chemical shift dispersion (~1 ppm) of all observable backbone amide proton resonances. In comparison, the full-length SNase displays a backbone amide proton resonance dispersion of ~4 ppm (28). This behavior of the MFH protein is in agreement with observations that C-terminal deletion mutants of SNase have at most residual structures (4, 23). The lack of a well-defined three-dimensional structure for MFH may be the reason why the expressed protein accumulates in inclusion bodies of *E. coli*. Nevertheless, the MFH protein is highly soluble and stays in solution upon solubilization of MFH into a denaturing buffer containing 6–8 M urea, even after urea is later removed by dialysis or other methods of buffer exchange.

Importantly, the solution behavior of isolated SFC120 and MFH is similar to that of their fusions with many other polypeptides of different origin and nature, which facilitates their use as carrier tags for the production of IDPs/IUPs. We have expressed SFC120 fused with affinity purification tags (such as poly-His), protease-specific cleavage sequences, and naturally occurring and designed polypeptides of various lengths. In all cases, the fusion protein was found in inclusion bodies and could be readily resolubilized after extraction and purification. The following examples will outline standard procedures and typical yields and will provide an understanding of some potential difficulties that can arise in the course of production.

22.3 CONSTRUCTION OF EXPRESSION VECTORS

The DNA sequence corresponding to the SFC120 protein was used to construct two expression plasmids, pTSN-6A (see item 1 below) and pMFH (item 2). The plasmid vector pTSN-6A contains SFC120 (Fig. 22.1) as the carrier protein, while pMFH encodes the methionine-free version of SFC120 to facilitate peptide purification upon release from the fusion protein through cyanogen bromide (CNBr) cleavage.

1. The DNA sequence of SFC120 was amplified from the *Staphylococcus aureus* genome by standard polymerase chain reaction (PCR) methods. The

```
sfc120      gca act tca act aaa aaa tta cat aaa gaa cct gcg act tta att
SFC120      Ala Thr Ser Thr Lys Lys Leu His Lys Glu Pro Ala Thr Leu Ile
mfh         gca act tca act aaa aaa tta cat aaa gaa cct gcg act tta att
MFH         Ala Thr Ser Thr Lys Lys Leu His Lys Glu Pro Ala Thr Leu Ile
            1               5                   10                  15

sfc120      aaa gcg att gat ggt gat acg gtt aaa tta atg tac aaa ggt caa
SFC120      Lys Ala Ile Asp Gly Asp Thr Val Lys Leu Met Tyr Lys Gly Gln
mfh         aaa gcg att gat ggt gat acg gtt aaa tta ttg tac aaa ggt caa
MFH         Lys Ala Ile Asp Gly Asp Thr Val Lys Leu Leu Tyr Lys Gly Gln
                            20                  25                  30

sfc120      cca atg aca ttc aga cta tta ttg gtt gat aca cct gaa aca aag
SFC120      Pro Met Thr Phe Arg Leu Leu Leu Val Asp Thr Pro Glu Thr Lys
mfh         cca ttg aca ttc aga cta tta ttg gtt gat aca cct gaa aca aag
MFH         Pro Leu Thr Phe Arg Leu Leu Leu Val Asp Thr Pro Glu Thr Lys
                            35                  40                  45

sfc120      cat cct aaa aaa ggt gta gag aaa tat ggt cct gaa gca agt gca
SFC120      His Pro Lys Lys Gly Val Glu Lys Tyr Gly Pro Glu Ala Ser Ala
mfh         cat cct aaa aaa ggt gta gag aaa tat ggt cct gaa gca agt gca
MFH         His Pro Lys Lys Gly Val Glu Lys Tyr Gly Pro Glu Ala Ser Ala
                            50                  55                  60

sfc120      ttt acg aaa aaa atg gta gaa aat gca aag aaa att gaa gtc gag
SFC120      Phe Thr Lys Lys Met Val Glu Asn Ala Lys Lys Ile Glu Val Glu
mfh         ttt acg aaa aaa ttg gta gaa aat gca aag aaa att gaa gtc gag
MFH         Phe Thr Lys Lys Leu Val Glu Asn Ala Lys Lys Ile Glu Val Glu
                            65                  70                  75

sfc120      ttt gac aaa ggt caa aga act gat aaa tat gga cgt ggc tta gcg
SFC120      Phe Asp Lys Gly Gln Arg Thr Asp Lys Tyr Gly Arg Gly Leu Ala
mfh         ttt gac aaa ggt caa aga act gat aaa tat gga cgt ggc tta gcg
MFH         Phe Asp Lys Gly Gln Arg Thr Asp Lys Tyr Gly Arg Gly Leu Ala
                            80                  85                  90

sfc120      tat att tat gct gat gga aaa atg gta aac gaa gct tta gtt cgt
SFC120      Tyr Ile Tyr Ala Asp Gly Lys Met Val Asn Glu Ala Leu Val Arg
mfh         tat att tat gct gat gga aaa ttg gta aac gaa gct tta gtt cgt
MFH         Tyr Ile Tyr Ala Asp Gly Lys Leu Val Asn Glu Ala Leu Val Arg
                           95                   100                 105

                                                        Eco RI
sfc120      caa ggc ttg gct aaa gtt gct tat gtt tac aaa cct gaa ttc atg
SFC120      Gln Gly Leu Ala Lys Val Ala Tyr Val Tyr Lys Pro Glu Phe Met
mfh         caa ggc ttg gct aaa gtt gct tat gtt tac aaa cct gaa ttg cat
MFH         Gln Gly Leu Ala Lys Val Ala Tyr Val Tyr Lys Pro Glu Leu His
                            110                 115                 120

                                   Bam HI
sfc120 (cloned sequence) tga gga tcc
SFC120 (cloned sequence) Stop

                                Eco RI                              Bam HI
mfh    cat cat cac cat cac gaa ttc atg (cloned sequence) tga gga tcc
MFH    His His His His His Glu Phe Met (cloned sequence) Stop
```

Figure 22.1 DNA and amino acid sequences of SFC120 and MFH carrier proteins. In bold are *EcoRI* and *BamHI* restriction sites used for subcloning an IDP of interest.

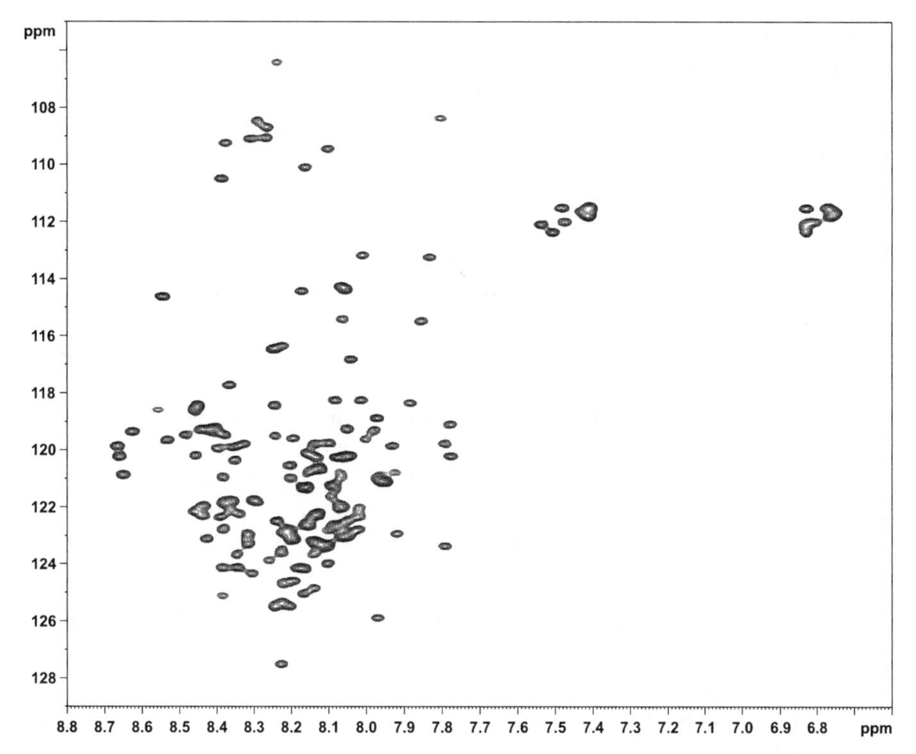

Figure 22.2 [^{15}N,^{1}H]-HSQC spectrum of a uniformly ^{15}N-enriched MFH-peptide construct in 5 mM Tris-HCl, 10% D_2O, pH 6.1, recorded at 298 K, 800 MHz.

restriction enzyme site of *NcoI* was generated at the 5′-end, and the two restriction sites of *EcoRI* and *BamHI* were generated in the 3′-end. The PCR product was double digested with *NcoI* and *BamHI*, and ligated into the pET15M vector, which was modified from the pET-15b vector (Novagen, EMD Biosciences, San Diego, CA) by removing the *EcoRI* site. The resultant plasmid was termed pTSN-6A. The gene of a target IDP/IUP with an appropriate cleavage site (e.g., Met) and a stop codon is inserted into pTSN-6A between the *EcoRI* and *BamHI* sites. A His-tag with six histidines can be placed at either the N-terminal or C-terminal side of the SFC120 carrier protein to simplify purification.

2. To generate MFH, four methionine residues, that is, Met26, Met32, Met65, and Met98, of SFC120 were changed into Leu by site-directed mutagenesis, which was carried out in a Perkin-Elmer (Waltham, MA) Thermocycler™ essentially by following the instruction manual for the QuikChange™ Site-Directed Mutagenesis Kit (Stratagene, La Jolla, CA). Initially, a plasmid with SFC120 was used as a template. The product from a previous PCR was used as a template in the next PCR. The site-directed

mutagenesis reaction was repeated until all four methionines were changed into leucine. The basic procedure utilizes a supercoiled, double-stranded DNA vector with two synthetic complementary oligonucleotide primers containing the desired mutation. The primers are annealed to opposite strands of the vector and are elongated during temperature cycling by means of *Pfu* DNA polymerase. On incorporation of the oligonucleotide primers, a mutated plasmid is generated. Following temperature cycling, the product is treated with *Dpn I* endonuclease to select for mutation-containing synthesized DNA. The *Dpn I* endonuclease is specific for methylated and hemimethylated DNA, and since DNA isolated from almost all *E. coli* strains is dam methylated, it digests parental DNA. The nicked vector DNA incorporating the desired mutations is then transformed into *E. coli* DH5α competent cells. After isolating the mutated plasmid and confirmation of the new sequence, the *EcoRI* and *BamHI* restriction sites were used to incorporate a piece of DNA in a form of two synthetic annealed complementary oligonucleotides encoding six consecutive histidine residues and a multiple cloning site (MCS). The inserted sequence is composed of six parts as follows:

Mfe I site → six histidines → Met → *EcoRI site* → MCS → *BamHI* site.

The resultant plasmid was termed pMFH, and, similar to pTSN-6A, an appropriate gene for IDP/IUP expression can be inserted into pMFH between the *EcoRI* and *BamHI* restriction sites.

22.4 PRODUCTION OF CDC42/RAC INTERACTIVE BINDING (CRIB) MOTIF POLYPEPTIDES

The CRIB domain is a highly conserved sequence motif of ~18 amino acid residues found in a number of cell-signaling kinases and effector proteins. It mediates the interactions of these kinases and effector proteins with Cdc42, an ancient, highly conserved, Rho-type small GTPase of the Ras superfamily serving as a molecular switch of intracellular signaling pathways (3, 18). Previous studies have shown that the consensus CRIB motif itself in the P21-associated kinase (PAK) or the effector molecule Wiskott-Aldrich syndrome protein (WASP) is insufficient for high-affinity binding to Cdc42 and residues C-terminal to the CRIB motif are required for high-affinity interactions (21, 27). Extended CRIB polypeptides were shown to form a complex with Cdc42 via an extended shallow surface (13, 15) and predicted to represent a somewhat disordered region with a propensity to form a structured interface upon complexation (12). This type of interaction is likely to be common among IDPs/IUPs, which acquire defined structures only upon binding to well-structured macromolecular partners. Considerable amounts of CRIB peptides are required for the in-depth characterization of these large peptides both *in vitro* (17, 26) and for cell-based assays (25).

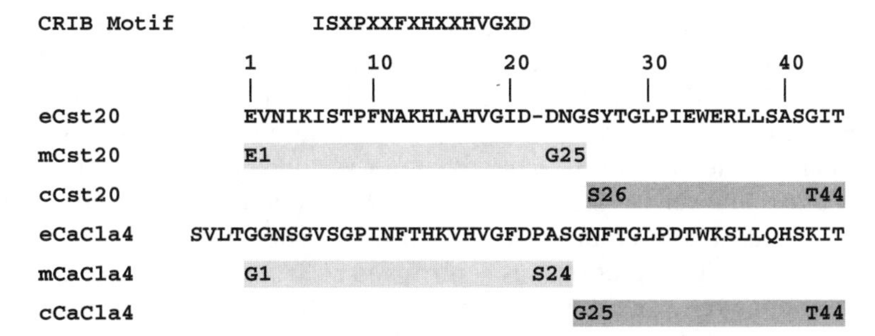

Figure 22.3 Amino acid sequences of recombinantly expressed CRIB fragments of the *Candida albicans* PAK kinases Cst20 and CaCla4.

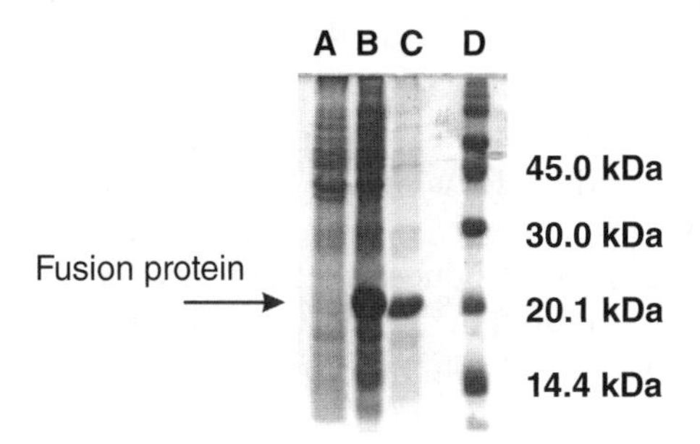

Figure 22.4 Expression and purification of eCst20. The panels show the SDS-PAGE of (A) total cell lysate before IPTG induction, (B) total cell lysate after IPTG induction, (C) fusion protein after ethanol precipitation step, and (D) molecular weight standard.

Figure 22.3 shows a number of CRIB fragments of the *Candida albicans* kinases of the PAK family Cst20 and CaCla4 chosen for recombinant expression. The extended CRIB fragments (or eCRIB peptides) consist of two subfragments, first the minimal, that is, the consensus, CRIB motif (mCRIB peptides), and a stretch of important residues as C-terminal extensions (cCRIB peptides). Both eCRIB peptides, that is, eCst20 and eCaCla4, exhibit nanomolar affinities for Cdc42, similar to those of the full-length kinases. In total, six target peptides, that is, eCst20, mCst20, cCst20, eCaCla4, mCaCla4, and cCaCla4, were chosen for expression and purification using the new carrier system.

A combination of expression/purification methods were used to obtain tens of milligrams of purified CRIB peptides. The extended eCst20 peptide was produced using the pTSN-6A vector (see section 22.3) as a fusion with SFC120 in the absence of a His-tag. High-level expression of the desired fusion protein is evident after the induction of the cell culture with isopropyl-β-thiogalactosidose (IPTG) (Fig. 22.4). The amount of fusion protein can be roughly estimated by sodium dodecyl sulfate-polyacrylamide gel electrophoresis (SDS-PAGE) as ~60% of the total protein content. Virtually pure fusion protein (>95%) was obtained after two rounds of ethanol precipitation (Fig. 22.4). In brief, cell pellets from 1L of culture were resuspended in 60–100mL of 20mM Tris, 100mM NaCl buffer, and 6M urea at pH 8.0 for 4h and then sonicated for 45s on ice. The solution was cleared by centrifugation at 7000 rpm for 20min. An equal volume of 100% ice-cold ethanol was added to the supernatant and the solution allowed to stand at 4°C for 2h. After centrifugation, another equal volume of cold ethanol was added to the supernatant and allowed to stand overnight. The solution was centrifuged at 8500rpm and the pellet containing the fusion protein was subjected to SDS-PAGE analysis. Generally, the pellet formed after the first ethanol addition contained only a small amount of fusion protein, whereas the majority of the fusion protein is precipitated after the second addition of ethanol. If necessary, the pellet was resuspended in 6M urea and applied to a Sep-Pak column (Waters) to remove any remaining impurities. The fusion protein can be lyophilized for further processing.

CNBr cleavage was used to release the eCst20 peptide from the fusion protein. The ethanol-precipitated fusion protein was dissolved in 70% trifluoroacetic acid (TFA) and CNBr added to a final molar ratio of 100:1, and the solution was allowed to stand for ~24h in the dark. CNBr cleavage in 0.1M HCl/6M GdnHCl (20) also produces excellent results. The samples were then diluted with water (×10), lyophilized to dryness, and purified by reversed-phase (RP)-HPLC on a Vydac C18 column (Grace, Deerfield, IL) using an acetonitrile-water gradient containing 0.1% TFA. The released peptide was lyophilized and its identity confirmed by electrospray mass spectrometry. A typical HPLC profile for eCst20 purification is shown in Figure 22.5. In addition to the peptide of interest, the sample contains a number of cleavage products of the carrier and contaminating proteins. Although the profile appears quite crowded, eCst20 can be readily separated from contaminants, and purified to a high degree by passing it through the RP-HPLC for the second time (Fig. 22.5).

The [^{15}N,^{1}H]-HSQC spectrum of ^{15}N-enriched eCst20 is typical of an unstructured polypeptide with poor resonance dispersion (Fig. 22.6A). Upon complexation with activated *Candida* Cdc42 (invisible in the [^{15}N,^{1}H]-HSQC spectra, due to lack of labeling), the resonance dispersion dramatically increases (17), indicating the formation of a defined structure upon binding (Fig. 22.6B).

To facilitate larger-scale productions, it is often desirable to utilize methionine-free version of the carrier protein (i.e., in the form of the pMFH vector; see section 22.3) fused with a His-tag for affinity purification. The procedure of fusion protein purification follows a standard protocol for His-tagged proteins under denaturing conditions. A low pH (<4.5) buffer is used for the elution of the fusion protein. Subsequent removal of denaturant is not necessary, and overnight CNBr cleavage can be performed in 0.1N HCl and in 6M GdnHCl (20). Figure 22.5C shows HPLC purification of CNBr-cleaved cCaCla4 on a Vydac C18 column using an acetonitrile-water gradient containing 0.1% TFA. The peptide is well separated from the MFH carrier and unprocessed fusion protein and can be used for physicochemical studies

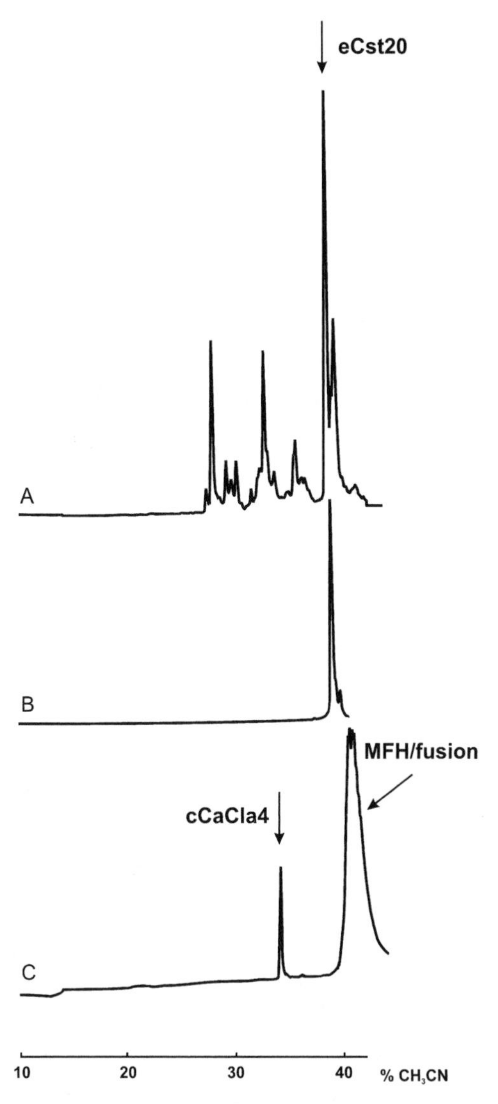

Figure 22.5 RP-HPLC purification of CRIB peptides after CNBr cleavage. (A) first round of HPLC purification of eCst20; (B) second round of HPLC purification of eCst20; (C) HPLC purification of cCaCla4. Arrows show positions of CRIB peptides and MFH/fusion proteins.

without further purification. In the event of partial overlap between the peptide of interest and MFH in terms of HPLC elution profiles, an RP column of slightly different selectivity can be used to reach a higher degree of peptide purity. We successfully utilized a pair of Vydac C18 RP silica columns (one based on polymeric and another on monomeric surface chemistry) to resolve and produce high yields of the cCst20 peptide.

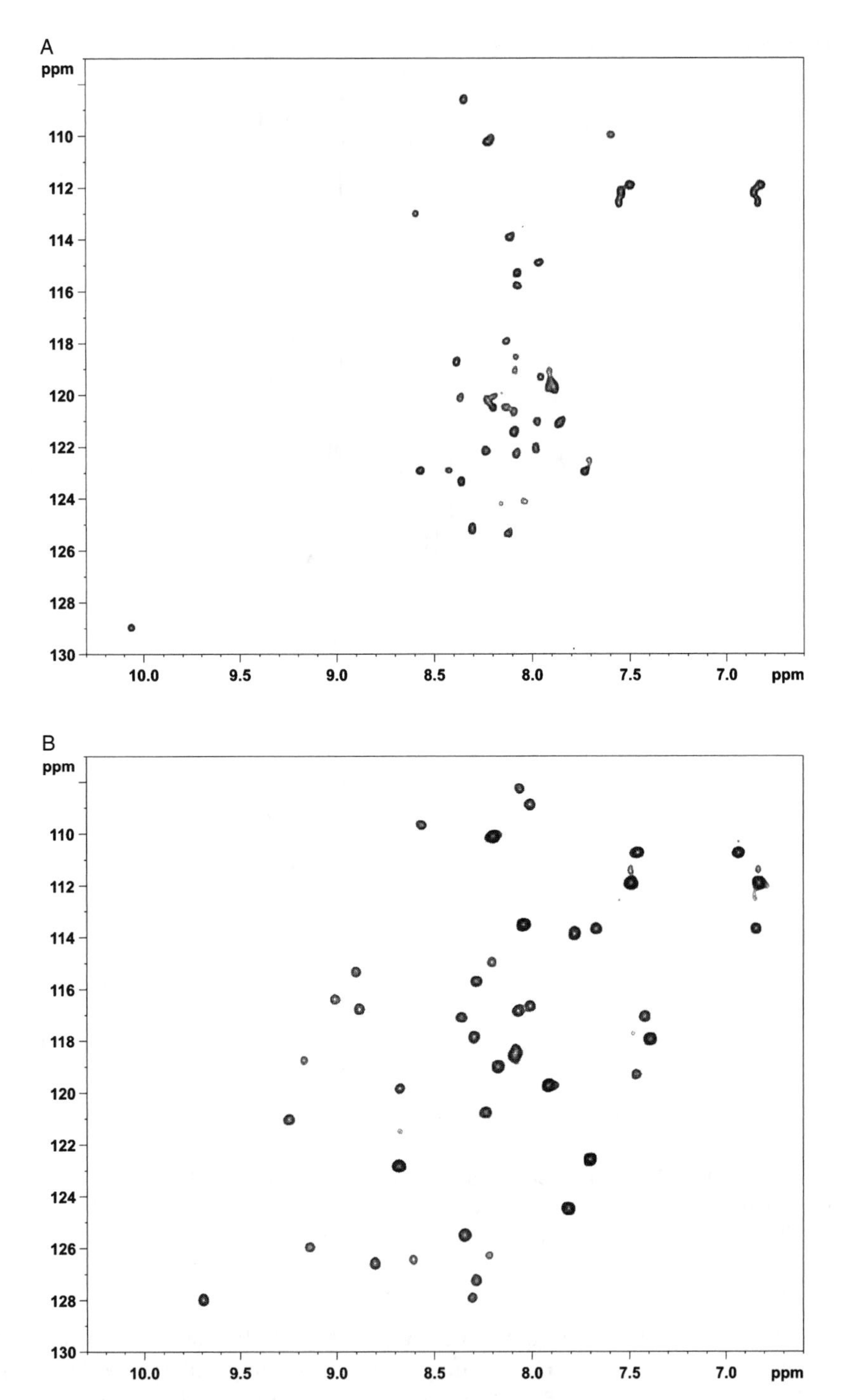

Figure 22.6 [^{15}N,^{1}H]-HSQC spectra of uniformly ^{15}N-enriched eCst20 peptide (A) alone and (B) in complex with Cdc42. The spectra were obtained at 298 K, 800 MHz in 50 mM sodium phosphate, 50 mM NaCl, 5 mM $MgCl_2$, and 90% H_2O/10% D_2O at pH 6.0.

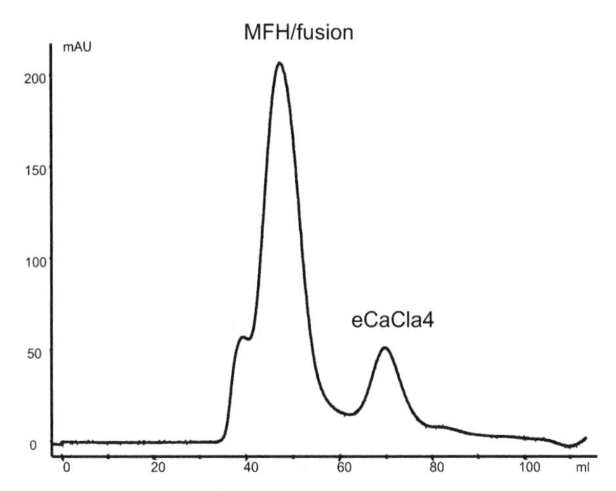

Figure 22.7 Isolation of eCaCla4 peptide on a HiPrep 16/60 Sephacryl S-100 high-resolution gel filtration column in 4 M urea and 0.1 N HCl at a flow rate of 1 mL/min.

HPLC profile of the eCaCla4 peptide overlapped with that of the MFH carrier, and RP purification was difficult as a result. In addition, we observed some precipitation out of the cleaved MFH-eCaCla4 solution on a desalting Sep-Pak C18 cartridge (Waters). In order to remove the remaining CNBr after cleavage reactions, we dialyzed the reaction mixture against 4 M urea/0.1 N HCl. To purify the eCaCla4 peptide, we chose to use size exclusion chromatography, which should separate the peptide from the uncleaved fusion protein and the carrier with molecular weights of ~5, ~19.5, and ~14.5 kDa, respectively. Since the carrier protein contained a His-tag, it was also possible to separate the peptide from the fusion and carrier proteins by Ni-agarose affinity chromatography, which would lead to a dilution of the peptide into an inconveniently large volume. The eCaCla4 peptide was therefore processed first using a HiPrep™ 16/60 Sephacryl™ S-100 high-resolution gel filtration column (Amersham Biosciences, Uppsala, Sweden) and further purified on HPLC. Figure 22.7 shows the fast protein liquid chromatography (FPLC) profile for the eCaCla4 purification.

The above-described examples demonstrate that the MFH/SFC120 carriers are highly expressed as solubilizable proteins compatible with a wide range of purification techniques, which can be used to isolate considerable amounts of a polypeptide of interest.

22.5 THE PROSEGMENT OF HUMAN CATHEPSIN B

The 62-residue propeptide of cathepsin B functions to regulate the degradative apparatus within the cell mediated by lysosomal cathepsin B (1). In the acidic environment of the lysosome, the interaction between the prosegment and the

```
cggagcaggccctctttccatcccgtgtcggatgagctggtcaactatgtcaacaaacgg
 R  S  R  P  S  F  H  P  V  S  D  E  L  V  N  Y  V  N  K  R
aataccacgtggcaggccgggcacaacttctacaacgtggacatgagctacttgaagagg
 N  T  T  W  Q  A  G  H  N  F  Y  N  V  D  M  S  Y  L  K  R
ctatgtggtaccttcctgggtgggcccaagccaccccagagagttatgtttaccgaggac
 L  C  G  T  F  L  G  G  P  K  P  P  Q  R  V  M  F  T  E  D
ctgaag
 L  K
```

Figure 22.8 DNA and amino acid sequences of human cathepsin B propeptide.

protease domain weakens, and the proregion becomes susceptible to proteolysis, leading to the release of active cathepsin B. Uncontrolled activation of cathepsin B (via release of the inhibitory propeptide) is implicated in a number of human diseases including cancers and inflammatory conditions (5, 14, 19, 29). The isolated cathepsin B propeptide is intrinsically unstructured (30) and appears to acquire some level of folding only in complex with cathepsin B (2).

The DNA of the human cathepsin B propeptide (see Fig. 22.8) was incorporated between *BamHI* and *EcoRI* restriction sites of the pHSN expression vector. This vector encodes a version of the SFC120 carrier protein with a single M65L mutation and a His-tag in the C-terminus. In addition, it incorporates a tetrapeptide sequence Phe-Asn-Pro-Arg before the cloning site for efficient thrombin cleavage (16). The construct was transformed into the *E. coli* strain BL21 (DE3) to overexpress the fusion protein. The cells were grown in Luria-Bertain (LB) medium supplemented with 50 μg/mL of ampicillin at 37°C and induced with 1 mM IPTG at the late exponential phase ($OD_{600nm} \approx 0.8$) for at least 4 h. Production of the ^{15}N-labeled fusion protein was performed in M9 minimal medium using ^{15}N-$(NH_4)_2SO_4$ as the sole nitrogen source. The harvested cells were resuspended in 50 mM phosphate, pH 8.0, 5 mM Tris, 50 mM NaCl, and 6 M urea for 3 h on ice. The suspension was cleared by centrifugation, and the supernatant containing the fusion protein was loaded on a Ni-NTA affinity agarose (Qiagen, Germantown, MD) column and eluted in the same buffer in the presence of 200 mM imidazole.

To release the propeptide, the fusion protein was cleaved by thrombin for 30 min to 2 h at room temperature. An aliquot from the reaction mixture was taken every 15 min to evaluate the cleavage process by SDS-PAGE. For bulk purification, the reaction was terminated by urea denaturation. The undigested fusion protein and the carrier protein were removed by Ni-NTA agarose. The flow-through was subjected to HPLC purification on an RP C18 Vydac semipreparative column at a flow rate of 5 mL/min. The ^{1}H-^{15}N HSQC spectrum of the propeptide is shown in Figure 22.9.

22.6 A BRIEF SUMMARY OF OTHER PRODUCTION EXAMPLES

The SFC120/MFH carrier has been used to produce a number of other polypeptides (Table 22.1), including a C-terminal fragment of hirudin and the

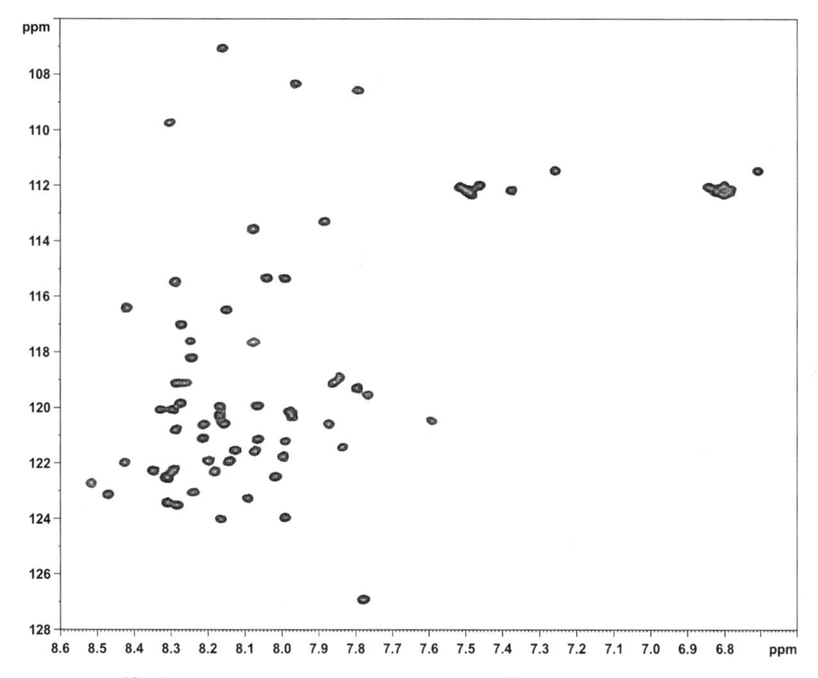

Figure 22.9 [^{15}N,^{1}H]-HSQC spectra of uniformly ^{15}N-enriched human cathepsin B propeptide. The spectra were obtained in 50 mM sodium acetate, pH 5.5, at 298 K, 800 MHz.

TABLE 22.1 Examples of the Expressed Peptides Using SFC120/MFH Carrier Proteins

Peptide	Size (AA)	Yield (mg/L)	Purity	Isotope Form	Medium
eCaCla4 (eCRIB)	48	>15	HPLC (>98%)	^{15}N	M_9
eCaCla4 (eCRIB)	48	~10	HPLC (>98%)	^{15}N/^{13}C/^{2}H	M_9
mCaCla4 (mCRIB)	22	>15	HPLC (>98%)	^{15}N	M_9
mSte20 (mCRIB)	22	>15	HPLC (>98%)	^{15}N	M_9
cCaCla4 (cCRIB)	22	>10	HPLC (>98%)	^{15}N	M_9
cSte20 (cCRIB)	22	>10	HPLC (>98%)	^{15}N	M_9
Hirudin$^{47–65}$	18	~20	HPLC (>98%)	Nonlabeled	LB
Propeptide of human cathepsin B	64	10	HPLC (>98%)	^{15}N	M_9
N-terminal fragment of parathyroid hormone	33	10	HPLC (>98%)	^{15}N	M_9
EphrinB peptide	33	>15	HPLC (>98%)	^{15}N	M_9

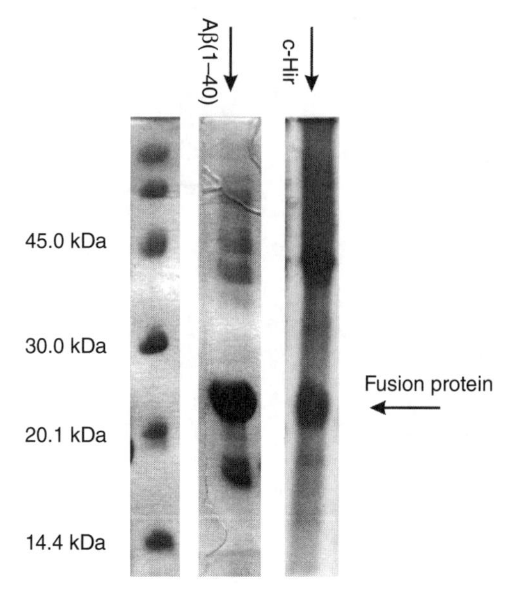

Figure 22.10 SDS-PAGE of *E. coli* cell pellets expressing Aβ (1–40) and a C-terminal fragment of hirudin (C-Hir).

amyloid β (1–40) (Aβ [1–40]) peptide. Hirudin, a protein of 65 residues from the European leech *Hirudo medicinalis*, is an extremely potent inhibitor of thrombin with a K_i at the subpicomolar level. Interaction between hirudin and thrombin involves an extended interface (22), which can be represented by a bivalency model (24). The N-terminal domain of hirudin is well folded, and it blocks the active site of thrombin. The C-terminal hirudin tail is largely unfolded in the free state, but it binds to the anion-binding exosite I of thrombin. Aβ (1–40) peptide is one of two major components of senile plaques and cerebrovascular deposits in Alzheimer's disease. Monomeric soluble form of Aβ (1–40) has a random conformation with some propensity for the formation of β-sheets (7). Yields of the fusion proteins for both of these polypeptides, that is, C-terminal fragment of hirudin and Aβ (1–40), were very high, >130 and 60 mg/L, respectively (Fig. 22.10). Production of the C-terminal fragment of hirudin followed standard procedures of fusion protein purification, peptide release by CNBr cleavage and HPLC separation, producing tens of milligrams of the peptide. However, processing of the fusion protein containing Aβ (1–40) illustrates a contrasting experience sometimes encountered for the production of IDPs/IUPs. The sequence of Aβ (1–40) contains a Met residue, and use of the preferred CNBr cleavage was not possible. To release Aβ (1–40) from the fusion protein, an enterokinase restriction site was inserted between the C-terminal His-tag of the carrier protein and the N-terminus of the peptide. Enterokinase is a serine protease with a proteolytic specificity for sequences -Asp-Asp-Asp-Asp-Lys-|-X for cleavage after Lys, which is largely independent of the nature of the P1′ (X) residues. Such a specificity of enterokinase makes it possible to remove the carrier protein while leaving no additional N-terminal residues on the designed polypeptide (6).

The fusion protein with Aβ (1–40) was desalted on a Sep-Pak C18 column by washing with 10% acetonitrile and 0.1% TFA, elution with 50% acetonitrile, 0.1% TFA, and freeze-drying. The powder of the fusion protein was resuspended in 10 mM Tris pH 6.8 and incubated with enterokinase at a concentration of 5 U per 1 mg of fusion protein at room temperature for 16 h until completion. The reaction mixture was desalted and freeze-dried for further purification on HPLC. Unfortunately, repeated attempts at HPLC purification did not produce any substantial amount of the purified Aβ (1–40) peptide. The Aβ (1–40) peptide was found to have aggregated/precipitated on the resin when a Ni-NTA agarose column was used to remove the remaining fusion and carrier proteins.

In conclusion, the described expression system is simple and appears to be universal, allowing adaptations to various purification procedures and solution properties of a particular polypeptide of interest. It can be used to produce unfolded proteins (IDPs/IUPs) with a yield comparable to and even higher than that for well-folded proteins. Chances of encountering "difficult" sequences are relatively low, and they are usually related to practical difficulties in handling unfolded peptides upon their separation from the carrier protein. Table 22.1 summarizes the small portion of sequences successfully expressed in isotopically labeled forms. Using SFC120/MHF carriers, we also prepared unfolded proteins up to ~80 residues long and did not observe significant drops in expression yields related to the size of the target IDP/IUP. From our experience, it is possible to scale up production by using bioreactors and/or, for shorter sequences, building expression vectors as tandem repeats to increase yields of production. In the latter case, treatment with CNBr releases several smaller peptides (separated by methionine residues in the fusion protein), which can also be used to generate equimolar mixtures of related peptides for ligand screening applications.

ACKNOWLEDGMENTS

We thank Dr. Ping Xu for recording the NMR spectra and Ping Wang for NMR characterization of the MFH carrier protein (Fig. 22.2).

REFERENCES

1. Barrett, A. J., and H. Kirschke. 1981. Methods Enzymol **80** (Pt C):535–61.
2. Cygler, M., J. Sivaraman, P. Grochulski, R. Coulombe, A. C. Storer, and J. S. Mort. 1996. Structure **4**:405–16.
3. Erickson, J. W., and R. A. Cerione. 2001. Curr Opin Cell Biol **13**:153–7.

4. Feng, Y., D. Liu, and J. Wang. 2003. J Mol Biol **330**:821–37.
5. Fox, T., E. de Miguel, J. S. Mort, and A. C. Storer. 1992. Biochemistry **31**:12571–6.
6. Hosfield, T., and Q. Lu. 1999. Anal Biochem **269**:10–6.
7. Hou, L., H. Shao, Y. Zhang, H. Li, N. K. Menon, E. B. Neuhaus, J. M. Brewer, I. J. Byeon, D. G. Ray, M. P. Vitek, T. Iwashita, R. A. Makula, A. B. Przybyla, and M. G. Zagorski. 2004. J Am Chem Soc **126**:1992–2005.
8. Jones, D. H., E. H. Ball, S. Sharpe, K. R Barber, and C. W. Grant. 2000. Biochemistry **39**:1870–8.
9. Koerner, T. J., J. E. Hill, A. M. Myers, and A. Tzagoloff. 1991. Methods Enzymol **194**:477–90.
10. Kuliopulos, A. and C. T. Walsh. 1994. J Am Chem Soc **116**:4599–607.
11. Majerle, A., J. Kidric, and R. Jerala. 2000. J Biomol NMR **18**:145–51.
12. Mohan, A., C. J. Oldfield, P. Radivojac, V. Vacic, M. S. Cortese, A. K. Dunker, and V. N. Uversky. 2006. J Mol Biol **362**:1043–59.
13. Morreale, A., M. Venkatesan, H. R. Mott, D. Owen, D. Nietlispach, P. N. Lowe, and E. D. Laue. 2000. Nat Struct Biol **7**:384–8.
14. Mort, J. S., and D. J. Buttle. 1997. Int J Biochem Cell Biol **29**:715–20.
15. Mott, H. R., D. Owen, D. Nietlispach, P. N. Lowe, E. Manser, L. Lim, and E. D. Laue. 1999. Nature **399**:384–8.
16. Ni, F., Y. Zhu, and H. A. Scheraga. 1995. J Mol Biol **252**:656–71.
17. Osborne, M. J., Z. Su, V. Sridaran, and F. Ni. 2003. J Biomol NMR **26**:317–26.
18. Pirone, D. M., D. E. Carter, and P. D. Burbelo. 2001. Trends Genet **17**:370–3.
19. Rao, J. S. 2003. Nat Rev Cancer **3**:489–501.
20. Rodriguez, J. C., L. Wong, and P. A. Jennings. 2003. Protein Expr Purif **28**:224–31.
21. Rudolph, M. G., P. Bayer, A. Abo, J. Kuhlmann, I. R. Vetter, and A. Wittinghofer. 1998. J Biol Chem **273**:18067–76.
22. Rydel, T. J., K. G. Ravichandran, A. Tulinsky, W. Bode, R. Huber, C. Roitsch, and J. W. Fenton. 1990. Science **249**:277–80.
23. Shortle, D., and A. K. Meeker. 1989. Biochemistry **28**:936–44.
24. Song, J., and F. Ni. 1998. Biochem Cell Biol **76**:177–88.
25. Su, Z., H. Li, Y. Li, and F. Ni. 2007. Chem Biol **14**:1273–82.
26. Su, Z., and F. Ni. 2003. Novel Fusion Proteins for the Production of Recombinant Peptides. United States Application No. 60/402,075, 10/524,053, Canadian Application No. 2,495,145, European Common Market Application No. 03783874.5, Patent Cooperation Treaty Application PCT/CA0301197.
27. Thompson, G., D. Owen, P. A. Chalk, and P. N. Lowe. 1998. Biochemistry **37**:7885–91.
28. Wang, J. F., A. P. Hinck, S. N. Loh, D. M. LeMaster, and J. L. Markley. 1992. Biochemistry **31**:921–36.
29. Yan, S., and B. F. Sloane. 2003. Biol Chem **384**:845–54.
30. Yu, Y., W. Vranken, N. Goudreau, E. de Miguel, M. C. Magny, J. S. Mort, R. Dupras, A. C. Storer, and F. Ni. 1998. FEBS Lett **429**:9–16.

23

LARGE-SCALE IDENTIFICATION OF INTRINSICALLY DISORDERED PROTEINS

VLADIMIR N. UVERSKY,[1,2] MARC S. CORTESE,[1] PETER TOMPA,[3] VERONIKA CSIZMOK,[3] AND A. KEITH DUNKER[1]

[1]*Institute for Intrinsically Disordered Protein Research, Center for Computational Biology and Bioinformatics, and Department of Biochemistry and Molecular Biology, Indiana University School of Medicine, Indianapolis, IN*

[2]*Institute for Biological Instrumentation, Russian Academy of Sciences, Pushchino, Moscow Region, Russia*

[3]*Institute of Enzymology, Biological Research Center, Hungarian Academy of Sciences, Budapest, Hungary*

ABSTRACT

A method to enrich cell extracts in totally unfolded proteins was investigated. A literature search revealed that 14 of 29 proteins isolated by their failure to precipitate during perchloric acid (PCA) or trichloroacetic acid (TCA) treatment were also shown experimentally to be totally disordered. A near 100,000-fold reduction in yield was observed after 5% or 9% PCA treatment of total soluble *Escherichia coli* protein. Despite this huge reduction, 158 and 142 spots were observed from the 5% and the 9% treated samples, respectively, on silver-stained two-dimensional (2D) sodium dodecyl sulfate polyacrylamide gel electrophoresis (SDS-PAGE) gels loaded with 10 μg of protein. Treatment with 1% PCA was less selective with more visible spots and a greater than threefold higher yield. A substantial yield of unprecipitated protein was

Instrumental Analysis of Intrinsically Disordered Proteins: Assessing Structure and Conformation, Edited by Vladimir Uversky and Sonia Longhi

obtained after 3% TCA treatment, suggesting that the common use of TCA precipitation prior to 2D gel analysis may result in loss of unstructured protein due to their failure to precipitate. Our preliminary analysis suggests that treating total protein extracts with 3–5% PCA and determining the identities of soluble proteins could be the starting point for uncovering unfoldomes (the complement of unstructured proteins in a given proteome). The 100,000-fold reduction in yield and concomitant reduction in number of proteins achieved by 5% PCA treatment produced a fraction suitable for analysis in its entirety using standard proteomic techniques. In this way, large numbers of totally unstructured proteins could be identified with minimal effort.

A method of different logic has been recently presented that not only allows the identification of intrinsically disordered proteins (IDPs) but at the same time also provides information about the structural state of an uncharacterized protein. The technique combines a native gel electrophoresis of heat-treated proteins with a second, denaturing gel containing 8 M urea. The heat treatment precipitates most globular proteins and leaves IDPs in solution, providing the first evidence for the lack of a well-defined structure. The subsequent 8 M urea electrophoresis is rationalized by the usual structural indifference of IDPs to chemical denaturation. As urea is uncharged and IDPs are just as "denatured" in 8 M urea as under native conditions, they run the same distance in the second dimension as in the first and end up along the diagonal. However, the heat-stable globular proteins unfold in 8 M urea, which make them run slower, and accumulate above the diagonal. This deviation in their running ensures the separation of IDPs from the globular proteins and enables the MS-based identification of IDPs.

23.1 INTRODUCTION

Intrinsically unstructured/disordered proteins (IDPs/IUPs) and protein domains lack a well-defined three-dimensional (3D) structure under physiological conditions. They are highly abundant in nature. Bioinformatics analyses of whole genomes using disorder predictors indicate that >50% of all eukaryotic genes encode proteins that contain lengthy disordered segments (>30 residues) (18, 76). Furthermore, many proteins in various genomes are likely to be wholly disordered; for example, >20% of eukaryotic proteins are expected to be mostly disordered (49). As the number of IDPs on various proteomes is very large, it makes sense to introduce the unfoldome and unfoldomics concepts. In fact, the suffix "-om-" is used to reflect totality. For example, all the genes in a given organism constitute its genome; all the proteins form its proteome; whereas the corresponding fields of studies are termed genomics and proteomics, respectively. Several "-omes" were proposed and accepted in the biology, including genome, proteome, interactome, metabolome, transcriptome, diseasome, toxicogenome, nutrigenome, cytome, oncoproteome, epitome, and glycome. It is believed that ome and omics concepts might add a new layer of knowledge, especially when a scientist is dealing with the data produced by the large-scale studies, including the high throughput experiments and the computational/bioinformatics analyses of the large data sets. The idea of unfoldomics is built on all that. Unfoldomics is a field that studies unfoldome. Unfoldome is attributed to a portion of proteome that includes a set of IDPs (also known as natively unfolded proteins, therefore unfoldome), their functions, structures, interactions, evolution, and so on. As IDPs are highly abundant in nature, have amazing structural variability, and possess a very wide variety of functions, we decided to name this realm of proteins unfoldome, thinking of the suffix "-om-" as a reflection of the totality of this phenomenon.

The above-mentioned high abundance of IDPs in various proteomes is determined by the fact that these proteins possess a number of crucial biological roles, as structural disorder imparts advantages in many nonconventional functions. These roles include regulation of cell division, transcription and translation, signal transduction, protein phosphorylation, storage of small molecules, chaperone action, and regulation of the self-assembly of large multiprotein complexes, such as the ribosome. These are complementary to the common catalysis and transport activities of proteins with well-defined, stable 3D structures (6, 13–17, 19, 20, 22, 36, 52, 59, 60, 62, 71, 75, 78–80). Although unbound IDPs are disordered in solution, they often perform their biological functions by binding to other biomolecules. This binding involves a disorder-to-order transition in which IDPs adopt a highly structured conformation upon binding to their biological partners (4, 21, 43–45, 47, 50, 74). In this way, IDPs play diverse roles in regulating the function of other biomolecules and in promoting the assembly of supramolecular complexes. Furthermore, because sites within their polypeptide chains are highly accessible, IDPs can undergo extensive posttranslational modifications, such as phosphorylation, acetylation, and/or ubiquitination (also sumoylation, neddylation), allowing for modulation of their biological activity or function.

Disorderedness is now linked to amino acid sequence. IDPs (and regions) exhibit low sequence complexity and are generally enriched in most polar and charged residues, and are depleted of hydrophobic residues (other than proline). These features are consistent with their inability to fold into globular structures and form the basis of computational tools for disorder prediction (11, 12, 16, 23, 52, 67). IDPs possess a number of very distinctive properties that are implemented for their discovery. This includes, but is not limited to, sensitivity to proteolysis (35), aberrant migration during SDS-PAGE (37), insensitivity to denaturing conditions (54), as well as definitive disorder characteristics visualized by CD spectropolarimetry, NMR spectroscopy, small-angle X-ray scattering, hydrodynamic measurement, fluorescence, Raman, and infrared spectroscopies (9, 53).

Structurally, IDPs range from completely unstructured polypeptides to extended partially structured forms to compact disordered ensembles containing substantial secondary structure (16, 17, 19, 62, 64). Many proteins contain mixtures of ordered and disordered regions. Here, we are focusing on proteins that are mostly, if not totally, disordered. The atypical conformational behavior of extended IDPs, with their "turn out" response to acidic pH, and high temperature and insensitivity to high concentrations of strong denaturants, is described in chapter 18. The reasons for this anomalous behavior are in the peculiarities of IDP amino acid sequences and in their lack of ordered 3D structure.

In this chapter, we show how the mentioned structural features of extended IDPs and their specific conformational behavior can be utilized in the large-scale identification of these important members of the protein kingdom. To this end, three methods for uncovering the unfoldomes are introduced. The first method is based on the finding that many proteins that fail to precipitate during perchloric acid or trichloroacetic acid treatment were IDPs (5). The second method utilizes the fact that IDPs possess high resistance toward the aggregation induced by the heat treatment (5, 8, 25). The third method utilizes heat treatment coupled with a novel 2D gel methodology (which is a combination of native and 8 M urea electrophoresis) to identify IDPs in cell extracts (8). It is anticipated that these methodologies, combined with the highly sensitive MS-based techniques, can be used for the detection and functional characterization of IDPs in various proteomes.

23.2 ENRICHMENT OF CELL EXTRACTS IN EXTENDED IDPS BY ACID TREATMENT

23.2.1 Basic Principles of the Approach

Importantly, extended IDPs, which are typically characterized by high percentages of charged residues, do not undergo large-scale structural changes at low pH. Thus, such proteins are likely to remain soluble under these extreme conditions (63, 68). For structured proteins, the protonation of negatively charged side chains leads to charge imbalances that disrupt salt bridges (10) and causes the disassociation of subunits (48) and cofactors (30), often leading to the random coil state (24, 26–28). Unlike intrinsically unfolded proteins, the acidic pH-induced random coils of structured proteins typically contain a large number of hydrophobic groups, and the exposure of these normally buried hydrophobic residues usually leads to aggregation and precipitation.

Furthermore, some structured proteins adopt a less open conformation in acid pH, often referred to as the molten globular or "A" state (24, 26, 58). This state may be "sticky." For example, proteins that rely on buried salt bridges are typically among those that aggregate and precipitate upon transition to the A state (57). The aggregation and precipitation of the A state can also depend on the anions in solution (1, 65, 66, 73). Interestingly, because of the influence of anions on A state stability, not all acids are capable of quantitatively precipitating such proteins (58). Note also that some structured proteins adopt nonsticky A states and thus remain in solution (55). Finally, some ordered proteins might be exceptionally stable and simply remain folded and soluble even at extremely low pH (57, 61).

Literature search generated a set of 29 proteins that had been isolated by treatment with PCA or TCA (5). Of these, 9 had not been structurally characterized, 6 were structured, and 14 were experimentally determined to be totally unstructured. This suggests that at least 50% of the proteins isolated by virtue of their resistance to PCA or TCA could be expected to be totally unstructured. This estimate could be higher if any of the nine uncharacterized proteins turn out to be unstructured.

To gain information on abundance of intrinsic disorder in acid-soluble proteins, their sequences were analyzed using two binary predictors of intrinsic disorder, charge-hydropathy plot (CH-plot) (67) and cumulative distribution function analysis (CDF) (49). Both of these methods perform binary classification of whole proteins as either mostly disordered or mostly ordered, where mostly ordered indicates proteins that contain more ordered residues than disordered residues and mostly disordered indicates proteins that contain more disordered residues than ordered residues (49).

Ordered and disordered proteins plotted in CH space can be separated to a significant degree by a linear boundary, with proteins located above this boundary line being disordered and with proteins below the boundary line being ordered (67). CDF analysis summarizes the per-residue disorder predictions by plotting Predictor of Native Disordered Regions (PONDR®) scores against their cumulative frequency, which allows ordered and disordered proteins to be distinguished based on the distribution of prediction scores (49). At any given point on the CDF curve, the ordinate gives the proportion of residues with a PONDR score less than or equal to the abscissa. CDF curves for PONDR VL-XT predictions begin at the point 0.0 and end at the point 1.1 because PONDR VL-XT predictions are defined only in the range 0–1, with values less than 0.5 indicating a propensity for order and values greater than or equal to 0.5 indicating a propensity for disorder. The optimal boundary that provided the most accurate order–disorder classification was determined, and it has been shown that seven boundary points located in the twelfth through the eighteenth bin provided the optimal separation of the ordered and disordered protein sets (49). For CDF analysis, order–disorder classification is based on whether a CDF curve is above or below a majority of boundary points (49).

Results of these analyses for the 24 proteins isolated with acid treatment and with known sequences are shown in Figure 23.1, which reflects an excellent correlation between experiment and prediction. The majority of proteins experimentally shown to be structured or unfolded were predicted to be ordered or intrinsically disordered, respectively, by both predictors. Additionally, three of four experimentally uncharacterized proteins were predicted to be wholly disordered by both classifiers. Thus, a combination of experimental and computational approaches suggested that ~70% of acid-

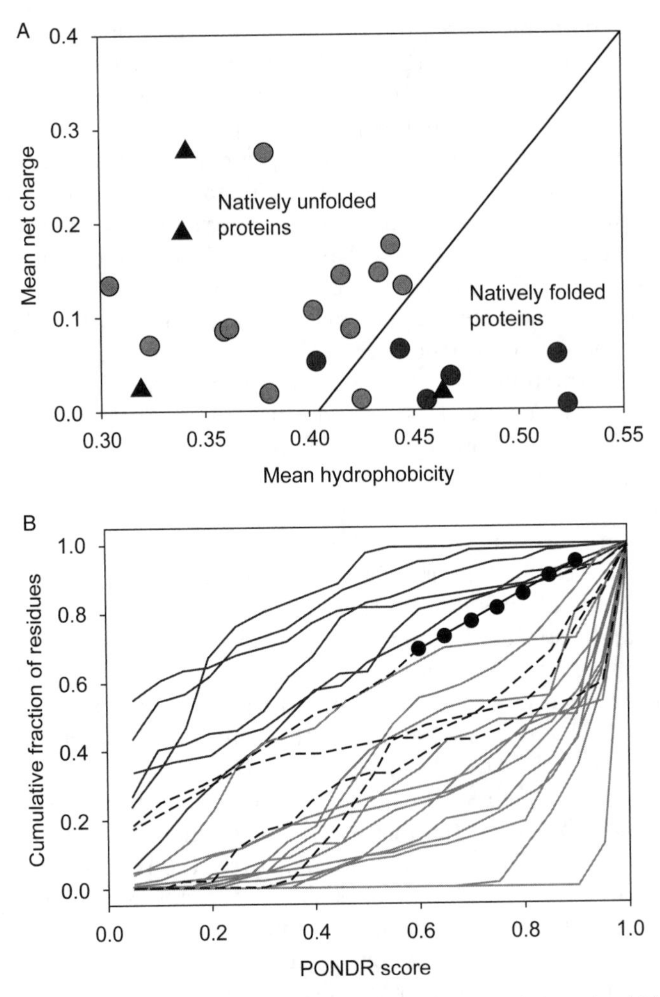

Figure 23.1 Disorder prediction of acid-soluble proteins. (A) CH plot (67). The black line represents the order–disorder boundary calculated according to Oldfield et al. (50). The light and dark gray circles correspond to proteins experimentally shown to be disordered and ordered, respectively. The black triangles represent four proteins for which structural information is not available. (B) CDF analysis (50). Data for 6 wholly ordered proteins (dark gray lines), 14 wholly disordered proteins (light gray lines), and 4 structurally noncharacterized proteins (black dashed lines), and the order–disorder boundary (black line with circles) are plotted. Reprinted with permission from Cortese et al. (5). Copyright 2009 American Chemical Society.

soluble proteins that were isolated based on their resistance to PCA or TCA could be expected to be totally unstructured (5).

Based on these observations, it was suggested that acid treatment should lead to substantial enrichment of intrinsically unfolded proteins in the soluble fraction. Therefore, it was proposed that the intrinsic properties of IUPs can be exploited to develop standard protocols to study them on a proteomic scale (5). That is, totally unstructured proteins should be separable from structured proteins by their intrinsic indifference to denaturing conditions that originates from their lack of tertiary structure. Initial studies to test this proposal are described in Reference 5 and are outlined below.

23.2.2 Acid Treatment-Based Enrichment of Cell Extracts in Extended IDPs

As noted above, PCA and TCA precipitation yielded several structured proteins that remained soluble (four and two, respectively), suggesting that neither acid is totally selective for IUPs (5).

Treatment with 1% PCA resulted in a reduction in total protein of ~30,000-fold when compared with the total soluble extract. For all higher PCA concentrations, the reduction was at least threefold greater (~80,800). Apparently, 3% PCA was sufficient to denature and precipitate all nonresistant proteins because higher PCA concentrations did not result in further yield reductions (5). Treatment with 3% TCA resulted in a yield similar to 1% PCA. However, an increased number and a different set of proteins were isolated when compared with 1%, 5%, or 9% PCA (Fig. 23.2A–C,E).

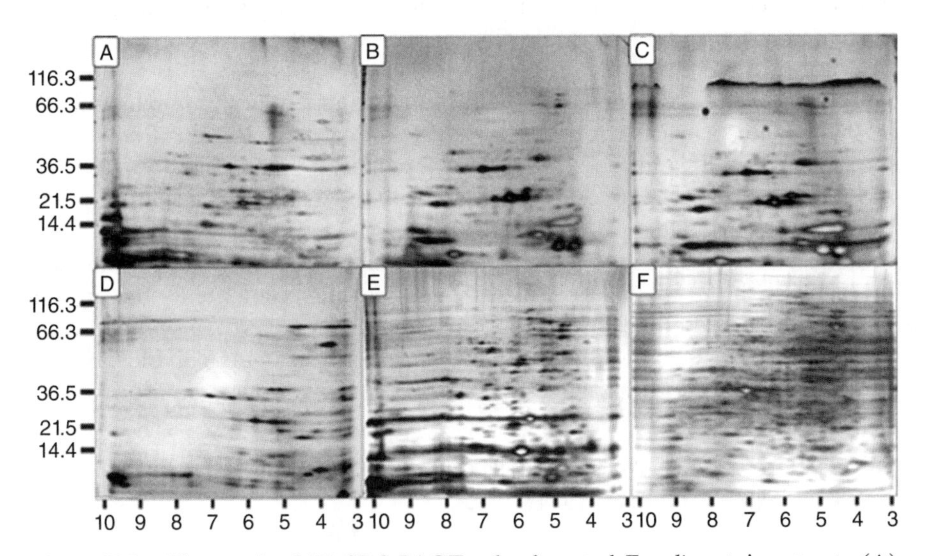

Figure 23.2 Silver-stained 2D SDS-PAGE gels of treated *E. coli* protein extracts: (A) 1% PCA soluble protein, (B) 5% PCA soluble protein, (C) 9% PCA soluble protein, (D) protein remaining soluble after heat treatment, (E) 3% TCA soluble protein, and (F) untreated total protein. All gels were loaded with ~10 μg of protein. Horizontal scales are approximate isoelectric pH. Vertical scales are molecular mass in kDa. Reprinted with permission from Cortese et al. (5). Copyright 2009 American Chemical Society.

The acid-soluble fractions from the *Escherichia coli* extracts were visualized using 2D SDS-PAGE. These extracts revealed that a substantial number of *E. coli* proteins were resistant to acid denaturation and concomitant precipitation (Fig. 23.2). The distribution and number of individual spots on the 5% and 9% PCA gels were similar, suggesting that, for the most part, the same subset of proteins was isolated using either condition. However, this correlation was not evident when comparing the 1% PCA with either the 5% or 9% PCA gels (Fig. 23.2A–C). This is consistent with the observed fourfold higher yield of extracted protein from the 1% PCA treatment. The less abundant proteins isolated using the two higher PCA concentrations were probably below the detection level on the 1% PCA gel, considering normalized gel loading and the number of high-density spots on the lower PCA percentage gel. For comparison, a 3% TCA extract was visualized (Fig. 23.2D). This treatment yielded a unique spot distribution pattern that was not comparable to the PCA gels. Another method of denaturing proteins is the application of heat. Boiling soluble extracts for 15 min resulted in the fewest number of spots, with a different spot distribution pattern compared with any of the acid gels (Fig. 23.2D). A high molecular mass band on the 9% PCA gel spanning the pH range from 3.5 to 8 may indicate that additional modification and cross-linking may have taken place with this treatment (Fig. 23.2E). Because of the reduction in number of proteins with each of these treatments, it would be difficult to correlate spot identities between denaturing treatments gels and untreated total protein gels (Fig. 23.2F) (i.e., only the most abundant acid-resistant proteins would be visible on total extract gels).

The ratio of unstructured to structured proteins known to remain soluble in PCA could provide a rough approximation of the relative ratio between the two classes in similar extractions. Applying the percentage (about 57%) of totally unstructured proteins that were isolated using PCA denaturation to the 158 proteins evident on the 5% PCA gel suggests that ~90 of the visible proteins could be expected to be totally unstructured. This number compares favorably with the 85–196 totally disordered proteins estimated to be present in the *E. coli* proteome (49). Therefore, treating total protein extracts with 3–5% PCA or TCA, and determining the identities of the soluble proteins could form the basis for uncovering unfoldomes in various organisms.

23.3 ENRICHMENT OF CELL EXTRACTS IN EXTENDED IDPS BY HEAT TREATMENT

23.3.1 Basic Principles of the Approach

Several studies indicated that IDPs are stable to heat denaturation and aggregation (see chapter 18). It is likely that resistance to thermal aggregation originates from the low mean hydrophobicity and high net charge characteristic of these proteins. It has been shown that the solubility and limited secondary structure of IDPs, such as p21, p27, α-synuclein, prothymosin α, and phosphodiesterase γ subunit, are virtually unaltered by heating to 90°C (3, 31, 43, 44, 46, 51, 68–70, 72, 77). Furthermore, heat treatment has been utilized for the recombinant IDP purification (2, 29, 32, 38, 77).

23.3.2 Large-Scale Analysis of IDPs Isolated from the Proteome of Mammalian Cells

The extracts of NIH3T3 mouse fibroblasts were heated at a variety of temperatures and analyzed by SDS-PAGE to determine the extent of protein precipitation under these conditions (25). This analysis revealed that the increase in the incubation temperature was accompanied by the decrease in the amount of soluble proteins. The most abundant proteins remaining soluble after the heat treatment were typically rather small. However, minor components in the range from 30 to 200 kDa (80 and 90°C) were also observed. Next, 2D gel electrophoresis was used to visualize the effects of heat treatment on cell extracts (Fig. 23.3), and the gel spots were picked and analyzed by MALDI-TOF/TOF MS. As a result, 375, 388, and 198 proteins (287, 304, and 124 nonredundant proteins) were identified from 584, 472, and 269 spots on 2D

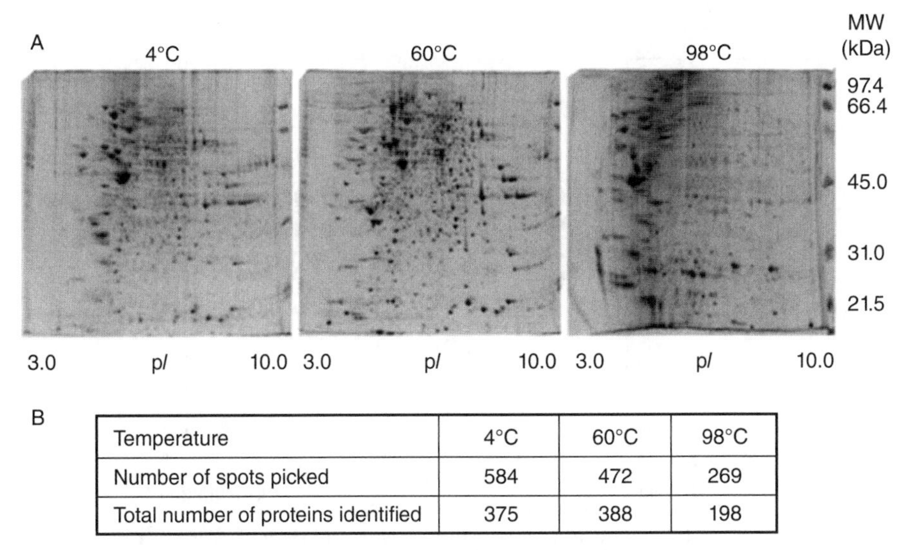

Temperature	4°C	60°C	98°C
Number of spots picked	584	472	269
Total number of proteins identified	375	388	198

Figure 23.3 2D gel analysis of proteins from untreated and heat-treated NIH3T3 cell extracts. (A) 2D gels for cell extracts incubated at 4°C or heat treated at 60°C for 10 min or 98°C for 1 h. Proteins (156 g) were loaded onto pH 310 ProteoGel IPG strips (Sigma-Aldrich, St. Louis, MO) (17 cm linear gradient) and resolved by isoelectric focusing (IEF) as outlined in the experimental procedures section. SDS-PAGE was performed with 10% gels. Spots were visualized by SYPRO (Sigma-Aldrich, St. Louis, MO) ruby staining. (B) Total number of gel spots was picked and proteins were identified for the 2D gels shown in (A). Reprinted with permission from Galea et al. (25). Copyright 2009 American Chemical Society.

gels obtained for cell extracts treated at 4, 60, and 98°C, respectively (Fig. 23.3B). These 287, 304, and 124 nonredundant proteins were further analyzed using a set of bioinformatics tools.

Figure 23.4 shows that heat treatment of NIH3T3 cell extracts resulted in a noticeable enrichment of proteins possessing high net charge (mean residue

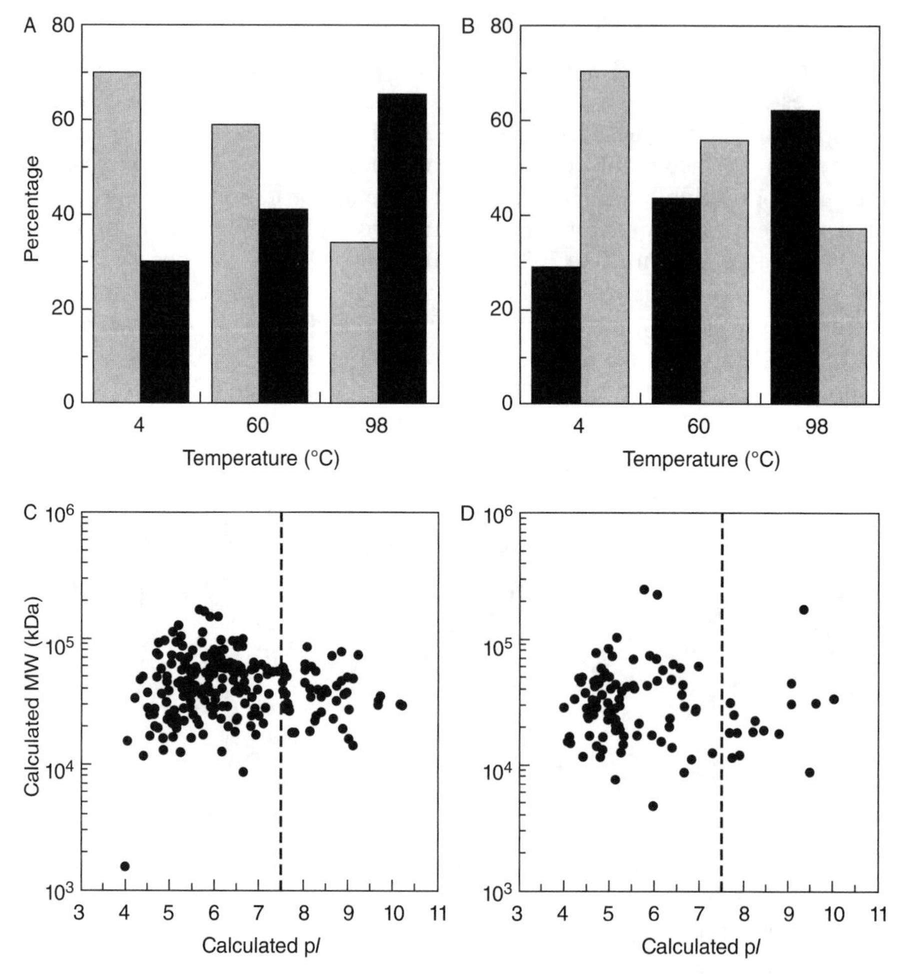

Figure 23.4 Heat treatment enriches proteins with high net charge and low hydrophobicity. (A) Percentage of proteins having a mean residue net charge less than (gray bars) or greater than (black bars) 0.025 and (B) percentage of proteins having a mean residue hydrophobicity of less than (black bars) or greater than (gray bars) 0.45 following incubation at 4°C or heat treatment at 60°C for 10 min or 98°C for 1 h. Calculated p*I* values were plotted against calculated molecular weight (on a logarithmic scale) for predicted IFPs, MPs, and IUPs for samples treated at 4°C (C) or 98°C (D). It should be noted that the actual net charge for some proteins will differ from the calculated value due to posttranslational modifications (e.g., phosphorylation). Reprinted with permission from Galea et al. (25). Copyright 2009 American Chemical Society.

net charge > 0.025) and/or low hydrophobicity (mean residue hydrophobicity < 0.45) (25). These features are characteristic of extended IDPs (67). At the next stage, these 287, 304, and 124 nonredundant proteins obtained from the cell extracts treated at 4, 60, and 98°C, respectively, were classified as IDPs, intrinsically folded proteins (IFPs), or mixed ordered/disordered proteins (MPs) based on the results of the disorder prediction by the PONDR VL-XT algorithm (http://www.pondr.com/). In this analysis, proteins having an average PONDR score >0.5 were classified as IDPs. Proteins having an average PONDR score of 0.32–0.5 and possessing a high mean net charge and low mean hydrophobicity were also classified as IDPs. Proteins having an average PONDR score <0.32 were classified as IFPs together with proteins having an average PONDR score of 0.32–0.5 and possessing a low mean net charge and high mean hydrophobicity. Proteins that did not meet the described above criteria for IUPs or IFPs were defined as MPs (25).

Figure 23.5 clearly shows that heat treatment resulted in an enrichment of IDPs and depletion of MPs and IFPs. In fact, although IDPs comprised only 11.8% of the proteins identified in the untreated cell extract (4°C), their relative population increased to 41.9% after the heat treatment at 98°C. However, MPs and IFPs, which comprised 42.8% and 45.4% of proteins in the untreated cell extract, were substantially depleted to 27.4% and 30.6%, respectively, after heat treatment at 98°C. Importantly, the number of IDPs identified in the cell extracts before and after heat treatment remained relatively unchanged, although there was a large decrease in the number of the identified MPs and IFPs (25).

Finally, the identified proteins were classified according to their reported biological function. In the untreated cell extracts, the largest functional group of proteins was involved in cellular metabolic processes, such as glycolysis, and in various biosynthetic pathways (excluding protein biosynthesis). Other functional categories included regulation, protein degradation, signaling, protein biosynthesis, structure, ribosomal proteins, heat shock proteins, chaperones, and transport. In contrast, the heat-treated extracts were enriched in proteins involved in regulation and structure, and depleted in metabolic proteins. Proteins involved in cell signaling and protein folding, and heat shock proteins were enriched to a lesser extent after the heat treatment (25). In conclusion, the heat treatment of mammalian cell extracts was shown to lead to enrichment in IDPs and in depletion in MPs and IFPs. The majority of the identified heat-resistant IDPs were cytoplasmic or nuclear proteins involved in cellular signaling or regulation.

23.4 2D ELECTROPHORESIS TECHNIQUE FOR DE NOVO RECOGNITION AND CHARACTERIZATION OF IDPS

23.4.1 Basic Principles of the Approach

This approach utilizes two specific features of the extended IDPs—their heat stability discussed above and their structural indifference to chemical dena-

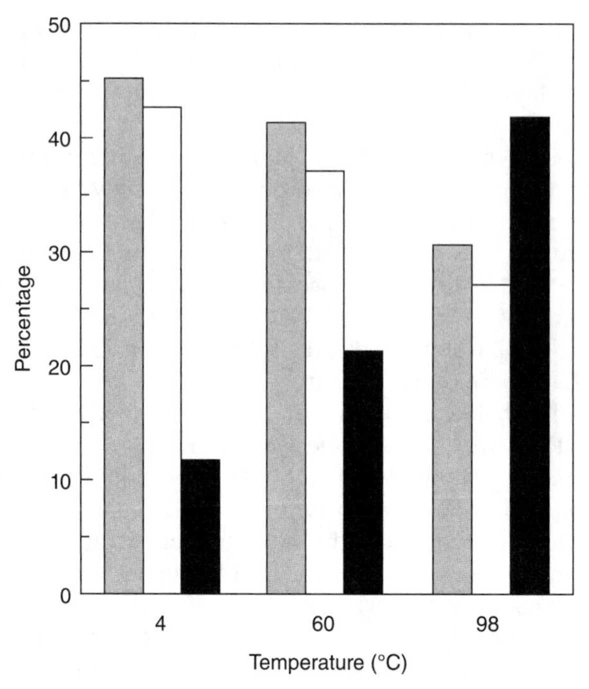

Figure 23.5 Heat treatment enriches IUPs (black bars), and depletes MPs (white bars) and IFPs (gray bars). NIH3T3 cell extracts were incubated at 4°C or heat treated at 60°C for 10 min or 98°C for 1 h. Reprinted with permission from Galea et al. (25). Copyright 2009 American Chemical Society.

turation (see chapter 21). This technique consists of the combination of native and 8 M urea electrophoresis of heat-treated proteins where the extended IDPs are expected to run into the diagonal, whereas globular proteins either precipitate upon heat treatment or unfold and run off the diagonal in the second dimension (8). Therefore, extended IUPs could be separated from folded globular proteins in a cellular extract by a novel 2D gel electrophoretic technique based on the combination of a native gel electrophoresis of heat-treated proteins followed by a second, denaturing gel containing 8 M urea (8). The rationales for this approach are considered below. Extended IDPs are often heat stable as demonstrated for Csd1 (29), MAP2 (32), α-synuclein (69, 77), its familial Parkinson's disease-related mutants (46), β- and γ-synucleins (70), stathmin (2), $p21^{Cip1}$ (43), prothymosin α (68), C-terminal domain of caldesmon (51), and phosphodiesterase γ (72). Therefore, heat treatment should lead to a dissent in initial separation of the extended IDPs from globular proteins, the vast majority of which are known to aggregate and precipitate at high temperatures. In the native gel, IUPs and rare heat-stable globular proteins will then be separated according to their charge/mass ratios. The second dimension in the proposed 2D gel electrophoretic approach relies on the insensitivity of extended IDPs to the denaturing action of high-urea concentrations. Since the extended IDPs are as unfolded in 8 M urea as under native conditions, they are expected to run the same distance in the second dimension and end up along the diagonal. Heat-stable globular proteins, however, will unfold in urea, slow down in the second direction, and accumulate above the diagonal. Because of this effective separation, extended IDPs could be amenable to subsequent MS identification (8).

The combination of these two dimensions was first proved by a collection of 10 experimentally confirmed IDPs (stathmin, MAP2c, Mypt1-[304–511], ERD10, α-casein, β-casein, α-synuclein, CSD1, Bob-1, and DARPP32) and 4 globular control proteins (fetuin, IPMDH, BSA, ovalbumin) (Fig. 23.6) (8). In this running, the sample was not boiled to keep all globular proteins in

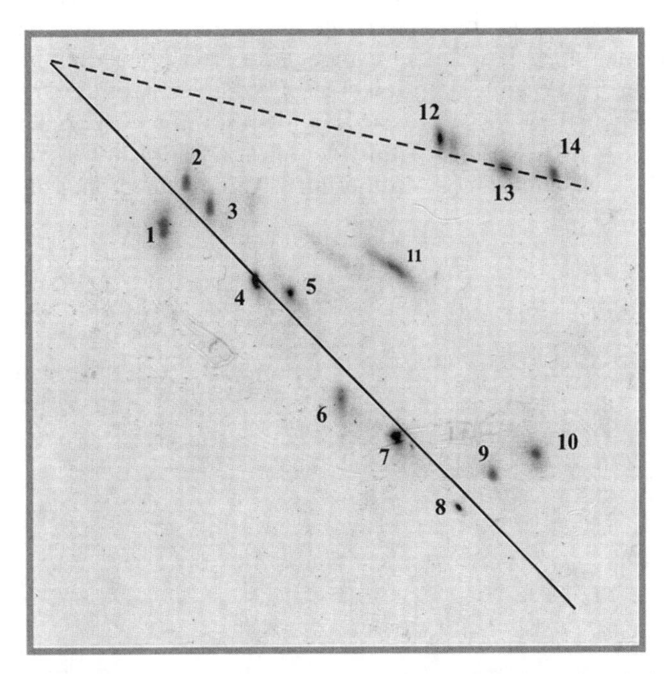

Figure 23.6 The native/8 M urea 2D electrophoresis of IUPs and globular proteins. A mixture of proteins (1 μg of each) was run on a 7.5% native gel in the first dimension (without heat treatment) and on a 7.5% gel containing 8 M urea in the second dimension in the large format (20 × 16 cm). The second gel was visualized by colloidal Coomassie staining. Individual proteins marked are as follows: IDPs: 1, stathmin; 2, MAP2c; 3, Mypt1-(304–511); 4, ERD10; 5, ß-casein; 6, NACP; 7, Csd1; 8, Bob-1; 9, DARPP32; and 10, α-casein. Globular proteins: 11, fetuin; 12, IPMDH; 13, BSA; and 14, ovalbumin. A *continuous line* marks the diagonal of the gel to where IUPs run. A *dashed line* marks the position of globular proteins.

solution and demonstrate the resolving power of the technique. Figure 23.6 shows that, indeed, IDPs ran at, or very near, the diagonal of the second (denaturing) gel, whereas globular proteins remained way above the diagonal. In fact, the ordered proteins ran mostly along a second line of much smaller slope. One IDP (α-casein) ran a little above the diagonal, likely due to significant residual structure present in caseins (34, 56). One globular protein (fetuin) ran closer to IDPs probably due to its high resistance to denaturation (40). Despite these two deviations, this experiment, overall, demonstrated that the 2D electrophoresis separates IDPs and globular proteins as predicted (8).

23.4.2 Identification of E. coli and Saccharomyces cerevisiae IDPs by Combining the 2D Technique with MS

Figure 23.7 represents the results of running heat-treated extracts of *E. coli* and *S. cerevisiae* in the large format gel. The comparison of the 2D pattern of the extracts from *E. coli* and *S. cerevisiae* reveals that more proteins were seen in the diagonal of the latter in agreement with predictions that the frequency of protein disorder increases with increasing complexity of the organisms (13, 18, 49, 76). Then, these gels were used for identification of spots at, and above, the diagonal by MS. Results of this identification are summarized in Tables 23.1 and 23.2, which also contain the intrinsic disorder propensity estimated

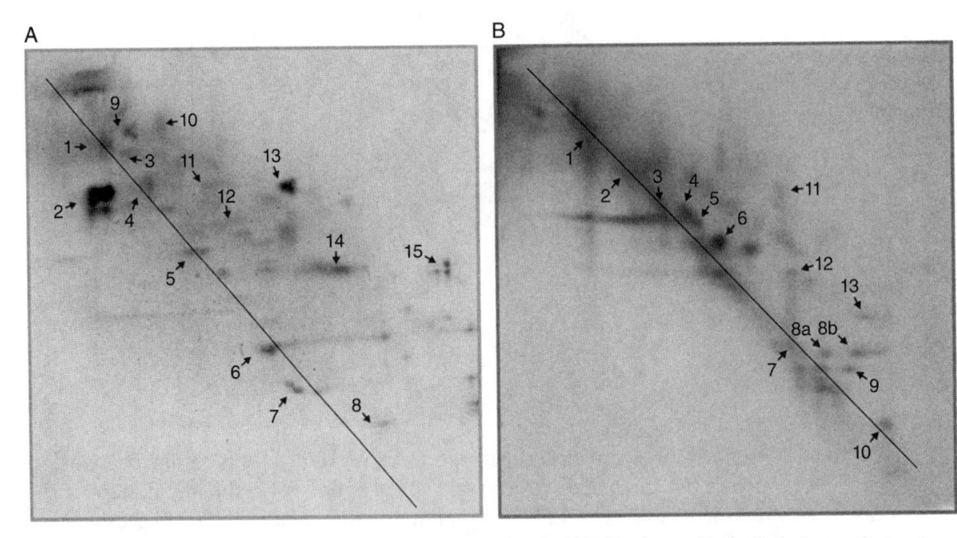

Figure 23.7 (A) Separation and identification of IDPs from *E. coli*. A large-format 7.5–15% gradient 2D gel was run with a heat-treated extract of *E. coli* strain BL21. The gel was visualized by colloidal Coomassie, and *dots* marked were cut out and sent for MS identification. (B) Separation and identification of IDPs from *S. cerevisiae*. A large-format 7.5–15% gradient 2D gel was run with a heat-treated extract of *S. cerevisiae*. The gel was visualized by colloidal Coomassie, and *dots* marked were cut out and sent for MS identification.

TABLE 23.1 Novel IUPs Identified in *E. coli*

Spot	GenBank™ Accession No.	Protein/Function	PONDR (%)	Charge/ Molecular Mass
1	15804734	GroES (10-kDa chaperonin)	43.3	−0.29
2	15804576	Ribosomal L7/L12 (L7/ L12)	57.02	−0.65
3	15799829	DnaKs	39.07	−0.4
4	15803789	Acetyl-CoA carboxylase, BCCP subunit, carrier of biotin (BCCP)	57.05	−0.54
5	15803918	Hypothetical protein YhgI (YhgI)	41.36	−0.71
6	15804517	Hypothetical protein (ORF1)	69.14	−0.83
7	26249319	Glycine cleavage complex H, carrier of aminomethyl moiety (GccH)	53.49	−1.37
8	15801211	Acyl carrier protein (ACP)	74.36	−1.74
9	15802070	Superoxide dismutase	20.21	−0.28
10	21238987	Aspartate 1-decarboxylase	22.22	−0.22
11	15804443	Hypothetical protein (ORF2)	15.73	−0.29
12	16131637	Thioredoxin	10.24	−0.29
13	16130343	PTS system, glucose-specific IIA component (PTSIIA)	34.91	−0.55
14	15803862	FKBP-type peptidyl-prolyl cis-trans isomerase, His-rich (FKBP isomerase)	41.33	−1.25
15	15800387	Flavodoxin 1	17.05	−1.22

The proteins identified on the diagonal (1–8) or above the diagonal (9–15) of the 2D gel of the *E. coli* extract are listed with their GenBank (gi) accession number. The sequences were submitted to disorder prediction by PONDR (VL-XT, default setting); the percentage of structural disorder thus estimated is also shown. The net charge/molecular mass (the latter in kDa) ratio was also calculated.

DnaKs, 70-kDa heat shock proteins, or chaperones Hsp70; FKBP, FK506-binding protein; ORF, open reading frame; PTS, phosphoenolypyruvate phosphotransferase system.

TABLE 23.2 Novel IUPs Identified in *S. cerevisiae*

Spot	GenBank™ Accession No.	Protein/Function	PONDR (%)	Charge/ Molecular Mass
1	6322325	Protein component of the small ribosomal subunit (sRib)	20.69	–0.10
2	6730103	Chain A, H48c yeast Cu(II),Zn-superoxide dismutase (H48c)	39.22	–0.32
3	6324768	TFIIA, large chain	63.29	–0.99
4	6324250	Actin-binding main tropomyosin (tropoM)	49.25	–1.06
5	6321606	Clathrin light chain	66.09	–1.17
6	6319315	Translation elongation factor eEF-1β (eEF1β)	42.23	–1.08
7	6320535	Centromere DNA-binding complex subunit D (centromere D)	55.15	–1.12
8	133071	60S acidic ribosomal protein P2-β (P2-β)	63.64	–1.35
9	133066	60S acidic ribosomal protein P2-α (P2-α)	70.75	–1.58
10	14318558	Ubiquinol-cytochrome c oxidoreductase subunit 6 (Ubi6)	51.70	–2.66
11	6321332	Myosin-2 light chain	19.46	–0.55
12	6320718	SUMO-1 homologue	49.50	–0.51
13	6319585	Yeast calmodulin	52.38	–1.30

The proteins identified on the diagonal (1–10) or above the diagonal (11–13) of the 2D gel of *S. cerevisiae* extract are listed with their GenBank (gi) accession number. The percentage of structural disorder estimated by PONDR prediction (VL-XT, default setting) is also shown. The net charge/molecular mass (the latter in kDa) ratio was also calculated. SUMO, small ubiquitin-like modifier.

by PONDR VL-XT for the identified proteins. Tables 23.1 and 23.2 show that the amount of predicted disorder in proteins located at the diagonal positions is very high (52.1 ± 14.1%), noticeably exceeding that of certain typical IDPs (e.g., α-synuclein 37.1% and α-casein, 41.15%) (8). Many of these proteins have never been structurally characterized. However, literature data were available for some of these proteins. For example, many ribosomal proteins are known to lack a well-defined structure when separated from ribosomal RNA (13, 67). Yeast ribosomal stalk proteins P1α and P2β were found to be mostly disordered by circular dichroism, nuclear magnetic resonance, and fluorescence spectroscopy, heat stability, and protease sensitivity (81). However, when these proteins were mixed together, they were shown to interact, and this interaction was accompanied by an increase in the global secondary structure content of both proteins (81). Similarly, *E. coli* ribosomal L7/L12 was shown to be a heat-stable, molten globule-like protein that acquired structure only upon dimerization (42). A similar behavior was reported for heat shock 10kDa protein 1 (chaperonin 10), also known as HSPE1 or GroES (33). *E. coli* acyl carrier protein (ACP) is stabilized by Mg^{2+} binding (Protein Data Bank code 1acp), but it is extremely flexible in solution (41), and its *Vibrio harveyi* homologue is fully disordered under physiological conditions (39). Some of the proteins have both disordered and ordered regions. For example, the structure of the 80–156 domain of the acetyl-coenzyme A (acetyl-CoA) carboxylase biotin carboxyl carrier protein (BCCP) subunit is known (Protein Data Bank code 1a6x), but the protein has a highly flexible N-terminal linker region (7). The majority of proteins above the diagonal were shown to be heat-stable enzymes (e.g., superoxide dismutase), which typically require a well-defined structure for function.

In conclusion, the structural analysis of the proteins identified by the 2D electrophoresis suggests that this method is suitable to identify the overall disordered status of a protein even if it contains a certain amount of secondary structural elements.

23.5 CONCLUDING REMARKS

Currently, there is a dramatic gap between the genome-based bioinformatics predictions, according to which significant fraction of any given proteome belong to the class of IDPs, and a relatively small set of experimentally characterized IDPs. In fact, only about 500 or so IDPs are structurally characterized now, which clearly represents a small fraction of this large class of proteins. Therefore, proteomic-scale identification and characterization of IDPs are clearly needed to fill this gap and advance our knowledge in this important field. IDPs, especially their extended forms, are characterized by several unique features that can be used for isolation of these proteins from the cell extracts. Described in this chapter, proteomic techniques utilize high resistance of these proteins against extreme pH and high temperature. The proposed approaches are the first steps toward the discovery of IDPs at the proteome level.

ACKNOWLEDGMENTS

This work was supported in part by the Program of the Russian Academy of Sciences for the "Molecular and cellular biology," and by grants R01

LM007688-01A1 and GM071714-01A2 from the National Institutes of Health. We gratefully acknowledge the support of the Indian University Purdue University at Indianapolis (IUPUI) Signature Centers Initiative.

REFERENCES

1. Ahmad, A., K. P. Madhusudanan, and V. Bhakuni. 2000. Trichloroacetic acid and trifluoroacetic acid-induced unfolding of cytochrome c: stabilization of a native-like folded intermediate(1). Biochim Biophys Acta **1480**:201–10.
2. Belmont, L. D., and T. J. Mitchison. 1996. Identification of a protein that interacts with tubulin dimers and increases the catastrophe rate of microtubules. Cell **84**:623–31.
3. Bienkiewicz, E. A., J. N. Adkins, and K. J. Lumb. 2002. Functional consequences of preorganized helical structure in the intrinsically disordered cell-cycle inhibitor p27(Kip1). Biochemistry **41**:752–9.
4. Cheng, Y., C. J. Oldfield, J. Meng, P. Romero, V. N. Uversky, and A. K. Dunker. 2007. Mining alpha-helix-forming molecular recognition features with cross species sequence alignments. Biochemistry **46**:13468–77.
5. Cortese, M. S., J. P. Baird, V. N. Uversky, and A. K. Dunker. 2005. Uncovering the unfoldome: enriching cell extracts for unstructured proteins by acid treatment. J Proteome Res **4**:1610–8.
6. Cortese, M. S., V. N. Uversky, and A. Keith Dunker. 2008. Intrinsic disorder in scaffold proteins: getting more from less. Prog Biophys Mol Biol **98**:85–106.
7. Cronan, J. E., Jr. 2002. Interchangeable enzyme modules. Functional replacement of the essential linker of the biotinylated subunit of acetyl-CoA carboxylase with a linker from the lipoylated subunit of pyruvate dehydrogenase. J Biol Chem **277**:22520–7.
8. Csizmok, V., E. Szollosi, P. Friedrich, and P. Tompa. 2006. A novel two-dimensional electrophoresis technique for the identification of intrinsically unstructured proteins. Mol Cell Proteomics **5**:265–73.
9. Daughdrill, G. W., G. J. Pielak, V. N. Uversky, M. S. Cortese, and A. K. Dunker. 2005. Natively disordered proteins, pp. 271–353. In J. Buchner and T. Kiefhaber (eds.), Handbook of Protein Folding. Wiley-VCH, Verlag GmbH & Co., Weinheim, Germany.
10. Dill, K. A., and D. Shortle. 1991. Denatured states of proteins. Annu Rev Biochem **60**:795–825.
11. Dosztanyi, Z., M. Sandor, P. Tompa, and I. Simon. 2007. Prediction of protein disorder at the domain level. Curr Protein Pept Sci **8**:161–71.
12. Dosztanyi, Z., and P. Tompa. 2008. Prediction of protein disorder. Methods Mol Biol **426**:103–15.
13. Dunker, A. K., C. J. Brown, J. D. Lawson, L. M. Iakoucheva, and Z. Obradovic. 2002. Intrinsic disorder and protein function. Biochemistry **41**:6573–82.
14. Dunker, A. K., C. J. Brown, and Z. Obradovic. 2002. Identification and functions of usefully disordered proteins. Adv Protein Chem **62**:25–49.

15. Dunker, A. K., M. S. Cortese, P. Romero, L. M. Iakoucheva, and V. N. Uversky. 2005. Flexible nets: the roles of intrinsic disorder in protein interaction networks. FEBS J **272**:5129–48.

16. Dunker, A. K., J. D. Lawson, C. J. Brown, R. M. Williams, P. Romero, J. S. Oh, C. J. Oldfield, A. M. Campen, C. M. Ratliff, K. W. Hipps, J. Ausio, M. S. Nissen, R. Reeves, C. Kang, C. R. Kissinger, R. W. Bailey, M. D. Griswold, W. Chiu, E. C. Garner, and Z. Obradovic. 2001. Intrinsically disordered protein. J Mol Graph Model **19**:26–59.

17. Dunker, A. K., and Z. Obradovic. 2001. The protein trinity—linking function and disorder. Nat Biotechnol **19**:805–6.

18. Dunker, A. K., Z. Obradovic, P. Romero, E. C. Garner, and C. J. Brown. 2000. Intrinsic protein disorder in complete genomes. Genome Inform Ser Workshop Genome Inform **11**:161–71.

19. Dunker, A. K., C. J. Oldfield, J. Meng, P. Romero, J. Y. Yang, J. W. Chen, V. Vacic, Z. Obradovic, and V. N. Uversky. 2008. The unfoldomics decade: an update on intrinsically disordered proteins. BMC Genomics **9** (Suppl 2):S1.

20. Dunker, A. K., I. Silman, V. N. Uversky, and J. L. Sussman. 2008. Function and structure of inherently disordered proteins. Curr Opin Struct Biol **18**:756–64.

21. Dyson, H. J., and P. E. Wright. 2002. Coupling of folding and binding for unstructured proteins. Curr Opin Struct Biol **12**:54–60.

22. Dyson, H. J., and P. E. Wright. 2005. Intrinsically unstructured proteins and their functions. Nat Rev Mol Cell Biol **6**:197–208.

23. Ferron, F., S. Longhi, B. Canard, and D. Karlin. 2006. A practical overview of protein disorder prediction methods. Proteins **65**:1–14.

24. Fink, A. L., L. J. Calciano, Y. Goto, T. Kurotsu, and D. R. Palleros. 1994. Classification of acid denaturation of proteins: intermediates and unfolded states. Biochemistry **33**:12504–11.

25. Galea, C. A., V. R. Pagala, J. C. Obenauer, C. G. Park, C. A. Slaughter, and R. W. Kriwacki. 2006. Proteomic studies of the intrinsically unstructured mammalian proteome. J Proteome Res **5**:2839–48.

26. Goto, Y., L. J. Calciano, and A. L. Fink. 1990. Acid-induced folding of proteins. Proc Natl Acad Sci U S A **87**:573–7.

27. Goto, Y., and A. L. Fink. 1990. Phase diagram for acidic conformational states of apomyoglobin. J Mol Biol **214**:803–5.

28. Goto, Y., N. Takahashi, and A. L. Fink. 1990. Mechanism of acid-induced folding of proteins. Biochemistry **29**:3480–8.

29. Hackel, M., T. Konno, and H. Hinz. 2000. A new alternative method to quantify residual structure in "unfolded" proteins. Biochim Biophys Acta **1479**:155–65.

30. Haines, D. C., I. F. Sevrioukova, and J. A. Peterson. 2000. The FMN-binding domain of cytochrome P450BM-3: resolution, reconstitution, and flavin analogue substitution. Biochemistry **39**:9419–29.

31. Hengst, L., V. Dulic, J. M. Slingerland, E. Lees, and S. I. Reed. 1994. A cell cycle-regulated inhibitor of cyclin-dependent kinases. Proc Natl Acad Sci U S A **91**:5291–5.

32. Hernandez, M. A., J. Avila, and J. M. Andreu. 1986. Physicochemical characterization of the heat-stable microtubule-associated protein MAP2. Eur J Biochem **154**:41–8.
33. Higurashi, T., K. Nosaka, T. Mizobata, J. Nagai, and Y. Kawata. 1999. Unfolding and refolding of *Escherichia coli* chaperonin GroES is expressed by a three-state model. J Mol Biol **291**:703–13.
34. Holt, C., and L. Sawyer. 1993 Caseins as rheomorphic proteins: interpretation of primary and secondary structures of the alpha(s1)-, beta- and kappa-caseins. J Chem Soc Faraday Trans **89**:2683–92.
35. Hubbard, S. J., R. J. Beynon, and J. M. Thornton. 1998. Assessment of conformational parameters as predictors of limited proteolytic sites in native protein structures. Protein Eng **11**:349–59.
36. Iakoucheva, L. M., C. J. Brown, J. D. Lawson, Z. Obradovic, and A. K. Dunker. 2002. Intrinsic disorder in cell-signaling and cancer-associated proteins. J Mol Biol **323**:573–84.
37. Iakoucheva, L. M., A. L. Kimzey, C. D. Masselon, R. D. Smith, A. K. Dunker, and E. J. Ackerman. 2001. Aberrant mobility phenomena of the DNA repair protein XPA. Protein Sci **10**:1353–62.
38. Kalthoff, C. 2003. A novel strategy for the purification of recombinantly expressed unstructured protein domains. J Chromatogr B Analyt Technol Biomed Life Sci **786**:247–54.
39. Keating, M. M., H. Gong, and D. M. Byers. 2002. Identification of a key residue in the conformational stability of acyl carrier protein. Biochim Biophys Acta **1601**:208–14.
40. Kim, T. D., H. J. Ryu, H. I. Cho, C. H. Yang, and J. Kim. 2000. Thermal behavior of proteins: heat-resistant proteins and their heat-induced secondary structural changes. Biochemistry **39**:14839–46.
41. Kim, Y., and J. H. Prestegard. 1989. A dynamic model for the structure of acyl carrier protein in solution. Biochemistry **28**:8792–7.
42. Kitaura, H., M. Kinomoto, and T. Yamada. 1999. Ribosomal protein L7 included in tuberculin purified protein derivative (PPD) is a major heat-resistant protein inducing strong delayed-type hypersensitivity. Scand J Immunol **50**:580–7.
43. Kriwacki, R. W., L. Hengst, L. Tennant, S. I. Reed, and P. E. Wright. 1996. Structural studies of p21Waf1/Cip1/Sdi1 in the free and Cdk2-bound state: conformational disorder mediates binding diversity. Proc Natl Acad Sci U S A **93**:11504–9.
44. Lacy, E. R., I. Filippov, W. S. Lewis, S. Otieno, L. Xiao, S. Weiss, L. Hengst, and R. W. Kriwacki. 2004. p27 binds cyclin-CDK complexes through a sequential mechanism involving binding-induced protein folding. Nat Struct Mol Biol **11**:358–64.
45. Lacy, E. R., Y. Wang, J. Post, A. Nourse, W. Webb, M. Mapelli, A. Musacchio, G. Siuzdak, and R. W. Kriwacki. 2005. Molecular basis for the specificity of p27 toward cyclin-dependent kinases that regulate cell division. J Mol Biol **349**:764–73.
46. Li, J., V. N. Uversky, and A. L. Fink. 2001. Effect of familial Parkinson's disease point mutations A30P and A53T on the structural properties, aggregation, and fibrillation of human alpha-synuclein. Biochemistry **40**:11604–13.

47. Mohan, A. 2006. MoRFs: A dataset of molecular recognition features. Master of Science, Indiana University, Indianapolis, IN.
48. Neumann, S., U. Matthey, G. Kaim, and P. Dimroth. 1998. Purification and properties of the F1F0 ATPase of Ilyobacter tartaricus, a sodium ion pump. J Bacteriol **180**:3312–6.
49. Oldfield, C. J., Y. Cheng, M. S. Cortese, C. J. Brown, V. N. Uversky, and A. K. Dunker. 2005. Comparing and combining predictors of mostly disordered proteins. Biochemistry **44**:1989–2000.
50. Oldfield, C. J., Y. Cheng, M. S. Cortese, P. Romero, V. N. Uversky, and A. K. Dunker. 2005. Coupled folding and binding with alpha-helix-forming molecular recognition elements. Biochemistry **44**:12454–70.
51. Permyakov, S. E., I. S. Millett, S. Doniach, E. A. Permyakov, and V. N. Uversky. 2003. Natively unfolded C-terminal domain of caldesmon remains substantially unstructured after the effective binding to calmodulin. Proteins **53**:855–62.
52. Radivojac, P., L. M. Iakoucheva, C. J. Oldfield, Z. Obradovic, V. N. Uversky, and A. K. Dunker. 2007. Intrinsic disorder and functional proteomics. Biophys J **92**:1439–56.
53. Receveur-Brechot, V., J. M. Bourhis, V. N. Uversky, B. Canard, and S. Longhi. 2006. Assessing protein disorder and induced folding. Proteins **62**:24–45.
54. Reeves, R., and M. S. Nissen. 1999. Purification and assays for high mobility group HMG-I(Y) protein function. Methods Enzymol **304**:155–88.
55. Roychaudhuri, R., G. Sarath, M. Zeece, and J. Markwell. 2004. Stability of the allergenic soybean Kunitz trypsin inhibitor. Biochim Biophys Acta **1699**:207–12.
56. Sawyer, L., and C. Holt. 1993. The secondary structure of milk proteins and their biological function. J Dairy Sci **76**:3062–78.
57. Schafer, K., U. Magnusson, F. Scheffel, A. Schiefner, M. O. Sandgren, K. Diederichs, W. Welte, A. Hulsmann, E. Schneider, and S. L. Mowbray. 2004. X-ray structures of the maltose-maltodextrin-binding protein of the thermoacidophilic bacterium Alicyclobacillus acidocaldarius provide insight into acid stability of proteins. J Mol Biol **335**:261–74.
58. Sivaraman, T., T. K. Kumar, G. Jayaraman, and C. Yu. 1997. The mechanism of 2,2,2-trichloroacetic acid-induced protein precipitation. J Protein Chem **16**:291–7.
59. Tompa, P. 2005. The interplay between structure and function in intrinsically unstructured proteins. FEBS Lett **579**:3346–54.
60. Tompa, P. 2002. Intrinsically unstructured proteins. Trends Biochem Sci **27**:527–33.
61. Tucker, D. L., N. Tucker, and T. Conway. 2002. Gene expression profiling of the pH response in *Escherichia coli*. J Bacteriol **184**:6551–8.
62. Uversky, V. N. 2002. Natively unfolded proteins: a point where biology waits for physics. Protein Sci **11**:739–56.
63. Uversky, V. N. 2003. A protein-chameleon: conformational plasticity of alpha-synuclein, a disordered protein involved in neurodegenerative disorders. J Biomol Struct Dyn **21**:211–34.
64. Uversky, V. N. 2002. What does it mean to be natively unfolded? Eur J Biochem **269**:2–12.

65. Uversky, V. N., and A. L. Fink. 1998. Structural effect of association on protein molecules in partially folded intermediates. Biochemistry (Mosc) **63**:456–62.

66. Uversky, V. N., and A. L. Fink. 1998. Structural properties of staphylococcal nuclease in oligomeric A-forms. Biochemistry (Mosc) **63**:463–9.

67. Uversky, V. N., J. R. Gillespie, and A. L. Fink. 2000. Why are "natively unfolded" proteins unstructured under physiologic conditions? Proteins **41**:415–27.

68. Uversky, V. N., J. R. Gillespie, I. S. Millett, A. V. Khodyakova, A. M. Vasiliev, T. V. Chernovskaya, R. N. Vasilenko, G. D. Kozlovskaya, D. A. Dolgikh, A. L. Fink, S. Doniach, and V. M. Abramov. 1999. Natively unfolded human prothymosin alpha adopts partially folded collapsed conformation at acidic pH. Biochemistry **38**:15009–16.

69. Uversky, V. N., J. Li, and A. L. Fink. 2001. Evidence for a partially folded intermediate in alpha-synuclein fibril formation. J Biol Chem **276**:10737–44.

70. Uversky, V. N., J. Li, P. Souillac, I. S. Millett, S. Doniach, R. Jakes, M. Goedert, and A. L. Fink. 2002. Biophysical properties of the synucleins and their propensities to fibrillate: inhibition of alpha-synuclein assembly by beta- and gamma-synucleins. J Biol Chem **277**:11970–8.

71. Uversky, V. N., C. J. Oldfield, and A. K. Dunker. 2005. Showing your ID: intrinsic disorder as an ID for recognition, regulation and cell signaling. J Mol Recognit **18**:343–84.

72. Uversky, V. N., S. E. Permyakov, V. E. Zagranichny, I. L. Rodionov, A. L. Fink, A. M. Cherskaya, L. A. Wasserman, and E. A. Permyakov. 2002. Effect of zinc and temperature on the conformation of the gamma subunit of retinal phosphodiesterase: a natively unfolded protein. J Proteome Res **1**:149–59.

73. Uversky, V. N., D. J. Segel, S. Doniach, and A. L. Fink. 1998. Association-induced folding of globular proteins. Proc Natl Acad Sci U S A **95**:5480–3.

74. Vacic, V., C. J. Oldfield, A. Mohan, P. Radivojac, M. S. Cortese, V. N. Uversky, and A. K. Dunker. 2007. Characterization of molecular recognition features, MoRFs, and their binding partners. J Proteome Res **6**:2351–66.

75. Vucetic, S., H. Xie, L. M. Iakoucheva, C. J. Oldfield, A. K. Dunker, Z. Obradovic, and V. N. Uversky. 2007. Functional anthology of intrinsic disorder. 2. Cellular components, domains, technical terms, developmental processes, and coding sequence diversities correlated with long disordered regions. J Proteome Res **6**:1899–916.

76. Ward, J. J., J. S. Sodhi, L. J. McGuffin, B. F. Buxton, and D. T. Jones. 2004. Prediction and functional analysis of native disorder in proteins from the three kingdoms of life. J Mol Biol **337**:635–45.

77. Weinreb, P. H., W. Zhen, A. W. Poon, K. A. Conway, and P. T. Lansbury, Jr. 1996. NACP, a protein implicated in Alzheimer's disease and learning, is natively unfolded. Biochemistry **35**:13709–15.

78. Wright, P. E., and H. J. Dyson. 1999. Intrinsically unstructured proteins: re-assessing the protein structure-function paradigm. J Mol Biol **293**:321–31.

79. Xie, H., S. Vucetic, L. M. Iakoucheva, C. J. Oldfield, A. K. Dunker, Z. Obradovic, and V. N. Uversky. 2007. Functional anthology of intrinsic disorder. 3. Ligands, post-translational modifications, and diseases associated with intrinsically disordered proteins. J Proteome Res **6**:1917–32.

80. Xie, H., S. Vucetic, L. M. Iakoucheva, C. J. Oldfield, A. K. Dunker, V. N. Uversky, and Z. Obradovic. 2007. Functional anthology of intrinsic disorder. 1. Biological processes and functions of proteins with long disordered regions. J Proteome Res **6**:1882–98.
81. Zurdo, J., C. Gonzalez, J. M. Sanz, M. Rico, M. Remacha, and J. P. Ballesta. 2000. Structural differences between *Saccharomyces cerevisiae* ribosomal stalk proteins P1 and P2 support their functional diversity. Biochemistry **39**:8935–43.

24

PURIFICATION OF INTRINSICALLY DISORDERED PROTEINS

AVIV PAZ,[1,2] TZVIYA ZEEV-BEN-MORDEHAI,[1,2] JOEL L. SUSSMAN,[1] AND ISRAEL SILMAN[2]

[1]*Department of Structural Biology, Weizmann Institute of Science, Rehovot, Israel*
[2]*Department of Neurobiology, Weizmann Institute of Science, Rehovot, Israel*

ABSTRACT

Since IDPs share physicochemical characteristics that differentiate them from globular proteins, the process of IDP purification can be highly efficient if one utilizes purification schemes that take advantage of these special characteristics. However, purification can be highly problematic when dealing with recombinant IDPs that are sensitive to the degradation machinery of the host cell in which they are being overexpressed. Herein, we survey some of the specialized procedures reported in the literature for purification of IDPs, elaborate on ways to stabilize IDPs in the course of purification, and focus on our experience in the purification of two highly protease-sensitive IDPs under denaturing conditions that inactivated the endogenous proteases of the host.

24.1 INTRODUCTION

In the process of purification of a recombinant protein, advantage is taken of its specific physicochemical attributes so as to separate it from the host's proteins. A survey of published procedures for purification of IDPs shows that

Instrumental Analysis of Intrinsically Disordered Proteins: Assessing Structure and Conformation, Edited by Vladimir Uversky and Sonia Longhi

most are similar to those adopted for well-folded globular proteins, even though IDPs differ from such proteins in many respects. Thus, in general, they display lower hydrophobicities and higher net charges (33), do not lose their activity upon heating, and lack substantial amounts of stable secondary and tertiary structures (29). Although the prevalence and functional importance of IDPs has only been realized in recent years (7), some scientists who studied them earlier utilized unconventional purification techniques, such as boiling, for the purification of heat-stable proteins (17) or of proteins emanating from thermophilic organisms (24). However, Kalthoff has recently argued that most procedures for the purification of recombinant proteins are tailored to meet the requirements of well-structured polypeptides, namely, purification under stabilizing, near-physiological conditions (16). The special attributes of the IDPs may thus be taken advantage of in their isolation and purification, especially when overexpression is carried out in *Escherichia coli*, in which the predicted frequency of functional proteins displaying a high degree of disorder is ~5% (25). In general, special care to avoid proteolytic degradation should be exercised when purifying IDPs, since they are readily available both to proteases (3) and to the cellular degradation machinery (32).

In summary, the "atypical" physicochemical attributes of the IDPs, taken together with their high sensitivity to proteolysis, demand implementation of "atypical" approaches to their purification.

24.2 IDPS: PROTEOLYSIS AND STABILIZATION

Disordered polypeptides often possess extended dimensions, with both high exposure to solvent and flexibility that render them highly sensitive to enzymes that introduce posttranslational modifications, such as acetylation, hydroxylation, methylation, phosphorylation, and ubiquitination, to endogenous proteases, and to the proteasomal degradation machinery (6, 15). In the case of the highly disordered cytoplasmic domain of the *Drosophila* adhesion protein, gliotactin, which we had overexpressed in *E. coli*, we were able to obtain the intact domain without taking any special precautions to avoid proteolysis, apart from the introduction of protease inhibitors during the lysis stage (35). In other cases, authors have reported that they were able to obtain a full-length IDP, accompanied by relatively small amounts of lower molecular weight species as the final product of their purification process, if protease inhibitors were present during lysis of the bacterial pellet, as was the case for the MARCKS proteins (1, 21). In contrast, high sensitivity to proteases was observed in the purification of the BH3 protein (14). In most cases, carrying out purification at a low temperature, in the presence of a suitable cocktail of protease inhibitors, indeed stabilized the IDP sufficiently to permit its purification in intact form. In other cases, such precautions were inadequate, and it was observed that the IDP degraded rapidly subsequent to cell lysis (26).

Presumably, the intracellular stability of such an IDP, prior to lysis, can be ascribed to compartmentalization that separates the IDP from the protease(s) (9).

One general approach to reducing the susceptibility to proteolytic degradation of an IDP, or, indeed, of any protein, is to increase its stability (22). IDPs can be stabilized by complexing them either with a binding partner or with a cofactor that is known to stabilize them. The interactions of three subunits of cytoplasmic dynein from *Drosophila melanogaster*, LC8, Tctex-1, and the N-terminal domain of IC74, which is known to be intrinsically disordered, protected the latter from digestion both by trypsin and by proteinase K (20). We have shown, using the 20S proteasome system, that formation of a complex of the intrinsically disordered cytoplasmic domain of human neuroligin 3 (hNL3-cyt), with either of two of its binding partners, PSD-95 and S-SCAM, partially protects it from degradation (32). Thus, inclusion of a suitable ligand in the lysis buffer and in the buffers used during purification, or coexpression of the IDP with a suitable partner, are both approaches that may afford protection against degradation during the course of purification. The latter option is attractive since, in general, IDPs have multiple partners (31), one or another of which may prove suitable for coexpression.

Another possible approach to the stabilization of IDPs involves their coexpression with chaperones. Chaperones are large protein complexes that assist other proteins or RNA molecules to achieve their functional structure, in an energy-dependent process, by preventing their misfolding and aggregation, and/or by aiding their assembly and transport (18). It was earlier suggested that chaperones may bind and sequester IDPs, in addition to their proposed function of promoting protein folding, as a means of protecting the IDPs against proteolytic digestion (5). It is of interest that many chaperones display intrinsic disorder (34), as might be anticipated for proteins capable of recognizing a wide variety of targets. It is also worth mentioning that some chaperones exert their protective function on IDP clients that are implicated in aggregation and amyloid formation, such as tau protein and α-synuclein (4, 27, 28). Since the half-lives of IDPs *in vivo* display only a weak correlation with their degree of disorder (30), Hegyi and Tompa investigated whether IDPs are preferentially associated with chaperones within the cell, since such an association might offer protection against proteolytic degradation (13). Although they found a negative correlation between chaperone binding and intrinsic disorder, this observation does not exclude coexpresssion of an IDP with an appropriate stabilizing chaperone in recombinant protein expression. The work of Gorovits and Horowitz (11), who studied the interaction of GroEL and GroES with both thermally and chemically unfolded dihydrofolate reductase (DHFR), shows the feasibility of this approach. Both the chemically and thermally denatured forms of DHFR were shown to bind to GroEL, but at different sites. Upon addition of GroES, the ternary complex formed and provided protection from proteolysis only for the more compact, thermally unfolded DHFR.

The various approaches presented above are all capable of producing a higher yield of the intact polypeptide for a given IDP. There are, however, two less subtle approaches that are especially suitable for purification of IDPs, since they take into account their special characteristics as compared with those of globular proteins. These are the heating/cooling strategy and purification in the presence of a denaturing agent. Both these approaches share four advantages: (1) They preclude the requirement for a separate cell lysis step; (2) in cases in which a His-tag is introduced into the target protein, they both greatly reduce the amount of contaminating proteins that may bind nonspecifically to the metal affinity column; (3) they both serve as very effective means of inactivating proteases; and (4) last but not least, since IDPs are both heat stable and almost devoid of secondary structure, both their thermal and their chemical denaturation is reversible.

24.3 PURIFICATION USING THERMAL DENATURATION

As already mentioned, boiling protein extracts as an approach to purifying heat-stable proteins is not a new approach (17). Thermal denaturation of globular proteins is often irreversible, since unfolding of the protein exposes the hydrophobic residues that are buried in the hydrophobic core in its native conformation; this, in turn, results in the formation of aggregates that often possess a high β-sheet content (2, 8, 19). Since IDPs have a low percentage of hydrophobic residues (33), and are heat resistant, such a procedure directly gets rid of many of the globular proteins, including the endogenous proteases. Thus, for example, Oka and colleagues used a single-step procedure for purification of a thermophilic leucine dehydrogenase overexpressed in *E. coli* cells. They heated the cells at 70°C for 30 min, and were able to achieve an amazing yield of 75 mg/g of wet cells of 95% homogeneous protein. The heating/cooling strategy was proposed by Kalthoff as a general approach to purification of IDPs (16), and he demonstrated its efficacy for two unstructured domains derived from the heat-stable proteins, epsin 1 and AP180. In his procedure, the bacterial pellet is resuspended in 10 mL per liter of culture medium of 3 mM β-mercaptoethanol/10 mM imidazole/300 mM NaCl/50 mM sodium phosphate and placed in a boiling water bath for 5 min under continuous agitation. The cells are then transferred to a −10°C NaCl/ice water bath for 5 min, followed by centrifugation for 15 min at 117,000 *g* to remove the cell debris and precipitate. In both cases, the proteins were constructed with His-tags that permitted their purification on a metal affinity column. Comparison of this heating/cooling procedure to a standard cell lysis protocol, using Triton X-100 and sonication, revealed that the yield of purified protein was increased by 56% for AP180 and by 106% for epsin 1. A caveat that Kalthoff himself pointed out is that the heating process may result in chemical modification of certain amino acids, for example, via the Maillard reaction (12).

24.4 PURIFICATION USING CHEMICAL DENATURATION

We are currently involved in structural and functional characterization of a family of transmembrane cellular adhesion proteins whose extracellular domains display high sequence homology to acetylcholinesterase, and are thus termed cholinesterase-like adhesion molecules (CLAMs). As mentioned above, overexpression in *E. coli*, followed by purification, under native conditions, of the cytoplasmic domain of one of the CLAMs, gliotactin, were both quite straightforward steps and resulted in a stable protein, Gli-cyt, that was subsequently shown to be an IDP (35). Attempts to similarly overexpress and purify the cytoplasmic domains of two other CLAMs, namely, those of the *Drosophila* protein, neurotactin, Nrt-cyt, and of the human NL3, hNL3-cyt, revealed that both displayed high protease sensitivity. Thus, even if lysis of the *E. coli* cells in which they were overexpressed was performed by sonication at 4°C, in the presence of a broad-range protease inhibitor cocktail (Sigma-Aldrich P8849 Protease Inhibitor Cocktail for use in purification of His-tagged proteins, Sigma-Aldrich, St. Louis, MO), purification of both proteins produced low yields. Although the amounts obtained were sufficient to establish that they were IDPs, both were contaminated with large amounts of degradation products. Furthermore, upon storage at 4°C, complete degradation occurred within about 1 week. To overcome proteolysis, we decided to perform both the lysis and most of the purification steps under denaturing conditions for both hNL3-cyt and Nrt-cyt. The two procedures are described below.

The protocol for hNL3-cyt started with lysis of the bacterial pellet from 1L of culture by resuspension in 30mL of 8M urea/100mM NaH_2PO_4/10mM Tris, pH 8.0 (buffer A), at 4°C. Lysis was followed by centrifugation (12,000*g*, 30min, 4°C) and filtration through a 0.22-μm Stericup filtration device (Millipore, Billerica, MA, USA). The filtered protein solution was loaded onto a 1-mL HisTrap column (GE Healthcare, Chalfont St. Giles, UK), pre-equilibrated in buffer A, and eluted using a linear pH 8.0–4.5 gradient of buffer A. Already at this stage, a high level of purification was achieved (Fig. 24.1, lane 2), as compared with the protein obtained using a standard lysis technique, which was both impure and degraded (Fig. 24.1, lane 1). Eluted fractions were analyzed by SDS-PAGE, and fractions containing hNL3-cyt were pooled and loaded onto a Sepharose Fast Flow (SP FF) cation exchange column (GE Healthcare). The protein was eluted using a linear gradient of 0–1M NaCl in 100mM MES, pH 6.0. Pure fractions were pooled and concentrated, with concomitant change of the buffer to 1mM EDTA/5mM DTT/250mM NaCl/100mM Tris, pH 8.5, using a Vivaspin concentration device (3000 MWCO, Sartorius, Göttingen, Germany), and further purified through a HiLoad 16/60 Superdex 75-pg column (GE Healthcare) pre-equilibrated with the same buffer. This protocol resulted in ~10mg of pure hNL3-cyt per liter of medium, which remained intact for at least 6 months when stored at 4°C (26).

Even though we switched to nondenaturing conditions at the stage of elution from the cation-exchange column elution, due to the high purity of the

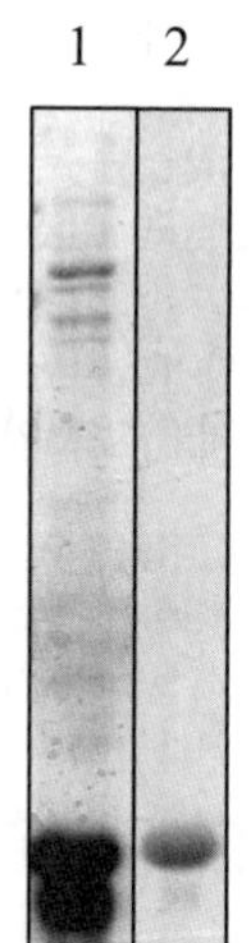

Figure 24.1 Comparison by SDS-PAGE of batches of hNL3-cyt overexpressed in *E. coli* and eluted from a HisTrap column after extraction under nondenaturing and denaturing conditions. Lane 1: Following lysis of the bacterial pellet by sonication in the absence of denaturant, hNL3-cyt was absorbed onto a HisTrap column in 10 mM imidazole/300 mM NaCl/50 mM NaH_2PO_4, pH 8.0, containing a broad-range protease inhibitor cocktail, and eluted in the same buffer, containing 250 mM imidazole. Lane 2: Eluant from a HisTrap column elution using denaturing conditions both for lysis of the bacterial pellet and, subsequently, its purification (see text for details).

protein, one can, if necessary, use a buffer containing 8 M urea also for size exclusion chromatography, thus further purifying the protein under protective denaturing conditions.

Figure 24.2 demonstrates the higher purity obtained for Nrt-cyt eluted from a HisTrap column under denaturing conditions as compared with nondenaturing ones (compare lane 2 with lane 1), as well as the final stable and purified preparation obtained (lane 3). Purification of Nrt-cyt involved three steps: Lysis was performed as described for hNL3-cyt, and the HisTrap column utilized the same buffers but with a segmented gradient of 5 column volumes (CVs) of 0–30% 8 M urea/100 mM NaH_2PO_4/10 mM Tris, pH 4.5 (buffer B), in buffer A, 20 CVs of 30–50% buffer B, and finally 20 CVs of 50–100% buffer B. Eluant fractions were analyzed by SDS-PAGE; those containing Nrt-cyt were pooled, and the buffer was converted to 4.5 M urea/1 M $(NH_4)_2SO_4$/60 mM NaH_2PO_4/6 mM Tris, pH 7.0, by addition of ~2 M $(NH_4)_2SO_4$ and appropriate dilution. The preparation was then loaded onto a 1-mL HiTrap Phenyl-Sepharose FF column (GE Healthcare). Nrt-cyt was eluted using a 20 CV gradient of descending salt concentration from 1 to 0 M $(NH_4)_2SO_4$, in 4.5 M urea/60 mM NaH_2PO_4/6 mM Tris, pH 7.0, concentrated, and finally loaded

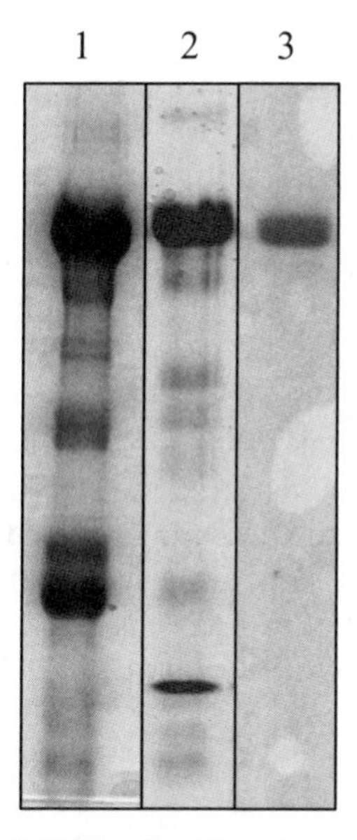

Figure 24.2 Comparison by SDS-PAGE of batches of Nrt-cyt overexpressed in *E. coli* and eluted from a HisTrap column after extraction under nondenaturing and denaturing conditions. Lane 1: Eluant from a HisTrap column of Nrt-cyt extracted and purified under nondenaturing conditions. Lane 2: Eluant from a HisTrap column of Nrt-cyt extracted and purified under denaturing conditions. Lane 3: Final purified product obtained under denaturing conditions (see text for details).

onto a Superdex 200 size exclusion column equilibrated in PBS, to yield ~1.5 mg of Nrt-cyt per liter of medium, whose purity on SDS-PAGE is shown in Figure 24.2, lane 3.

Purification under denaturing conditions was recently used for the bovine viral diarrhea virus core protein, which is an IDP. But this was only done so as to increase the purity of the protein in the final stages of purification (subsequent to lysis and GST-tag purification, both under nondenaturing conditions) (23). As mentioned above, our protocol for purification under denaturing conditions was devised based on our knowledge that the target protein was an IDP. It is of interest that a similar protocol was developed for lysis and purification of secreted bacterial proteins (10). Regardless of their initial solubility, or of their fold, recombinant fusion proteins were extracted from bacterial cells with guanidium chloride, purified under denaturing conditions by metal affinity chromatography, refolded, and further purified. The 12 proteins so purified were all monomeric and displayed unique far-UV circular dichroism spectra with observed secondary structure signals for each protein.

The advantages of purifying IDPs under conditions of chemical denaturation, especially utilizing urea (since the chaotropic salt, guanidinium hydrochloride, is not compatible with ion-exchange columns), are the following: (1)

Protease inhibition that can, if necessary, be maintained throughout the purification process. Since 8M urea is compatible with many resins, several successive purification steps can be performed in its presence, thus maintaining inhibition of residual proteases until they are completely removed. (2) A higher degree of purification for His-tag-labeled proteins by metal affinity chromatography under denaturing conditions, relative to purification under nondenaturing conditions, since smaller amounts of contaminating proteins bind nonspecifically to the chelating resin. (3) Since the urea buffer has a low ionic strength, it is possible to skip a desalting/dialysis step, and to directly load the eluate from the metal affinity column onto an ion exchanger.

In the above, we have demonstrated that the unique physicochemical characteristics of IDPs can be taken advantage of in development of protocols for their purification that, at the same time, overcome their high susceptibility to proteolysis. Two protocols that can serve as paradigms were presented for the disordered cytoplasmic domains of two CLAMs, namely, hNL3-cyt and Nrt-cyt.

ACKNOWLEDGMENTS

This work was supported by grants from Autism Speaks, the Divadol Foundation, the Israel Science Foundation, the Nalvyco Foundation, the Neuman Foundation, a research grant from Mr. Erwin Pearl, the Benoziyo Center for Neuroscience, the European Commission Sixth Framework Research and Technological Development Programme "SPINE2-COMPLEXES" Project, under contract number LSHG-CT-2006-031220, and the "Teach-SG" Project, under contract number ISSG-CT-2007-037198 (I. S. and J. L. S.). J. L. S. is the Morton and Gladys Pickman Professor of Structural Biology.

REFERENCES

1. Arbuzova, A., A. A. Schmitz, and G. Vergeres. 2002. Cross-talk unfolded: MARCKS proteins. Biochem J **362**:1–12.
2. Aune, K. C., A. Salahuddin, M. H. Zarlengo, and C. Tanford. 1967. Evidence for residual structure in acid- and heat-denatured proteins. J Biol Chem **242**: 4486–9.
3. Denning, D. P., V. Uversky, S. S. Patel, A. L. Fink, and M. Rexach. 2002. The *Saccharomyces cerevisiae* nucleoporin Nup2p is a natively unfolded protein. J Biol Chem **277**:33447–55.
4. Dou, F., W. J. Netzer, K. Tanemura, F. Li, F. U. Hartl, A. Takashima, G. K. Gouras, P. Greengard, and H. Xu. 2003. Chaperones increase association of tau protein with microtubules. Proc Natl Acad Sci U S A **100**:721–6.
5. Dunker, A. K., C. J. Brown, J. D. Lawson, L. M. Iakoucheva, and Z. Obradovic. 2002. Intrinsic disorder and protein function. Biochemistry **41**:6573–82.

6. Dunker, A. K., J. D. Lawson, C. J. Brown, R. M. Williams, P. Romero, J. S. Oh, C. J. Oldfield, A. M. Campen, C. M. Ratliff, K. W. Hipps, J. Ausio, M. S. Nissen, R. Reeves, C. Kang, C. R. Kissinger, R. W. Bailey, M. D. Griswold, W. Chiu, E. C. Garner, and Z. Obradovic. 2001. Intrinsically disordered protein. J Mol Graph Model **19**:26–59.
7. Dyson, H. J., and P. E. Wright. 2005. Intrinsically unstructured proteins and their functions. Nat Rev Mol Cell Biol **6**:197–208.
8. Elwell, M., and J. Schellman. 1975. Phage T4 lysozyme. Physical properties and reversible unfolding. Biochim Biophys Acta **386**:309–23.
9. Frankel, A. D., and P. S. Kim. 1991. Modular structure of transcription factors: implications for gene regulation. Cell **65**:717–9.
10. Geisbrecht, B. V., S. Bouyain, and M. Pop. 2006. An optimized system for expression and purification of secreted bacterial proteins. Protein Expr Purif **46**:23–32.
11. Gorovits, B. M., and P. M. Horowitz. 1997. Conditions of forming protein complexes with GroEL can influence the mechanism of chaperonin-assisted refolding. J Biol Chem **272**:32–5.
12. Grandhee, S. K., and V. M. Monnier. 1991. Mechanism of formation of the Maillard protein cross-link pentosidine. Glucose, fructose, and ascorbate as pentosidine precursors. J Biol Chem **266**:11649–53.
13. Hegyi, H., and P. Tompa. 2008. Intrinsically disordered proteins display no preference for chaperone binding in vivo. PLoS Comput Biol **4**:e1000017.
14. Hinds, M. G., C. Smits, R. Fredericks-Short, J. M. Risk, M. Bailey, D. C. Huang, and C. L. Day. 2007. Bim, Bad and Bmf: intrinsically unstructured BH3-only proteins that undergo a localized conformational change upon binding to prosurvival Bcl-2 targets. Cell Death Differ **14**:128–36.
15. Iakoucheva, L. M., C. J. Brown, J. D. Lawson, Z. Obradovic, and A. K. Dunker. 2002. Intrinsic disorder in cell-signaling and cancer-associated proteins. J Mol Biol **323**:573–84.
16. Kalthoff, C. 2003. A novel strategy for the purification of recombinantly expressed unstructured protein domains. J Chromatogr B **786**:247–54.
17. Khandelwal, R. L., and S. M. Zinman. 1978. Purification and properties of a heat-stable protein inhibitor of phosphoprotein phosphatase from rabbit liver. J Biol Chem **253**:560–5.
18. Korcsmaros, T., I. A. Kovacs, M. S. Szalay, and P. Csermely. 2007. Molecular chaperones: the modular evolution of cellular networks. J Biosci **32**:441–6.
19. Lowe, P. A., U. Aebi, C. Gross, and R. R. Burgess. 1981. In vitro thermal inactivation of a temperature-sensitive sigma subunit mutant (rpoD800) of *Escherichia coli* RNA polymerase proceeds by aggregation. J Biol Chem **256**:2010–5.
20. Makokha, M., M. Hare, M. Li, T. Hays, and E. Barbar. 2002. Interactions of cytoplasmic dynein light chains Tctex-1 and LC8 with the intermediate chain IC74. Biochemistry **41**:4302–11.
21. Manenti, S., O. Sorokine, A. Van Dorsselaer, and H. Taniguchi. 1992. Affinity purification and characterization of myristoylated alanine-rich protein kinase C substrate (MARCKS) from bovine brain. Comparison of the cytoplasmic and the membrane-bound forms. J Biol Chem **267**:22310–5.

22. Markus, G. 1965. Protein substrate conformation and proteolysis. Proc Natl Acad Sci U S A **54**:253–8.
23. Murray, C. L., J. Marcotrigiano, and C. M. Rice. 2008. Bovine viral diarrhea virus core is an intrinsically disordered protein that binds RNA. J Virol **82**:1294–304.
24. Oka, M., Y. S. Yang, S. Nagata, N. Esaki, H. Tanaka, and K. Soda. 1989. Overproduction of thermostable leucine dehydrogenase of Bacillus stearothermophilus and its one-step purification from recombinant cells of *Escherichia coli*. Biotechnol Appl Biochem **11**:307–11.
25. Oldfield, C. J., Y. Cheng, M. S. Cortese, C. J. Brown, V. N. Uversky, and A. K. Dunker. 2005. Comparing and combining predictors of mostly disordered proteins. Biochemistry **44**:1989–2000.
26. Paz, A., T. Zeev-Ben-Mordehai, M. Lundqvist, E. Sherman, E. Mylonas, L. Weiner, G. Haran, D. I. Svergun, F. A. Mulder, J. L. Sussman, and I. Silman. 2008. Biophysical characterization of the unstructured cytoplasmic domain of the human neuronal adhesion protein neuroligin 3. Biophys J **95**:1928–44.
27. Rekas, A., C. G. Adda, J. Andrew Aquilina, K. J. Barnham, M. Sunde, D. Galatis, N. A. Williamson, C. L. Masters, R. F. Anders, C. V. Robinson, R. Cappai, and J. A. Carver. 2004. Interaction of the molecular chaperone alphaB-crystallin with alpha-synuclein: effects on amyloid fibril formation and chaperone activity. J Mol Biol **340**:1167–83.
28. Thorn, D. C., S. Meehan, M. Sunde, A. Rekas, S. L. Gras, C. E. MacPhee, C. M. Dobson, M. R. Wilson, and J. A. Carver. 2005. Amyloid fibril formation by bovine milk kappa-casein and its inhibition by the molecular chaperones alphaS- and beta-casein. Biochemistry **44**:17027–36.
29. Tompa, P. 2002. Intrinsically unstructured proteins. Trends Biochem Sci **27**:527–33.
30. Tompa, P., J. Prilusky, I. Silman, and J. L. Sussman. 2007. Structural disorder serves as a weak signal for intracellular protein degradation. Proteins **71**:903–9.
31. Tompa, P., C. Szasz, and L. Buday. 2005. Structural disorder throws new light on moonlighting. Trends Biochem Sci **30**:484–9.
32. Tsvetkov, P., G. Asher, A. Paz, N. Reuven, J. L. Sussman, I. Silman, and Y. Shaul. 2007. Operational definition of intrinsically unstructured protein sequences based on susceptibility to the 20S proteasome. Proteins **70**:1357–66.
33. Uversky, V. N. 2002. What does it mean to be natively unfolded? Eur J Biochem **269**:2–12.
34. Uversky, V. N., C. J. Oldfield, and A. K. Dunker. 2005. Showing your ID: intrinsic disorder as an ID for recognition, regulation and cell signaling. J Mol Recognit **18**:343–84.
35. Zeev-Ben-Mordehai, T., E. H. Rydberg, A. Solomon, L. Toker, V. J. Auld, I. Silman, S. Botti, and J. L. Sussman. 2003. The intracellular domain of the *Drosophila* cholinesterase-like neural adhesion protein, gliotactin, is natively unfolded. Proteins **53**:758–67.

INDEX

Note: *f* indicates a figure; *t* indicates a table.

Instrumental Analysis of Intrinsically Disordered Proteins: Assessing Structure and Conformation, Edited by Vladimir Uversky and Sonia Longhi